THE CHEMISTRY OF
MOLECULAR IMAGING

THE CHEMISTRY OF MOLECULAR IMAGING

Edited by

NICHOLAS LONG
WING-TAK WONG

Library of Congress Cataloging-in-Publication Data:

The chemistry of molecular imaging / edited by Nicholas Long, Wing-Tak Wong.
      pages   cm
   Summary: "This book investigates the chemistry of molecular imaging and helps to educate non-chemists already involved in the area of molecular imaging. It addresses all the major modalities and techniques, such as MRI, positron emission tomography, single photon emission computed tomography, ultrasound, and fluorescence/optical imaging"– Provided by publisher.
      Includes bibliographical references and index.
      ISBN 978-1-118-09327-6 (hardback)
1. Imaging systems.   2. Imaging systems in medicine.   3. Molecular probes.   I. Long, Nicholas J., 1965– editor.   II. Wong, Wing-Tak, editor.
   TK8315.C476 2015
   616.07′540154–dc23

                                                                                                2014032169

Printed in the United States of America

10 9 8 7 6 5 4 3 2 1

CONTENTS

PREFACE

Since the emergence of molecular imaging over 30 years ago, it has arguably become one of the most rapidly growing fields of scientific research, spanning multiple disciplines such as medicine, pharmacology, chemistry, cell biology, and biomedical engineering. In contrast to conventional biomedical imaging by microscopy, where investigations are generally performed on excised tissues, molecular imaging includes *in vivo* techniques that provide visual and quantitative information on normal or pathological processes at the cellular or sub-cellular level. Most pertinently, these techniques have been designed to be noninvasive to the subject body. As a result, molecular imaging has revolutionised how one can monitor and characterise complex, dynamically changing molecular pathways within the living organisms, thus spurring its rapid development.

Nowadays, molecular imaging via Positron Emission Tomography (PET), Single-Photon Emission Computed Tomography (SPECT), Computed Tomography (CT) or Computed Axial Tomography (CAT), Magnetic Resonance Imaging (MRI), Optical Imaging, and Ultrasound (US) have already been regularly applied in the clinical environment as diagnostic tools. Each of these imaging modalities brings its own advantages and disadvantages, utilising a specific wavelength of electromagnetic or sound wave for excitation or detection. Due to their specific properties, imaging targets can range from cell surface markers or genes and their related products to a particular cellular or pathological process. Once the imaging target is identified, an appropriate imaging contrast agent is needed to be selected. This agent, mostly in the form of a small molecule, protein, or antibody, can bind to or enter the target environment upon injection into the subject body. Traditionally, imaging contrast agents are likely to be specific to one particular imaging mode, but in recent years, multimodality probes have been increasingly common as scientists strive for improvements in efficiency. For the first time in the area, we present a book dedicated to the *chemistry* of molecular imaging—other excellent texts and monographs describe the principles of biomedical imaging, focusing on the physics and mathematics behind the techniques.

Consisting of 16 chapters, this book is designed to provide (i) an in-depth discussion on the chemistry of various imaging contrast agents, probes, and biomarkers being applied in different imaging modalities and (ii) the methodology in which the agent becomes bound to its intended target and how it acts within *in vitro* and *in vivo* environments.

Following a general introduction to molecular imaging and the various imaging modes in Chapter 1, Chapter 2 lays out the principles with which imaging contrast agents achieve their labelling or bioconjugated status. In Chapters 3 to 7, radioactive isotopes employed in the nuclear medicine imaging techniques PET or SPECT are discussed, while agents for MRI are examined in Chapters 8 to 10. This is followed by the discussion of organic molecules, metal complexes, and nanoparticles being utilised in optical imaging, comprising Chapters 11 to 14. Chapter 15 details the applications of microbubbles in ultrasound, MRI, and more, and finally, the last two chapters of the book investigate the nature and properties of multimodality imaging contrast agents. Throughout this book, we hope to construct a comprehensive picture of imaging chemistry, with examples and illustrations, thus affording the readers a thorough understanding of the art of imaging contrast agent design.

We thank all the authors for the preparation of their individual contributions that really *make* this book, and their patience with the project. We hope that readers do enjoy the book and that it will prove useful and stimulating for their own research, helping to reinforce this burgeoning and exciting area of scientific discovery.

London and Hong Kong, June 2014   NICHOLAS J. LONG
                                                                                                      WING-TAK WONG

LIST OF CONTRIBUTORS

Silvio Aime, Departmento di Chimica I.F.M. and Centro di Imaging Moleculare, Universita di Torino, Torino, Italy

Octavia A. Blackburn, Chemistry Research Laboratory, University of Oxford, Oxford, UK

Weibo Cai, Departments of Radiology and Medical Physics, University of Wisconsin-Madison, Madison, WI, USA

April M. Chow, Laboratory of Biomedical Imaging and Signal Processing and Department of Electrical and Electronic Engineering, The University of Hong Kong, Pokfulam, Hong Kong SAR, China

Michael P. Coogan, Department of Chemistry, Lancaster University, Lancaster, UK

Lina Cui, Molecular Imaging Program at Stanford, Departments of Radiology and Chemistry, School of Medicine, Stanford University, Stanford, CA, USA

Jonathan R. Dilworth, Department of Chemistry, University of Oxford, Oxford, UK

Frédéric Dollé, CEA, I2BM, Service Hospitalier Frédéric Joliot, Orsay, France

Amanda L. Eckermann, Departments of Chemistry, Molecular Biosciences, Neurobiology, Biomedical Engineering and Radiology, Northwestern University, Evanston, IL, USA

Osasere M. Evbuomwan, Department of Chemistry, University of Texas at Dallas, Richardson, TX, USA

Stephen Faulkner, Chemistry Research Laboratory, University of Oxford, Oxford, UK

Wei Feng, Department of Chemistry, Fudan University, Shanghai, China

Juan Gallo, Department of Chemistry, Imperial College London, London, UK

Reinier Hernandez, Department of Medical Physics, University of Wisconsin-Madison, Madison, WI, USA

Hao Hong, Department of Radiology, University of Wisconsin-Madison, Madison, WI, USA

Koichi Kato, Department of Molecular Imaging, National Centre of Neurology and Psychiatry, Kodaira, Tokyo, Japan

Ga-Lai Law, Department of Applied Biology and Chemical Technology, Hong Kong Polytechnic University, Hung Hom, Kowloon, Hong Kong SAR, China

Michael Hon-Wah Lam, Department of Biology and Chemistry, City University of Hong Kong, Kowloon, Hong Kong SAR, China

Bengt Långström, Department of Biochemistry and Organic Chemistry, Uppsala University, Uppsala, Sweden; Neuropsychopharmacology Unit, Centre for Pharmacology and Therapeutics, Division of Experimental Medicine,

Imperial College London, London, UK; Department of Nuclear Medicine, PET & Cyclotron Unit, Odense University Hospital, University of Southern Denmark, Institute of Clinical Research, Odense, Denmark

Chi-Sing Lee, Laboratory of Chemical Genomics, School of Chemical Biology and Biotechnology, Peking University Shenzhen Graduate School, Shenzhen University Town, Xili, Shenzhen, China

Fuyou Li, Department of Chemistry, Fudan University, Shanghai, China

Nicholas J. Long, Department of Chemistry, Imperial College London, London, UK

Daniel J. Mastarone, Departments of Chemistry, Molecular Biosciences, Neurobiology, Biomedical Engineering and Radiology, Northwestern University, Evanston, IL, USA

Thomas J. Meade, Departments of Chemistry, Molecular Biosciences, Neurobiology, Biomedical Engineering and Radiology, Northwestern University, Evanston, IL, USA

Philip W. Miller, Department of Chemistry, Imperial College London, London, UK

Tapas R. Nayak, Department of Radiology, University of Wisconsin-Madison, Madison, WI, USA

Chris Orvig, Medicinal Inorganic Chemistry Group, Department of Chemistry, University of British Columbia, Vancouver, BC, Canada

Sofia I. Pascu, Department of Chemistry, University of Bath, Bath, UK

Simon J. A. Pope, School of Chemistry, Cardiff University, Cardiff, UK

Eric W. Price, Medicinal Inorganic Chemistry Group, Department of Chemistry, University of British Columbia, Vancouver, BC, Canada

Jianghong Rao, Molecular Imaging Program at Stanford, Departments of Radiology and Chemistry, School of Medicine, Stanford University, Stanford, CA, USA

Dirk Roeda, CEA, I2BM, Service Hospitalier Frédéric Joliot, Orsay, France

A. Dean Sherry, Advanced Imaging Research Center, UT Southwestern Medical Center, Dallas, TX, USA; Department of Chemistry, University of Texas at Dallas, Richardson, TX , USA

Yun Sun, Department of Chemistry, Fudan University, Shanghai, China

Enzo Terreno, Departmento di Chimica I.F.M. and Centro di Imaging Moleculare, Universita di Torino, Torino, Italy

Ka-Leung Wong, Department of Chemistry, Ho Sin Hang Campus, Hong Kong Baptist University, Kowloon Tong, Kowloon, Hong Kong SAR, China

Wing-Tak Wong, Department of Applied Biology and Chemical Technology, Hong Kong Polytechnic University, Hung Hom, Kowloon, Hong Kong SAR, China

Ed X. Wu, Laboratory of Biomedical Imaging and Signal Processing and Department of Electrical and Electronic Engineering, The University of Hong Kong, Pokfulam, Hong Kong SAR, China

Jing Zhou, Department of Chemistry, Fudan University, Shanghai, China

1

AN INTRODUCTION TO MOLECULAR IMAGING

GA-LAI LAW AND WING-TAK WONG

Department of Applied Biology and Chemical Technology, Hong Kong Polytechnic University, Hung Hom, Kowloon, Hong Kong SAR, China

1.1 INTRODUCTION

The aim of this book is to introduce the concepts of different imaging techniques that are employed for diagnostics and therapy and the role that chemistry has played in their evolution. The book provides a general introduction to the area of molecular imaging, giving an account of the role of molecular design and its importance in modern-day techniques, with an in-depth introduction of some of the probes and methodologies employed. This first chapter introduces the different types of imaging modalities currently at the forefront of imaging and illustrates some basic concepts underlying these techniques. It acts as a simplified background to set the scene for the following chapters, which will discuss the chemical properties of molecules and the role they play in different imaging modalities. For the interested readers, other textbooks are referenced that will provide more detailed information regarding the different techniques reviewed.

In life everything is incessantly changing. There is constant evolution in life sciences, evolution in the way problems arise, and evolution in the way they are solved. Diagnostics and therapy are both important, but as Einstein said, "intellectuals solve problems, geniuses prevent them." The key challenge still remains to unravel the hidden knowledge within life sciences, which constantly challenges us with new diseases and mechanistic mutation of biological systems and pathways [1]. Again, as stated by Einstein, "once we accept our limits, we go beyond them."

Molecular imaging aims to detect and monitor mechanistic processes in cells, tissues, or living organisms with the use of instruments and contrast mechanisms without perturbing their living system. Ultimately, it is a field that utilises molecular building blocks to bring solutions to problems by specialised imaging techniques that have matured into a large integrated field enveloped within various branches of science (Figure 1.1) [2]. In the area of modern-day imaging where technology is at its pinnacle, molecular design still holds a dominant role in the forefront of molecular imaging.

In the past, developments in contrast agents, probes, and dyes have brought about an era of creativity where new techniques, materials, and designs have flourished to form a concrete foundation resulting in today's achievements in diagnosis and therapy (Figure 1.2). The construction of better chemical molecules will continue to help us develop a more comprehensive picture of learning about life science. Figure 1.3 depicts a timeline in the development of the field [1–3].

The Chemistry of Molecular Imaging, First Edition. Edited by Nicholas Long and Wing-Tak Wong.
© 2015 John Wiley & Sons, Inc. Published 2015 by John Wiley & Sons, Inc.

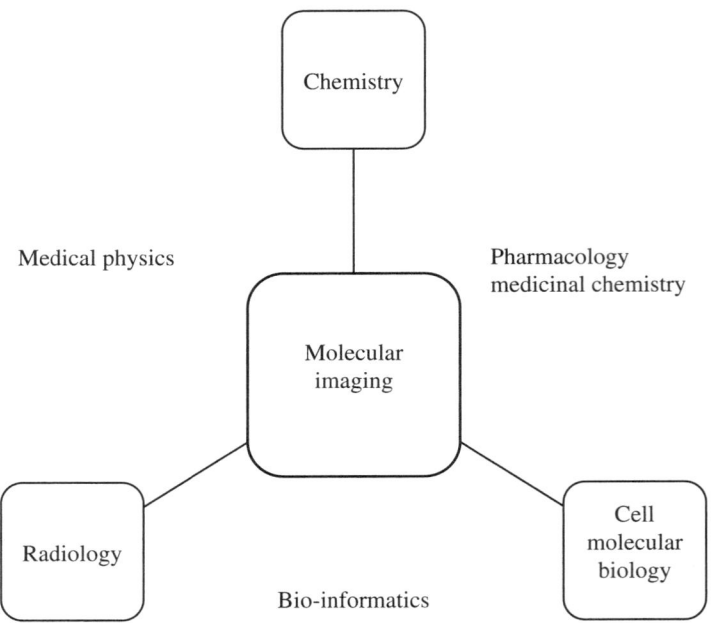

FIGURE 1.1   Types of multidisciplinary fields related to molecular imaging.

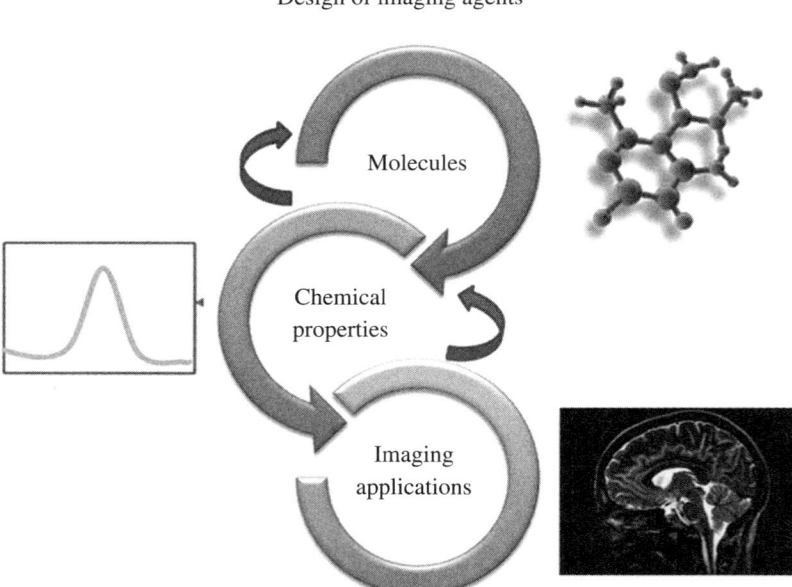

FIGURE 1.2   Diagram showing the links in the design rationale of imaging agents.

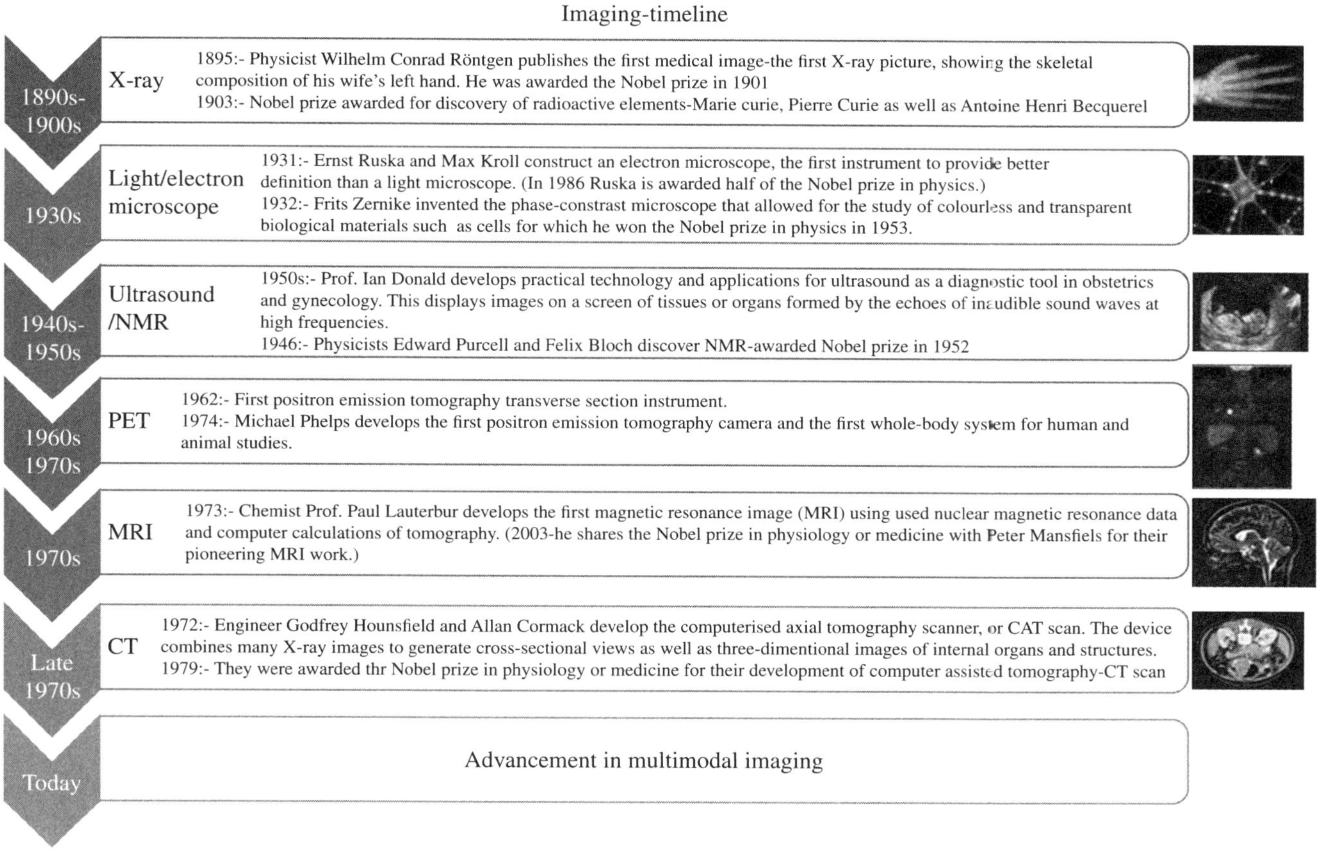

Imaging-timeline

1890s-1900s	X-ray	1895:- Physicist Wilhelm Conrad Röntgen publishes the first medical image-the first X-ray picture, showing the skeletal composition of his wife's left hand. He was awarded the Nobel prize in 1901 1903:- Nobel prize awarded for discovery of radioactive elements-Marie curie, Pierre Curie as well as Antoine Henri Becquerel	
1930s	Light/electron microscope	1931:- Ernst Ruska and Max Kroll construct an electron microscope, the first instrument to provide better definition than a light microscope. (In 1986 Ruska is awarded half of the Nobel prize in physics.) 1932:- Frits Zernike invented the phase-contrast microscope that allowed for the study of colourless and transparent biological materials such as cells for which he won the Nobel prize in physics in 1953.	
1940s-1950s	Ultrasound /NMR	1950s:- Prof. Ian Donald develops practical technology and applications for ultrasound as a diagnostic tool in obstetrics and gynecology. This displays images on a screen of tissues or organs formed by the echoes of inaudible sound waves at high frequencies. 1946:- Physicists Edward Purcell and Felix Bloch discover NMR-awarded Nobel prize in 1952	
1960s 1970s	PET	1962:- First positron emission tomography transverse section instrument. 1974:- Michael Phelps develops the first positron emission tomography camera and the first whole-body system for human and animal studies.	
1970s	MRI	1973:- Chemist Prof. Paul Lauterbur develops the first magnetic resonance image (MRI) using used nuclear magnetic resonance data and computer calculations of tomography. (2003-he shares the Nobel prize in physiology or medicine with Peter Mansfiels for their pioneering MRI work.)	
Late 1970s	CT	1972:- Engineer Godfrey Hounsfield and Allan Cormack develop the computerised axial tomography scanner, or CAT scan. The device combines many X-ray images to generate cross-sectional views as well as three-dimentional images of internal organs and structures. 1979:- They were awarded thr Nobel prize in physiology or medicine for their development of computer assisted tomography-CT scan	
Today		Advancement in multimodal imaging	

FIGURE 1.3   An approximate timeline showing the development of the different imaging modalities [1–3].

Fluorodeoxyglucose

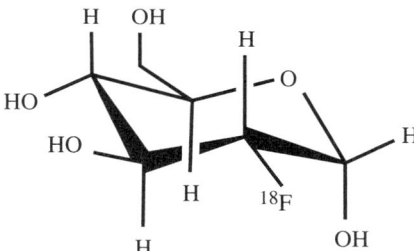

FIGURE 1.4   ¹⁸FDG, a typical contrast agent used in PET.

1.2 WHAT IS POSITRON EMISSION TOMOGRAPHY (PET)?

Positron Emission Tomography (PET) is a nuclear medicine tomographic modality and one of the most sensitive methods for quantitative measurement of physiologic processes *in vivo* [4]. This technique utilises positron-emitting radionuclides and requires the use of radiotracers that decay and produce two 511 keV γ-rays resulting from the annihilation of a positron and an electron. One of the most commonly used molecules is ^{18}F-labelled fluorodeoxyglucose (^{18}FDG), which has radioactive fluorine and is readily taken up by tumours (Figure 1.4) [5].

1.2.1 Basic Principles

In PET, a neutron-deficient isotope causes positron annihilation to produce two 511 keV γ-rays, which are simultaneously emitted when a positron from a nuclear disintegration annihilates in tissue. PET imaging, unlike MRI, ultrasound, and optical imaging, does not require any external sources for probing or excitation; instead, the source is generated from radioisotopes and emitted from but not transmitted through an object/patient, as in CT imaging [4–7]. Radionuclides are incorporated as part of a small metabolically active molecule to generate radiotracers such as [18]FDG, which are then intravenously injected into patients at trace dosage for PET imaging. [18]FDG is a favourable radiotracer because it is inhibited from metabolic degradation before it decays due to the fluorine at the $2'$ position in the molecule. Upon decay, the fluorine is converted into [18]O. There is generally a short period of time before accumulation of radiotracers into the targeted organs or tissues that are being examined, so it is important for radiotracers to have a suitable half-life—some commonly used radionuclei have very short half-lives. Some common radionuclides used in PET are 11-C (half-life ~20 min), 13-N (~10 min), 15-O (~2 min) and 18-F (~110 min). These are produced by a cyclotron, whereas 82-Rb (76 s), which is used in clinical cardiac PET, is produced by a generator [8–9].

When a radioisotope undergoes positron emission decay (positive β-decay), it emits a positron that travels through the tissue for a short distance (~<2 mm) whilst decelerating by the loss of its kinetic energy until it collides with an electron. This results in back-to-back annihilation of γ-ray photons, which move in opposite directions and are emitted nearly 180 degrees apart before being detected by scintillators and a photomultiplier tube. This type of coincidence is a true coincidence event; to detect this, the detectors are designed like a ring that surrounds the patient during the scanning procedure. Several parallel rings form the complete detection panel of the PET system in a cylindrical geometry (Figure 1.5).

PET has relatively high sensitivity in detecting molecular species ($10^{-11} – 10^{-12}$ M), even though not all annihilation photons are used for image reconstruction because not all coincidences are true coincidences. A coincidence event is assigned to a line of response where the two relevant detectors are joined (detectors opposite to each other); this allows for positional information to be located from the detected radiation without any physical collimators. This is known as electronic collimation. There are four types of coincidence events in PET: true, scattered, random, and multiple (Figure 1.6). Only true coincidence, which is the simultaneous detection of two emissions from a single annihilation event, is useful. No other events are detected within this coincidence time-window.

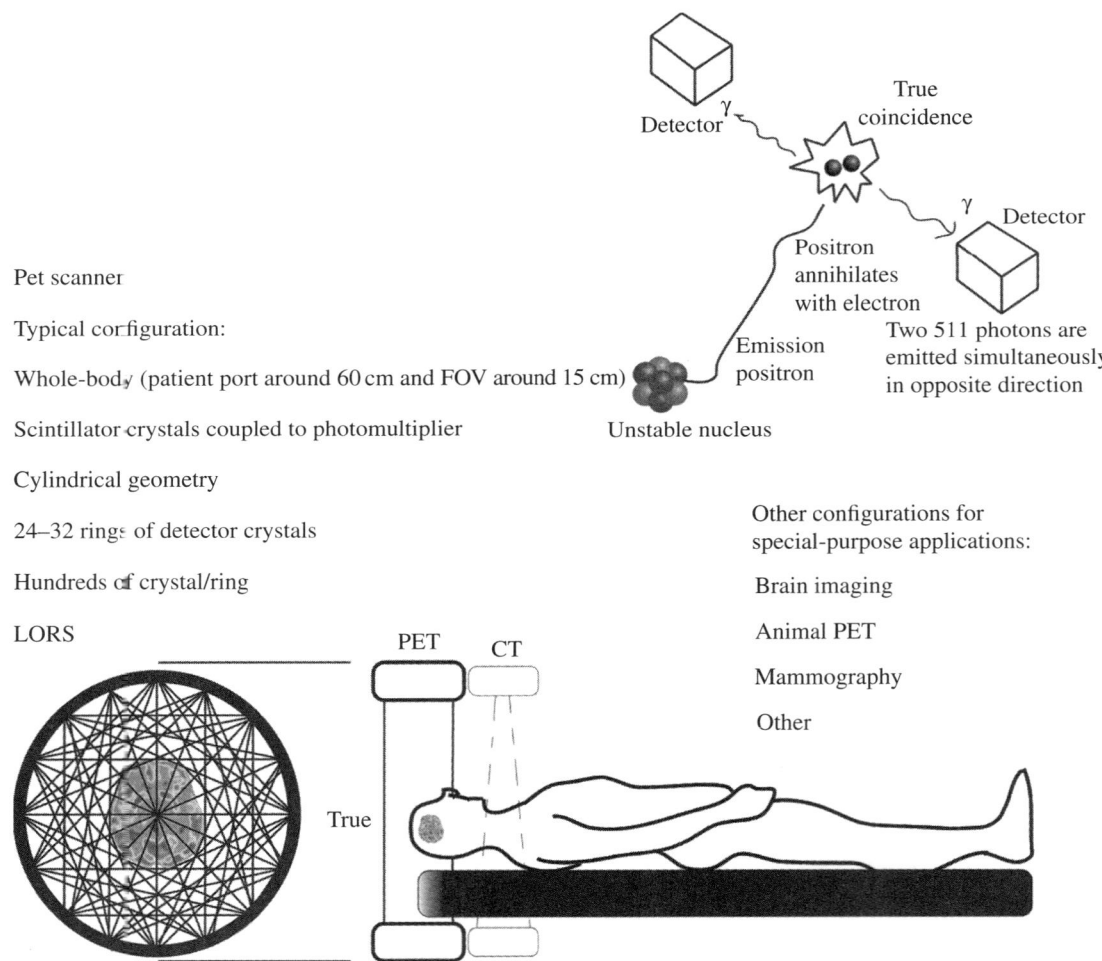

FIGURE 1.5 Typical configuration of a PET scanner.

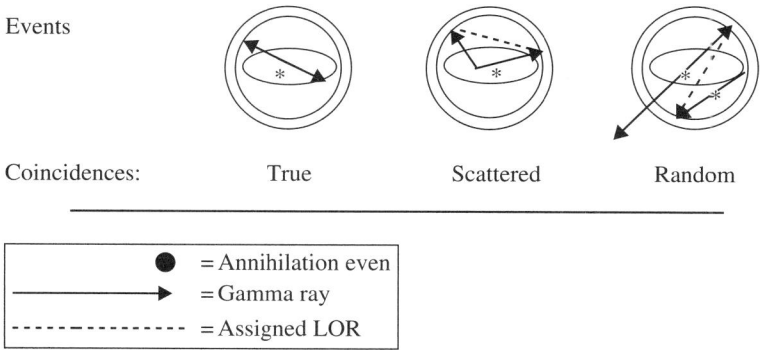

FIGURE 1.6 Different types of coincidence events.

Scattered coincidence occurs when one or both photons from a single event are scattered and both are detected; however, one of the photons must have undergone at least one Compton scattering event prior to detection. This type of event adds a background to the true coincidence event and causes overestimation of the isotope concentration as well as decreasing image contrast. In Compton scattering, a photon interacts with an electron in the absorber material, resulting in an increase in the kinetic energy of the electron as well as a change in direction in the photon. The energy of the photon after interaction is defined as:

$$E' = \frac{E}{1 + \left(\dfrac{E}{m_0 c^2}\right)(1 - \cos\theta)} \tag{1.1}$$

where E is the energy of the incident photon, E' is the energy of the scattered photon, $m_0 c^2$ is the rest mass of the electron, and θ is the scattering angle [10]. From Equation 1.1, it can be seen that fairly large deflections can occur with just a small loss of energy; for example, for 511 keV photons, a Compton scattering event results in a deflection of over 25 degrees but results in just a 10% loss in the photon energy. Random coincidence is the simultaneous detection of emission from more than one decay event. It occurs when two photons not arising from the same annihilation event are incident on the detectors within the coincidence time-window of the system. This contributes to statistical noise in data as well as overestimation of isotope concentration [8].

Multiple coincidences occur when more than two photons are detected by different detectors within the coincidence resolving time. This type of event either causes event mis-positioning or rejection because it is not possible to determine the line of response to which the event should be assigned. Coincidence events are grouped together to produce projection images called sonograms. Acquisition of PET images is not a simple process because data corrections are required for scattered, random coincidences as well as for the effects of attenuation, because the data acquired from the PET camera are given as projections. The measured projections are different from the projections assumed in image reconstruction [9]. Reconstruction of images from projections is computationally burdensome. Data reconstruction and correction are usually carried out by analytical or iterative methods. Analytical methods are simple, fast, and usually have predictable linear behaviour. However, such methods are not very flexible and have problems associated with noise resolution and image properties and do not allow for quantitative imaging. On the other hand, iterative methods allow for quantitative imaging but require long calculation times as well as amplification of the background noise; it thus requires counts to be low to reduce the projection noise.

1.2.2 Advantages and Limitations

PET is a highly sensitive and popular technique in preclinical and clinical imaging. It is a very important diagnostic technique because disease processes such as cancer often begin with functional changes at the cellular level. There are many radioactive tracers with various half-lives applicable for different preclinical and clinical applications. The half-lives are often very short and therefore must be injected immediately after production. Due to the mechanism of decay being the same for all different radioactive tracers, it is only possible to trace one molecular species in a given imaging experiment or clinical scan where only true coincidence events are used.

Tracers can be designed to be target-specific to tumours and allow study of metabolic activities such as bone metabolism and bone metastasis that are common in a lot of cancers. Thus this technique can be used to monitor disease processes and patients' responses to therapy. However, one of the major limitations of PET is its poor spatial resolution. It is also limited by pixel sampling rate, quantity of the radioactive source, and blurring in the phosphor screens of the detector rings. However, the use of electronic collimation over physical collimation helps to improve sensitivity and uniformity of the emitting source response function.

1.3   WHAT IS SINGLE PHOTON EMISSION COMPUTED TOMOGRAPHY (SPECT)?

SPECT is another nuclear imaging technique for imaging molecules, metabolisms, and biochemical functions of organs and cells, and like PET, the use of radioisotopes is required. As its name suggests, it involves the emission of a single γ-ray per nuclear disintegration, which is measured directly, unlike in PET, where the positrons are emitted to produce the γ-rays. Numerous single γ-rays are detected by rotating gamma cameras to reconstruct an image of the origin of the γ-rays, which identifies the location of the radioisotope. Thus, specific radio-ligands are used to incorporate typical radioisotopes such as [99mTc] to target to areas of interest [11]. An example of a radiopharmaceutical commonly used in cardiac imaging is [9mTc]-tetrofosmin, also known as 'Myoview' (Figure 1.7) [12].

1.3.1   Basic Principles

Gamma rays from radioactive nuclides, apart from radioactive decay, also produce other forms of radiation such as alpha and beta. A gamma ray results from the relaxation of an excited daughter nucleus to a lower energy state after a nucleus emits an α or β particle [13]. An example of this is technetium-99, which is a commonly used radioisotope in radiopharmaceuticals produced from molybdenum-99. As shown in Figure 1.8, the excited nuclear state is more stable than average excited states after a β-decay where a β particle is released, thus the daughter nucleus forms a metastable excited state resulting in the

Technetium ([99mTc]) tetrofosmin

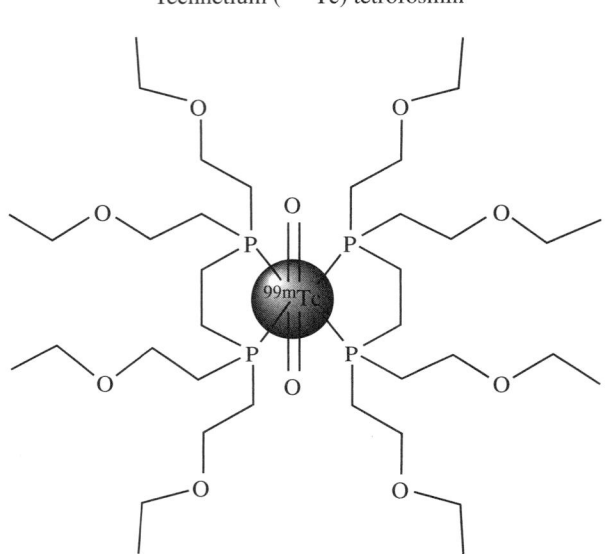

FIGURE 1.7   Myoview: A typical contrast agent used in SPECT.

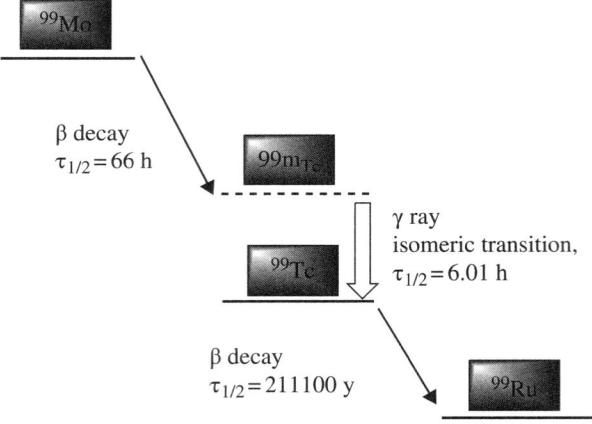

FIGURE 1.8   A schematic for the formation of [99mTc].

nuclear isomer, ^{99m}Tc. This isomer has a short half-life of ~6 hours before it goes to ^{99}Tc by isomeric transition, a radioactive decay process from an excited metastable state that results in γ-ray emission [13, 14].

In SPECT, these γ-ray emissions are detected by a 360^0 rotating photon detector array around the body known as the gamma camera, which can acquire multiple 3D projections at multiple angles. Sodium iodide or solid-state cadmium-zinc-telluride detectors, which provide spatial resolution of 1–2 mm, are usually used. Images are formed with the information given on the position and concentration of the radionuclide biodistribution in two dimensions [14]. However, due to the attenuation effects of γ-ray emission as it is transmitted from the injected tracers inside the body, mathematical reconstruction algorithms have been developed to improve resolution. Some other common radionuclides used in SPECT in addition to ^{99m}Tc are ^{111}In (half-life 2.8 days), ^{123}I (13.2 h), and ^{125}I (59.5 days). Due to the different half-lives, dual tracers can be used to give simultaneous imaging because the γ-ray emissions have different energies [14, 15].

A gamma detector is made up of a few cameras that are placed opposite to one another to form a cylindrical detector that allows rotation around an axis centre. Due to its multiple camera heads, it only needs to rotate 120–180 degrees to collect data around the entire body. The gamma camera is made up of three basic layers, the first of which contains a collimator, a special lens that only allows entry of γ-rays that are perpendicular to the plane of the camera. The other two layers consist of a crystal and detectors. The crystal is usually a thallium-activated sodium iodide [NaI(Tl)] detector crystal, which, when absorbing γ-rays, would scintillate to cause a light signal to be detected (Figure 1.9) [15].

The data are collected as a planar matrix of values that correspond to the number of gamma counts and can be processed to give planar scintigrams for constructing 2D images. Typically, each row across the matrix represents an intensity displayed across a single projection, whereas the successive rows represent successive projection angles. There are different techniques to reconstruct tomographic images that are different for 2D cross-sectional images and for 3D images. A common reconstruction method is the simple back-projection method, which generates 2D cross-sectional images of activity from a slice within the detected object, using the projection profiles obtained for that slice. However, there is a flaw in this method of data reconstruction: The final SPECT images have poorer spatial resolution than the raw 2D images used to produce them. For better spatial resolution, other processing techniques such as direct Fourier transform reconstruction as well as data filtering can be used. Filtered back-projection is a favourable method for data reconstruction. For tomographic 3D images, 3D reconstruction algorithms can also be used to visualise the 3D biodistribution of the radiotracers.

The resolution and sensitivity of SPECT is dependent on the pinhole of the collimator multiple-pinhole and multiple-solid state detector systems are often used to allow for lower radiation dosages and shorter scan times, hence improving the sensitivity and resolution of this imaging modality [16, 17].

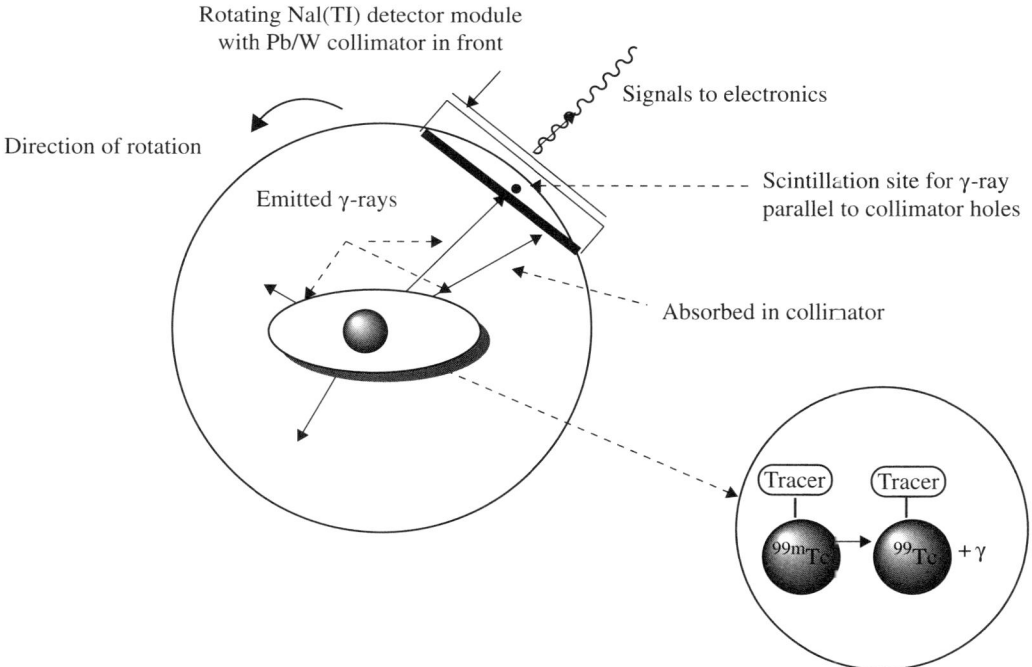

FIGURE 1.9 Schematic diagram showing the basic principles of SPECT.

1.3.2 Advantages and Limitations

One of the major advantages of SPECT over other imaging modalities is that it provides improved contrast between regions of different functions, thus allowing improved detection of abnormal physiological functions. SPECT also gives better spatial localisation as well as improved quantification. It offers greater accessibility because it uses radioisotopes with longer half-lives and, unlike PET, does not require any cyclotrons to generate these radioisotopes. Due to the selection of radioisotopes available, it allows simultaneous imaging because different radioisotopes produce different γ-rays of different energies (Table 1.1). This is a unique advantage of SPECT because multiple energy windows can be used for concurrent imaging of different functions and metabolic processes.

However, this technique also has disadvantages: It is often necessary to use long scanning times, which can cause discomfort to patients. Artefacts can also be easily generated by numerous uncontrollable factors such as patient movement and uneven distribution of the radiotracer.

1.4 WHAT IS COMPUTED TOMOGRAPHY (CT) OR COMPUTED AXIAL TOMOGRAPHY (CAT)?

Computed tomography (CT) or computed axial tomography (CAT) is a diagnostic technique that uses special X-ray equipment to obtain cross-sectional pictures of the body [18]. This technique allows detailed imaging of organs, bones, and tissues and is often used in conjunction with other diagnostic methods such as MRI and PET (Figure 1.10) [19].

1.4.1 Basic Principles

X-rays are a form of electromagnetic radiation with a wavelength in the range of 0.01 to 10 nm that can transverse along the cross-section of an object in straight lines. The wavelength is attenuated by the object through which it passes but is still detectable outside the object [20]. In CT imaging, the cross-section is probed with X-rays from various directions with the use of rotating X-ray equipment where both the low energy X-ray source and the detector are rotated at 360 degrees around the patient. The acquired attenuated signals are then recorded and converted into projections of the linear attenuation coefficient distribution of the cross-section to produce volumetric data. Charged Coupled Device (CCD) detectors are used to carry out photo-transduction of incoming X-rays to produce images. The contrast in CT images relies on the

TABLE 1.1   Types of Radioisotopes Used for Different Studies by SPECT Imaging.

Study	Radioisotope	Emission energy (keV)	Half-life	Radiopharmaceutical	Activity (MBq)
Bone scan	technetium-99 m	140	6 hours	Phosphonates / Bisphosphonates	800
Myocardial perfusion scan	technetium-99 m	140	6 hours	tetrofosmin; Sestamibi	700
Brain scan	technetium-99 m	140	6 hours	HMPAO; ECD	555–1110
Tumour scan	iodine-123	159	13 hours	MIBG	400
White cell scan	indium-111 & technetium-99 m	171 & 245	67 hours	in vitro labelled leucocytes	18

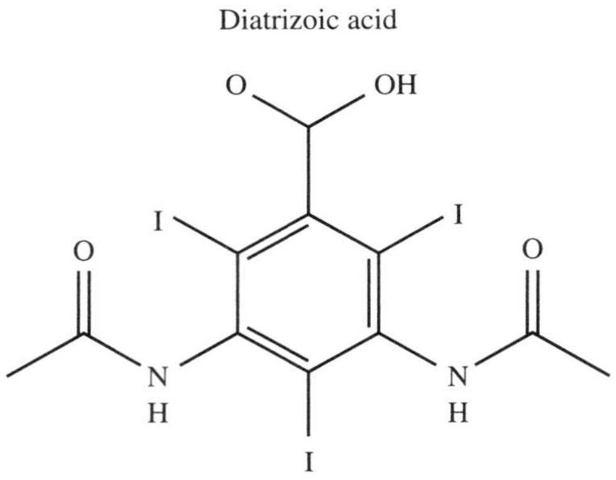

Diatrizoic acid

FIGURE 1.10   A typical contrast agent used in CT-Gastrografin.

intrinsic structural and absorption differences in the properties through which the X-rays travel, such as tissues, organs, bone, fat, water, and air [20].

In CT imaging, cross-sectional images of body organs and tissues are obtained at very high resolution, thus allowing differentiation among different types of tissue. The images are reconstructed from computers using the Fourier transform, a mathematical operation that reconstructs the cross-section to form an image of a slice through the body with the focal point centred at the position of the X-ray beam. This is the most basic image that can be obtained. Other more advanced image reconstruction, such as 3D images, can also be produced by taking multiple scans at short intervals and then stacking the slices together. Dynamic spatial reconstruction (DSR) is also possible by using approximately 30 X-ray tubes to produce images that show changes through time; thus, dynamic changes in structure and function can be monitored [20, 21]. CT imaging can provide views of soft tissue, bone, muscle, and blood vessels without compromising clarity. Other imaging techniques are much more limited in the types of images they can provide, so CT is commonly used for diagnostic purposes [21]. With the use of contrast agents such as iodinated contrast agents, imaging of tumours is also possible, providing information on size and localisation, thus aiding planning in treatment, whether for surgery or radiotherapy.

The use of higher energy X-rays can also improve resolution but can be problematic especially in the health of human patients because it increases ionizing radiation damage. Normally, higher X-ray CT machines are used for animal studies. As mentioned previously, CT images are generated by creating slices through the body; this is recorded as areas of varying CT intensities and represented in pixels. These pixels are represented by a number that is quantitative to the amount of X-ray beam absorbed by the tissues at each individual point in the body.

The image is created according to the density of the tissue from a matrix of pixels that are converted to CT numbers known as Hounsfield numbers (HU, Eq. 1.2). This scale is defined in Hounsfield units (HU) and shown in Figure 1.11. The denser the tissues are, the higher the value of the numbers, thus a greyscale is created. Different windows and levels are created from these numbers to produce an image. Different size windows affect the quality of the image by defining the range as well as the upper and lower limits. For example, a large window shows the major structures, whereas a small window shows finer details that are often used to discern tissues of similar density. The Hounsfield number at the centre of the window is referenced at the level and is used to define the range for the window associated with the type of structure of interest. To enhance CT images, high density contrast agents as well as multi-slice detector geometries that allow whole-body imaging are used; these help to reduce scanning times and patient discomfort as well as artefacts.

$$HU = \frac{\mu_X - \mu_{Water}}{\mu_{Water} - \mu_X} \times 1000 \qquad (1.2)$$

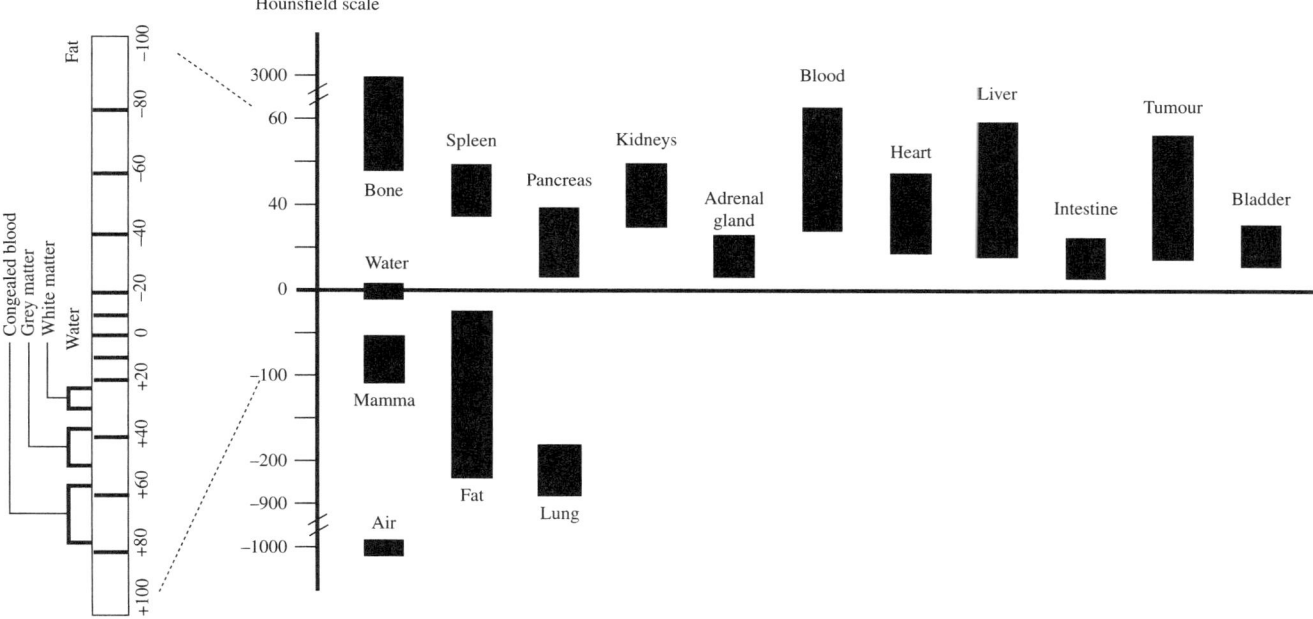

FIGURE 1.11   The Hounsfield Scale and the equation for the Hounsfield Unit (Eq. 1.2).

1.4.2 Advantages and Limitations

CT imaging provides fast and high spatial resolution images that allow fine visualization of anatomical details. As a diagnostic technique, it is often used to detect or confirm cancerous tumours by providing detailed information such as the size and location of the tumour to help planning for radiation therapy or surgery. One of the major advantages of CT is that it could be combined with other imaging modalities such as PET and MRI to provide other information such as dynamic and metabolic data (Figure 1.12). However, repeated CT imaging carries health risks to the patient due to exposure to non-negligible radiation dosage.

Another problem is that the scanning procedure can be quite time-consuming and may require up to an hour to complete. This could cause discomfort to patients in some cases. In general, CT imaging is a pain-free procedure, and it is often incorporated into other techniques to create powerful multimodal imaging modalities to give improved sensitivity and resolution for diagnosis, especially in cancer. The improvement in images due to the use of combined techniques such as CT and PET are shown in Figure 1.13. It is apparent that the combined use of CT and PET provides more information on tumours, such as their location and size as well as growth and metabolic activity of tissues [23].

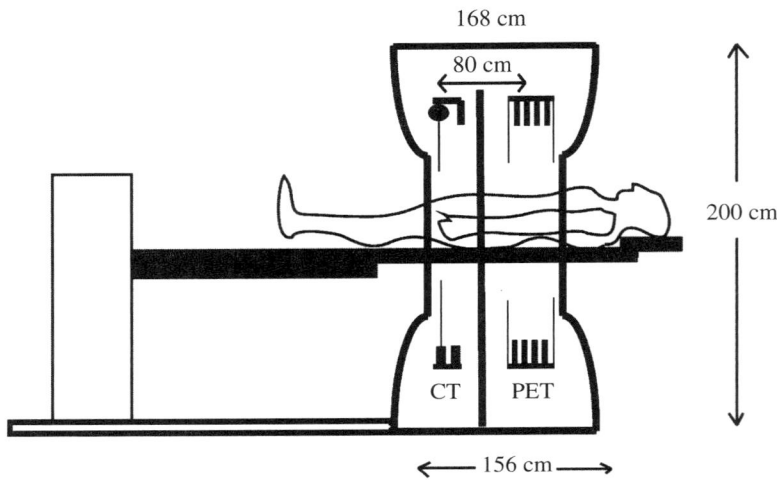

Dual-modality imaging range

FIGURE 1.12 Diagram showing a typical CT-PET instrument [22].

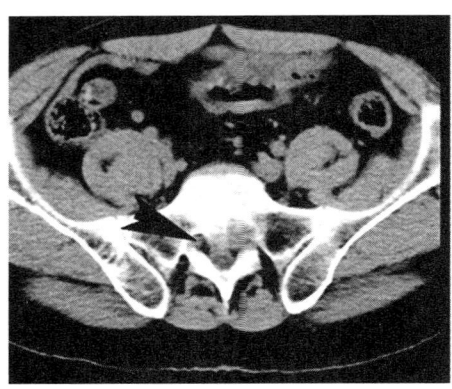 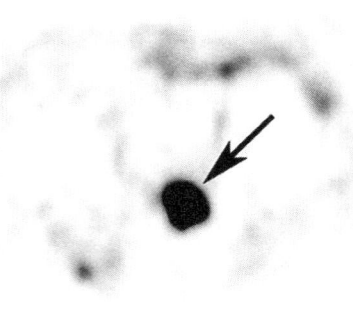 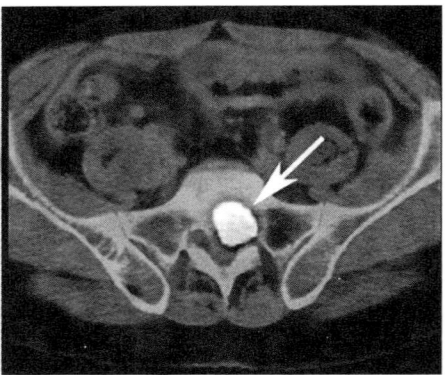

FIGURE 1.13 Left: Image from a CT scan; Middle: Image from a PET scan; Right: Image from a CT-PET scan [23]. (*See insert for colour representation of the figure.*)

Gadoteric acid

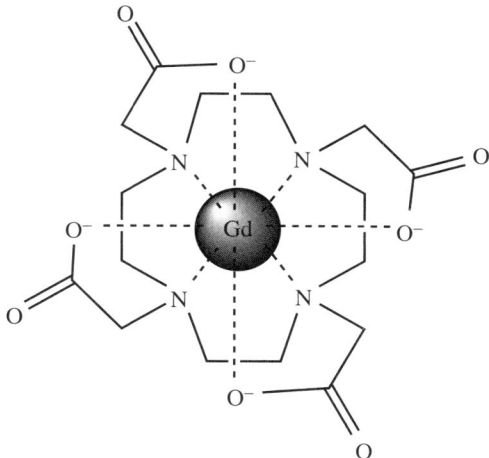

FIGURE 1.14   Dotarem: A typical contrast agent used in MRI.

1.5   WHAT IS MAGNETIC RESONANCE IMAGING (MRI)?

Magnetic resonance imaging (MRI) is an imaging technique based on the principles of nuclear magnetic resonance (NMR), which provides microscopic chemical and physical information about molecules. Instead of obtaining information about chemical shifts and coupling constants, MRI gives spatial distribution of the intensity of water proton signals in the body [24]. MRI measures the relaxation of free hydrogen nuclei when they realign to their original state in the direction of the magnetic field after having been excited by a radio frequency (Rf). Different image contrasts can be achieved by using different pulse sequences or by changing the imaging parameters such as longitudinal relaxation time (T_1) and transverse relaxation time (T_2). Contrast agents are also used to improve the quality of the image (Figure 1.14) [25].

1.5.1   Basic Principles

Approximately 63% of the human body is primarily fat and water, which are comprised of many hydrogen atoms. Thus, MRI focuses mostly on NMR signals from hydrogen nuclei. Nuclei are charged particles that have characteristic motion or precession that produces a small magnetic moment. In the presence of a magnetic field, the nuclei would move about it in a phenomenon known as the Larmor precession. The frequency of Larmor precession is proportional to the applied magnetic field strength as defined by the Larmor frequency ω_0, $\Rightarrow \omega_0 = \gamma B_0$, where γ is the gyromagnetic ratio and B_0 is the strength of the applied magnetic field. The gyromagnetic ratio is a nuclei-specific constant, for hydrogen, $\gamma = 42.6$ MHz/Tesla. A strong uniform magnetic field of 1.5 or 3 Tesla is generally used in a typical human scanner [24–26]. MRI measures the relaxation of free hydrogen nuclei after they have been excited by a radio frequency. The electric field of the radio frequency creates a new magnetic field, B_1, which induces protons away from the original field, B_0. The nuclei spins acquire enough energy to tilt/flip and precess. This 'flip' is time- and power-dependent on the B_1 field and hence the RF pulse. When the RF field is off, the protons are able to realign to their original state in the direction of the magnetic field, B_0, by T_1 and T_2 relaxation. In a strong magnetic field, the hydrogen nuclei spin is aligned in a direction parallel to the field. This process is illustrated in Figure 1.15.

   The signal recorded in MRI is the energy given or lost after relaxation of the nuclei from the RF excitation. This signal is the "spin echo," which is composed of multiple frequencies, reflecting different positions along the magnetic field gradient. Fourier transform is used to process these frequencies where the magnitude of the signal at each frequency is proportional to the hydrogen density at that location, thus allowing images to be constructed. Hence, spatial information in MRI is encoded in the frequency of the signal, which is dependent on the local value of the magnetic field. The generation of MRI images requires the combination of both spatial and intensity information. The signal intensity is mostly affected by the T_1 and T_2 relaxation parameters. although in general, the overall quality of the image is strongly dependent on hardware design such as the transmitting and receiving coil design.

   In order to understand how the contrasts of the images are generated, it is important to note that each proton has a unique T_1 and T_2, which are parameters that can be easily altered [27–29]. Proton relaxation is a process of realigning

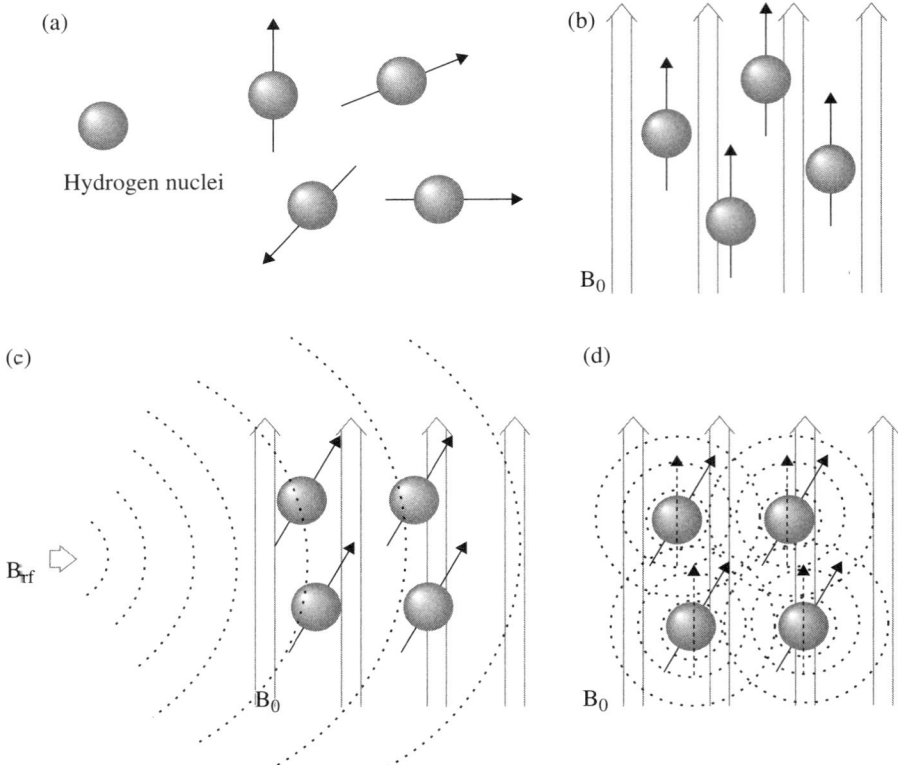

FIGURE 1.15 (a) Hydrogen nuclei are randomly aligned in the absence of any strong magnetic fields. (b) All hydrogen nuclei, in the presence of a strong magnetic field B_0, are aligned in parallel with the magnetic field to create a net magnetic moment, M, which is parallel to B_0. (c) A radio-frequency pulse B_{rf} is applied perpendicularly to the magnetic field B_0. This pulse, with a frequency equal to the Larmor frequency, causes the net magnetic moment of the nuclei M to tilt away from B_0. (d) The RF pulse stops and the net momentum of the nuclei realigns back in parallel to B_0 by relaxation, at the same time the nuclei lose energy to give an RF signal.

with B_0. There are two different types of relaxation, T_1 and T_2, the longitudinal relaxation time and the transverse relaxation time respectively. The T_1 and T_2 relaxation times define the way the protons revert back to their resting states after the initial RF pulse. As protons relax, they realign along B_0 by T_1 where T_1 is the recovery of magnetisation along the longitudinal axis. They also lose phase coherence by T_2, which is the decay of magnetisation along the transverse axis. These two parameters are the most significant in providing image contrast in MRI. The time constant of T_1 is tissue-dependent. Signal strength decreases in time with a loss of phase coherence of the spins. This decrease occurs at a time constant T_2, which is always less than T_1 (Figure 1.16). During T_1 and T_2 relaxation, the nuclei lose energy by emitting their own RF signals; however, only transverse magnetisation produces a signal. This signal is referred to as the free-induction decay (FID) response signal. The FID response signal is measured by a conductive field coil placed around the object being imaged, and the FID decays at a rate given by the tissue relaxation parameter known as T_2^*. The measurement of the FID signal gives images that have different weightings depending on the T_1, T_2, and T_2^*. These signals are processed or reconstructed to obtain greyscale contrast 3D images [24]. The different properties between T_1 and T_2 are shown in Table 1.2.

Although the human body contains a high percentage of water, the signal intensity is not just dependent on the amount of water at the location, and experimentally it is hard to change the proton density in tissue to look at small changes. Thus, chemical contrast agents are used to change the characteristics of tissue by altering the magnetic relaxation times of T_1 and T_2, which normally amplifies the contrast, and these agents are classed as T_1 and T_2 agents.

Ferromagnetic contrast agents alter the contrast by changing the T_2^* of the water molecules around the ferromagnetic contrast agent by distorting the B_0 magnetic field around the ferromagnetic material in the contrast agent. These contrast agents are typically iron nanoparticles with bio-organic compatible substrates. Paramagnetic contrast agents, which are much more commonly used, alter the contrast by producing time-varying magnetic fields that promote T_1 relaxation of water molecules. The time-varying magnetic fields come from both rotational motion of the contrast agent and electron spin flips associated with the unpaired electrons in the paramagnetic material in the contrast agent. This is why gadolinium agents are the most favourable because the f-element Gd ion has the maximum seven unpaired electrons [29].

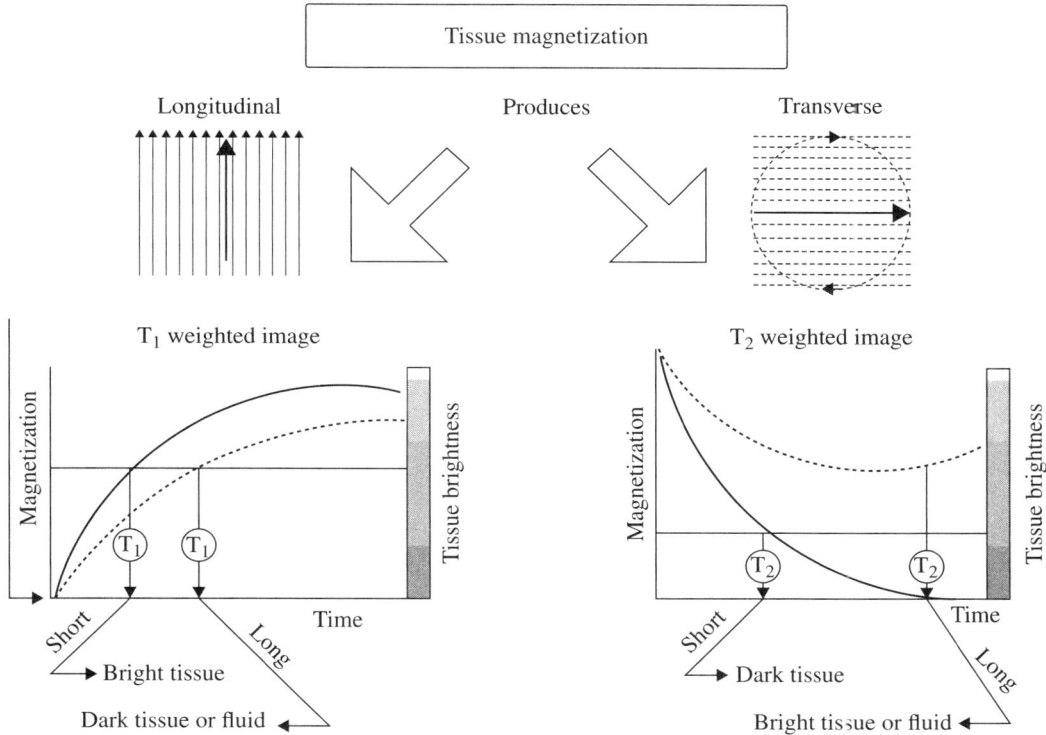

FIGURE 1.16   Comparison of T_1 and T_2 properties.

TABLE 1.2   Properties of T_1 and T_2 Relaxation.

T_1 is called the "spin-lattice" relaxation time	T_2 is called the "spin-spin" relaxation time
Recovery of magnetisation along the longitudinal axis: longitudinal relaxation time	Decay of magnetisation along the transverse axis: transverse relaxation time
Measurement of the "lattice" around the spin	Measurement of the "spins" around the spin
Rotational correlation time " τr " $\sim 1/T_1$	Translational correlation " τc " $\sim 1/T_2$
Time to rotate one radian (10–12 sec)	Time to diffuse one diameter (10–11 sec)
$= (4\pi)\, \eta\, \mathbf{a}3\, /\, 3kT$, where a is a radius, η is viscosity	$= (6\pi)\, \eta\, D\, /\, kT$, where D is diffusion, η is viscosity
Free protons rotate rapidly and have short τr	Free protons move rapidly and have short τc. This is observed as a long T_2.
This is observed as a long T_1.	
Bound protons hit barriers and have long τr. This translates into a short T_1.	Hindered protons move slowly and have long τc. This translates into a short T_2.

T_1 and T_2 may be shortened considerably in the presence of a paramagnetic contrast agent. However, a compromise is necessary because the shortening of T_1 would lead to an increase in signal intensity, while the shortening of T_2 would produce broader lines with decreased intensity. This reflects a nonlinear relationship between signal intensity and concentration of the contrast agent. At low concentrations, an increase in the concentration of the contrast agent would cause an increase in signal intensity until the optimal concentration is reached due to effects on T_1. Further increase in the concentration would reduce the intensity of the signal because of the effects on T_2. The relationship between T_1 and T_2 dictates the design of contrast agents, which must have a relatively greater effect on T_1 than on T_2, as well as the use of pulse sequences that emphasise changes in T_1.

Gadolinium-based contrast agents are usually used in MRI because of their excellent paramagnetic properties and biological tolerance. However, there are some concerns regarding the problem of toxicity. Problems are related to trans-metallation, which is the exchange of the metal in the contrast agent with a metal ion in solution (Eq. 1.3) [28].

$$ML + M' \rightleftarrows M'L + M \tag{1.3}$$

Toxicity, specificity, and relaxivity are three important critieria in the design of contrast agents. Many of these paramagnetic metals are toxic as free ions. Thus, the design and study of contrast agents is very important in MRI because these agents are injected in large doses especially compared with the quantities used in nuclear imaging techniques. For example, Gd-based contrast agents are injected intravenously at a dosage of 0.1 mmol/kg. Relaxivity is the ability of a magnetic contrast agent to increase the relaxation rates of the surrounding water proton spins, which depends on the molecular structure and kinetics of the complex. The relaxivity (r_1 or r_2) of a contrast agent in water is the change in $1/T_1$ or $1/T_2$ of water per concentration of contrast agent and is dependent on parameters such as temperature and magnetic field. The relationship between T_1, r_1, and the concentration of the paramagnetic material is given in Equation (1.4) where M is a magnetic ion [29].

$$1/T_{1\,(Measured)} = 1/T_{1\,(Water)} + r_1[M] \tag{1.4}$$

Because most T_1 contrast agents are unable to cross the intact blood-brain barrier, they are not used for brain imaging. Studies have shown that T_2-weighted imaging is more sensitive for detecting brain pathology as well as for imaging edema. However, the use of functional MRI (fMRI), a branch of the MRI modality, is more commonly used to measure brain activity in response to specified stimuli. This technique allows study of the functions of the living brain in a noninvasive manner and is clinically used in the treatment of brain tumours [29].

1.5.2 Advantages and Limitations

MRI provides excellent tissue contrast and higher spatial resolution and is one of the best techniques for showing anatomical detail. It is very useful in early-stage diagnosis of diseases, especially for brain tumours, and in providing information on the biochemistry and metabolism of tissues. It does not require any ionising radiation, and simultaneous extraction of physiologic and anatomic information is possible. Spatial resolution can be increased by increasing the acquisition time without endangering patients, compared with radioisotope imaging techniques. fMRI allows noninvasive imaging of the brain without the use of any external contrast agents, but the images are of lower spatial resolution. Advances in technology design allows for a smaller receiver coil radius and high magnetic field strengths to improve signal to noise ratio and resolution. However, this gives rise to other technical challenges, such as artefacts and problems of physiological effects on human patients. Figure 1.17 schematically shows the basics of an MRI machine.

In terms of disadvantages, MRI is several orders of magnitude less sensitive than nuclear imaging modalities because only a small percentage of protons are able to absorb the radiofrequency energy to generate a data signal. Reliable signal amplification strategies as well as good contrast agents are thus required. A good contrast agent can improve the sensitivity of the signals because it alters the relaxation times of tissues in the area but only in regions in which the contrast agents are concentrated. MRI also requires a much larger dosage of contrast agents than, for example, radiotracers in nuclear imaging, making toxicity of the imaging agents a major area of concern in molecular design and development. Rigorous tests and clinical trials are thus required in order for MRI contrast agents to be approved by the FDA.

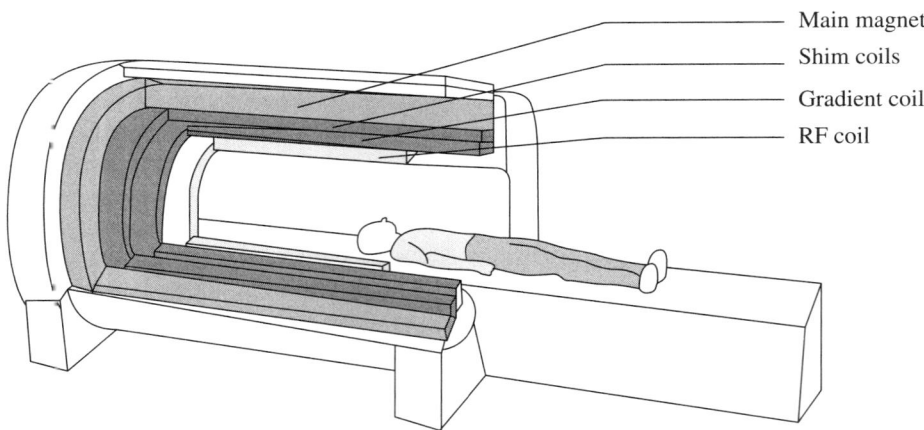

FIGURE 1.17   Components of a typical MRI scanner.

1.6 WHAT IS OPTICAL IMAGING?

Optical imaging techniques exploit different properties of light, such as absorption, emission, reflectance, scattering, polarisation, coherence, and fluorescence, as a source of contrast. Such contrast is created by the interaction of photons with different tissue or cellular components. Fluorescent and luminescent properties of light are generally used in *in vivo* studies. Therefore, this section will focus on fluorescence imaging. Although other optical imaging systems such as bioluminescence imaging, optical coherence tomography, photoacoustic microscopy, and tissue spectroscopy are not mentioned here (but are briefly discussed in Chapter 11), this does not mean that they play a less significant role in the development of optical imaging.

Fluorescence-based techniques are extremely valuable for studying cellular structures, functions, and pathways as well as molecular interactions in biological systems. These techniques utilise microscopy systems to observe and image at the microscopic and macroscopic levels, as well as molecular labels/dyes for contrast enhancement [30]. There are numerous types of labels and dyes for optical imaging; one example of a classic dye is shown in Figure 1.18.

1.6.1 Basic Principles

Fluorescence results from a process that occurs when certain molecules absorb light. These molecules are generally polyaromatic hydrocarbons or heterocycles known as fluorophores. They absorb energy at a particular wavelength and emit energy at a different but specific wavelength. This can be explained by a simple electronic state Jablonski diagram (Figure 1.19).

The electronic states of most organic molecules are either singlet states (S_1), when all electrons in the molecule are spin-paired, or triplet states (T_1), when one set of electron spins is unpaired. Upon excitation, the molecules move to a temporary excited state S_1, which can relax by numerous decay mechanisms to go back to its original ground state S_0. There are numerous decay processes; the two important types are nonradiative decay, where energy is lost by vibronic motions, or by radiative decay, in which energy is lost by fluorescence. Fluorescence is an important property that is manipulated in fluorescence microscopy. It can be quantified by monitoring the emission quantum yield and the emission lifetime (decay), which dictate the selection criteria for designing fluorophores. The fluorescence quantum efficiency of a fluorophore is the ratio of fluorescence photons emitted to the photons absorbed [31]. It is given as a quantum yield percentage where the quantum efficiency of emission is calculated using standard samples (which have a fixed and known fluorescence quantum yield value) according to Equation (1.5) [32]:

Cy5

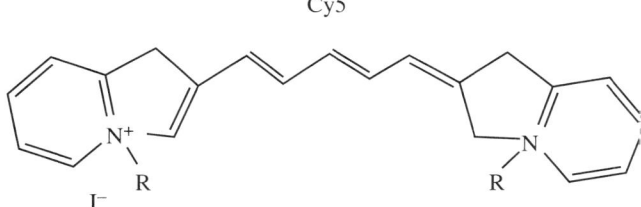

FIGURE 1.18   Cy5, a typical dye used in optical imaging.

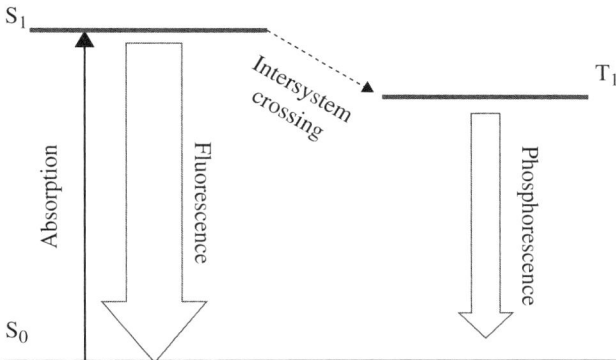

FIGURE 1.19   Jablonski diagram showing the basic principles of fluorescence.

$$\Phi_x = \Phi_{ST}\left[A_X/A_{ST}\right]\left[\eta^2_x/\eta^2_{ST}\right] \qquad (1.5)$$

where the subscripts ST and X denote standard and samples respectively, Φ is the fluorescence quantum yield, A the integrated area of emission band, and η the refractive index of the solvent. Φ can be used to assess the brightness of the fluorophore, which is the product of the quantum yield and the extinction coefficient of the fluorophore.

The advent of traditional fluorescence microscopy has initiated an important era in the study of living cells. Since then, many more creative engineering and sophisticated designs of microscopes have emerged, providing new and better optical imaging techniques for both academic and clinical studies. Simultaneously, it has also given rise to a new research area in the development of biological fluorophores, such as fluorescent proteins and GFP, which have provided insight into cellular structures and functions. Fluorophores improve or define image contrast, which is the difference between the highest and

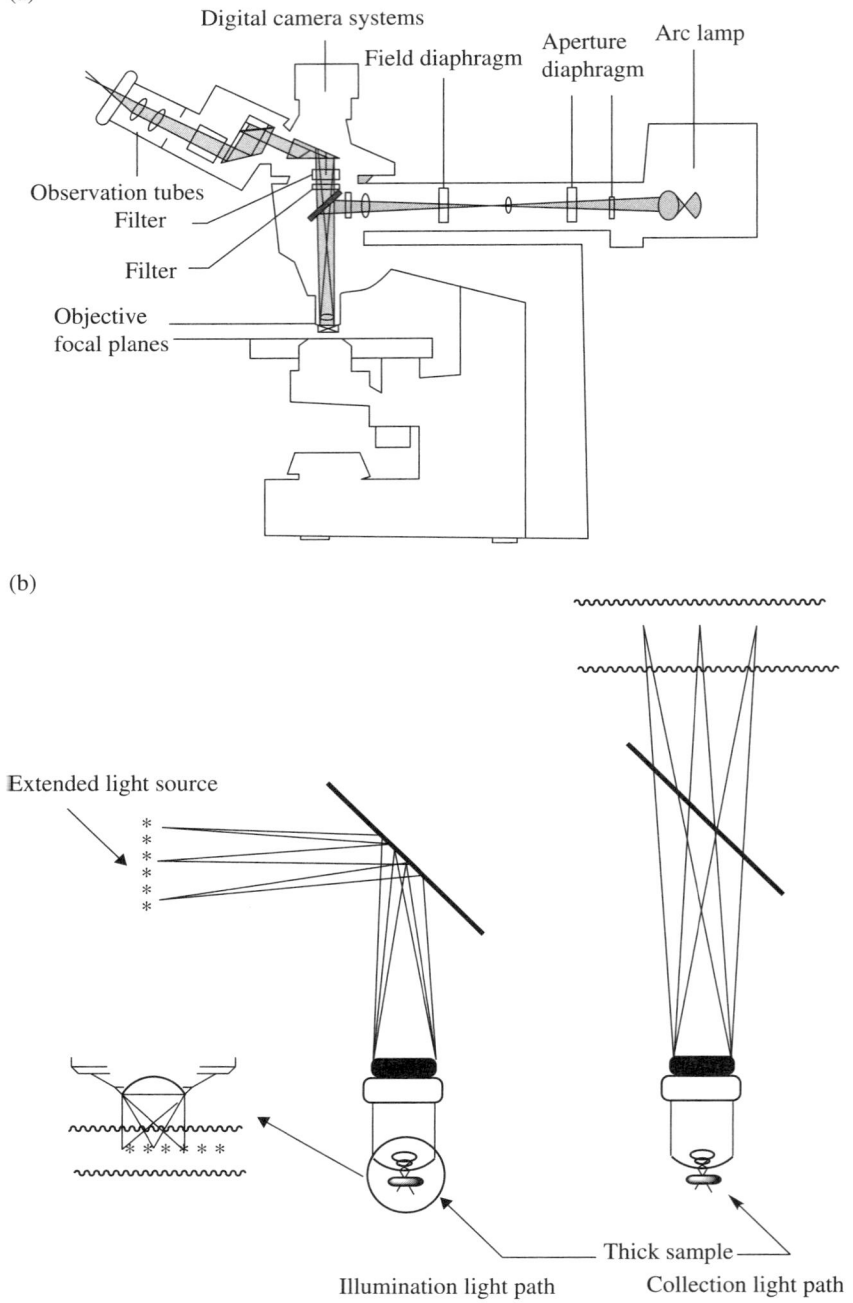

FIGURE 1.29 (a) A conventional fluorescence microscope and (b) the light path in a fluorescence microscope.

lowest signal intensity of two points in an image. This is important because without contrast, signal differences between background and the resolution of the optical lens cannot be differentiated [33]. All fluorescence imaging systems consist of the following key elements: excitation source; light delivery optics such as mirror; light collection and filtration optics; and light detection, amplification, and digitisation systems. There are numerous types of fluorescence microscope systems available; however, only two of the simplest and most common microscopes, conventional and confocal, will be discussed.

1.6.2 Conventional or Wide-Field Fluorescence Microscopy

In conventional fluorescence microscopy, also known as wide-field fluorescence microscopy, the excitation source excites the entire sample, which is lit up laterally and vertically (Figure 1.20). This causes interference and produces stray light, which decreases the resolution of the image. The excitation source is usually a mercury lamp that gives a window of various wavelengths at various intensities, the strongest excitation wavelengths being in the UV region, which are unsuitable for *in vivo* excitation. The window is quite broad, ranging from the UV to the visible red regions. Emission filters are necessary for wavelength selection.

Because of the mercury lamp excitation source, it is generally advantageous for imaging probes to have high molar extinction coefficients and high quantum yielding, especially for excitation in the continuous window between 450–540 nm where there is much lower intensity [34].

1.6.3 Confocal Microscopy or Confocal Laser Scanning Microscopy

In contrast, confocal microscopy uses point illumination where only a part or a point of the sample is excited at any one time. This technique utilises a pinhole in an optically conjugate plane in front of the detector to eliminate out-of-focus information and produce better quality images compared with wide-field images. In theory, only light from the focused focal plane reaches the detector. This is due to the attenuation of the light intensity, which rapidly falls off above and below the plane of focus as the beam converges and diverges. This reduces excitation of things that are out of the focal plane, hence eliminating a lot of unwanted background signals because these are deflected. Any out-of-focus light that enters the photo-detector normally has intensity that is too weak to be detected. Moreover, any point of light that is in the focal plane but not at the focal point will be blocked by the pinhole screen. This method, illustrated in Figure 1.21, is known as optical sectioning. Optical sectioning is affected by the size of the pinhole: The smaller the pinhole, the thinner the slice would become; but this is not definite because there are other influencing factors such as the wavelength of the light, numerical aperture of the lens, as well as reflecting index of the medium.

Three-dimensional images can be made from scanning many thin sections through the sample to create numerous optical sections that can be stacked together to produce an image. All these properties enable confocal microscopes to obtain better resolution. Generally, to obtain higher resolution images, a laser is used as the excitation source because it provides discrete wavelengths with very high intensities as well as a point light source of illumination. Depending on the laser system used, the desired wavelength can be selected, enabling a wider range of fluorophore probes/labels to be utilised [35–36].

By using laser sources, near-infrared fluorescence imaging is also possible. This is a less well developed technique that allows for deeper tissue imaging because at the excitation region of 650–900 nm, it allows for maximal tissue penetration but minimal autofluorescence. The near infrared region is also the region with the lowest absorption for haemoglobin, which is found in the blood and is responsible for absorbing the majority of visible light [37]. Optical imaging helps to increase the knowledge of entire biological pathways and accelerate a systems-wide understanding of biological complexity. In optical imaging high-affinity imaging agents/labels with appropriate pharmacokinetics are essential for imaging at the molecular level. This is because it is almost impossible to distinguish all cells, regardless of whether they are cancerous, from one another by *in vivo* imaging without labelling. There are numerous commercially available fluorescent labels to enhance the quality of optical imaging. These can be broadly categorised into genetic reporters, injectable imaging agents, and exogenous cell trackers [36–39].

1.6.4 Advantages and Limitations

The advantages and disadvantages of optical imaging methods are summarised in Table 1.3.

Although optical imaging methods are highly sensitive and relatively low cost, they have low spatial resolution (~1 mm) and poor depth penetration that is limited to several millimetres of tissues. Studies have shown that there is a 10-fold loss in photon intensity for every centimetre of tissue depth that the light penetrates, which leads to problems in signal quantification. Other typical problems are associated with absorption and light scattering. Thus, the biggest problem and challenge remains in manipulating this technique so that it can be used for opaque animal and hence, human studies. Sophisticated designs in microscopy and the use of different wavelength excitation sources, such as near infrared, have partially ameliorated this technique. Yet there are still too many limitations for it to be useful for clinical purposes, although it has been increasingly used for *in vivo* mechanistic and cellular studies.

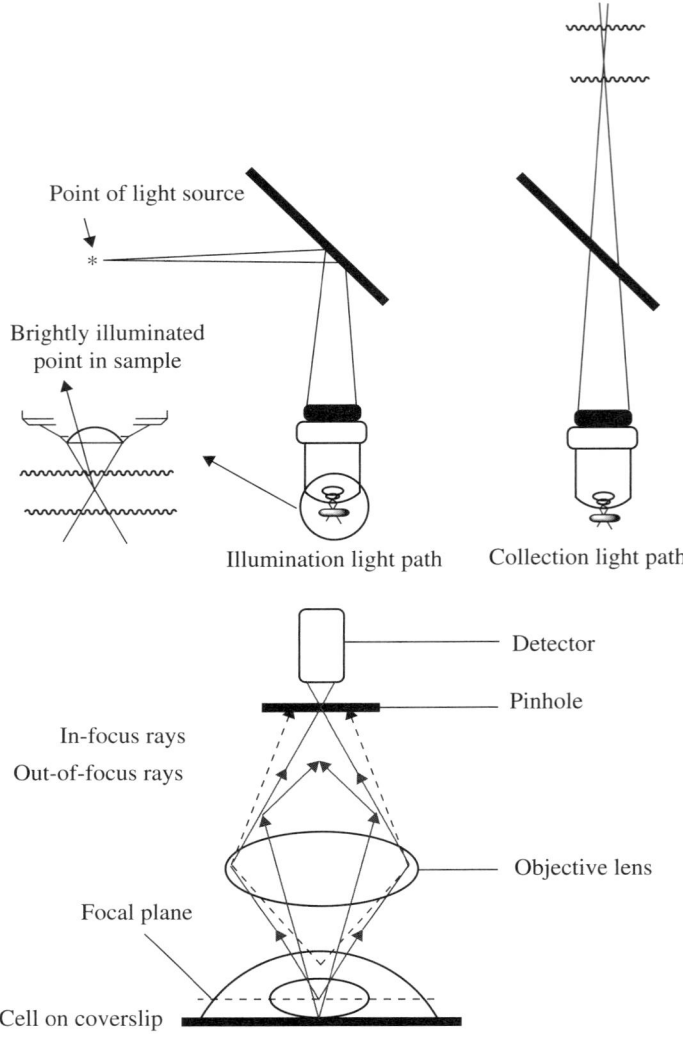

Point of light source

Brightly illuminated
point in sample

Illumination light path   Collection light path

Detector

Pinhole

In-focus rays
Out-of-focus rays

Objective lens

Focal plane

Cell on coverslip

FIGURE 1.21   The light path in a confocal microscope.

TABLE 1.3   Different Types of Optical Imaging Methods.

Imaging Methods	Maximum Depth	Target	Time	Primary Small Animal Use	Clinical Potential
Confocal microscopy	300 um	Physiological, Molecular	Sec/Min	Gene expression, reporter enzyme targeting, high sensitivity, and use in quantitative translational research	Experimental
Two-photon microscopy	800 um	Physiological, Molecular	Sec/Min	Gene expression imaging, multiple probes simultaneously	Experimental
Fluorescence reflectance imaging	5–7 mm	Physiological, Molecular	Sec/Min	Gene expression activatable, detects fluorochromes in live and dead cells	Yes
Diffuse optical tomography	15–20 mm	Physiological, Molecular	Sec		Yes
Fluorescence molecular tomography	0–15 cm	Physiological, Molecular	Sec	Quantitative imaging of targeted fluorochrome reporters in deep tissues	No
Bioluminescence imaging	2–3 cm	Molecular	Min	Gene expression, cell tracking, quick and easy	Experimental

1.7 WHAT IS ULTRASOUND (US)?

Ultrasound waves are longitudinal sound waves that oscillate back and forth. Sound waves travel at a fixed velocity and oscillate by compression and decompression of the medium through which they are travelling or transmitting [40]. In US imaging, sound waves are emitted as pulses that are partly reflected and transmitted from a boundary between two tissue structures and are detected as echoes. The reflections of these waves are dependent on the different acoustical impedance between the two tissues: The larger the difference is, the stronger the signal of the echo would be. The echo is caused by the depth of the tissue interface and is the measured time response for the echo to travel back (Figure 1.22) [42, 43].

1.7.1 Basic Principles

The ultrasound used in diagnostic applications has frequencies (1–12 MHz) higher than typical human hearing frequency ranges (15 000 ~ 20 000 Hz) [43]. Unlike sound waves with lower frequencies that could diffract around corners, ultrasound travels more in a straight line, like electromagnetic light beams, and thus will be reflected. They are however, still longitudinal waves (Figure 1.23).

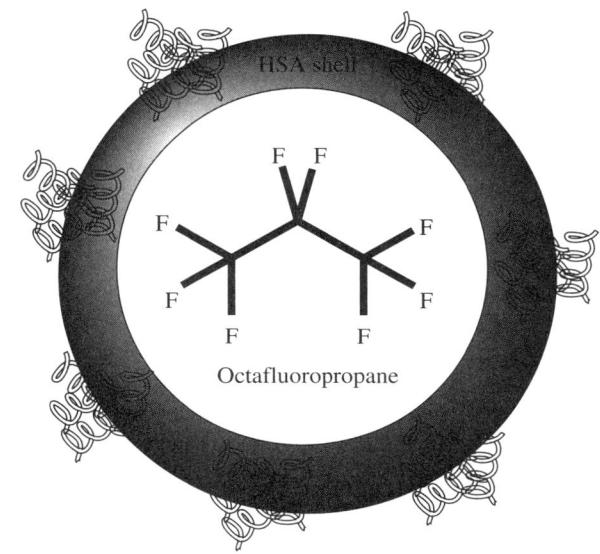

FIGURE 1.22 Microbubble, a typical contrast agent used in US.

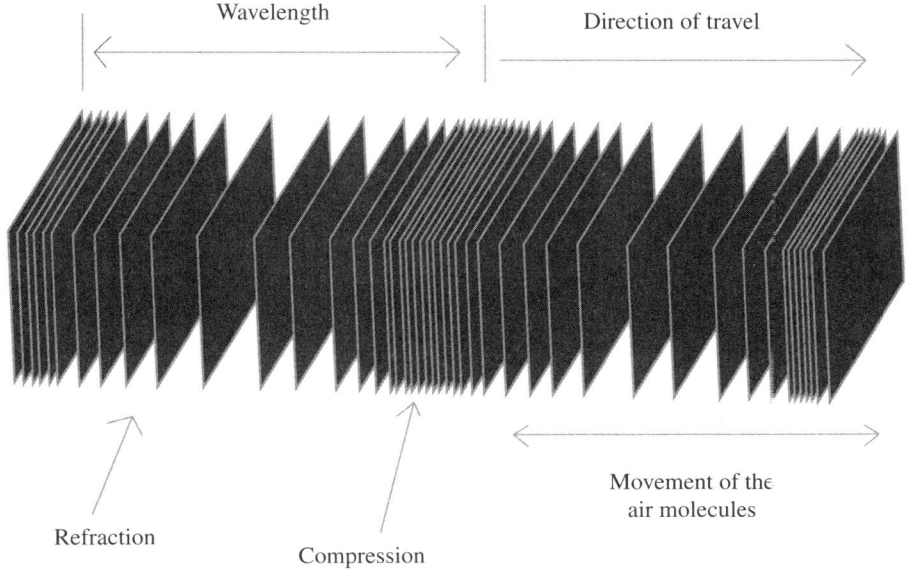

FIGURE 1.23 Basic properties of sound waves.

Because of this property, such waves can be used in diagnostic applications where they will be reflected by small objects. Yet it is also for this reason that US does not penetrate very deeply; this is a serious limitation for an imaging tool. Some general properties of sound waves will be mentioned briefly before discussing the use of US in imaging [44]. Wavelength λ is inversely related to the frequency f by the sound velocity c: Thus the velocity equals the wavelength times the number of oscillations per second (Eq. 1.6).

$$\lambda = c / f \Rightarrow c = \lambda f \tag{1.6}$$

Therefore, at a given temperature in a given material/medium, sound velocity is constant. Sound velocity varies due to the medium/material through which it is transmitting, and it is this property that is utilised in US imaging. This also means that simple gaseous media are problematic, because the sound waves cannot propagate easily through the medium. Therefore, ultrasound is unsuitable for imaging certain parts of the body, for example, the bowel, which is filled with air and organs that are obscured by the bowel. In general ultrasound imaging, only amplitude information is used in the reflected signal generated by an alternating current applied across piezoelectric crystals. These crystals are used in ultrasound probes to generate echoing signals to produce vibrations by compression and decompression. They also act as receiver of the reflected ultrasound. In ultrasound imaging, millions of pulses and echoes are transmitted and received every second. For each pulse emitted, the reflected signal is sampled multiple times. Different tissue structures reflect different amounts of emitted energy to produce signals with different amplitudes caused by the different depths of the structures. There are two different types of amplitudes: those from the transmitting pulses and those of the incoming pulses or signals that are a result of reflections produced from the sound waves hitting a surface structure. The energy of the amplitude of the reflected signals, as well as the incident, is known as the reflection coefficient, whereas the energy of the amplitude of the incident pulse and the transmitted pulse are called the transmission coefficient. These signals are affected by the difference in acoustic impedance of the different materials they are travelling through. The acoustic impedance of a medium is given by the equation $Z = c \times \rho$ and is defined as the speed of sound in a material × the density [45].

When ultrasound signals hit a surface, not all the signals are reflected directly back to the transmitter; often many are lost due to scattering via the nature of the reflecting surface. When ultrasound is scattered in multiple directions, the reflecting surfaces are known as scatterers. There are two types of scatterers: irregular and regular. These are dependent on the types of surfaces that the sound waves are subjected to (Figure 1.24). An irregular scatterer reflects only a small portion of the incoming sound wave back to the detecting probe. A regular scatterer reflects a larger portion of the sound waves back and is caused by reflecting surfaces that are perpendicular to the ultrasound beam. In general, the reflecting surface affects the

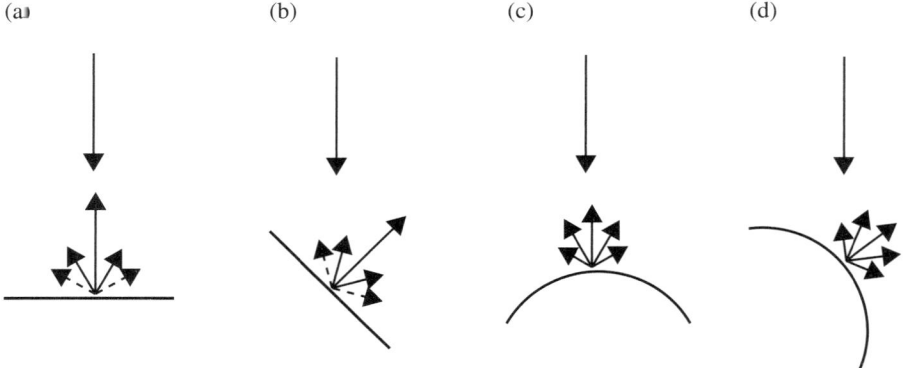

FIGURE 1.24   (a) An ideal surface, where most of the energy is reflected back to the transducer (high amplitude echo). (b) An ideal surface but at an angle of 45°, which will reflect most of the energy away from the surface (very low amplitude echo). (c) A curved surface that is a scatterer because it spreads out energy in all directions (low amplitude signal). (d) A curved surface that is perpendicular to the US beam; it is also a scatterer but more energy is reflected back to the beam.

size and direction of the signals, which in turn affects the type of scatters produced. It is important to note that the amplitude of the reflected signal, that is, the energy, is dependent on both the direction of the reflected signals and the reflection coefficient [44].

In ultrasound, the waves are attenuated when they are reflected and scattered in the tissue as well as when they are reflected and passed back to the probe. Typically, there is a 10% loss of total energy. The total energy decreases with penetration because there is an increase in energy lost the deeper the sound waves travel into the tissues because the sound waves are absorbed. Absorption in tissues is the most important cause of attenuation.

Absorption is extremely important in ultrasound for two reasons: depth penetration and safety concerns, the latter of which is the major limiting factor for its use as an imaging modality. The heat produced from absorption changes the temperature of the surrounding tissues, making it a limitation in ultrasound equipment due to safety issues for patients. The other is penetration issues because the attenuation of the US waves increases with depth. There are many factors that affect absorption such as the density of tissue and the frequency of the ultrasound beam. Good absorption generally results from high tissue density and sound frequency; hence penetration can be increased by increasing the transmitted energy. However, there are side effects such as tissue damage due to the extreme heat generated.

There are other types of ultrasound such as 3D ultrasound and Doppler ultrasound imaging based on the Doppler Effect. 3D imaging in general allows higher resolution imaging and thus provides more detailed information. This is commonly used to assess foetus development, as well as for biopsies. Doppler ultrasound is generally used for the study of the rate of blood flow through the heart and arteries [46].

There are many different ways to store ultrasound data, either in their full waveform as RF data that consist of both amplitude and frequency data or in pulse form where only data of the amplitude are collected and is less demanding on the storage systems. The signals can also be stored by taking the spectrum of frequencies from the reflected ultrasound pulse, which is then represented as a numerical value per image pixel. This method of storage is commonly used in Doppler imaging.

1.7.2 Advantages and Limitations

Ultrasound imaging is virtually noninvasive; it is used in a variety of clinical settings, especially in obstetrics and gynaecology, cardiology, and cancer detection. One of its most important uses is in studying and monitoring foetal development. No radiation is used in ultrasound imaging and the procedure can be performed much faster than X-rays and other radiographic techniques. Its major limitation is its poor penetration: US waves have difficulty penetrating the bone because they attenuate when passing deeper into the body. The body is also acoustically homogeneous because it contains around 70% water, thus it is difficult to discriminate the interface between the tissue and blood, but real-time evaluation of blood flow is possible. In general, ultrasound generates more heat as the frequency increases, so the ultrasonic frequency has to be carefully monitored. The signal intensities can be enhanced by the intravenous injection of contrast agents such as microbubbles at very low dosage, allowing the technique to remain minimally invasive.

Nevertheless, there are still safety concerns because the local temperature of tissue increases because heat is developed when the tissue or water absorbs the ultrasound energy. This local heat can also cause formation of cavitation.

Additionally, microbubbles have low circulation residence times and are easily taken up in certain locations—for example, the liver and the spleen—and can be destructed, inducing local microvasculature ruptures. Image enhancement can generally be improved by having high acoustic power output, but this again has to be compromised with the contrast agents, because high mechanical indices as well as low ultrasound frequencies tend to cause the microbubbles to burst. However, such properties can be beneficial for therapeutic purposes: Some studies have shown that the destructive nature of the microbubbles can be exploited for drug targeting and delivery. Despite concerns over the technique, US imaging is the most efficient technique with regard to cost, time, and safety compared to other imaging modalities such as MRI, PET, and SPECT, making it one of the most frequently used diagnostic techniques. Recent developments in ultrasound imaging have improved the resolution as well as the technique itself, allowing it to be incorporated into other methods such as photoacoustic imaging, a technique that uses the properties of both light and sound.

1.8   CONCLUSIONS

This chapter gives an overview of some of the most common imaging modalities and their basic principles. Their advantages and disadvantages are summarised in Figure 1.25, and a more detailed comparison is shown in Table 1.4. Each imaging modality has its own strengths and weaknesses for a particular area—for example, some techniques are more suited for cellular, molecular, or anatomic imaging—and in fact, many of these techniques are complementary. These detection systems differ in cost, availability, technical expertise needed, sensitivity, accuracy, and signal detection efficiency. Thus, the key issues depend on the type of research questions being addressed.

However, advancement in imaging technology has enabled the creation of multimodal imaging platforms, which has solved some of the dilemmas regarding the decision on what techniques should be used. At the same time, multimodal imaging helps to resolve time and cost issues. Multimodal imaging modalities are an attractive solution, especially in the area of diagnosis and in monitoring therapeutic responses; CT-MRI and CT-PET are already commonly used. The developments of new techniques are continuously being explored, so it is likely that multimodal imaging modalities will continue to be an area of interest, especially in the next decade. However, beneath this surface excitement around the advancement of new technology, a key influencing element in imaging is still in the area of developing molecular design and understanding their chemical properties. The demand for better imaging agents and probes parallels the development of new instrumentation. The need to overcome and comply with the endless hurdles concerning biological barriers and new mutagenic developments in diseases requires a constant need to review amplification strategies as well as to develop new molecular designs for better probes and contrast agents. There will be a continued search for creating specialised and specific probes propagated by endless, and as yet unknown, questions in areas such as oncology, physiology, and pathology.

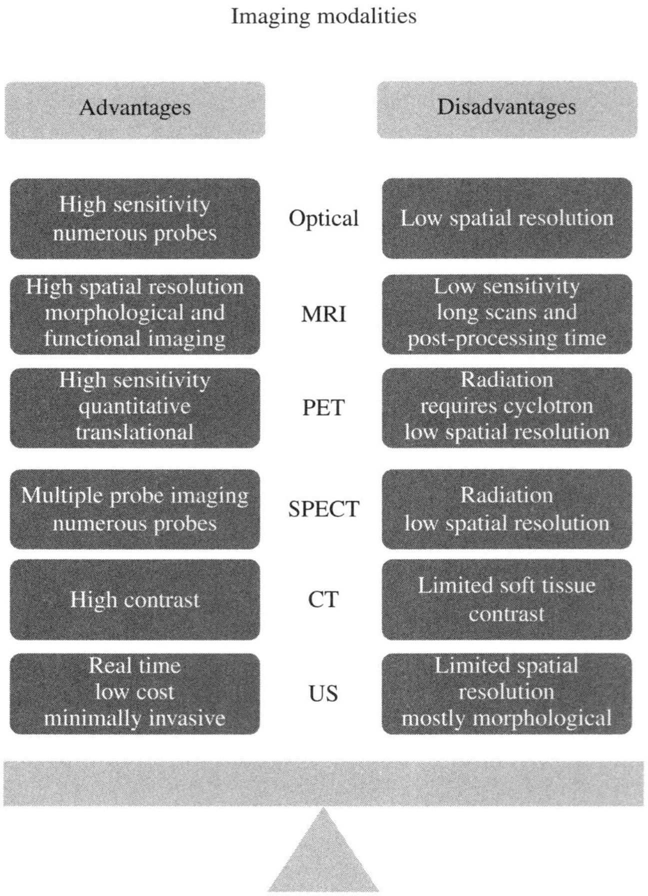

Imaging modalities

Advantages		Disadvantages
High sensitivity numerous probes	Optical	Low spatial resolution
High spatial resolution morphological and functional imaging	MRI	Low sensitivity long scans and post-processing time
High sensitivity quantitative translational	PET	Radiation requires cyclotron low spatial resolution
Multiple probe imaging numerous probes	SPECT	Radiation low spatial resolution
High contrast	CT	Limited soft tissue contrast
Real time low cost minimally invasive	US	Limited spatial resolution mostly morphological

FIGURE 1.25   Pros and cons of different imaging modalities.

TABLE 1.4 Comparison of the Different Imaging Modalities

Imaging technique	Optical	MRI	PET	SPECT	X-ray CT	Ultrasound	Use of current imaging modalities
Type of electromagnetic radiation energy used in image generation	Visible to near infrared	Radiowaves	High energy γ rays	Lower energy γ rays	X-rays	High frequency sounds	Molecular Metabolism
Spatial resolution	15–1000 nm	4–100 um ~1 mm-fMRI	1–2 mm	1–2 mm	12–50 um 50–200 um	50–500 um	
Depth	< 1 cm	No limit	No limit	No limit	No limit	mm to cm	Physiology
Scan time	Seconds	Minutes to hours seconds to minutes-fMRI	Minutes		Minutes	Seconds	
Type of molecular probe and quantity of probes used	Activatable, direct or indirect micrograms to milligrams	Activatable, direct or indirect micrograms to miligrams	Radiolabeled, director indirect nanograms	Radiolabeled, direct or indirect nanograms		Limited activatable, direct micrograms to milligrams	Anatomy
Imaging agents- Contrast agents and molecular probes (Most of the imaging agents can be found from the Molecular Imaging and contrast Agent Database (MICAD))	5-Carboxy-fluorescein, fluorescein, FITC, Alexa Fluor 488 and 647, 680, 750, Oregon Green 488 Cy5, Cy5.5 Cy7 Rhodamin X, Rhodamin Green, Luciferin, GFP IR-783, IR-786, IRDye78, IRDye800CW NIR2, Quantum dot, Single walled carbon nanotubes, Ln based molecular dyes	Gadolinium based contrast agents- Magnevist, Dotarem, Prohance, Gadovist, Optimark, Eovist, Vasovist/Ablavar. $Tm3+$, ^{23}C, ^{29}F, , Manganese ferrite, Iron oxides	Technetium-99m: used mostly for bone and heart scans ^{28}F-fluoromisonidazole/^{64}Cu-ATSM-identify hypoxia in tissues, gallium-attaches to area of inflammation ^{28}F-fluoride-imaging of new bone formation ^{28}F-fluorodeoxyglucose (FDG) images the metabolic activity of tissues. Can identify cancerous cells.	99mTc 186Re, 188Re 111In, 123I, 125I, 131I, 170Tm, 177Lu,	Iodine contrast agents-highlights blood vessels as well as tissues of various organs Barium-images of the abdomen and pelvis Gastrografin	Microbubbles/nanobubbles Optison microbubbles-perfluoropropane gas encapsulated by a serum albumin shell Dyes and Glod	
Key use	Visualisation of cell structures	Anatomical imaging and functional imaging of brain activity-fMRI	Metabolic imaging		Lung and bone tumour imaging	Vascular imaging	

REFERENCES

[1]   H. R. Herschman, *Science* **302**, 605–608 (2003).

[2]   G. L. ten Kate, E. J. G. Sijbrands, R. Valkema, F. J. ten Cate, S. B. Feinstein, A. F. W. van der Steen, M. J. A. P. Daemen and A. F. L. Schinkel. *J. Nuclear Cardiology* **17**, 897–912 (2011).

[3]   F. Dalagija, A. Mornjaković and I. Sefić, *Acta Medica Academica* **35**, 35–39 (2006).

[4]   M. D. Seemann, *Eur. J. Med. Res.* **28**, 241–246 (2004).

[5]   T. Ido, C. N. Wan V. Casella, J. S. Fowler, A. P. Wolf, M. Reivich and D.E. Kuhl, *J. Label. Compd. Radiopharm.* **14**, 175–182 (1978).

[6]   T. G. Tukington, *J. Nucl. Med. Technol.* **29**, 4–11 (2001).

[7]   M. M. Ter-Pogossian, M. E. Phelps, E. J. Hoffman and N. A. Mullani, *Radiology* **114**, 89–98 (1975).

[8]   N. Blow, *Nat. Meth.* **6**, 465–469 (2009).

[9]   O. Belohlavek, E Bombardieri, R. Hicks and Y. Sasaki, *A Guide to Clinical PET in Oncology: Improving Clinical Management of Cancer Patients*, 2008, Chapter **1**, 1–8.

[10] S. M. Ametamey, M. Honer and P. A. Schubiger, *Chem. Rev.* **108**, 1501–1516 (2008).

[11] A. K. Buck, S. Nekolla, S. Ziegler, A. Beer, B. J. Krause, K. Herrmann, K. Scheidhauer, H. J. Wester, E. J. Rummeny, M. Schwaiger and A. Drzezga, *J. Nucl. Med.* **49**, 1305–1319 (2008).

[12] K. Schwochau, *Angew. Chem. Int. Ed. Engl.* **33**, 2258–2267 (1994).

[13] G. F. Knoll, *Proceedings of the IEEE* **71**, 320–329 (1983).

[14] R. J. Jaszczak, *Physics in Medicine and Biology* **51**, R99–R115 (2006).

[15] M. M. Khalil, J. L. Tremoleda, T. B. Bayomy and W. Gsell, *Int. J. Mol. Imag.* 1–15 (2011).

[16] W. C. Lavely, S. Goetze, K. P. Friedman, J. P. Leal, Z. Zhang, E. Garret-Mayer, A. P. Dackiw, R. P. Tufano, M. A. Zeiger and H. A. Ziessman, *J. Nucl. Med.* **48**, 1084–1089 (2007).

[17] N. J. Dougall, S. Bruggink and K. P. Ebmeier, *Am. J. Geriatr. Psychiatry* **12**, 554–570 (2004).

[18] A. G. Filler, *Int. J. Neur.* **1**, (2010).

[19] E. C. Lasser, C. C. Berry, L. B. Talner, L. C. Santini, E. K. Lang, F. H. Gerber and H. O. Stolberg, *N. Engl. J. Med.* **317**, 845–849 (1987).

[20] D. J. Brenner and E. J. Hall, *N. Engl. J. Med.* **357**, 2277–2284 (2007).

[21] G. T. Herman, *Fundamentals of Computerized Tomography: Image Reconstruction from Projection*. 2nd ed. Springer, 2009.

[22] D. W. Townsend, *Annals Acad. Med.* **33**, 133–145 (2004).

[23] http://reference.medscape.com/features/slideshow/lung-cancer-staging

[24] R. V. Damadian, *Science* **171**, 1151–1153 (1971).

[25] K. J. Murphy, J. A. Brunberg and R. H. Cohan, *AJR Am J Roentgenol.* **167**, 847–849 (1996).

[26] W. R. Hendee and C. J. Morgan, *West. J. Med.* **141**, 491–500 (1984).

[27] D. Le Bihan, E. Breton, D. Lallemand, P. Grenier, E. Cabanis and M. Laval-Jeantet, *Radiology* **161**, 401–407 (1986).

[28] S. Aime, M. Botta, M. Fasano and E. Terreno, *Chem. Soc. Rev.* **27** 19–29 (1998).

[29] W. B. Amos, J. G. White and M. Fordham, *Appl. Opt.* **26**, 3239–3243 (1987).

[30] J. R. Lakowicz, *Principles of Fluorescence Spectroscopy*. 3rd ed. Springer, 2006.

[31] A. T. R. Williams, S. A. Winfield and J. N. Miller, *Analyst* **108**, 1067–1071 (1983).

[32] D. R. Sandison, D. W. Piston, R. M. Williams and W. W. Webb, *Appl. Opt.* **34**, 3576–3588 (1995).

[33] W. B. Amos and J G. White, *Biol. Cell* **95**, 335–342 (2003).

[34] J. B. Pawley, *Handbook of Biological Microscopy*. Springer, 2006.

[35] C. Preza, D. L. Snyder and J.-A. Conchello, *J. Opt. Soc. Am. A* **16**, 2185–2199 (1999).

[36] J. B. de Monvel, S. Le Calvez and M. Ulfendahl, *Biophysical* **80**, 2455–2470 (2001).

[37] J. W. Lichtman and J.-A. Conchello, *Nat. Meth.* **2**, 910–919 (2005).

[38] C. Kaminski and J. R. Soc. *Interface* **6**, S1–S2 (2009).

[39] F. Helmchen, *Exp. Phys.* 738–745 (2011).

[40] P. N. T. Wells, *Phys. Med. Biol.* **51**, R83–R98, (2006).

[41] P. N. T. Wells, *Rep. Prog. Phys.* **62**, 671–722 (1999).

[42] M. S. Hughes, G. M. Lanza, J. N. Marsh and S. A. Wickline, *MedicaMundi* **47**, 66–73 (2003).

[43] D. Cosgrove, *Eur. J. Radiol.* **60**, 324–330 (2006).

[44] K. G. Baker, V. J. Robertson and F. A. Duck, *Phys. Ther.* **81**, 1351–1358 (2001).

[45] R. S. C. Cobbold, *Foundations of Biomedical Ultrasound*. Oxford University Press, 2007.

2

CHEMICAL METHODOLOGY FOR LABELLING AND BIOCONJUGATION

LINA CUI AND JIANGHONG RAO

Molecular Imaging Program at Stanford, Departments of Radiology and Chemistry, School of Medicine, Stanford University, Stanford, CA, USA

2.1 INTRODUCTION

In molecular imaging, bioconjugation chemistry is crucial in the synthesis of imaging probes, as well as in the coupling of various imaging modalities with small molecules, macromolecules, or nanostructures. Preparation of imaging probes with short half-lives can benefit from the reactions of fast kinetics, such as various so-called 'click' reactions [1], while labelling of target molecules using imaging probes in living cells or organisms relies not only on the kinetics of the reaction, but the compatibility in different contexts or media [2–5]. Modification of target molecules, such as proteins, at selected residues enhances the precision of observed biological events probed by these target molecules [6–8]. Traditional bioconjugation methods for commonly seen functional groups, such as amines, thiols, hydroxyls, carboxyls, and phosphates, have been exhaustively discussed in Hermanson's *Bioconjugate Techniques* [9]. For instance, the amino group, a very prevalent functionality in various molecules or constructs, provides us with readily available conjugation sites via reactions with a plethora of reagents, such as carboxylic acids, frequently activated by succinimides [10], carbodiimides [11], or phosphonium/uronium reagents [12], isothiocyanates [13], acyl/sulfonyl halides [14], squarates [15], epoxides [16], and phosphoramidites [17]. The thiol group, another class of ubiquitous functional handle, is most commonly modified via halide substitution [18], epoxy/aziridine ring opening [19], thiol-maleimide reactions [20], disulfide bond formation [21], Michael addition [22, 23], and radical thiol-ene reactions [24]. Although these reactions can be very efficient in flasks, their use in cell media or living systems is prohibited due to the pervasiveness of amino and thiol groups.

This chapter will focus on highly efficient coupling reactions and their applications in various reaction media, as well as newly developed bioconjugation strategies that can be useful in the introduction of molecular imaging probes. We will highlight both chemical and biochemical conjugation approaches, specifically those that can occur in aqueous phase. Chemistry for specific imaging modalities will be covered in subsequent chapters.

2.2 CHEMICAL METHODS

2.2.1 Through Reactions with Aldehydes or Ketones

The studies of reactions between an aldehyde or a ketone and a hydrazine/hydrazide (producing a hydrazone) or an aminooxy group (producing an oxime) can be traced back to a century ago [9, 25]. These reactions have been used successfully in protein and peptide modification [26] or bioconjugation [27, 28], dendrimer synthesis [29], modification of mammalian cell surface [30, 31], as well as labelling of proteins inside bacterial cells [32]. Typically, to achieve sufficient coupling, these

The Chemistry of Molecular Imaging, First Edition. Edited by Nicholas Long and Wing-Tak Wong.
© 2015 John Wiley & Sons, Inc. Published 2015 by John Wiley & Sons, Inc.

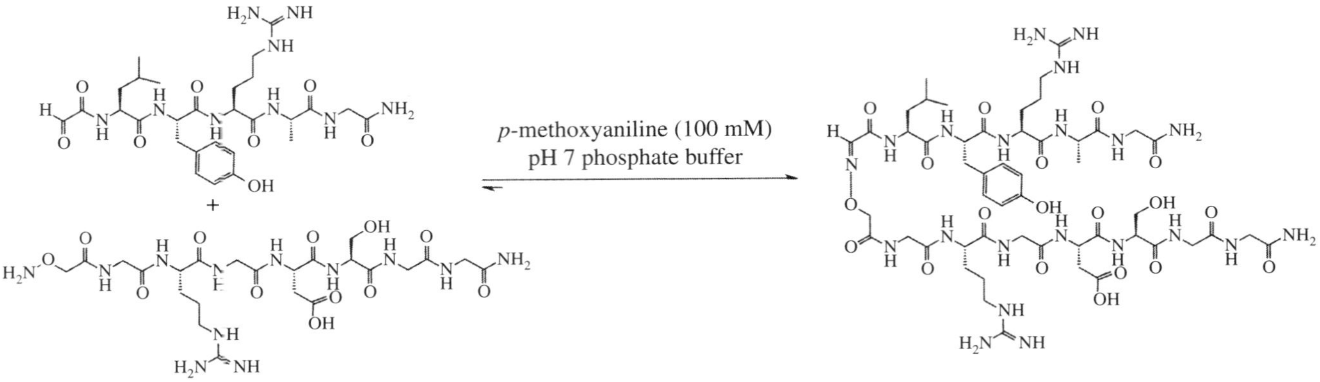

SCHEME 2.1   Oxime ligation between glyoxylyl-LYRAG and aminooxyacetyl-GRGDSGG (1 mM) at pH 7.0.

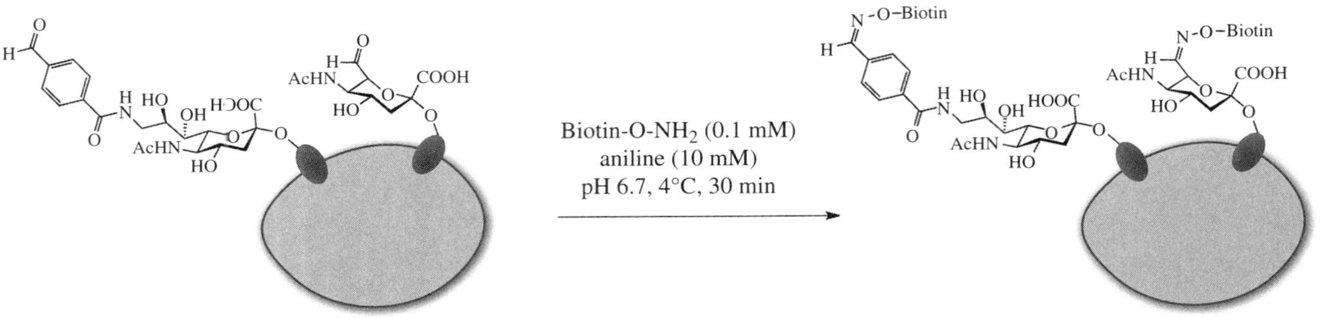

SCHEME 2.2   Oxime ligation on cell surface with sialic acid derivatives, generated by metabolic labelling or by chemical treatment with sodium periodate.

reactions are performed under acidic conditions (pH 4–5) at high concentrations with a large excess of one of the reagents. Built on prior work from Schiff, Cordes, and Jencks [33–35], Dawson and co-workers have found that addition of aniline as a nucleophilic catalyst (millimolar) greatly facilitated the oxime formation reactions at much lower concentrations at pH 7 at an observed rate constant of 0.061 $M^{-1}s^{-1}$ (Scheme 2.1) [36, 37].

One prominent example of the oxime ligation was carried on living cells, wherein probes were introduced to cell surfaces through reaction with aldehydes generated by oxidation of the glycans of cell surface glycoproteins (Scheme 2.2) [38].

The aniline-accelerated oxime ligation chemistry has been utilised in the synthesis of head-group functionalised phospholipids [39], chemoselective surface immobilisation of proteins [40], and incorporation of glycans or glycopeptides to gold nanoparticles [41]. The near-neutral conditions used in the ligation is of particular importance, while molecular probes need to be introduced to materials containing acid-degradable bonds, such as acetalated dextran (Ac-DEX) particles developed in the Fréchet's group [10]. Alkoxyamine-functionalised peptides and fluorescent molecules were conjugated to the Ac-DEX particles efficiently through reaction of alkoxyamine and aldehyde of reducing end of the polysaccharide at pH 7.4.

Similarly, hydrazone formation between the carbonyl group and hydrazine or hydrazide can also be accelerated in the presence of aniline [42, 43]. The utilities of aniline-catalysed hydrazone ligation have been demonstrated in a few examples – labelling of unprotected peptide with Alexa Fluor 488 [42], modification of CdSe-ZnS core-shell quantum dots (Q dots, or QDs) with peptides [44, 45] and immobilisation of aldehyde-modified antibodies to hydrazine-functionalised surface supports [46]. Generally, hydrazone bonds are not as stable as oxime linkages [47]. Nonetheless, the degradability of a hydrazone bond allows faster hydrolysis under acidic conditions than at physiological pH, making it a promising trigger-cleavable linkage in drug delivery [12, 13, 16, 17, 48].

2.2.2   Through Reactions with Azides

Azides are one of the most frequently used chemical tags for bioconjugation or labelling, due to the fact that they are not endogenously available in biological systems, very stable under physiological conditions, and react very efficiently and selectively once activated under various reaction conditions, such as the Staudinger ligation, Huisgen azide-alkyne cycloaddition, and azide-cyclooctyne reaction.

2.2.2.1 Staudinger Ligation The Staudinger reaction occurs rapidly between an azide and a triphenylphosphine to form an aza-ylide, which hydrolyzes spontaneously to generate a primary amine (Figure 2.1a) [49]. Bertozzi and co-workers redesigned the molecule by introducing a well-positioned electrophilic trap in the phosphine structure, which could capture the nucleophilic aza-ylide intermediate and form an amide bond upon hydrolysis (Figure 2.1b) [50]. The mild reaction conditions allowed the introduction of a biotin molecule to living Jurkat cell surface, which was metabolically engineered to present azido groups.

Following the original work on Staudinger ligation reactions, the Bertozzi group reported studies on 'traceless' Staudinger ligation to form an amide bond between the two components on the phosphine and azide side, without other additional atoms in the linkage (Scheme 2.3) [51]. Although only preliminary work was done to show the formation of amide bonds, the work certainly shone light on its application in peptide coupling and modification of cellular components such as proteins and glycans.

At about the same time, Raine and co-workers reported their successful coupling of two amino acids using a two-step traceless Staudinger ligation (Scheme 2.4) [52]. The first step involved a trans-thioesterification of amino acid thioester with phosphinothiol, and the second step resembled that used in Bertozzi's "traceless" Staudinger ligation.

Both nontraceless and traceless Staudinger ligations exhibit rapid kinetics and can be performed in living systems due to abiotic nature of their reagents. The ligation reactions have found their widespread use in the modification of glycans, proteins, lipids, and DNA molecules, in the context of small molecules, polymers, and solid supports. Past and recent examples

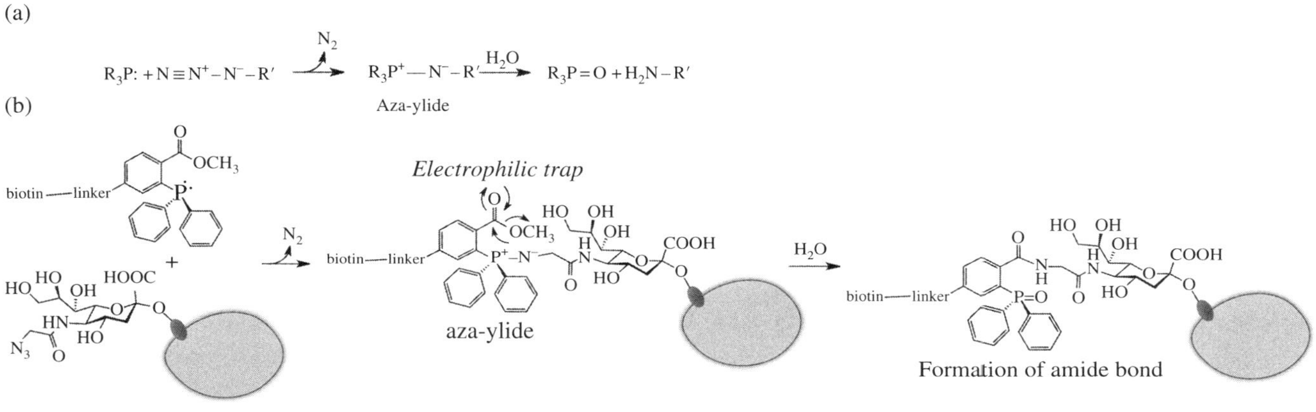

FIGURE 2.1 (a) Traditional Staudinger reaction. (b) Staudinger ligation used to label azide-containing cell surface glycans with biotin.

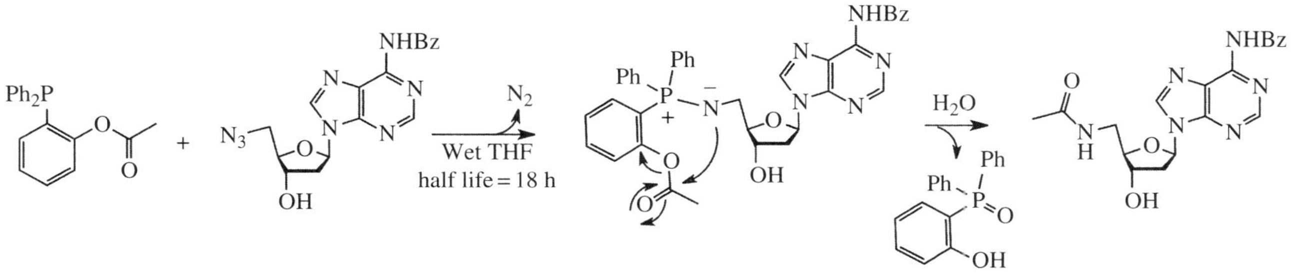

SCHEME 2.3 Traceless Staudinger ligation to form an amide bond.

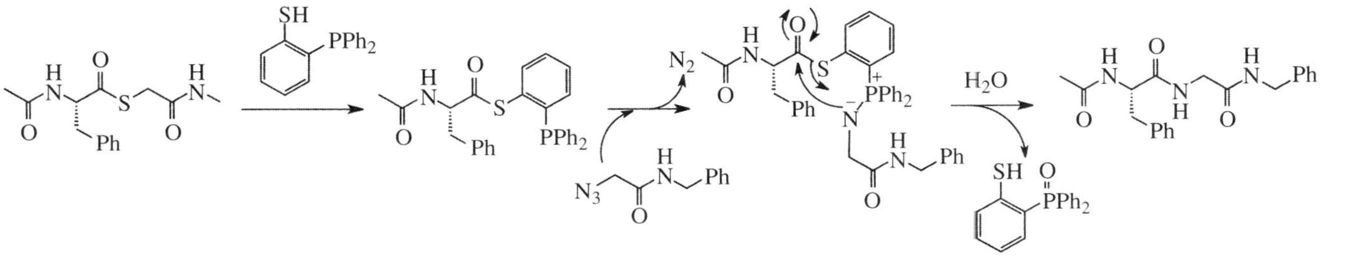

SCHEME 2.4 Traceless Staudinger ligation to form amide bond between two amino acids.

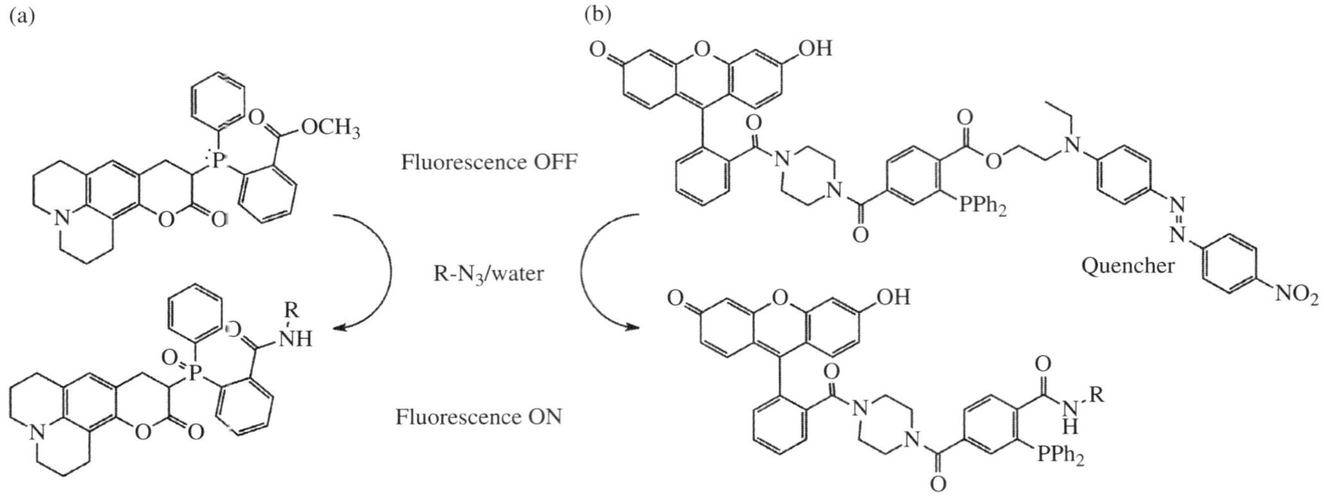

(a)   (b)

FIGURE 2.2   Fluorescence switch upon Staudinger ligation. Fluorescent was quenched by an electron lone pair from phosphorus (a) and a quencher (b).

utilising Bertozzi-Staudinger ligation have been reviewed extensively elsewhere [53, 54]. Here, we will focus on the application of this reaction in the introduction of molecular imaging probes.

Using the original Staudinger ligation approach, phosphine-conjugated fluorescent labels, Cy5.5, fluorescein, and rhodamine, were introduced to living cells by visualised cell surface glycans after metabolic labelling with azido groups [55]. Besides the direct fluorescent labelling of the azide-modified biomolecules, the Bertozzi group also developed 'switch-on' probes upon Staudinger ligation. One of the probes was a coumarin-phosphine conjugate, where the fluorescence of coumarin was quenched by the electron lone pair of the phosphine phosphorus (Figure 2.2**a**) [56]. Upon reaction with azido-protein, coumarin was attached to the protein and fluorescence was turned on. Another example employed a dye-quencher system, wherein a fluorescein analogue and Disperse Red 1 were attached to the same phosphine molecule (Figure 2.2**b**) [57]. Once Staudinger ligation occurred on the azidoglycan-expressing cell surface and was obtained after metabolic labelling, the quencher molecule Disperse Red 1 was cleaved off, turning fluorescence on.

Bioluminescence imaging of these engineered cell surface azido-glycans was evaluated by feeding phosphine-linked luciferin to luciferase-expressing cell lines [58]. Staudinger ligation released the free luciferin, which diffused into cells, where it served as a luciferase substrate leading to light emission. Although only visualisation of engineered cell surface glycan was reported, these strategies may find their way in the imaging of other azide-modified molecules in cells and whole animals.

Despite the widespread use of Staudinger ligation *in vitro* and in living cells, only a handful of reports are available for its application in medical imaging *in vivo*. Following Bertozzi's initial nontraceless Staudinger ligation work, a recent report applied intravenous injection of NeutrAvidin, labelled with either a far-red fluorophore DyLight649 or DOTA-^{111}In, after the same metabolic engineering and Staudinger ligation as that in Bertozzi's work [59]. The researchers found significant azido-labelled *N*-acetyl-mannosamine-dependent increase in tissue contrast, which was detected using optical imaging and single-photon-emission computed tomography (SPECT). Although this case showed great promise for Staudinger ligation in nuclear imaging, another group could not reproduce the successful use of this reaction in live animals [60]. In this study, researchers attempted to introduce radiolabels to tumours in a two-step approach, first pretargeting the tumour with azido-antibody followed by radiolabelling of antibody with phosphine derivatives via Staudinger ligation. The advantage of the approach is obvious, because it could reduce the radiation of a directly radiolabelled antibody, which has longer circulation half-life than radiolabelled phosphine probes, such as [89]Zr- and [67/68]Ga-labelled desferrioxamine-phosphines, ^{177}Lu-DOTA-phosphine and ^{123}I-cubyl phosphine probes. However, *in vitro* Staudinger ligation between the azido-antibody and phosphine probes was not efficient in the presence of serum, and a side product of phosphine was identified; no *in vivo* Staudinger ligation product was observed in the mouse model. Clearly, extension of the reaction in living system would require further investigation and optimisation.

2.2.2.2 *Azide-Alkyne Cycloaddition*   Huisgen 1,3-dipolar cycloadditions are exergonic reactions to condense two unsaturated molecules to yield various five-membered heterocycles [14]. Among them, azide-alkyne cycloaddition is arguably the mostly explored and used reaction; it was originally performed at elevated temperatures, giving a mixture of 1,4- and 1,5-cycloaddition isomers. The Sharpless group and the Meldal group independently discovered that copper (I) could greatly accelerate the reaction rate (by seven orders of magnitude compared with reactions without copper) at lowered temperatures leading to products with higher regioselectivity (1,4-isomer only) and often nearly quantitative yields (Scheme 2.5) [61, 62], making the azide-alkyne cycloaddition one of the most popularly used conjugation methods.

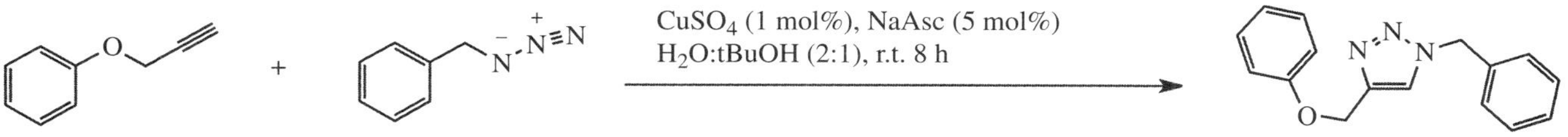

SCHEME 2.5   Copper (I)-catalysed azide-alkyne cycloaddition (CuAAC) to give only 1,4-isomer.

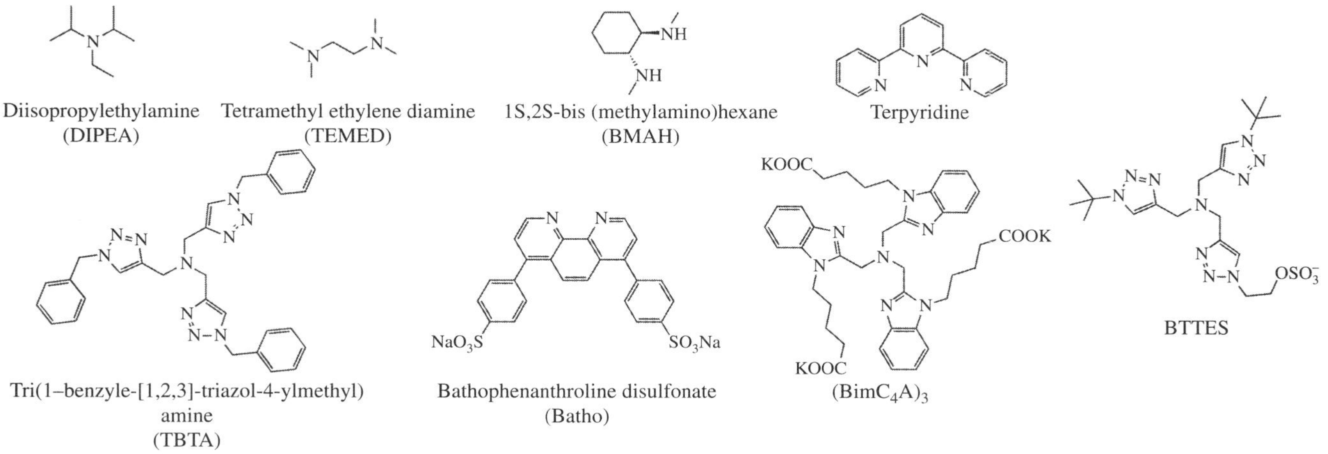

Diisopropylethylamine (DIPEA)

Tetramethyl ethylene diamine (TEMED)

1S,2S-bis (methylamino)hexane (BMAH)

Terpyridine

Tri(1–benzyle-[1,2,3]-triazol-4-ylmethyl) amine (TBTA)

Bathophenanthroline disulfonate (Batho)

(BimC₄A)₃

BTTES

FIGURE 2.3   Commonly used ligands in CuAAC.

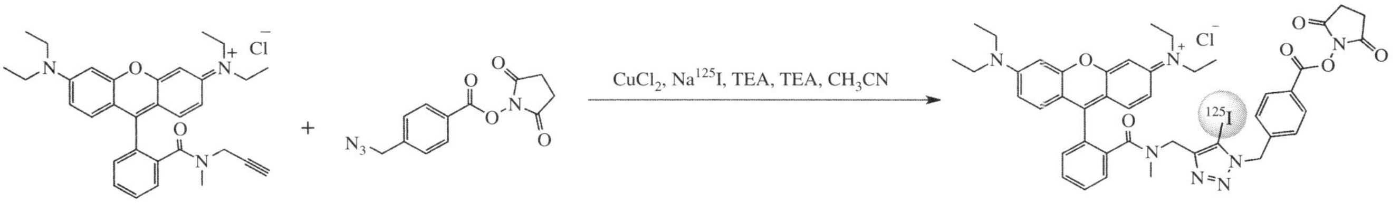

SCHEME 2.6   Incorporation of ^{125}I-label via CuAAC.

Addition of ligands (Figure 2.3), although not required in most cases, was found to accelerate the reaction with lowered copper concentration, presumably by stabilising copper (I), protecting it from oxidation and disproportionation, and thereby enhancing the catalytic activity of the copper (I) ion [63–67].

In the past a few years, the CuAAC has been studied and reviewed extensively, for its mechanistic analysis [68–70], reactivity and general applications [71–76], application in small molecule synthesis [77, 78], polymer and material sciences [79–83], synthesis of glycopolymers [20, 84], synthesis of peptidomimetics [85, 86], dendrimer construction [87, 88], preparation of ligand-bound liposomes, modification of nucleic acids [89–92], modification of solid surfaces [93–96], modification of polymeric nanoparticles [18, 97, 98], modification of carbon nanotubes [99], modification of virus particles [63, 100], and bioconjugation of peptide or proteins [101, 102].

To prepare molecular imaging probes, CuAAC has been used to introduce F-18 to small molecule inhibitors of target proteins [103–105], folates [106, 107], peptides or glycopeptides [108–114], proteins (via azido FDG) [115], and oligonucleotides [116–118], C-11[119] and Tc-99m can also be introduced to carrier molecules effortlessly [120–122]. Interestingly, in a recent report, I-125 was incorporated to the triazole ring during the cyclisation reaction when CuI was used as a copper source, leading to an easy one-pot synthesis of multimodal imaging probe (Scheme 2.6) [123]. A few MRI probes were also synthesised using CuAAC, which usually coupled chelators or ligands of contrast agents to a carrier molecule, followed by the addition of contrast metal salts [124–126]. In a recent report, pre-labelled Gd chelate was successfully introduced to dendrimers using CuAAC [127]. Surface modification of iron oxide nanoparticles has been very efficient with CuAAC as seen in the cases of introduction of biotin, fluorochrome, or steroid moieties [128], targeting groups [109], cell-penetrating ligands [129, 130], and antibodies [131]. Similarly, grafting of PEG to self-assembled polyelectrolyte nanoparticles was made promptly with CuAAC [132].

Most often, the CuAAC reaction has been performed with CuSO₄/ascorbate in either water or water/organic solvents or with CuI in organic solvents, such as THF, CH₃CN, DMF, DMSO, DCM, toluene, with or without ligands, at room temperature or

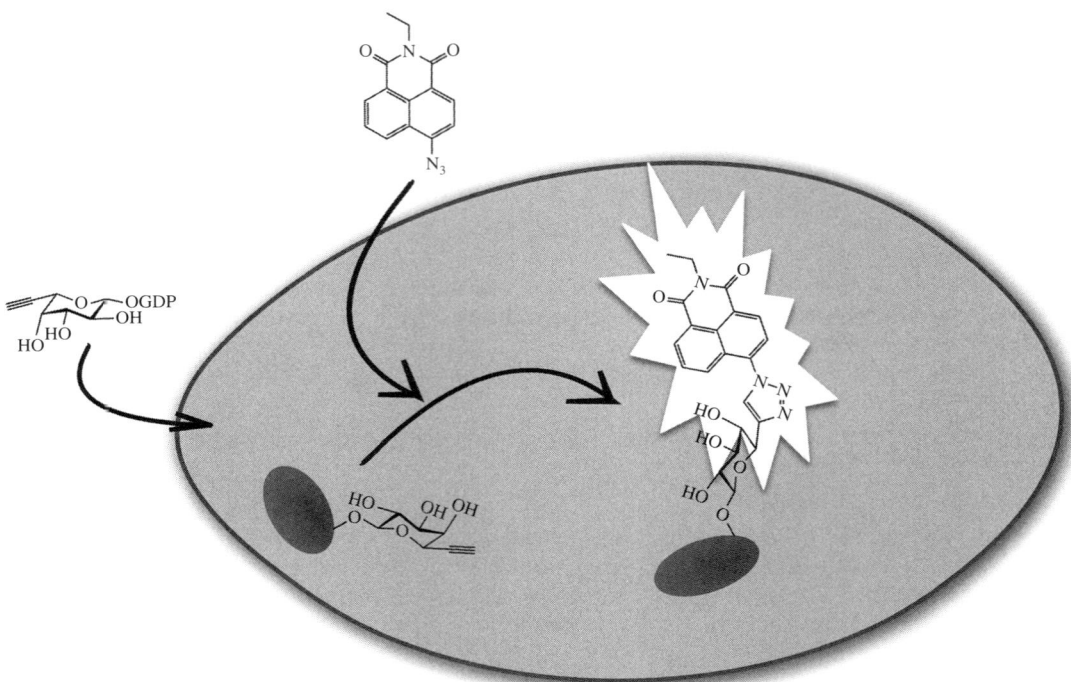

FIGURE 2.4 Fluorescent labelling of fucosylated glycans in cells via CuAAC. Propargyl or azide-modified fucoside was first incorporated to proteins inside cells via glycosylation pathways, and fucosylated proteins could then be imaged using azide or propargyl-modified fluorogenic probe 1,8-naphthalimide derivative, which turned on upon reaction with modified fucosylated proteins.

sometimes at slightly elevated temperatures [74]. However, the high concentration of copper ion used in the reactions becomes a concern in causing adverse side reactions or toxicity in living systems [133], therefore limiting the usage of CuAAC in living cells or organisms. Nonetheless, a few groups have reported the use of regular CuAAC in fixed mammalian cells. Wong and co-workers applied the reaction to probe fucosylated glycans in cells [134]. Fucosylation sites could be identified after incorporation of alkyne or azide modified GDP-L-fucose to glycoproteins by the addition of an azide or alkyne modified cell-penetrating fluorogenic probe in the presence of millimolar $CuSO_4$ and ascorbate (Figure 2.4).

In one report by Jao et al., CuAAC was performed in cells to image RNA transcription and turnover [135]. Incorporation of alkyne-modified uridine into newly transcribed RNA can be visualised by treating the cells with fluorescent azides. Dynamic labelling of DNA was achieved using alkyne-modified deoxyuridine [136]. Jao et al. also demonstrated the use of CuAAC in imaging incorporation of propargylcholine, an analogue of the most abundant head group choline, into phospholipids [137]. In another example, cells were metabolically labelled with either alkyne or azides (using artificial amino acids or carbohydrate) on the cell surface, and semiconducting polymer dots were then attached to cell surface through CuAAC using millimolar $CuSO_4$ and ascorbate [138].

Recently, after screening of a small library of analogues of TBTA, a commonly used ligand to assist CuAAC, Wu and coworkers have identified one tighter binding ligand, therein termed as BTTES, which required only micromolar copper for efficient labelling in living cells within minutes, without apparent toxicity after days (Figure 2.3) [67]. Using the CuAAC-BTTES system, *in vivo* imaging of fucosylated glycans during zebrafish early embryogenesis was achieved for the first time. This approach may open a new era for *in vivo* imaging of living systems using CuAAC, which was not possible to achieve previously.

2.2.2.3 Cyclooctyne-Azide Cycloaddition

In order to render the azide-alkyne cycloaddition biocompatible inside cells without disturbing normal cellular functions, efforts have been made to optimise the reaction. Ju and co-workers found use of electron-deficient internal or terminal alkynes allowed the azide-alkyne cycloaddition to complete under biological conditions in hours without any catalysts, such as copper [139]. However, under the same conditions, no reaction occurred for the alkynes without any neighbouring electron-withdrawing group. Although not perfect, this approach made it promising to achieve biocompatible azide-alkyne cycloaddition.

Besides tuning the electronic effects, release of strain energy can be an alternative approach. Inspired by the discovery by Wittig and Krebs that cyclooctyne, the smallest of the stable cycloalkynes, and phenyl azide can undergo explosion-like

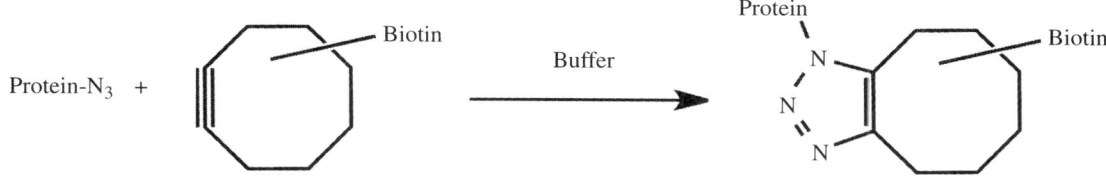

SCHEME 2.7   Strain-promoted azide-alkyne cycloaddition.

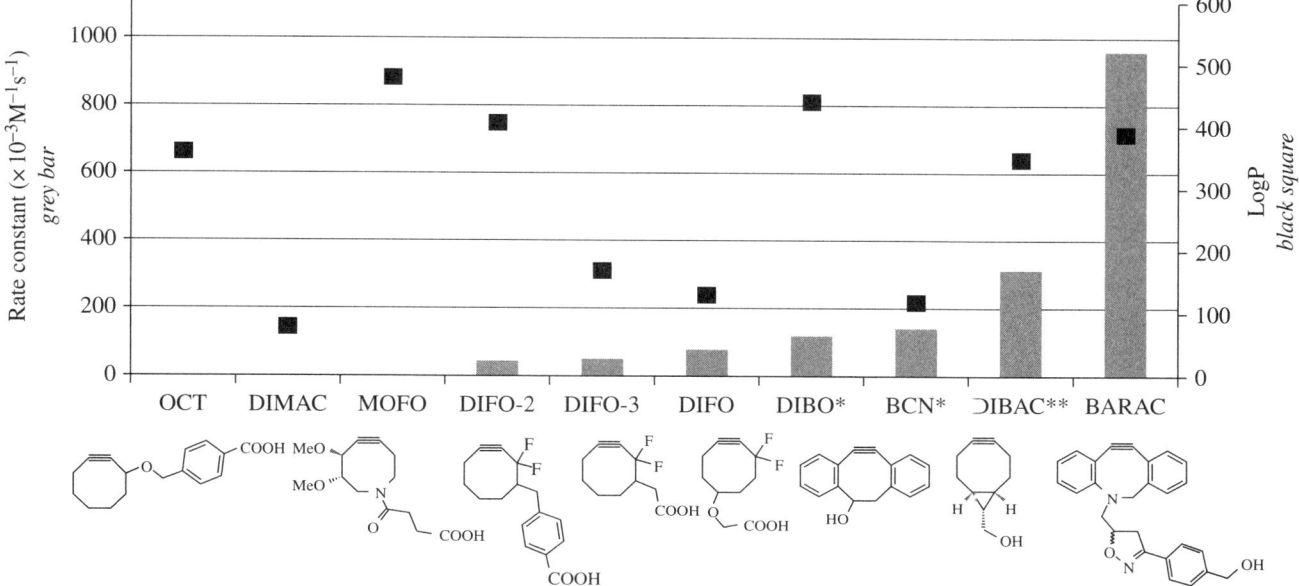

FIGURE 2.5   Reactivity (second-order rate constant) and lipophilicity (logP) of various cyclooctynes. Data was plotted from values in references. Reactions were in acetonitrile, acetonitrile-water (*) and methanol (**).

reaction to give triazole as a single product [140], Bertozzi and co-workers developed the 'copper-free' strain-promoted azide-alkyne cycloaddition as a bioconjugation method for use under physiological conditions (Scheme 2.7) [141, 142]. Azido-proteins either in buffers or on a living cell surface could be labelled efficiently in 1 h under ambient conditions, and the reaction was more efficient than Staudinger ligation (around two-fold).

To enhance the efficiency of the cyclooctyne-azide reaction, electron-withdrawing groups were introduced into cyclooctyne, leading to their revolutionary cyclooctyne – difluorinated cyclooctyne (DIFO) (Figure 2.5) [143]. The combined effects of ring-strain release and favourable electronic properties greatly accelerated the cyclooctyne-azide cycloaddition to be comparable with CuAAC in protein labelling. Its second-order rate constant was measured to be 7.6×10^{-2} M^{-1}s^{-1} (with benzyl azide), which is 17 to 63 times faster than the Staudinger ligation or previous strain-promoted cycloadditions. Labelling of azido-glycan on the live cell surface was observed after 1 minute reaction with the DIFO fluorescent probe. Besides the highly sensitive azide labelling, DIFO-azide cycloaddition did not show cellular cytotoxicity as assessed by morphology and propidium iodide staining. These features allowed the dynamic imaging of glycan internalisation and sub-cellular partitioning in living cells, and its compatibility with glycan trafficking in the time examined was verified by two-colour imaging using DIFO tethered fluorescent probes. In zebrafish embryos, the DIFO fluorescent probe was applied to label the metabolically generated azido-glycans, fluorescence was observed after 1 minute reaction with DIFO probe, and the intensity increased with reaction time [144]. Using multicolour DIFO fluorescent probes during zebrafish embryogenesis, spatiotemporal glycan expression was visualised to reveal its differences in the cell-surface expression, intracellular trafficking, and tissue distribution in the embryo. The initial breakthrough in imaging glycans in zebrafish embryo by cyclooctyne-azide cycloaddition led to successful studies of glycan expression and trafficking in living *Caenorhabditis elegans* [145], in zebrafish during early embryogenesis [146, 147], and in live mice [148].

Although very efficient and biocompatible, the use of DIFO may be limited due to its time-consuming synthesis. The Bertozzi group reported a more tractable synthetic route with a 20-fold increase in overall yields, and the products were comparable to the originally reported DIFO in terms of reaction efficiency and low cytotoxicity [150]. Boons and co-workers reported the use of dibenzocyclooctynes (DIBO) to detect azide, and they exhibited comparable reactivity to DIFO. Without electron withdrawing groups, the dibenzo system achieved similar kinetics as DIFO, probably through increased strain energy. Researchers from the Netherlands have reported a bicyclo[6.1.0]nonyne (BCN) as a novel ring-strained alkyne [151]. The synthesis of BCN conjugates was highly straightforward from cyclopropanation of 1,5-cyclooctadiene, and the BCN probes showed around twofold better kinetics compared with DIBO in the cycloaddition reaction with azides. It has been used in the labelling of proteins and glycans, and in the three-dimensional visualisation of living melanoma [151]. The so-called BARAC, biarylazacyclooctynone, introduced amide in the cyclooctyne ring adding even more strain energy through its resonant structure (Figure 2.5) [152]. The easily synthesised (seven-step) BRACO, in fact, was around threefold more sensitive to the azido group than DIFO, making it promising for in vivo labelling. Interestingly, a fluorogenic cyclooctyne was prepared based on the parent structure of BARAC and gave 10-fold enhancement in fluorescence quantum yield upon triazole formation with organic azides. This may find its use in real-time imaging of azide-labelled biomolecules [153]. To overcome the possible bioavailability problem caused by the hydrophobic nature of DIFO, hydrophilic azacyclooctyne, therein termed DIMAC, was synthesised in nine steps from glucose (Figure 2.5) [154]. The reactivity of DIMAC was similar to that of the nonfluorinated cyclooctynes. Examples using monofluorinated cyclooctyne (MOFO) [142] and dibenzoazacyclooctyne (DIBAC) [155] were also examined. Besides reactivity, lipophilicity of these compounds is another parameter for selection of proper reagents (Figure 2.5), because lipophilic compounds have poor solubility in water, and may have hydrophobic interactions with proteins in the reaction mixture such as that in cells [149]. Along with the frequently used Alexa Fluor dyes and FITC in Bertozzi's and other groups, BODIPY-cyclooctyne conjugates [156] and luminescent quantum dot conjugates [157] were also applied for labelling in live cells.

Besides in vivo labelling, the highly efficient 'copper-free' azide-alkyne cycloaddition can be used simply as a bioconjugation strategy ex vivo, such as in situ cross-linking of photodegradable polymeric networks using bifunctional, fluorinated cyclooctynes [158], cross-linking of communication-mediating (COM) domains of nonribosomal peptide synthetase (NRPS) proteins to study their protein-protein interactions [159], as well as dendrimer synthesis [88, 160]. Monofluorocyclooctyne functionalisation of radionuclide chelators, such as DOTA and NOTA, has been used to introduce radiolabels such as In-111 to azide-modified peptides via this copper-free click chemistry [161, 162]. F-18 radiolabels were also introduced to bombesin, a 14-amino acid neuropeptide that binds to gastrin-releasing peptide receptor, for diagnosis and imaging of cancer [163].

Similar to the cyclooctyne-azide click chemistry, strain-promoted cycloadditions of cyclic nitrones with cyclooctynes are also very efficient, or even faster in some cases than the reactions of azides [164, 165] and can also be useful tools in bioconjugation.

2.2.3 Through Reactions with Alkenes

2.2.3.1 Diels-Alder Reaction with Tetrazine   1,2,4,5-Tetrazine or s-tetrazine can react with electron rich dienophiles at room temperature via inverse electron-demand Diels-Alder reactions, followed by a retro-[4+2] cycloaddition to release N_2 as the only byproduct [166, 167]. Later, Sauer studied the kinetics of inverse electron-demand Diels-Alder reactions between electron-deficient tetrazines with various linear and cyclic dienophiles, among which the highly strained trans-cyclooctene was the fastest (seven orders of magnitude faster than the cis-cyclooctene) (Scheme 2.8) [168]. In protic solvents, the 4,5-dihydropyridizine rapidly rearranges to its isomer.

However, the tetrazines in Sauer's work underwent rapid hydrolysis in water. Fox and co-workers replaced the strongly electron-withdrawing ester or trifluoromethyl groups with aromatic groups, achieving more stable tetrazine derivatives in aqueous media, and made this extremely efficient reaction promising for bioconjugation (Scheme 2.8) [169]. The reactions are orthogonal to a variety of functionalities and proceed in most organic solvents, water, cell media, or cell lysates in high yields. For instance, in methanol/water mixture, the reaction between 3,6-di-(2-pyridyl)-s-tetrazine and trans-cyclooctene was extremely fast at a second-order rate constant of 2000 $M^{-1}s^{-1}$ with quantitative yield, while only micromolar reagents were needed. Recently Fox et al. reported a group of even more reactive trans-cyclooctenes [170]. Use of the

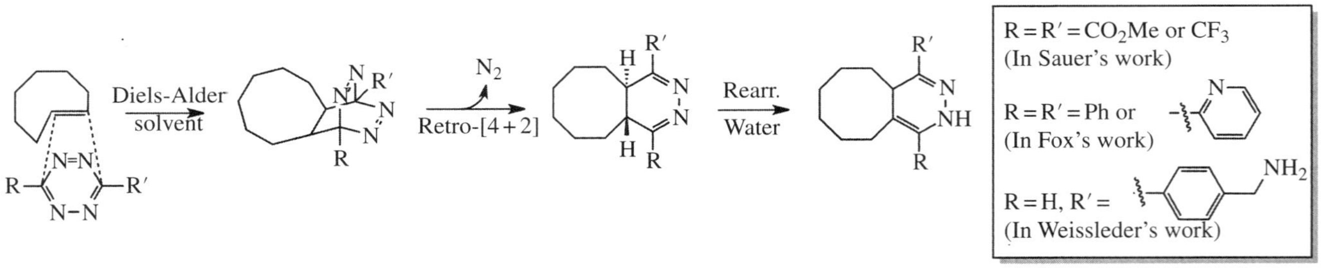

SCHEME 2.8   Strain-promoted Diels-Alder reaction between tetrazine and trans-cyclooctene (TCO).

tetrazine-*trans*-cyclooctene reaction as a ligation method can be very practical in the sense that the starting tetrazine derivative can be synthesised in several steps from commercially available materials, and *trans*-cyclooctene derivative can be generated from its *cis*-cyclooctene analogue using a one-step photochemical reaction reported by the Fox group [171].

Weissleder and co-workers discovered a more stable monosubstituted tetrazine derivative for biomedical application, after noticing some degradation of 3,6-di-(2-pyridyl)-s-tetrazine (Scheme 2.8). Importantly, they found that a dye-labelled tetrazine showed even faster kinetics while reacting with *trans*-cyclooctene modified antibody, at a second order rate constant of 6000 $M^{-1}s^{-1}$ at 37°C. The utility of the reaction for live-cell labelling and in serum-containing media was demonstrated using a pre-targeted labelling protocol. A549 lung cancer cells overexpressing EGFR were pretargeted by anti-EGFR antibody cetuximab labelled with *trans*-cyclooctene and rhodamine, followed by subsequent ligation with a NIR dye (VT680) labelled tetrazine. Efficient coupling was achieved within 30 min with a submicromolar tetrazine probe, which co-localised with the rhodamine signal, showing the specificity of the tetrazine-probe for *trans*-cyclooctene tethered to an antibody (Scheme 2.9). Interestingly, the extent of the reaction is proportional to the number of *trans*-cyclooctenes on the antibody, as shown by flow cytometry and confocal imaging. Similarly, magnetofluorescent nanoparticles attached to tetrazine could also be introduced to live cell surface using the pretargeted labelling protocol [172]. Compared to the traditional preassembled antibody-nanoparticle approach, there are several obvious advantages: It is easier to conjugate antibodies with *trans*-cyclooctene than with nanoparticles; maximisation of the nanoparticles on an antibody is possible, while higher loading of nanomaterials on antibodies tends to lead to aggregates and precipitates; an antibody binds to cell surface antigens before the conjugation with nanoparticles, decreasing the chances of blocked binding sites by bulky nanoparticles. This antibody pretargeted labelling protocol was also used to label intracellular molecules [173].

Weissleder and co-workers also explored the reaction between tetrazine and norbornene as a new bioconjugation strategy (Scheme 2.10) [174]. To solve the stability issue of tetrazines in water [175], they incorporated a benzylamine into tetrazine, making it very stable in a buffer and moderately stable in serum. Although the kinetics of this reaction was not as good as the *trans*-cyclooctene-tetrazine ligation, it showed a second-order rate constant of 1.9 $M^{-1}s^{-1}$ and 1.6 $M^{-1}s^{-1}$ in an aqueous buffer and serum respectively. The reaction was sufficiently fast in live cell imaging using the aforementioned pretargeted labelling protocol. Efficient coupling was achieved within 30 minutes with the tetrazine probe at the micromolar scale, and the tetrazine-probe showed specificity for norbornene attached on antibody.

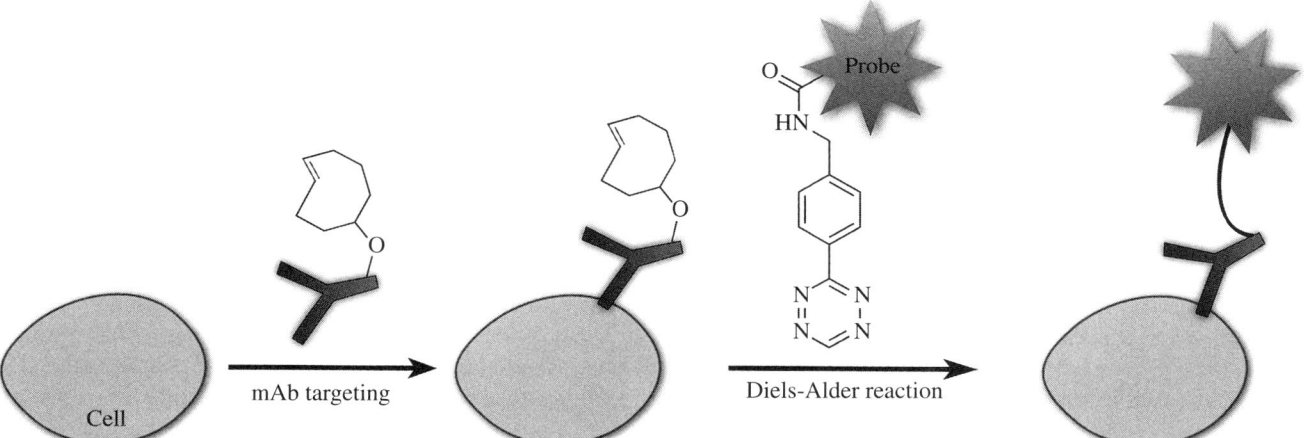

SCHEME 2.9 Live cell pre-targeting. Step one, *trans*-cyclooctene-modified monoclonal anti-EGFR antibody was targeted to EGFR on cell surface. Step two, the pre-targeted cells were labelled with tetrazine-modified fluorophores or other imaging probes.

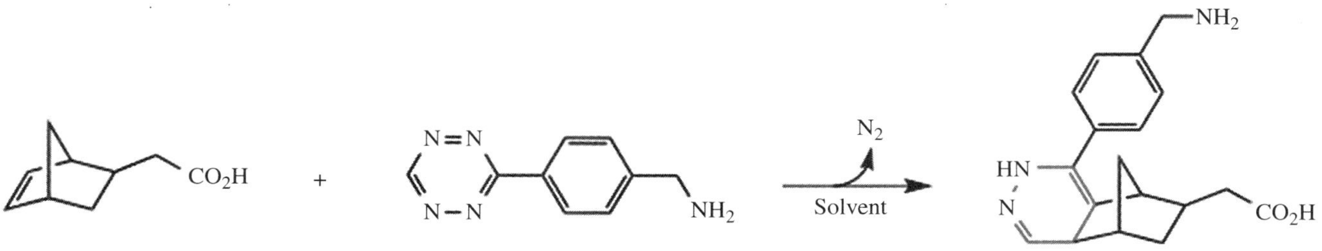

SCHEME 2.10 Conjugation reaction between tetrazine and norbornene.

Since the initial reports of the two types of tetrazine Diels-Alder reactions in bioconjugation, there have been many reports on the development and biomedical applications of the reactions [176], such as *in vitro* labelling of quantum dots [177] and nucleic acids [178, 179], modification of polymer [180], labelling of live cell surface antigens using small dye molecules or nanoparticles [172, 174, 181], imaging of intracellular small molecules [173, 182], and *in vivo* imaging [183–186]. The first application of the tetrazine-*trans*-cyclooctene ligation for *in vivo* imaging was performed using a pretargeted labelling protocol (similar to Scheme 2.9) [183]. Anti-TAG72 (TAG72 – a biomarker overexpressed in a wide range of solid tumours) monoclonal antibody CC49 was first tethered to *trans*-cyclooctene, and DOTA was conjugated with the Fox type tetrazine. In a PBS buffer, the cycloaddition finished within 3 minutes at 1.67 μM and the second-order rate constant was found to be 13090 M^{-1}s^{-1} at 37°C, very promising for *in vivo* labelling. Faster kinetics than the original report by Fox were expected, because a large increase of the reaction rate in aqueous media was seen in many other cases. Very low nonspecific binding of DOTA(In-111)-tetrazine conjugate to unmodified CC49 or other media constituents was observed, and the probes showed decent stability in live mice (75% remained after 24 hrs). *In vivo* administration of the *trans*-cyclooctene-modified CC49, followed one day later with injection of DOTA(In-111)-tetrazine, resulted in significant localisation of radioactivity in the tumour, as imaged by SPECT/CT of live mice three hours later. Only a background signal was observed in tumours in the mice administered with unmodified CC49 and DOTA(In-111)-tetrazine or with *trans*-cyclooctene-modified nonspecific antibody to TAG72 and DOTA(In-111)-tetrazine.

The Fox group verified the applicability of the tetrazine-*trans*-cyclooctene ligation to introduce an F-18 label [187]. Weissleder and co-workers took it one step further and modified inhibitor AZD2281 [188] of poly-ADP-ribose-polymerase 1(PARP-1) with F-18 (Scheme 2.11) [189]. This is advantageous because direct F-18 labelling of the AZD2281 is challenging and usually leads to low radiochemical yields and purity. F-18-AZD2281 showed targeted accumulation in mice with breast cancer tumour xenografts that overexpressed PARP-1 [184]. Using the tetrazine-*trans*-cyclooctene ligation, F-18 was also attached to RGD peptide with high efficiency (5 min > 90% yield). The F-18-RGD conjugate showed superior tumour uptake *in vivo*, with specificity for α$_v$β$_3$ integrin [185]. Tetrazine-norbornene cycloaddition was used to construct antibody-Cu-64 or antibody-Zr-89 conjugates *in vitro* by tethering anti-HER2-antibody and chelator (DOTA or DFO) to norbornene and tetrazine respectively followed by the cycloaddition and radiometallation [186]. *In vivo* PET imaging and biodistribution studies showed significant and specific uptake of the conjugates in HER2-positive xenografts with little background noise.

2.2.3.2 Radical Thiol-Ene Reaction
Reactions between alkenes and thiols can occur either through free-radical addition of thiol to alkenes (termed thiol-ene reaction) or through Michael addition of thiol to electrophilic carbon-carbon double bonds (Scheme 2.12). While application of the Michael addition has been explored widely in the past, here we will only focus on the radical type thiol-ene reaction for its rapid advances in recent years [24].

Thiol-ene reactions are considered to be a type of 'click chemistry,' owing to the high reactivity of the reactants upon addition of low concentrations of catalysts with or without UV irradiation. Their high yields within short periods under ambient conditions and readily installable thiols or alkenes to various molecules make them one of the most frequently used reactions in polymer sciences, such as modification of substrate surface, fabrication of photolithography and microdevices, and formation of nanostructured networks [190]. Nearly all sterically favourable alkenes can undergo thiol-ene reactions, while electron-rich (vinyl) and strain-involved alkenes (norbornene) react faster than electron-poor alkenes.

Recently, thiol-ene chemistry has been explored as a conjugation method to modify various molecules, such as polymers [191, 192], dendrimers [193, 194], peptides or proteins [195], and nanoparticles [196–198]. The robustness of thiol-ene reactions was demonstrated in the synthesis of a fourth-generation dendrimer (G4); all of the 48 peripheral alkene groups could be modified nearly quantitatively in 30 minutes using the photoinitiator 2,2-dimethoxy-2-phenylacetophenone (DMPA) under UV irradiation (Scheme 2.13) [194].

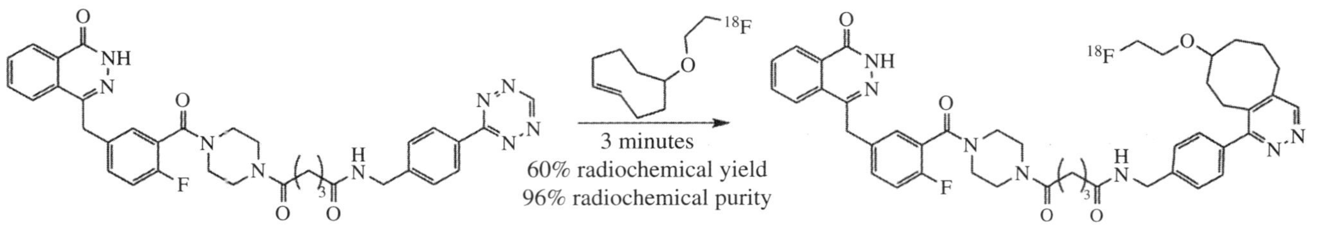

SCHEME 2.11 F-18 labelling of PARP-1 inhibitor AZD2281 using tetrazine-*trans*-cyclooctene ligation.

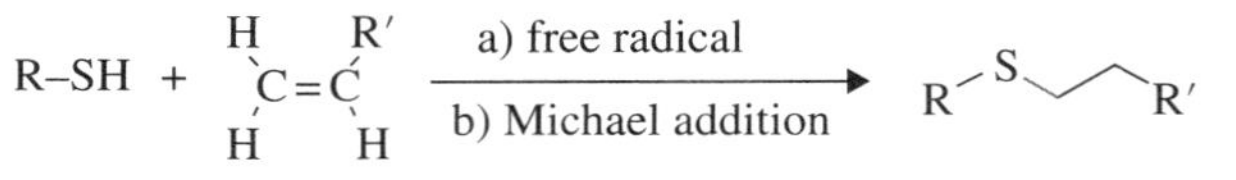

SCHEME 2.12 Thiol-ene coupling via either free radical (a) or Michael addition reactions (b).

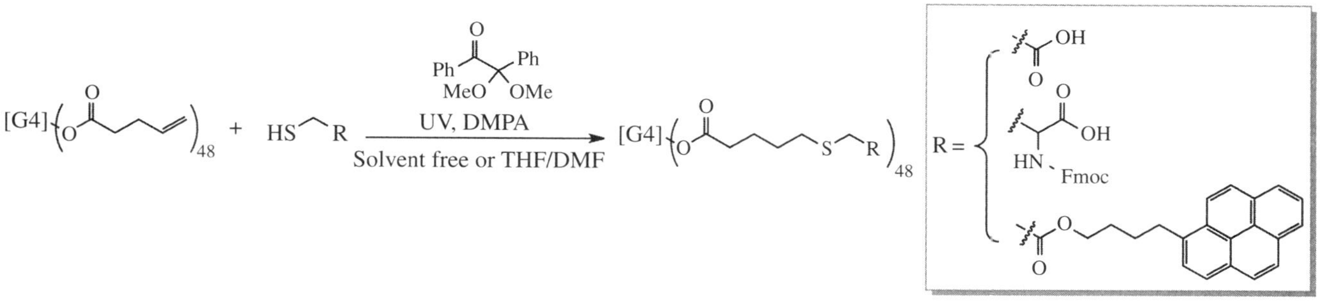

SCHEME 2.13 Modification of dendrimer G4-ene₄₈ using thiol-ene radical reaction.

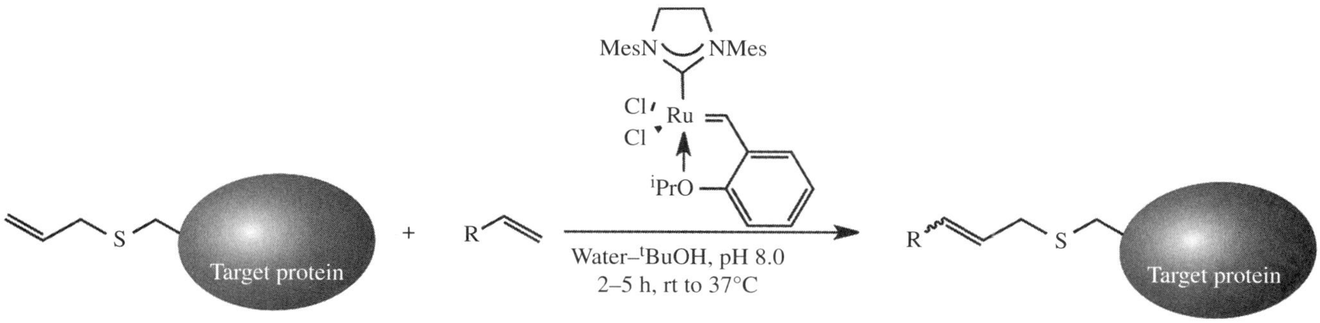

SCHEME 2.14 Protein modification using olefin cross metathesis.

Recently, thiol-ene chemistry has been used on the cell surface of living cells, which metabolically present methacryloyl groups [199]. After UV exposure for 10 minutes in the presence of a photoinitiator (Irgacure 2959 or BiMA), thiol-PEG could be introduced to the cell surface. Although no cytotoxicity was reported in the paper, thiol-ene chemistry, at its current stage, does not seem to be an optimal chemistry for use in cells or living systems.

Interestingly, electron-deficient alkynes, such as alkynoic amides, esters, and alkynones, could react with cysteine-containing peptides or proteins in aqueous media to form reversible modifications [200]. Modification of unprotected peptides and proteins occurred via the formation of a vinyl sulphide linkage, which could be cleaved off by addition of thiols under mild conditions.

2.2.3.3 Cross Metathesis Olefin metathesis is a great tool for building molecular architectures in organic synthesis [201–203]. Among various types of olefin metathesis, cross-metathesis is not as straightforward or efficient as ring-closing metathesis and ring-opening metathesis polymerisation due to the lack of the entropic driving force existing in RCM and side reactions arising from self-metathesis. Development of metathesis catalysts that can be used in aqueous media greatly facilitated theirs application to modify biomolecules in water or water-organic solvent mixtures [204–206]. Protein modification using cross metathesis was first realised by the Davis group, after chemical introduction of alkene groups to cysteine thiols on proteins (Scheme 2.14) [207]. Addition of a magnesium salt in the aqueous-alcohol media in the presence of a ruthenium catalyst (Hoveyda-Grubbs second generation) allowed the reactions to be completed at pH 8 at 37°C with high efficiency, and various groups, such as alcohols, carbohydrates, and oligoethyleneglycol, could be introduced to a protein surface. It was also noticed that allyl chalcogenides generally enhance the rate of alkene metathesis reactions, and allyl selenides were found to be exceptionally reactive olefin metathesis substrates, enabling a broad range of protein modifications [208]. Although not a direct conjugation strategy, ring-opening metathesis polymerisation could generate hydroxypyridonate-type gadolinium chelators, built in polymer backbones [209]. These chelates exhibited extremely large molecular relaxivities ($r_1 > 100$ mM^{-1}s^{-1}), presumably due to the longer rotational correlation times estimated from the molecular weights of linear polymers.

2.2.4 Cross-Coupling Reactions

Cross-coupling reactions, such as those in synthesis of natural products and construction of polymers [210, 211], have been applied extensively in organic synthesis. In molecular imaging, cross-coupling reactions are also used in the preparation of radiotracers [212]. However, their use in conjugation or labelling of biomolecules is limited by low conversion rates and harsh conditions with the use of transition metal catalysts, which may denature the biomolecules. Nonetheless, by carefully modifying reaction conditions, a few groups have extended the cross-coupling reactions in protein labelling or modification.

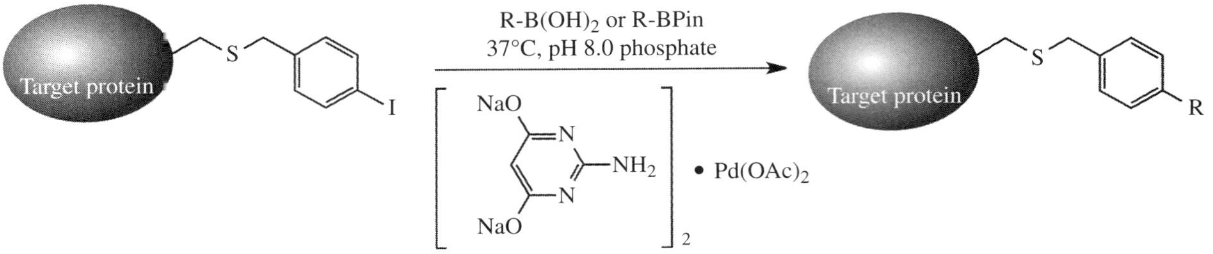

SCHEME 2.15   Protein modification using the Suzuki-Miyaura cross-coupling reaction.

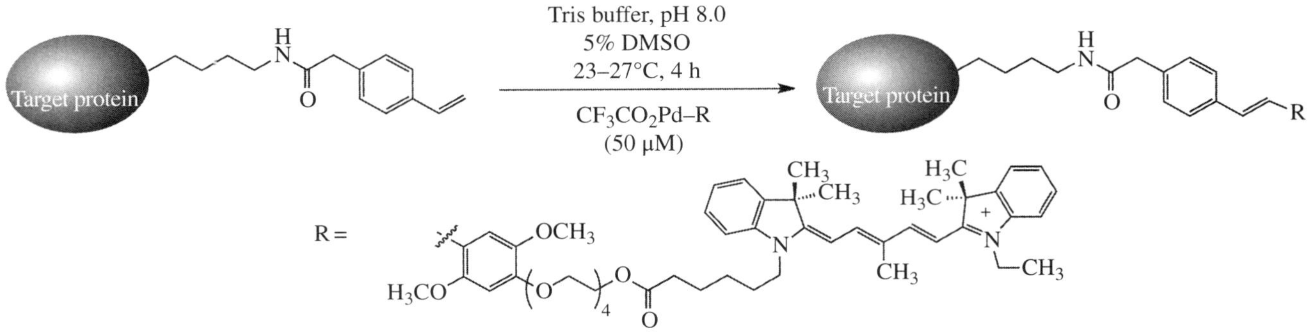

SCHEME 2.16   Protein modification using Heck-type coupling reaction.

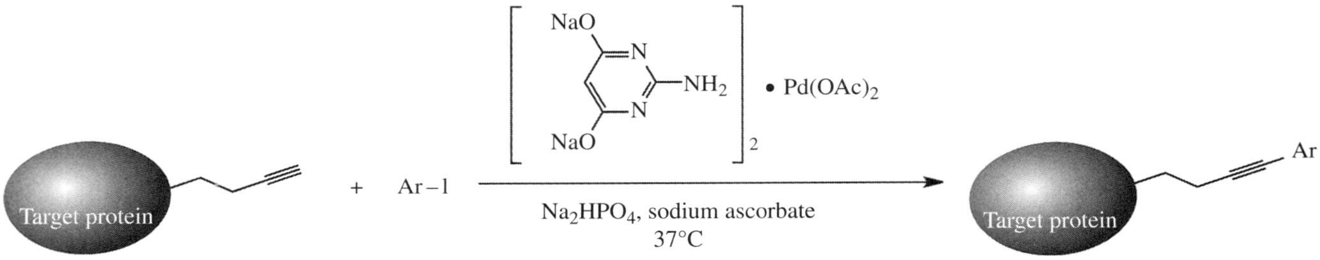

SCHEME 2.17   Protein modification using a Sonogashira coupling reaction.

For instance, Davis and co-workers, using the Suzuki-Miyaura cross-coupling reaction, have made it possible to modify a protein with an aryl or lipid group and carbohydrate molecule (Scheme 2.15) [213, 214].

The reaction employed a new ligand for palladium, which allowed completion of the reaction within one hour at 37°C, making it promising to modify proteins via genetically or chemically incorporated cross-coupling partners.

Myers and co-workers developed a protocol to label proteins using a Heck-type coupling under mild conditions (Scheme 2.16) [215, 216].

These reactions add new tools for labelling proteins and other biomolecules in aqueous media; however, their utility in the context of live cells or organisms has not been investigated. Recently, a copper-free Sonogashira coupling was developed for protein labelling in aqueous media and bacterial cells (Scheme 2.17) [217, 218]. Homopropargylglycine-modified protein could be labelled with an array of molecules such as fluorophores and fluorinated compounds with good to excellent yields.

2.3   SITE-SPECIFIC MODIFICATION OF PROTEINS OR PEPTIDES

2.3.1   N-terminal Cysteine

2.3.1.1   Native Chemical Ligation (NCL)   Kent and co-workers found that a peptide-α-thioester and a peptide with N-terminal cysteine could ligate and form a native amide bond between the two peptide fragments, namely native chemical ligation [219]. Regardless of the presence of any internal cysteine residues, N-terminal cysteine residue of one peptide/protein can undergo *trans*-thioesterification with a thioester attached to another peptide/protein, followed by a S⟶N acyl transfer to form an amide bond at the ligation site (Scheme 2.18). To improve the reactivity of the peptide-α-thioester, mostly

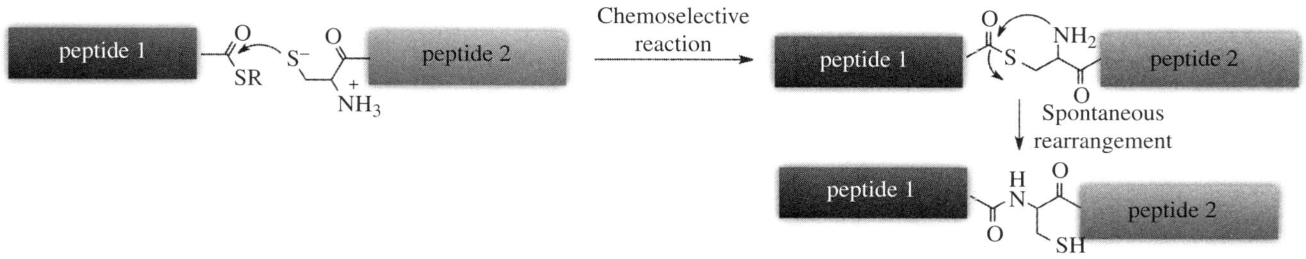

SCHEME 2.18 Native chemical ligation of two peptide fragments.

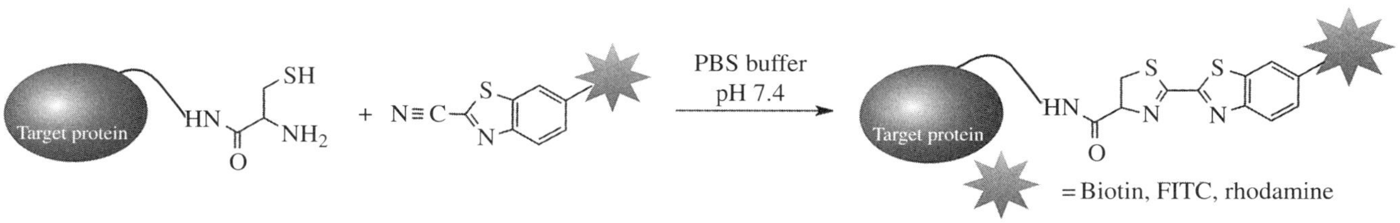

SCHEME 2.19 Protein modification using the CBT-cysteine condensation reaction.

alkyl esters, thiol additives are often used to perform *in situ trans*-thioesterification before cysteine is utilised [220]. Addition of a water-soluble thiol, (4-carboxylmethyl) thiophenol (MPAA), was able to complete the NCL reaction within several hours [221]. The utility of the NCL to ligate peptide-α-thioester has been expanded from N-terminal cysteine-peptide to peptides ended with glycine alanine [222], phenyl alanine [223, 224], valine [225], lysine [226], leucine [227], and glutamine [228], all of which require a second step for desulfurisation after the original NCL.

Besides its initial application in peptide and protein synthesis [229–231], NCL has become a useful tool in the synthesis and ligation of peptide nucleic acids [232, 233], preparation of lipid-GFP conjugates [234], construction of glycopeptides [235], synthesis of peptide or protein dendrimers [236], hydrogel cross-linking [237], construction of protein-based micelles, and conjugation of peptide to solid supports [239], amongst others [240].

2.3.1.2 *CBT Condensation Chemistry*

A condensation reaction between D-cysteine and 2-cyanobenzothiazole (CBT) was first used in the synthesis of D-luciferin, a common firefly luciferase substrate [241]. Rao and co-workers found the reaction could proceed rapidly under physiological pH conditions in water. This finding led to a new biocompatible conjugation method (Scheme 2.19) [242]. CBT derivatives only react with 1,2- or 1,3-aminothiol-containing molecules to form stable condensation products, while reactions with the thiol group alone can only lead to the formation of unstable or reversible products. Therefore, peptide or protein possessing N-terminal cysteine can be modified to introduce various functionalities, such as fluorescent molecules, biotin, and F-18 with high selectivity and efficiency (at room temperature within 1 hour). The CBT condensation reaction also allowed successful labelling of cell surface proteins while N-terminal cysteine residues were present. This efficient conjugation reaction led to self-condensation and assembly of imaging molecules, which possess both CBT or its analogues and a free cysteine residue on the same molecule, in living cells [243–245].

Besides the generation of N-terminal cysteine using recombinant proteins followed by enzymatic cleavage [242], Chin and co-workers recently reported ways of genetically introducing 1,2-aminothiol for site-specific protein labelling [246], expanding the utilities of the CBT condensation for site-specific labelling of proteins.

2.3.2 Aromatic Residues

Modification of native protein residues other than cysteine is less common. The Francis group has developed a few conjugation protocols for the modification specifically on tyrosine and tryptophan [247]. Using the efficient diazonium-phenol coupling reaction, they were able to modify the tyrosine residues on MS2 virus capsid protein with small functional groups or polymer selectively with high yields (Scheme 2.20) [248]. Tyrosine residues on proteins were also modified selectively in aqueous media using a Mannich type reaction [249], via π-allylpalladium complexes [250], or via oxidative modification using cerium(IV) ammonium nitrate [251].

Methods for selective modification of the tryptophan residue on proteins using rhodium carbenoids at near neutral pHs were also studied in the Francis group (Scheme 2.21) [252, 253].

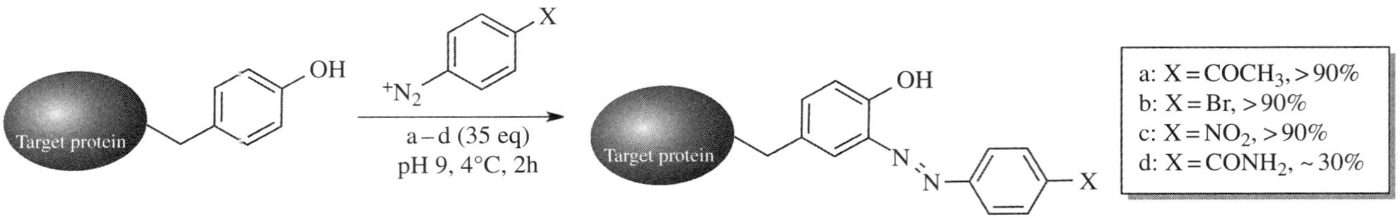

SCHEME 2.20 Modification of tyrosine residues on proteins via diazonium-phenol coupling.

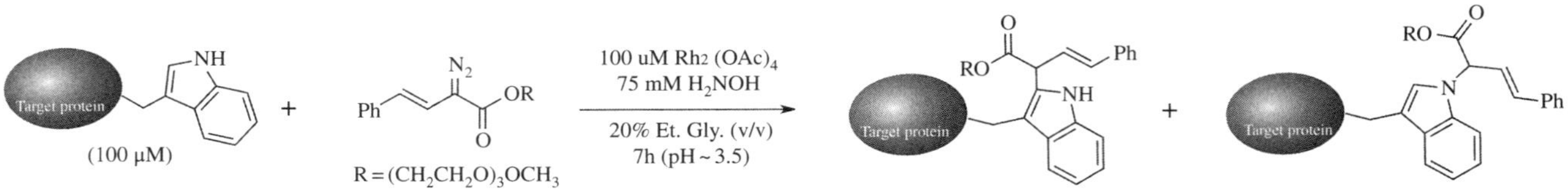

SCHEME 2.21 Modification of tryptophan residues on protein using rhodium carbenoids.

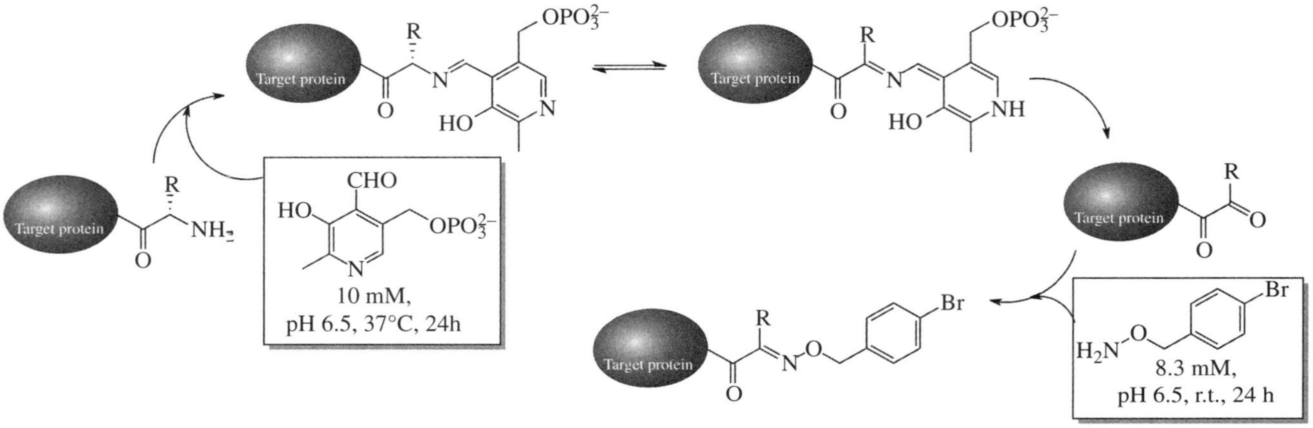

SCHEME 2.22 Site-selective modification of a protein at its N-terminus where the terminal residue is glycine (R = H), aspartic acid (R = CH$_2$COOH), valine (R = CH(CH$_3$)$_2$), or methionine (R = CH$_2$CH$_2$SCH$_3$).

2.3.3 N-terminus of Protein

Mild transamination reactions were discovered for site-specific modification of N-termini of proteins (Scheme 2.22). The protein is first treated with an aldehyde overnight at slightly elevated temperature to generate a reactive ketone or aldehyde group at its N-terminus, which can then be further modified via hydrazone or oxime ligation [254]. The following N-terminal residues are preferred: aspartic acid, glycine, valine, and methionine. On the other hand, serine, threonine, cysteine, tryptophan, and proline are found incompatible with the use of aldehydes in the reaction.

2.3.4 C-terminus of Protein

One extension of NCL is expressed protein ligation (EPL), which utilised a protein-splicing mechanism by using intein instead of thioester in one of the peptide components [255, 256]. The target protein can be expressed as a fusion protein with intein, a protein subunit that is cleaved off via self-catalysed rearrangement by forming a thioester during protein splicing (Scheme 2.23). The target protein fusion can then be modified in a similar way as that in NCL. Although EPL is more efficient in the synthesis or ligation of proteins of a large molecular weight (greater than 15 KDa), a thiol additive such as thiophenol is necessary to facilitate the ligation.

For example, a murine leptin was fused to GyrA intein, which underwent thiolysis first to give an α-thioester before introduction of other functionalities such as imaging probes or other reactive groups [257]. Intein-based expressed protein ligation has its advantage in terms of site specificity, but its efficiency may not be comparable to that of aniline-accelerated oxime/hydrazone ligations. Therefore, to achieve highly efficient site-specific labelling of proteins, combined labelling techniques may be necessary. Using a modified intein-mediated ligation approach, Rao and co-workers constructed a QD-based

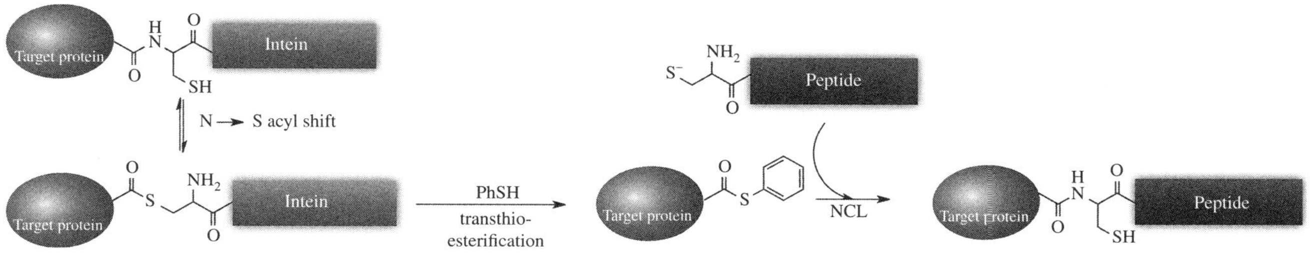

SCHEME 2.23 Synthesis of protein-peptide conjugate via expressed protein ligation.

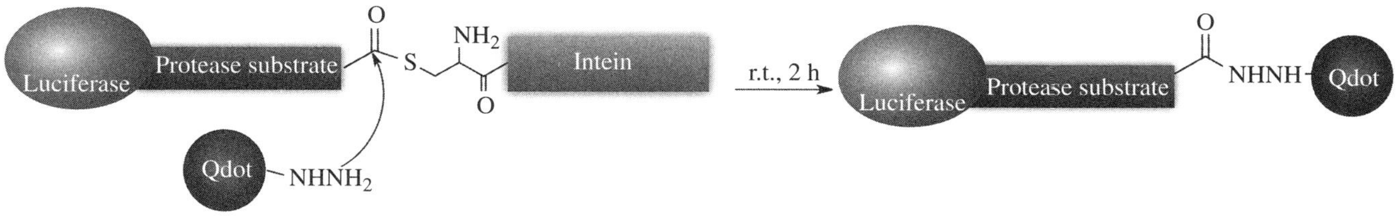

SCHEME 2.24 Construction of quantum dot-protein conjugates via a modified intein-mediated ligation.

nanosensor to detect matrix metalloproteinase (MMP) activity [258]. The attachment of luciferase and protease substrate fusion protein to QDs was achieved through replacement of intein fused to the target protein by hydrazides displayed on the surface of QDs (Scheme 2.24). The reaction using hydrazide-QD was very efficient, as all the QDs were modified with luciferase in 2 hrs at room temperature.

2.3.5 Introduction of Chemical Tags for Site-Specific Labelling on Peptides or Proteins

2.3.5.1 Introduction of Small Functional Group Tags

Introduction of Ketone or Aldehyde Tags To achieve site-specific labelling of proteins, approaches have been developed to introduce aldehydes/ketones and aminooxy or hydrazide functionalities to proteins through genetic encoding, enzymatic labelling, and chemical reactions [259, 260]. Sodium periodate oxidation has been used frequently in the past to generate aldehyde functionalities on glycans or N-terminal serines or threonines [9]. The newly generated aldehyde groups on the proteins or glycoproteins can then be labelled selectively via oxime/hydrozone ligation to introduce functional molecules, such as PEG, fluorophores, and other proteins or peptides [261, 262]. Introduction of ketone groups using a cell's own machinery has been well utilised by the Bertozzi group in labelling cell surface proteins [30]. Using a metabolic labelling approach, monosaccharides with modified functional handles could enter the metabolic pathway to generate glycoproteins bearing these handles, including, but not limited to, keto, azido, alkynyl, alkanyl, crotonoyl, and thiol groups [263–266]. It has been proven as a valuable approach to labelling cell surface proteins, specifically at the sialic acid residues of the glycan, in live mammalian cells and in living organisms.

Schultz and co-workers demonstrated the first genetic encoding of keto amino acids *p*-acetyl-L-phenylalanine and *m*-acetyl-L-phenylalanine into proteins expressed in *E. coli* via amber suppression method, and applied these groups for selective labelling of proteins with fluorescent dyes both *in vitro* and in living *E. coli* [267–270]. In mammalian cells, Ye et al. incorporated unnatural amino acids *p*-acetyl-L-phenylalanine and *p*-benzoyl-L-phenylalanine (Bzp) into chemokine receptor CCR5 and rhodopsin at three specific sites with high efficiency, allowing their modification with different reagents [271]. More recently, Bertozzi and co-workers incorporated aldehyde tags via genetic introduction of formylglycine to the target protein, allowing further site-specific modification of the protein through reactions of aldehyde, such as oxime/hydrazone ligation [259, 272]. In an improved procedure of site-directed spin labelling, genetically encoded unnatural amino acid *p*-acetyl-L-phenylalanine was introduced to mutants of T4 lysozyme, allowing further labelling with hydroxylamine reagents [273]. Robust protein farnesyltransferase was able to incorporate an aldehyde-containing substrate analogue to a protein, which was subsequently immobilised onto aminooxy-functionalised agarose beads or labelled with a fluorophore [274]. An F-18 radiolabel was introduced to a mouse hormone protein leptin via the oxime ligation, wherein the aminooxy group was incorporated at the C-terminus of leptin using expressed protein ligation, followed by aniline-catalysed oximation reaction with [F-18]fluorobenzaldehyde. After modification, the hormone was biologically active both *in vitro* and *in vivo* and was applied to positron emission tomography (PET) in ob/ob mice [257]. In a different approach, the aminooxy group was efficiently introduced to protein through chemical substitution of C-terminal thioacid [275] by a 1,2-bis(oxyamino)ethane under mild conditions (pH 7.4 buffer) [276].

Introduction of Azide or Alkyne Tags   Similarly, many efforts have been made to introduce azide or alkyne, another pair of popularly used chemical tags, into proteins. Besides Bertozzi's metabolic incorporation of azide or alkyne groups to cell surface proteins using glycan synthetic pathways, Tirrell and co-workers also demonstrated the use of azido-amino acids to introduce azide groups to the cell surface of *E. coli* using its metabolic pathways [133]. The Schultz group again incorporated *p*-azidophenylalanine and *p*-propargyloxyphenylalanine to proteins using orthogonal TyrRS/tRNA$_{CUA}$ pairs that genetically encode these unnatural amino acids in yeast [277]. A similar approach was performed to introduce alkynyl groups to proteins in *E. coli* [278]. Using a genetic engineering approach, Cazalis et al. generated a thrombomodulin mutant with a C-terminal azido-methionine in *E. coli*, allowing site-specific PEGylation of the thrombomodulin mutant via Staudinger ligation while maintaining the enzymatic activity [279]. Davis and co-workers expressed a mutant of TIM barrel protein SSβG bearing an azidohomoalanine chemical tag in a methionine auxotrophic strain of *E. coli* and proved the use of the chemical tag by formation of glycoprotein conjugate using CuAAC [280, 281]. Alkyne-incorporated SSβG was also generated using the same method. Using a similar approach, Finn and colleagues introduced azide- and alkyne-containing unnatural amino acids to virus-like particles, while the incorporation of the functional handles did not affect the particles' self-assembly [282]. Accessibility of the newly introduced azido or alkynyl groups was verified by labelling the virus-like particles using alkynyl- or azido-functionalised fluorophores, biotin, and DOTA(Gd). Recently, chemical approaches to introduce azide groups to proteins using aqueous diazotransfer reactions were also developed [283].

Introduction of Alkene Tags   A number of methods have been developed to incorporate alkene groups into proteins either *in vitro* or in bacteria or yeast. Using methionyl-tRNA synthetase, Tirrell and co-workers could replace the methionine residues with alkene-containing methionine analogues, such as homoanaline, in proteins expressed in a methionine auxotrophic strain of *E. coli* [284, 285]. The Schultz group has genetically encoded O-allyltyrosine and phenylselenocysteine in *E. coli* with genetically engineered tRNA/aminoacyl-tRNA synthetase [286, 287]. The same group also introduced several alkene-containing unnatural amino acids into proteins in yeast using tRNA/aminoacyl-tRNA synthetase pairs [288]. Use of pyrrolysyl-tRNA synthetase has been exploited to incorporate the alkene-containing nonnatural amino acid, 6-N-allyloxycarbonyl-L-lysine into proteins site-specifically [289]. Chemical methods have also been developed to introduce alkene groups to proteins using reactions of O-mesitylenesulfonylhydroxylamine with cysteine thiol groups on protein, followed by replacement with a thiol nucleophile containing alkene [207].

2.3.5.2   Introduction of Peptide Tags
Peptide tags can be introduced to a target protein via genetic approaches using a plasmid encoding fusion of the target protein and a peptide tag, which either have high affinity to other molecular parts, or can be a substrate for a particular enzyme for further modification [5–7].

Tetracysteine Tag   In search of substituents for large fluorescent proteins [290, 291], the Tsien group has developed an arsenic-modified fluorescein derivative (FlAsH) system (Figure 2.6) [292]. Genetic introduction of four cysteine residues at the i, i + 1, i + 4, and i + 5 positions on a β hairpin of a protein promoted their binding with FlAsH, making the protein of interest fluorescent. Benefiting from the small size of the tetracysteine tag (CCXXCC) and small molecule fluorophores,

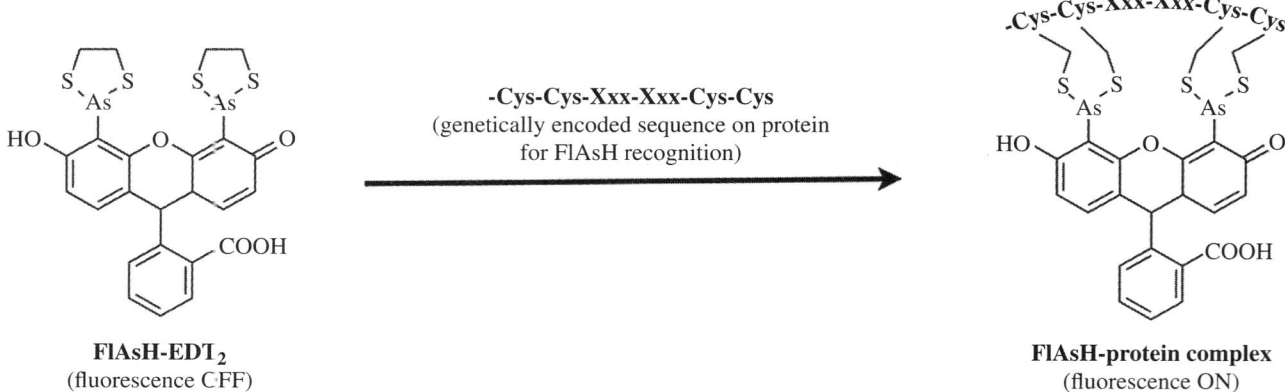

FIGURE 2.6   Protein fluorescent labelling via a tetracysteine tag using FlAsH.

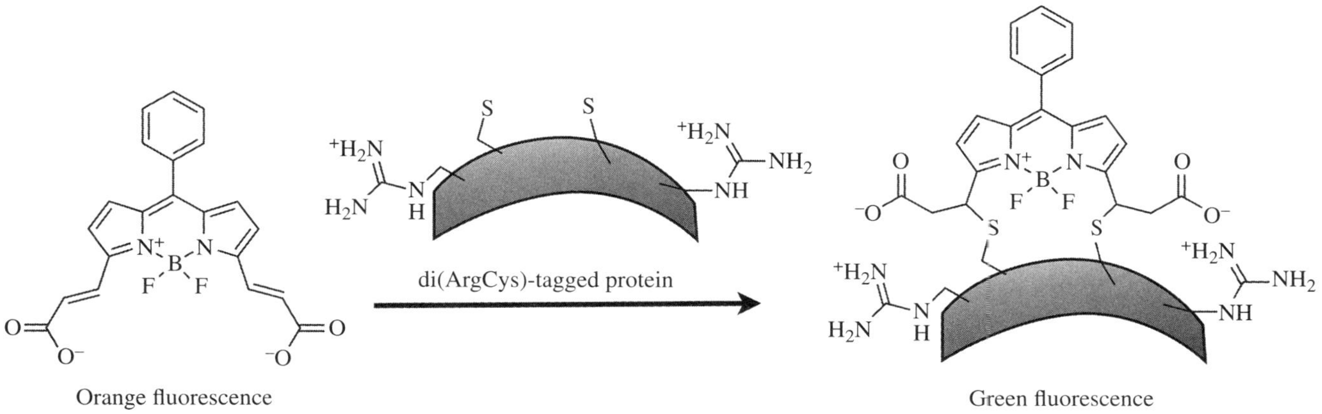

FIGURE 2.7 Protein fluorescent labelling via di(ArgCys) tag using BODIPY.

protein imaging in live cells has been greatly facilitated using FlAsH and other later-developed bisarsenical fluorophores, such as the red-fluorescent ReAsH [293–298], especially when protein labelling is prohibited by the large size of fluorescent proteins [299]. However, concerns should also be kept in mind while using the biarsenical fluorophores, such as background noise caused by nonspecific binding to endogenous cysteine groups and potential arsenical toxicity.

Fluorette or Dye-Binding Peptide Tag Using phage display, Nolan and co-workers developed a set of fluorophore-binding peptides (fluorettes) for Texas Red with high affinity (binding constants in the picomolar range) [300, 301]. The peptide aptamers with stably folded structures interact with overlapping regions of Texas Red without interference with the fluorescence of Texas Red. One of them could bind to x-rhod calcium sensors, which are structurally similar to Texas Red. Fusion proteins containing this peptide could bind to the x-rhod calcium sensors in living cells.

Peptides or proteins containing a tetraserine motif (SSPGSS) can bind to a fluorogenic rhodamine-derived bis-boronic acid with high affinity (in sub-micromolar range), and the binding to the tetraserine-containing peptide sequence was over four orders of magnitude higher than that with a simple monosaccharide. Imaging of endogenous proteins containing the SSPGSS motif was attempted in living cells, and fluorescence was turned on only at the SSPGSS-rich cell interior [302]. Recently, a peptide tag containing two pairs of Arg-Cys was designed to label protein with a BODIPY dye covalently via Michael addition in live cells (Figure 2.7) [303]. This strategy is comparable to the tetracysteine-biarsenic approach, whilst exhibiting large spectral change and low cytotoxicity.

Metal Ion-Binding Peptide Tag Metal ion-binding peptide sequences, such as oligohistidine (His-tags) [304] and lanthanide-binding tags [305], have been widely used in protein purification, immobilisation, as well as detection [306]. Inspired by these *in vitro* applications, recent years have seen many examples of metal ion-binding peptide tags in imaging of proteins in the context of cells [307–311]. For instance, proteins (ligand-gated ion channel and G protein-coupled receptor) fused with polyhistidine were visualised within seconds in living cells by addition of a chromophore containing a metal-ion-chelating nitrilotriacetate (NTA) moiety [307].

Biotin Ligase and Lipoic Acid Ligase Peptide Substrate Tags Ting and co-workers reported the introduction of a bioorthogonal tag to cell surface proteins using an approach combining genetic and enzymatic methods (Figure 2.8). First, a peptide sequence, an *E. coli* biotin ligase (BirA) substrate [312], is fused to the target protein genetically, then a biotin [313, 314] or biotin-mimicking molecule [315], in which two amide nitrogens of biotin are replaced with carbons to produce a ketone moiety, is introduced to the fusion protein via biotin ligase. The biotin- or ketone-possessing biotin-mimicry on the protein can then be labelled with fluorescent probes via streptavidin binding or oxime/hydrazine ligation. Besides direct protein staining, the biotin ligase-mediated site-specific labelling was also used to detect protein-protein interactions *in vitro* and in cells, when proteins of interest were fused to BirA and BirA's peptide substrate, respectively [316]. Although this proximity biotinylation for protein-protein interaction studies has its advantages, such as high signal to background ratio and sensitivity, the introduction of BirA may alter the protein's behaviour while interacting with other proteins.

In an analogous approach, the Ting lab explored the applicability of lipoic acid ligase (LplA) in place of biotin ligase (Figure 2.8) [317, 318]. LplA could catalyse the reaction between lipoic acid or its azidoalkyl carboxylic acid analogue (discovered through screening) and the lysine side chain of ligase peptide substrate to form an amide bond. The azido group can therefore be introduced to protein site-specifically when the protein is fused to LplA's peptide substrate. The Ting lab

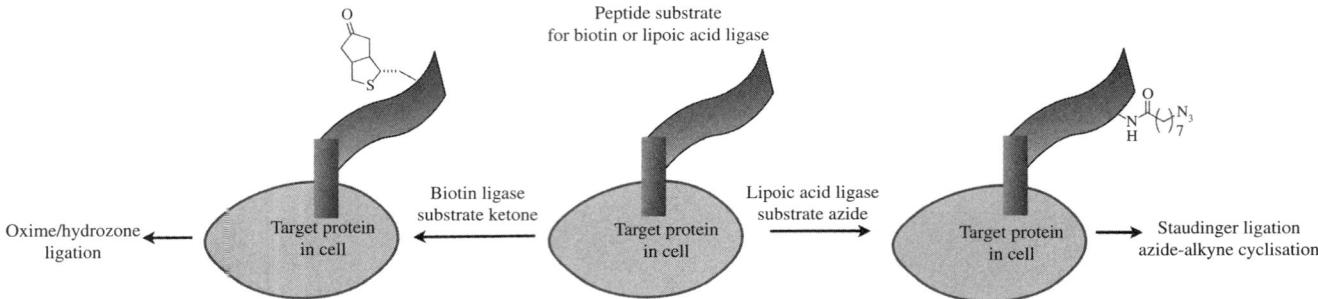

FIGURE 2.8   Protein labelling using biotin ligase or lipoic acid ligase peptide substrate tags.

demonstrated the accessibility of the azido group using its reaction with a cyclooctyne-conjugated fluorescent probe, and the azido-modified cell surface fusion protein was visualised in live cells. The orthogonal use of LplA and biotin ligase was also verified. One interesting application of this LplA-mediated labelling was illustrated in a study of protein-protein interaction [319]. A photo-crosslinker, aryl azide, was introduced to one protein fused to LplA's peptide substrate, and the interacting protein can be cross-linked upon UV irradiation.

Sortase Peptide Substrate Tag   Sortase, a class of bacterial transpeptidase, is in charge of anchoring cell surface proteins to the cell wall by catalysing the cleavage of the threonine-glycine bond on the conserved LPXTG peptide substrate of its target protein, followed by formation of a new bond between threonine and amino group on the cell wall. Mao et al. first demonstrated the utilities of sortase in introduction of proteins, peptides, and small molecules to protein with LPXTG motif [320]. Since then, sortase-mediated ligation has been applied in protein labelling with functional ligands, protein imaging in living cells [321, 322], protein immobilisation, protein purification, and construction of neoglycoconjugates [323, 324].

2.3.5.3   Introduction of Protein Tags

scFV tag   During the early time of developing surrogates to fluorescent proteins, the high affinity and specificity of the antibody-hapten interaction has been applied in protein labelling strategies. Single-chain antibodies (scFV) targeting specified sites in living cells could arrest cell-permeable hapten-fluorophore conjugates through their highly affinity interactions (nanomolar range) and were able to image specific subcellular localisation expressing the scFV, such as the endoplasmic reticulum, Golgi, and plasma membrane of living mammalian cells [325]. Chemical probes with various spectral and indicator properties could be used in this approach.

FKBP12:F36V Tag   Nolan and co-workers developed a protein labelling approach using high-affinity interaction between an FKBP12 mutant (F36V) and a synthetic, engineered ligand SLF′ [326]. A protein fused with FKBP12:F36V tag could be labelled noncovalently by SLF′-fluorophore conjugates in live mammalian cells, and the level of staining was proportional to the expression level of the fusion protein. After SLF′-fluorophore labelling, β-galactosidase-FKBP12:F36V fusion protein lost its enzymatic activity by 90% when fluorophore assisted laser inactivation was applied.

SNAP-tag or hAGT Tag   Human O_6-alkylguanine-DNA-alkyltransferase (hAGT) is a 20 kDa DNA repair protein that dealkylates O6-alkylguanine in damaged DNA. 'SNAP-tag' uses O_6-guanine-modified fluorophore to covalently label proteins in living cells (Scheme 2.25) [327]. This represents the first example using fusion protein technology to label target proteins covalently and has been applied successfully in protein labelling in live mammalian cells, super-resolution imaging of intracellular proteins [328], cell surface protein labelling for protein-protein interaction studies and virus-cell interaction studies [329, 330], two-step protein immobilisation on solid surface [331], and quantum dots modification [332].

To image the reactive oxygen species in cells, SNAP-tag fusion proteins on the surface or interior of living cells were labelled with boronate-capped dyes that could turn on in response to changes in local peroxide species levels [333]. A tumour-targeting anti-EGFR antibody fused to the SNAP-tag allowed specific attachment of O_6-guanine-modified photosensitiser at the target sites, increasing the phototoxicity in tumour cells [334]. Similarly, CLIP-tag, an AGT variant using O_2-benzylcytosine derivatives as substrate, has also been developed as an orthogonal analogue of SNAP-tag [335]. SNAP and CLIP fusion proteins can be labelled with different imaging probes specifically and simultaneously in living cells.

HaloTag   Bacterial haloalkane dehalogenases can replace halides from aliphatic hydrocarbons by nucleophilic attack from aspartate in the enzyme to form a covalent ester bond with the hydrocarbon substrate (Scheme 2.26) [336]. Protein fused

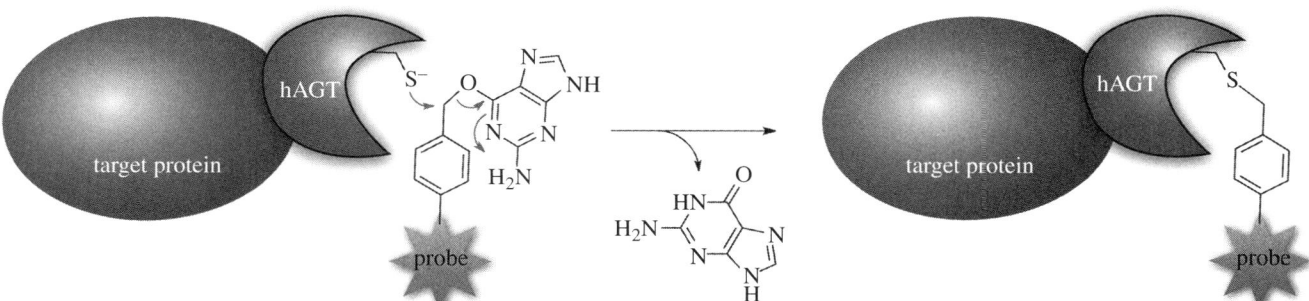

SCHEME 2.25 Protein labelling via hAGT fusion protein.

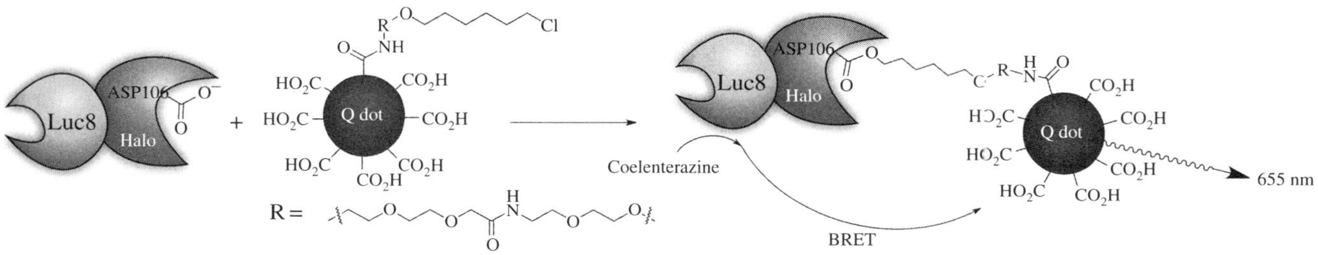

SCHEME 2.26 HaloTag used to conjugate luciferase and quantum dots.

with haloalkane dehalogenases (HaloTag, around 33 kDa) can therefore be modified with different functionalities, such as fluorophores, affinity handles, and solid support, selectively using synthetic ligands that bear haloalkane substrates. Los et al. designed a modular system for protein labelling using HaloTag, which allowed modification of proteins in buffers and in fixed or living cells and applied the protocol in imaging of subcellular protein (NF-κB proteins) translocation and protein-protein or protein-DNA interactions [337]. Visualisation of cell membrane–bound β-integrin molecules was achieved using HaloTag fusion [338]. The HaloTag catalysed labelling occurs very rapidly in buffer, and a second-order rate constant was obtained to be 2.7×10^6 M^{-1}s^{-1}, in the same speed scale as the binding between streptavidin and biotin (8.5×10^6 M^{-1}s^{-1}) using nanomolar protein and ligand. In living cells, sufficient labelling was obtained at micromolar of ligand within 15 min.

Using the HaloTag labelling protocol, the Rao group constructed quantum dot-luciferase conjugates using luciferase-HaloTag fusion protein and quantum dots modified with haloalkane chains [339]. Quantum dots can also be fixed to a living cell surface when the HaloTag is introduced to cell membrane bound protein genetically [340]. The Moerner group and collaborators demonstrated localisation of α-tubulin in cells transfected to express HaloTag-α-tubulin [341]. After addition of the HaloTag ligand, a chloroalkylated photoactivatable dye DCDHF, cells were visualised using single-molecule super-resolution imaging. Protein localisation in live *Caulobacter crescentus* bacteria was also visualised using transgenic organisms expressing HaloTag fusion proteins. The utility of molecular imaging using HaloTag in cells has also been shown in the imaging of peroxisome dynamics in various cultured mammalian cells using C-terminal peroxisomal matrix protein targeting signalling protein fused to HaloTag [342]. Real-time single-molecule imaging of *trans*-translation entry process was realised by anchoring ribosomal protein to solid surface via HaloTag labelling [343].

In tumour models intraperitoneally implanted with cancer cells with highly expressed HaloTags, *in vivo* spectral fluorescence imaging of the tumour cells was obtained via endoscopy after administration of various green to near IR fluorophore-conjugated HaloTag ligands, demonstrating a valuable tool for cancer research in animals [344]. In other work, tumours with a high expression of α subunit hypoxia-inducible factor (HIFα) were visualised in live animals, which relied on the construction of a near-infrared fluorescently labelled HIFα-targeted fusion protein via HaloTag labelling [345]. Recently, DOTA(Gd) was introduced to protein via HaloTag labelling, and increased relaxivity (r_1 and T_1-weighted images) was observed compared to unbound DOTA(Gd) [346]. The best DOTA(Gd)-protein conjugate exhibited sixfold enhancement in r_1 compared to its unbound state. The increased relaxivity is probably caused by the receptor-induced magnetisation enhancement effect while the motion of a contrast agent is coupled to that of a target protein.

eDHFR tag   Utilising the tight and specific binding between *E. coli* dihydrofolate reductase (eDHFR) and trimpethoprim (TMP) ($K_D < 1$ nM) [347], the Cornish and Sheetz groups developed a robust protein labelling method using an eDHFR tag (Scheme 2.27) [348]. Protein fused with eDHFR could be labelled with near stoichiometric concentration of TMP-modified

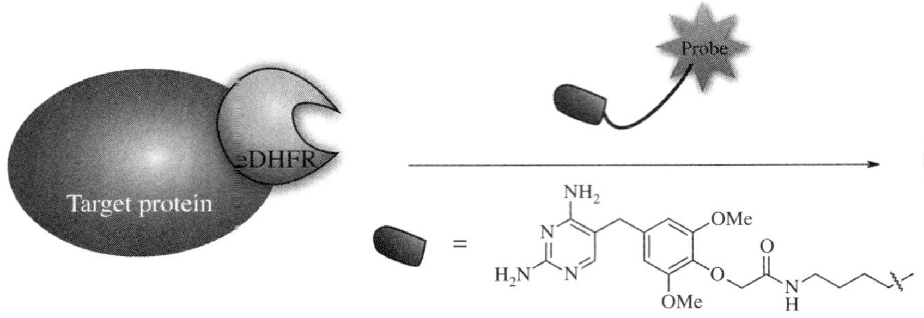

SCHEME 2.27 Labelling of target protein via eDHFR fusion.

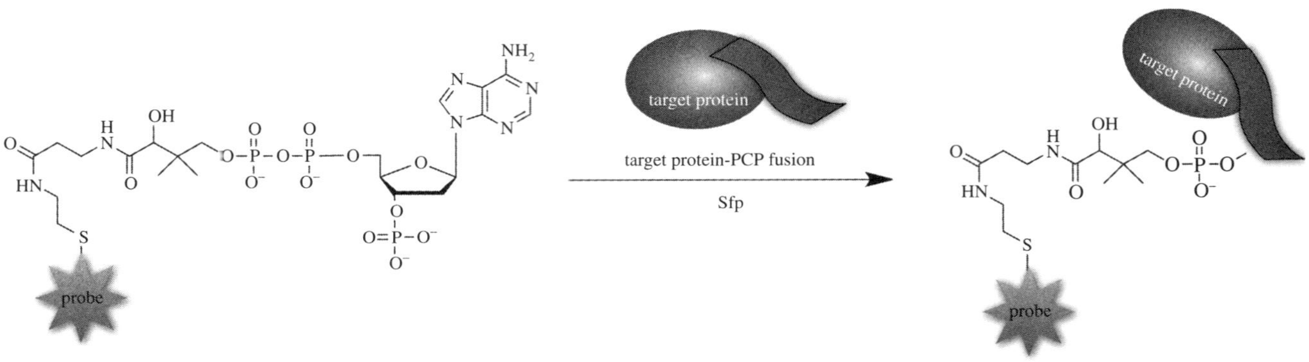

SCHEME 2.28 Protein labelling via PPTase-PCP tag.

probe with fast kinetics (association within minutes) and minimal background noise in living mammalian cells. The eDHFR approach can be an alternative for GFP for protein labelling with a few benefits, such as smaller size of eDHFR (~18 kDa), and easier modification of spectral properties of imaging probes. The system has been used to label the fusion proteins with many different fluorophores, such as fluorescein-based green and red dyes, BODIPY Texas Red [348], far-red photoswitching Atto-655 [349], two-photon fluorophore BC575 [350], and luminescent terbium probe [351], for various applications including super-resolution imaging [349] and time-resolved imaging in living cells [351]. Although the dissociation half-life of eDHFR-TMP complexes is over tens of minutes, covalent modification is preferred in imaging approaches such as single-molecule tracking and pulse-chase labelling. The Cornish and Sheetz groups applied the proximity-induced reactivity in inhibitor design and rendered the covalent bond formation between eDHFR fusion protein and TMP probe by installing a Cys residue on eDHFR in the contacting position with an arylamide modified TMP probe [352].

PPTase-PCP Tag Walsh and co-workers found that a post-translational modification enzyme, phosphopantetheinyl transferase (PPTase, Sfp) could catalyse the covalent modification of a target protein fused to a peptide carrier protein (PCP) moiety, excised from nonribosomal peptide synthetase, by a phosphopantetheinyl-modified small molecule (Scheme 2.28) [353]. The labelling of PCP fusion protein by PPTase was highly specific and efficient in cell lysates and was compatible with various small molecule probes and proteins.

Cutinase Tag Bonasio et al. reported the use of a fugal enzyme cutinase as a protein tag to label integrin lymphocyte function-associated antigen-1 on the surface of living cells, whilst not interfering with the conformation or function of the integrin [354]. Various imaging probes, including quantum dots, could be introduced to the target protein covalently via a catalysed reaction of p-nitrophenyl phosphonate-conjugates by cutinase.

β-Lactamase E166NTEM Tag β-Lactamases are small bacterial enzymes that hydrolyze β-lactam structures that have been seen as reporter enzymes for gene expression in mammalian cells [355, 356], as diagnostic and therapeutic targets of bacterial infection [357–360], as well as in prodrug activation [361]. A mutant of β-lactamase TEM-1, E166NTEM, was found to accumulate the enzyme-substrate intermediate with very slow regeneration of the enzyme. Utilising this property of E166NTEM, Mizukami et al. demonstrated fusion protein between epidermal growth factor receptor (EGFR), and E166NTEM could be labelled specifically and covalently by β-lactam probes, which were either β-lactam-dye conjugate or β-lactam bridged dye-quencher system [362].

2.4 CONCLUSIONS

As this chapter summarises, a rich collection of chemical and biochemical reactions currently exists in the literature for labelling a wide variety of functional groups on target molecules specifically or nonspecifically at a range of settings from *in vitro* to live cells and even live animals. Selection of proper bioconjugation strategies is thus important to successful labelling of target molecules for imaging. While doing so, one should consider the compatibility of the reactions with the target molecule and its reaction medium, as well as the efficiency of the conjugation reaction. Labelling target molecules with imaging moieties of a short half-life requires high reaction yields and fast reaction kinetics. Site-specific labelling chemistry gives uniform products but often requires introduction of a chemical tag to the target molecule through genetic encoding or chemoenzymatic reactions. These chemical tags range from small functional groups (a few atoms) to large proteins (several to tens of kDa), the size of which should also be taken into consideration to avoid perturbations on the normal biological functions of the target molecule. Bioorthogonal reactions allow direct labelling in biological media with high specificity and efficiency and thus are in high demand, but only a few are available. During the final compilation of the book, significant progress in conjugation chemistry has been made toward clinical applications. For example, strain-promoted cycloadditions, such as TCO-tetrazine reaction and cyclooctyne-azide cycloaddition, have been applied in tumour pre-targeting using nanoparticles [363, 364]. The Rao lab has applied the CBT condensation in *in situ* self-assembly of nanoparticles in living mice, leading its way toward early cancer detection [365] and therapy response monitoring [366–368]. Ongoing efforts of many chemists and biologists will lead to the discovery of more efficient, site-specific, and biocompatible biochemical methods and further enlarge the repertoire of labelling strategies for the development of molecular imaging probes.

REFERENCES

[1] C. Mamat, T. Ramendaa and F. R. Wuest, *Mini-Rev. Org. Chem.* **6**, 21–34 (2009).

[2] E. M. Sletten and C. R. Bertozzi, *Angew. Chem. Int. Ed.* **48**, 6974–6998 (2009).

[3] T. Kurpiers and H. D. Mootz, *Angew. Chem. Int. Ed.* **48**, 1729–1731 (2009).

[4] T. K. Tiefenbrunn and P. E. Dawson, *Biopolymers* **94**, 95–106 (2010).

[5] A. Dragulescu-Andrasi and J. H. Rao, *ChemBioChem* **8**, 1099–1101 (2007).

[6] K. M. Marks and G. P. Nolan, *Nat. Methods* **3**, 591–596 (2006).

[7] C. R. Jing and V. W. Cornish, *Acc. Chem. Res.* **44**, 784–792 (2011).

[8] H. Wang and X. Y. Chen, *Frontiers in Bioscience-Landmark* **13**, 1716–1732 (2008).

[9] G. T. Hermanson *Bioconjugate Techniques,* 2nd ed.. Academic Press, New York, 2008.

[10] T. T. Beaudette, J. A. Cohen, E. M. Bachelder, K. E. Broaders, J. L. Cohen, E. G. Engleman and J. M. J. Fréchet, *J. Am. Chem. Soc.* **131**, 10360–10361 (2009).

[11] W. C. W. Chan and S. M. Nie, *Science* **281**, 2016–2018 (1998).

[12] S. O. Doronina, B. E. Toki, M. Y. Torgov, B. A. Mendelsohn, C. G. Cerveny, D. F. Chace, R. L. DeBlanc, R. P. Gearing, T. D. Bovee, C. B. Siegall, J. A. Francisco, A. F. Wahl, D. L. Meyer and P. D. Senter, *Nat. Biotechnol.* **21**, 778–784 (2003).

[13] Y. Bae, N. Nishiyama, S. Fukushima, H. Koyama, M. Yasuhiro and K. Kataoka, *Bioconjugate Chem.* **16**, 122–130 (2005).

[14] R. Huisgen, *1,3-Dipolar Cycloaddition Chemistry.* Wiley, New York, 1984.

[15] D. R. Bundle, J. R. Rich, S. Jacques, H. N. Yu, M. Nitz and C. C. Ling, *Angew. Chem. Int. Ed.* **44**, 7725–7729 (2005).

[16] C. C. Lee, E. R. Gillies, M. E. Fox, S. J. Guillaudeu, J. M. J. Frechet, E. E. Dy and F. C. Szoka, *Proc. Natl. Acad. Sci.* **103**, 16649–16654 (2006).

[17] O. L. P. De Jesus, H. R. Ihre, L. Gagne, J. M. J. Frechet and F. C. Szoka, *Bioconjugate Chem.* **13**, 453–461 (2002).

[18] R. Franke, C. Doll and J. Eichler, *Tetrahedron Lett.* **46**, 4479–4482 (2005).

[19] D. Solomon, P. I. Kitov, E. Paszkiewicz, G. A. Grant, J. M. Sadowska and D. R. Bundle, *Org. Lett.* **7**, 4369–4372 (2005).

[20] L. Cui, P. I. Kitov, G. C. Completo, J. C. Paulson and D. R. Bundle, *Bioconjugate Chem.* **22**, 546–550 (2011).

[21] D. Witt, *Synthesis* **16**, 2491–2509 (2008).

[22] A. J. Chmura, M. S. Orton and C. F. Meares, *Proc. Natl. Acad. Sci.* **98**, 8480–8484 (2001).

[23] S. Dziadek, S. Jacques and D. R. Bundle, *Chem. Eur. J.* **14**, 5908–5917 (2008).

[24] C. E. Hoyle and C. N. Bowman, *Angew. Chem., Int. Ed.* **49**, 1540–1573 (2010).

[25] E. Barrett and A. Lapworth, *J. Chem. Soc.* **93**, 85–93 (1908).

[26] H. Heitzman and F. M. Richards, *Proc. Natl. Acad. Sci.* **71**, 3537–3541 (1974).

[27] H. F. Gaertner, K. Rose, R. Cotton, D. Timms, R. Camble and R. E. Offord, *Bioconjugate Chem.* **3**, 262–268 (1992).

[28] T. D. Pallin and J. P. Tam, *Chem. Commun.* **19**, 2021–2022 (1995).

[29] J. Shao and J. P. Tam, *J. Am. Chem. Soc.* **117**, 3893–3899 (1995).

[30] L. K. Mahal, K. J. Yarema and C. R. Bertozzi, *Science* **276**, 1125–1128 (1997).

[31] G. A. Lemieux, K. J. Yarema, C. L. Jacobs and C. R. Bertozzi, *J. Am. Chem. Soc.* **121**, 4278–4279 (1999).

[32] J. Rayo, N. Amara, P. Krief and M. M. Meijler, *J. Am. Chem. Soc.* **133**, 7469–7475 (2011).

[33] T. T. Tidwell, *Angew. Chem. Int. Ed.* **47**, 1016–1020 (2008).

[34] E. H. Cordes and W. P. Jencks, *J. Am. Chem. Soc.* **84**, 826–831 (1962).

[35] E. H. Cordes and W. P. Jencks, *J. Am. Chem. Soc.* **84**, 832–837 (1962).

[36] A. Dirksen, T. M. Hackeng and P. E. Dawson, *Angew. Chem Int. Ed.* **45**, 7581–7584 (2006).

[37] A. Dirksen, S. Dirksen, T. M. Hackeng and P. E. Dawson, *J. Am. Chem. Soc.* **128**, 15602–15603 (2006).

[38] Y. Zeng, T. N. C. Ramya, A. Dirksen, P. E. Dawson and J. C. Paulson, *Nat. Methods* **6**, 207–209 (2009).

[39] T. Furuta, M. Mochizuki, M. Ito, T. Takahashi, T. Suzuki and T. Kan, *Org. Lett.* **10**, 4847–4850 (2008).

[40] E. H. M. Lempers, B. A. Helms, M. Merkx, E. W. Meijer, *ChemBioChem* **10**, 658–662 (2009).

[41] M. B. Thygesen, K. K. Sorensen, E. Clo and K. J. Jensen, *Chem. Commun.* **42**, 6367–6369 (2009).

[42] A. Dirksen and P. E. Dawson, *Bioconjugate Chem.* **19**, 2543–2548 (2008).

[43] A. Dirksen, S. Yegneswaran and P. E. Dawson, *Angew. Chem. Int. Ed.* **49**, 2023–2027 (2010).

[44] J. B. Blanco-Canosa, I. L. Medintz, D. Farrell, H. Mattoussi and P. E. Dawson, *J. Am. Chem. Soc.* **132**, 10027–10033 (2010).

[45] G. Iyer, F. Pinaud, J. M. Xu, Y. Ebenstein, J. Li, J. Chang, M. Dahan and S. Weiss, *Bioconjugate Chem.* **22**, 1006–1011 (2011).

[46] J. Y. Byeon, F. T. Limpoco and R. C. Bailey, *Langmuir* **26**, 15430–15435 (2010).

[47] J. Kalia and R. T. Raines, *Angew. Chem. Int. Ed.* **47**, 7523–7526 (2008).

[48] M. Hruby, C. Konak and K. Ulbrich, *J. Controlled Release* **103**, 137–148 (2005).

[49] H. Staudinger and J. Meyer, *Helv. Chim. Acta* **2**, 635–646 (1919).

[50] E. Saxon and C. R. Bertozzi, *Science* **287**, 2007–2010 (2000).

[51] E. Saxon, J. I. Armstrong and C. R. Bertozzi, *Org. Lett.* **2**, 2141–2143 (2000).

[52] B. L. Nilsson, L. L. Kiessling and R. T. Raines, *Org. Lett.* **2**, 1939–1941 (2000).

[53] S. S. van Berkel, M. B. van Eldijk and J. C. M. van Hest, *Angew. Chem. Int. Ed.* **50**, 8806–8827 (2011).

[54] M. Kohn and R. Breinbauer, *Angew. Chem. Int. Ed.* **43**, 3106–3116 (2004).

[55] P. V. Chang, J. A. Prescher, M. J. Hangauer and C. R. Bertozzi, *J. Am. Chem. Soc.* **129**, 8400–8401 (2007).

[56] G. A. Lemieux, C. L. de Graffenried and C. R. Bertozzi, *J. Am. Chem. Soc.* **125**, 4708–4709 (2003).

[57] M. J. Hangauer and C. R. Bertozzi, *Angew. Chem. Int. Ed.* **47**, 2394–2397 (2008).

[58] A. S. Cohen, E. A. Dubikovskaya, J. S. Rush and C. R. Bertozzi, *J. Am. Chem. Soc.* **132**, 8563–8565 (2010).

[59] A. A. Neves, H. Stockmann, R. R. Harmston, H. J. Pryor, I. S. Alam, H. Ireland-Zecchini, D. Y. Lewis, S. K. Lyons, F. J. Leeper and K. M. Brindle, *FASEB J.* **25**, 2528–2537 (2011).

[60] D. J. Vugts, A. Vervoort, M. Stigter-van Walsum, G. W. M. Visser, M. S. Robillard, R. M. Versteegen, R. C. M. Vulders, J. D. M. Herscheid and G. van Dongen, *Bioconjugate Chem.* **22**, 2072–2081 (2011).

[61] C. W. Tornøe, C. Christensen and M. Meldal, *J. Org. Chem.* **67**, 3057–3064 (2002).

[62] V. V. Rostovtsev, L. G. Green, V. V. Fokin and K. B. Sharpless, *Angew. Chem. Int. Ed.* **41**, 2596–2599 (2002).

[63] Q. Wang, T. R. Chan, R. Hilgraf, V. V. Fokin, K. B. Sharpless and M. G. Finn, *J. Am. Chem. Soc.* **125**, 3192–3193 (2003).

[64] T. R. Chan, R. Hilgraf, K. B. Sharpless and V. V. Fokin, *Org. Lett.* **6**, 2853–2855 (2004).

[65] V. O. Rodionov, S. I. Presolski, S. Gardinier, Y. H. Lim and M. G. Finn, J. Am. Chem. Soc. **129**, 12696–12704 (2007).

[66] S. I. Presolski, V. Hong, S. H. Cho and M. G. Finn, *J. Am. Chem. Soc.* **132**, 14570–14576 (2010).

[67] D. S. del Amo, W. Wang, H. Jiang, C. Besanceney, A. C. Yan, M. Levy, Y. Liu, F. L. Marlow and P. Wu, *J. Am. Chem. Soc.* **132**, 16893–16899 (2010).

[68] V. O. Rodionov, V. V. Fokin and M. G. Finn, *Angew. Chem. Int. Ed.* **44**, 2210–2215 (2005).

[69] V. O. Rodionov, S. I. Presolski, D. D. Diaz, V. V. Fokin and M. G. Finn, *J. Am. Chem. Soc.* **129**, 12705–12712 (2007).

[70] V. D. Bock, H. Hiemstra and J. H. van Maarseveen, *Eur. J. Org. Chem.* **1**, 51–68 (2006).

[71] Q. Wang, T. R. Chan, R. Hilgraf, V. V. Fokin, K. B. Sharpless and M. G. Finn, *J. Am. Chem. Soc.* **125**, 3192–3193 (2003).

[72] P. Wu and V. V. Fokin, *Aldrichim. Acta* **40**, 7–17 (2007).

[73] J. E. Moses and A. D. Moorhouse, *Chem. Soc. Rev.* **36**, 1249–1262 (2007).

[74] M. Meldal and C. W. Tornoe, *Chem. Rev.* **108**, 2952–3015 (2008).

[75] J. E. Hein and V. V. Fokin, *Chem. Soc. Rev.* **39**, 1302–1315 (2010).

[76] H. C. Kolb, M. G. Finn and K. B. Sharpless, *Angew. Chem., Int. Ed.* **40**, 2004–2021 (2001).

[77] J. Hu, J. R. Lu and Y. Ju, *Chem. Asian J.* **6**, 2636–2647 (2011).

[78] W. C. Qu, M. P. Kung, C. Hou, S. Oya and H. F. Kung, *J. Med. Chem.* **50**, 3380–3387 (2007).

[79] W. H. Binder and R. Sachsenhofer, *Macromol. Rapid Commun.* **28**, 15–54 (2007).

[80] M. van Dijk, D. T. S. Rijkers, R. M. J. Liskamp, C. F. van Nostrum and W. E. Hennink, *Bioconjugate Chem.* **20**, 2001–2016 (2009).

[81] J. A. Johnson, M. G. Finn, J. T. Koberstein and N. J. Turro, *Macromol. Rapid Commun.* **29**, 1052–1072 (2008).

[82] E. S. Read and S. P. Armes, *Chem. Commun.* 3021–3035 (2007).

[83] O. Altintas and U. Tunca, *Chem. Asian J.* **6**, 2584–2591 (2011).

[84] S. Slavin, J. Burns, D. M. Haddleton and C. R. Becer, *Eur. Polym. J.* **47**, 435–446 (2011).

[85] Y. L. Angell and K. Burgess, *Chem. Soc. Rev.* **36**, 1674–1689 (2007).

[86] J. M. Holub and K. Kirshenbaum, *Chem. Soc. Rev.* **39**, 1325–1337 (2010).

[87] B. Helms, J. L. Mynar, C. J. Hawker and J. M. J. Frechet, *J. Am. Chem. Soc.* **126**, 15020–15021 (2004).

[88] P. A. Ledin, F. Friscourt, J. Guo and G. J. Boons, *Chem. Eur. J.* **17**, 839–846 (2011).

[89] P. M. E. Gramlich, C. T. Wirges, A. Manetto and T. Carell, *Angew. Chem., Int. Ed.* **47**, 8350–8358 (2008).

[90] T. Yamada, C. G. Peng, S. Matsuda, H. Addepalli, K. N. Jayaprakash, M. R. Alam, K. Mills, M. A. Maier, K. Charisse, M. Sekine, M. Manoharan and K. G. Rajeev, *J. Org. Chem.* **76**, 1198–1211 (2011).

[91] E. Paredes and S. R. Das, *ChemBioChem* **12**, 125–131 (2011).

[92] H. J. Liu, T. Torring, M. D. Dong, C. B. Rosen, F. Besenbacher and K. V. Gothelf, *J. Am. Chem. Soc.* **132**, 18054–18056 (2010).

[93] Y. Li and C. Z. Cai, *Chem. Asian J.* **6**, 2592–2605 (2011).

[94] N. K. Devaraj and J. P. Collman, *QSAR Comb. Sci.* **26**, 1253–1260 (2007).

[95] M. A. Watson, J. Lyskawa, C. Zobrist, D. Fournier, M. Jimenez, M. Traisnel, L. Gengembre and P. Woisel, *Langmuir* **26**, 15920–15924 (2010).

[96] S. Y. Ku, K. T. Wong and A. J. Bard, *J. Am. Chem. Soc.* **130**, 2392–2393 (2008).

[97] R. K. O'Reilly, M. J. Joralemon, K. L. Wooley and C. J. Hawker, *Chem. Mater.* **17**, 5976–5988 (2005).

[98] L. Cui, J. A. Cohen, K. E. Broaders, T. T. Beaudette and J. M. J. Frechet, *Bioconjugate Chem.* **22**, 949–957 (2011).

[99] J. Y. Liu, Z. H. Nie, Y. Gao, A. Adronov and H. M. Li, *J. Polym. Sci. Pol. Chem.* **46**, 7187–7199 (2008).

[100] N. F. Steinmetz, V. Hong, E. D. Spoerke, P. Lu, K. Breitenkamp, M. G. Finn and M. Manchester, *J. Am. Chem. Soc.* **131**, 17093–17095 (2009).

[101] M. Colombo and A. Bianchi, *Molecules* **15**, 178–197 (2010).

[102] X. C. Li, *Chem. Asian J.* **6**, 2606–2616 (2011).

[103] D. Kobus, Y. Giesen, R. Ullrich, H. Backes and B. Neumaier, *Appl. Radiat. Isot.* **67**, 1977–1984 (2009).

[104] H. J. Breyholz, S. Wagner, A. Faust, B. Riemann, C. Holtke, S. Hermann, O. Schober, M. Schafers and K. Kopka, *ChemMedChem* **5**, 777–789 (2010).

[105] F. Pisaneschi, Q. D. Nguyen, E. Shamsaei, M. Glaser, E. Robins, M. Kaliszczak, G. Smith, A. C. Spivey, and E. O. Aboagye, *Bioorg. Med. Chem.* **18**, 6634–6645 (2010).

[106] T. L. Ross, M. Honer, P. Y. H. Lam, T. L. Mindt, V. Groehn, R. Schibli, P. A. Schubiger and S. M. Ametamey, *Bioconjugate Chem.* **19**, 2462–2470 (2008).

[107] T. L. Mindt, C. Muller, F. Stuker, J. F. Salazar, A. Hohn, T. Mueggler, M. Rudin and R. Schibli, *Bioconjugate Chem.* **20**, 1940–1949 (2009).

[108] H. S. Gill and J. Marik, *Nat. Protoc.* **6**, 1718–1725 (2011).

[109] Z. B. Li, Z. Wu, K. Chen, F. T. Chin and X. Chen, *Bioconjugate Chem.* **18**, 1987–1994 (2007).

[110] S. H. Hausner, J. Marik, M. K. J. Gagnon and J. L. Sutcliffe, *J. Med. Chem.* **51**, 5901–5904 (2008).

[111] J. A. H. Inkster, B. Guerin, T. J. Ruth and M. J. Adam, *J. Labelled Compd. Radiopharm.* **51**, 444–452 (2008).

[112] M. Lei, M. Tian and H. Zhang, *Current Medical Imaging Reviews* **6**, 33–41 (2010).

[113] S. Maschauer, J. Einsiedel, R. Haubner, C. Hocke, M. Ocker, H. Hubner, T. Kuwert, P. Gmeiner and O. Prante, *Angew. Chem. Int. Ed.* **49**, 976–979 (2010).

[114] D. E. Olberg and O. K. Hjelstuen, *Curr. Top. Med. Chem.* **10**, 1669–1679 (2010).

[115] O. Boutureira, F. D'Hooge, M. Fernandez-Gonzalez, G. J. L. Bernardes, M. Sanchez-Navarro, J. R. Koeppe and B. G. Davis, *Chem. Commun.* **46**, 8142–8144 (2010).

[116] J. A. H. Inkster, M. J. Adam, T. Storr and T. J. Ruth, *Nucleosides Nucleotides Nucleic Acids* **28**, 1131–1143 (2009).

[117] F. Mercier, J. Paris, G. Kaisin, D. Thonon, J. Flagothier, N. Teller, C. Lemaire and A. Luxen, *Bioconjugate Chem.* **22**, 108–114 (2011).

[118] J. Schulz, D. Vimont, T. Bordenave, D. James, J. M. Escudier, M. Allard, M. Szlosek-Pinaud and E. Fouquet, *Chem. Eur. J.* **17**, 3096–3100 (2011).

[119] P. J. H. Scott, *Angew. Chem., Int. Ed.* **48**, 6001–6004 (2009).

[120] S. Celen, J. Cleynhens, C. Deroose, T. de Groot, A. Ibrahimi, R. Gijsbers, Z. Debyser, L. Mortelmans, A. Verbruggen and G. Bormans, *Bioorg. Med. Chem.* **17**, 5117–5125 (2009).

[121] G. Gasser, K. Jager, M. Zenker, R. Bergmann, J. Steinbach, H. Stephan and N. Metzler-Nolte, *J. Inorg. Biochem.* **104**, 1133–1140 (2010).

[122] E. M. Kim, M. H. Joung, C. M. Lee, H. J. Jeong, S. T. Lim, M. H. Sohn and D. W. Kim, *Bioorg. Med. Chem. Lett.* **20**, 4240–4243 (2010).

[123] R. Yan, E. El-Emir, V. Rajkumar, M. Robson, A. P. Jathoul, R. B. Pedley and E. Arstad, *Angew. Chem., Int. Ed.* **50**, 6793–6795 (2011).

[124] I. Dijkgraaf, A. Y. Rijnders, A. Soede, A. C. Dechesne, G. W. van Esse, A. J. Brouwer, F. H. M. Corstens, O. C. Boerman, D. T. S. Rijkers and R. M. J. Liskamp, *Org. Biomol. Chem.* **5**, 935–944 (2007).

[125] M. Q. Tan, X. M. Wu, E. K. Jeong, Q. J. Chen and Z. R. Lu, *Biomacromolecules* **11**, 754–761 (2010).

[126] J. M. Bryson, W. J. Chu, J. H. Lee and T. M. Reineke, *Bioconjugate Chem.* **19**, 1505–1509 (2008).

[127] F. Fernandez-Trillo, J. Pacheco-Torres, J. Correa, P. Ballesteros, P. Lopez-Larrubia, S. Cerdan, R. Riguera and E. Fernandez-Megia, *Biomacromolecules* **12**, 2902–2907 (2011).

[128] E. Y. Sun, L. Josephson and R. Weissleder, *Molecular Imaging* **5**, 122–128 (2006).

[129] S. Santra, C. Kaittanis, J. Grimm and J. M. Perez, *Small* **5**, 1862–1868 (2009).

[130] A. L. Martin, L. M. Bernas, B. K. Rutt, P. J. Foster and E. R. Gillies, *Bioconjugate Chem.* **19**, 2375–2384 (2008).

[131] D. L. J. Thorek, D. R. Elias and A. Tsourkas, *Molecular Imaging* **8**, 221–229 (2009).

[132] K. Y. Pu, K. Li and B. Liu, *Adv. Funct. Mater.* **20**, 2770–2777 (2010).

[133] A. J. Link and D. A. Tirrell, *J. Am. Chem. Soc.* **125**, 11164–11165 (2003).

[134] M. Sawa, T. L. Hsu, T. Itoh, M. Sugiyama, S. R. Hanson, P. K. Vogt and C. H. Wong, *Proc. Natl. Acad. Sci. U. S. A.* **103**, 12371–12376 (2006).

[135] C. Y. Jao and A. Salic, *Proc. Natl. Acad. Sci.* **105**, 15779–15784 (2008).

[136] A. B. Neef and N. W. Luedtke, *Proc. Natl. Acad. Sci.* **108**, 20404–20409 (2011).

[137] C. Y. Jao, M. Roth, R. Welti and A. Salic, *Proc. Natl. Acad. Sci.* **106**, 15332–15337 (2009).

[138] C. F. Wu, Y. H. Jin, T. Schneider, D. R. Burnham, P. B. Smith and D. T. Chiu, *Angew. Chem. Int. Ed.* **49**, 9436–9440 (2010).

[139] Z. M. Li, T. S. Seo and J. Y. Ju, *Tetrahedron Lett.* **45**, 3143–3146 (2004).

[140] G. Wittig and A. Krebs, *Chem. Ber.-Recl.* **94**, 3260–3275 (1961).

[141] N. J. Agard, J. A. Prescher and C. R. Bertozzi, *J. Am. Chem. Soc.* **126**, 15046–15047 (2004).

[142] N. J. Agard, J. M. Baskin, J. A. Prescher, A. Lo and C. R. Bertozzi, *ACS Chem. Biol.* **1**, 644–648 (2006).

[143] J. M. Baskin, J. A. Prescher, S. T. Laughlin, N. J. Agard, P. V. Chang, I. A. Miller, A. Lo, J. A. Codelli and C. R. Bertozzi, *Proc. Natl. Acad. Sci.* **104**, 16793–16797 (2007).

[144] S. T. Laughlin, J. M. Baskin, S. L. Amacher and C. R. Bertozzi, *Science* **320**, 664–667 (2008).

[145] S. T. Laughlin and C. R. Bertozzi, *ACS Chem. Biol.* **4**, 1068–1072 (2009).

[146] J. M. Baskin, K. W. Dehnert, S. T. Laughlin, S. L. Amacher and C. R. Bertozzi, *Proc. Natl. Acad. Sci.* **107**, 10360–10365 (2010).

[147] K. W. Dehnert, B. J. Beahm, T. T. Huynh, J. M. Baskin, S. T. Laughlin, W. Wang, P. Wu, S. L. Amacher and C. R. Bertozzi, *ACS Chem. Biol.* **6**, 547–552 (2011).

[148] P. V. Chang, J. A. Prescher, E. M. Sletten, J. M. Baskin, I. A. Miller, N. J. Agard, A. Lo and C. R. Bertozzi, *Proc. Natl. Acad. Sci.* **107**, 1821–1826 (2010).

[149] M. F. Debets, S. S. Van Berkel, J. Dommerholt, A. J. Dirks, F. Rutjes and F. L. Van Delft, *Acc. Chem. Res.* **44**, 805–815 (2011).

[150] J. A. Codelli, J. M. Baskin, N. J. Agard and C. R. Bertozzi, *J. Am. Chem. Soc.* **130**, 11486–11493 (2008).

[151] J. Dommerholt, S. Schmidt, R. Temming, L. J. A. Hendriks, F. Rutjes, J. C. M. van Hest, D. J. Lefeber, P. Friedl, F. L. van Delft, *Angew. Chem., Int. Ed.* **49**, 9422–9425 (2010).

[152] J. C. Jewett, E. M. Sletten, C. R. Bertozzi, *J. Am. Chem. Soc.* **132**, 3688–3690 (2010).

[153] J. C. Jewett and C. R. Bertozzi, *Org. Lett.* **13**, 5937–5939 (2011).

[154] E. M. Sletten and C. R. Bertozzi, *Org. Lett.* **10**, 3097–3099 (2008).

[155] M. F. Debets, S. S. van Berkel, S. Schoffelen, F. Rutjes, J. C. M. van Hest and F. L. van Delft, *Chem. Commun.* **46**, 97–99 (2010).

[156] K. E. Beatty, J. Szychowski, J. D. Fisk and D. A. Tirrell, *ChemBioChem* **12**, 2137–2139 (2011).

[157] A. Bernardin, A. Cazet, L. Guyon, P. Delannoy, F. Vinet, D. Bonnaffe and I. Texier, *Bioconjugate Chem.* **21**, 583–588 (2010).

[158] J. A. Johnson, J. M. Baskin, C. R. Bertozzi, J. T. Koberstein and N. J. Turro, *Chem. Commun.* 3064–3066 (2008).

[159] G. H. Hur, J. L. Meier, J. Baskin, J. A. Codelli, C. R. Bertozzi, M. A. Marahiel and M. D. Burkart, *Chem. Biol.* **16**, 372–381 (2009).

[160] C. Ornelas, J. Broichhagen and M. Weck, *J. Am. Chem. Soc.* **132**, 3923–3931 (2010).

[161] M. E. Martin, S. G. Parameswarappa, M. S. O'Dorisio, F. C. Pigge and M. K. Schultz, *Bioorg. Med. Chem. Lett.* **20**, 4805–4807 (2010).

[162] N. J. Baumhover, M. E. Martin, S. G. Parameswarappa, K. C. Kloepping, M. S. O'Dorisio, F. C. Pigge and M. K. Schultz, *Bioorg. Med. Chem. Lett.* **21**, 5757–5761 (2011).

[163] L. S. Campbell-Verduyn, L. Mirfeizi, A. K. Schoonen, R. A. Dierckx, P. H. Elsinga and B. L. Feringa, *Angew. Chem., Int. Ed.* **50**, 11117–11120 (2011).

[164] C. S. McKay, J. A. Blake, J. Cheng, D. C. Danielson and J. P. Pezacki, *Chem. Commun.* **47**, 10040–10042 (2011).

[165] X. H. Ning, R. P. Temming, J. Dommerholt, J. Guo, D. B. Ania, M. F. Debets, M. A. Wolfert, G. J. Boons and F. L. van Delft, *Angew. Chem., Int. Ed.* **49**, 3065–3068 (2010).

[166] R. A. Carboni and R. V. Lindsey, *J. Am. Chem. Soc.* **81**, 4342–4346 (1959).

[167] D. L. Boger, *Chem. Rev.* **86**, 781–793 (1986).

[168] F. Thalhammer, U. Wallfahrer and J. Sauer, *Tetrahedron Lett.* **31**, 6851–6854 (1990).

[169] M. L. Blackman, M. Royzen and J. M. Fox, *J. Am. Chem. Soc.* **130**, 13518–13519 (2008).

[170] M. T. Taylor, M. L. Blackman, O. Dmitrenko and J. M. Fox, *J. Am. Chem. Soc.* **133**, 9646–9649 (2011).

[171] M. Royzen, G. P. A. Yap and J. M. Fox, *J. Am. Chem. Soc.* **130**, 3760–3761 (2008).

[172] J. B. Haun, N. K. Devaraj, S. A. Hilderbrand, H. Lee and R. Weissleder, *Nat. Nanotechnol.* **5**, 660–665 (2010).

[173] J. B. Haun, N. K. Devaraj, B. S. Marinelli, H. Lee and R. Weissleder, *ACS Nano* **5**, 3204–3213 (2011).

[174] N. K. Devaraj, R. Weissleder and S. A. Hilderbrand, *Bioconjugate Chem.* **19**, 2297–2299 (2008).

[175] M. R. Karver, R. Weissleder and S. A. Hilderbrand, *Bioconjugate Chem.* **22**, 2263–2270 (2011).

[176] N. K. Devaraj and R. Weissleder, *Acc. Chem. Res.* **44**, 816–827 (2011).

[177] H. S. Han, N. K. Devaraj, J. Lee, S. A. Hilderbrand, R. Weissleder and M. G. Bawendi, *J. Am. Chem. Soc.* **132**, 7838–7839 (2010).

[178] J. Schoch, M. Wiessler and A. Jaschke, *J. Am. Chem. Soc.* **132**, 8846–8847 (2010).

[179] J. Schoch, S. Ameta and A. Jaschke, *Chem. Commun.* **47**, 12536–12537 (2011).

[180] C. F. Hansen, P. Espeel, M. M. Stamenovic, I. A. Barker, A. P. Dove, F. E. Du Prez and R. K. O'Reilly, *J. Am. Chem. Soc.* **133**, 13828–13831 (2011).

[181] N. K. Devaraj, R. Upadhyay, J. B. Hatin, S. A. Hilderbrand and R. Weissleder, *Angew. Chem., Int. Ed.* **48**, 7013–7016 (2009).

[182] N. K. Devaraj, S. Hilderbrand, R. Upadhyay, R. Mazitschek and R. Weissleder, *Angew. Chem., Int. Ed.* **49**, 2869–2872 (2010).

[183] R. Rossin, P. R. Verkerk, S. M. van den Bosch, R. C. M. Vulders, I. Verel and J. Lub, M. S. Robillard, *Angew. Chem., Int. Ed.* **49**, 3375–3378 (2010).

[184] T. Reiner, E. J. Keliher, S. Earley, B. Marinelli and R. Weissleder, *Angew. Chem., Int. Ed.* **50**, 1922–1925 (2011).

[185] R. Selvaraj, S. L. Liu, M. Hassink, C. W. Huang, L. P. Yap, R. Park, J. M. Fox, Z. B. Li and P. S. Conti, *Bioorg. Med. Chem. Lett.* **21**, 5011–5014 (2011).

[186] B. M. Zeglis, P. Mohindra, G. I. Weissmann, V. Divilov, S. A. Hilderbrand, R. Weissleder and J. S. Lewis, *Bioconjugate Chem.* **22**, 2048–2059 (2011).

[187] Z. B. Li, H. C. Cai, M. Hassink, M. L. Blackman, R. C. D. Brown, P. S. Conti and J. M. Fox, *Chem. Commun.* **46**, 8043–8045 (2010).

[188] T. Reiner, S. Earley, A. Turetsky and R. Weissleder, *ChemBioChem* **11**, 2374–2377 (2010).

[189] E. J. Keliher, T. Reiner, A. Turetsky, S. A. Hilderbrand and R. Weissleder, *ChemMedChem* **6**, 424–427 (2011).

[190] C. E. Hoyle, T. Y. Lee and T. Roper, *J.Polym. Sci. Pol. Chem.* **42**, 5301–5338 (2004).

[191] A. Gress, A. Völkel and H. Schlaad, *Macromolecules* **40**, 7928–7933 (2007).

[192] R. L. A. David and J. A. Kornfield, *Macromolecules* **41**, 1151–1161 (2008).

[193] C. D. Heidecke and T. K. Lindhorst, *Chem. Eur. J.* **13**, 9056–9067 (2007).

[194] K. L. Killops, L. M. Campos and C. J. Hawker, *J. Am. Chem. Soc.* **130**, 5062–5064 (2008).

[195] A. Dondoni, A. Massi, P. Nanni and A. Roda, *Chem. Eur. J.* **15**, 11444–11449 (2009).

[196] A. van der Ende, T. Croce, S. Hamilton, V. Sathiyakumar and E. Harth, *Soft Matter* **5**, 1417–1425 (2009).

[197] K. Hayashi, K. Ono, H. Suzuki, M. Sawada, M. Moriya, W. Sakamoto and T. Yogo, *Chem. Mater.* **22**, 3768–3772 (2010).

[198] A. Pfaff, A. Schallon, T. M. Ruhland, A. P. Majewski, H. Schmalz, R. Freitag and A. H. E. Muller, *Biomacromolecules* **12**, 3805–3811 (2011).

[199] Y. Iwasaki and H. Matsuno, *Macromol. Biosci.* **11**, 1478–1483 (2011).

[200] H. Y. Shiu, T. C. Chan, C. M. Ho, Y. Liu, M. K. Wong and C. M. Che, *Chem. Eur. J.* **15**, 3839–3850 (2009).

[201] R. H. Grubbs, *Handbook of Metathesis.* Wiley-VCH: Weinheim, Germany, 2003.

[202] A. H. Hoveyda and A. R. Zhugralin, *Nature* **450**, 243–251 (2007).

[203] S. J. Connon and S. Blechert, *Angew. Chem. Int. Ed.* **42**, 1900–1923 (2003).

[204] J. B. Binder and R. T. Raines, *Curr. Opin. Chem. Biol.* **12**, 767–773 (2008).

[205] S. Zaman, O. J. Curnow and A. D. Abell, *Aust. J. Chem.* **62**, 91–100 (2009).

[206] Y. Y. A. Lin, J. M. Chalker and B. G. Davis, *ChemBioChem* **10**, 959–969 (2009).

[207] Y. A. Lin, J. M. Chalker, N. Floyd, G. J. L. Bernardes and B. G. Davis, *J. Am. Chem. Soc.* **130**, 9642–9643 (2008).

[208] Y. A. Lin, J. M. Chalker and B. G. Davis, *J. Am. Chem. Soc.* **132**, 16805–16811 (2010).

[209] M. J. Allen, R. T. Raines and L. L. Kiessling, *J. Am. Chem. Soc.* **128**, 6534–6535 (2006).

[210] R. Skoda-Foldes and L. Kollar, *Chem. Rev.* **103**, 4095–4129 (2003).

[211] T. Yamamoto, *J. Organomet. Chem.* **653**, 195–199 (2002).

[212] M. Pretze, P. Grosse-Gehling and C. Mamat, *Molecules* **16**, 1129–1165 (2011).

[213] N. Miyaura and A. Suzuki, *Chem. Rev.* **95**, 2457–2483 (1995).

[214] J. M. Chalker, C. S. C. Wood and B. G. Davis, *J. Am. Chem. Soc.* **131**, 16346–16347 (2009).

[215] I. P. Beletskaya and A. V. Cheprakov, *Chem. Rev.* **100**, 3009–3066 (2000).

[216] R. L. Simmons, R. T. Yu and A. G. Myers, *J. Am. Chem. Soc.* **133**, 15870–15873 (2011).

[217] R. Chinchilla and C. Najera, *Chem. Rev.* **107**, 874–922 (2007).

[218] N. Li, R. K. V. Lm, S. Edwardraja and Q. Lin, *J. Am. Chem. Soc.* **133**, 15316–15319 (2011).

[219] P. E. Dawson, T. W. Muir, I. Clarklewis and S. B. H. Kent, *Science* **266**, 776–779 (1994).

[220] P. E. Dawson, M. J. Churchill, M. R. Ghadiri and S. B. H. Kent, *J. Am. Chem. Soc.* **119**, 4325–4329 (1997).

[221] E. C. B. Johnson and S. B. H. Kent, *J. Am. Chem. Soc.* **128**, 6640–6646 (2006).

[222] L. E. Canne, S. J. Bark and S. B. H. Kent, *J. Am. Chem. Soc.* **118**, 5891–5896 (1996).

[223] L. Z. Yan and P. E. Dawson, *J. Am. Chem. Soc.* **123**, 526–533 (2001).

[224] D. Crich and A. Banerjee, *J. Am. Chem. Soc.* **129**, 10064–10065 (2007).

[225] C. Haase, H. Rohde and O. Seitz, *Angew. Chem. Int. Ed.* **47**, 6807–6810 (2008).

[226] R. L. Yang, K. K. Pasunooti, F. P. Li, X. W. Liu and C. F. Liu, *J. Am. Chem. Soc.* **131**, 13592–13593 (2009).

[227] Z. Harpaz, P. Siman, K. S. A. Kumar and A. Brik, *ChemBioChem* **11**, 1232–1235 (2010).

[228] B. Wu, J. H. Chen, J. D. Warren, G. Chen, Z. H. Hua and S. J. Danishefsky, *Angew. Chem. Int. Ed.* **45**, 4116–4125 (2006).

[229] P. E. Dawson and S. B. H. Kent, *Annu. Rev. Biochem.* **69**, 923–960 (2000).

[230] W. Y. Lu, M. A. Qasim and S. B. H. Kent, *J. Am. Chem. Soc.* **118**, 8518–8523 (1996).

[231] S. Y. Shang, Z. P. Tan and S. J. Danishefsky, *Proc. Natl. Acad. Sci.* **108**, 5986–5989 (2011).

[232] C. Dose and O. Seitz, *Org. Lett.* **7**, 4365–4368 (2005).

[233] S. Ficht, C. Dose and O. Seitz, *ChemBioChem* **6**, 2098–2103 (2005).

[234] M. J. Grogan, Y. Kaizuka, R. M. Conrad, J. T. Groves and C. R. Bertozzi, *J. Am. Chem. Soc.* **127**, 14383–14387 (2005).

[235] S. Ingale, T. Buskas and G. J. Boons, *Org. Lett.* **8**, 5785–5788 (2006).

[236] I. van Baal, H. Malda, S. A. Synowsky, J. L. J. van Dongen, T. M. Hackeng, M. Merkx and E. W. Meijer, *Angew. Chem. Int. Ed.* **44**, 5052–5057 (2005).

[237] B. H. Hu, J. Su and P. B. Messersmith, *Biomacromolecules* **10**, 2194–2200 (2009).

[238] S. W. A. Reulen, P. Y. W. Dankers, P. H. H. Bomans, E. W. Meijer and M. Merkx, *J. Am. Chem. Soc.* **131**, 7304–7312 (2009).

[239] E. Wieczerzak, R. Hamel, V. Chabot, V. Aimez, M. Grandbois, P. G. Charette and E. Escherl, *Biopolymers* **90**, 415–420 (2008).

[240] D. S. Y. Yeo, R. Srinivasan, G. Y. J. Chen and S. Q. Yao, *Chem. Eur. J.* **10**, 4664–4672 (2004).

[241] E. H. White, F. McCapra, G. F. Field, and W. D. McElroy, *J. Am. Chem. Soc.* **83**, 2402–2403 (1961).

[242] H. J. Ren, F. Xiao, K. Zhan, Y. P. Kim, H. X. Xie and Z. Y. Xia, J. Rao, *Angew. Chem. Int. Ed.* **48**, 9658–9662 (2009).

[243] G. L. Liang, H. J. Ren and J. H. Rao, *Nature Chem.* **2**, 54–60 (2010).

[244] G. L. Liang, J. Ronald, Y. X. Chen, D. J. Ye, P. Pandit, M. L. Ma, B. Rutt and J. H. Rao, *Angew. Chem. Int. Ed.* **50**, 6283–6286 (2011).

[245] D. J. Ye, G. L. Liang, M. L. Ma and J. H. Rao, *Angew. Chem. Int. Ed.* **50**, 2275–2279 (2011).

[246] D. P. Nguyen, T. Elliott, M. Holt, T. W. Muir and J. W. Chin, *J. Am. Chem. Soc.* **133**, 11418–11421 (2011).

[247] J. M. Antos and M. B. Francis, *Curr. Opin. Chem. Biol.* **10**, 253–262 (2006).

[248] T. L. Schlick, Z. B. Ding, E. W. Kovacs and M. B. Francis, *J. Am. Chem. Soc.* **127**, 3718–3723 (2005).

[249] N. S. Joshi, L. R. Whitaker and M. B. Francis, *J. Am. Chem. Soc.* **126**, 15942–15943 (2004).

[250] S. D. Tilley and M. B. Francis, *J. Am. Chem. Soc.* **128**, 1080–1081 (2006).

[251] K. L. Seim, A. C. Obermeyer and M. B. Francis, *J. Am. Chem. Soc.* **133**, 16970–16976 (2011).

[252] J. M. Antos and M. B. Francis, *J. Am. Chem. Soc.* **126**, 10256–10257 (2004).

[253] J. M. Antos, J. M. McFarland, A. T. Iavarone and M. B. Francis, *J. Am. Chem. Soc.* **131**, 6301–6308 (2009).

[254] J. M. Gilmore, R. A. Scheck, A. P. Esser-Kahn, N. S. Joshi and M. B. Francis, *Angew. Chem. Int. Ed.* **45**, 5307–5311 (2006).

[255] V. Muralidharan and T. W. Muir, *Nat. Methods* **3**, 429–438 (2006).

[256] T. W. Muir, D. Sondhi and P. A. Cole, *Proc. Natl. Acad. Sci.* **95**, 6705–6710 (1998).

[257] R. R. Flavell, P. Kothari, M. Bar-Dagan, M. Synan, S. Vallabhajosula, J. M. Friedman, T. W. Muir and G. Ceccarini, *J. Am. Chem. Soc.* **130**, 9106–9112 (2008).

[258] Z. Y. Xia, Y. Xing, M. K. So, A. L. Koh, R. Sinclair and J. H. Rao, *Anal. Chem.* **80**, 8649–8655 (2008).

[259] I. S. Carrico, B. L. Carlson and C. R. Bertozzi, *Nat. Chem. Biol.* **3**, 321–322 (2007).

[260] Z. Y. Hao, S. L. Hong, X. Chen and P. R. Chen, *Acc. Chem. Res.* **44**, 742–751 (2011).

[261] K. F. Geoghegan and J. G. Stroh, *Bioconjugate Chem.* **3**, 138–146 (1992).

[262] H. F. Gaertner and R. E. Offord, *Bioconjugate Chem.* **7**, 38–44 (1996).

[263] O. T. Keppler, R. Horstkorte, M. Pawlita, C. Schmidts and W. Reutter, *Glycobiology* **11**, 11R–18R (2001).

[264] J. A. Prescher, D. H. Dube and C. R. Bertozzi, *Nature* **430**, 873–877 (2004).

[265] S. G. Sampathkumar, M. B. Jones and K. J. Yarema, *Nat. Protoc.* **1**, 1840–1851 (2006).

[266] T. L. Hsu, S. R. Hanson, K. Kishikawa, S. K. Wang, M. Sawa and C. H. Wong, *Proc. Natl. Acad. Sci.* **104**, 2614–2619 (2007).

[267] L. Wang, A. Brock, B. Herberich and P. G. Schultz, *Science* **292**, 498–500 (2001).

[268] L. Wang, Z. W. Zhang, A. Brock and P. G. Schultz, *Proc. Natl. Acad. Sci.* **100**, 56–61 (2003).

[269] Z. W. Zhang, B. A. C. Smith, L. Wang, A. Brock, C. Cho and P. G. Schultz, *Biochemistry* **42**, 6735–6746 (2003).

[270] H. T. Liu, L. Wang, A. Brock, C. H. Wong and P. G. Schultz, *J. Am. Chem. Soc.* **125**, 1702–1703 (2003).

[271] S. X. Ye, C. Kohrer, T. Huber, M. Kazmi, P. Sachdev, E. C. Y. Yan, A. Bhagat, U. L. RajBhandary and T. P. Sakmar, *J. Biol. Chem.* **283**, 1525–1533 (2008).

[272] P. Wu, W. Q. Shui, B. L. Carlson, N. Hu, D. Rabuka, J. Lee and C. R. Bertozzi, *Proc. Natl. Acad. Sci.* **106**, 3000–3005 (2009).

[273] M. R. Fleissner, E. M. Brustad, T. Kalai, C. Altenbach, D. Cascio, F. B. Peters, K. Hideg, S. Peuker, P. G. Schultz and W. L. Hubbell, *Proc. Natl. Acad. Sci.* **106**, 21637–21642 (2009).

[274] M. Rashidian, J. K. Dozier, S. Lenevich and M. D. Distefano, *Chem. Commun.* **46**, 8998–9000 (2010).

[275] X. H. Zhang, F. P. Li, X. W. Lu and C. F. Liu, *Bioconjugate Chem.* **20**, 197–200 (2009).

[276] L. Yi, H. Y. Sun, Y. W. Wu, G. Triola, H. Waldmann and R. S. Goody, *Angew. Chem. Int. Ed.* **49**, 9417–9421 (2010).

[277] A. Deiters, T. A. Cropp, M. Mukherji, J. W. Chin, J. C. Anderson and P. G. Schultz, *J. Am. Chem. Soc.* **125**, 11782–11783 (2003).

[278] A. Deiters and P. G. Schultz, *Bioorg. Med. Chem. Lett.* **15**, 1521–1524 (2005).

[279] C. S. Cazalis, C. A. Haller, L. Sease-Cargo and E. L. Chaikof, *Bioconjugate Chem.* **15**, 1005–1009 (2004).

[280] S. I. Van Kasteren, H. B. Kramer, D. P. Gamblin and B. G. Davis, *Nat. Protoc.* **2**, 3185–3194 (2007).

[281] S. I. van Kasteren, H. B. Kramer, H. H. Jensen, S. J. Campbell, J. Kirkpatrick, N. J. Oldham, D. C. Anthony and B. G. Davis, *Nature* **446**, 1105–1109 (2007).

[282] E. Strable, D. E. Prasuhn, A. K. Udit, S. Brown, A. J. Link, J. T. Ngo, G. Lander, J. Quispe, C. S Potter, B. Carragher, D. A. Tirrell and M. G. Finn, *Bioconjugate Chem.* **19**, 866–875 (2008).

[283] S. Schoffelen, M. B. van Eldijk, B. Rooijakkers, R. Raijmakers, A. J. R. Heck and J. C. M. van Hest, *Chem. Sci.* **2**, 701–705 (2011).

[284] J. C. M. van Hest and D. A. Tirrell, *FEBS Lett.* **428**, 68–70 (1998).

[285] J. C. M. van Hest, K. L. Kick and D. A. Tirrell, *J. Am. Chem. Soc.* **122**, 1282–1288 (2000).

[286] Z. W. Zhang, L. Wang, A. Brock and P. G. Schultz, *Angew. Chem. Int. Ed.* **41**, 2840–2842 (2002).

[287] J. Wang, S. M. Schiller and P. G. Schultz, *Angew. Chem. Int. Ed.* **46**, 6849–6851 (2007).

[288] H. W. Ai, W. J. Shen, E. Brustad and P. G. Schultz, *Angew. Chem. Int. Ed.* **49**, 935–937 (2010).

[289] T. Yanagisawa, R. Ishii, R. Fukunaga, T. Kobayashi, K. Sakamoto and S. Yokoyama, *Chem. Biol.* **15**, 1187–1197 (2008).

[290] M. Chalfie, Y. Tu, G. Euskirchen, W. W. Ward and D. C. Prasher, *Science* **263**, 802–805 (1994).

[291] R. Heim, D. C. Prasher and R. Y. Tsien, *Proc. Natl. Acad. Sci.* **91**, 12501–12504 (1994).

[292] B. A. Griffin, S. R. Adams and R. Y. Tsien, *Science* **281**, 269–272 (1998).

[293] S. R. Adams, R. E. Campbell, L. A. Gross, B. R. Martin, G. K. Walkup, Y. Yao, J. Llopis and R. Y. Tsien, *J. Am. Chem. Soc.* **124**, 6063–6076 (2002).

[294] M. Andresen, R. Schmitz-Salue and S. Jakobs, *Mol. Biol. Cell* **15**, 5616–5622 (2004).

[295] B. R. Martin, B. N. G. Giepmans, S. R. Adams and R. Y. Tsien, *Nat. Biotechnol.* **23**, 1308–1314 (2005).

[296] S. R. Adams and R. Y. Tsien, *Nat. Protoc.* **3**, 1527–1534 (2008).

[297] C. Hoffmann, G. Gaietta, A. Zurn, S. R. Adams, S. Terrillon, M. H. Ellisman, R. Y. Tsien and M. J. Lohse, *Nat. Protoc.* **5**, 1666–1677 (2010).

[298] C. F. Pereira, P. C. Ellenberg, K. L. Jones, T. L. Fernandez, R. P. Smyth, D. J. Hawkes, M. Hijnen, V. Vivet-Boudou, R. Marquet, I. Johnson and J. Mak, *PLoS One* **6**, e17016 (2011).

[299] G. Gaietta, T. J. Deerinck, S. R. Adams, J. Bouwer, O. Tour, D. W. Laird, G. E. Sosinsky, R. Y. Tsien and M. H. Ellisman, *Science* **296**, 503–507 (2002).

[300] M. N. Rozinov and G. P. Nolan, *Chem. Biol.* **5**, 713–728 (1998).

[301] K. M. Marks, M. Rosinov and G. P. Nolan, *Chem. Biol.* **11**, 347–356 (2004).

[302] T. L. Halo, J. Appelbaum, E. M. Hobert, D. M. Balkin and A. Schepartz, *J. Am. Chem. Soc.* **131**, 438–439 (2009).

[303] J. J. Lee, S. C. Lee, D. T. Zhai, Y. H. Ahn, H. Y. Yeo, Y. L. Tan and Y. T. Chang, *Chem. Commun.* **47**, 4508–4510 (2011).

[304] E. Hochuli, H. Dobeli and A. Schacher, *J. Chromatogr.* **411**, 177–184 (1987).

[305] K. J. Franz, M. Nitz and B. Imperiali, *ChemBioChem* **4**, 265–271 (2003).

[306] H. Block, B. Maertens, A. Spriestersbach, N. Brinker, J. Kubicek, R. Fabis, J. Labahn and F. Schafer, In *Guide to Protein Purification, Second Edition*, R. R. Burgess and M. P. Deutscher, Eds., 2009; Vol. **463**, pp. 439–473.

[307] E. G. Guignet, R. Hovius and H. Vogel, *Nat. Biotechnol.* **22**, 440–444 (2004).

[308] C. R. Goldsmith, J. Jaworski, M. Sheng and S. J. Lippard, *J. Am. Chem. Soc.* **128**, 418–419 (2006).

[309] A. Ojida, K. Honda, D. Shinmi, S. Kiyonaka, Y. Mori and I. Hamachi, *J. Am. Chem. Soc.* **128**, 10452–10459 (2006).

[310] S. Lata, M. Gavutis, R. Tampe and J. Piehler, *J. Am. Chem. Soc.* **128**, 2365–2372 (2006).

[311] C. T. Hauser and R. Y. Tsien, *Proc. Natl. Acad. Sci.* **104**, 3693–3697 (2007).

[312] I. Chen, Y. A. Choi and A. Y. Ting, *J. Am. Chem. Soc.* **129**, 6619–6625 (2007).

[313] M. Howarth and A. Y. Ting, *Nat. Protoc.* **3**, 534–545 (2008).

[314] M. Howarth, K. Takao, Y. Hayashi and A. Y. Ting, *Proc. Natl. Acad. Sci.* **102**, 7583–7588 (2005).

[315] I. Chen, M. Howarth, W. Y. Lin and A. Y. Ting, *Nat. Methods* **2**, 99–104 (2005).

[316] M. Fernandez-Suarez, T. S. Chen and A. Y. Ting, *J. Am. Chem. Soc.* **130**, 9251–9253 (2008).

[317] S. Puthenveetil, D. S. Liu, K. A. White, S. Thompson and A. Y. Ting, *J. Am. Chem. Soc.* **131**, 16430–16438 (2009).

[318] M. Fernandez-Suarez, H. Baruah, L. Martinez-Hernandez, K. T. Xie, J. M. Baskin, C. R. Bertozzi and A. Y. Ting, *Nat. Biotechnol.* **25**, 1483–1487 (2007).

[319] H. Baruah, S. Puthenveetil, Y. A. Choi, S. Shah and A. Y. Ting, *Angew. Chem. Int. Ed.* **47**, 7018–7021 (2008).

[320] H. Y. Mao, S. A. Hart, A. Schink and B. A. Pollok, *J. Am. Chem. Soc.* **126**, 2670–2671 (2004).

[321] M. W. Popp, J. M. Antos, G. M. Grotenbreg, E. Spooner and H. L. Ploegh, *Nat. Chem. Biol.* **3**, 707–708 (2007).

[322] T. Yamamoto and T. Nagamune, *Chem. Commun.* 1022–1024 (2009).

[323] S. Tsukiji and T. Nagamune, *ChemBioChem* **10**, 787–798 (2009).

[324] T. Proft, *Biotechnol. Lett.* **32**, 1–10 (2010).

[325] J. Farinas and A. S. Verkman, *J. Biol. Chem.* **274**, 7603–7606 (1999).

[326] K. M. Marks, P. D. Braun and G. P. Nolan, *Proc. Natl. Acad. Sci.* **101**, 9982–9987 (2004).

[327] A. Keppler, S. Gendreizig, T. Gronemeyer, H. Pick, H. Vogel and K. Johnsson, *Nat. Biotechnol.* **21**, 86–89 (2003).

[328] T. Klein, A. Loschberger, S. Proppert, S. Wolter, S. V. van de Linde and M. Sauer, *Nat. Methods* **8**, 7–9 (2011).

[329] D. Maurel, L. Comps-Agrar, C. Brock, M. L. Rives, E. Bourrier, M. A. Ayoub, H. Bazin, N. Tinel, T. Durroux, L. Prezeau, E. Trinquet and J. P. Pin, *Nat. Methods* **5**, 561–567 (2008).

[330] M. Eckhardt, M. Anders, W. Muranyi, M. Heilemann, J. Krijnse-Locker and B. Muller, *PLoS One* **6**, e22007 (2011).

[331] L. Iversen, N. Cherouati, T. Berthing, D. Stamou and K. L. Martinez, *Langmuir* **24**, 6375–6381 (2008).

[332] A. Petershans, D. Wedlich and L. Fruk, *Chem. Commun.* **47**, 10671–10673 (2011).

[333] D. Srikun, A. E. Albers, C. I. Nam, A. T. Iavaron and C. J. Chang, *J. Am. Chem. Soc.* **132**, 4455–4465 (2010).

[334] A. F. Hussain, F. Kampmeier, V. von Felbert, H. F. Merk, M. K. Tur and S. Barth, *Bioconjugate Chem.* **22**, 2487–2495 (2011).

[335] A. Gautier, A. Juillerat, C. Heinis, I. R. Correa, M. Kindermann, F. Beaufils and K. Johnsson, *Chem. Biol.* **15**, 128–136 (2008).

[336] D. B. Janssen, *Curr. Opin. Chem. Biol.* **8**, 150–159 (2004).

[337] G. V. Los, L. P. Encell, M. G. McDougall, D. D. Hartzell, N. Karassina, C. Zimprich, M. G. Wood, R. Learish, R. F. Ohane, M. Urh, D. Simpson, J. Mendez, K. Zimmerman, P. Otto, G. Vidugiris, J. Zhu, A. Darzins, D. H. Klaubert, R. F. Bulleit and K. V. Wood, *ACS Chem. Biol.* **3**, 373–382 (2008).

[338] J. Schroder, H. Eenink, M. Dyba and G. V. Los, *Biophys. J.* **96**, L1–L3 (2009).

[339] Y. Zhang, M. K. So, A. M. Loening, H. Q. Yao, S. S. Gambhir and J. H. Rao, *Angew. Chem. Int. Ed.* **45**, 4936–4940 (2006).

[340] M. K. So, H. Q. Yao and J. H. Rao, *Biochem. Biophys. Res. Commun.* **374**, 419–423 (2008).

[341] H. L. D. Lee, S. J. Lord, S. Iwanaga, K. Zhan, H. X. Xie, J. C. Williams, H. Wang, G. R. Bowman, E. D. Goley, L. Shapiro, R. J. Twieg, J. H. Rao and W. E. Moerner, *J. Am. Chem. Soc.* **132**, 15099–15101 (2010).

[342] S. J. Huybrechts, P. P. Van Veldhoven, C. Brees, G. P. Mannaerts, G. V. Los and M. Fransen, *Traffic* **10**, 1722–1733 (2009).

[343] Z. P. Zhou, Y. Shimizu, H. Tadakuma, H. Taguchi, K. Ito and T. Ueda, *J. Biochem.* **149**, 609–613 (2011).

[344] N. Kosaka, M. Ogawa, P. L. Choyke, N. Karassina, C. Corona, M. McDougall, D. T. Lynch, C. C. Hoyt, R. M. Levenson, G. V. Los and H. Kobayashi, *Bioconjugate Chem.* **20**, 1367–1374 (2009).

[345] T. Kuchimaru, T. Kadonosono, S. Tanaka, T. Ushiki, M. Hiraoka and S. Kizaka-Kondoh, *PLoS One* **5**, e15736 (2010).

[346] R. C. Strauch, D. J. Mastarone, P. A. Sukerkar, Y. Song, J. J. Ipsaro and T. J. Meade, *J. Am. Chem. Soc.* **133**, 16346–16349 (2011).

[347] D. P. Baccanari, S. Daluge and R. W. King, *Biochemistry* **21**, 5068–5075 (1982).

[348] L. W. Miller, Y. F. Cai, M. P. Sheetz and V. W. Cornish, *Nat. Methods* **2**, 255–257 (2005).

[349] R. Wombacher, M. Heidbreder, S. van de Linde, M. P. Sheetz, M. Heilemann, V. W. Cornish and M. Sauer, *Nat. Methods* **7**, 717–719 (2010).

[350] S. S. Gallagher, C. R. Jing, D. S. Peterka, M. Konate, R. Wombacher, L. J. Kaufman, R. Yuste and V. W. Cornish, *ChemBioChem* **11**, 782–784 (2010).

[351] H. E. Rajapakse, D. R. Reddy, S. Mohandessi, N. G. Butlin and L. W. Miller, *Angew. Chem., Int. Ed.* **48**, 4990–4992 (2009).

[352] S. S. Gallagher, J. E. Sable, M. P. Sheetz and V. W. Cornish, *ACS Chem. Biol.* **4**, 547–556 (2009).

[353] J. Yin, F. Liu, X. H. Li and C. T. Walsh, *J. Am. Chem. Soc.* **126**, 7754–7755 (2004).

[354] R. Bonasio, C. V. Carman, E. Kim, P. T. Sage, K. R. Love, T. R. Mempel, T. A. Springer and U. H. von Andrian, *Proc. Natl. Acad. Sci.* **104**, 14753–14758 (2007).

[355] J. T. Moore, S. T. Davis and I. K. Dev, *Anal. Biochem.* **247**, 203–209 (1997).

[356] G. Zlokarnik, P. A. Negulescu, T. E. Knapp, L. Mere, N. Burres, L. X. Feng, M. Whitney, K. Roemer and R. Y. Tsien, *Science* **279**, 84–88 (1998).

[357] W. Z. Gao, B. G. Xing, R. Y. Tsien and J. H. Rao, *J. Am. Chem. Soc.* **125**, 11146–11147 (2003).

[358] B. Xing, A. Khanamiryan and J. H. Rao, *J. Am. Chem. Soc.* **127**, 4158–4159 (2005).

[359] H. Yao, M. K. So and J. Rao, *Angew. Chem. Int. Ed.* **46**, 7031–7034 (2007).

[360] S. M. Drawz and R. A. Bonomo, *Clin. Microbiol. Rev.* **23**, 160–201 (2010).

[361] H. P. Svensson, J. F. Kadow, V. M. Vrudhula, P. M. Wallace and P. D. Senter, *Bioconjugate Chem.* **3**, 176–181 (1992).

[362] S. Mizukami, S. Watanabe, Y. Hori and K. Kikuchi, *J. Am. Chem. Soc.* **131**, 5016–5017 (2009).

[363] F. Emmetiere, C. Irwin, N. T. Viola-Villegas, V. Longo, S. M. Cheal, P. Zanzonico, N. Pillarsetty, W. A. Weber, J. S. Lewis and T. Reiner, *Bioconjugate Chem.* **24**, 1784–1789 (2013).

[364] S. B. Lee, H. L. Kim, H.-J. Jeong, S. T. Lim, M.-H. Sohn and D. W. Kim, *Angew. Chem. Int. Ed.* **52**, 10549–10552 (2013).

[365] A. Dragulescu-Andrasi, S. Kothapalli, G. A. Tikhomirov, J. Rao and S. S. Gambhir, *J. Am. Chem. Soc.* **135**, 11015–11022 (2013).

[366] B. Shen, J. Jeon, M. Palner, D. Ye, A. J. Shuhendler, F. T. Chin and J. Rao, *Angew. Chem. Int. Ed.* **52**, 10511–10514 (2013).

[367] D. Ye, A. J. Shuhendler, L. Cui, L. Tong, S. S. Tee, G. A. Tikhomirov, D. W. Felsher and J. Rao. Bioorthogonal cyclization-mediated in situ self-assembly of small-molecule probes for imaging caspase activity in vivo. *Nature Chem.* **6**, 519–526 (2014).

[368] D. Ye, A. J. Shuhendler, P. Pandit, K. D. Brewer, S. S. Tee, L. Cui, G. A. Tikhomirov, B. Rut and J. Rao, *Chemical Science*, **5**, 3845–3852 (2014).

3

RECENT DEVELOPMENTS IN THE CHEMISTRY OF [^{18}F]FLUORIDE FOR PET

DIRK ROEDA AND FRÉDÉRIC DOLLÉ

CEA, I2BM, Service Hospitalier Frédéric Joliot, Orsay, France

3.1 INTRODUCTION

Molecular imaging in medicine and biology provides noninvasive views inside tissues of a living organism in order to obtain information at the molecular level. In this endeavour, positron emission tomography (PET) has always tried to stay as close as possible to the 'tracer principle' of George de Hevesy (Nobel prize 1943), which states that 'a radioactive atom might be used as a "representative" tracer of stable atoms of the same element whenever and wherever it accompanied them in biological systems.' The element fluorine, although as inorganic fluoride probably an essential element for humans, is not known to play any role in human physiological processes in an organically bound form. Why then does the positron emitting radioisotope fluorine-18 occupy such a prominent place in PET and associated radiochemistry [1], while *a priori* its principal alternative carbon-11 has the potential to label any carbon-containing compound? There are several answers to this question, but the convenience of the 110-minute physical half-life is preponderant. This time span is long enough to allow for relatively elaborate radiochemical processing and short enough to ensure that the radiation dose to a patient is not too high. It also makes distribution of radiofluorinated products possible over considerable geographical distances. Indeed, if fluorine-18 had the half-life of carbon-11 (20 minutes), it most probably would not have the status it has today nor would carbon-11. A true tracer, in the sense of de Hevesy, is however possible with fluorine-18 in the field of medicinal drugs and related compounds that show a fluorine atom in their structure. The numbers of this type of substance are growing continuously, and a good deal of ^{18}F-PET studies have been done with radiotracers based on it, notably in *in vivo* (neuro)receptor studies. An example is the central benzodiazepine antagonist [^{18}F]flumazenil (**1**) (Figure 3.1) [2].

But ^{18}F-PET does not stop there. A C–F entity in an organic molecule resembles sizewise a C–H or a C–OH group, and the latter two can sometimes be replaced by a fluorine atom without changing too much the *in vivo* behaviour of the molecule, especially where it concerns mechanisms of transport into tissue cells, and this in spite of the considerably greater electronegativity of fluorine. Three of the most successful PET radiopharmaceuticals are based on this design principle, namely 2-[^{18}F]fluoro-2-deoxy-D-glucose (**2**) (C–H for C-^{18}F in 2-deoxy-D-glucose or C–OH for C-^{18}F in D-glucose), 3'-[^{18}F]fluoro-3'-deoxythymidine (**3**) (C–H for C-^{18}F in 3'-deoxythymidine or C–OH for C-^{18}F in thymidine) and 6-[^{18}F]fluoro-L-DOPA (**4**) (aromatic C–H for C-^{18}F in L-DOPA). These tracers are used to measure glucose metabolism, cell proliferation and dopamine storage respectively.

The third option in fluorine-18 radiopharmaceutical design, in addition to the above true labelling and mimicry of H or OH, is to provide the molecule of interest with a prosthesis that can accommodate the fluorine-18 atom. This technique, called prosthetic labelling, is first of all applied in the labelling of macromolecules (e.g., proteins, peptides, oligonucleotides) but also in the derivation of small molecules such as in [^{18}F]LBT-999 (**6**) (dopamine transporter ligand) [3, 4] and in particular by the replacement of a methoxy- by a [^{18}F]fluoroalkoxy group [5] like in [^{18}F]DPA-714 (**7**) (TSPO ligand) [6–8]

The Chemistry of Molecular Imaging, First Edition. Edited by Nicholas Long and Wing-Tak Wong.
© 2015 John Wiley & Sons, Inc. Published 2015 by John Wiley & Sons, Inc.

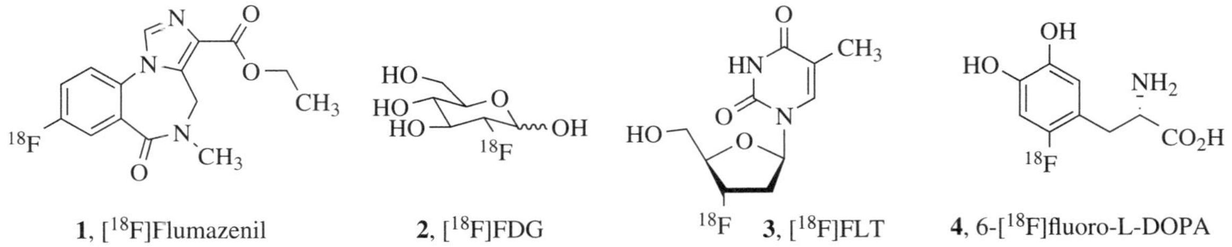

1, [18F]Flumazenil   2, [18F]FDG   18F 3, [18F]FLT   4, 6-[18F]fluoro-L-DOPA

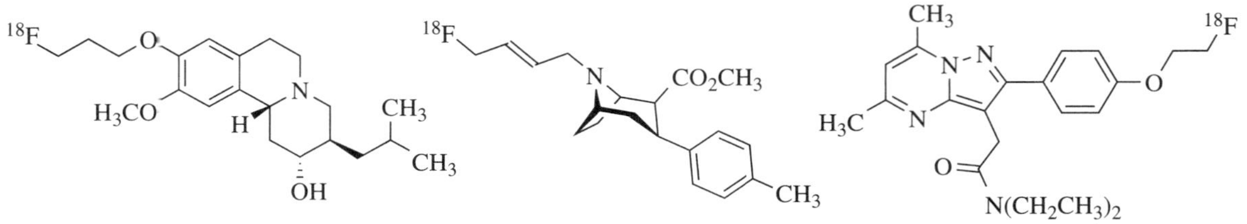

5, [18F]AV-133   6, [18F]LBT-999   7, [18F]DPA-714

FIGURE 3.1   Three categories of radiopharmaceutical design: True labelling (**1**), H- or OH mimicking by a [18F]fluorine atom (**2–4**) and prosthetic labelling (**5–7**).

and [18F]AV-133 (**5**) (vesicular monoamine transporter ligand) [9]. The latter compound shows that an OH for F swap can change receptor affinity considerably because the corresponding alcohol has a 100 times smaller affinity than [18F]AV-133 and can be tolerated as a contaminant in the radiopharmaceutical preparation [10].

3.2   FLUORINE-18: THE STARTING MATERIAL

Fluorine-18 is conveniently produced with a particle accelerator, normally a cyclotron, by bombarding an appropriate target with a charged-particle beam. The target can be liquid water or oxygen gas, highly enriched in the isotope ^{18}O, and the incident particles are protons with an energy usually between 10 and 20 MeV, inducing the nuclear reaction $^{18}O(p,n)^{18}F$ [11, 12]. The radioisotope is recovered as an aqueous solution of [18F]fluoride. A typical radioactivity level produced is about 1 Ci having a specific radioactivity (SRA) of 2 to 3 Ci/µmol corresponding to 0.3 to 0.5 µmol of fluoride. Obviously an important dilution with non-radioactive fluoride occurs because the theoretical SRA of carrier-free fluorine-18 is 1712 Ci/µmol (0.6 nmol/Ci). A considerable source of carrier fluoride can be radiolysis of commonly used polytetrafluoroethylene (PTFE) transport lines [13, 14]. SRAs of more than 100 Ci/µmol have also been reported, but it should be noted that these extremely high values may lead to rapid radiolysis of a radiofluorinated compound [15]. Neutron irradiation of a lithium carbonate target in a nuclear reactor can be an alternative way of making [18F]fluoride [16–18], but this method is seldom used. Fluorine-18 can also be produced by irradiation of a neon gas target with deuterons by the reaction $^{20}Ne(d,\alpha)^{18}F$. This method is used to make [18F]F$_2$ for electrophilic radiofluorination, for example, the synthesis of 6-[18F]fluoro-L-DOPA (**4**), but the drawback is that the carrier F$_2$ must be added to extract the radioactivity from the target holder [19]. Electrophilic [18F]F$_2$ is also made from an [18O]O$_2$ target, equally with added carrier. In this chapter we will not discuss electrophilic radiofluorination [20, 21] in detail but will focus on the much more current nucleophilic radiochemistry with [18F]fluoride [22] illustrated with recent developments.

3.3   REACTIVE [18F]FLUORIDE

Radiofluorination with [18F]fluoride implies nucleophilic substitution reactions. In these, the [18F]fluoride anion attacks the molecule to be labelled at an atom, normally carbon, that bears a suitable leaving group, which is expulsed while being replaced by the radioactive fluorine atom. The carbon atom is either aliphatic [23] or aromatic [24–26]. Cyclotron-produced [18F]fluoride comes as an aqueous solution. A fluoride anion in aqueous media is surrounded by a close shell of water dipoles that effectively hinders nucleophilic action in most cases. For a reaction to take place, the protective water shell must be broken up, which is not easy where the medium is water. Interestingly, nature has found a way to do this by the enzyme 5′-fluoro-5′-deoxyadenosine synthase isolated from *Streptomyces cattleya*. It was used in ^{18}F-chemistry to synthesise some

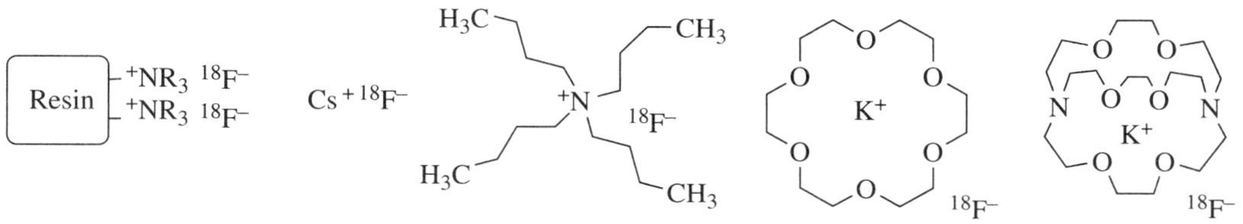

FIGURE 3.2 Various forms of [¹⁸F]fluoride, activated by dehydration.

radiofluorinated nucleotides and sugars and fluoroacetic acid [27–34, 34, 35]. Besides this special case, there are more examples where hydration of the fluoride is not prohibitive for nucleophilic substitution. This is when the elements silicon, boron, or aluminium carry the leaving group [36]. The driving force of these reactions is thought to be the favourable high strength of the fluorine-silicon, -boron, and -aluminium bonds so that the reaction can take place in an aqueous medium. In line with this, aluminium species leaching from borosilicate glassware were shown to have a negative impact on [¹⁸F] fluoride reactivity [37]. As far as radiofluorination on a carbon centre is concerned, the [¹⁸F]fluoride normally is rigorously dried and activated. One method has been to trap it on an anion exchange resin (**8**, Figure 3.2) and use the [¹⁸F]fluoride-loaded resin as a reagent [18, 38]. This method is not commonly used today. However, a strong anion exchanger in the form of a commercially available solid phase extraction (SPE) cartridge is currently used for initial trapping of the [¹⁸F]fluoride from the target water. It is then eluted with a mixture of acetonitrile (MeCN) and water containing a suitable base. The eluate is evaporated to dryness at elevated temperature; often this azeotropic process is repeated several times after adding MeCN. The base can be, for instance, cesium carbonate or –hydroxide, leaving the fluoride anion beside the relatively large cesium counter ion in a more or less naked reactive form (**9**). More frequently, instead of the cesium cation, a bulky tetraalkylammonium ion, usually tetrabutyl (**10**), is applied by using the corresponding hydroxide or (bi)carbonate as base. Most popular is potassium carbonate in conjunction with the cage-ligand kryptofix-222 (**12**), the latter encapsulating the potassium cation and promoting [¹⁸F]fluoride in its naked reactive form. Sometimes kryptofix-222 is replaced by 18-crown-6 (**11**) in cases where the substrate is too reactive and would react with the nitrogen atoms of kryptofix-222 [39–42].

However, in many cases drying does not need to be pushed to extremes; small amounts of water are often tolerated and sometimes deliberately added after drying to diminish loss of [¹⁸F]fluoride through adsorption to the reaction vessel wall [39, 43, 44]. The tolerance for some water may be useful because it offers a possibility to simplify the drying procedure, which is much sought after in the application of microfluidics in PET chemistry. This recent development consists of the use of reactor devices that mix reactants via microchannels and microvessels (down to sub-millimetre scale) in very small amounts allowing miniaturisation of the process and making it more efficient due to the high surface-to-volume ratio [45–47]. In this context, an older method [48] of [¹⁸F]fluoride isolation was recently revived and improved. [¹⁸F]Fluoride was extracted from the cyclotron target on a cartridge containing a water-wettable macroporous co-polymer loaded with the long-chain phase-transfer catalyst N-tetradecyltrimethylammonium bicarbonate. After purging with a nitrogen gas stream, the radioactivity could be eluted with 1 mL of dry MeCN because the tetraalkylammonium [¹⁸F]fluoride and subsequent radiochemistry could be performed in this eluate, which contained about 5000 ppm (~200 μmol) of water. Evaporation to dryness is circumvented because this method is compatible with closed systems such as microchannel devices [49]. Reversed-phase cartridges have also been used [50]. In a similar vein [¹⁸F]fluoride, trapped and 'dried' on the usual anion-exchange SPE cartridge, can, as an alternative to the conventional MeCN/aqueous potassium carbonate mixtures, be eluted with wet MeCN (e.g., 5000 ppm of water) provided with a strong organic base. The formed hydroxide anions push the [¹⁸F]fluoride off the exchange sites. Again, nucleophilic substitution reactions are possible directly in the eluate. Instead of water, traces of alcohols such as methanol can be used [51]. A similar way of [¹⁸F]fluoride recovery consists of elution with tetrabutylammonium mesylate or 1-butyl-3-methylimidazolium triflate in methanol, but in this case the solvent is evaporated before further chemistry [52]. A completely different approach, avoiding the evaporation step, is the electrochemical isolation of [¹⁸F]fluoride from an aqueous solution on a positively charged electrode [53] and its subsequent release, after rinsing the electrode with dry solvent, into a solution of kryptofix-222 in MeCN by reversing the voltage [54, 55]. This procedure has now been miniaturised in a microchannel device for microfluidic radiochemistry [56]. Notwithstanding the tolerance of water in a number of nucleophilic radiofluorination reactions, the search for highly reactive anhydrous [¹⁸F]fluoride continues. The idea of carrying over [¹⁸F]fluoride as an intermediate volatile product is well known. An example is the use of the volatile trimethylsilyl [¹⁸F]fluoride, which can be easily made from [¹⁸F]fluoride and (CH₃)₃SiCl in aqueous MeCN and then decomposed again by reaction with tetraethylammonium hydroxide liberating [¹⁸F] fluoride [57–60]. A similar approach was recently reported in which conventionally prepared kryptofix-222/K[¹⁸F]F reacted with phenyl triflate to give volatile trifluoromethanesulphonyl [¹⁸F]fluoride. It was carried by a gas stream into a solution of tributylammonium azide to produce highly reactive tributylammonium [¹⁸F]fluoride [61].

3.4 THE RADIOFLUORINATION

3.4.1 Electrophilic Reactions

Reactive [¹⁸F]fluoride is practically always applied in a nucleophilic substitution reaction. Before looking at this in more detail, we should mention recent progress in the conversion of [¹⁸F]fluoride into electrophilic fluorine-18 with a SRA intermediate between that of [¹⁸F]fluoride and the habitual low value of [¹⁸F]F$_2$. Gaseous methyl [¹⁸F]fluoride, made by nucleophilic substitution on methyl iodide, was made to react with a small amount of F$_2$ in an electric arc to give [¹⁸F] F$_2$ by a radical exchange process [62]. This 'high SRA' [¹⁸F]F$_2$ was converted into [¹⁸F]selectfluor bis(triflate), which was successfully used in electrophilic aromatic model destannylations [63]. High SRA [¹⁸F]F$_2$ itself has been used in the synthesis of 6-[¹⁸F]fluoro-L-DOPA (4) [64]. Xenon [¹⁸F]fluoride of relatively high SRA was made by an exchange reaction on a small amount of XeF$_2$ in a microfluidic reactor and used in electrophilic model reactions [65]. Very recently a palladium-based method was proposed to convert no-carrier-added [¹⁸F]fluoride into an electrophilic form for aromatic radiofluorination [66] (Scheme 3.1).

[¹⁸F]Fluoride is captured (by reaction with complex 13) to give a specially designed Pd(IV) complex (14). The oxidation state IV is high for palladium, and nucleophilic attack on the fluorine atom is possible. The palladium atom together with its other ligands acts as a leaving group recovering the more comfortable oxidation state Pd(II). In the presented method, this nucleophilic attack is performed by another palladium(II) complex 15 carrying the aromatic precursor, resulting in a transfer of the [¹⁸F]fluoride onto the Pd atom of the precursor complex 15 giving 16, in its turn in the Pd(IV) state. The latter gives a fast reductive elimination to the desired [¹⁸F]aryl fluoride 17. Various relatively complex radiofluorinated compounds, for example, [¹⁸F]fluorodeoxyestrone, were labelled in radiochemical yields of more than 30%. Further experience should show whether this is perhaps the long-sought-after grail of no-carrier-added electrophilic radiofluorination.

3.4.2 Nucleophilic Reactions

The vast majority of radiofluorination reactions are nucleophilic substitution reactions with a carbon atom as the reaction centre similar to the above CH$_3$[¹⁸F]F synthesis. Other reaction centres that are being explored are silicon, boron, and aluminium. These centres are especially finding application in bioconjugation, which is not surprising because these elements lend themselves primarily to prosthetic labelling because they normally are foreign to biomolecules. Reactions can often be carried out in aqueous media, which is an advantage in the chemical manipulation of proteins. The substitution reactions at a carbon centre are usually done in an organic solvent, most often of an aprotic dipolar character such as MeCN, N,N-dimethylformamide (DMF), or dimethylsulphoxide (DMSO). Carbon-centred nucleophilic radiofluorination can be divided into an aliphatic [23] and an aromatic [24–26, 67] field.

3.4.2.1 Nucleophilic Aliphatic Substitution

Leaving Groups In the aliphatic field, the leaving group is often a sulphonate group such as, in decreasing nucleofugacity, triflate, mesylate, or tosylate with the highest incidence of the last one because of the stability of tosylate esters. The reaction takes place with inversion of configuration, which is important for the design of the precursor when the reaction centre is an asymmetric carbon such as in the radiosynthesis of all four optically pure stereoisomers of 4-[¹⁸F]fluoroglutamines [41]. For the frequently encountered radiofluorinated sugar derivatives, of which [¹⁸F]FDG (2) [68] and [¹⁸F]FLT (3) [23, 69] (Scheme 3.2) are emblematic, this is naturally also an important issue.

The two-step radiosynthesis of [¹⁸F]FMAU (23) is a more recent example [70] (Scheme 3.3). The Boc protecting group on the pyrimidine ring of 22 is necessary to avoid neighbouring group participation of the pyrimidine carbonyl oxygen atom at the 2-position (as in 20, Scheme 3.2) through electron-pair donation by the 3-nitrogen, which could result in an unwanted

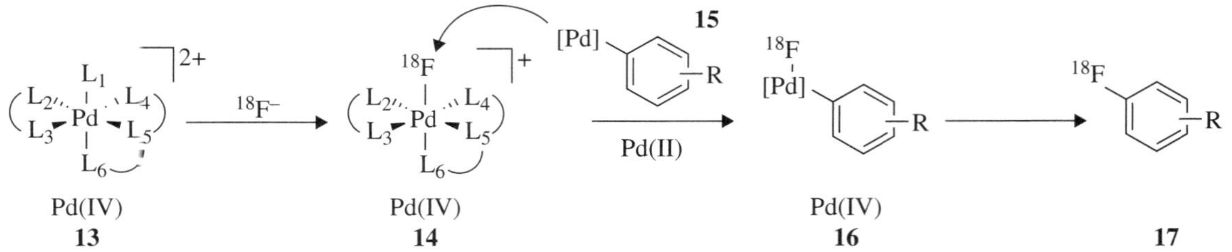

SCHEME 3.1 Schematic representation of creating electrophilic [¹⁸F]fluorine as a Pd(IV) complex (14) reacting with another Pd(II) complex (15) bearing the aromatic precursor.

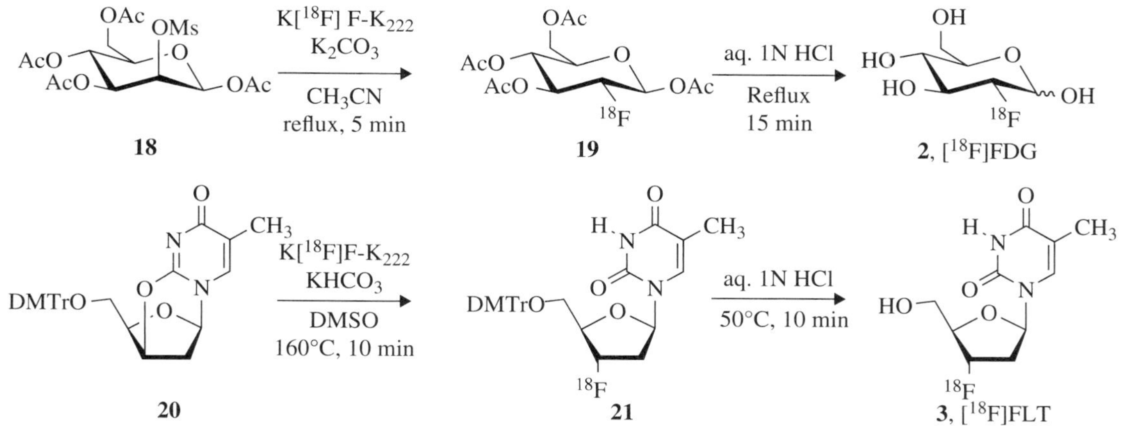

SCHEME 3.2 Radiosynthesis of [^{18}F]FDG (**2**) and [^{18}F]FLT (**3**) illustrating inversion of configuration in these secondary SN2 substitutions (from precursor **18** (respectively **20**) to intermediate **19** (respectively **21**)).

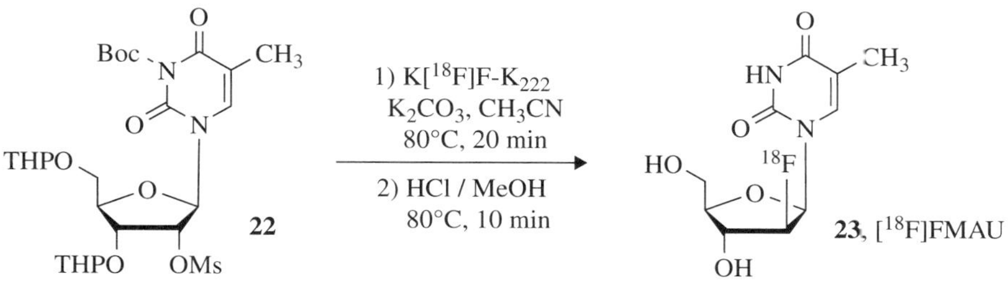

SCHEME 3.3 Two-step synthesis of [^{18}F]FMAU (**23**) with the pyrimidine moiety in place, replacing the habitual three-step procedure, which included final [^{18}F]sugar pyrimidine coupling.

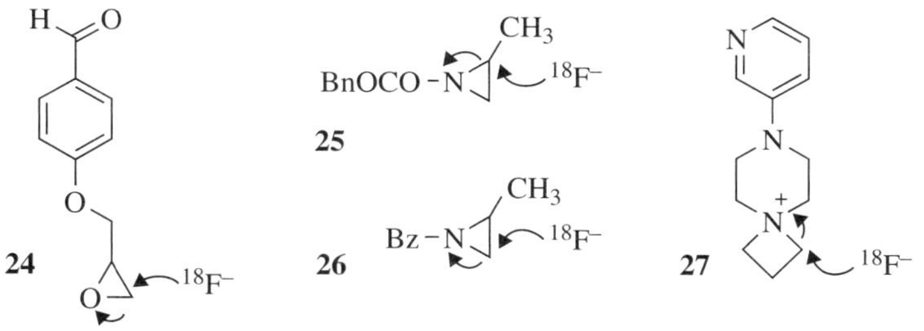

FIGURE 3.3 Radiofluorination by ring-opening reactions.

labelled product with the ^{18}F in the 'down' position. The yields were very low, but possibly worthwhile because the alternative is the time-consuming three-step procedure, habitual for this type of molecule, in which the pyrimidine moiety is attached after the fluorine-18 introduction. The low yield is probably due to steric hindrance. Another reason could be a side reaction, common in this type of chemistry, which is the elimination of methylsulphonic acid by base attack, including that by [^{18}F] fluoride, on the hydrogen at the 3'-position, which also deactivates the [^{18}F]fluoride. Note that corresponding purines with the more reactive triflate as leaving group have given reasonable yields, although here the 2'-position seems sterically hindered to some extent [71].

The halogens chlorine [72, 73], bromine [10, 74, 75], and iodine [76–79] are also used as leaving groups but less than the sulphonates. Exchange reactions with aliphatic fluorine have only been reported with polyfluorinated carbon centres [80–86]. Aliphatic radiofluorination based on ring opening, in which the leaving group remains attached to the molecule, [23, 87–94] remains a topical subject. The examples shown in Figure 3.3 feature the regioselective opening of an epoxide ring (**24**) to give 4-(3-[^{18}F]fluoro-2-hydroxypropoxy)benzaldehyde for peptide labelling [95], the opening of 2-methylaziridine derivatives (**25** and **26**) regiocontrolled by the choice of the *N*-protective group [96], and the ring opening of an azetidinium salt **27** to give a 2-[^{18}F]fluoroethylpiperazine [97].

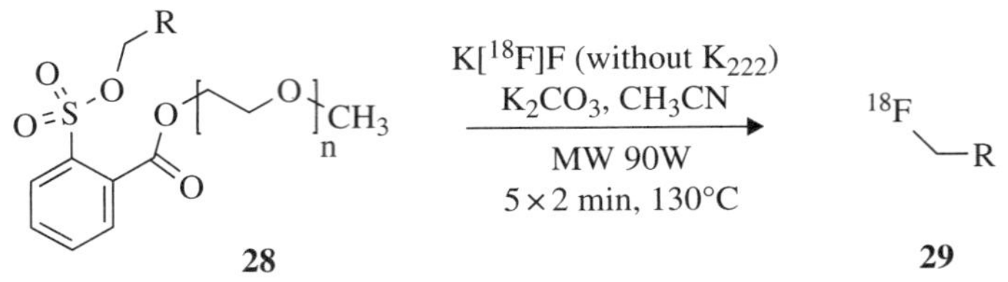

SCHEME 3.4   Synthesis of a fluorine-18-labelled aliphatic compound **29** by the use of a nucleophilic assisting leaving group. Compound **28** complexes by its PEG arm the counter cation, positioning the latter so that it can stabilize the transition state.

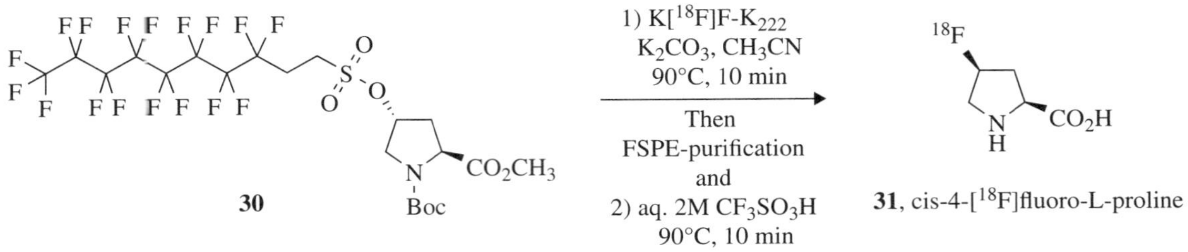

SCHEME 3.5   Synthesis of cis-4-[^{18}F]fluoro-L-proline (**31**) from a polyfluorinated sulphonate precursor **30**.

A new concept is represented by the so-called nucleophile assisting leaving groups (NALGs). These are primary arene-sulphonates that bear at the *ortho* position an entity that has chelating properties toward metal cations, for example, a PEG motif (**28**, Scheme 3.4). While the [^{18}F]fluoride attacks the sulphonate-bearing aliphatic carbon, its accompanying metal cation is complexed, allowing it to stabilise the negative charges on both the incoming fluoride and the leaving sulphonate unit during the transition state. No phase transfer kryptand is needed, and the rate of the reaction is considerably enhanced relative to the corresponding tosylate under the same reaction conditions [98].

The application of sulphonate esters having a long polyfluorinated alkyl chain attached to the sulphur atom is a new and remarkable trend [99]. These sulphonates make very good leaving groups because of the electron-attractive power of the fluorine atoms. More important even is the fact that the polyfluorinated state of the starting material facilitates separation from the non-polyfluorinated product by so-called fluorous solid-phase extraction (FSPE) based on the high mutual affinity between a polyfluorinated compound and a fluorous solid phase [100]. Such a separation could potentially replace an HPLC purification. For example, cis-4-[^{18}F]fluoro-L-proline (**31**) was synthesised from the corresponding 1H,1H,2H,2H-perfluorodecane-1-sulphonate precursor (**30**) and purified by FSPE in somewhat higher yields than with the tosylate (Scheme 3.5) [101]. Note that the first two carbon atoms of the chain are not fluorinated. Indeed, in model studies benzyl perfluorobutanesulphonate or -perfluorooctanesulphonate were reported not to give the desired product [99]. Some evidence was recently presented pointing to the possibility of an attack by fluoride on the sulphur atom, provoking expulsion of perfluoroalkene having eliminated a fluoride ion [102]. On the other hand, the *in situ* prepared perfluorobutanesulphonyl precursor of cis-4-[^{18}F]fluoro-L-proline (**31**) was reported to give the desired product [103, 104].

Polyfluorinated long-chain sulphonate precursors for [^{18}F]FDG have been attached to a resin with the chain acting as a linker [105, 106]. The leaving group and also unreacted or modified precursor remain on the resin, which facilitates purification. The presence of the fluorine atoms greatly enhances the reactivity, which was higher than that of the corresponding triflate. Leaching of non-radioactive fluoride from the chain, for example, by exchange with [^{18}F]fluoride, was claimed to be negligible. The issue of ^{18}F/^{19}F exchange is something to be kept in mind when the SRA is important because the latter is reduced by this phenomenon. While there does not seem to be any exchange with compounds such as perfluoro-*n*-hexane or perfluoro-*n*-hexylbenzene [107] and only a little bit in triflates [108], it is very important in tresylates (CF$_3$CH$_2$SO$_3$R) in which ^{18}F/^{19}F exchange seriously competes with nucleophilic displacement of the tresylate group [86, 109]. It was reported that nucleophilic radiofluorination with 1H,1H,2H,2H-perfluorooctanesulphonates resulted in a ten- to hundred-fold decrease in SRA [101].

Solvents   The solvents applied in both aliphatic and aromatic radiofluorination are usually of the polar aprotic kind, such as MeCN, DMSO, or DMF. Although these solvents may not be optimal in terms of relative stabilisation of transition state and starting products, they have the merit of promoting solubility of the fluoride and the precursor and not presenting loose protons that can deactivate the [^{18}F]fluoride anion. This last concept has lately been moderated by two developments in aliphatic radiofluorination, namely bulky alcohols and ionic liquids. It was shown that bulky tertiary alcohols like *t*-butanol or *t*-amyl alcohol can act as solvents in nucleophilic aliphatic radiofluorination, in spite of the hydroxy group, generally reputed for deactivation of fluoride in the same way as water [52, 110–117]. Most information on this method is about primary mesylates. Sometimes the yields are better than in aprotic solvents, and elimination reactions are suppressed [52, 114]. It is believed that the tertiary alcohols solvate the [¹⁸F]fluoride in a much less tight way than water does, so that nucleophilicity is not inhibited. Also they may enhance the nucleofugacity of the sulphonate leaving group by hydrogen bonds to the latter's oxygen atoms [118]. A comparative study on benzylic chloride and mesylate model compounds showed that tertiary alcohols perform less well than MeCN and DMF for this reaction. The same study also confirmed the current opinion that they do not work for aromatic substitution [118], although a heteroaromatic substitution in *t*-butanol has been published [119]. The important radiopharmaceutical [¹⁸F]FLT (**3**) has been produced in *t*-butanol, but the question whether elimination, notorious for this molecule [120], had been suppressed or not was not addressed [111, 113].

Ionic liquids are salt-like compounds constituted by an organic cation like 1-butyl-3-methylimidazolium (bmim) and an anion that can be inorganic (e.g., BF_4^- or SbF_6^-) or organic, such as a sulphonate. They are in the liquid state at ambient temperature and can serve as solvents, usually with a co-solvent (MeCN), in nucleophilic aliphatic radiofluorination [121–125]. Large amounts of water are tolerated, offering the advantage that a drying procedure of the [¹⁸F]fluoride is not necessary. The presence of water seems to reduce elimination, for example, in 1-mesyloxy-2-(naphth-2-yl)ethane, a substrate that normally is very susceptible to elimination by fluoride [121, 122]. [¹⁸F]FDG (**2**) and [¹⁸F]FLT (**3**) have been made successfully in ionic liquid systems [124, 125]. The use of *t*-butanol as a co-solvent in an ionic liquid does not seem advantageous; [121] however, incorporation of a tertiary alcohol moiety in the organic cation of an ionic liquid may produce a synergistic effect [126].

3.4.2.2 Nucleophilic Homoaromatic Substitution

Homoaromatic nucleophilic radiofluorination is in general more difficult than its aliphatic counterpart, and higher temperatures are usually required. The solvent should be polar and aprotic, like MeCN, DMSO, or DMF, and no divergences to more aqueous systems, as for the aliphatic case above, have been reported. A representative example is shown in Scheme 3.6.

While the aliphatic substitution has a one-step mechanism with a single transition state (**38**, Scheme 3.7b), the aromatic substitution passes via a normally negatively charged intermediate like **35** (Scheme 3.7a), the so-called Meisenheimer complex, resulting from the addition of the [¹⁸F]fluoride anion to the substrate molecule, in this case **34**.

The formation of the Meisenheimer complex **35** is the rate-determining step. The subsequent loss of the leaving group is fast. The most current leaving groups used in this reaction are the nitro group, the halogens (reactivity order: F >> Cl > Br >>> I, which is the inverse of aliphatic reactions [127]), and the trimethylammonium group. The latter has the particular

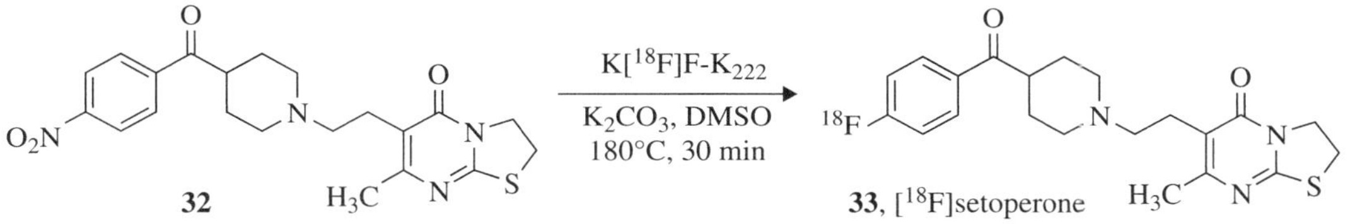

SCHEME 3.6 Representative example of a homoaromatic radiofluorination. Synthesis of the serotoninergic 5-HT$_2$ ligand [¹⁸F]setoperone (**33**).

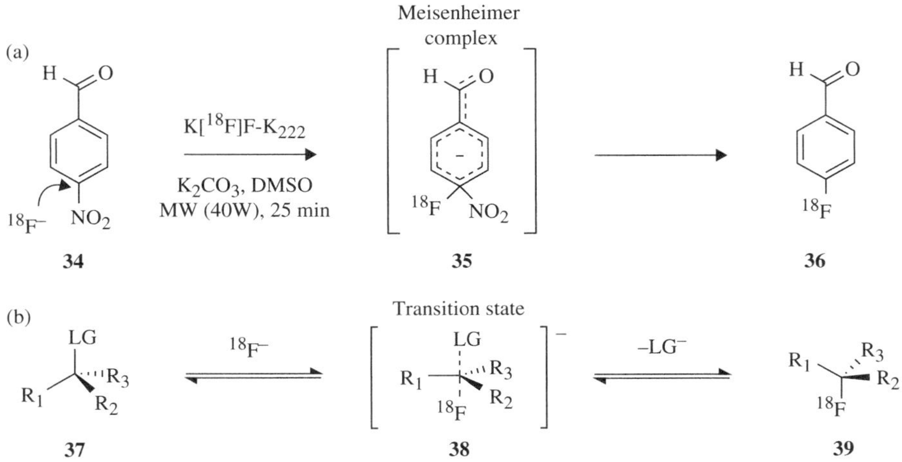

SCHEME 3.7 (a) Two-step reaction mechanism of an aromatic radiofluorination as illustrated by the synthesis of *p*-[¹⁸F]fluorobenzaldehyde (**36**), showing the intermediate Meisenheimer complex **35**. (b) One-step reaction mechanism of an aliphatic radiofluorination (from **37** to **39** via **38**).

advantage, besides its high nucleofugacity, of giving an electrical charge to the precursor molecule that facilitates its separation from the electrically neutral product. On the other hand, it may suffer from a side-reaction producing [¹⁸F]methyl fluoride by [¹⁸F]fluoride attack on one of the methyls of the trimethylammonium group [128]. In order to be formed at a reasonable rate, the Meisenheimer complex needs (resonance) stabilisation by the presence of auxiliary electron-withdrawing groups on the ring, preferably at an *ortho* or *para* position relative to the leaving group, such as the carbonyl groups in **32** (Scheme 3.6) and **34** (Scheme 3.7). Electron withdrawing groups can be nitro, cyano, aldehyde, ketone, ester, and carboxylic acid functions. More recently, a benzothiazol-2-yl moiety [129, 130], a 1,3,4-oxadiazol-2-yl [131] as well as a triphenylphosphonium group [132] *para* to a nitro group were found to have sufficient activating power to allow radiofluorination. The activating group is either an integrated part of the desired radiolabelled structure or is changed or removed afterwards [26]. In the first case, the radiosynthesis can sometimes be performed in one step such as with [¹⁸F]setoperone (**33**) [133, 134] or *p*-[¹⁸F]MPPF (**40**) [135] (Figure 3.4). With the latter, the activating group is an amide function, which usually gives relatively low yields. The presence of activating groups at an *ortho* or *para* position is mostly but not always a guarantee of good yields. The efflux protein inhibitor 1-[¹⁸F]fluoroelacridar (**41**) was obtained in <2% yield despite the presence of both an *ortho* carbonyl and a *para* amide function [136].

The formyl group in *p*-[¹⁸F]fluorobenzaldehyde (**36**) (Scheme 3.7) is a representative example of an activating group that can be transformed after having served in the labelling reaction. *p*-[¹⁸F]Fluorobenzaldehyde (**36**) is a common starting building block for more complex molecules in which the carbonyl group serves as a handle for further derivation [25], for example, the transformation of *p*-[¹⁸F]fluorobenzaldehyde (**36**) into *p*-[¹⁸F]fluorobenzyl azide for further use in Huisgen 1,3-cycloaddition (see below) for peptide labelling [137]. A similar strategy, with the fluorination *ortho* to the formyl group, is regularly applied in various syntheses of 6-[¹⁸F]fluoro-L-DOPA (**4**) [138]. A formyl group can be removed from an aromatic position, using Wilkinson's catalyst, after having served as an activating group in radiofluorination. This was the initial strategy in the radiosynthesis of certain 4-(4-[¹⁸F]fluorobenzyl)piperidinyl compounds such as **46** via **43** [139] (Scheme 3.8),

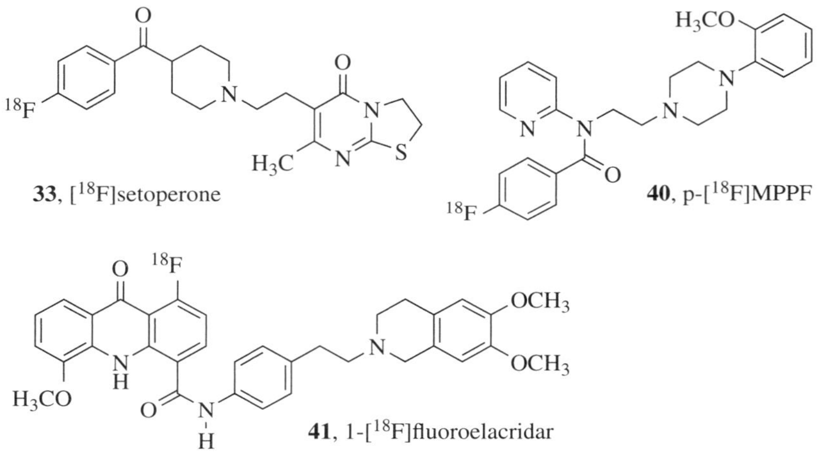

33, [¹⁸F]setoperone

40, p-[¹⁸F]MPPF

41, 1-[¹⁸F]fluoroelacridar

FIGURE 3.4 Examples of [¹⁸F]molecules featuring an activating group so that direct radiofluorination was possible.

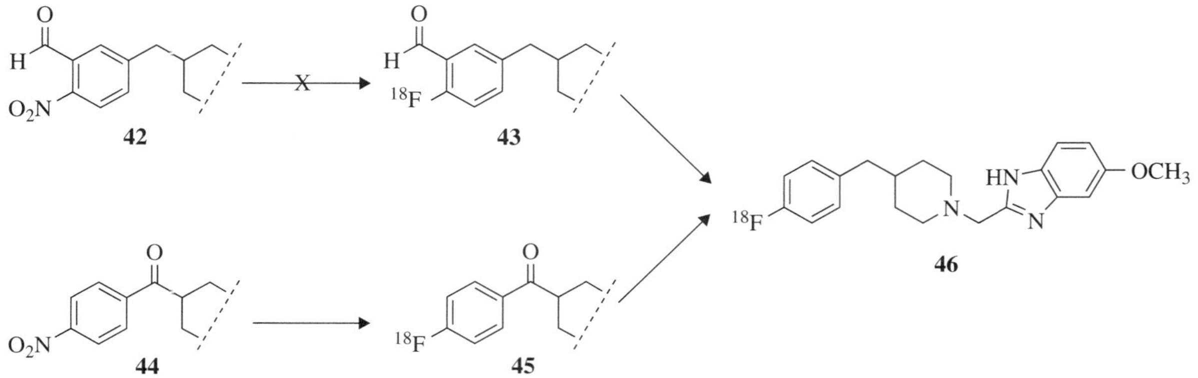

42   **43**   **46**

44   **45**

SCHEME 3.8 Synthesis of certain ([¹⁸F]fluorobenzyl)piperidinyl compounds such as **46** with a *para* or *meta* carbonyl function as a temporary activating group that is reduced after radiofluorination.

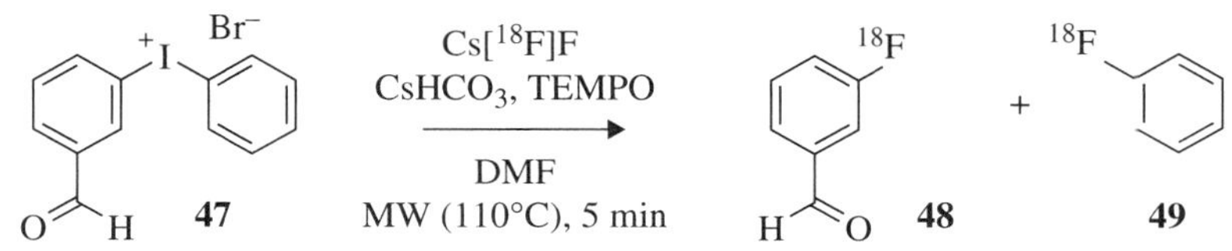

SCHEME 3.9 Synthesis of *meta*-[¹⁸F]fluorobenzaldehyde (**48**) via an iodonium precursor **47**.

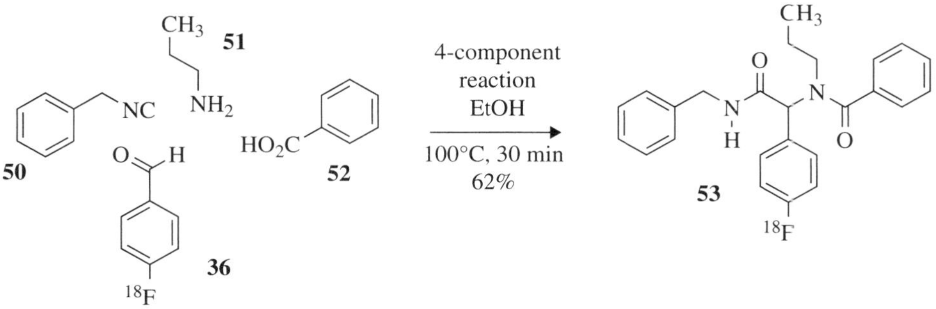

SCHEME 3.10 A convergent synthesis involving four components (**36**, **50**, **51**, and **52**) applied to the radiosynthesis with fluorine-18 of **53**.

which probably failed because of oxidation of the methylene group of **42** under the fluorination conditions. Therefore, an alternative 4-(4-nitrobenzoyl)piperidinyl precursor **44** was chosen. The activating benzoyl carbonyl group in **45** is then reduced after the fluorination. With an activating group at a *meta* position, the radiochemical yields are considerably lower, but sometimes useful activities can still be obtained especially with microwave heating [127, 140–143].

Meta-[¹⁸F]fluorobenzaldehyde (**48**) was recently made from an appropriate diaryliodonium precursor **47** (Scheme 3.9) [144] with a yield of 80%. It was subsequently converted into *meta*-[¹⁸F]fluorobenzyl bromide, which was needed for the synthesis of a labelled tyrosine kinase inhibitor containing a *meta*-[¹⁸F]fluorobenzylether moiety [145]. The leaving group in this positively charged precursor **47** is an aryl iodide, the incoming [¹⁸F]fluoride being predominantly directed toward the most electron deficient aryl system, in this case the benzene ring bearing the formyl group, leading to **48** rather than to **49**. This orientation was reinforced in the presence of a radical scavenger, and the yield was also dependent on the choice of the counter ion.

This reaction type provides a general strategy for the radiofluorination of arenes, especially the electron-rich ones, which are otherwise not accessible with habitual procedures of aromatic nucleophilic fluorination, but also electron-deficient ones as in the above example [146–159]. Apart from electronic effects of the substituents, a steric effect also exists, named the *ortho* effect, which implies that an *ortho* substituent on one of the rings, for example, a methyl group, directs the fluorination toward that ring [146, 147, 155, 156]. Aromatic moieties other than benzenes, such as 2-thienyl, have been used to direct the fluorination to the other aromatic ring [154, 157]. Although the number of published applications is slowly growing, the use of the method remains modest because of complicated precursor syntheses that do not always lead to very stable compounds.

Multicomponent chemistry is a convergent synthetic approach in which three or more substrates react simultaneously in one step. This could be a very useful tool in radiofluorination in order to access products with the fluorine label in positions that would normally not be considered as feasible. In a first validation of this concept, the easily accessible *para*-[¹⁸F]fluoroacetaldehyde (**36**) and substituted derivatives as well as *para*-[¹⁸F]fluorobenzoic acid were condensed in four different types of convergent three- or four-component reactions, optimised for the specific requirements of stoichiometry and time in PET chemistry, leading to labelled compounds that would not be easily accessible via traditional late-stage bimolecular condensation involving a [¹⁸F]fluoroaromatic compound or [¹⁸F]fluoride itself [160] (Scheme 3.10). A multicomponent version of fluoroalkylations of aromatic compounds was also proposed lately [161]. A multicomponent *N*-[¹⁸F]fluoroalkylation has been shown useful in the synthesis of [¹⁸F]fluoroethylcholine [162].

3.4.2.3 Nucleophilic Heteroaromatic Substitution

Scheme 3.11 depicts the synthesis of the nicotinic acetylcholine receptor ligand [¹⁸F]F-A-85380 (**56**) [163], which is a heteroaromatic radiofluorination. When an aromatic ring contains one or more heteroatoms, nucleophilic radiofluorination is enhanced relative to the homoaromatic counterpart [67, 164]. Most important is the pyridine ring, in which the ring nitrogen exerts a similar, if not slightly greater activating effect on the α and γ positions as the aldehyde group in benzaldehyde [165].

No further activating groups on the pyridine ring are needed to make the reaction proceed, as can also be noted in the syntheses of the TSPO [166], ligand 6[¹⁸F]fluoro-PBR28 (**57**) [167, 168], and the [¹⁸F]epibatidine derivatives **61**–**64** [169–172]

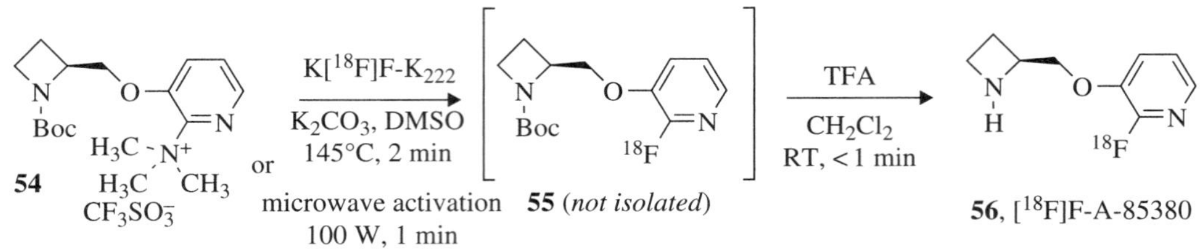

SCHEME 3.11 Synthesis of the nicotinic acetylcholine receptor ligand [¹⁸F]F-A-85380 (**56**) from the Me₃N⁺-pyridine precursor **54**, via non-isolated **55** as an example of a heteroaromatic radiofluorination.

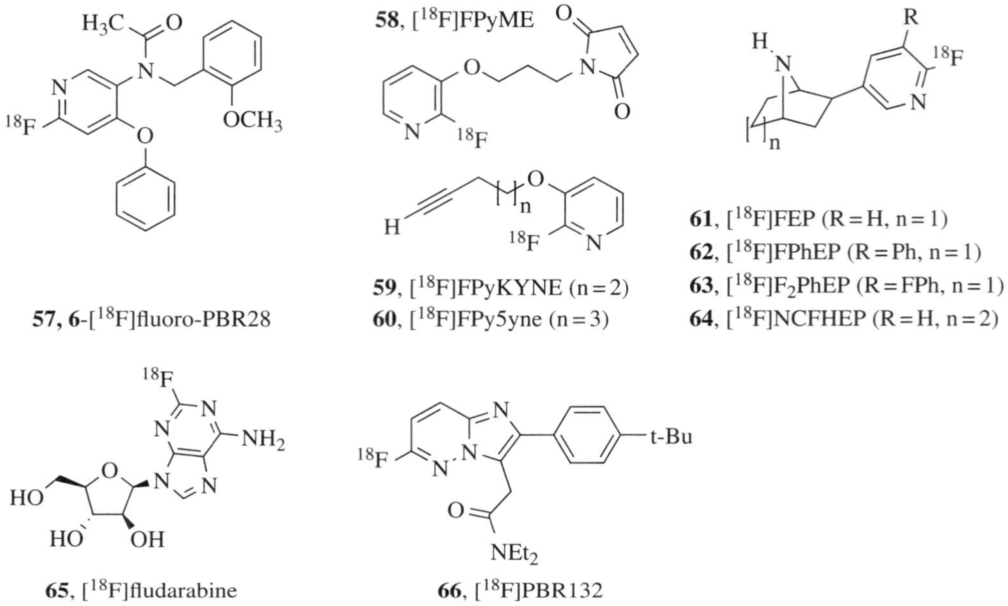

FIGURE 3.5 Various radioligands (**57**, **61–66**) and prosthetic reagents (**58–60**) as examples of radiolabelled compounds obtained by heteroaromatic radiofluorination.

(Figure 3.5). The activating effect by the nitrogen atom is hardly perturbed by the presence of deactivating groups such as methoxy or methyl *ortho* to the leaving group [165, 173]. This is also reflected in the syntheses of the prosthetic agents [¹⁸F]FPyKYNE (**59**) [174], [¹⁸F]FPy5yne (**60**) [175, 176], and [¹⁸F]FPyMe (**58**) [177].

The scope of the radiofluorination of pyridine itself was studied; it was found that nitro and trimethylammonium groups perform better than the halogens [178], that the α position is somewhat more reactive than the γ position, and that the β position is completely unreactive [179]. However, in analogy to the above example of *meta*-[¹⁸F]fluorobenzaldehyde (**48**), *meta*-[¹⁸F]fluoropyridine can be produced from an appropriate iodonium salt [152]. The β-position is the most stable one for *in vivo* defluorination, and the γ position is the most labile, in fact 4-[¹⁸F]fluoropyridine rapidly hydrolyses in water to 4-[¹⁸F]fluoro-2-pyridone. The α-position is practically always chosen for radiofluorination because the chemistry is the easiest and the fluorine in this position is relatively stable *in vivo*. In aromatic substitutions, fluorine is the best leaving group of the halogens, and indeed ¹⁹F/¹⁸F exchange with 2-fluoropyridines is very easy with high yields using small substrate amounts, resulting in very reasonable SRAs in the order of 10 GBq/μmol [173]. Examples of radiofluorination of heteroaromatic systems with two nitrogen atoms in the ring are [¹⁸F]fludarabine (**65**) [180] (2-position of a pyrimidine moiety) and [¹⁸F] PBR132 (**66**) [181] (3-position of a pyridazine moiety).

2-Chloro- and 2-bromo-1,3-thiazoles were recently shown to react easily with [¹⁸F]fluoride in DMSO to give the corresponding 2-[¹⁸F]fluoro-1,3-thiazoles [182]. Their reactivity is comparable to those of 2-chloro- and 2-bromopyridine. The 2-[¹⁸F]fluoro-1,3-thiazole moiety presents a potential alternative to the 2-[¹⁸F]fluoropyridine entity in prosthetic group design. In contrast to the 1,3-thiazole ring itself, which is susceptible to oxidative ring opening *in vivo*, the 2-fluoro-1,3-thiazole ring is metabolically much more stable. A first application has been the synthesis of a high-affinity metabotropic glutamate receptor subtype 5 radioligand (**68**) [183] (Scheme 3.12). Note that the fluorine atom on the benzene ring is susceptible to considerable exchange giving rise to **69**, in spite of its non-activated position, a reminder of the fact that fluorine is the

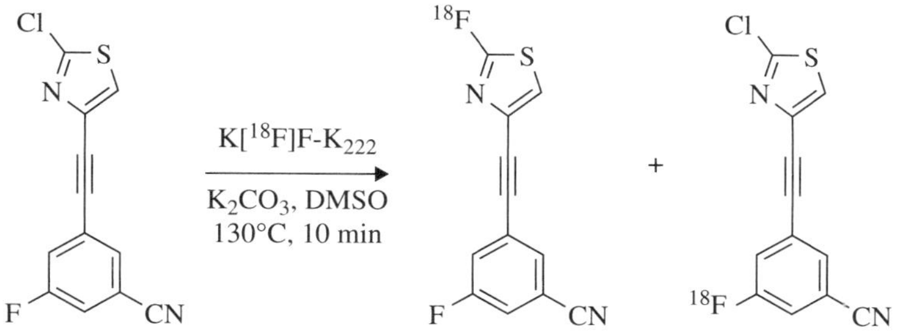

SCHEME 3.12 Radiolabelling of a high-affinity metabotropic glutamate receptor subtype 5 radioligand **68** from the corresponding 2-chloro-thiazole **67**. A radioactive side product (**69**) is formed by fluorine exchange on the benzene ring.

best leaving group of the halogens in aromatic substitution, which, for example, has been recently applied in the fluorination of the aromatic amino acids L-DOPA, tyrosine, and phenylalanine [184].

3.5 LABELLING OF LARGE BIOLOGICAL MOLECULES

3.5.1 The Prosthetic Group Concept

When large biological molecules, such as peptides, proteins, antibodies, or oligonucleotides, are to be labelled with fluorine-18, true labelling is not an issue because these molecules do not possess fluorine atoms of their own. If a polypeptide structure contains one or more aromatic amino acid residues, it can in principle be labelled by an electrophilic H for ^{18}F substitution [185], but this method is seldom used because of lack of specificity and side reactions. Instead, it is current practice to call on an intermediary chemical entity, a so-called prosthetic group, that can be, often specifically, attached to the substrate molecule and also accommodate the fluorine-18 atom. Most frequently, the prosthetic agent is first labelled and then attached to the macromolecule, so that the often harsh reaction conditions of radiofluorination do not need to be inflicted upon the latter. Alternatively, the prosthetic entity can first be coupled to the macromolecule after which the adduct serves as the labelling substrate. A large number of fluorine-18 prosthetic reagents have been published and have been compiled in a recent review [186]. Figure 3.6 shows some representative examples: [^{18}F]SFB (**70**), [^{18}F]FBAM (**71**), [^{18}F]FBnBrA (**72**), and [^{18}F]F-PEG-SH (**73**) [187–192]. The chemistry of the coupling between the prosthetic entity and the biomolecule is varied and includes acylation, S-alkylation, oxime or hydrazone formation, click chemistry, photoconjugation, glycosylation, amidation, amidination, and thiourea formation. Some more recent developments are discussed hereafter.

The prosthetic group carrying the fluorine-18 atom often has an appreciable size and usually is attached at a well-defined position of the macromolecule, where it is thought not to perturb the latter's function, for example, at the terminal amino group where it concerns a protein or peptide. However, random labelling of ε-amino groups of lysine residues in a polypeptide chain by acetaldehyde under reductive alkylation conditions is known to give minimal disturbance of the protein's structure and function because of the small size of the added ethyl group. [^{11}C]Formaldehyde was used in the past to a similar end in PET chemistry [193, 194]. Now [^{18}F]fluoroacetaldehyde (**76**) has been used successfully for the labelling of recombinant human interleukin-1 receptor antagonist, which contains nine lysine residues, in aqueous medium using sodium cyanoborohydride as reducing agent [195]. The easy two-step one-pot preparation of [^{18}F]fluoroacetaldehyde (**76**) is interesting and worth a closer look (Scheme 3.13) [196]. 1-[^{18}F]fluoro-tosyloxyethane (**75**) is produced in the habitual way (**74**, K[^{18}F]F/K$_{222}$/K$_2$CO$_3$, MeCN, or DMSO, 90°C, 8 minutes). The MeCN is then evaporated and DMSO is added, followed by heating at 150°C for 4 minutes and at 130°C for another 4 minutes while [^{18}F]fluoroacetaldehyde (**76**) is distilling out. This is the

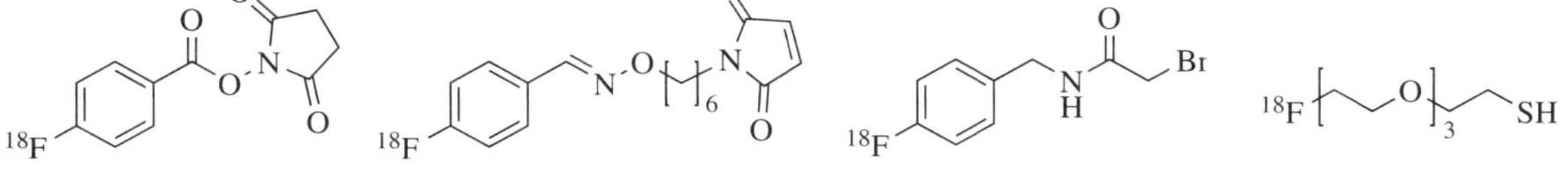

FIGURE 3.6 Some representative examples of [^{18}F]prosthetic reagents (**70–73**).

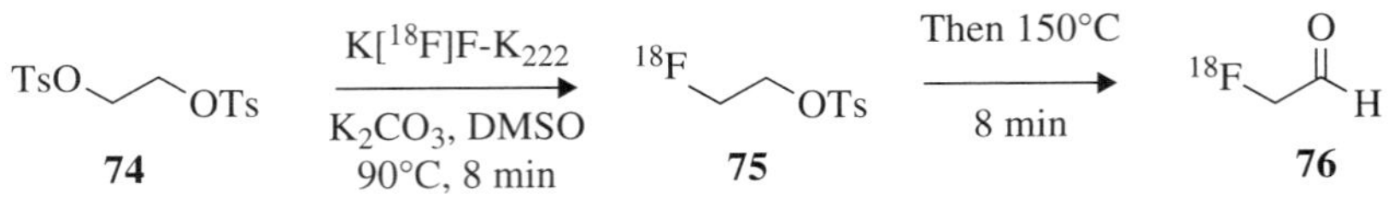

SCHEME 3.13   Two-step one-pot radiosynthesis of [^{18}F]fluoroacetaldehyde (**76**) by the Kornblum oxidation of tosylate **75** by DMSO.

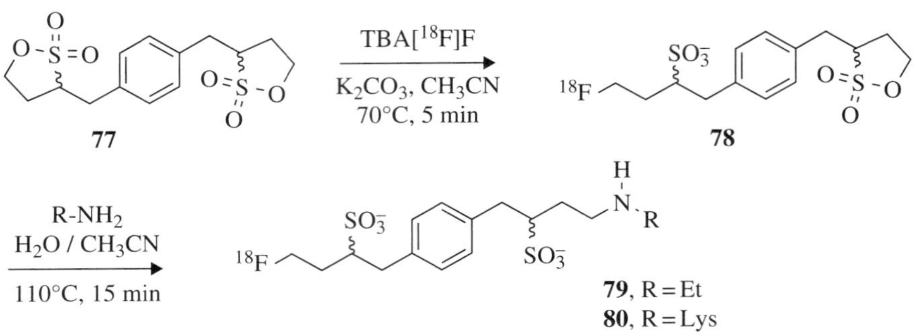

SCHEME 3.14   The use of two successive sulton ring opening reactions as radiolabelling step and bioconjugation step, resulting in relatively polar products (**79**, **80**).

Kornblum oxidation of a tosylate by DMSO in the presence of K_2CO_3. Note that this reaction potentially can cause unwanted precursor degradation in ordinary aliphatic radiofluorination in DMSO at temperatures ≥150°C.

Addition of a prosthetic group to a biomolecule is likely to cause an unwanted increase of lipophilicity. A way to counter this can be the integration of a polyethyleneglycol (PEG) linker in the macromolecule/prosthetic group construct. An elegant alternative has been proposed using the ring opening of a sulton by [^{18}F]fluoride creating at the same time an anionic moiety on the prosthetic agent. This increases the hydrophilicity and facilitates separation from the much less polar precursor [197] (Scheme 3.14). Thus the *bis*-sulton **77** is radiofluorinated at one of the two sulton rings to give **78**, which is rapidly separated from its precursor by a SPE method. The product **78** was then aminated with ethylamine or lysine by opening of the second sulton ring. Products **79** and **80** could again be easily purified by SPE. A one-pot procedure without the intermediate purification is also possible. This method should be extendable to macromolecule labelling.

3.5.2   Prosthetic Entities with Fluorine-18 Bound to Aluminium, Boron, or Silicon

Radiometals play an important role in nuclear medicine imaging. Well-known examples are copper-64, gallium-68, and yttrium-86 for PET and technetium-99m and indium-111 for SPECT. As far as macromolecules are concerned, these metals are invariably attached in some oxidised form through a prosthetic polydental chelating entity [198]. This labelling strategy is now also available for fluorine-18 by complexing the latter with aluminium as the $Al[^{18}F]F^{2+}$ cation [36, 199–204] (**81**, Figure 3.7). The chelates that have been used are of the NOTA or NODA type, and X-ray structure determinations of the latter have been performed [203, 204]. The procedure is simple: The reaction takes place in aqueous media and consists of adding $AlCl_3$ to the peptide-chelate adduct followed by the [^{18}F]fluoride followed by heating at 100°C for 15 minutes or less. Radiochemical yields of up to 85% have been reported [200], and the well-known peptides octreotide [201] and RGD2 were successfully labelled and evaluated. This chemistry should lend itself to a kit-like approach. Stereoisomers at the level of the orientation of the covalent Al-F bond in the complex seem to be possible and separable with HPLC [200]. New NODA derivatives that function at lower temperatures for the labelling step have recently been proposed [205].

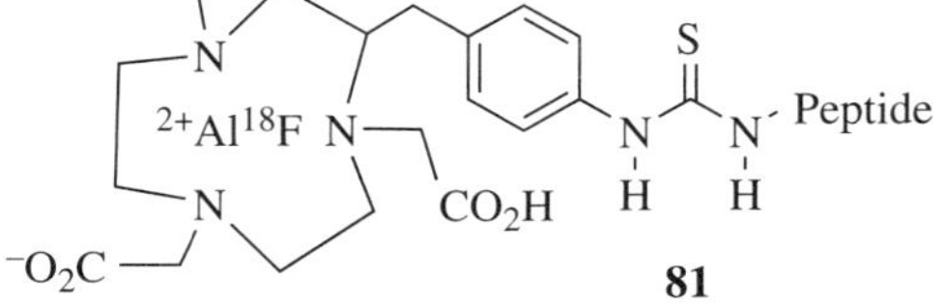

FIGURE 3.7   An aluminium-fluorine-18 complex cation bound by a prosthetic NOTA ring.

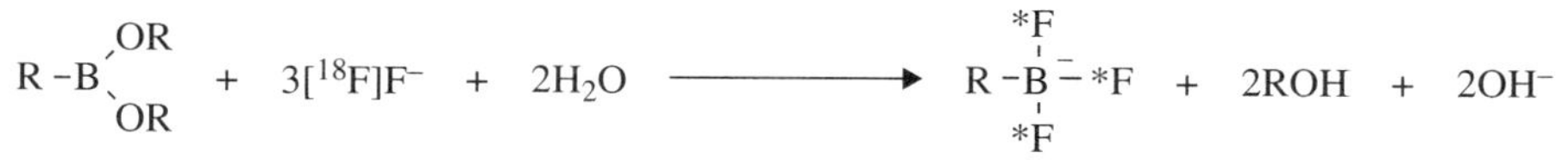

SCHEME 3.15 Tri[¹⁸F]fluoroborate anion formation from the corresponding boronic esters **82**. The label in **83** is designated by *F and not by ¹⁸F to avoid the wrong impression that an individual molecule should necessarily bear three radioactive fluorine atoms at the time.

Boron is situated just above aluminium in the periodic table and shows similar binding properties to fluorine. However, it is used in a different way than aluminium in radiofluorination [36], [206–213]. Arylboronic esters such as the pinacol diester, but also free arylboronic acids, react at ambient temperature in aqueous media with fluoride to give anionic complexes **83**, in which all three available positions on the boron atom are occupied with fluorine atoms (Scheme 3.15), and which are stable toward hydrolysis [207–209]. For labelling with fluorine-18, carrier fluoride is often added in order to make sure that three fluorine atoms are introduced to each product molecule. The detrimental effect of this on the SRA is limited by the use of as little precursor as possible in a small volume (a high concentration counteracts the relatively low rate constant of the reaction) combined with the fact that three fluorine atoms enter each product molecule. With an excess of fluoride, the fluorination efficiency can be quantitative [206]. The entry of three fluorine atoms per molecule on the one hand triples the initial SRA but on the other hand may render the reaction less efficient when using no-carrier-added [¹⁸F]fluoride. However, even with added carrier fluoride, SRAs of more than 1 Ci/μmol are attainable [207]. Boron-fluorine chemistry looks highly promising in prosthetic labelling because of its simplicity and its performance in aqueous media. Recent applications of this methodology have been the labelling of an aryltrifluoroborate derivative of marimastate, a matrix metalloproteinase kinase inhibitor, also in a kit-like procedure [211, 212], and a bodipy dye for dual PET and fluorescence imaging [213].

A third newcomer in prosthetic fluorine-18 labelling is silicon-fluorine bond formation [36, 210]. Although this bond, like the B-F and the Al-F, is a strong one, it is relatively sensitive to hydrolysis in aqueous media. Bulky substituents on the silicon atom effectively protect against this [214] and therefore, the prosthetic entity often takes the form of the aryl-*bis*-*t*-butylsilanyl motive (**84**, Figure 3.8). Leaving groups that have been employed are OEt, OH, H, and F. The reaction can be carried out under mild conditions and in aqueous media, although organic solvents such as MeCN or DMSO have been applied more often thus far. Although unusual in fluorine-18 PET chemistry because of SRA considerations, the ¹⁸F/¹⁹F exchange on a fluoro-silicon precursor (**85**, X=F) has proved to be very useful. This exchange is very rapid and efficient, probably proceeding through a penta coordinate intermediate. Therefore, very small amounts of precursor, down to 10 nmol or less, are sufficient, which leads to SRAs high enough for receptor studies [215, 216].

p-[¹⁸F]Fluorobenzaldehyde (**36**) is useful in both bioconjugation [186] and the radiosynthesis of smaller molecules [25], and so is its silylated analogue *p*-(di-*tert*-butyl[¹⁸F]fluorosilyl)benzaldehyde (**86**), made by ¹⁸F/¹⁹F exchange (Figure 3.8). It can either be labelled as such and then conjugated to a peptide [215] by oxime formation, or it can be labelled when already attached to the peptide [217, 218]. In the latter case, a kit-like approach should be possible. Similar prosthetic agents for ¹⁸F/¹⁹F exchange, differing by having another functionality instead of aldehyde, have been used, that is, thiol [219, 220] (coupled with maleimide derivated rat serum albumin (RSA) [220]), isothiocyanato (thiourea formation, RSA, apotransferrin, bovine IgG [219, 221]), isocyanato, *N*-maleimido (with SH derived RSA), carboxyl and its active esters [219], and azidomethyl [222]. The silylated amino acid *p*-(di-*tert*-butylfluorosilyl)phenylalanine (**87**, Figure 3.8) was synthesised enantioselectively recently, incorporated in Tyr3-octreoate derivatives, and labelled with fluorine-18 by ¹⁸F/¹⁹F exchange [223]. The bulky alkyl groups on the silicon atom contribute to an increase in lipophilicity of the biomolecule to be labelled; this may be a problem. Therefore, a positively charged compound containing both the *p*-(di-*tert*-butyl[¹⁸F]fluorophenylsilane motive and a tetraalkylammonium moiety was proposed as a lead compound for the development of a more hydrophilic prosthetic group [224]. Another approach has been the insertion of a hydrophilic moiety in the peptide, for example, PEG [218]. Although fluorine seems to be the preferred leaving group, H, OH, and OEt are also encountered [225–229]. Acid often accelerates the

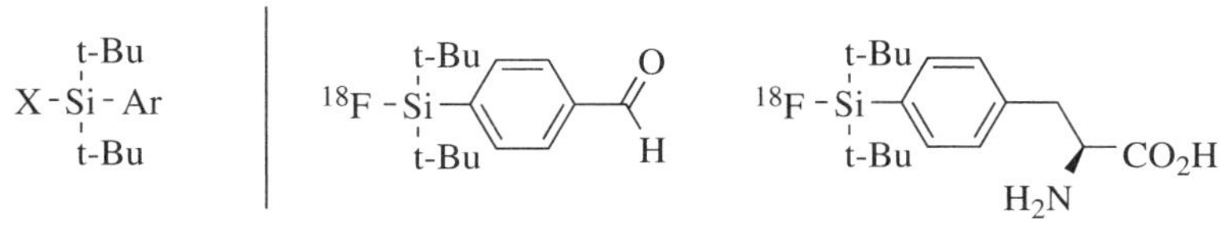

FIGURE 3.8 General structure of most prosthetic [¹⁸F]fluorosilyl compounds (**84**) exemplified by *p*-(di-*tert*-butyl[¹⁸F]fluorosilyl)benzaldehyde (**86**) and *p*-(di-*tert*-butylfluorosilyl)phenylalanine (**87**), as well as general structure of their precursors (**85**).

reaction with these leaving groups, especially with OH and OEt. An example with hydride as leaving group is the derivatisation of a nucleoside with a -Ph(t-Bu)₂SiH moiety with subsequent incorporation into an oligonucleotide, followed by labelling at 165°C in DMSO [228]. The fluorosilicon approach is most often used in bioconjugation, but also small molecules have been made, for example, silicon-based derivatives of the D₂ ligand fallypride (¹⁸F/¹⁹F exchange, 2.4 Ci/μmol) [216] and of the hypoxia marker misonidazole (OEt and Oi-Pr leaving groups) [227]. Finally, labelled alkyl tetrafluorosilicates present an alternative to the prosthetic groups with bulky substituents discussed above [206]. They were made from the corresponding triethoxysilyl precursors, and carrier fluoride was added to ensure efficient incorporation. The labelled fluorosilicates proved to be moderately stable in aqueous media.

3.5.3   Click Chemistry

Organic azides (**90**) react with terminal alkynes (**89**) in a 1,3-dipolar cycloaddition to give 1,4-disubstituted 1,2,3-triazoles (**88**) (Scheme 3.16). This so-called Huisgen cycloaddition is Cu(I) catalysed, which orientates the substituents specifically in the 1,4 direction. It is an example of a 'click reaction,' which is a modular synthetic approach using always a same efficient condensation reaction to join a pair of variable chemical moieties. Click chemistry has become highly fashionable in ¹⁸F-PET chemistry, especially in bioconjugation because of the mild reaction conditions in water or aqueous organic mixtures, the high yields, and the *in vivo* stability of the triazole entity, which mimics an amide or peptide bond, and the relative inertness of the azide and alkyne functions toward other reactions in the applied conditions [230–232]. The fluorine-18 label can be either on the azide or on the alkyne partner, of which 2-[¹⁸F]fluoroethyl azide and 4-[¹⁸F]fluoro-1-butyne are the simplest representatives. These can be easily prepared from corresponding sulphonate precursors and, if desired, isolated by distillation [101], [233–242]. More complex ¹⁸F-labelled click components that can contribute to an increased hydrophilicity are known too, for example, pyridines such as compounds **58**, **59,** and **60** [174–177], reagents derived from 2-[¹⁸F]fluoro-2-deoxy-D-glucose (**2**) [243] or containing PEG [244–247]. An illustration of the efficiency of the Huisgen click reaction is a recent oligonucleotide labelling using azido([¹⁸F]fluoromethyl)benzenes and only 20 nmol of precursor [248].

The use of copper in the above click reaction sometimes presents a problem because of the potential risk of contamination of an injection with this toxic metal. A copper-free variation on the above azide cycloaddition was therefore developed, in which the driving force of the reaction is energy relief in a strained azadibenzocyclo-octyne (**91**) when the azide (**90**) adds to the triple bond to give **92** (Scheme 3.16). Again the radioactively labelled part can be attached to either of the two reaction partners. The alkyne partner can be labelled at the ring nitrogen in the form of, for example, a 3-(p-[¹⁸F]fluorobenzoylamino) propionyl group (easily accessible via [¹⁸F]SFB (**70**)) [249] or a 3-(6-[¹⁸F]fluorohexanoylamino)propionyl group [250]. In this case the biomolecule to be labelled bears the azido function, for example, Tyr³-octreotate [250]. Bombesin was labelled the other way around with various azides of varying lipophilicity [251]. This click construction can be made in either aqueous or alcoholic media and seems metabolically stable. [249] Other 1-3 dipoles potentially can be useful too, as the recently proposed nitrone function, which adds to alkenes without catalyst [252].

Another catalyst-free bioconjugation method that is drawing attention is the inverse electron demand Diels-Alder cycloaddition of a *trans*-cyclo-octane (**94**) and an aryl-substituted tetrazine derivative (**93**) [253] (Scheme 3.17). This reaction, leading to **95**, is practically immediate and can be carried out in aqueous media. The *trans*-cyclo-octene **94**, carrying a fluorine-18 labelled chain, adds with expulsion of dinitrogen to the tetrazine **93**, carrying the biomolecule. The driving force is again strain relief. *Trans*-cyclo-octenes are conveniently generated from the corresponding *cis* compounds by photochemical means. This method has been successfully applied in the labelling of a PARP-1 inhibitor, which is a small molecule [254], and in the labelling of a cyclic RGD peptide [255]. A similar reaction is the photoactivated addition of a dipolarophilic alkene (**97**) to an aryl substituted tetrazole (**96**), which was applied to peptide bioconjugation with a ¹⁸F-moiety attached to the tetrazole and the peptide to the alkene, leading to product **98** [256] (Scheme 3.17).

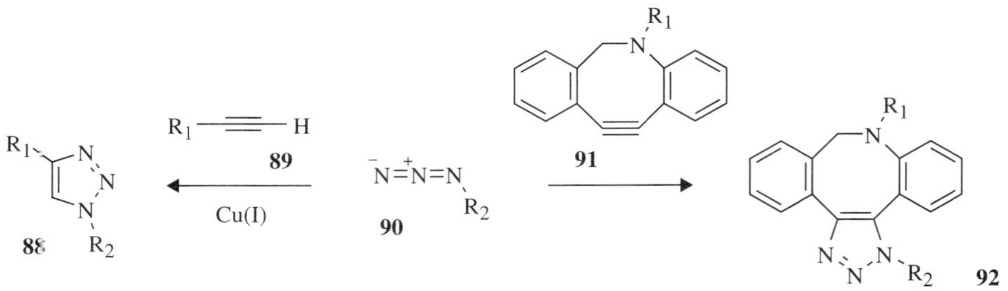

SCHEME 3.16   The Huisgen 1,3-dipolar cycloaddition (on the left) and a copper-free variation using a strained azadibenzocyclo-octyne (on the right) in radiofluorination. The radioactive part can be either on R₁ or on R₂.

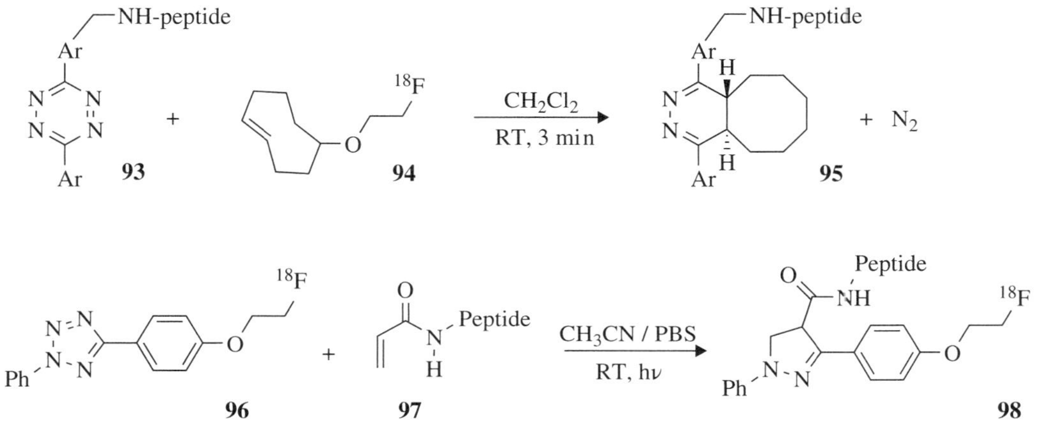

SCHEME 3.17 Two prosthetic radiofluorinations, one using the inverse electron demand Diels-Alder cycloaddition of a *trans*-cyclo-octane (**94**) to an aryl-substituted tetrazine derivative (**93**) and the other the photoactivated addition of a dipolarophilic alkene (**97**) to an aryl substituted tetrazole (**96**). Ar = Aryl.

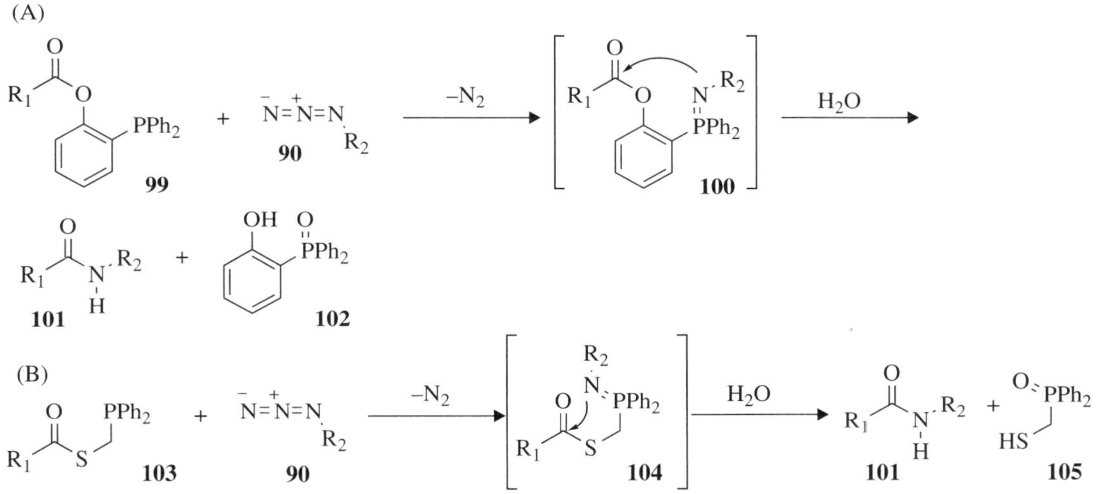

SCHEME 3.18 Two variants (A and B) of the Staudinger ligation used in prosthetic radiofluorination. The radioactive part can be either on R_1 or on R_2.

The Staudinger ligation [257–260] also employs an organic azide (**90**) that, as in the Huisgen 1,3-cycloaddition, can either be the derivatised macromolecule or the fluorine-18 bearing counterpart (Scheme 3.18). It reacts with a diphenyl-arylphosphine (**99**) or a diphenylalkylphosphine (**103**), resulting in an amide bond between the two structures to be linked (**101**). The reactive groups involved are bio-orthogonal, and the phosphine part is eliminated from the intermediate (**100**, **104**) as a phosphine oxide (**102**, **105**), leaving no trace in the product.

Although thought of as a tool in fluorine-18 bioconjugation, so far only model reactions with relatively small entities have been presented. Indeed, the method is equally suitable for the construction of small radioligands, such as, for example, the 4-quinolone derivative **109** from precursor **106** (Scheme 3.19) [257]. When the radiolabel is on the azide side as in this example, 2-[18F]fluoroethyl azide (**108**, generated from the tosylate **107**) has been the radiosynthon of choice. With the Staudinger reaction it gives access to N-(2-[18F]fluoroethyl)amides, which are difficult to make by a direct SN_2 fluorination because the corresponding precursors tend to give an intramolecular cyclisation with the amide moiety to 4,5-dihydro-oxazoles. It provides an alternative to coupling reactions involving 2-[18F]fluoroethylamine [261]. The N-H group would also need protection in a direct fluorination, but not with the Staudinger ligation.

Other model reactions involving 2-[18F]fluoroethyl azide (**108**) have also been designed according to variant B in Scheme 3.18 [259]. Indeed, structure **103** looks more amenable to derivation with complex molecules than **99**. When the

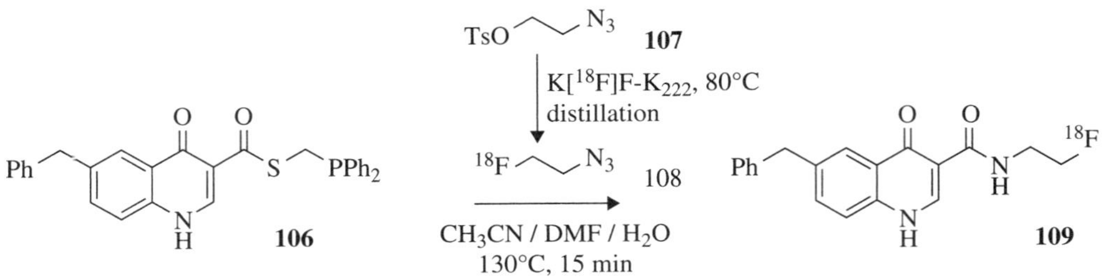

SCHEME 3.19   Example of the synthesis of an N-(2-[¹⁸F]fluoroethyl)amide (**109**) using the Staudinger ligation.

azide function is envisaged on the biomolecule, variant A of Scheme 3.18 has been chosen so far with $R_1 = p$-[¹⁸F]fluorophenyl, p-(2-[¹⁸F]fluoroethyl)phenyl or 5-[¹⁸F]fluoropentyl [258, 260, 262]. The radiochemistry of these phosphine compounds is not always straightforward, and reaction conditions such as solvents have to be selected carefully.

3.6   CONCLUSIONS

Fluorine-18 chemistry for PET has now reached a certain degree of maturity. Notably nucleophilic aliphatic and aromatic substitution have been consolidated to robust and reliable methodologies serving as a stepping stone for further development. The advent of microfluidic synthesis devices is promoting a nucleophilic radiofluorination in which drying is not necessarily a prerequisite. In bioconjugation, new click-like procedures are following each other at a great pace while alumina, boron, and silicon claim their place next to carbon as the reaction centre. In electrophilic substitution, after a long status quo, new ideas have begun to emerge on how to keep the SRA high. Thus, the field of fluorine-18 chemistry continues to invite significant creativity and thus should be assured of a bright future.

REFERENCES

[1]   S. Banister, D. Roeda, F. Dollé and M. Kassiou, *Curr. Radiopharm.* **3**, 68–80 (2010).

[2]   N. N. Ryzhikov, N. Seneca, R. N. Krasikova, N. A. Gomzina, E. Shchukin, O. S. Fedorova, D. A. Vassiliev, B. Gulyas, H. Hall, I. Savic and C. Halldin, *Nucl. Med. Biol.* **32**, 109–116 (2005).

[3]   F. Dollé, F. Hinnen, P. Emond, S. Mavel, Z. Mincheva, W. Saba, M.-A. Schöllhorn-Peyronneau, H. Valette, L. Garreau, S. Chalon, C. Halldin, J. Helfenbein, J. Legaillard, J.-C. Madelmont, J.-B. Deloye, M. Bottlaender and D. Guilloteau, *J. Label Compounds Radiopharm.* **49**, 687–698 (2006).

[4]   F. Dollé, J. Helfenbein, F. Hinnen, S. Mavel, Z. Mincheva, W. Saba, M.-A. Schöllhorn-Peyronneau, H. Valette, L. Garreau, S. Chalon, C. Halldin, J.-C. Madelmont, J.-B. Deloye, M. Bottlaender, J. Legaillard, D. Guilloteau and P. Emond, *J. Label. Compounds Radiopharm.* **50**, 716–723 (2007).

[5]   M. R. Zhang and K. Suzuki, *Curr. Topics Med. Chem.* **7**, 1817–1828 (2007).

[6]   M. L. James, R. R. Fulton, J. Vercoullie, D. J. Henderson, L. Garreau, S. Chalon, F. Dollé, S. Selleri, D. Guilloteau and M. Kassiou, *J. Nucl. Med.* **49**, 814–822 (2008).

[7]   A. Damont, F. Hinnen, B. Kuhnast, M.-A. Schöllhorn-Peyronneau, M. L. James, C. Luus, B. Tavitian, M. Kassiou and F. Dollé, *J. Label. Compounds Radiopharm.* **51**, 286–292 (2008).

[8]   B. Kuhnast, A. Damont, F. Hinnen, T. Catarina, S. Demphel, S. Le Helleix, C. Coulon, S. Goutal, P. Gervais and F. Dollé, *Appl. Radiat. Isot.* **70**, 439–477 (2012).

[9]   M. R. Kilbourn, B. Hockley, L. Lee, C. Hou, R. Goswami, D. E. Ponde, M-P. Kung and H. F. Kung, *Nucl. Med. Biol.* **34**, 233–237 (2007).

[10] L. Zhu, Y. J. Liu, K. Plossl, B. Lieberman, J. Y. Liu and H. F. Kung, *Nucl. Med. Biol.* **37**, 133–141 (2010).

[11] E. Hess, S. Takács, B. Scholten, F. Tárkányi, H. H. Coenen and S. M. Qaim, *Radiochim. Acta* **89**, 357–362 (2001).

[12] T. J. Ruth, K. R. Buckley, K. S. Chun, E. T. Hurtado, S. Jivan and S. Zeisler, *Appl. Radiat. Isot.* **55**, 457–461 (2001).

[13] F. Füchtner, S. Preusche, P. Mäding, J. Zessin and J. Steinbach, *Nuklearmedizin* **47**, 116–119 (2008).

[14] M. S. Berridge, S. M. Apana and J. M. Hersh, *J. Label Compounds Radiopharm.* **52**, 543–548 (2009).

[15] P. J. H. Scott, B. C. Hockley, H. F. Kung, R. Manchanda, W. Zhang and M. R. Kilbourn, *Appl. Radiat. Isot.* **67**, 88–94 (2009).

[16] J. P. de Kleijn, H. J. Meeuwissen and B. van Zanten, *Radiochem. Radioanal. Letters* **23**, 139–143 (1975).

[17] J. P. de Kleijn, R. F. Ariaansz and B. van Zanten, *Radiochem. Radioanal. Letters* **28**, 257–262 (1977).

[18] J. P. de Kleijn, J. W. Seetz, J. F. Zawierko and B. van Zanten, *Int. J. Appl. Radiat. Isot.* **28**, 591–594 (1977).

[19] M. Guillaume, A. Luxen, B. Nebeling, M. Argentini, J. Clark and V. W. Pike, *Appl. Radiat. Isot.* **42**, 749–762 (1991).

[20] F. Dollé, D. Roeda, B. Kuhnast and M-C. Lasne, in A. Tressaud and G. Haufe, eds., *Fluorine and Health: Molecular Imaging, Biomedical Materials and Pharmaceuticals.* Elsevier, Amsterdam-Boston-Heidelberg-London-New York-Oxford-Paris-San Diego-San Francisco-Singapore-Sydney-Tokyo, 2008; pp 3–65.

[21] M.-C. Lasne, C. Perrio, J. Rouden, L. Barré, D. Roeda, F. Dollé and C. Crouzel, in W. Krause, ed., *Topics in Current Chemistry*, Springer-Verlag, Berlin-Heidelberg, Vol. **222**, 2002, pp 201–258.

[22] L. Cai, S. Lu and V. W. Pike, *Eur. J. Org. Chem.* 2853–2873 (2008).

[23] D. Roeda and F. Dollé, *Curr. Radiopharm.* **3**, 81–108 (2010).

[24] H. H. Coenen and J. Ermert, *Curr. Radiopharm.* **3**, 163–173 (2010).

[25] J. Ermert and H. H. Coenen, *Curr. Radiopharm.* **3**, 127–160 (2010).

[26] J. Ermert and H. H. Coenen, *Curr. Radiopharm.* **3**, 109–126 (2010).

[27] D. O'Hagan, C. Schaffrath, S. L. Cobb, J. T. G. Hamilton and C. D. Murphy, *Nature* **416**, 279–279 (2002).

[28] L. Martarello, C. Schaffrath, H. Deng, A. D. Gee, A. Lockhart and D. O'Hagan, *J. Label. Compounds. Radiopharm.* **46**, 1181–1189 (2003).

[29] H. Deng, S. L. Cobb, A. D. Gee, A. Lockhart, L. Martarello, R. P. McGlinchey, D. O'Hagan and M. Onega, *Chem. Comm.* 652–654 (2006).

[30] D. O'Hagan, *J. Fluor. Chem.* **127**, 1479–1483 (2006).

[31] X. F. Zhu, D. A. Robinson, A. R. McEwan, D. O'Hagan and J. H. Naismith, *J. Am. Chem. Soc.* **129**, 14597–14604 (2007).

[32] M. Winkler, J. Domarkas, L. F. Schweiger and D. O'Hagan, *Angew. Chem. Int. Ed.* **47**, 10141–10143 (2008).

[33] H. Deng, S. M. Cross, R. P. McGlinchey, J. T. G. Hamilton and D. O'Hagan, *Chem. Biol.* **15**, 1268–1276 (2008).

[34] X. G. Li, J. Domarkasc and D. O'Hagan, *Chem. Comm.* **46**, 7819–7821 (2010).

[35] M. Onega, J. Domarkas, H. Deng, L. F. Schweiger, T. A. D. Smith, A. E. Welch, C. Plisson, A. D. Gee and D. O'Hagan, *Chem. Comm.* **46**, 139–141 (2010).

[36] G. E. Smith, H. L. Sladen, S. C. G. Biagini and P. J. Blower, *Dalton Trans.* **40**, 6196–6205 (2011).

[37] A. Svadberg, A. Clarke, K. Dyrstad, I. Martinsen and O. K. Hjelstuen, *Appl. Radiat. Isot.* **69**, 289–294 (2011).

[38] S. A. Toorongian, G. K. Mulholland, D. M. Jewett, M. A. Bachelor and M. R. Kilbourn, *Nucl. Med. Biol.* **17**, 273–279 (1990).

[39] E. Briard and V. W. Pike, *J. Label. Compounds Radiopharm.* **47**, 217–232 (2004).

[40] E. Briard, S. S. Zoghbi, F. G. Siméon, M. Imaizumi, J. P. Gourley, H. U. Shetty, S. Y. Lu, M. Fujita, R. B. Innis and V. W. Pike, *J. Med. Chem.* **52**, 688–699 (2009).

[41] W. C. Qu, Z. H. Zha, K. Ploessl, B. P. Lieberman, L. Zhu, D. R. Wise, C. B. Thompson and H. F. Kung, *J. Am. Chem. Soc.* **133**, 1122–1133 (2011).

[42] Z. H. Zha, L. Zhu, J. Y. Liu, F. H. Du, H. M. Gan, J. P. Qiao and H. F. Kung, *Nucl. Med. Biol.* **38**, 501–508 (2011).

[43] M. R. Kilbourn, J. W. Brodack, D. Y. Chi, C. S. Dence, P. A. Jerabek, J. A. Katzenellenbogen, T. E. Patrick and M. J. Welch, *J. Label. Compounds Radiopharm.* **23**, 1174–1176 (1986).

[44] M. L. Kornguth, T. R. DeGrado, J. E. Holden and S. J. Gatley, *J. Label. Compounds Radiopharm.* **25**, 369–381 (1988).

[45] P. W. Miller, A. J. deMello and A. D. Gee, *Curr. Radiopharm.* **3**, 254–262 (2010).

[46] G. Pascali, G. Mazzone, G. Saccomanni, C. Manera and P. A. Salvadori, *Nucl. Med. Biol.* **37**, 547–555 (2010).

[47] R. Bejot, A. M. Elizarov, E. Ball, J. Z. Zhang, R. Miraghaie, H. C. Kolb and V. Gouverneur, *J. Label. Compounds Radiopharm.* **54**, 117–122 (2011).

[48] D. M. Jewett, S. A. Toorongian, G. K. Mulholland, G. L. Watkins and M. R. Kilbourn, *Appl. Radiat. Isot.* **39**, 1109–1111 (1988).

[49] J. Aerts, S. Voccia, C. Lemaire, F. Giacomelli, D. Goblet, D. Thonon, A. Plenevaux, G. Warnock and A. Luxen, *Tetra. Lett.* **51**, 64–66 (2010).

[50] B. Y. Yang, J. M. Jeong, Y. S. Lee, D. S. Lee, J. K. Chung and M. C. Lee, *Tetrahedron* **67**, 2427–2433 (2011).

[51] C. F. Lemaire, J. J. Aerts, S. Voccia, L. C. Libert, F. Mercier, D. Goblet, A. R. Plenevaux and A. J. Luxen, *Angew. Chem. Int. Ed.* **49**, 3161–3164 (2010).

[52] J. W. Seo, B. S. Lee, S. J. Lee, S. J. Oh and D. Y. Chi, *Bull. Korean Chem. Soc.* **32**, 71–76 (2011).

[53] D. Alexoff, D. J. Schlyer and A. P. Wolf, *Appl. Radiat. Isot.* **40**, 1–6 (1989).

[54] K. Hamacher, T. Hirschfelder and H. H. Coenen, *Appl. Radiat. Isot.* **56**, 519–523 (2002).

[55] G. Reischl, W. Ehrlichmann and H.-J. Machulla, *J. Radioanal. Nucl. Chem.* **254**, 29–31 (2002).

[56] H. Saiki, R. Iwata, H. Nakanishi, R. Wong, Y. Ishikawa, S. Furumoto, R. Yamahara, K. Sakamoto and E. Ozeki, *Appl. Radiat. Isot.* **68**, 1703–1708 (2010).

[57] M. S. Rosenthal, A. L. Bosch, R. J. Nickles and S. J. Gatley, *Int. J. Appl. Radiat. Isot.* **36**, 318–319 (1985).

[58] L. G. Hutchins, A. L. Bosch, M. S. Rosenthal, R. J. Nickles and S. J. Gatley, *Int. J. Appl. Radiat. Isot.* **36**, 375–378 (1985).

[59] S. J. Gatley, *Appl. Radiat. Isot.* **40**, 541–544 (1989).

[60] G. K. Mulholland, *Appl. Radiat. Isot.* **42**, 1003–1008 (1991).

[61] J. Tewson, *J. Label. Compounds Radiopharm.* **54** (Suppl. 1), S451 (2011).

[62] J. Bergman and O. Solin, *Nucl. Med. Biol.* **24**, 677–683 (1997).

[63] H. Teare, E. G. Robins, A. Kirjavainen, S. Forsback, G. Sandford, O. Solin, S. K. Luthra and V. Gouverneur, *Angew. Chem. Int. Ed.* **49**, 6821–6824 (2010).

[64] S. Forsback, O. Eskola, M. Haaparanta, J. Bergman and O. Solin, *Radiochim. Acta* **96**, 845–848 (2008).

[65] S. Y. Lu and V. W. Pike, *J. Fluor. Chem.* **131**, 1032–1038 (2010).

[66] E. Lee, A. S. Kamlet, D. C. Powers, C. N. Neumann, G. B. Boursalian, T. Furuya, D. C. Choi, J. M. Hooker and T. Ritter, *Science* **334**, 639–642 (2011).

[67] F. Dollé, *Curr. Pharmaceutical Design* **11**, 3221–3235 (2005).

[68] K. Hamacher, H. H. Coenen and G. Stöcklin, *J. Nucl. Med.* **27**, 235–238 (1986).

[69] A. Blocher, M. Kuntzsch, R. Wei and H. J. Machulla, *J. Radioanal. Nucl. Chem.* **251**, 55–58 (2002).

[70] N. Turkman, J. G. Gelovani and M. M. Alauddin, *J. Label. Compounds Radiopharm.* **53**, 782–786 (2010).

[71] M. M. Alauddin, J. D. Fissekis and P. S. Conti, *J. Label. Compounds Radiopharm.* **46**, 805–814 (2003).

[72] E. D. Hostetler, S. Sanabria-Bohorquez, H. Fan, Z. Z. Zeng, L. Gantert, M. Williams, P. Miller, S. O'Malley, M. Kameda, M. Ando, N. Sato, S. Ozaki, S. Tokita, H. Ohta, D. Williams, C. Sur, J. J. Cook, H. D. Burns and R. Hargreaves, *Neuroimage* **54**, 2635–2642 (2011).

[73] T. Bourdier, R. Shepherd, P. Berghofer, T. Jackson, C. J. R. Fookes, D. Denoyer, D. S. Dorow, I. Greguric, M.-C. Grégoire, R. Hicks and A. Katsifis, *J. Med. Chem.* **54**, 1860–1870 (2011).

[74] K. Kersemans, J. Mertens and V. Caveliers, *J. Label. Compounds Radiopharm.* **53**, 58–62 (2010).

[75] N. Turkman, A. Pal, W. P. Tong, J. G. Gelovani and M. M. Alauddin, *J. Label. Compounds Radiopharm.* **54**, 233–238 (2011).

[76] T. R. DeGrado, S. Y. Wang and D. C. Rockey, *J. Nucl. Med.* **41**, 1727–1736 (2000).

[77] J. Bergman, O. Eskola, P. Lehikoinen and O. Solin, *Appl. Radiat. Isot.* **54**, 927–933 (2001).

[78] P. Kumar, L. I. Wiebe and A. J. B. McEwan, *J. Label. Compounds Radiopharm.* **48** (Suppl. 1), S203 (2005).

[79] F. Wüest, T. Kniess, M. Kretzschmar and R. Bergmann, *Bioorg. Med. Chem. Letters* **15**, 1303–1306 (2005).

[80] M. R. Kilbourn, M. Pavia and V. Gregor, *Appl. Radiat. Isot.* **41**, 823–828 (1990).

[81] A. Hammadi and C. Crouzel, *J. Label. Compounds Radiopharm.* **33**, 703–710 (1993).

[82] M. R. Satter, C. C. Martin, T. R. Oakes, B. F. Christian and R. J. Nickles, *Appl. Radiat. Isot.* **45**, 1093–1100 (1994).

[83] P. Johnström and S. Stone-Elander, *J. Label. Compounds Radiopharm.* **36**, 537–547 (1995).

[84] P. S. Johnström and S. Stone-Elander, *Appl. Radiat. Isot.* **47**, 401–407 (1996).

[85] T. Viljanen, P. Lehikoinen and O. Solin, *J. Label. Compounds Radiopharm.* **46** (Suppl. 1), S219 (2003).

[86] M. Suehiro, G. B. Yang, G. Torchon, E. Ackerstaff, J. Humm, J. Koutcher and O. Ouerfelli, *Bioorg. Med. Chem.* **19**, 2287–2297 (2011).

[87] U. Roehn, J. Becaud, L. J. Mu, A. Srinivasan, T. Stellfeld, A. Fitzner, K. Graham, L. Dinkelborg, A. P. Schubiger and S. M. Ametamey, *J. Fluor. Chem.* **130**, 902–912 (2009).

[88] A. K. Podichetty, S. Wagner, S. Schröer, A. Faust, M. Schäfers, O. Schober, K. Kopka and G. Haufe, *J. Med. Chem.* **52**, 3484–3495 (2009).

[89] N. Vasdev, E. M. van Oosten, K. A. Stephenson, N. Zadikian, A. K. Yudin, A. J. Lough, S. Houle and A. A. Wilson, *Tetrahedron Letters* **50**, 544–547 (2009).

[90] N. Ahmed, G. Garcia, H. Ali and J. E. van Lier, *Steroids* **74**, 42–50 (2009).

[91] N. Jarkas, R. J. Voll, L. Williams, V. M. Camp and M. M. Goodman, *J. Med. Chem.* **53**, 6603–6607 (2010).

[92] R. Bejot, V. Kersemans, C. Kelly, L. Carroll, R. C. King and V. Gouverneur, *Nucl. Med. Biol.* **37**, 565–575 (2010).

[93] W. P. Yu, L. Williams, V. M. Camp, J. J. Olson and M. M. Goodman, *Bioorg. Med. Chem. Letters* **20**, 2140–2143 (2010).

[94] W. P. Yu, J. McConathy, L. Williams, V. M. Camp, E. J. Malveaux, Z. B. Zhang, J. J. Olson and M. M. Goodman, *J. Med. Chem.* **53**, 876–886 (2010).

[95] R. Schirrmacher, F. Lucas, E. Schirrmacher, B. Wängler and C. Wängler, *Tetra. Lett.* **52**, 1973–1976 (2011).

[96] E. M. van Oosten, M. Gerken, P. Hazendonk, R. Shank, S. Houle, A. A. Wilson and N. Vasdev, *Tetra. Lett.* **52**, 4114–4116 (2011).

[97] P. Grosse-Gehling, F. R. Wuest, T. Peppel, M. Köckerling and C. Mamat, *Radiochim. Acta* **99**, 365–373 (2011).

[98] S. Y. Lu, S. D. Lepore, S. Y. Li, D. Mondal, P. C. Cohn, A. K. Bhunia and V. W. Pike, *J. Org. Chem.* **74**, 5290–5296 (2009).

[99] E. Blom, F. Karimi and B. Långström, *J. Label. Compounds Radiopharm.* **53**, 24–30 (2010).

[100] W. Zhang, *Chem. Rev.* **109**, 749–795 (2009).

[101] R. Bejot, T. Fowler, L. Carroll, S. Boldon, J. E. Moore, J. Declerck and V. Gouverneur, *Angew. Chem. Int. Ed.* **48**, 586–589 (2009).

[102] D. Roeda, B. Kuhnast, A. Damont, E. Cerutti and F. Dollé, *J. Label. Compounds Radiopharm.* **54** (Suppl. 1), S504 (2011).

[103] M. Jelinski, K. Hamacher and H. H. Coenen, *J. Label. Compounds Radiopharm.* **44** (Suppl. 1), S151–S153 (2001).

[104] M. Jelinski, Markierungsverfahren zur Synthese 4-[18F]Fluorprolyl-haltiger Peptide. Thesis University of Cologne 2002, Berichte des Forschungszentrums Jülich ; 4008, ISSN 0944-2952.

[105] L. J. Brown, D. R. Bouvet, S. Champion, A. M. Gibson, Y. L. Hu, A. Jackson, I. Khan, N. C. Ma, N. Millot, H. Wadsworth and R. C. D. Brown, *Angew. Chem. Int. Ed.* **46**, 941–944 (2007).

[106] L. J. Brown, N. C. Ma, D. R. Bouvet, S. Champion, A. M. Gibson, Y. L. Hu, A. Jackson, I. Khan, N. Millot, A. C. Topley, H. Wadsworth, D. Wynn and R. C. D. Brown, *Org. Biomol. Chem.* **7**, 564–575 (2009).

[107] E. Blom, F. Karimi and B. Långström, *J. Label. Compounds Radiopharm.* **52**, 504–511 (2009).

[108] T. J. Tewson, M. J. Welch and M. E. Raichle, *J. Nucl. Med.* **19**, 1339–1345 (1978).

[109] F. I. Aigbirhio, V. W. Pike, S. L. Waters and R. J. N. Tanner, *J. Fluor. Chem.* **70**, 279–287 (1995).

[110] D. W. Kim, D. S. Ahn, Y. H. Oh, S. Lee, H. S. Kil, S. J. Oh, S. J. Lee, J. S. Kim, J. S. Ryu, D. H. Moon and D. Y. Chi, *J. Am. Chem. Soc.* **128**, 16394–16397 (2006).

[111] S. J. Lee, S. J. Oh, D. Y. Chi, H. S. Kil, E. N. Kim, J. S. Ryu and D. H. Moon, *Eur. J. Nucl. Med. Mol. Imag.* **34**, 1406–1409 (2007).

[112] D. W. Kim, H. J. Jeong, S. T. Lim, M. H. Sohn and J. A. Katzenellenbogen, D. Y. Chi, *J. Org. Chem.* **73**, 957–962 (2008).

[113] S. J. Lee, S. J. Oh, D. Y. Chi, B. S. Lee, J. S. Ryu and D. H. Moon, *J. Label. Compounds Radiopharm.* **51**, 80–82 (2008).

[114] K. Sachin, E. M. Kim, S. J. Cheong, H. J. Jeong, S. T. Lim, M. H. Sohn and D. W. Kim, *Bioconj. Chem.* **21**, 2282–2288 (2010).

[115] J. H. Lee, H. B. Zhou, C. S. Dence, K. E. Carlson, M. J. Welch and J. A. Katzenellenbogen, *Bioconj. Chem.* **21**, 1096–1104 (2010).

[116] K. Sachin, H. J. Jeong, S. T. Lim, M. H. Sohn, D. Y. Chi and D. W. Kim, *Tetrahedron* **67**, 1763–1767 (2011).

[117] S. J. Lee, S. J. Oh, W. Y. Moon, M. S. Choi, J. S. Kim, D. Y. Chi, D. H. Moon and J. S. Ryu, *Nucl. Med. Biol.* **38**, 593–597 (2011).

[118] T. Koivula, J. Šimeček, J. Jalomäki, K. Helariutta and A. J. Airaksinen, *Radiochim. Acta* **99**, 293–300 (2011).

[119] D. E. Olberg, J. M. Arukwe, D. Grace, O. K. Hjelstuen, M. Solbakken, G. M. Kindberg and A. Cuthbertson, *J. Med. Chem.* **53**, 1732–1740 (2010).

[120] M. Suehiro, S. Vallabhajosula, S. J. Goldsmith and D. J. Ballon, *Appl. Radiat. Isot.* **65**, 1350–1358 (2007).

[121] D. W. Kim, C. E. Song and D. Y. Chi, *J. Am. Chem. Soc.* **124**, 10278–10279 (2002).

[122] D. W. Kim, Y. S. Choe and D. Y. Chi, *Nucl. Med. Biol.* **30**, 345–350 (2003).

[123] D. W. Kim, Y. S. Choe and D. Y. Chi, *J. Label. Compounds Radiopharm.* **46** (Suppl. 1), S89 (2003).

[124] H. W. Kim, J. M. Jeong, Y. S. Lee, D. Y. Chi, K. H. Chung, D. S. Lee, J. K. Chung and M. C. Lee, *Appl. Radiat. Isot.* **61**, 1241–1246 (2004).

[125] B. S. Moon, K. C. Lee, G. I. An, D. Y. Chi, S. D. Yang, C. W. Choi, S. M. Lim and K. S. Chun, *J. Label. Compounds Radiopharm.* **49**, 287–293 (2006).

[126] S. S. Shinde, Y. S. Lee and D. Y. Chi, *Org. Letters* **10**, 733–735 (2008).

[127] N. Guo, D. Alagille, G. Tamagnan, R. R. Price and R. M. Baldwin, *Appl. Radiat. Isot.* **66**, 1396–1402 (2008).

[128] H. R. Sun and S. G. DiMagno, *J. Fluor. Chem.* **128**, 806–812 (2007).

[129] K. Serdons, T. Verduyckt, D. Vanderghinste, J. Cleynhens, P. Borghgraef, P. Vermaelen, C. Terwinghe, F. Van Leuven, K. Van Laere, H. Kung, G. Bormans and A. Verbruggen, *Bioorg. Med. Chem. Letters* **19**, 602–605 (2009).

[130] K. Serdons, C. Terwinghe, P. Vermaelen, K. Van Laere, H. Kung, L. Mortelmans, G. Bormans, and A. Verbruggen, *J. Med. Chem.* **52**, 1428–1437 (2009).

[131] W. Deuther-Conrad, S. Fischer, A. Hiller, E. O. Nielsen, D. B. Timmermann, J. Steinbach, O. Sabri, D. Peters and P. Brust, *Eur. J. Nucl. Med. Mol. Imag.* **36**, 791–800 (2009).

[132] T. M. Shoup, D. R. Elmaleh, A. L. Brownell, A. Zhu, J. L. Guerrero, and A. J. Fischman, *Mol. Imag. Biol.* **13**, 511–517 (2011).

[133] C. Crouzel, M. Venet, T. Irie, G. Sanz and C. Boullais, *J. Label. Compounds Radiopharm.* **25**, 403–414 (1988).

[134] D. Roeda, B. Kuhnast, A. Hammadi and F. Dollé, *J. Label. Compounds Radiopharm.* **50**, 848–866 (2007).

[135] D. LeBars, C. Lemaire, N. Ginovart, A. Plenevaux, J. Aerts, C. Brihaye, W. Hassoun, V. Leviel, P. Mekhsian, D. Weissmann, J. F. Pujol, A. Luxen and D. Comar, *Nucl. Med. Biol.* **25**, 343–350 (1998).

[136] B. Dörner, C. Kuntner, J. P. Bankstahl, T. Wanek, M. Bankstahl, J. Stanek, J. Mullauer, F. Bauer, S. Mairinger, W. Löscher, D. W. Miller, P. Chiba, M. Müller, T. Erker and O. Langer, *Bioorg. Med. Chem.* **19**, 2190–2198 (2011).

[137] D. Thonon, C. Kech, J. Paris, C. Lemaire and A. Luxen, *Bioconj. Chem.* **20**, 817–823 (2009).

[138] B. Shen, W. Ehrlichmann, M. Uebele, H. J. Machulla and G. Reischl, *Appl. Radiat. Isot.* **67**, 1650–1653 (2009).

[139] R. Labas, G. Gilbert, O. Nicole, M. Dhilly, A. Abbas, O. Tirel, A. Buisson, J. Henry, L. Barré, D. Debruyne and F. Sobrio, *Eur. J. Med. Chem.* **46**, 2295–2309 (2011).

[140] G. Massaweh, E Schirrmacher, C. la Fougere, M. Kovacevic, C. Wängler, D. Jolly, P. Gravel, A. J. Reader and A. Thiel, *Nucl. Med. Biol.* **36**, 721–727 (2009).

[141] K. S. Mandap, T. Ido, Y. Kiyono, M. Kobayashi, T. G. Lohith, T. Mori, S. Kasamatsu, T. Kudo, H. Okazawa and Y. Fujibayashi, *Nucl. Med. Biol.* **36**, 403–409 (2009).

[142] N. Lazarova, F. G. Siméon, J. L. Musachio, S. Y. Lu and V. W. Pike, *J. Label. Compounds Radiopharm.* **50**, 463–465 (2007).

[143] N. Vasdev, P. N. Dorff, J. P. O'Neil, F. T. Chin, S. Hanrahan and H. F. VanBrocklin, *Bioorg. Med. Chem.* **19**, 2959–2965 (2011).

[144] F. Basuli, H. T. Wu and G. L. Griffiths, *J. Label. Compounds Radiopharm.* **54**, 224–228 (2011).

[145] F. Basuli, H. T. Wu, C. H. Li, Z. D. Shi, A. Sulima and G. L. Griffiths, *J. Label. Compounds Radiopharm.* **54**, 633–636 (2011).

[146] A. Shah, V. W. Pike and D. A. Widdowson, *J. Chem. Soc. Perkin Trans. 1.* 2043–2046 (1998).

[147] E. D. Hostetler, S. D. Jonson, M. J. Welch and J. A. Katzenellenbogen, *J. Org. Chem.* **64**, 178–185 (1999).

[148] F. R. Wüst and T Kniess, *J. Label. Compounds Radiopharm.* **46**, 699–713 (2003).

[149] F. R. Wüst and T Kniess, *J. Label. Compounds Radiopharm.* **47**, 457–468 (2004).

[150] J. Ermert, C. Hocke, T. Ludwig, R. Gail and H. H. Coenen, *J. Label. Compounds Radiopharm.* **47**, 429–441 (2004).

[151] E. G. Robins, F. Brady and S. K. Luthra, *J. Label. Compounds Radiopharm.* **48** (Suppl. 1), S145 (2005).

[152] M. A. Carroll, J. Nairne and J. L. Woodcraft, *J. Label. Compounds Radiopharm.* **50**, 452–454 (2007).

[153] M. R. Zhang, K. Kumata and K. A. Suzuki, *Tetra. Lett.* **48**, 8632–8635 (2007).

[154] T. L. Ross, J. Ermert, C. Hocke and H. H. Coenen, *J. Amer. Chem. Soc.* **129**, 8018–8025 (2007).

[155] B. C. Lee, C. S. Dence, H. B. Zhou, E. E. Parent, M. J. Welch and J. A. Katzenellenbogen, *Nucl. Med. Biol.* **36**, 147–153 (2009).

[156] J. H. Chun, S. Y. Lu, Y. S. Lee and V. W. Pike, *J. Org. Chem.* **75**, 3332–3338 (2010).

[157] J. H. Chun, S. Lu and V. W. Pike, *Eur. J. Org. Chem.* 4439–4447 (2011).

[158] B. C. Lee, J. S. Kim, B. S. Kim, J. Y. Son, S. K. Hong, H. S. Park, B. S. Moon, J. H. Jung, J. M. Jeong and S. E. Kim, *Bioorg. Med. Chem.* **19**, 2980–2990 (2011).

[159] T. L. Ross, J. Ermert and H. H. Coenen, *Molecules* **16**, 7621–7626 (2011).

[160] L. Li, M. N. Hopkinson, R. L. Yona, R. Bejot, A. D. Gee and V. Gouverneur, *Chem. Sci.* **2**, 123–131 (2011).

[161] M. Placzek, P. LaBeaume, L. Harris, P. Ng, M. Daniels, A. Kallmerten and G. B. Jones, *Tetra. Lett.* **52**, 332–335 (2011).

[162] M. Asti, D. Farioli, M. Iori, C. Guidotti, A. Versari and D. Salvo, *Nucl. Med. Biol.* **37**, 309–315 (2010).

[163] F. Dollé, L. Dolci, H. Valette, F. Hinnen, F. Vaufrey, I. Guenther, C. Fuseau, C. Coulon, M. Bottlaender and C. Crouzel, *J. Med. Chem.* **42**, 2251–2259 (1999).

[164] F. Dollé, in Ernst Schering Research Foundation, ed., *PET chemistry: The driving force in molecular imaging*, Springer Verlag, Berlin-Heidelberg, 2007, Vol. **62**, pp. 113–157.

[165] N. Malik, C. Solbach, W. Voelter and H. J. Machulla, *J. Radioanal. Nucl. Chem.* **283**, 757–764 (2010).

[166] F. Dollé, C. Luus, A. Reynolds and M. Kassiou, *Curr. Med. Chem.* **16**, 2899–2923 (2009).

[167] A. Damont, R. Boisgard, B. Kuhnast, F. Lemée, G. Raggiri, A. M. Scarf, E. Da Pozzo, S. Selleri, C. Martini, B. Tavitian, M. Kassiou and F. Dollé, *Bioorg. Med. Chem. Letters* **21**, 4819–4822 (2011).

[168] D. Roeda, B. Kuhnast, A. Damont and F. Dollé, *J. Fluor. Chem.* **134**, 107–114 (2012).

[169] L. Dolci, F. Dollé, H. Valette, F. Vaufrey, C. Fuseau, M. Bottlaender and C. Crouzel, *Bioorg. Med. Chem.* **7**, 467–479 (1999).

[170] G. Roger, W. Saba, H. Valette, F. Hinnen, C. Coulon, M. Ottaviani, M. Bottlaender and F. Dollé, *Bioorg. Med. Chem.* **14**, 3848–3858 (2006).

[171] G. Roger, F. Hinnen, H. Valette, W. Saba, M. Bottlaender and F. Dollé, *J. Label. Compounds Radiopharm.* **49**, 489–504 (2006).

[172] J. T. Patt, W. Deuther-Conrad, P. Brust, M. Patt, O. Sabri and J. Steinbach, *J. Label. Compounds Radiopharm.* **48** (Suppl. 1), S90 (2005).

[173] N. Malik, W. Voelter, H. J. Machulla and C. Solbach, *J. Radioanal. Nucl. Chem.* **287**, 287–292 (2011).

[174] B. Kuhnast, F. Hinnen, B. Tavitian and F. Dollé, *J. Label. Compounds Radiopharm.* **51**, 336–342 (2008).

[175] J. A. H. Inkster, B. Guerin, T. J. Ruth and M. J. Adam, *J. Label. Compounds Radiopharm.* **51**, 444–452 (2008).

[176] J. A. H. Inkster, M. J. Adam, T. Storr and T. J. Ruth, *Nucleosides Nucleotides Nucleic Acids* **28**, 1131–1143 (2009).

[177] B. de Bruin, B. Kuhnast, F. Hinnen, L. Yaouancq, M. Amessou, L. Johannes, A. Samson, R. Boisgard, B. Tavitian and F. Dollé, *Bioconj. Chem.* **16**, 406–420 (2005).

[178] L. Dolci, F. Dollé, S. Jubeau, F. Vaufrey and C. Crouzel, *J. Label. Compounds Radiopharm.* **42**, 975–985 (1999).

[179] M. Karramkam, F. Hinnen, F. Vaufrey and F. Dollé, *J. Label. Compounds Radiopharm.* **46**, 979–992 (2003).

[180] P. Marchand, C. Lorilleux, G. Gilbert, F. Gourand, F. Sobrio, D. Peyronnet, M. Dhilly and L. Barré, *ACS Med Chem Letters* **1**, 240–243 (2010).

[181] T. Pham, C. Fookes, X. Liu, I. Greguric, T. Bourdier and A. Katsifis, *J. Label. Compounds Radiopharm.* **50** (Suppl. 1), S204 (2007).

[182] F. G. Siméon, M. T. Wendahl and V. W. Pike, *Tetra. Lett.* **51**, 6034–6036 (2010).

[183] F. G. Siméon, M. T. Wendahl and V. W. Pike, *J. Med. Chem.* **54**, 901–908 (2011).

[184] F. M. Wagner, J. Ermert and H. H. Coenen, *J. Nucl. Med.* **50**, 1724–1729 (2009).

[185] M. Ogawa, K. Hatano, S. Oishi, Y. Kawasumi, N. Fujii, M. Kawaguchi, R. Doi, M. Imamura, M. Yamamoto, K. Ajito, T. Mukai, H. Saji and K. Ito, *Nucl. Med. Biol.* **30**, 1–9 (2003).

[186] B. Kuhnast and F. Dollé, *Curr. Radiopharm.* **3**, 174–201 (2010).

[187] S. M. Okarvi, *Eur. J. Nucl. Med.* **28**, 929–938 (2001).

[188] F. Wuest, M. Berndt, R. Bergmann, J. van den Hoff and J. Pietzsch, *Bioconj. Chem.* **19** 1202–1210 (2008).

[189] F. Dollé, F. Hinnen, F. Vaufrey, B. Tavitian and C. Crouzel, *J. Label. Compounds Radiopharm.* **39**, 319–330 (1997).

[190] B. Tavitian, S. Terrazzino, B. Kuhnast, S. Marzabal, O. Stettler, F. Dollé, J.-R. Deverre, A. Jobert, F. Hinnen, B. Bendriem, C. Crouzel and L. Di Giamberardino, *Nature Med.* **4**, 467–471 (1998).

[191] B. Kuhnast, F. Dollé, S. Terrazzino, B. Rousseau, C. Loc'h, F. Vaufrey, F. Hinnen, I. Doignon, F. Pillon, C. David, C. Crouzel and B. Tavitian, *Bioconj. Chem.* **11**, 627–636 (2000).

[192] M. Glaser, H. Karlsen, M. Solbakken, J. Arukwe, F. Brady, S. K. Luthra and A. Cuthbertson, *Bioconj. Chem.* **15**, 1447–1453 (2004).

[193] M. G. Straatmann and M. J. Welch, *J. Nucl. Med.* **16**, 425–428 (1975).

[194] G. Berger, M. Mazière, C. Prenant, J. Sastre and D. Comar, *Int J Appl Radiat Isot.* **35**, 81–83 (1984).

[195] C. Prenant, C. Cawthorne, M. Fairclough, N. Rothwell and H. Boutin, *Appl. Radiat. Isot.* **68**, 1721–1727 (2010).

[196] C. Prenant, J. Gillies, J. Bailey, G. Chimon, N. Smith, G. C. Jayson and J. Zweit, *J. Label. Compounds Radiopharm.* **51**, 262–267 (2008).

[197] S. Schmitt, C. Bouteiller, L. Barré and C. Perrio, *Chem. Comm.* **47** 11465–11467 (2011).

[198] D. Roeda, B. Kuhnast and F. Dollé, in Xiaoyuan Chen ed., *Recent Advances of Bioconjugate Chemistry in Molecular Imaging*, Research Signpost, Kerala 2008, pp. 115–153.

[199] W. J. McBride, R. M. Sharkey, H. Karacay, C. A. D'Souza, E. A. Rossi, P. Laverman, C. H. Chang, O. C. Boerman and D. M. Goldenberg, *J. Nucl. Med.* **50**, 991–998 (2009).

[200] W. J. McBride, C. A. D'Souza, R. M. Sharkey, H. Karacay, E. A. Rossi, C. H. Chang and D. M. Goldenberg, *Bioconj. Chem.* **21**, 1331–1340 (2010).

[201] P. Laverman, W. J. McBride, R. M. Sharkey, A. Eek, L. Joosten, W. J. G. Oyen, D. M. Goldenberg and O. C. Boerman, *J. Nucl. Med.* **51**, 454–461 (2010).

[202] S. L. Liu, H. G. Liu, H. Jiang, Y. Xu, H. Zhang and Z. Cheng, *Eur. J. Nucl. Med. Mol. Imag.* **38**, 1732–1741 (2011).

[203] D. Shetty, S. Y. Choi, J. M. Jeong, J. Y. Lee, L. Hoigebazar, Y. S. Lee, D. S. Lee, J. K. Chung, M. C. Lee and Y. K. Chung, *Chem. Comm.* **47**, 9732–9734 (2011).

[204] C. A. D'Souza, W. J. McBride, R. M. Sharkey, L. J. Todaro and D. M. Goldenberg, *Bioconj. Chem.* **22**, 1793–1803 (2011).

[205] W. J. McBride, C. A. D'Souza, R. M. Sharkey and D. M. Goldenberg, *Appl. Radiat. Isot.* **70**, 200–204 (2012).

[206] R. Ting, M. J. Adam, T. J. Ruth and D. M. Perrin, *J. Am. Chem. Soc.* **127**, 13094–13095 (2005).

[207] R. Ting, C. Harwig, U. auf dem Keller, S. McCormick and P. Austin, *J. Am. Chem. Soc.* **130**, 12045–12055 (2008).

[208] R. Ting, J. Lo, M. J. Adam, T. J. Ruth and D. M. Perrin, *J. Fluor. Chem.* **129**, 349–358 (2008).

[209] C. W. Harwig, R. Ting, M. J. Adam, T. J. Ruth and D. M. Perrin, *Tetra. Lett.* **49**, 3152–3156 (2008).

[210] L. Mu, P. A. Schubiger and S. M. Ametamey, *Curr. Radiopharm.* **3**, 224–242 (2010).

[211] U. auf dem Keller, C. L. Bellac, Y. Li, Y. M. Lou, P. F. Lange, R. Ting, C. Harwig, R. Kappelhoff, S. Dedhar, M. J. Adam, T. J. Ruth, F. Benard, D. M. Perrin and C. M. Overall, *Cancer Res.* **70**, 7562–7569 (2010).

[212] Y. Li, R. Ting, C. W. Harwig, U. A. D. Keller, C. L. Bellac, P. F. Lange, J. A. H. Inkster, P. Schaffer, M. J. Adam, T. J. Ruth, C. M. Overall and D. M. Perrin, *Medchemcomm.* **2**, 942–949 (2011).

[213] Z. Li, T. P. Lin, S. Liu, C. W. Huang, T. W. Hudnall, F. P. Gabbai and P. S. Conti, *Chem. Comm.* **47**, 9324–9326 (2011).

[214] A. Höhne, L. Yu, L. J. Mu, M. Reiher, U. Voigtmann, U. Klar, K. Graham, P. A. Schubiger and S. M. Ametamey, *Chem. Eur. J.* **15**, 3736–3743 (2009).

[215] E. Schirrmacher, B. Wängler, M. Cypryk, G. Bradtmöller, M. Schäfer, M. Eisenhut, K. Jurkschat and R. Schirrmacher, *Bioconj. Chem.* **18**, 2085–2089 (2007).

[216] L. Iovkova-Berends, C. Wängler, T. Zöller, G. Höfner, K. T. Wanner, C. Rensch, P. Bartenstein, A. Kostikov, R. Schirrmacher, K. Jurkschat and B. Wängler, *Molecules* **16**, 7458–7479 (2011).

[217] R. Schirrmacher, G. Bradtmöller, E. Schirrmacher, O. Thews, J. Tillmanns, T. Siessmeier, H. G. Buchholz, P. Bartenstein, B. Waengler, C. M. Niemeyer and K. Jurkschat, *Angew. Chem. Int. Ed.* **45**, 6047–6050 (2006).

[218] C. Wängler, B. Waser, A. Alke, L. Iovkova, H. G. Buchholz, S. Niedermoser, K. Jurkschat, C. Fottner, P. Bartenstein, R. Schirrmacher, J. C. Reubi, H. J. Wester and B. Wängler, *Bioconj. Chem.* **21**, 2289–2296 (2010).

[219] L. Iovkova, B. Wängler, E. Schirrmacher, R. Schirrmacher, G. Quandt, G. Boening, M. Schurmann and K. Jurkschat, *Chem. Eur. J.* **15**, 2140–2147 (2009).

[220] B. Wängler, G. Quandt, L. Iovkova, E. Schirrmacher, C. Wängler, G. Boening, M. Hacker, M. Schmoeckel, K. Jurkschat, P. Bartenstein and R. Schirrmacher, *Bioconj. Chem.* **20**, 317–321 (2009).

[221] P. Rosa-Neto, B. Waengler, L. Iovkova, G. Boening, A. Reader, K. Jurkschat and E. Schirrmacher, *Chembiochem.* **10**, 1321–1324 (2009).

[222] L. F. Tietze and K. Schmuck, *Synlett* 1697–1700 (2011).

[223] L. Iovkova, D. Konning, B. Wängler, R. Schirrmacher, S. Schoof, H. D. Arndt and K. Jurkschat, *Eur. J. Inorg. Chem.* 2238–2246 (2011).

[224] A. P. Kostikov, L. Iovkova, J. Chin, E. Schirrmacher, B. Wängler, C. Wängler, K. Jurkschat, G. Cosa and R. Schirrmacher, *J. Fluor. Chem.* **132**, 27–34 (2011).

[225] L. J. Mu, A. Höhne, R. A. Schubiger, S. M. Ametamey, K. Graham, J. E. Cyr, L. Dinkelborg, T. Stellfeld, A. Srinivasan, U. Voigtmann and U. Klar, *Angew. Chem. Int. Ed.* **47**, 4922–4925 (2008).

[226] A. Hoehne, L. Mu, M. Honer, P. A. Schubiger, S. M. Ametamey, K. Graham, T. Stellfeld, S. Borkowski, D. Berndorff, U. Klar, U. Voigtmann, J. E. Cyr, M. Friebe, L. Dinkelborg and A. Srinivasan, *Bioconj. Chem.* **19**, 1871–1879 (2008).

[227] P. Bohn, A. Deyine, R. Azzouz, L. Bailly, C. Fiol-Petit, L. Bischoff, C. Fruit, F. Marsais and P. Vera, *Nucl. Med. Biol.* **36**, 895–905 (2009).

[228] J. Schulz, D. Vimont, T. Bordenave, D. James, J. M. Escudier, M. Allard, M. Szlosek-Pinaud and E. Fouquet, *Chem. Eur. J.* **17**, 3096–3100 (2011).

[229] E. Balentova, C. Collet, S. Lamandé-Langle, F. Chrétien, D. Thonon, J. Aerts, C. Lemaire, A. Luxen and Y. Chapleur, *J. Fluor. Chem.* **132**, 250–257 (2011).

[230] M. Glaser and E. G. Robins, *J. Label. Compounds Radiopharm.* **52**, 407–414 (2009).

[231] T. L. Ross, *Curr. Radiopharm.* **3**, 202–223 (2010).

[232] J. C. Walsh and H. C. Kolb, *Chimia*, **64**, 29–33 (2010).

[233] J. Marik and J. L. Sutcliffe, *Tetrahedron Letters* **47**, 6681–6684 (2006).

[234] M. Glaser and E. Årstad, *Bioconj. Chem.* **18**, 989–993 (2007).

[235] G. Smith, M. Glaser, M. Perumal, Q. D. Nguyen, B. Shan, E. Årstad and E. O. Aboagye, *J. Med. Chem.* **51**, 8057–8067 (2008).

[236] D. H. Kim, Y. S. Choe, J. Y. Choi, Y. Choi, K. H. Lee and B. T. Kim, *Bioconj. Chem.* **20**, 1139–1145 (2009).

[237] D. Kobus, Y. Giesen, R. Ullrich, H. Backes and B. Neumaier, *Appl. Radiat. Isot.* **67**, 1977–1984 (2009).

[238] D. H. Kim, Y. S. Choe and B. T. Kim, *Appl. Radiat. Isot.* **68**, 329–333 (2010).

[239] J. McConathy, D. Zhou, S. E. Shockley, L. A. Jones, E. A. Griffin, H. Lee, S. J. Adams and R. H. Mach, *Mol. Imag.* **9**, 329–342 (2010).

[240] F. Pisaneschi, Q. D. Nguyen, E. Shamsaei, M. Glaser, E. Robins, M. Kaliszczak, G. Smith, A. C. Spivey and E. O. Aboagye, *Bioorg. Med. Chem.* **18**, 6634–6645 (2010).

[241] L. Iddon, J. Leyton, B. Indrevoll, M. Glaser, E. G. Robins, A. J. T. George, A. Cuthbertson, S. K. Luthra and E. O. Aboagye, *Bioorg. Med. Chem. Letters* **21**, 3122–3127 (2011).

[242] U. Ackermann, G. O'Keefe, S. T. Lee, A. Rigopoulos, G. Cartwright, J. I. Sachinidis, A. M. Scott and H. J. Tochon-Danguy, *J. Label. Compounds Radiopharm.* **54**, 260–266 (2011).

[243] S. Maschauer and O. Prante, *Carbohydrate Res.* **344**, 753–761 (2009).

[244] N. K. Devaraj, E. J. Keliher, G. M. Thurber, M. Nahrendorf and R. Weissleder, *Bioconj. Chem.* **20**, 397–401 (2009).

[245] H. S. Gill, J. N. Tinianow, A. Ogasawara, J. E. Flores, A. N. Vanderbilt, H. Raab, J. M. Scheer, R. Vandlen, S. P. Williams and J. Marik, *J. Med. Chem.* **52**, 5816–5825 (2009).

[246] H. S. Gill and J. Marik, *Nature Protocols* **6**, 1718–1725 (2011).

[247] K. Michel, K. Büther, M. P.Law, S. Wagner, O. Schober, S. Hermann, M. Schäfers, B. Riemann, C. Höltke and K. Kopka, *J. Med. Chem.* **54**, 939–948 (2011).

[248] T. Kuboyama, M. Nakahara, M. Yoshino, Y. L. Cui, T. Sako, Y. Wada, T. Imanishi, S. Obika, Y. Watanabe, M. Suzuki and H. Doi, *Bioorg. Med. Chem.* **19**, 249–255 (2011).

[249] V. Bouvet, M. Wuest and F. Wuest, *Org. Biomol. Chem.* **9**, 7393–7399 (2011).

[250] S. Arumugam, J. Chin, R. Schirrmacher, V.V. Popik and A. P. Kostikov, *Bioorg. Med. Chem. Letters* **21**, 6987–6991 (2011).

[251] L. S. Campbell-Verduyn, L. Mirfeizi, A. K. Schoonen, R. A. Dierckx, P. H. Elsinga and B. L. Feringa, *Angew. Chem. Int. Ed.* **50**, 11117–11120 (2011).

[252] B. D. Zlatopolskiy, R. Kandler, F. M. Mottaghy and B. Neumaier, *Appl. Radiat. Isot.* **70**, 184–192 (2012).

[253] Z. B. Li, H. C. Cai, M. Hassink, M. L. Blackman, R. C. D. Brown, P. S. Conti and J. M. Fox, *Chem. Comm.* **46**, 8043–8045 (2010).

[254] E. J. Keliher, T. Reiner, A. Turetsky, S. A. Hilderbrand and R. Weissleder, *ChemMedChem.* **6**, 424–427 (2011).

[255] R. Selvaraj, S. L. Liu, M. Hassink, C. W. Huang, L. P. Yap, R. Park, J. M. Fox, Z. B. Li and P. S. Conti, *Bioorg. Med. Chem. Letters* **21**, 5011–5014 (2011).

[256] D. Thonon, E. Goukens, G. Kaisin, J. Paris, J. Flagothier and A. Luxen, *Tetrahedron* **67**, 5572–5576 (2011).

[257] A. Gaeta, J. Woodcraft, S. Plant, J. Goggi, P. Jones, M. Battle, W. Trigg, S. K. Luthra and M. Glaser, *Bioorg. Med. Chem. Letters* **20**, 4649–4652 (2010).

[258] M. Pretze, F. Wuest, T. Peppel, M. Köckerling and C. Mamat, *Tetrahedron Letters* **51**, 6410–6414 (2010).

[259] L. Carroll, S. Boldon, R. Bejot, J. E. Moore, J. Declerck and V. Gouverneur, *Org. Biomol. Chem.* **9**, 136–140 (2011).

[260] C. Mamat, M. Franke, T. Peppel, M. Köckerling and J. Steinbach, *Tetrahedron* **67**, 4521–4529 (2011).

[261] I. F. Antunes, H. J. Haisma, P. H. Elsinga, R. A. Dierckx and E. F. J. de Vries, *Bioconj. Chem.* **21**, 911–920 (2010).

[262] C. Mamat, M. Pretze, J. Steinbach and F. Wuest, *J. Label. Compounds Radiopharm.* **52** (Suppl. 1), S142 (2009).

4

CARBON-11, NITROGEN-13, AND OXYGEN-15 CHEMISTRY: AN INTRODUCTION TO CHEMISTRY WITH SHORT-LIVED RADIOISOTOPES

PHILIP W. MILLER

Department of Chemistry, Imperial College London, London, UK

KOICHI KATO

Department of Molecular Imaging, National Centre of Neurology and Psychiatry, Kodaira, Tokyo, Japan

BENGT LÅNGSTRÖM

Department of Biochemistry and Organic Chemistry, Uppsala University, Uppsala, Sweden
Neuropsychopharmacology Unit, Centre for Pharmacology and Therapeutics, Division of Experimental Medicine, Imperial College London, London, UK
Department of Nuclear Medicine, PET & Cyclotron Unit, Odense University Hospital, University of Southern Denmark, Institute of Clinical Research, Odense, Denmark

4.1 INTRODUCTION

Carbon-11, nitrogen-13, and oxygen-15, with their respective half-lives of 20.4, 10.0, and 2.0 min., present significant chemical challenges for the production of tracer molecules for PET imaging [1–3]. Despite their short half-lives all three isotopes are important and versatile isotopes for the synthesis of PET tracers. The presence of stable carbon-12, nitrogen-14, and oxygen-16 in natural products and drug compounds make the corresponding PET isotopes obvious choices for the preparation of equivalent labelled compounds that would display virtually identical chemical and biological behaviour. Although the half-lives of these PET isotopes preclude time-consuming multistep syntheses, a range of clever radiochemistry has been developed over many years for introducing these isotopes into tracer molecules. Of these isotopes, C-11 has received the most attention owing to its longer half-life, which enables a wider range of chemical transformations, and access to more diverse chemical structures. N-13 and O-15, in comparison, are typically used to prepare simple molecules such as $^{13}NH_3$, $^{15}O_2$, and $[^{15}O]H_2O$. The aim of this chapter is twofold, first to introduce the field of C-11, N-13, and O-15 chemistry to the reader who may be new to the field, and second, to provide an up-to-date account of their chemistry for those who are more familiar with the field. The application of radiolabelled tracers prepared from these isotopes will not be discussed in much detail.

There are many challenges facing the chemist in the preparation of PET tracers using short-lived isotopes. Time is the most obvious parameter and places severe limits on the possible types of chemistry. A general rule of two to three half-lives is loosely applied to PET tracer synthesis from the end of cyclotron bombardment (EOB). For example, the production of an O-15 labelled compound would have to be complete within a short 6 min. 'window' to ensure a viable production method. For C-11 labelling protocols, a typical 40 min. reaction window will generally permit a maximum of only two or three discrete chemical transformations. Consequently, much of the chemistry development for synthesis with short-lived isotopes is centred on the preparation of simple and reactive precursor molecules for the introduction of the isotope quickly and typically within one chemical step.

The Chemistry of Molecular Imaging, First Edition. Edited by Nicholas Long and Wing-Tak Wong.
© 2015 John Wiley & Sons, Inc. Published 2015 by John Wiley & Sons, Inc.

The production of a PET tracer is initiated with the preparation of a simple precursor molecule in a particle accelerator called a cyclotron. The cyclotron directs a beam of high-energy protons or deuterons into a target chamber that contains the target material suitable for required radioisotope production (Table 4.1).

This simple radiolabelled product is then transferred from the cyclotron target to a 'hotcell' for further chemical reactions to convert it into the desired radiolabelled product. Nowadays there is a greater degree of computer automation for tracer production in order to reduce the risk of radiation exposure to the user to improve good manufacturing processes (GMP). Many sophisticated systems have now been developed for tracer synthesis that integrate computer controlled valves, pumps, heaters, and robotics to process and manipulate radioactive samples safely within the confines of the hotcell (Figure 4.1a) whilst rapid preparative HPLC techniques are used as standard for purification and analysis prior to PET scanning (Figure 4.1b and 4.1c).

In spite of the challenges, a combination of innovative chemistry and engineering has enabled the routine production of a wide range PET tracers and radiolabelling precursors. One key example is the production of the important C-11 reagent [^{11}C]methyl iodide (^{11}CH$_3$I), which can be produced rapidly in reaction times of <20 min. from cyclotron-generated ^{11}CO$_2$ via a gas phase catalytic hydrogenation reaction to give ^{11}CH$_4$, followed by a free radical iodination. As outlined below, ^{11}CH$_3$I is currently used to label a wide range of PET tracers.

TABLE 4.1   Characteristics of the Major Short-lived PET Radioisotopes C-11, N-13, and O-15.

Radioisotope	Half-life, $t_{1/2}$ (min)	Nuclear Reaction	Target	Major Precursor	Decay Product
^{11}C	20.4	^{14}N$(p, \alpha)^{11}$C	N$_2$(+O$_2$)	[^{11}C]CO$_2$	^{11}B
			N$_2$(+H$_2$)	[^{11}C]CH$_4$	
^{13}N	9.97	^{16}O$(p,\alpha)^{13}$N	H$_2$O	[^{13}N]NO$_x$	^{13}C
			H$_2$O+EtOH	[^{13}N]NH$_3$	
^{15}O	2.04	^{15}N$(d,n)^{15}$O	N$_2$(+O$_2$)	[^{15}O]O$_2$	^{15}N

(a)

(b)

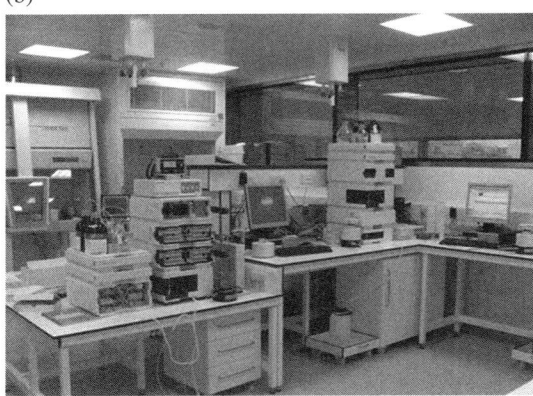

(c)

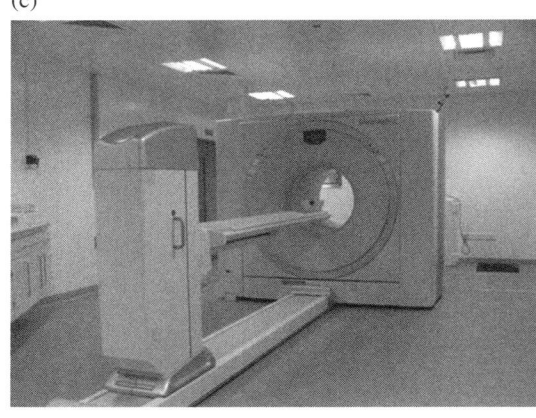

FIGURE 4.1   (a) A typical lead shielded hotcell used to perform radiolabelling reactions. (b) Radiochemistry quality control laboratory. (c) Combined PET-CT scanner.

Although the time issues and safe handling of short-lived radioactive PET isotopes may be obvious, there are added difficulties when dealing with the extremely small mass of radioisotopes produced. Typical reactions with PET isotopes are performed on the picomolar to nanomolar scale. Reacting and processing such small amounts of material have led to the development of miniaturised reaction systems and integration of microfluidic technology for radiolabelling reactions. Additionally, these very low amounts of radioisotopes result in a stoichiometric imbalance with the 'cold' reagent precursors with which they are reacting. The vast stoichiometric excess ($\sim 1 \times 10^3 - 1 \times 10^5$ fold) of cold precursor results in pseudo-first-order reaction kinetics with respect to the radioisotope concentration. This, in fact, can be advantageous for certain reactions and result in an acceleration of the labelling process; however, such excess reagents may be difficult to remove during the purification process. The efficiency of a labelling process is judged by both its radiochemical yield (RCY) and specific activity (SA) of the final labelled compound. RCY is a function of both the chemical yield and half-life of the radioisotope and is expressed as a fraction of the radioactivity present after a radiochemical separation. RCY is quoted as being either decay corrected, taking into account the radioactive decay that occurred between two different times, and non-decay corrected, which does not account for radioactive decay. High RCYs, although desirable, are not always essential for a viable tracer production. The specific activity is a measure of the radioactivity per unit mass of the labelled compound and is commonly expressed as GBq/µmol or Ci/µmol. Inevitably, some isotopic dilution with the naturally occurring isotope occurs during the labelling process, which means that theoretical SA maxima are never reached even for carrier-free methods. Specific activities are much lower for carrier-added synthesis methods, such as those used for the production of $^{18}F[F_2]$ in electrophilic fluorinations (see Chapter 3), to due to a direct result of isotopic dilution from the added carrier. Specific activities of PET-labelled products are typically in the order of 50-500 GBq/µmol ($\sim$1–15 Ci/µmol), are generally required to give a good quality PET data.

4.2 CARBON-11 CHEMISTRY

The carbon-11 isotope is most widely produced by the proton bombardment of nitrogen-14 ($^{14}N(p,\alpha)^{11}C$) in a gas phase cyclotron target. $^{11}CO_2$ and $^{11}CH_4$ are the two most commonly used ^{11}C 'primary' precursors and are formed when a small percentage of either oxygen or hydrogen is present in the target gases. The vast majority of ^{11}C labelled PET tracers are made from these two simple precursors; consequently, there is considerable effort placed into converting $^{11}CO_2$ and $^{11}CH_4$ into more reactive molecules for labelling. Scheme 4.1 summarises some of the major reactive 'secondary' precursors that are derived from $^{11}CO_2$ and $^{11}CH_4$.

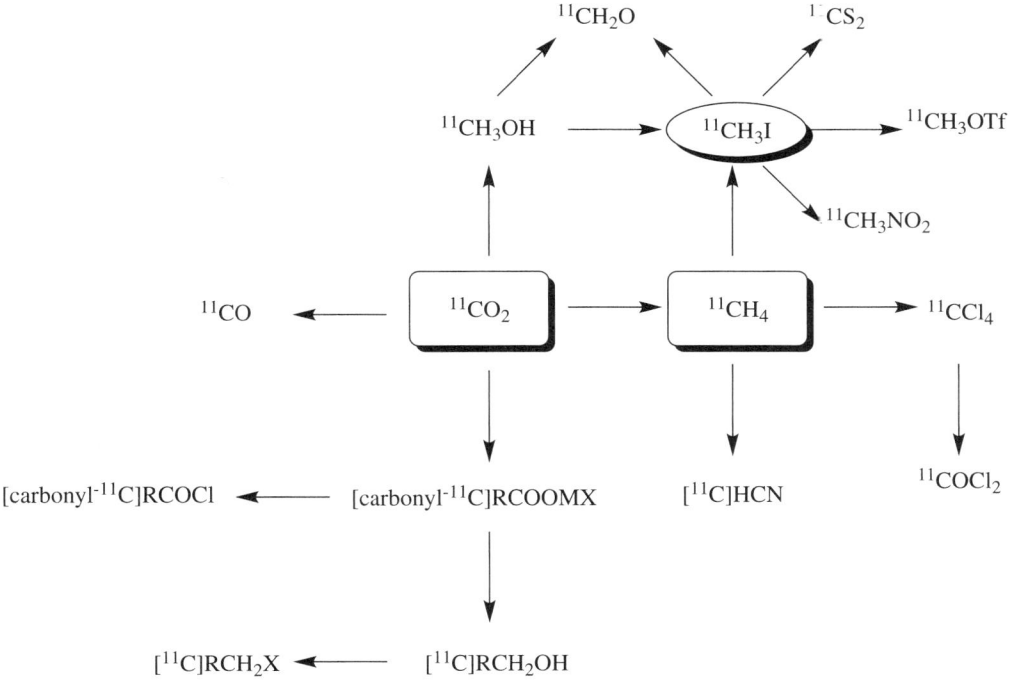

SCHEME 4.1   The major ^{11}C-precursors used in the synthesis of ^{11}C-labelled compounds produced from either $^{11}CO_2$ or $^{11}CH_4$.

4.2.1　Synthesis with [^{11}C]Carbon Dioxide

^{11}CO$_2$ is the most widely produced in target C-11 primary precursor. Although it is often converted to other reactive secondary precursors (Scheme 4.1), ^{11}CO$_2$ is a versatile reagent in its own right and has been recently exploited for the rapid labelling of a range of useful functional groups and target molecules. The main advantage of using ^{11}CO$_2$ directly is that labelling can be performed much faster and without the need for time-consuming secondary precursor production. One of the drawbacks of its direct reaction is the potential risk of contamination with atmospheric ^{12}CO$_2$, which can diminish specific radioactivities; however, with due care, excessive contamination with atmospheric CO$_2$ may be minimised. Carbon dioxide is widely known to rapidly react with organometallic reagents such as Grignard and organolithium reagents. The reaction of ^{11}CO$_2$ with Grignard reagents can be used to prepare a range of [^{11}C]carboxyl acids from the corresponding Grignards following hydrolysis. [Carbonyl-^{11}C]amides may be formed directly from [^{11}C]carboxymagnesium halides using conventional heating [4, 5] or microwave assisted heating [6] (Scheme 4.2).

[Carbonyl-^{11}C]acetate, which is used in the evaluation of myocardial oxygen metabolism [7] and diagnosis of prostate cancer [8], can be conveniently produced via [^{11}C]carboxylation of methyl magnesium bromide [9–11] following a quenching and neutralisation step. [^{11}C]carboxylic acids may be converted into more reactive acid chloride species via reaction with chlorinating agents such as thionyl chloride or phthaloyl dichloride. This route has been used for the synthesis of [carbonyl-^{11}C]amides and has been effectively applied to labelling the important 5HT$_{1A}$ receptor ligand WAY100635 (Scheme 4.3) [12, 13] and to the oncology biomarker BAY 59-8862, a derivative of taxane [14].

Recently, boronic acid esters have been used for metal-mediated ^{11}CO$_2$ carboxylation reactions. In contrast to organomagnesium and organolithium reagents, boronic acid esters are stable toward air and moisture and, importantly, have a high degree of functional group tolerance. Riss et al. [15] demonstrated a highly effective way of performing CuI-mediated carboxylations of boronic acid esters using a combination of TMEDA, KF, and crypt-222 (Scheme 4.4). A range of [^{11}C] carboxylic acids was obtained in high RCY that would be otherwise difficult to synthesise using traditional organometallic

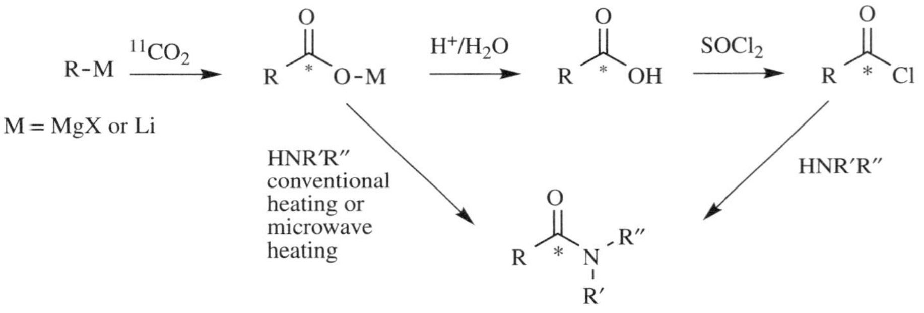

SCHEME 4.2　Reaction of ^{11}CO$_2$ with organometallic reagents to form [carbonyl-^{11}C]carboxylic acids, [carbonyl-^{11}C]acid chlorides, and [carbonyl-^{11}C]amides. [*] indicates C-11 labelling position.

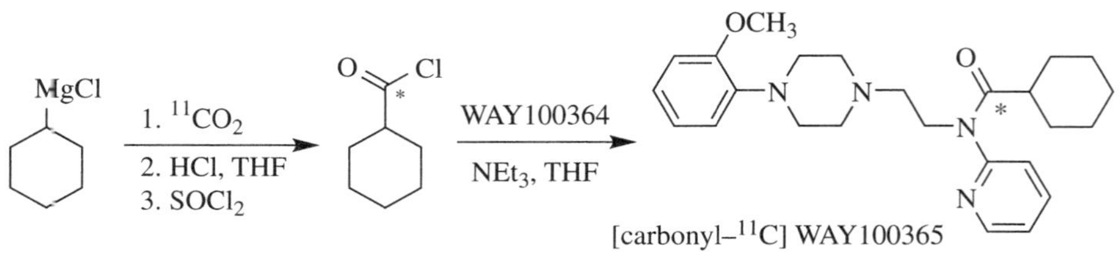

SCHEME 4.3　One-pot synthesis of [carbonyl-^{11}C]WAY100635 via reaction of [carbonyl-^{11}C]cyclohexyl acid chloride with WAY100634. [*] indicates C-11 labelling position.

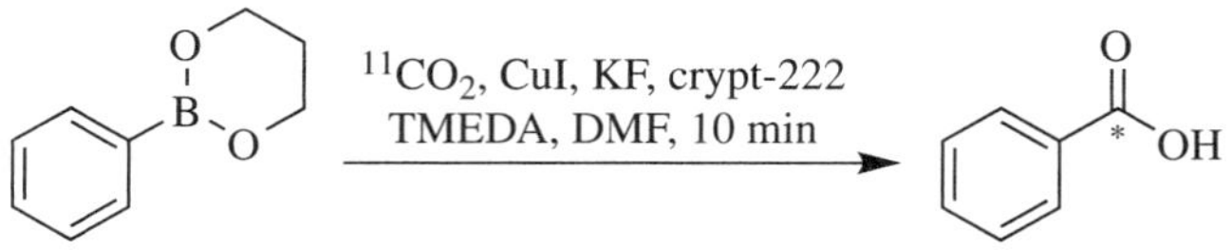

SCHEME 4.4　Model reaction of phenyl boronic ester with ^{11}CO$_2$ to form [carbonyl-^{11}C]benzoic acid in high RCY (99%). [*] indicates C-11 labelling position.

carboxylations. Furthermore, the carboxylic acids could be converted to esters or amides with good RCYs (>40%) and within short timeframes.

Ureas are common functional groups found in many drug-like molecules and therefore present valuable targets for C-11 labelling in the carbonyl position. Labelling of ureas has been previously achieved using C-11 phosgene and ^{11}CO (see below); however, the use of ^{11}CO$_2$ for C-11 carbonyl urea synthesis is potentially more advantageous because of the availability of ^{11}CO$_2$. A route to unsymmetrical carbonyl C-11 labelled ureas using ^{11}CO$_2$ has been recently developed using triphenylphosphinine precursors [16]. Initially, [^{11}C]phenylisocyanate is formed via the reaction of triphenylphosphinimine and ^{11}CO$_2$, following reaction of the [^{11}C]phenylisocyanate with a range of amines a series of small molecule [*carbonyl*-^{11}C] ureas were obtained. The highest RCYs were achieved when basic primary amines were used (45-49%), but disappointingly low yields were obtained when aniline was used as the amine source. This method may be useful for the preparation of a wide range of [^{11}C]ureas by variation of the phosphinimine and amine precursors; however, it has wider limitations due to the sensitivity of phosphinimine reagents.

Recently, the groups of Wilson [17, 18] and Fowler [19] have reported the use ^{11}CO$_2$ trapping reagents for the direct fixation of CO$_2$ to form labelled [*carbonyl* C-11] carbamate molecules. Both methods rely on the use of a strong amine base, DBU (1,8-diazabicycloundec-7-ene) or BEMP (2-*tert*-butylimino-2-diethylamino-1,3-dimethylperhydro-1,3,2-diazaphosphorine), to sequester and trap the ^{11}CO$_2$ from its nitrogen carrier gas stream. Exceptionally good trapping efficiencies were obtained when model alkyl primary or secondary amines were used; however, yields diminished when aniline derivatives were employed. Both groups were able to label PET imaging agents with high specific activities (>100 GBq µmol^{-1}).

4.2.2 [^{11}C]methylation Reactions

Of the secondary labelling shown in Figure 4.1, [^{11}C]methyl iodide is by far the most popular and commonly used C-11 precursor molecule. It is a highly versatile reagent capable of efficiently labelling a wide range of organic compounds. Reliable synthesis protocols and commercially available production units have further increased its popularity for C-11 labelling. There are two common synthetic routes to generate ^{11}CH$_3$I: the so-called 'wet method,' which is performed in an organic solvent and involves LiAlH$_4$ reduction of ^{11}CO$_2$ followed by reaction with hydroiodic acid [20], or the 'gas phase' free radical iodination reaction of ^{11}CH$_4$ at high temperature [21, 22]. The gas phase method has proven to be most popular because it of its ease of production and reliability. Today, several types of commercial units are available for ^{11}CH$_3$I production. In recent years, [^{11}C]methyl triflate, ^{11}CH$_3$OTf, has become an important alternative to ^{11}CH$_3$I because of its much greater reactivity [23], Typically, ^{11}CH$_3$OTf will be used when ^{11}CH$_3$I fails to give satisfactory yields [24–28]. [^{11}C]methyl triflate can be prepared by passing gaseous ^{11}CH$_3$I through a column of silver triflate at 200 °C [29]. Other ^{11}C-alkylating agents have been developed, such as [^{11}C]ethyliodide, [^{11}C]propyliodide, [^{11}C]butyliodide, and [^{11}C]benzyliodide; however, these have not proven to be as popular as the methylating reagents [30, 31].

Nucleophilic substitution is the standard route by which ^{11}CH$_3$ is reacted with the precursor molecule to form a tracer. *N*-, *O*-, and *S*- methylation reactions of amines, alcohols, and thiols labelled primary or secondary amines, ethers, or thioethers are common. Technically, this method is straightforward and usually involves passing a gas stream of ^{11}CH$_3$I or ^{11}CH$_3$OTf through a basic solution of the precursor; the solution may occasionally need to be heated to improve reaction times. Numerous ^{11}C-methylation procedures have been reported [32], and there are many technical variations of this method using various automated systems or captive solvent 'loop' methods [33]. As mentioned above, all PET labelling reactions have a vast stoichiometric excess of cold precursor reagent compared to the isotopically labelled precursor regent. This can be beneficial for the labelling of amine precursors. Oversubstitution, typical of 'cold' stoichiometric scale amine-alkylation reactions, is never observed at the concentrations used in typical PET labelling reactions. Such labelling reactions always result in exclusive labelling of only one C-11 methyl per molecule.

Methylation reactions are commonly used for the production of many of the key ^{11}C-tracers (Scheme 4.5) including Pittsburgh Compound B (PIB) [28, 34–36], raclopride (Scheme 4.5) [33, 37, 38], [^{11}C]*N*-methylspiperone [39], (NMSP) [40], [^{11}C]*N*-methylpiperidin-4-yl propinoate (PMP) [41], [^{11}C]flumazenil [42], and [^{11}C]carfentanil [43–45].

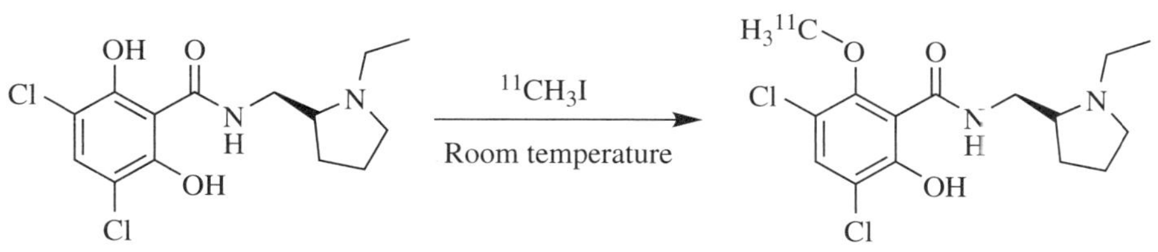

SCHEME 4.5 Formation of [^{11}C]raclopride (**5**) via *O*-[^{11}C]methylation reaction using ^{11}CH$_3$I .

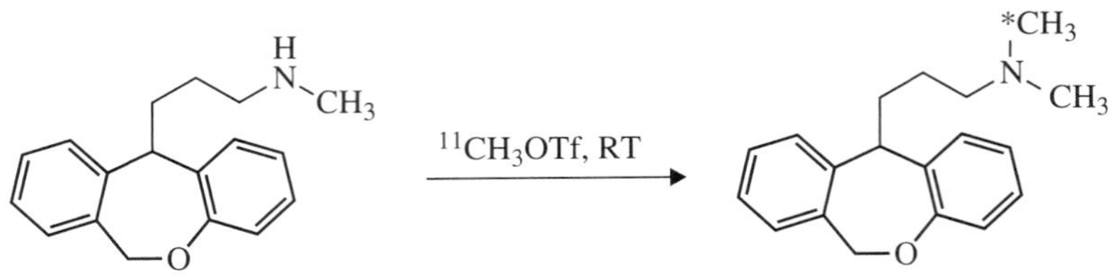

SCHEME 4.6   N-[^{11}C]methylation of nordoxepin using ^{11}CH$_3$OTf to form [^{11}C]doxepin (**10**).

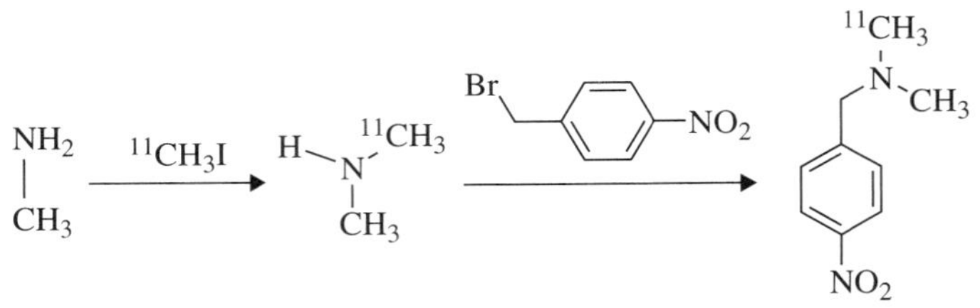

SCHEME 4.7   Indirect C-11 methylation reaction of [^{11}C]dimethylamine compounds.

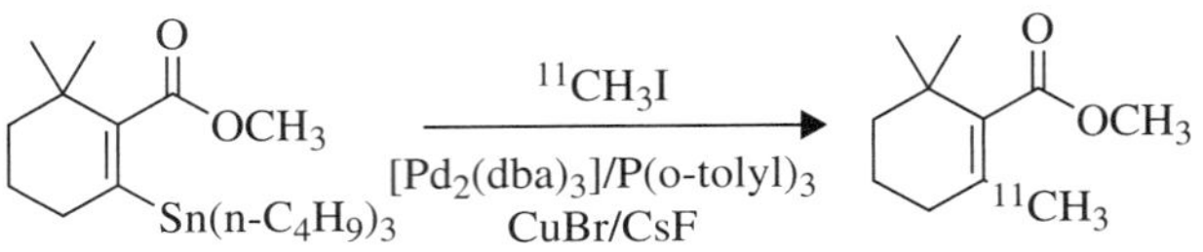

SCHEME 4.8   Stille cross-coupling for the formation of ^{11}C labelling compounds using ^{11}CH$_3$I.

By combining automated loop and SPE (solid phase extraction) methods and using the more reactive [^{11}C]methyl triflate it is possible to improve labelling efficiencies. This has been exemplified for the preparation of [^{11}C]doxepin (Scheme 4.6) [46, 47]. Such [^{11}C]methylation loop reactions have now been applied to the synthesis of [^{11}C]carfentanil [48, 49], [^{11}C]-L-[methyl]methionine [50], [^{11}C]gefitinib [51], and [^{11}C]flumazenil [52].

Although C-11 methylation reactions are generally performed directly, that is, in one step (C-11 methyl precursor plus target molecule precursor), there are occasionally advantages to labelling in sequential steps or via so-called 'indirect' methods. Indirect routes, which avoid direct reaction with C-11 labelling agents, can be beneficial if there are multiple methylation locations on the precursor molecule or if there are purification issues and/or difficulties sourcing suitable target precursors. An example is the labelling of dimethylamine groups that are prevalent in several tracer molecules (cf. [^{11}C] doxepin above), which are therefore a valuable target for labelling strategies. Although dimethylamine groups can be labelled directly with ^{11}CH$_3$I or ^{11}CH$_3$OTf, the indirect method using [^{11}C]dimethylamine and bromide precursors has been demonstrated to improve precursor handling and [^{11}C]dimethylamine product purification (Scheme 4.7) [53].

Transition metal catalysed cross-coupling reactions that have found wide application in synthetic organic chemistry have now been applied to radiolabelling protocols. In recent years, palladium-mediated Stille- and Suzuki-type reactions for C-^{11}CH$_3$ bond formation have found application for C-11 tracer synthesis. Palladium-mediated Stille-type couplings of orga-nostannanes with [^{11}C]methyl iodide have been the most widely investigated of this type. Organostannanes have wide functional group tolerances and are easily separated from radiolabelled species after reaction using preparative HPLC; how-ever, there are concerns about their toxicity, which may prohibit these methods for clinical use. One recent study [54] reported C-11 methylation of a series of alkenyltributylstannanes giving methylated alkenes and the further application of this method to the synthesis of ^{11}CH$_3$ labelled analogues of 1-methylalkene (Scheme 4.8).

Under optimised reaction conditions (solvent/base/Pd precursor/phosphine ligand) reactions could be carried out in 5 min. at 60 °C. The order of reagent addition was found to be critical to give reproducible yields and required that C-11 methyl iodide be added to the palladium complex, followed by the stannane precursor. This is presumably to better facilitate the oxidative addition step of methyl iodide to the palladium catalyst.

The [^{11}C]diaryl alkyne M-MTEB ([^{11}C]3-methyl-5-[(2-methyl-1,3-thiazol-4-yl)ethynyl] benzonitrile) radioligand, for imaging metabotropic glutamate receptors (mGluR5), has been synthesised via Suzuki and Stille palladium-mediated cross-coupling reactions [55] (Scheme 4.9). The Suzuki method, which uses microwave radiation, was found to give superior RCY in shorter reaction times over the Stille method.

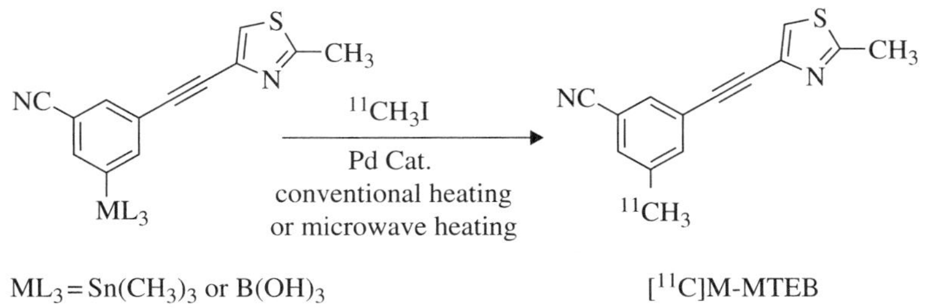

SCHEME 4.9 Radiosynthesis of [^{11}C]M-MTEB ([^{11}C]3-methyl-5-[(2-methyl-1,3-thiazol-4-yl)ethynyl]benzonitrile) via Suzuki or Stille reactions.

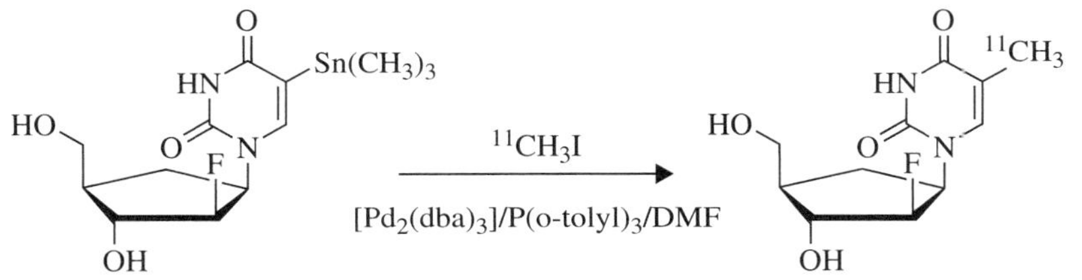

SCHEME 4.10 Synthesis of [^{11}C]FMAU via palladium mediated ^{11}CH$_3$I Stille cross-coupling.

The palladium source and phosphine ligand used in ^{11}CH$_3$I Stille cross-coupling reactions has been shown to have a marked effect on RCYs of obtained products. A recent study showed that the bulky P(o-tolyl)$_3$ phosphine ligand and the palladium(0) complex Pd$_2$(dba)$_3$ proved the best combination for the formation of 1-(2′-deoxy-2′-fluoro-β-D-arabinofuranosyl)-[methyl-^{11}C]thymine ([^{11}C]FMAU) [56] via a Stille coupling (Scheme 4.10). P(o-tolyl)$_3$ is thought to be a most effective ligand due to its large cone angle, which enhances the transmetallation step of the cycle by releasing steric strain. The reaction was also dependent on solvent with DMF proving best for both by acting as a trapping and reaction medium for ^{11}CH$_3$I.

The palladium-mediated Stille reaction of ^{11}CH$_3$I has been exploited for the synthesis of other biological molecules, such as [^{11}C]prostaglandins [57, 58] and the selective serotonin transporter (5-HTT) radioligand N,N-dimethyl-2-(2-amino-[^{11}C]4-methylphenylthio)benzylamine ([^{11}C]MADAM) [59]. The desire for alternative synthetic methods for ^{11}C-C coupling other than the Stille reaction, because of the formation of toxic tin by-products and purification difficulties, has led to the development of palladium-mediated ^{11}CH$_3$-C Suzuki coupling reactions. Suzuki-type coupling of ^{11}CH$_3$I to aryl boronic acids or esters is an arguably better alternative to the Stille reaction. Test reactions using a range of simple arylhalides with different functional groups have been carried out [60] with good RCYs (49-92%) using Pd(dppf)Cl$_2$ in the timeframe of several minutes at temperatures of 100 °C with microwave irradiation. Again, the order of reagent addition for these coupling reactions was found to be crucial in obtaining consistent and high RCY. ^{11}CH$_3$I must be added first to the palladium catalyst forming the oxidative addition product, followed by the aryl boronic species and base; similar observations were also made for the palladium-mediated terminal alkyne Sonogashira coupling reaction with ^{11}CH$_3$I [61]. A key disadvantage of these Pd-mediated C-11 methylation reactions is the necessary preparation of bespoke organostannane or organoborane precursors for tracer synthesis; this can be complex, time consuming, and expensive, which can inhibit the wider application of these types of labelling reactions.

4.2.2.1 Conversion of ^{11}CH$_3$I to Other Reagents

An alternative to traditional nucleophilic C-11 methylation and Pd-mediated reactions, where ^{11}CH$_3$I behaves as an electrophile, is conversion of ^{11}CH$_3$I into a nucleophile of the type ^{11}CH$_3$-M. The development of a generic C-11 methyl transfer reagent would circumvent the need to prepare organostannane or organoborane precursors for Pd-mediated reactions and enable more general labelling of aryl and vinyl halides with ^{11}CH$_3$. This has been demonstrated to some extent with fluoride-activated mono-organotin [^{11}C]methyl transfer reagent, [^{11}C]monomethylstannate (^{11}CH$_3$SnF$_2$[N(TMS)$_2$]$_2$) (Scheme 4.11) [62]. The conversion of ^{11}CH$_3$I into the [^{11}C]methyl tin reagent was achieved in quantitative yields and then coupled with an arylhalide under ligand-free palladium-catalysed conditions at the optimised conditions of 120 °C for 5 min. in dioxane. This reaction was applied to the [^{11}C]methylation of a series of bromoquinolines, which gave good RCYs ranging from 41–78% depending on the substrate used.

Other tin-based methyl transfer reagents include 5-[^{11}C]methyl-1-aza-5-stanna-bicyclo[3.3.3]undecane [63], which has been used to couple ^{11}CH$_3$ to a range of aryl and vinyl halides (Scheme 4.12). The preparation of this methyl transfer reagent requires the generation of ^{11}CH$_3$Li from ^{11}CH$_3$I and n-butyllithium followed by reaction with 5-chloro-1-aza-5-stanna-bicyclo[3.3.3]

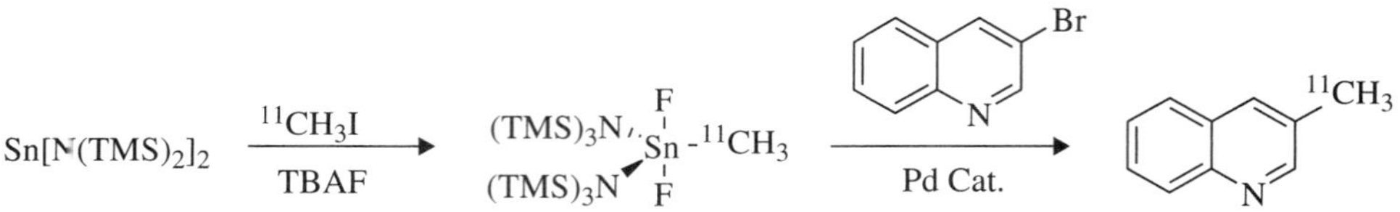

SCHEME 4.11 Synthesis of [^{11}C]monomethylstannate ($^{11}CH_3SnF_2[N(TMS)_2]_2$) and reaction with 3-bromoquinoline to give [^{11}C]3-methylquinoline.

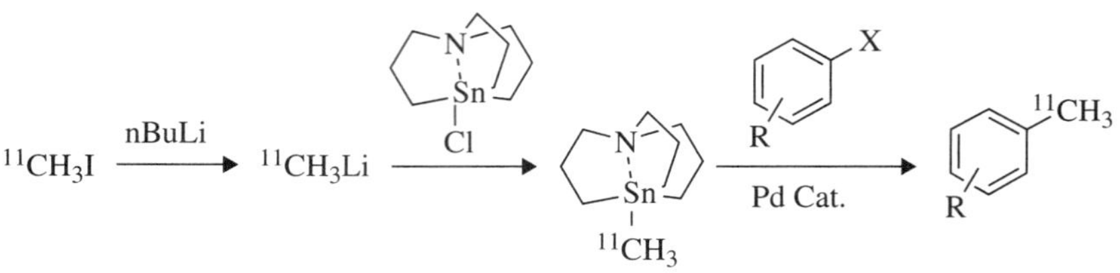

SCHEME 4.12 Preparation of the [^{11}C]methyl transfer reagent 5-[^{11}C]methyl-1-aza-5-stanna-bicyclo[3.3.3]undecane and Stille cross-coupling reaction with aryl or vinyl halides.

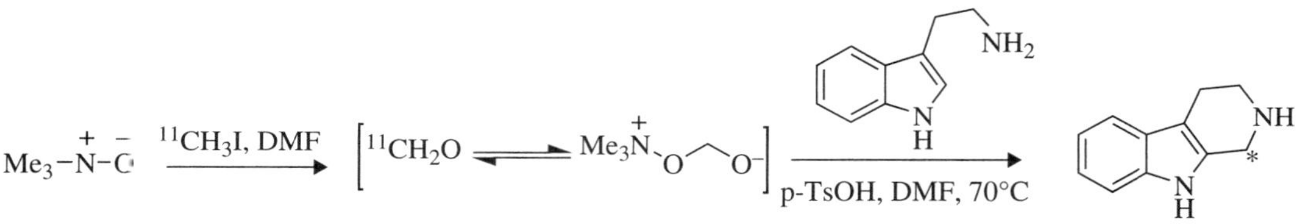

SCHEME 4.13 Preparation of $^{11}CH_2O$ from $^{11}CH_3I$ and subsequent reaction with tryptamine to form [$_{11}$C]-2,3,4,9-tetrahydro-1H-b-carboline. Ts = toluenesulfonyl. [*] indicates labelling position.

undecane. Coupling reactions of aryl and vinyl halides were found to give best RCY when carried out in the presence of a palladium allyl catalyst at temperatures above 100 °C. However, RCYs are highly dependent on the arylhalide substrate. One drawback of this system is the difficulty associated with the preparation of $^{11}CH_3Li$ which may be responsible for inconsistent RCYs.

Although methyl iodide has proven to be an exceptionally important labelling precursor, it is limited to labelling on the periphery of target molecules. For the majority of precursor compounds, a reactive nucleophilic functional group, such as an amine or alcohol, is required in the structure. In order to label a wider variety of atomic positions and functional groups within the skeletal or ring structure of a molecule, more diverse and reactive secondary precursor molecules are needed. Although is C-11 methyl iodide is a secondary precursor, its production is now so routine, fast, and reliable at many PET centres that it can in fact be used as starting point to form other reactive secondary labelling precursors. Hooker et al. recently reported a convenient synthesis of C-11 formaldehyde from C-11 methyl iodide [64]. Conventionally, C-11 formaldehyde is produced either by the partial reduction of $^{11}CO_2$ [65] or by the oxidation of C-11 methanol; [66] both routes result in a mixture of C-11 labelled products. By using trimethylamine N-oxide (TMAO) as an oxidant, it was possible to achieve instantaneous conversion of $^{11}CH_3I$ to $^{11}CH_2O$ in high radiochemical yields (>80%) (Scheme 4.13), thus providing a particularly convenient and high yielding route that should be easy to adopt by other PET centres. Furthermore, Hooker et al. were able to demonstrate that $^{11}CH_2O$ could be used to label [^{11}C]-2,3,4,9-tetrahydro-1H-b-carboline in the ring structure.

The rapid gas phase conversion of methyl iodide to the versatile electrophilic reagent C-11 carbon disulfide has been recently reported by Miller et al. [67] Despite carbon disulfide being a widely known reagent in the chemical industry its C-11 radiochemistry had not been previously investigated up to this point. It was reported that $^{11}CH_3I$ could be readily converted to $^{11}CS_2$ in high RCYs within a short reaction time (<10 min., from end of $^{11}CH_3I$ production) via a high temperature (400 °C) gas phase reaction with P_2S_5. $^{11}CS_2$ could be conveniently trapped in acetonitrile at room temperature, making it easy to handle and process. Reaction with primary, secondary, and aromatic amines proceeded very efficiently to form C-11 dithiocarbamates (Scheme 4.14) in quantitative yields and within 5 min. in reaction times. Subsequent reaction of the C-11 dithiocarbamates with alkylhalides afforded quantitative conversion to the dithiocarbamate esters, while reaction of the benzyl dithiocarbamate with $POCl_3$ resulted in desulfurisation and formation of the corresponding isothiocyanate.

C-11 nitromethane is another reagent that can be quickly generated from $^{11}CH_3I$. It is typically prepared via passing a gas stream of $^{11}CH_3I$ through a column of $AgNO_2$ that is heated to 100 °C. The conversion is fast, complete in a few minutes, and produces $^{11}CH_3NO_2$ with conversions of about 90% (Scheme 4.15). An additional column of $NaHCO_3$ attached to the $AgNO_2$

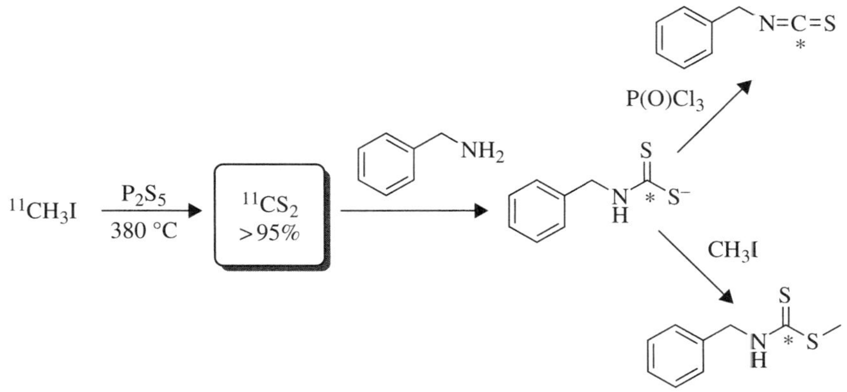

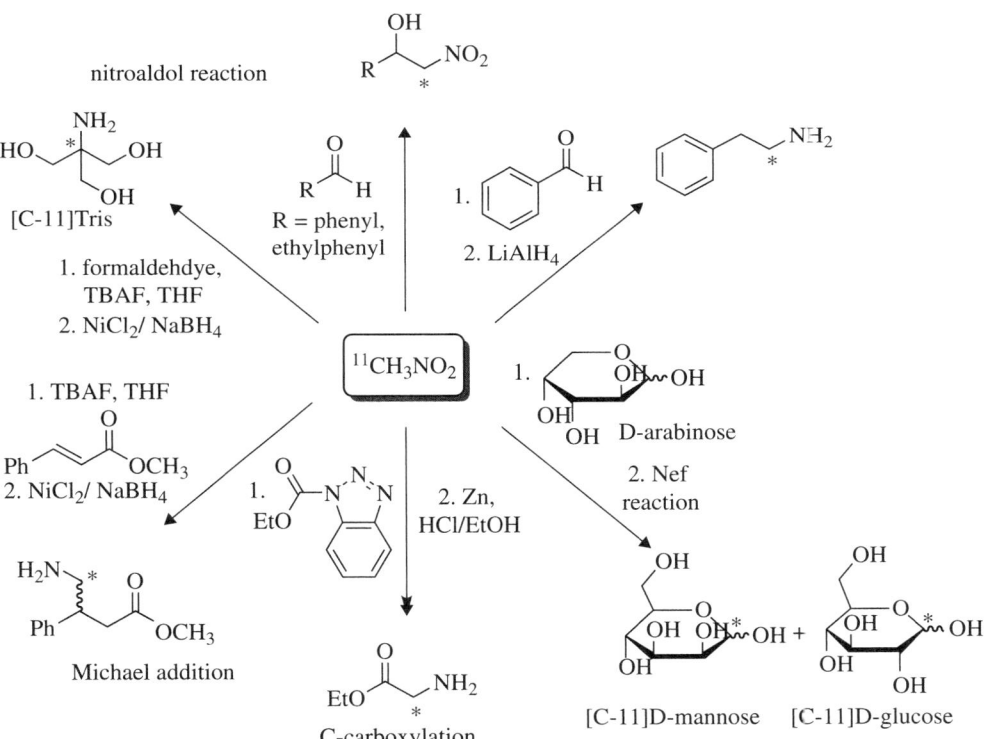

SCHEME 4.14 Preparation of $^{11}CS_2$ from $^{11}CH_3I$ and subsequent reactions with benzylamine to form a C-11 dithiocarbamate, C-11 dithiocarbamate ester and an C-11 isothiocyanate. [*] indicates labelling position.

$$^{11}CH_3I_{(g)} \xrightarrow[\text{100 °C}]{\text{AgNO}_2} {}^{11}CH_3NO_2$$

SCHEME 4.15 Conversion of $^{11}CH_3I$ to $^{11}CH_3NO_2$ via reaction with $AgNO_2$.

SCHEME 4.16 C-11 radiolabelling reactions using $^{11}CH_3NO_2$.

outlet has been found to improve the quality of $^{11}CH_3NO_2$ by removing nitrogen oxide compounds that are the result of $AgNO_2$ pyrolysis [68].

Although $^{11}CH_3NO_2$ is a much less reactive molecule than the other $^{11}CH_3I$-derived precursors described above, it has been applied to the synthesis of a number of labelled compounds (Scheme 4.16). It is a versatile reagent that has not been widely exploited in the field. The condensation reaction of $^{11}CH_3NO_2$ with D-arabinose has been used to form the epimeric mixtures of [C-11]D-nitroalcohols, which can then be subsequently converted to [C-11]D-glucose and [C-11]D-mannose via the Nef reaction (oxidation of nitro compounds to carbonyls). [C-11]D-mannose was formed as the major product; both products could be separated, and isolation could be achieved via preparative HPLC in moderate RCYs <20% [69]. $^{11}CH_3NO_2$ has also been used to label the neurotransmitter phenethylamine; reaction of $^{11}CH_3NO_2$ with benzaldehyde to generate [C-11]beta-nitrostyrene, followed by a reduction with $LiAlH_4$, was reported to give [C-11]phenethylamine in moderate RCYs. More recently, Kato et al. have exploited $^{11}CH_3NO_2$ to prepare a range of labelled nitro and amine compounds. The fluoride-assisted Michael addition of $^{11}CH_3NO_2$ to the α,β-unsaturated compound, p-chlorocinnamate, followed by a $NiCl_2$/$NaBH_4$ reduction step gave a reasonable 36% RCY of the amine within an acceptable timeframe. The nitroaldol reaction of $^{11}CH_3NO_2$ with formaldehyde and other aldehydes has also been used by Kato et al. to label a range of nitro compounds and the resulting amines following a reduction step [68, 70]. Excellent RCYs (68%) were obtained for the synthesis of the labelled amino-triol compound [C-11]Tris (Figure 4.16). C-carboxylation of $^{11}CH_3NO_2$ has also proven to be a useful strategy for the synthesis for nitroacetate compounds. [C-11]ethyl nitroacetate, an interesting C-11 intermediate, was prepared in good RCY (75%) via the reaction of $^{11}CH_3NO_2$ with 1-ethoxycarbonylbenzotriazole and quantitatively converted to the [C-11]glycine ethyl ester using Zn powder (Scheme 4.16) [71].

4.2.3   [^{11}C]Phosgene Reactions

C-11 phosgene ($^{11}CCCl_2$) is a highly reactive gaseous small molecule labelling precursor that can be used to efficiently label ureas, carbamates, and carbonates in the carbonyl position [72]. In some respects, $^{11}COCl_2$ is an ideal C-11 labelling synthon because of its reactivity and potential to form a wide range of compounds; however, its routine synthesis can be problematic, which has resulted in only a handful of groups worldwide developing its chemistry. [C-11]phosgene is generally prepared via the chlorination reaction of $^{11}CH_4$ to form $^{11}CCl_4$ followed by an oxidation reaction over iron or copper catalysts at high temperature [73, 74]. Recently, a new method of $^{11}COCl_2$ has been reported for the room temperature conversion of $^{11}CCl_4$ to $^{11}COCl_2$ using a working-environmental gas detection tube (Kitagawa gas detector tube) [75]. This method involves passing a stream $^{11}CCl_4$ through a glass tube filled with I_2O_5 and fuming H_2SO_4. Consistently high and reproducible RCYs of $^{11}COCl_2$ were reported with the added benefits of room temperature conversion and a simpler experimental setup. Another recent report describes the preparation of $^{11}COCl_2$ using two quartz columns [76]. Initially, $^{11}CH_4$ is reacted with Cl_2 at 510 °C to form $^{11}CCl_2$, followed by removal of Cl_2 by reaction with antimony the $^{11}CCl_4$ gas stream that is heated to 750 °C in the second empty quartz tube, giving 30–35% RCY of $^{11}COCl_2$. With regard to radiolabelling target molecules C-11, phosgene has been used for the rapid preparation of C-11 labelled ureas [77–79], carbamates, carbamoyl chlorides [80], amides [80], and uric acids [81, 82] (Scheme 4.17).

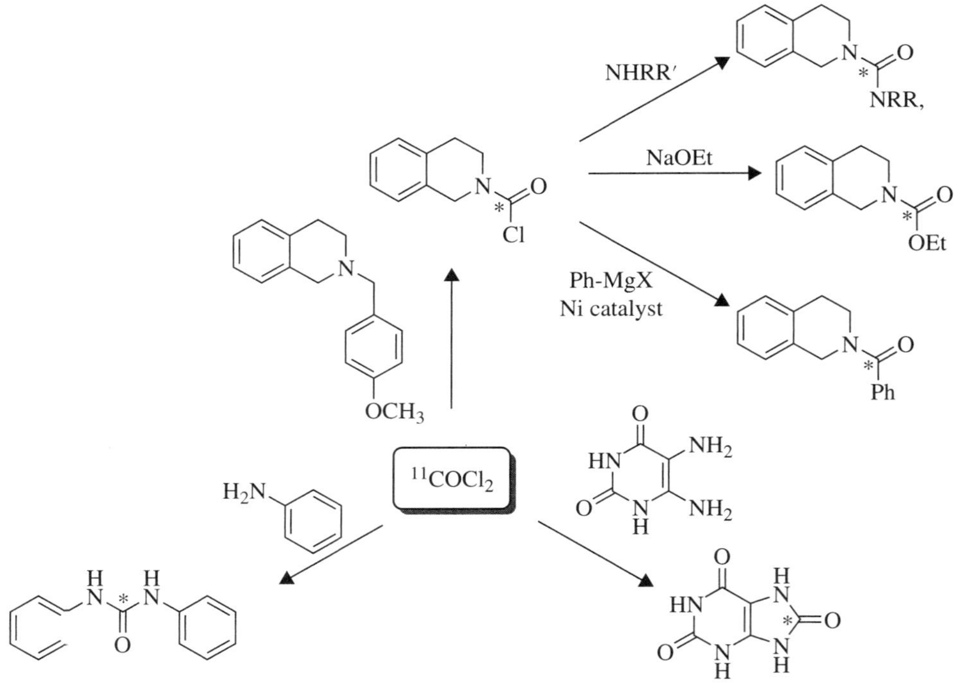

SCHEME 4.17   Various reactions [^{11}C]phosgene to form C-11 labelled carbonyl compounds.

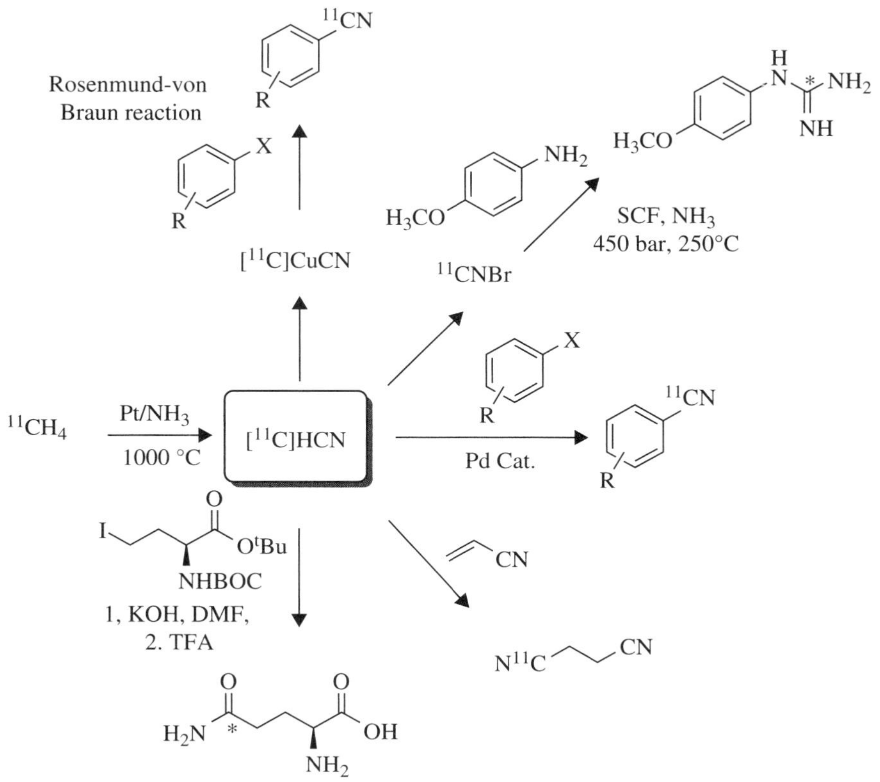

SCHEME 4.18 Synthesis of [^{11}C]HCN and selected [^{11}C]cyanation reactions.

4.2.4   [^{11}C]Cyanation Reactions

[^{11}C]HCN is a useful synthon for the introduction of nitrile groups into a range of tracer compounds. Nitrile labelling is important because the CN group is found in many organic compounds and natural products. [^{11}C]HCN is typically prepared via the reaction of ^{11}CH$_4$ with NH$_3$ over a platinum catalyst [83] and can be directly reacted with precursor molecules, as in the synthesis of labelled amino acids [84] or converted to ^{11}CNBr [85, 86] or [^{11}C]CuCN [87] for further reaction. [^{11}C]CuCN can be reacted with aryl halides via the Rosenmund von Braun reaction [87–89]. Palladium-mediated reactions of [^{11}C]HCN with aryl iodides have also been used to introduce the ^{11}CN group into target molecules [90, 91] (Scheme 4.18).

The [^{11}C]cyano functional group provides further opportunity for conversion to other products such as carboxylic acids, amides [91], tetrazoles [87], amidines [89], and amino acids [84, 92]. C-11 cyanogen bromide has been applied to the synthesis of ^{11}C-labelled guanidines [85]. The first step is the reaction of ^{11}CNBr with amines to rapidly form C-11 cyanamides, followed by a high pressure, high temperature supercritical fluid (SCF) synthesis method to afford the corresponding C-11 guainidines (Scheme 4.18). Recently, a convenient synthesis of [^{11}C]glutamine, a potential marker for imaging tumour metabolism, has been reported by reacting [^{11}C]HCN with 4-iodo-2-amino-butanoic ester in the presence of KOH base followed by hydrolysis with trifluoroacetic acid [92] (Scheme 4.18).

4.2.5   [^{11}C]Carbonylation Reactions

C-11 carbon monoxide is a highly versatile reagent that has been demonstrated to form a wide range of C-11 carbonyl labelled compounds [93]. ^{11}CO can be prepared quickly and with high radiochemical yield by either the high temperature reduction of ^{11}CO$_2$ over zinc [94] or molybdenum [95]. The molybdenum route is more popular because it is more reliable and requires less maintenance. Palladium-mediated C-11 carbonylation reactions have been most widely exploited and have been used to effectively label imides, ketones, carboxylic acids, esters, amides, and acrylamides [96–99]. The palladium-catalysed carbonylation of aryl and vinyl halides used for many of these labelling reactions, originally developed by Heck, proceeds via three characteristic steps in the catalytic cycle: oxidative addition of an aryl halide species to the *in situ* palladium(0) catalyst, insertion-migration of carbon monoxide to form a Pd-acyl, and nucleophilic attack followed by reductive elimination to form the product. Carbon monoxide has low solubility in many organic solvents that can result in poor reactivity unless the pressure

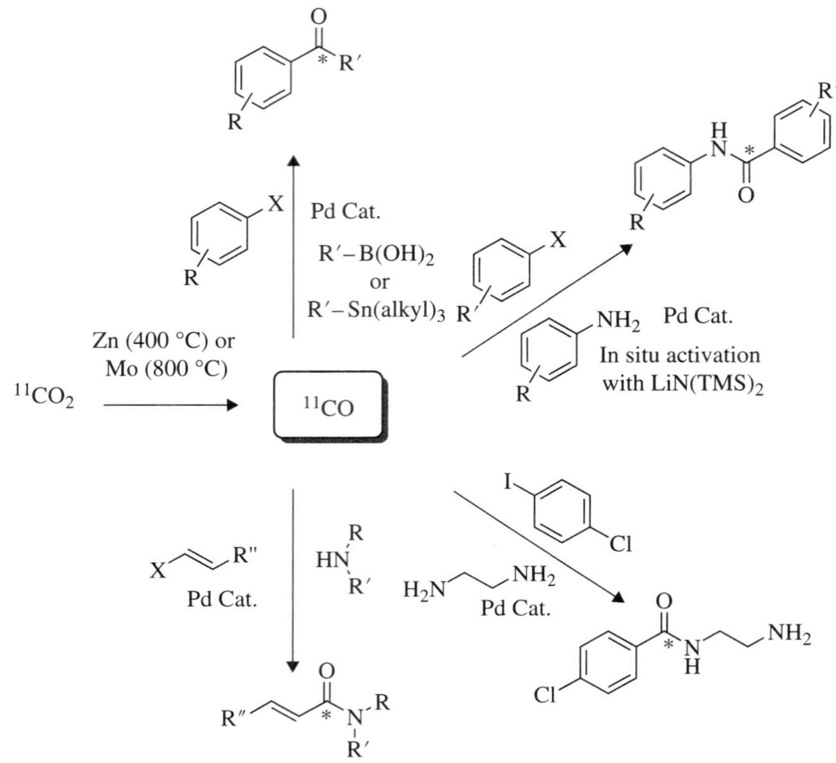

SCHEME 4.19   Selected palladium-mediated [^{11}C]carbonylation reactions using high pressure micro-autoclave system.

in the system is increased; the CO insertion step can therefore be rate limiting at low pressures. For labelling reactions using ^{11}CO, the high isotopic dilution of ^{11}CO results in very low partial pressures of ^{11}CO, thus inhibiting the ^{11}CO insertion step of the catalytic cycle. Several strategies in recent years have been developed to improve the reactivity of ^{11}CO by enhancing the transfer of ^{11}CO into the solution phase to improve reactivity: (1) Recirculation of the ^{11}CO gas through the reaction mixture; [99] (2) Increasing the pressure of ^{11}CO using a high-pressure HPLC micro-autoclave reactor systems; [96, 100] (3) Chemical trapping agents to enhance solubility; [101–104] and (4) microfluidic reactors to enhance mass transport between gas and liquid phases [105, 105].

The high pressure HPLC C-11 carbonylation system has been applied to a wide range of reactions, including Suzuki and Stille C-11 carbonylation reactions using boronic acid [107, 108] and alkyl tin precursors [109] for the preparation of [*carbonyl*-^{11}C]biaryl or aryl-benzyl ketones [97] (Scheme 4.19). The synthesis of a wide range [*carbonyl*-^{11}C]amides and esters has been prepared using the high pressure autoclave system via ^{11}CO palladium-mediated carboxyaminations [98, 110–114]. Further investigations of these [^{11}C]carboxyamination reactions showed that *in situ* activation using lithium bis(trimethylsilyl) amide [112] and 1,2,2.6,6,-pentamethylpiperidine [113] enhanced RCYs for less reactive amines such as aniline.

Two low pressure palladium-mediated C-11 carbonylation methods have been developed that rely on a pre-trapping and solubilising step prior to Pd-mediated insertion. The first of these uses a BH$_3$.THF solution to form BH$_3$.^{11}CO; the borane acts as Lewis acid by accepting a lone pair of electrons from the ^{11}CO forming the adduct and solubilising the ^{11}CO [104]. In a typical procedure. ^{11}CO/He gas stream was bubbled through a solution of BH$_3$.THF to give BH$_3$.^{11}CO, addition of the cross-coupling regents. palladium catalyst, aryl halide, and suitable amine or alcohol nucleophiles gave reasonable yields of [*carbonyl*-^{11}C]amide or [*carbonyl*-^{11}C]esters (Scheme 4.20).

More recently, copper tris(pyrazolyl)borate complexes (Scheme 4.21) have been used as highly efficient ^{11}CO trapping agents and have proven to give higher trapping efficiencies and to be technically simpler to use than the borane trapping methods for low pressure Pd-mediated carbonylation reactions [101, 103, 115, 116]. The tris(pyrazolyl)borate (Tp) ligands (also referred to as scorpionates) enforce strict tridentate coordination geometry on the central copper ion while leaving a vacant coordination site to permit ^{11}CO binding to the copper.

The choice of Tp ligand was found to be important for enhancing trapping efficiencies. The tris(3,5-dimethylpyrazolyl)-borate (Tp*) was an excellent ligand because the methyl groups on the pyrazol rings enhance the back bonding between the Cu and CO, stabilising this bond, and improve complex solubility in organic solvents compared to the non-methylated

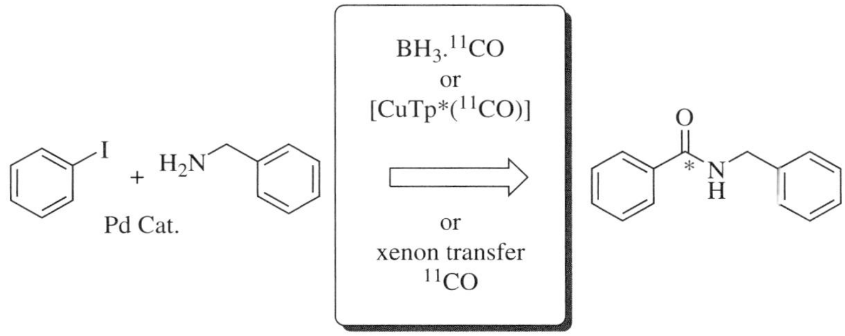

SCHEME 4.20 Low pressure ^{11}CO reactions. Synthesis of the model compound [*carbonyl*-^{11}C]*N*-benzylbenzamide via palladium-mediated carbonylation using the BH$_3$.^{11}CO adduct, [Cu(Tp*)^{11}CO] complex or ^{11}CO xenon transfer method.

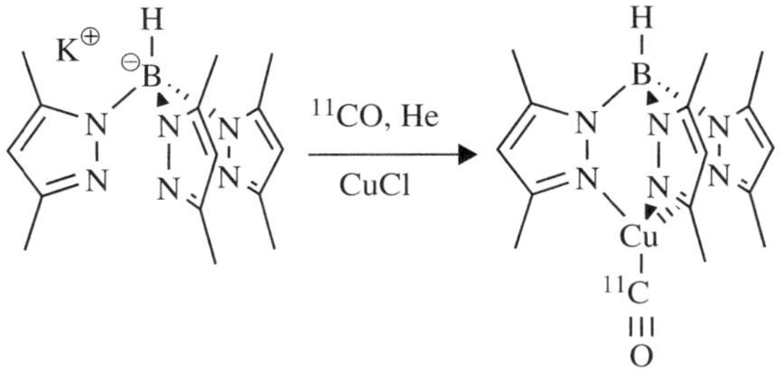

SCHEME 4.21 Formation of [Cu(Tp*)^{11}CO] via the complexation reaction of potassium tris(3,5-dimethylpyrazolyl)-borate (Tp*), CuCl and ^{11}CO.

analogue. Excellent RCY of model amides were achieved (Scheme 4.20); however, the yields were found to be highly dependent on the Pd catalyst system [116]. A third low pressure ^{11}C carbonylation method has now been recently developed by Eriksson et al. [117] that exploits the high solubility of xenon gas in organic solvents to efficiently transfer ^{11}CO to a small volume reaction vial. Because xenon is highly soluble in organic solvents such as THF and toluene typically used for metal-mediated coupling reactions, it can be used to transfer ^{11}CO into small volumes of reagent solution without a build-up of gas pressures. The method has been demonstrated to quantitatively transfer ^{11}CO from a silica trap into solvent volumes of less than 1 ml without significant pressure build-up and has been used to perform several model Pd- and Rh-coupling reactions for the formation of labelled amides, ureas, and esters. This method is particularly interesting because it avoids the use of additional trapping reagents, thus making it a chemically much simpler protocol for ^{11}CO labelling (Scheme 4.20).

Microfluidic systems have been used for performing low pressure palladium mediated C-11 carbonylation reactions of aryl halides. One system exploits a silica-supported palladium catalyst that was packed into PTFE tubing, effectively acting as a mini-packed bed reactor [105] (Figure 4.2). A high surface area-to-volume ratio is generated within this device owing to the large surface area of the silica supported catalyst material. This improves the interfacial gas-liquid contact between the ^{11}CO/He gas stream and the aryl halide/amine coupling reagents in solution. A series of different C-11 amide systems could be effectively labelled in good to excellent RCY. Advantageously, the method is technically straightforward but requires a low temperature ^{11}CO trapping step prior to reaction in the microfluidic device. The micro-tube reactor loops could be reused for successive labelling reactions for the synthesis of the same labelled compound, changing to a different substrate using the same reactor resulted in mixtures of labelled products. More recently, a glass fabricated device has been used to perform ^{11}CO carbonylation labelling reactions. The device contains two inlet ports for gas and liquid reagents, a 5 m long reaction channel and an outlet port. The microchannels were chemically etched, have a semicircular cross-sectional profile, and are 220 μm wide and 100 μm deep. The main reaction channel, which is 5 m long, occupies most of the 90 x 15 mm footprint area. The long residence channel was necessary to give sufficient residence times of the gas and liquid reagents on the device under gas-liquid annular type flows. A typical ^{11}CO labelling process, involving a ^{11}CO pre-trapping stage, reaction on the device, and a flushing step is complete in less than 15 min. A range of simple amide and ester molecules were labelled with ^{11}CO in good RCYs using [PdCl$_2$(Xantphos)] as catalyst.

Besides palladium-mediated reactions, rhodium-mediated carbonylation reactions have also been used for the synthesis of a range of C-11 carbonyl containing compounds, including malonates [118], hydroxyureas [119], carbamates [120], and diphenyl ureas [120]. In the presence of an azide, the rhodium catalyst enables the insertion of ^{11}CO forming Rh-complexed

(a)

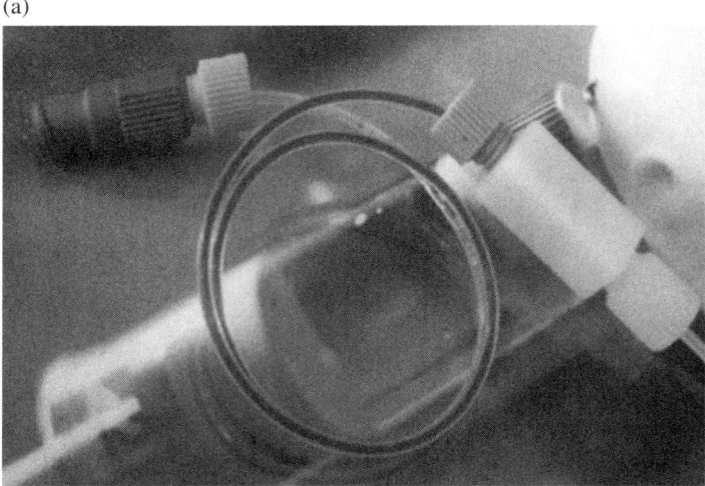

(b)

FIGURE 4.2 (a) Microtube reactor packed with a silica-supported palladium phosphine catalyst. (b) Glass fabricated microfluidic device used for gas-liquid phase ¹¹CO carbonylation reactions (Figure reproduced with permission from Ref. [106]).

C-11 isocyanate or C-11 ketene intermediates that undergo either reaction with a suitable nucleophile. [*Carbonyl*-¹¹C] diphenyl ureas and [*carbonyl*-¹¹C]ethylphenyl carbamates respectively [120] have been prepared in this way using the high pressure micro autoclave system (Scheme 4.22).

In a typical procedure a mixture of Rh(I) catalyst (typically [RhCl(COD)] and triphenylphosphine) and diazo precursor were injected into the micro-autoclave reactor previously charged with ¹¹CO/He gas. The selected nucleophile (alcohol or amine) is then added after a short time. A Rh-carbenoid complex is initially formed on reaction with the diazo precursors with concomitant loss of N₂. ¹¹CO migratory insertion of ¹¹CO results in the rhodium-ketenyl complex and following nucleophilic attack by an alcohol or amine give [*carbonyl*-¹¹C]carbamates or ureas respectively. One drawback of this reaction system is the use of diazo precursors that may not be possible to generate for the labelling of more complex molecules.

Free radical photo-initiated reactions of carbon monoxide provide an alternative route to the traditional metal catalysed reactions for the introduction of CO into organic molecules. Recently, these types of photo-initiated carbonylation have been used to prepare C-11 labelled carboxylic acids [121, 122], esters [122–124] and amides [125] (Scheme 4.23). In comparison to the Pd- and Rh-mediated reactions described above, these photo-initiated reactions can be used to label aliphatic molecules with ¹¹CO, which cannot be readily achieved via transition metal reactions owing to beta-hydride elimination processes. The photo-initiated C-11 carbonylations were performed in a modified micro-autoclave system equipped with a sapphire window to allow penetration of UV light [121]. The system was used to prepare a wide range of [*carbonyl*-¹¹C] carboxylic acids, esters, or amides from the corresponding alkyliodides and nucleophile (water, alcohol, or amine). The reactor was typically pressurised to 40 mPa with ¹¹CO/He gas and then irradiated with UV light (280-400 nm). It was necessary to have a photo-sensitiser in the reaction mixture (benzophenone or acetone) and polar solvents such as DMF or DMSO to give higher RCYs. Conversions of ¹¹CO were also found to be dependent on agitation of the solution, concentration of alkyl iodide, and pressure and intensity of UV irradiation.

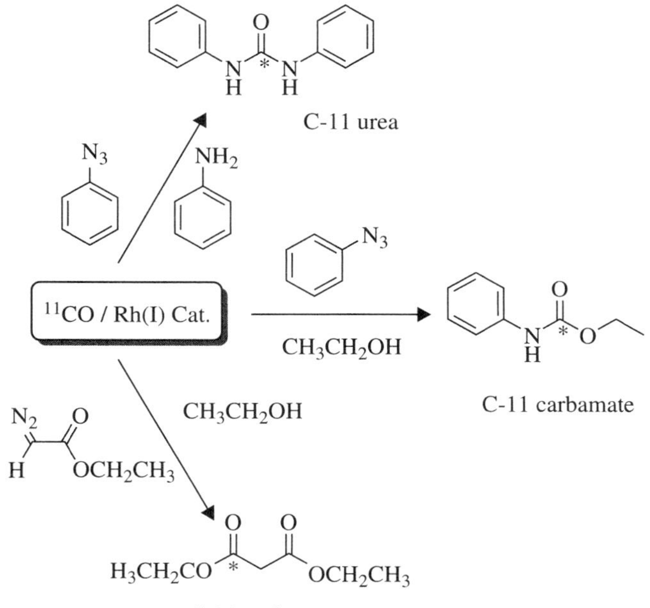

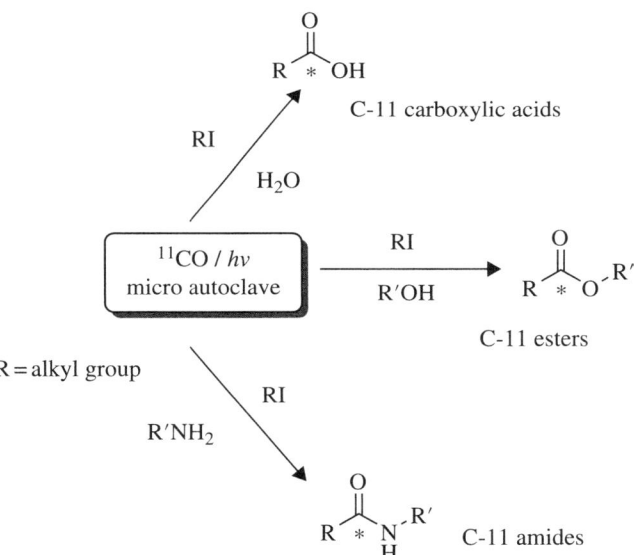

SCHEME 4.22   Selected rhodium-mediated [^{11}C]carbonylation reactions forming [*carbonyl*-^{11}C]urea, carbamate, and malonate compounds.

SCHEME 4.23   Synthesis of [*carbonyl*-^{11}C]-labelled aliphatic carboxylic acids, esters, and amides in high pressure photo-initiated carbonylation system.

4.3   NITROGEN-13 CHEMISTRY

Nitrogen-13 is produced by the nuclear reaction ^{16}O(p,α)^{13}N via the proton irradiation of H$_2$O creating a mixture of [^{13}N] ammonia, [^{13}N]nitrogen, [^{13}N]nitrite, and [^{13}N]nitrate [126, 127]. [^{13}N]nitrite and [^{13}N]nitrate are formed predominately; however, [^{13}N]ammonia is the most desirable ^{13}N product because of its direct application in PET myocardial perfusion studies [128] and its potential for further chemical reactions. The synthesis of [^{13}N]ammonia is typically achieved outside the target via the reduction of ^{13}NO$_x$s using Devarda's alloy [129–131] or TiCl$_3$; [132] however, in-target [^{13}N]ammonia production is also accessible. One reported method uses an ethanol/hydrogen gas in-target [133, 134], while another reports high specific activity of ^{13}NH$_3$ production by proton irradiation of H$_2$O in the presence of 10 mM ethanol saturated with oxygen gas [131] (Scheme 4.24).

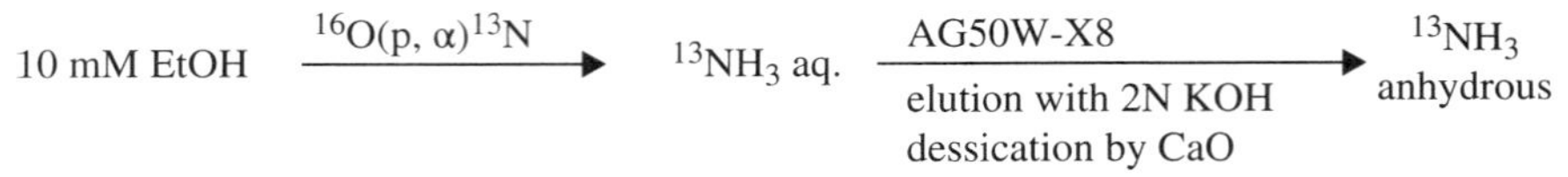

SCHEME 4.24   In-target production route of $^{13}NH_3$.

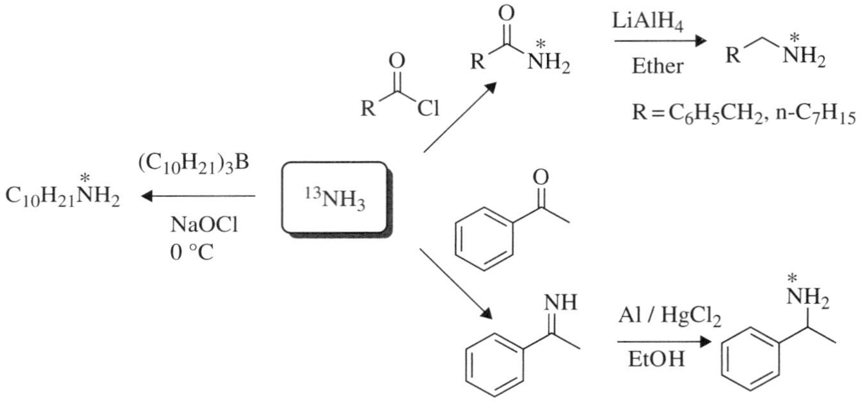

SCHEME 4.25   Synthesis of N-13 amines from $^{13}NH_3$.

Although the in-target $^{13}NH_3$ production generating higher specific radioactivities for PET imaging is a more recent development, N-13 labelling chemistry using [^{13}N]ammonia has been studied since the 1970s using low specific activity carrier added methods. The short 10-minute half-life of nitrogen-13 has clearly limited the range of N-13 chemistry with the majority of N-13 chemistry being dominated by $^{13}NH_3$ production for myocardial blood flow measurements [128, 135–137]. Despite the obvious time challenges, N-13 labelling routes have been developed for amino acid synthesis using enzymes [138–143], acid chloride substitution reactions for amide synthesis [139, 144], and nitrophenyl carbamate synthesis (Scheme 4.25) [145].

4.3.1   Synthesis of ^{13}N-labelled Amines

N-13 labelled amines are of interest for improving PET myocardial perfusion imaging and representing their metabolism. The preparation of N-13 labelled amines has been achieved in good radiochemical yields via the reduction of N-13 amides using LiAlH$_4$, for example, [^{13}N]phenetylamine and [^{13}N]octylamine have been prepared from their respective N-13 amides in excellent RCYs (60–70%) [146]. The synthesis of [^{13}N]phenetylamine via Hoffman rearrangement of the corresponding amide is also possible, but RCYs were poor owing to the suspected decomposition of the product by excess of NaOBr in the reaction mixture [147]. The RCY was improved by using LiAlH$_4$ but a longer reaction time (5–20 min.) and tedious work-up were unsuitable for the routine N-13 labelling. The synthesis of [^{13}N]amphetamine was accomplished by the reduction of the corresponding imine prepared from phenylacetone and [^{13}N]ammonia [148]. In this process, a mixture of aluminium and HgCl$_2$ was used as the reductant. However, carrier ammonia is required to avoid the formation of secondary amine, which results in significantly decreased specific radioactivity. Organoborane precursors have recently been used for the preparation of N-13 labelled organic amines. The amination reaction of tri-decylborane with [^{13}N]ammonia in the presence of NaOCl resulted in the rapid formation of [1-^{13}N]aminodecane in 40–60% RCY (Figure 4-23) [149].

4.3.2   Enzymatic Synthesis of ^{13}N-labelled Amino Acid

Radiolabelled amino acids are attractive molecules for PET studies because of their role in better understanding myocardium and tumour metabolism. The labelling of an α-amino group or amide group with N-13 has been achieved using enzymatic reactions. A key step in this labelling process was the immobilisation of enzymes onto a surface in order to avoid contamination of the final labelled product [150]. Enzymes such as L-glutamic acid dehydrogenase, L-glutamine synthetase, L-glutamic-pyruvic acid transaminase, and L-aspartase were immobilised to CNBr-activated Sepharose [140] and used for the enzymatic reaction of $^{13}NH_3$ with amino acid precursors. The synthesis of ^{13}N-labelled L-glutamate was achieved via the reductive amination with α-ketoglutarate and $^{13}NH_3$ while the N-13 labelled L-glutamine was obtained by the amidation of L-glutamate with $^{13}NH_3$. Nitrogen-13 labelled L-aspartate was prepared by fumarate amination. Reactions are summarised in Scheme 4.26 and were performed under the no-carrier-added conditions using a semi-automated procedure [140, 142, 150, 151].

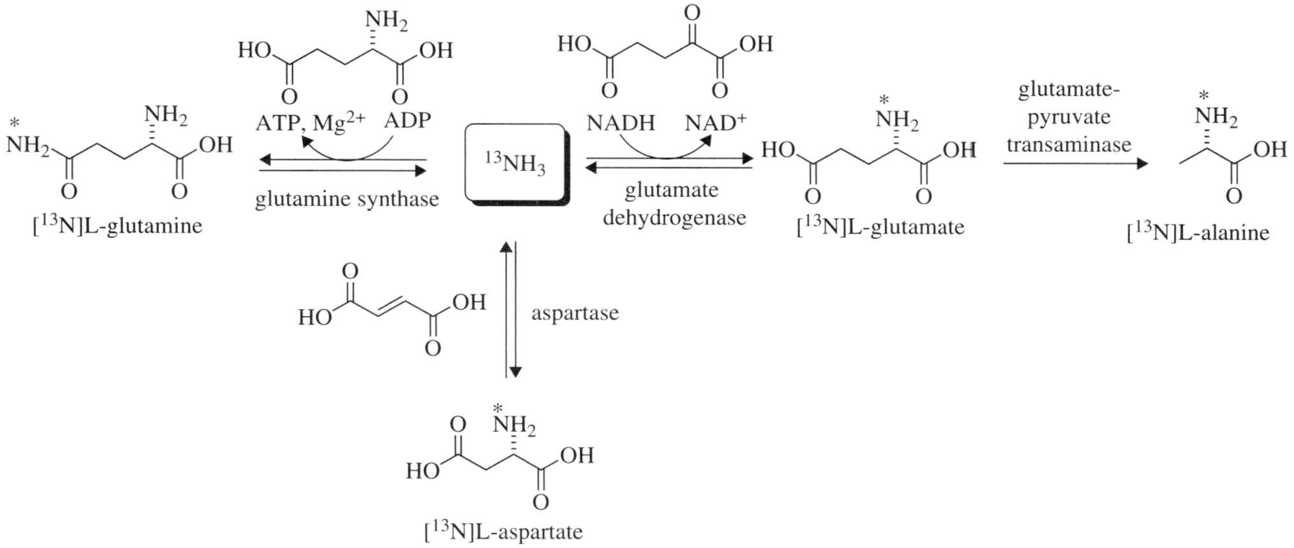

SCHEME 4.26   Summary of the enzymatic synthesis of N-13 labelled amino acids.

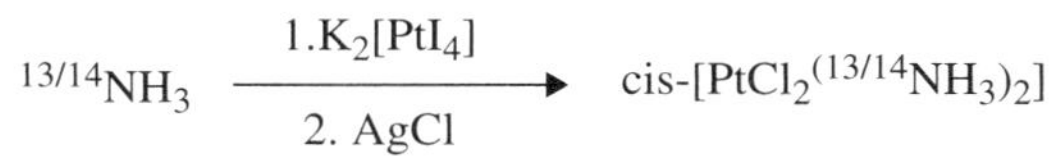

SCHEME 4.27   Synthesis of N-13 labelled cisplatin.

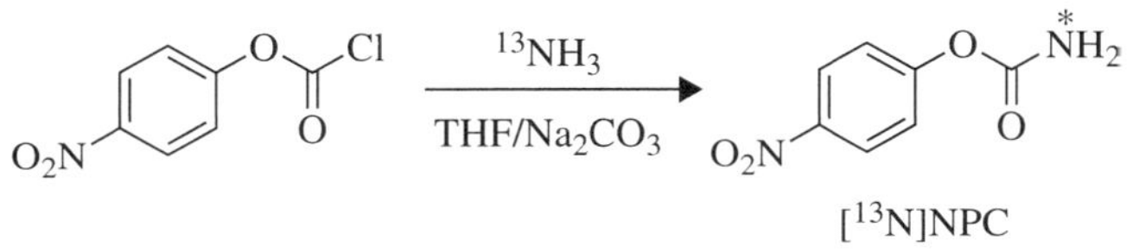

SCHEME 4.28   Synthesis of N-13 labelled NPC.

4.3.3   Synthesis of [^{13}N]cisplatin

Cisplatin, cis-[PtCl$_2$(NH$_3$)$_2$], has been used for the treatment of a range of cancers, including brain tumours. N-13 labelling of the widely used anti-cancer treatment cisplatin has been achieved using ^{13}NH$_3$ in order to better understand its blood-brain-barrier penetration for improving brain tumour treatment. N-13 labelled cisplatin was prepared by the reaction of K$_2$PtI$_4$ with [^{13}N]ammonia, forming cis-[PtI$_2$($^{13/14}$NH$_3$)$_2$] [152, 153], following ligand exchange of iodide to chloride; using silver chloride, cis-[PtCl$_2$($^{13/14}$NH$_3$)$_2$] was obtained in good RCY (Scheme 4.27).

Holschbach et al. [154] reported a more refined synthesis using solid phase extraction (SPE) technology. A strong anion exchange (SAX) cartridge was used for the [^{13}N]ammonia introduction step, and a cation exchange (SCX) cartridge was used to remove cationic silver species after ligand exchange. A decay-corrected RCY of 80% was obtained with 30 MBq/mmol specific activity.

4.3.4   Synthesis of [^{13}N] Carbamates and Ureas

The preparation of the N-13 labelled compounds performed using carrier-added methods results in specific radioactivities that are much lower than the carrier-free methods typically used for ^{11}C and ^{18}F labelling. High specific radioactivities of labelled compounds are generally required for brain-receptor studies in order to avoid saturating the receptor sites with cold compound. No-carrier-added production methods are therefore essential for N-13 if it is to be used for these types of studies in PET. An example of a no-carrier-added reaction for the production of a high specific activity compound is the synthesis of p-nitrophenyl [^{13}N]carbamate ([^{13}N]NPC) from [^{13}N]ammonia and p-nitrophenyl chloroformate [145] (Scheme 4.28). The high reactivity of chloroformate precursor was key to the success of this reaction; however, chloroformate analogues were found to be difficult to handle owing to their high sensitivity to moisture.

Recently, a more practical method for N-13 carbamate labelling was developed via the *in situ* preparation of chloroformate analogues from stable precursors [155]. A mixture of *p*-nitrophenol and commercial triphosgene was employed for the preparation of *p*-nitrophenyl chloroformate in a one-pot process that was found to be more practical. Careful control of the quantities of triphosgene and appropriate base, usually *i*-PrNEt, was found to be important to achieve high RCY (Table 4.2).

This strategy was used to prepare a range of urea analogues where isocyanate intermediates were formed from amines and triphosgene. This method has recently been applied to the synthesis of N-13 labelled carbamazepine [155] and thalidomide [156] (Scheme 4.29).

4.3.5 Other ^{13}N-labelling Reactions

4.3.5.1 *Preparation and Reaction of [^{13}N]nitrogen Dioxide, [^{13}N]nitrous Acid, and [^{13}N]nitrite* Although much less popular than ^{13}NH$_3$, the precursors ^{13}NO$_2$, ^{13}NO$_2^-$, and [^{13}N]HNO$_2$ have been used for N-13 labelling reactions. The first reported preparation of ^{13}NO$_2$ was via the oxidation of [^{13}N]ammonia using gallium and cobalt oxides (Scheme 4.30) [157].

TABLE 4.2   One-pot Radiosynthesis of ^{13}N-labelled Urea and Carbamate Analogues.

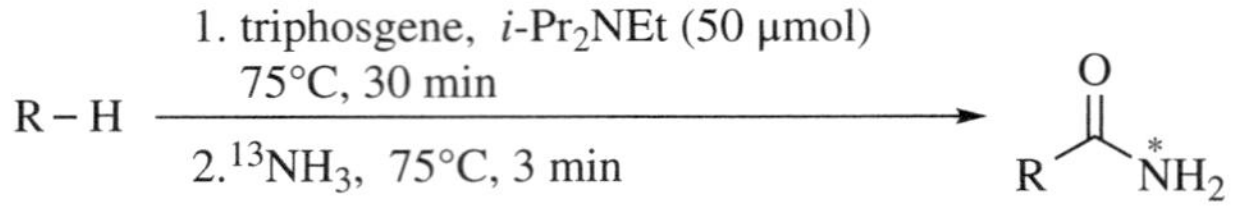

$$R-H \xrightarrow[\text{2.}^{13}NH_3,\ 75°C,\ 3\ min]{\begin{array}{c}\text{1. triphosgene, } i\text{-Pr}_2\text{NEt (50 µmol)}\\ 75°C,\ 30\ min\end{array}} \underset{R}{\overset{O}{\underset{}{\|}}}\!\!-\!\!\overset{*}{N}H_2$$

Entry	RH	Triphosgene (µmol)	Radiochemical Yields (%)[a] Based on ^{13}NH$_3$
1	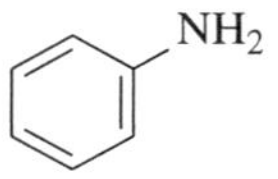 (aniline) NH$_2$	3.3	0–70 ($n = 5$)
2		2.5	78
3	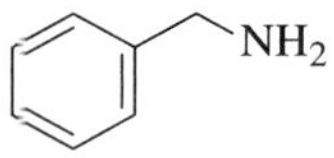 (benzylamine) NH$_2$	2.5	85
4	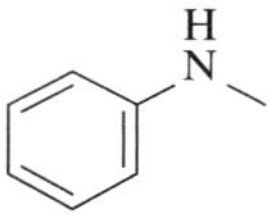 (N-methylaniline) H, N	3.3	0–36 ($n=5$)
5		2.5	52
6	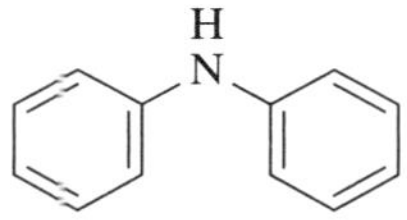 (diphenylamine) H, N	2.5	34
7	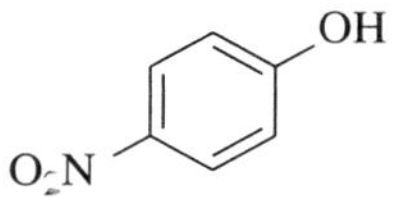 (4-nitrophenol) OH, O$_2$N	3.3	0–62 ($n=5$)
8		2.5	84
9	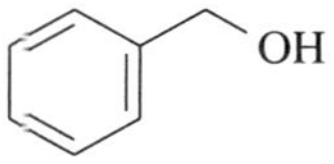 (benzyl alcohol) OH	2.5	90

[a]Radiochemical yield was determined by analytical HPLC. All results are the mean ($n=3$) with a maximum range of ±10%. Radioactive products were identified using authentic non-radioactive samples.

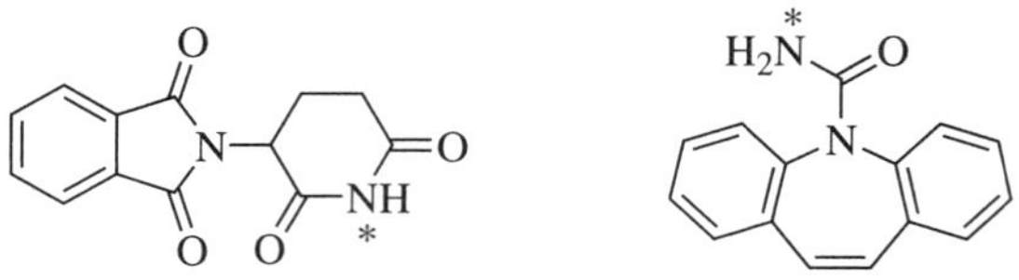

[^{13}N] thalidomide   [^{13}N] carbamazepine

SCHEME 4.29   Examples of N-13 labelled compounds.

$$4 \ ^{13/14}NH_3 + 5O_2 \xrightarrow{Ga_2O_3, \ CaO} 4 \ ^{13/14}NO + 6H_2O \quad (1)$$

$$2 \ ^{13/14}NO + O_2 \longrightarrow 2 \ ^{13/14}NO_2 \quad (2)$$

$$^{13/14}NO + {}^{13/14}NO_2 + H_2O \longrightarrow 2 \ [^{13/14}N]HNO_2 \quad (3)$$

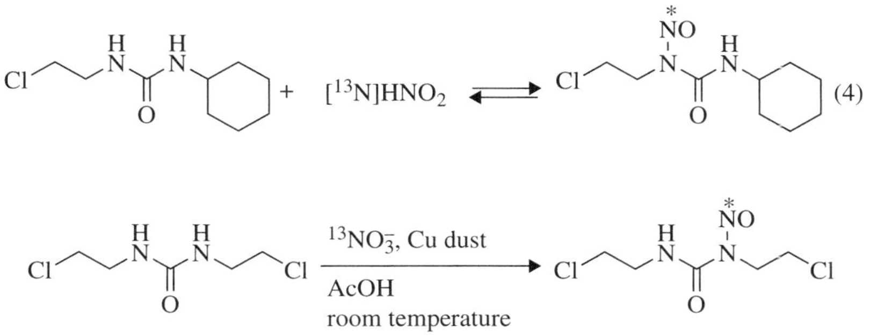

SCHEME 4.30   Synthesis of [^{13}N]nitrous acid and the ^{13}N-nitrosation of ureas.

$$^{13}NO_2^- + 2HSO_3^- + H^+ \longrightarrow HO[^{13}N](SO_3)_2^{2-} + H_2O \quad (1)$$

$$HO[^{13}N](SO_3)_2^{2-} + 2H_2O \longrightarrow {}^{13}NH_3OH^+ + 2SO_4^{2-} + H^+ \quad (2)$$

SCHEME 4.31   Preparation of [^{13}N]hydroxamine via reaction of [^{13}N]nitrite with sulphurous acid.

Reaction with water resulted in the formation of [^{13}N]nitrous acid; however, reactions with ureas via nitrosation resulted in total synthesis times that were unacceptably long (60-65 min.) and in low specific radioactivities. An improved method was later reported, as mentioned above, using the in-target ^{16}O(p,α)^{13}N nuclear reaction that yielded a mixture of [^{13}N]nitrite and [^{13}N]nitrate. The reduction of this mixture using Cu dust gave high radiochemical purity [^{13}N]nitrite and resulted in an improved yield of the nitrosation reaction of bis-(2-chloroethyl)-urea to give bis-(2-chloroethyl)-[^{13}N]nitrosourea, which was used in cancer chemotherapy programmes (Scheme 4.30) [158]. Although the synthesis time was reduced to 15-20 min., the method was carrier-added, which resulted in low specific radioactivities. Additionally, the removal of insoluble material was found to be tedious work under radiolabelling reaction conditions. Recently, Llop et al. [159] reported an improved online preparation of [^{13}N]nitrite where the reduction of [^{13}N]nitrate was effectively achieved by simply passing the mixture through a column containing cadmium on sand. The only radiochemical impurity was found to be [^{13}N]ammonia, which did not affect subsequent reactions.

4.3.5.2   *Synthesis of Hydroxyl[^{13}N]amine*
Hydroxylamine forms a variety of characteristic derivatives such as hydroxamic acids, oximes, and amidoximes. Such derivatives have proven useful in understanding the structure and function of many biologically active molecules. [^{13}N]hydroxylamine has been prepared from [^{13}N]nitrite [160] via with sulphurous acid (Scheme 4.31), although low specific activities were obtained and [^{13}N]nitrate was a major radiolabelling impurity in the reaction from the initial nuclear reaction.

4.3.5.3   *Synthesis of Nitrosothiols, Nitrosamines, and Diazo Compounds*
S-nitrosoglutathione is a known platelet aggregation inhibitor and a potentially useful vascular imaging agent. [^{13}N]S-nitrosoglutathione has recently been efficiently prepared via the no-carrier added reaction of [^{13}N]nitrite with a free thiol group under acidic conditions (Scheme 4.32) [159]. A fully automated method has been developed for the radiosynthesis of [^{13}N]S-nitrosothiols using a trapping [^{13}N]nitrite and nitrosation reaction on anion exchange resin [161]. A similar anion exchange resin technique has also proven useful for the preparation of a range of [^{13}N]nitrosoamines (Scheme 4.32) [162].

The aromatic diazo moiety is a structural component of the Congo Red dye, a widely used marker to identify beta-amyloid (Aβ) aggregation in the brains of people who have suffered from Alzheimer's disease. C-11 and F-18 labelled Aβ marker analogues have received considerable attention over the past decade as a diagnostic tool for Alzheimer's disease. N-13 labelling of the diazo groups in Congo Red derivatives also presents as a viable route for the preparation of PET tracers for amyloid imaging. One route to labelling such diazo compounds recently developed by Llop et al. [163] reports the reaction of primary aromatic amines with [^{13}N]nitrite to give a [^{13}N]diazonium salt followed by reaction with an aromatic amine or phenol to generate diazo compounds in good radiochemical yields (Scheme 4.32).

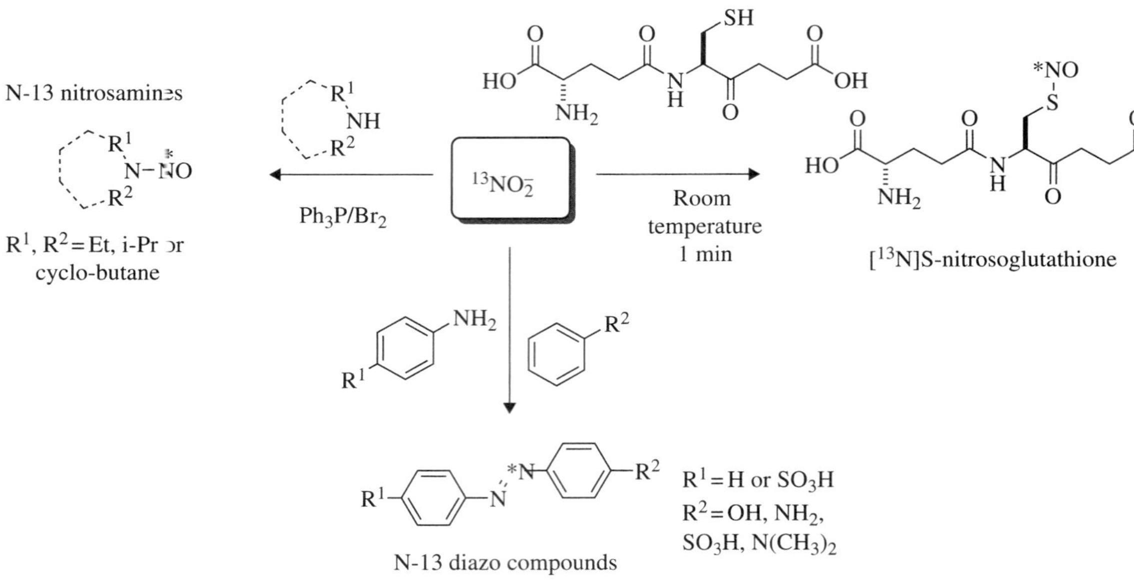

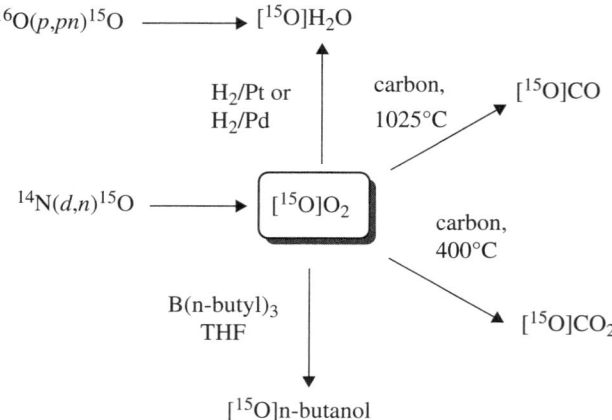

SCHEME 4.32   [^{13}N]nitrite a useful precursor for labelling nitrosothiols, nitrosamines and diazo compounds.

4.4   OXYGEN-15 CHEMISTRY

Oxygen-15 is generally produced in target via the bombardment of nitrogen gas with deuterons ^{14}N(d,n)^{15}O. A small percentage of carrier oxygen is required in the target gas that results in low specific activities of [^{15}O]O$_2$. Oxygen-15 has the shortest half-life ($t_{1/2}$ = 2.0 min) of the commonly used PET isotopes, which seriously prohibits its use in multistep chemical syntheses. The majority of O-15 chemistry is therefore based on the production of small molecules either directly within the cyclotron target or via one chemical transformation using high temperature gas phase methods. Despite the time challenges of using O-15, some of the earliest PET scans in human were achieved using [^{15}O]O$_2$, [^{15}O]CO, [^{15}O]CO$_2$, and [^{15}O]H$_2$O. Oxygen-15 labelled water has been especially important for the measurement of regional cerebral blood flow [164–167]. The production of [^{15}O]H$_2$O for these studies has been achieved in several ways, most simply by the inhalation of [^{15}O]CO$_2$, which is instantaneously converted into [^{15}O]H$_2$O in the lungs by carbonic anhydrase enzyme [168] and then transported around the body in the bloodstream. [^{15}O]H$_2$O can be produced directly via the high temperature reduction ^{15}O$_2$ over platinum [169] or palladium [170–172] catalysts (Scheme 4.33).

Alternatively, [^{15}O]H$_2$O can be produced in target by bombardment of natural water with protons [173, 174] (Scheme 4.33) yielding [^{15}O]H$_2$O that can be administered intravenously. Early investigations of blood flow studies in brain and other organ by PET relied on [^{15}O]O$_2$ or [^{15}O]H$_2$O before other the PET isotopes became available [175–177]. [^{15}O]CO can also be used to measure regional blood volume and can be produced rapidly via the reduction of [^{15}O]O$_2$ over carbon beads at high temperature [169, 171, 178]. The number of other more complicated molecules labelled with O-15 is very limited.

^{16}O(p,pn)^{15}O $\longrightarrow$ [^{15}O]H$_2$O

H$_2$/Pt or H$_2$/Pd   carbon, 1025°C   [^{15}O]CO

^{14}N(d,n)^{15}O $\longrightarrow$ [^{15}O]O$_2$

carbon, 400°C   [^{15}O]CO$_2$

B(n-butyl)$_3$ THF

[^{15}O]n-butanol

SCHEME 4.33   Synthesis of major oxygen-15 compounds for PET.

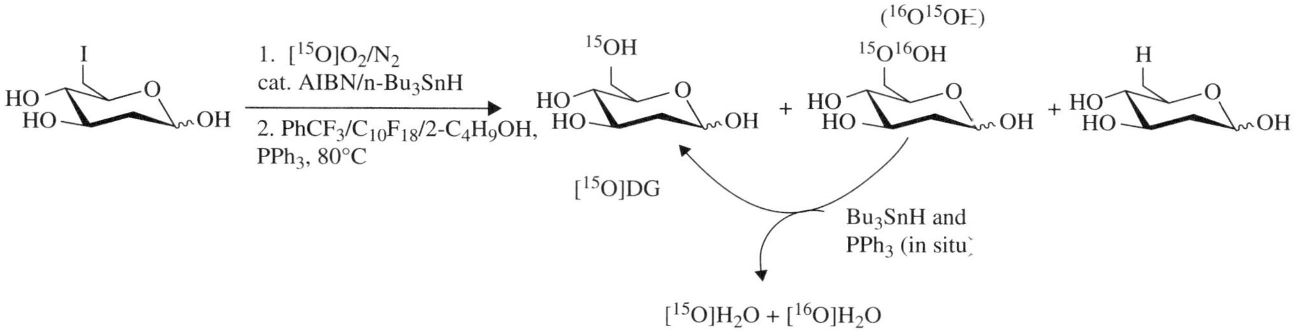

SCHEME 4.34 Free radical synthesis of 6-[^{15}O]-2-deoxy-D-glucose ([^{15}O]DG) from iodinated sugar and [^{15}O]O$_2$.

[^{15}O]Butanol, prepared via the reaction of [^{15}O]O$_2$ tributyl borane [179, 180], has proven to be superior to [^{15}O]H$_2$O in measuring cerebral blood flows and performing neuroactivation studies [181]. Another recent example reports the practical one-step preparation of ^{15}O labelled deoxyglucose O6-[^{15}O]-2-deoxy-D-glucose ([^{15}O]DG) within a 7 min. reaction time (Scheme 4.34) [182].

The synthesis was carried out using the iodinated precursor under mild reaction conditions that importantly removed the need for protecting the hydroxyl groups on the sugar. By circumventing a deprotection step, it was then possible to reduce the overall reaction time from hours to minutes. A specially adapted hot-air jacket reaction vessel equipped with a sintered glass bottom introduced the O-15 gas into the liquid reagents as fine bubbles. Typical decay-corrected radiochemical purity of the labelled [^{15}O]DG was found to be >70%. [^{15}O]DG was used to perform sequential [^{15}O]DG-[^{15}O]H$_2$O-[^{18}F]FDG measurements and PET images showing accumulation in the heart, kidneys, and bladders of animal models.

4.5 CONCLUSIONS

Over the past two decades there has been a diverse range of chemistry applied to the synthesis of C-11, N-13, and O-15 radio-labelled compounds for PET. As new chemistry is adopted from the mainstream chemical literature and specifically applied to radiolabelling, new possibilities have emerged for labelling different functional groups and labelling in different atomic positions within target molecules. The application of transition metal cross-coupling reactions and carbonylation methods are prime examples. Equally, the development of new technologies (e.g., microfluidics, microwave, and UV-photochemistry) and the improvement in automated procedures have had an important impact on the safety and the reliability of labelling procedures. The application of technology and its commercialisation is an important driver in this regard and can have an important influence on developing new radiochemistry. An example of this is the new radiochemistry that has emerged around ^{11}CH$_2$O, ^{11}CH$_3$NO$_2$, and ^{11}CS$_2$, which is a direct result of the now routine production of C-11 methyl iodide. The development of chemistry with C-11, N-13, and O-15 is ultimately driven by their application in PET imaging and the resultant need to radiolabel an ever-wider range of compounds. There are still many challenges in this area with regards to widening reaction scope, the development of novel precursor synthesis, reducing reaction times and improving radiosynthesis reliability.

REFERENCES

[1] P. W. Miller, N. J. Long, R. Vilar and A. D. Gee, *Angew. Chem. Int. Ed.* **47**, 8998–9033 (2008).

[2] J. S. Fowler and A. P. Wolf, *Acc. Chem. Res.* **30**, 181–188 (1997).

[3] S. M. Ametamey, M. Honer and P. A. Schubiger, *Chem. Rev.* **108**, 1501–1516 (2008).

[4] C. Aubert, C. HuardPerrio and M. C. Lasne, *J. Chem. Soc., Perkin Trans. 1* 2837–2842 (1997).

[5] C. Perrio-Huard, C. Aubert and M. C. Lasne, *J. Chem. Soc., Perkin Trans. 1* 311–316 (2000).

[6] S. Y. Lu, J. S. Hong and V. W. Pike, *J. Labelled Compd. Radiopharm.* **46**, 1249–1259 (2003).

[7] J. van den Hoff, W. Burchert, A. R. Borner, H. Fricke, G. Kuhnel, G. J. Meyer, D. Otto, E. Weckesser, H. G. Wolpers and W. H. Knapp, *J. Nuc. Med.* **42**, 1174–1182 (2001).

[8] D. E. Ponde, C. S. Dence, N. Oyama, J. Kim, Y. C. Tai, R. Laforest, B. A. Siegel and M. J. Welch, *J. Nuc. Med.* **48**, 420–428 (2007).

[9] R. J. Davenport, K. Dowsett and V. W. Pike, *Appl. Radiat. Isot.* **48**, 1117–1120 (1997).

[10] D. Le Bars, M. Malleval, F. Bonnefoi and C. Tourvieille, *J. Labelled Compd. Radiopharm.* **49**, 263–267 (2006).

[11] D. Soloviev and C. Tamburella, *Appl. Radiat. Isot.* **64**, 995–1000 (2006).

[12] D. R. Hwang, N. R. Simpson, J. Montoya, J. J. Mann and M. Laruelle, *Nuc. Med. Biol.* **26**, 815–819 (1999).

[13] J. A. McCarron, D. R. Turton, V. W. Pike and K. G. Poole, *J. Labelled Compd. Radiopharm.* **38**, 941–953 (1996).

[14] P. Mading, J. Zessin, U. Pleiss, F. Fuchtner and F. Wuest, *J. Labelled Compd. Radiopharm.* **49**, 357–365 (2006).

[15] P. J. Riss, S. Lu, S. Telu, F. I. Aigbirhio and V. W. Pike, *Angew. Chem. Int. Ed.* **51**, 2698–2702 (2012).

[16] E. W. Van Tilburg, A. D. Windhorst, M. Van der Mey and J. D. M. Herscheid, *J. Labelled Compd. Radiopharm.* **49**, 321–330 (2006).

[17] A. A. Wilson, A. Garcia, S. Houle and N. Vasdev, *Org. Bio. Chem.* **8**, 428–432 (2010).

[18] A. A. Wilson, A. Garcia, S. Houle, O. Sadovski and N. Vasdev, *Chem. Eur. J.* **17**, 259–264 (2011).

[19] J. M. Hooker, A. T. Reibel, S. M. Hill, M. J. Schueller and J. S. Fowler, *Angew. Chem. Int. Ed.* **48**, 3482–3485 (2009).

[20] B. Langstrom, G. Antoni, P. Gullberg, C. Halldin, P. Malmborg, K. Nagren, A. Rimland and H. Svard, *J. Nuc. Med.* **28**, 1037–1040 (1987).

[21] P. Larsen, J. Ulin, K. Dahlstrom and M. Jensen, *Appl. Radiat. Isot.* **48**, 153–157 (1997).

[22] A. M. Spence, H. M. Graham, M. Muzi, S. D. Freeman, J. M. Link, J. R. Grierson, F. Osullivan, D. Stein, G. L. Abbott and K. A. Krohn, *J. Nuc. Med.* **38**, 617–624 (1997).

[23] D. M. Jewett, *Appl. Radiat. Isot.* **43**, 1383–1385 (1992).

[24] C. Thominiaux, F. Dolle, M. L. James, Y. Bramoulle, H. Boutin, L. Besret, M. C. Gregoire, H. Valette, M. Bottlaender, B. Tavitian, P. Hantraye, S. Selleri and M. Kassiou, *Appl. Radiat. Isot.* **64**, 570–573 (2006).

[25] F. Dolle, P. Emond, S. Mavel, S. Demphel, F. Hinnen, Z. Mincheva, W. Saba, H. Valette, S. Chalon, C. Halldin, M. Helfenbein, J. Legaillard, J. C. Madelmont, J. B. Deloye, M. Bottlaender and D. Guilloteau, *Biorg. Med. Chem.* **14**, 1115–1125 (2006).

[26] A. A. Wilson, A. Garcia, A. Chestakova, H. Kung and S. Houle, *J. Labelled Compd. Radiopharm.* **47**, 679–682 (2004).

[27] K. Kawamura and K. Ishiwata, *Ann. Nuc. Med.* **18**, 165–168 (2004).

[28] C. Solbach, M. Uebele, G. Reischl and H. J. Machulla, *Appl. Radiat. Isot.* **62**, 591–595 (2005).

[29] D. M. Jewett, *Applied Radiation and Isotopes* **43**, 1383–1385 (1992).

[30] B. Langstrom, G. Antoni, P. Gullberg, C. Halldin, K. Nagren, A. Rimland and H. Svard, *Appl. Radiat. Isot.* **37**, 1141–1145 (1986).

[31] G. Antoni and B. Langstrom, *J. Labelled Compd. Radiopharm.* **24**, 125–143 (1987).

[32] R. Bolton, *J. Labelled Compd. Radiopharm.* **44**, 701–736 (2001).

[33] A. A. Wilson, A. Garcia, L. Jin and S. Houle, *Nuc. Med. Biol.* **27**, 529–532 (2000).

[34] W. E. Klunk, H. Engler, A. Nordberg, Y. M. Wang, G. Blomqvist, D. P. Holt, M. Bergstrom, I. Savitcheva, G. F. Huang, S. Estrada, B. Ausen, M. L. Debnath, J. Barletta, J. C. Price, J. Sandell, B. J. Lopresti, A. Wall, P. Koivisto, G. Antoni, C. A. Mathis and B. Langstrom, *Ann. Neurol.* **55**, 306–319 (2004).

[35] A. Nordberg, *Lancet Neurol.* **3**, 519–527 (2004).

[36] C. A. Mathis, B. J. Lopresti and W. E. Klunk, *Nuc. Med. Biol.* **34**, 809–822 (2007).

[37] X. S. Fei, B. H. Mock, T. R. DeGrado, J. Q. Wang, B. E. Glick-Wilson, M. L. Sullivan, G. D. Hutchins and Q. H. Zheng, *Synth. Commun.* **34**, 1897–1907 (2004).

[38] O. Langer, K. Nagren, F. Dolle, C. Lundkvist, J. Sandell, C. G. Swahn, F. Vaufrey, C. Crouzel, B. Maziere and C. Halldin, *J. Labelled Compd. Radiopharm.* **42**, 1183–1193 (1999).

[39] K. Suzuki, O. Inoue, K. Tamate and F. Mikado, *Appl. Radiat. Isot.* **41**, 593–599 (1990).

[40] M. Laruelle, *J. Cerebr. Blood F. Met.* **20**, 423–451 (2000).

[41] D. E. Kuhl, R. A. Koeppe, S. Minoshima, S. E. Snyder, E. P. Ficaro, N. L. Foster, K. A. Frey and M. R. Kilbourn, *Neurology* **52**, 691–699 (1999).

[42] V. W. Pike, C. Halldin, C. Crouzel, L. Barre, D. J. Nutt, S. Osman, F. Shah, D. R. Turton and S. L. Waters, *Nuc. Med. Biol.* **20**, 503–525 (1993).

[43] J. R. Lever, *Curr. Pharm. Design* **13**, 33–49 (2007).

[44] H. T. Ravert, B. Bencherif, I. Madar and J. J. Frost, *Curr. Pharm. Design* **10**, 759–768 (2004).

[45] D. J. Scott, C. S. Stohler, R. A. Koeppe and J. K. Zubieta, *Synapse* **61**, 707–714 (2007).

[46] R. Iwata, C. Pascai, A. Bogni, Y. Miyake, K. Yanai and T. Ido, *Appl. Radiat. Isot.* **55**, 17–22 (2001).

[47] R. Iwata, C. Pascai, A. Bogni, K. Yanai, M. Kato, T. Ido and K. Ishiwata, *J. Labelled Compd. Radiopharm.* **45**, 271–280 (2002).

[48] A. R. Studenov, S. Jivan, M. J. Adam, T. J. Ruth and K. R. Buckley, *Appl. Radiat. Isot.* **61**, 1195–1201 (2004).

[49] A. R. Studenov, S. Jivan, K. R. Buckley and M. J. Adam, *J. Labelled Compd. Radiopharm.* **46**, 837–842 (2003).

[50] V. Gomez, J. D. Gispert, V. Amador and J. Llop, *J. Labelled Compd. Radiopharm.* **51**, 83–86 (2008).

[51] D. P. Holt, H. T. Ravert, R. F. Dannals and M. G. Pomper, *J. Labelled Compd. Radiopharm.* **49**, 883–888 (2006).

[52] M. C. Cleij, J. C. Clark, J. C. Baron and F. I. Aigbirhio, *J. Labelled Compd. Radiopharm.* **50**, 19–24 (2007).

[53] O. Jacobson and E. Mishani, *Appl. Radiat. Isot.* **66**, 188–193 (2008).

[54] T. Hosoya, K. Sumi, H. Doi, M. Wakao and M. Suzuki, *Org. Bio. Chem.* **4**, 410–415 (2006).

[55] T. G. Hamill, S. Krause, C. Ryan, C. Bonnefous, S. Govek, T. J. Seiders, N. D. P. Cosford, J. Roppe, T. Kamenecka, S. Patel, R. E. Gibson, S. Sanabria, K. Riffel, W. S. Eng, C. King, X. Q. Yang, M. D. Green, S. S. O'Malley, R. Hargreaves and H. D. Burns, *Synapse* **56**, 205–216 (2005).

[56] L. Samuelsson and B. Langstrom, *J. Labelled Compd. Radiopharm.* **46**, 263–272 (2003).

[57] M. Bjorkman, H. Doi, B. Resul, M. Suzuki, R. Noyori, Y. Watanabe and B. Langstrom, *J. Labelled Compd. Radiopharm.* **43**, 1327–1334 (2000).

[58] M. Suzuki, H. Doi, K. Kato, M. Bjorkman, B. Langstrom, Y. Watanabe and R. Noyori, *Tetrahedron* **56**, 8263–8273 (2000).

[59] J. Tarkiainen, J. Vercouille, P. Emond, J. Sandell, J. Hiltunen, Y. Frangin, D. Guilloteau and C. Halldin, *J. Labelled Compd. Radiopharm.* **44**, 1013–1023 (2001).

[60] E. D. Hostetler, G. E. Terry and H. D. Burns, *J. Labelled Compd. Radiopharm.* **48**, 629–634 (2005).

[61] F. Wuest, J. Zessin and B. Johannsen, *J. Labelled Compd. Radiopharm.* **46**, 333–342 (2003).

[62] M. Huiban, A. Huet, L. Barre, F. Sobrio, E. Fouquet and C. Perrio, *Chem. Commun.* 97–99 (2006).

[63] T. Forngren, L. Samuelsson and B. Langstrom, *J. Labelled Compd. Radiopharm.* **47**, 71–78 (2004).

[64] J. M. Hooker, M. Schonberger, H. Schieferstein and J. S. Fowler, *Angew Chem Int Ed Engl* **47**, 5989–5992 (2008).

[65] M. W. Nader, S. K. Zeisler, A. Theobald and F. Oberdorfer, *Appl. Radiat. Isot.* **49**, 1599–1603 (1998).

[66] D. Roeda and F. Dollé, *J. Labelled Compd. Radiopharm.* **46**, 456–458 (2003).

[67] P. W. Miller and D. Bender, *Chem. Eur. J.* **18**, 433–436 (2012).

[68] K. Kato, M. R. Zhang, K. Minegishi, N. Nengaki, M. Takei and K. Suzuki, *J. Labelled Compd. Radiopharm.* **54**, 140–144 (2011).

[69] K. O. Schoeps, B. Langstrom, S. Stoneelander and C. Halldin, *Appl. Radiat. Isot.* **42**, 877–883 (1991).

[70] K. Kato, S. A. Gustavsson and B. Langstrom, *Tetrahedron Lett.* **49**, 5837–5839 (2008).

[71] K. Kato, M. R. Zhang and K. Suzuki, *Mol. Biosyst.* **4**, 53–55 (2008).

[72] D. Roeda and F. Dolle, *Curr. Top. Med. Chem.* **10**, 1680–1700 (2010).

[73] K. I. Nishijima, Y. Kuge, K. Seki, K. Ohkura, N. Motoki, K. Nagatsu, A. Tanaka, E. Tsukamoto and N. Tamaki, *Nuc. Med. Biol.* **29**, 345–350 (2002).

[74] P. Landais and C. Crouzel, *Appl. Radiat. Isot.* **38**, 297–300 (1987).

[75] M. Ogawa, Y. Takada, H. Suzuki, K. Nemoto and T. Fukumura, *Nuc. Med. Biol.* **37**, 73–76 (2010).

[76] Y. Bramoulle, D. Roeda and F. Dolle, *Tetrahedron Lett.* **51**, 313–316 (2010).

[77] F. Dolle, L. Martarello, Y. Bramoulle, M. Bottlaender and A. D. Gee, *J. Labelled Compd. Radiopharm.* **48**, 501–513 (2005).

[78] P. Lidstrom, T. A. Bonasera, M. Marquez-M, S. Nilsson, M. Bergstrom and B. Langstrom, *Steroids* **63**, 228–234 (1998).

[79] D. Roeda, B. Tavitian, C. Coulon, F. David, F. Dolle, C. Fuseau, A. Jobert and C. Crouzel, *Biorg. Med. Chem.* **5**, 397–403 (1997).

[80] L. Lemoucheux, J. Rouden, M. Ibazizene, F. Sobrio and M. C. Lasne, *J. Org. Chem.* **68**, 7289–7297 (2003).

[81] K. Yashio, Y. Katayama, T. Takashima, N. Ishiguro, H. Doi, M. Suzuki, Y. Wada, I. Tamai and Y. Watanabe, *Biorg. Med. Chem. Lett.* **22**, 115–119 (2012).

[82] C. Asakawa, M. Ogawa, K. Kumata, M. Fujinaga, T. Yamasaki, L. Xie, J. Yui, K. Kawamura, T. Fukumura and M. R. Zhang, *Biorg. Med. Chem. Lett.* **21**, 7017–7020 (2011).

[83] R. Iwata, T. Ido, T. Takahashi, H. Nakanishi and S. Iida, *Appl. Radiat. Isot.* **38**, 97–102 (1987).

[84] N. M. Gillings and A. D. Gee, *J. Labelled Compd. Radiopharm.* **4**, 909–920 (2001).

[85] G. B. Jacobson, G. Westerberg, K. E. Markides and B. Langstrom, *J. Am. Chem. Soc.* **118**, 6863–6872 (1996).

[86] G. Westerberg and B. Langstrom, *Appl. Radiat. Isot.* **48**, 459–461 (1997).

[87] M. Ponchant, F. Hinnen, S. Demphel and C. Crouzel, *Appl. Radiat. Isot.* **48**, 755–762 (1997).

[88] W. B. Mathews, J. A. Monn, H. T. Ravert, D. P. Holt, D. D. Schoepp and R. F. Dannals, *J. Labelled Compd. Radiopharm.* **49**, 829–834 (2006).

[89] F. Simeon, F. Sobrio, F. Gourand and L. Barre, *J. Chem. Soc., Perkin Trans. 1* 690–694 (2001).

[90] Y. Andersson and B. Langstrom, *J. Chem. Soc., Perkin Trans. 1* 1395–1400 (1994).

[91] Y. Andersson, M. Bergstrom and B. Langstrom, *Appl. Radiat. Isot.* **45**, 707–714 (1994).

[92] W. Qu, S. Oya, B. P. Lieberman, K. Ploessl, L. Wang, D. R. Wise, C. R. Divgi, L. P. Chodosh, C. B. Thompson and H. F. Kung, *J. Nuc. Med.* **53**, 98–105 (2012).

[93] B. Langstom, O. Itsenko and O. Rahman, *J. Labelled Compd. Radiopharm.* **50**, 794–810 (2007).

[94] J. C. Clark and P. D. Buckingham, *Short-lived Radioactive Gases for Medical Use*, Butterworths, London, 231 (1975).

[95] S. K. Zeisler, M. Nader, A. Theobald and F. Oberdorfer, *Appl. Radiat. Isot.* **48**, 1091–1095 (1997).

[96] T. Kihlberg, B. Langstrom and F. T., J. Eriksson, *Methods and Apparatus for Production and Use of [^{11}C]Carbon Monoxide in Labelling Synthesis.* International patent application, submitted on July 9, 2005, PCT/IB2005/001939 (2005).

[97] F. Karimi, J. Barletta and B. Langstrom, *Eur. J. Org. Chem.* 2374–2378 (2005).

[98] J. Eriksson, O. Aberg and B. Langstrom, *Eur. J. Org. Chem.* 455–461 (2007).

[99] P. Lidstrom, T. Kihlberg and B. Langstrom, *J. Chem. Soc., Perkin Trans. 1* 2701–2706 (1997).

[100] E. D. Hostetler and H. D. Burns, *Nuc. Med. Biol.* **29**, 845–48 (2002).

[101] S. Kealey, P. W. Miller, N. J. Long, C. Plisson, L. Martarello and A. D. Gee, *Chem. Commun.* 3696–3698 (2009).

[102] C. R. Child, S. Kealey, H. Jones, P. W. Miller, A. J. P. White, A. D. Gee and N. J. Long, *Dalton Trans.* **40**, 6210–6215 (2011).

[103] L. E. Jennings, S. Kealey, P. W. Miller, A. D. Gee and N. J. Long, *J. Labelled Compd. Radiopharm.* **54**, 135–139 (2011).

[104] H. Audrain, L. Martarello, A. Gee and D. Bender, *Chem. Commun.* 558–559 (2004).

[105] P. W. Miller, N. J. Long, A. J. de Mello, R. Vilar, H. Audrain, D. Bender, J. Passchier and A. Gee, *Angew. Chem. Int. Ed.* **46**, 2875–2878 (2007).

[106] P. W. Miller, H. Audrain, D. Bender, A. J. deMello, A. D. Gee, N. J. Long and R. Vilar, *Chem. Eur. J.* **17**, 460–463 (2011).

[107] O. Rahman, J. Llop and B. Langstrom, *Eur. J. Org. Chem.* 2674–2678 (2004).

[108] O. Rahman, T. Kihlberg and B. Langstrom, *Eur. J. Org. Chem.* 474–478 (2004).

[109] M. H. Al-Qahtani and V. W. Pike, *J. Chem. Soc., Perkin Trans. 1* 1033–1036 (2000).

[110] T. Kihlberg and B. Langstrom, *J. Org. Chem.* **64**, 9201–9205 (1999).

[111] F. Karimi and B. Långström, *J. Chem. Soc., Perkin Trans. 1,* 2111–2116 (2002).

[112] F. Karimi and B. Langstrom, *Org. Bio. Chem.* **1**, 541–546 (2003).

[113] F. Karimi and B. Langstrom, *Eur. J. Org. Chem.* 2132–2137 (2003).

[114] O. Rahman, T. Kihlberg and B. Långström, *J. Org. Chem.* **68**, 3558– (2003).

[115] S. Kealey, C. Plisson, T. L. Collier, N. J. Long, S. M. Husbands, L. Martarello and A. D. Gee, *Org. Bio. Chem.* **9**, 3313–3319 (2011).

[116] G. Buscemi, P. W. Miller, S. Kealey, A. D. Gee, N. J. Long, J. Passchier and R. Vilar, *Org. Bio. Chem.* **9**, 3499–3503 (2011).

[117] J. Eriksson, J. van den Hoek, A. D. Windhorst, *J. Labelled Compd. Radiopharm.* **55**, 223–228 (2012).

[118] J. Barletta, F. Karimi, H. Doi and B. Langstrom, *J. Labelled Compd. Radiopharm.* **49**, 801–809 (2006).

[119] J. Barletta, F. Karimi and B. Langstrom, *J. Labelled Compd. Radiopharm.* **49**, 429–436 (2006).

[120] H. Doi, J. Barletta, M. Suzuki, R. Noyori, Y. Watanabe and B. Langstrom, *Org. Bio. Chem.* **2**, 3063–3066 (2004).

[121] O. Itsenko and B. Langstrom, *J. Org. Chem.* **70**, 2244–2249 (2005).

[122] O. Itsenko and B. Langstrom, *Org. Lett.* **7**, 4661–4664 (2005).

[123] O. Itsenko, T. Kihlberg and B. Langstrom, *Eur. J. Org. Chem.* 3830–3834 (2005).

[124] O. Itsenko, D. Norberg, T. Rasmussen, B. Langstrom and C. Chatgilialoglu, *J. Am. Chem. Soc.* **129**, 9020–9031 (2007).

[125] O. Itsenko, T. Kihlberg and B. Langstrom, *J. Org. Chem.* **69**, 4356–4360 (2004).

[126] R. J. Nickles, S. J. Gatley, R. D. Hichwa, D. J. Simpkin and J. L. Martin, *Int. J. Appl. Radiat. Is.* **29**, 225–227 (1978).

[127] N. J. Parks and K. A. Krohn, *Int. J. Appl. Radiat. Is.* **29**, 754–757 (1978).

[128] H. R. Schelbert, M. E. Phelps, S. C. Huang, N. S. Macdonald, H. Hansen and D. E. Kuhl, *Circulation* **63**, 1259–1272 (1981).

[129] W. Vaalburg, J. A. A. Kamphuis, H. D. Beerlingvandermolen, S. Reiffers, A. Rijskamp and M. G. Woldring, *Int. J. Appl. Radiat. Is.* **26**, 316–318 (1975).

[130] S. J. Gatley and C. Shea, *Appl. Radiat. Isot.* **42**, 793–796 (1991).

[131] K. Suzuki and Y. Yoshida, *Appl. Radiat. Isot.* **50**, 497–503 (1999).

[132] H. Krizek, N. Lembares, Dinwoodi.R, I. Gloria, K. A. Lathrop and P. V. Harper, *J. Nuc. Med.* **14**, 629–630 (1973).

[133] B. Wieland, G. Bida, H. Padgett, G. Hendry, E. Zippi, G. Kabalka, J. L. Morelle, R. Verbruggen and M. Ghyoot, *Appl. Radiat. Isot.* **42**, 1095–1098 (1991).

[134] M. S. Berridge and B. J. Landmeier, *Appl. Radiat. Isot.* **44**, 1433–1441 (1993).

[135] G. Porenta, J. Czernin and H. R. Schelbert, *Positron Emission Tomography of the Heart,* , S. R. Bergmann, B. E. Sobel, (Eds.). Futura Publishers, Kisco, New York, 153–183 (1992).

[136] T. Schepis, O. Gaemperli, V. Treyer, I. Valenta, C. Burger, P. Koepfli, M. Namdar, I. Adachi, H. Alkadhi and P. A. Kaufmann, *J. Nuc. Med.* **48**, 1783–1789 (2007).

[137] W. G. Kuhle, G. Porenta, S. C. Huang, D. Buxton, S. S. Gambhir, H. Hansen, M. E. Phelps and H. R. Schelbert, *Circulation* **86**, 1004–1017 (1992).

[138] A. S. Gelbard, E. Nieves, S. Filcdericco and K. C. Rosenspire, *J. Labelled Compd. Radiopharm.* **23**, 1055–1055 (1986).

[139] K. Suzuki and K. Tamate, *Int. J. Appl. Radiat. Is.* **35**, 771–777 (1984).

[140] J. R. Barrio, F. J. Baumgartner, E. Henze, M. S. Stauber, J. E. Egbert, N. S. Macdonald, H. R. Schelbert, M. E. Phelps and F. T. Liu, *J. Nuc. Med.* **24**, 937–944 (1983).

[141] A. J. L. Cooper and A. S. Gelbard, *Anal. Biochem.* **111**, 42–48 (1981).

[142] A. S. Gelbard, R. S. Benua, R. E. Reiman, J. M. McDonald, J. J. Vomero and J. S. Laughlin, *J. Nuc. Med.* **21**, 988–991 (1980).

[143] A. S. Gelbard, L. P. Clarke, J. M. McDonald, W. G. Monahan, R. S. Tilbury, T. Y. T. Kuo and J. S. Laughlin, *Radiology* **116**, 127–132 (1975).

[144] T. Irie, O. Inoue, K. Suzuki and T. Tominaga, *Int. J. Appl. Radiat. Is.* **36**, 345–347 (1985).

[145] K. Suzuki, Y. Yoshida, N. Shikano and A. Kubodera, *Appl. Radiat. Isot.* **50**, 1033–1038 (1999).

[146] T. Tominaga, O. Inoue, K. Suzuki, T. Yamasaki and M. Hirobe, *Appl. Radiat. Isot.* **37**, 1209–1212 (1986).

[147] T. Tominaga, O. Inoue, T. Irie, K. Suzuki, T. Yamasaki and M. Hirobe, *Int. J. Appl. Radiat. Is.* **36**, 555–557 (1985).

[148] R. D. Finn, D. R. Christman and A. P. Wolf, *J. Labelled Compd. Radiopharm.* **18**, 909–913 (1981).

[149] P. J. Kothari, R. D. Finn, G. W. Kabalka, M. M. Vora, T. E. Boothe and A. M. Emran, *Appl. Radiat. Isot.* **37**, 469–470 (1986).

[150] M. B. Cohen, L. Spolter, C. C. Chang, N. S. Macdonald, J. Takahashi and D. D. Bobinet, *J. Nuc. Med.* **15**, 1192–1195 (1974).

[151] A. S. Gelbard, R. S. Benua, J. S. Laughlin, G. Rosen, R. E. Reiman and J. M. McDonald, *J. Nuc. Med.* **20**, 782–784 (1979).

[152] M. T. Haber, A. J. L. Cooper, K. C. Rosenspire, J. Z. Ginos and D. A. Rottenberg, *J. Labelled Compd. Radiopharm.* **22**, 509–516 (1985).

[153] B. De Spiegeleer, G. Slegers, C. Vandecasteele, W. Van Den Bossche, K. Schelstraete, A. Claeys and P. De Moerloose, *J. Nuc. Med.* **27**, 399–403 (1986).

[154] M. Holschbach, W. Hamkens, A. Steinbach, K. Hamacher and G. Stocklin, *Appl. Radiat. Isot.* **48**, 739–744 (1997).

[155] K. Kumata, M. Takei, M. Ogawa, K. Kato, K. Suzuki and M. R. Zhang, *J. Labelled Compd. Radiopharm.* **52**, 166–172 (2009).

[156] K. Kumata, M. Takei, M. Ogawa, J. J. Yui, A. Hatori, K. Suzuki and M. R. Zhang, *J. Labelled Compd. Radiopharm.* **53**, 53–57 (2010).

[157] W. A. Pettit, R. H. Mortara, G. A. Digenis and M. F. Reed, *J. Med. Chem.* **18**, 1029–1031 (1975).

[158] W. A. Pettit, R. S. Tilbury, G. A. Digenis and R. H. Mortara, *J. Labelled Compd. Radiopharm.* **13**, 119–122 (1977).

[159] J. Llop, V. Gomez-Vallejo, M. Bosque, G. Quincoces and I. Penuelas, *Applied Radiation and Isotopes* **67**, 95–99 (2009).

[160] D. S. Kaseman, A. J. L. Cooper, A. Meister, A. S. Gelbard and R. E. Reiman, *J. Labelled Compd. Radiopharm.* **21**, 803–814 (1984).

[161] V. Gomez-Vallejo, K. Kato, I. Oliden, J. Calvo, Z. Baz, J. I. Borrell and J. Llop, *Tetrahedron Lett.* **51**, 2990–2993 (2010).

[162] V. Gomez-Vallejo, K. Kato, M. Hanyu, K. Minegishi, J. I. Borrell and J. Llop, *Biorg. Med. Chem. Lett.* **19**, 1913–1915 (2009).

[163] V. Gomez-Vallejo, J. I. Borrell and J. Llop, *Eur. J. Med. Chem.* **45**, 5318–5323 (2010).

[164] S. C. Jones, J. H. Greenberg, R. Dann, G. D. Robinson, M. Kushner, A. Alavi and M. Reivich, *J. Cerebr. Blood F. Met.* **5**, 566–575 (1985).

[165] B. Weber, G. Westera, V. Treyer, C. Burger, N. Khan and A. Buck, *J. Nuc. Med.* **45**, 1344–1350 (2004).

[166] F. Schneider, R. E. Gur, L. H. Mozley, R. J. Smith, P. D. Mozley, D. M. Censits, A. Alavi and R. C. Gur, *Psychiat. Res-Neuroim.* **61**, 265–283 (1995).

[167] P. Herscovitch, M. E. Raichle, M. R. Kilbourn and M. J. Welch, *J. Cerebr. Blood F. Met.* **7**, 527–542 (1987).

[168] I. Kanno, A. A. Lammertsma, J. D. Heather, J. M. Gibbs, C. G. Rhodes, J. C. Clark and T. Jones, *J. Cerebr. Blood F. Met.* **4**, 224–234 (1984).

[169] M. S. Berridge, A. H. Terris and E. H. Cassidy, *Appl. Radiat. Isot.* **41**, 1173–1175 (1990).

[170] M. Sajjad, J. S. Liow and J. Moreno-Cantu, *Appl. Radiat. Isot.* **52**, 205–210 (2000).

[171] J. C. Clark, C. Crouzel, G. J. Meyer and K. Strijckmans, *Appl. Radiat. Isot.* **38**, 597–600 (1987).

[172] B. M. Palmer, M. Sajjad and D. A. Rottenberg, *Nuc. Med. Biol.* **22**, 241–249 (1995).

[173] G. K. Mulholland, M. R. Kilbourn and J. J. Moskwa, *Appl. Radiat. Isot.* **41**, 1193–1199 (1990).

[174] J. VanNaemen, M. Monclus, P. Damhaut, A. Luxen and S. Goldman, *Nuc. Med. Biol.* **23**, 413–416 (1996).

[175] B. D. Ahluwalia, C. A. Hales, G. L. Brownell and H. Kazemi, *RSNA and AAPM*, Chicago, (Nov. 1973).

[176] G. L. Brownell, C. A. Burnham, D. A. Chester, J. A. Correia, J. E. Correll, B. J. Hoop, J. Parker and R. Subramanyam, *Workshop on Reconstruction Tomography, 1976. Publ. Reconstruction Tomography in Diagnostic Radiology and Nuclear Medicine*, Eds. M. M. Ter-Pogossian, M. E. Phelps, G. L. Brownell, University Park Press, Baltimore, 293–307 (1977).

[177] M. E. Raichle, *Scientific American* **270**, 58–64 (1994).

[178] E. Rostrup, G. M. Knudsen, I. Law, S. Holm, H. B. W. Larsson and O. B. Paulson, *Neuroimage* **24**, 1–11 (2005).

[179] G. W. Kabalka, R. M. Lambrecht, M. Sajjad, J. S. Fowler, S. A. Kunda, G. W. McCollum and R. Macgregor, *Int. J. Appl. Radiat. Is.* **36**, 853–855 (1985).

[180] S. M. Moerlein, G. G. Gaehle, K. R. Lechner, R. K. Bera and M. J. Welch, *Appl. Radiat. Isot.* **44**, 1213–1218 (1993).

[181] J. R. Votaw, T. R. Henry, T. M. Shoup, J. M. Hoffman, J. L. Woodard and M. M. Goodman, *J. Cerebr. Blood F. Met.* **19**, 982–989 (1999).

[182] H. Yorimitsu, Y. Murakami, H. Takamatsu, S. Nishimura and E. Nakamura, *Angew. Chem. Int Ed.* **44**, 2708–2711 (2005).

5

THE CHEMISTRY OF INORGANIC NUCLIDES (^{86}Y, ^{68}Ga, ^{64}Cu, ^{89}Zr, ^{124}I)

Eric W. Price and Chris Orvig

Medicinal Inorganic Chemistry Group, Department of Chemistry, University of British Columbia, Vancouver, BC, Canada

5.1 INTRODUCTION: INORGANIC NUCLIDE-BASED RADIOPHARMACEUTICALS

The chemistry of molecular imaging agents based on radiometals is intimately tied to several core chemistry disciplines, such as aqueous metal ion chemistry, coordination chemistry, chelator design/synthesis, and bioconjugation chemistry with various biomolecular vectors. Biomolecular vectors (biovectors) are biological molecules such as peptides and antibodies that have high affinity and specificity for specific receptors or tissues. Radiometal-based radiopharmaceuticals have almost exclusively relied on bifunctional chelators (BFC) as a core to provide site-specific delivery of a tightly bound radioactive metal ion to biological targets. The targets are specified by a variety of biovectors that can be conjugated (attached) to the BFC agent. A BFC is simply a chelating agent that possesses the functionality to coordinate both a metal ion and a biovector. To this end, a successful radiopharmaceutical agent using radiometals must incorporate a high stability chelator with a robust conjugation site to provide a covalent linkage to an established biovector with high target affinity. The radioactive isotope used in a radiopharmaceutical is selected based on its decay properties (half-life, emission type and energy, branching ratios) and is matched with the chelator, biovector, biological target, and purpose of the agent. The radiohalogen ^{124}I is not utilised in the same manner as radiometals and is typically incorporated into radiopharmaceutical agents using direct covalent linkages, often replacing a hydrogen atom. The chemistry of radioiodine isotopes is very similar to non-radioactive iodine chemistry.

The chemistry of radioactive "hot" metal ions is identical to regular non-radioactive "cold" metal ions; however, radiochemistry is typically performed under extremely dilute conditions. It is also important to note that several of the elements being discussed have multiple radioactive isotopes that are useful for diagnostic or therapeutic purposes ($^{86/90}$Y, $^{123/124/125/131}$I, $^{67/68}$Ga, $^{60/61/62/64}$Cu), and all isotopes of a given element have identical chemistry [1–6]. A radiopharmaceutical agent based on one specific isotope will have identical chemistry and biological behaviour when synthesised with any other isotopes of that same element [1–6]. Assuming that the radioactive decay properties of the other isotopes of the element are useful for the specific radiopharmaceutical agent in question, substitution is seamless and can yield multiple applications from one molecular scaffold.

Positron emission tomography (PET) imaging is a very accurate and quantitative imaging technique, so exploration of positron emitting isotopes is of great significance. The archetypical PET isotope is ^{18}F, which has a very high positron (β^+) abundance of 96% and a low energy β^+ emission of 640 keV (short mean free path from decay location providing high resolution and accuracy). Many of the more exotic PET isotopes discussed here have lower positron abundances (low branching ratios, ~20-60% decay by β^+) and higher positron energies, which decreases the accuracy of data collection and requires longer image acquisition times and/or higher activity injected doses (Table 5.1) [7].

For detection to occur, the positron must be sufficiently slowed down after emission for it to meet an electron and annihilate. The size of the spherical radius that a positron travels from its source is dependent on its energy, and higher energy

---

The Chemistry of Molecular Imaging, First Edition. Edited by Nicholas Long and Wing-Tak Wong.
© 2015 John Wiley & Sons, Inc. Published 2015 by John Wiley & Sons, Inc.

TABLE 5.1 Relevant Properties of Selected PET Imaging Isotopes, EC = Electron Capture; Some Low Abundance Positrons have been Omitted for Clarity [10, 11].

Isotope	$t_{1/2}$ (h)	Decay Mode	E_β^+ (keV)	Production Method
^{60}Cu	0.4	β^+ (93%)	3920, 3000	cyclotron, ^{60}Ni(p,n)^{60}Cu
		EC (7%)	2000	
^{61}Cu	3.3	β^+ (62%)	1220, 1150	cyclotron, ^{61}Ni(p,n)^{61}Cu
		EC (38%)	940, 560	
^{62}Cu	0.16	β^+ (98%)	2910	^{62}Zn/^{62}Cu generator
		EC (2%)		
^{64}Cu	12.7	β^+ (19%)	656	cyclotron, ^{64}Ni(p,n)^{64}Cu
		EC (41%)		
		β^- (40%)		
^{66}Ga	9.5	β^+ (60%)	4150, 935	cyclotron, ^{63}Cu(α,nγ)^{66}Ga
		EC (10%)		
^{68}Ga	1.1	β^+ (90%)	1880	^{68}Ge/^{68}Ga generator
		EC (10%)		
^{86}Y	14.7	β^+ (33%)	1221	cyclotron, ^{86}Sr(p,n)^{86}Y
		EC (66%)		
^{89}Zr	78.5	β^+ (23%)	897	cyclotron, ^{89}Y(p,n)^{89}Zr
		EC (77%)		
^{124}I	100.2	β^+ (23%)	2138, 1535	cyclotron, ^{124}Te(p,n)^{124}I
		EC (77%)		

positrons travel a larger radius and therefore decrease spatial resolution. In general, lower energy β^+ and γ emissions provide better image quality. The nuclides ^{86}Y, ^{89}Zr, and ^{124}I emit a large amount of γ rays relative to the amount of positrons (poor branching ratios). These additional γ emissions can both complicate PET imaging by interfering with the detection of coincident 511 keV γ rays that originate from β^+ emission/annihilation and increase the radioactive dose accumulated by patients [8, 9]. Despite these shortcomings, the PET nuclides discussed here have a multitude of chemical and physical properties that make them attractive for imaging purposes. Radionuclides are typically produced by proton (p,n) or deuteron (d,2n) bombardment via a cyclotron, neutron bombardment via a nuclear reactor (n,xp), or elution from a generator system (the parent nuclide in generators must be produced via cyclotron or reactor). The most common production methods of various isotopes are displayed in Table 5.1 [10, 11]. Several of the most promising inorganic PET nuclides (excluding ^{18}F, Chapter 3) will be discussed in this section; however, there are many other unconventional PET nuclides that have received less attention and have been discussed elsewhere [12]. As recently noted, there are many discrepancies in nuclear decay properties reported in the literature, thus the half-lives, branching ratios, and positron energies reported here should be considered approximate [12, 13]

The preparation of imaging agents based on ^{18}F and radioiodine typically involve complex radiolabelling syntheses and are technically demanding (see Chapter 3). With the short half-life of ^{18}F (110 minutes), this provides significant logistical challenges. On the contrary, radiopharmaceutical preparations based on radiometals can be very simple to radiolabel, because they are often suitable for making kit formulations. These kit formulations simply require the addition of a pure and high specific activity radiometal to a buffered solution of BFC-biovector conjugate, incubation at an appropriate temperature to allow for quantitative radiometallation, and finally purification. This is possible because the majority of synthesis work is accomplished in the laboratory before radiolabelling is performed, thus these kit formulations are designed with a simple "shake and bake" preparation style that requires minimal processing and training for clinical deployment. The obvious drawback to BFC-radiometal preparations is that the radiometal-chelate complex is relatively large and often charged, which means that it may not be compatible with traditional agents such as small molecule drugs and neurotransmitters. The small molecule medicinal agents can have a hydrogen atom (typically aromatic) substituted by ^{18}F or a radioiodine isotope with minimal biological impact, allowing for imaging of the molecule's native and essentially unadulterated function. Radiohalogens can also be labelled onto a prosthetic group, which is typically comprised of an aromatic moiety that undergoes facile radiohalogenation and an appropriate bioconjugation tether for attachment to a biovector. The use of prosthetic groups to utilise radioiodine is a comparable modality to radiometal-based BFC systems, where the radioactive moiety is a separate chemical entity and is conjugated to a biovector.

Gallium(III), yttrium(III), and zirconium(IV) ions are resistant to redox reactions under biological conditions (aqueous, ~pH 7.4), which is important for complex stability; however, they are sensitive to hydrolysis. Changing the oxidation state

TABLE 5.2 Relevant Properties of Metal Cations: [a]**Ionic Radius in Picometers [15], CN = Coordination Number,** [b,c] **in Water as Hydrated Ions [16, 17].**

Metal	Ionic Radius[a] (CN)	pK_a[b]	$k_{exchange}$ (Water)[c], s[-1]
Cu(II)	57 (4)	7.53	2×10^8
	65 (5)		
	73 (6)		
Ga(III)	62 (6)	2.6	7.6×10^2
Y(III)	102 (8)	7.7	1.3×10^7
	108 (9)		
Zr(IV)	72 (6)	0.22	-
	84 (8)		
	89 (9)		

of a metal ion can drastically change its coordination properties with a chelator, as evidenced by the activity of bacterial iron binding siderophores such as enterobactin and desferal [14]. Bacteria excrete siderophores to bind iron(III) from the surrounding environment with exceptionally high binding affinity and then release the metal when inside the bacteria via reduction to iron(II) [14]. Copper is known to be redox active and has potential to be reduced and subsequently decomplexed *in vivo*. As described by the hard-soft acid-base theory (HSAB), copper(II) is a borderline soft metal ion, whereas gallium(III), yttrium(III), and particularly zirconium(IV) are very hard acidic metal ions and prone to hydrolysis and hydroxide formation (Table 5.2).

Under less acidic conditions the metal ions Ga(III) and Zr(IV) will readily form insoluble hydroxide species that do not coordinate with chelators. When these acidic metal ions are complexed by strong chelators, the formation of hydroxides is retarded and complexes can remain stable over a wide pH range (some even in 3-6 M HCl). Cu(II) complexes are often unstable with respect to metal ion decomplexation, because Cu(II) has a d^9 electronic configuration and is prone to distortions (e.g., axial Jahn-Teller distortions in six coordinate octahedral complexes). Cu(II) has a very high water exchange rate of 2×10^8 s[-1] (Table 5.2) and shows a high degree of lability [15–17]. These inherent electronic instabilities toward redox chemistry and geometric distortions in Cu(II) complexes require caution and extra vigilance in chelator selection. If a chelator-metal ion complex can be crystallised, its solid-state structure can be obtained by X-ray crystallography in order to study its geometry and bonding. Solid-state structures provide important information on the coordination number, geometry, and specific donor atoms used in a complex; however, the solid-state structure is often not representative of the solution-phase behaviour of a complex [18, 19].

5.2 RADIOPHARMACEUTICAL DESIGN

The modular design of BFC systems allows for a theoretically limitless number of biovectors (peptides, nucleotides, antibodies, nanoparticles, etc.) to be conjugated, providing site-specific molecular targeting to a constantly expanding number of disease states and biochemical processes. The radiometal-chelate complex has a large impact on the pharmacokinetics of the BFC-biovector conjugate, with many radiometal complexes being very hydrophilic and providing rapid renal excretion [1–6]. BFCs have an attractive level of modularity built into their design, because they contain several core modules that can be swapped (with varying degrees of difficulty) to provide vastly different applications, yielding a rich set of tools for nuclear medicine practitioners. These modules are discussed as: (1) the *radiometal*, which can be changed to tune the radiation type (γ for SPECT, β^+ for PET, and β^-, α, or Auger electrons for therapy); (2) the *chelator*, which can be picked to optimise binding and stability with a chosen radiometal; (3) the *bioconjugation site*, which can be changed to various functionalisable handles for different types of linkages, including spacer groups that can be added to ensure minimal impact on receptor targeting (often short polymers or peptides); and (4) the *biovectors*, which can be chosen as molecular targeting groups for nearly any biochemical process or tissue type (Figure 5.1). As an example of BFC modularity, theranostics are radiopharmaceutical agents that share the same molecular scaffold (same chelator and biovector), but provide diagnostic information when using an appropriate SPECT/PET isotope such as ^{68}Ga/^{111}In/^{89}Zr/^{86}Y and provide a therapeutic effect when using β^- emitters such as ^{90}Y/^{177}Lu [20].

Although each radiometal has different preferences for ligand donor atoms (N, O, S), coordination number, and complex geometry, there are many key design considerations that can be applied universally [21]. Ligand synthesis should be relatively simple and avoid stereoisomers and non-enantio/diastereospecific reactions, because different isomers can often

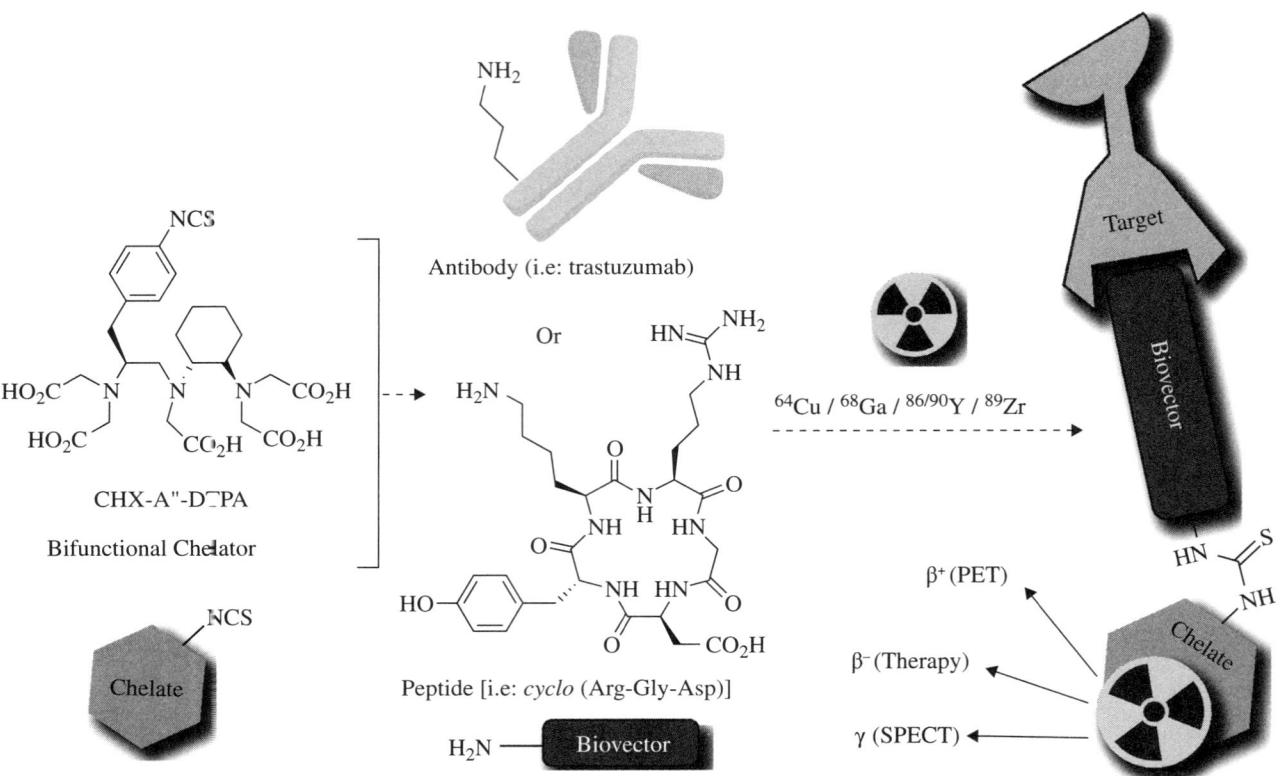

FIGURE 5.1   Illustration of a BFC-based radiopharmaceutical agent conjugated to a biological targeting group (biovector).

exhibit different biodistribution profiles and binding activity. Effort should be channelled into designing a chelate system that is modular and easily modified for different linker types so that different bioconjugation techniques can be used. The biodistribution profile of a compound and its interaction with a conjugated biovector can be tuned by changing the polarity and charge of the chelate (the degree of polarity can be assessed by octanol-water partition coefficients (log P)), thus a modular design containing easily modified synthons is important for optimising these properties. It is most common to conjugate a BFC to a biovector prior to radiolabelling with a radiometal, which is an ideal method for clinic-ready kit formulations. There are many excellent and extensive review articles available that elaborate on different aspects of these inorganic PET isotopes and their incorporation into radiopharmaceuticals, as well as many of the other concepts discussed in this chapter [12, 18, 19, 21–33].

5.3   RADIOPHARMACEUTICAL STABILITY

The purpose of a bifunctional chelating agent is to sequester a radiometal ion with such high thermodynamic stability and kinetic inertness that it is not released through any mechanism under physiological conditions (decomplexation, acid-catalysed dissociation, trans-chelation by serum proteins, protonation, trans-chelation by bone, complex adsorption to bone, hydrolysis, etc.). The result of radiometal loss from a radiopharmaceutical *in vivo* is non-targeted distribution of the radiometal to various compartments in the body, depending on the biological interactions of the specific radiometal. Each metal ion has its own unique properties to contend with when attempting to incorporate it into an imaging/therapeutic agent, such as its aqueous chemistry, redox chemistry, and affinity for native biological chelators.

When evaluating chelators for use in radiopharmaceuticals, kinetic inertness *in vivo* is a much more important factor than absolute thermodynamic stability of the metal-chelate complex (Figure 5.2). Thermodynamic stability constants ($K_{ML} = [ML]/[M][L]$) can be calculated from experiments such as potentiometric and/or spectrophotometric titrations. These thermodynamic stability values can be useful as preliminary comparisons of the efficiencies of various chelators for a particular metal ion [17, 34]. Stability constants give a number for the direction and magnitude of the equilibrium in a metal-chelate coordination reaction; however, they give no rate information, thus the kinetics of dissociation must be probed through other methods (Figure 5.2) [35, 36].

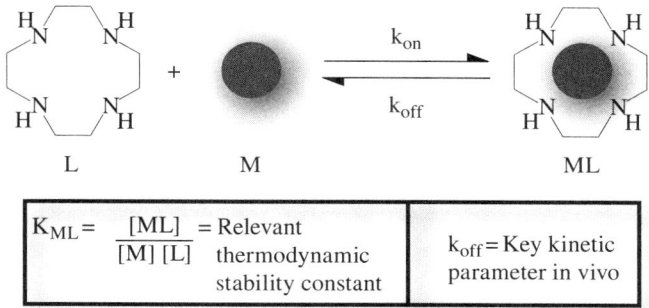

$$K_{ML} = \frac{[ML]}{[M][L]} = \text{Relevant thermodynamic stability constant}$$

$$k_{off} = \text{Key kinetic parameter in vivo}$$

FIGURE 5.2 Cartoon representation of basic thermodynamic and kinetic properties of metal-chelate complexes, with k_{off} being the most relevant measure for kinetic inertness *in vivo*.

This is a very important factor because the rate of dissociation *in vivo* is what governs the kinetic inertness of a radiometal complex and therefore the ultimate suitability of a metal-chelate complex for medical applications. Another important thermodynamic parameter, perhaps even more important than K_{ML} values, is the pM value (-log[M]$_{Free}$) [37, 38]. The pM value is the negative log of the concentration of free metal ion left uncomplexed by a given chelator. It is calculated under a specific set of conditions. The pM value takes into account several factors that are not included in the K_{ML} value, such as ligand basicity, metal ion hydrolysis, and various other parameters, but it is condition-dependent. When a radiopharmaceutical is introduced into biological circulation, the extreme dilution may dramatically increase the rate of dissociation, despite a very high thermodynamic stability (K_{ML}, pM). One common method employed to access this rate information is acid dissociation experiments, typically performed in HCl/DCl solutions (often 5-6 M) and monitored via ultraviolet-visible spectroscopy (UV-Vis) or by nuclear magnetic resonance (NMR) in acidified D_2O over extended periods of time [35, 36]. A comprehensive table of K_{ML} and pM values has not been included, because it has been extensively discussed elsewhere [18, 37, 38]. Although K_{ML} and pM values are very useful, *in vivo* behaviour is ultimately not well predicted using these parameters.

Macrocyclic chelators are generally more kinetically inert than acyclic chelators, even when their thermodynamic stabilities have been determined to be very similar [39–43]. Macrocyclic chelators possess constrained geometries and partially pre-organised metal ion binding sites, which decrease the entropic penalty paid on metal ion coordination. When a metal ion is bound, it forces a rigid and constrained geometry on an otherwise freely moving acyclic chelator, so there is a significant decrease in entropy. The already constrained and pre-organised structures of macrocycles result in a less significant decrease in entropy on binding, therefore adding a powerful thermodynamic driving force toward complex formation. This phenomenon is referred to as the 'macrocycle effect,' and is an energetic incentive that macrocycles have over acyclic chelators, which is in addition to the chelate effect from which they both benefit [44]. Acyclic chelators, however, have much faster coordination kinetics than do macrocycles, and when properly designed are often able to quantitatively coordinate a radiometal in ~10 minutes at room temperature, whereas macrocycles often require heating above 80 °C and longer reaction times (30-90 minutes) [45–47]. Fast room temperature labelling becomes a crucial property when working with heat-sensitive molecules such as antibodies and their derivatives, or when working with short half-life isotopes such as ^{68}Ga and ^{62}Cu.

Without adequate complex stability, radioactive metal ions may be absconded by blood serum proteins such as transferrin or serum albumin, decreasing selectivity and contrast from the intended agent, and in extreme cases rendering it useless. Without adequate radiopharmaceutical uptake, target specificity, and stability, background activity in non-target tissue will be high and will complicate the collection and interpretation of imaging data, such as delineation of cancerous lesions. Many different *in vitro* stability assays can be performed to estimate *in vivo* stability; however, the ultimate determinant for the usefulness of metal-chelate complexes in biological systems is obtained from *in vivo* analysis by biodistribution and/or imaging studies. Each metal ion will have its own unique biological fate when freed from a radiopharmaceutical *in vivo*, where they are typically distributed and managed by transport proteins and other biological systems (Figure 5.3).

Translation from test tubes into living organisms presents a multitude of complications. The human body contains many endogenous ligands (naturally produced chelators/transport proteins/enzymes, etc.) that play vital roles in the metabolism and homeostasis of metal ions. When evaluating a new radiopharmaceutical agent *in vivo*, the stability can be monitored by watching for abnormal accumulation of the free nuclide in various organs (where each radiometal ion has been previously determined to accumulate). Biodistribution studies of radiometal-chelate complexes can be used to evaluate nonspecific organ uptake (bone, kidneys, liver, lungs etc.) and ensure that the radiometal remains chelator bound and is excreted intact (typically through the urine and/or faeces). Another negative consequence of radiometal loss *in vivo* is target receptor saturation (Figure 5.3). The concentration of cell-bound receptors in tumours is low, and if there is a significant concentration of

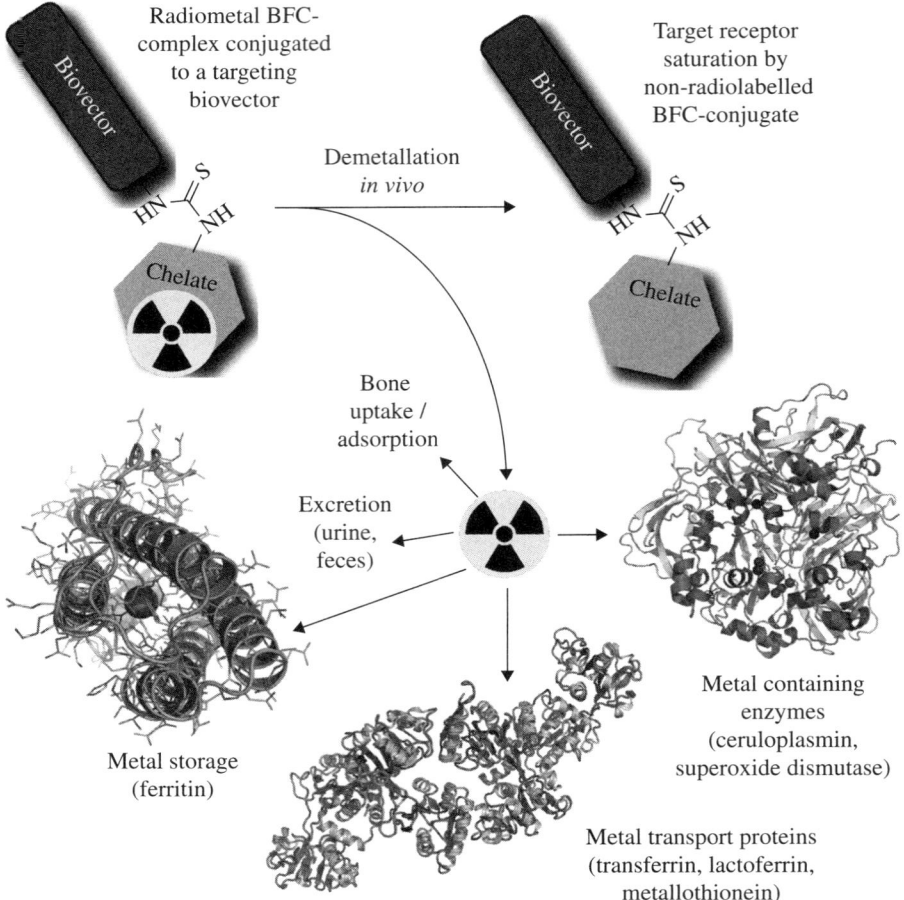

FIGURE 5.3 Some of the possible biological fates and consequences of BFC radiodemetallation *in vivo* (solid-state structures of ferritin H-chain homopolymer PDB file 1FHA, ceruloplasmin PDB file 2J5W, and *apo*-transferrin PDB file 2HAV shown).

demetallated radiopharmaceutical in circulation (now non-radioactive), the target receptors can become saturated and then block any further binding by the circulating population that is still radioactive [23]. The radiopharmaceutical must therefore have a high apparent specific activity (high quantity of radioactivity per mass unit of compound) and be remarkably stable and inert. Small molecule agents containing covalently bound radiohalogens such as [124]I are usually challenged by enzymatic degradation, and once metabolised, the radioiodine is either excreted/accumulated as a metabolite of the parent radiopharmaceutical or as free radioiodide (which typically localises in the thyroid and stomach).

5.4   [86]YTTRIUM RADIOMETAL ION PROPERTIES

Yttrium is the largest of the metals discussed in this chapter and typically forms eight or nine coordinate chelate complexes, with the most common geometries being square antiprismatic and monocapped square antiprismatic. Y(III) is the least acidic metal ion discussed in this section (pKa = 7.7) and, with a high affinity for hydroxyl ions, above pH 3 has a tendency to form insoluble $Y(OH)_3$ species. Yttrium is redox stable in the 3+ oxidation state and is a hard metal ion with a preference for hard donors such as carboxylate-oxygens and amine-nitrogens. [90]Y is a β^- emitter used for therapy, and although it is possible to perform biodistribution, imaging, and dosimetry studies with its bremsstrahlung X-rays, the spatial resolution obtained is poor [48]. [90]Y also emits a very low abundance of positrons (0.003%), which can be used to collect PET imaging data that are more accurate than those from ^{90}Y bremsstrahlung imaging [48]. Because ^{90}Y has no significant γ or β^+ emissions, radio-pharmaceuticals that incorporate it require a 'matched isotope pair' to be used as a surrogate for imaging and dosimetry. Yttrium has been most commonly investigated as its therapeutic isotope ^{90}Y; however, [86]Y offers an attractive PET imaging isotope that shares identical chemistry and therefore can be seamlessly substituted for ^{90}Y into existing radiopharmaceutical preparations as a 'matched isotope pair' [1–6]. [86]Y is a cyclotron-produced isotope that has a significant positron abundance of 33% and a half-life of 14.7 hours (Table 5.1) that in comparison to a 64.2-hour half-life for 90Y is relatively short. 111It has a closer matched half-life of 67.2 hours and is most typically used as a SPECT imaging surrogate ('matched isotope pair') for performing dosimetry with 90Y-based therapeutics. Despite this mismatch in half-life, 86Y is chemically identical to 90Y and is therefore a biologically equivalent surrogate, in contrast to 111In, which can have completely different properties in its complexes and their biodistributions [1–6]. 86Y is also a PET isotope, which is generally preferred to SPECT in terms of spatial resolution and quantitative dosimetry [48]. 86Y has a relatively high-energy positron emission of 1221 keV, and along with its low branching ratio of 33% and high-energy γ emissions, it is less ideal for PET imaging than are other long-lived alternatives such as 89Zr and 64Cu, with their more favourable emission profiles. 86Y is typically cyclotron-produced via the nuclear reaction 86Sr(p,n)86Y, and purified by ion exchange chromatography or electrolysis [49, 50].

5.4.1 Clinical Trials Based on 86Yttrium

The long half-lives of 86/90Y match well with the long biological half-lives of antibodies, which take two to three days to fully penetrate tumours and can remain circulating *in vivo* for weeks [26]. Antibodies are typically metabolised in the liver, which will subject conjugated radiometal-chelate complexes to harsh conditions [26]. Antibody conjugates with long circulation times and prominent hepatobiliary metabolism make the need for high thermodynamic and kinetic stability even more important than for conjugates of relatively fast clearing small biomolecules such as peptides (i.e., RGD, octreotide) [26]. Unlike magnetic resonance imaging (MRI), contrast agents (Chapter 8) that utilise an open coordination site for water exchange, it is best that the coordination sphere of the quasi-lanthanide Y be fully saturated for maximum stability when used in radiopharmaceuticals. There is currently one Y-based radiopharmaceutical FDA-approved for clinical use, 90Y-ibritumomab tiuxetan, trade name Zevalin® [51]. 90Y-Zevalin is a radioimmunotherapeutic (RIT) for non-Hodgkins lymphoma that utilises the BFC tiuxetan conjugated to the antibody ibritumomab [51]. Tiuxetan is a modified version of the chelator DTPA, whose carbon backbone contains an isothiocyanatobenzyl and a methyl group (Figure 5.4). Zevalin is also approved for use with 111In for imaging and dosimetry applications [52]. Zevalin could be used with 86Y for PET imaging/dosimetry, because different isotopes of the same element have identical chemical properties, but 86Y-Zevalin is currently not approved [1–6]. The success of Zevalin in the clinic for treating non-Hodgkins lymphoma is well established [52]. Clinical trials are currently underway to investigate the co-administration of Zevalin with the non-radiolabelled antibody rituximab to potentially enhance treatment success [53].

There are several Y-based radiopharmaceuticals in clinical trials; however, none of them utilise the positron-emitting isotope 86Y. As previously mentioned, the chemistry of 86Y and 90Y are identical (with the exception of possible differences in specific activity and concentration), so all 90Y-based therapeutics could potentially be used for 86Y PET imaging agents, should the properties of a given radiopharmaceutical agent warrant its use in PET imaging (i.e., for dosimetry as a theranostic pair) [1–6]. To this end, 90Y-based radiopharmaceutical agents that are currently in clinical trials will be discussed. Although BFC systems are most common, nanoparticles are gaining traction in the literature, and glass microspheres (i.e., TheraSpheres) that incorporate 90Y are currently in clinical trials [54]. They are injected into arteries that flow into liver tumours, where they become lodged in capillaries and bombard liver cancer cells with high doses of localised radiation [54].

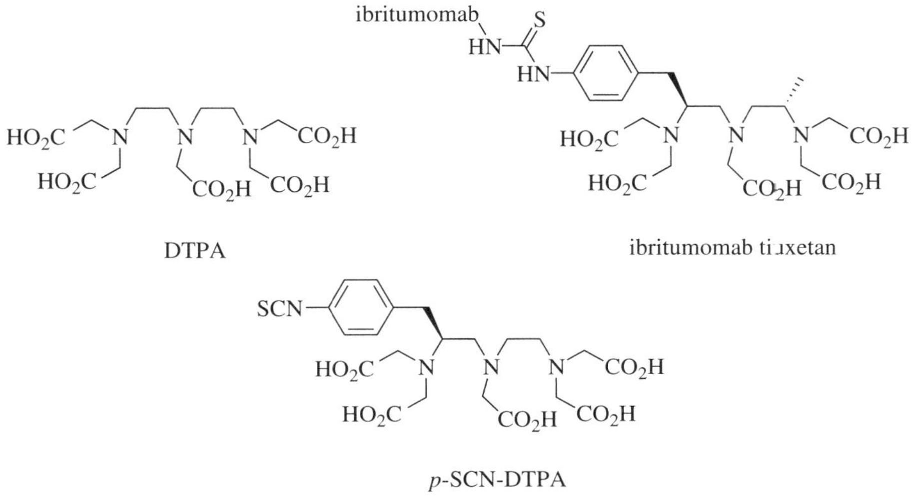

FIGURE 5.4 Commonly used Y(III) chelator DTPA, the BFC precursor *p*-SCN-DTPA, and the FDA-approved BFC system ibritumomab tiuxetan (trade name Zevalin when used with 90Y).

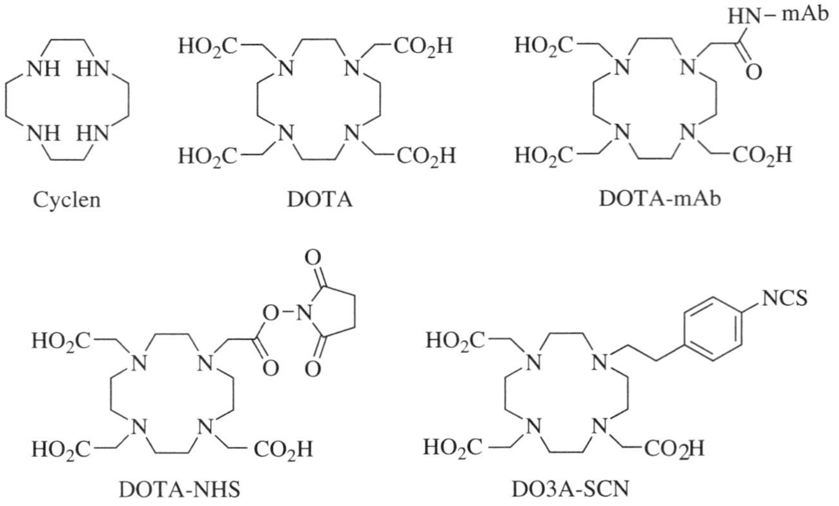

FIGURE 5.5 Chemical structure of the cyclen-based chelator DOTA, and the commonly used N-bifunctional DOTA derivatives DOTA-mAb (where mAb = monoclonal antibody), DOTA-NHS (N-hydroxysuccinimide ester), and DO3A-SCN.

These microspheres may not be appropriate for ^{86}Y PET imaging because their mode of targeting is nonspecific; however, ^{86}Y may be a useful surrogate for dosimetry of these microspheres to assess radioactive dose to non-target organs and confirm tumour localisation. A variety of theranostic nanoparticles, found in the literature, incorporate combinations of MRI contrast agents, PET/SPECT nuclides, fluorescent markers, and biovectors [55].

Of more relevance to ^{86}Y PET imaging are bifunctional chelating agents such as ^{90}Y-DOTA, which have been conjugated to various biovectors (Figure 5.5).

^{90}Y-DOTA has been conjugated to the anti-carcinoembryonic antigen (CEA) antibody M5A and is in clinical trials for the relatively nonspecific treatment of a variety of tumours (including colon, breast, lung, and medullar thyroid) [53]. The ^{90}Y-DOTA BFC system has been in clinical trials for use with many other antibodies as well, including chimeric monoclonal antibody cG250 for treatment of renal cancer, anti-CD45 monoclonal antibody BC8 for acute myeloid leukaemia, anti-CD22 monoclonal antibody LL2 (epratuzumab) for lymphoma or leukaemia, monoclonal antibody CC49 for ovarian cancer, humanised MN-14 (anti-CEA) antibody for relapsed or refractory small cell lung cancer, and humanised monoclonal antibody BrE-3 for metastatic breast cancer [53]. A bioconjugate containing the cyclic peptide octreotide, ^{90}Y-DOTA-TOC, is in clinical trials for treatment of neuroendocrine tumours (also being used with ^{111}In and ^{68}Ga for imaging/dosimetry) [53, 56]. ^{86}Y might be useful as a surrogate for ^{90}Y in all of these agents for accurate pre-therapy imaging and dosimetry. Some specific examples of ^{86}Y-antibody conjugates that have been recently reported but are not in clinical trials are CHX-A″-DTPA conjugated to cetuximab panitumumab for imaging human malignant mesothelioma tumours [1, 57], and bevacizumab for imaging of tumour angiogenesis [2].

These examples of ^{90}Y-based BFC systems in clinical trials should emphasise the importance of a strong and well-developed chelate system as the foundation of a radiometal-based radiopharmaceutical, because it can then be conjugated to a variety of biovectors and applied to a large number of imaging and treatment modalities. Once a suitable chelate system is matched with a metal ion and found to have favourable coordination and stability properties, it can then be conjugated to a theoretically limitless number of biovectors. A quick browse of the current clinical trial offerings [53] demonstrates this quite clearly, because a majority of investigated agents are based on the chelators DTPA and DOTA. The chelators DTPA and DOTA are ubiquitous in radiometal and coordination chemistry and have been used to complex many metal ions, such as indium, gallium, copper, zirconium, and lutetium. The versatile metal ion binding properties of these chelators means that radiopharmaceutical preparations that use them can be evaluated with many different nuclides, although not all metal ions bind with the same affinity and stability.

5.4.2   Recent ^{86}Yttrium Work

As evidenced from the popularity of DOTA, the high thermodynamic stability and kinetic inertness of macrocyclic chelators is ideal for *in vivo* radiopharmaceutical applications (they benefit from the macrocycle effect); however, sensitive biological molecules such as antibodies (and their derivatives, such as affibodies, diabodies, nanobodies) are not ideally compatible with the high labelling temperatures typically required for macrocycles [58, 59]. DTPA is an acyclic chelator with much

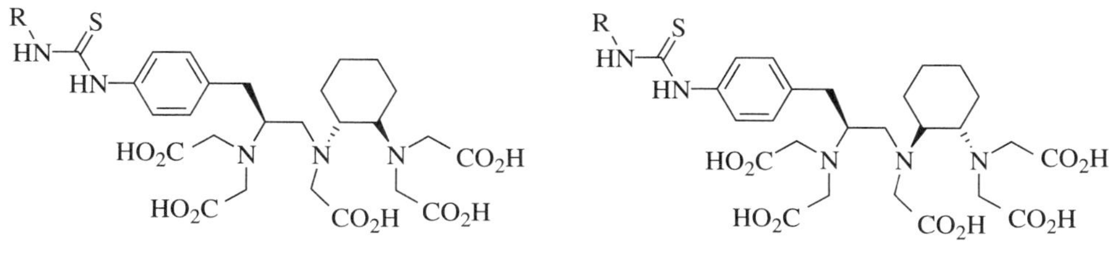

FIGURE 5.6 Isomers of the modified DTPA chelator CHX-DTPA, with CHX-A″-DTPA being the most stable isomer and having improved kinetic inertness *in vivo* over DTPA (**R** = biovector) [42].

faster reaction kinetics that can label some radiometals quantitatively at room temperature over a short period of time; however, it is not nearly as stable *in vivo* as its macrocyclic counterparts [42]. This shortcoming could be improved through design of novel acyclic chelators or by modification of well-known acyclic chelators such as the DTPA derivative CHX-A″-DTPA (Figure 5.6) [42].

Although DTPA and DOTA are the most commonly investigated chelate systems for Y(III), the chelator CHX-A″-DTPA shows significantly improved stability versus DTPA, but it still may not be as stable as DOTA [42]. The cyclohexyl backbone of CHX-A″-DTPA makes the chelator more rigid and imposes a degree of preorganisation on the metal ion binding site [42]. Conjugation through functionalisation in the chelate backbone (Figure 5.6) does not interfere with the metal ion coordination sphere and has been shown to yield much more stable complexes *in vitro* and *in vivo* than does amide conjugation to one of the carboxylate arms [42, 60]. Due to the success of antibody biovectors, developing acyclic chelators with fast metal ion coordination kinetics, as well as high stability and kinetic inertness comparable to that of macrocycles like DOTA, would be ideal.

An important lesson learned from the example of CHX-A″-DTPA (Figure 5.6) is that the CHX-B″-DTPA isomer is much less stable *in vivo* with Y(III), although *in vitro* stability assays suggested they were identical [42, 61]. If CHX-DTPA is used as a racemic mixture it would result in significant decomposition and radio-demetallation *in vivo* [42, 61]. The metal-chelate portion of a radiopharmaceutical ideally does not interact with the target receptor, and often there is a spacer group placed between the biovector and the metal-chelate complex to minimise any interference. Considering this, one might assume that having different isomers of the metal-chelate complex should not affect receptor binding or biodistribution, because it is not the radiometal complex that interacts with the target receptor. This proposition assumes that there is sufficient space between the radiometal complex and biovector and that the chemical properties such as coordination number, geometry, and polarity are the same between isomers. This is an oversimplified view and although it appears to be true according to many *in vitro* assays, it turns out to be inaccurate *in vivo*, as demonstrated by the different *in vivo* stabilities of CHX-A″-DTPA versus CHX-B″-DTPA [42, 61]. Multiple isomers of a BFC-metal complex can exhibit different pharmacokinetic properties, potentially leading to differential rates of decomplexation depending on organ distribution (Figure 5.2) [45, 62]. These examples demonstrate why the stereochemistry of chelators and their metal ion complexes should be fully investigated. It should be noted that the studies performed above, comparing the different isomers of CHX-DTPA, were performed with isotopes of Y(III) [42, 61], thus the results may vary with different radiometals, although the same principles apply.

The site of bioconjugation on the BFC must be chosen carefully because it can compromise the metal ion coordination sphere. For example, the carboxylate arm functionalised DOTA derivatives interfere with metal ion coordination by blocking one of the coordinating carboxylate arms (although the amide carbonyl groups can still weakly coordinate, Figure 5.5) [45]. Several DOTA derivatives have been synthesised that alleviate this problem through carbon backbone and side-arm functionalisation, for example, DOTAGA, DOTASA, and various isothiocyanate derivatives (Figure 5.7) [45, 63, 64].

These DOTA-BFC derivatives retain full denticity (octadentate), as well as the same thermodynamic stability and kinetic inertness as DOTA [45, 63, 64]. Side-arm and backbone functionalisation provide sites for biovector conjugation that do not disturb radiometal coordination; however, as with DOTA, radiometallation conditions still require heating and therefore are not ideal for antibody conjugations [58, 59]. The rigid pre-organised metal ion binding sites in macrocycles provide very favourable bonding interactions (macrocycle effect) [44], and often result in fewer isomeric species being present in solution [63]. The chelator must be carefully matched with the specific radiometal; as an example of chelate-radiometal specificity, DOTA is the most kinetically inert chelator for Y(III); [42] however, the trend is reversed for the similarly sized metal ion Lu(III), where DOTA is actually found to be less inert to acid decomplexation than is the acyclic chelator CHX-A″-DTPA [40]. This example stresses the importance of carefully matching the chelator with the metal ion and that stability trends evaluated with a specific metal ion and a set of chelators are not universal and often do not translate to other metal ions.

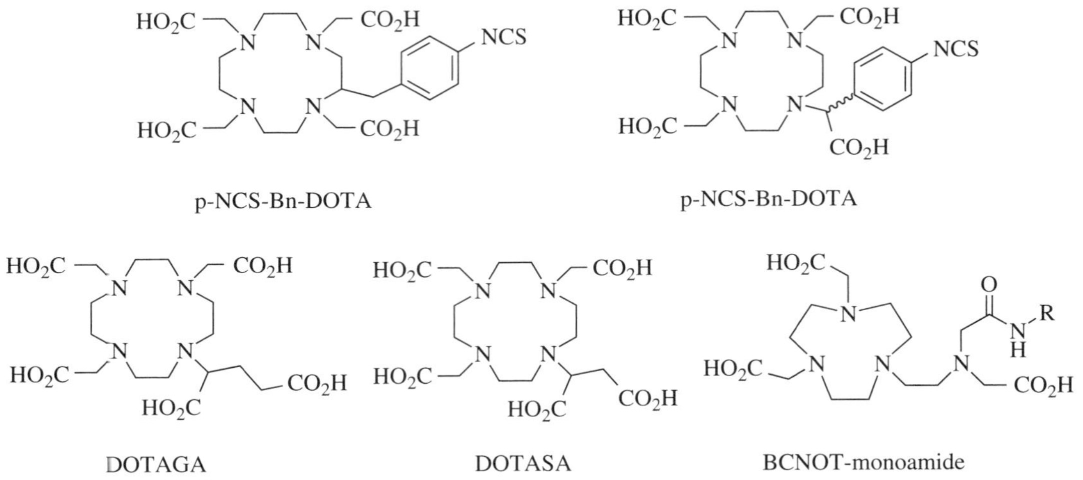

FIGURE 5.7 Bifunctional DOTA derivatives that retain full octadentate coordination, and the novel NOTA-BFC BCNOT-monoamide (NETA) that expands its coordination sphere from 6 to 7 coordinate (8 including amide oxygen, **R** = biovector) [45, 63–65].

Every metal ion will coordinate with a given ligand differently, and therefore can form different isomers in solution. These isomers often have a temperature-dependent fluxional behaviour, which can be studied by variable temperature (VT) NMR [63, 65]. VT NMR experiments have been performed with In(III) and Y(III) complexes of DOTA [63, 65], which demonstrate this fluxional behaviour in solution, as well as the difference in coordination environment between the two metal complexes. Chelators that exhibit the lowest degree of isomerism with a given metal ion are generally preferred, because they tend to be more inert [45, 62]. The modified NOTA chelator BCNOTA (NETA, BCNOT-monoamide as the BFC derivative, Figure 5.7) is promising for use with Y(III), because it has been reported to have labelling kinetics as fast as acyclic chelators, as well as a high degree of stability and rigidity imparted by the macrocyclic framework [66].

Because ^{90}Y in practice is strictly a β⁻ emitter, it must be used in conjunction with a SPECT or PET isotope such as ^{111}In, ^{86}Y, or ^{89}Zr in order to perform imaging studies and collect quantitative location data (typically combined with CT/MRI data). This provides crucial information on organ uptake kinetics and radiochemical complex stability, identifies dose-limiting organs (often kidneys, bone, or liver depending on mode of excretion and metabolism), and allows for dose estimates to various organs to be calculated (dosimetry) [67]. This allows clinicians to calculate the amount of ^{90}Y to administer to a patient while minimising toxicity to the bone marrow and kidney, often the most problematic and radiosensitive organs to ^{90}Y [67]. The combined use of two isotopes for imaging and therapy is referred to as a matched isotope pair, or a theranostic agent [20]. The goal of internal radiation therapy is to maximise the amount of radioactive dose delivered to a biological target (i.e., tumours), minimise localisation in non-target tissue, spare organs from damage, and have fast blood clearance and excretion of any non-target-bound radiopharmaceutical. This highlights the importance of accurate dosimetry, which relies on a close match in chemical and biological properties in a BFC-based radiopharmaceutical when coordinated with two different isotopes [1–6]. Problems arise here because, as previously discussed, different metal ions have different preferences for coordination number, geometry, and donor types, and often exhibit different behaviour *in vivo* [1–6]. These differences can result in a lack of bioequivalence, which causes different degrees of organ uptake and decreases the accuracy and predictive power of dosimetry techniques [1–6]. For these reasons, the matched isotope pair of ^{86}Y and ^{90}Y is ideal, because they form chemically identical chelate complexes [1–6].

When chelate complexes are coupled to peptides, they can significantly affect biodistribution and receptor binding, but when looking at large biovectors such as antibodies, the change is relatively small, because the antibody is massive (~150 kDa for an intact antibody) relative to the chelate complex [68]. Work performed with a 1B4M-DTPA antiTac monoclonal antibody conjugate demonstrated that the differences in biodistribution between ^{111}In and ^{90}Y complexes was between 10-15%, with bone marrow uptake being underestimated by ^{111}In biodistribution data [68]. The story changes when working with smaller peptide biovectors (bombesin, RGD, octreotide, etc.); in these cases, different metal ions and coordination modes can impart significantly different properties to the complex and result in larger differences in tissue distributions and receptor binding affinities [67, 69].

Using a matched isotope pair such as ^{111}In (γ) and ^{90}Y (β⁻) is not ideal if the two complexes are not sufficiently similar in their coordination spheres and properties. The availability of ^{86}Y as a positron emitter allows for high-resolution PET imaging and dosimetry, while retaining identical chemistry to ^{90}Y and providing a true matched isotope pair [1–6]. PET imaging and dosimetry performed with ^{86}Y-CHX-A''-DTPA-trastuzumab as a surrogate for ^{90}Y therapy has demonstrated its superior accuracy and effectiveness [3]. Studies comparing the accuracy of dosimetry performed with [86]Y/^{111}In-DOTA-TOC and ^{111}In-DTPA-octreotide as surrogates for ^{90}Y-DOTA-TOC therapy revealed that ^{111}In-DTPA-octreotide and ^{111}In-DOTA-TOC had significantly different organ uptake when compared to the analogous [86/90]Y complexes [67, 69]. The chemically identical option [86]Y-DOTA-TOC was the most suitable, with [86]Y being an ideal surrogate for [90]Y [67, 69]. One minor shortcoming of [86]Y is that the half-life of 14.7 hours, although long enough for most applications, does not enable image acquisition beyond 48-72 hours (unlike ^{111}In, $t_{1/2} = 67.9$ h). Another potential problem with [86]Y is production and limited availability. PET imaging is generally considered to be superior to SPECT; however, PET cameras are not as clinically ubiquitous as are SPECT cameras. If equipment is available and half-lives are appropriately matched, PET imaging with isotopes such as [124]I, ^{68}Ga, [86]Y, and [89]Zr would be preferred to SPECT imaging with isotopes such as [123/125]I or [111]In. The principle that differences in bioequivalence are observed for chelators containing different radiometals and/or different chelate isomers is universally applicable to all of the radiometals discussed in this section (^{64}Cu/^{68}Ga/[89]Zr).

5.4.3 Stability of [86]Yttrium-Based Radiopharmaceuticals

The metal ions Ga(III) and Fe(III) are bound with very high affinity by the blood serum protein transferrin; however, larger cationic metal ions such as Y(III) and the lanthanides are not bound as strongly [18, 38, 70–74]. The larger lanthanides such as neodymium and praseodymium are bound even more poorly and are often not able to occupy both binding sites of transferrin [71]. The metal ions yttrium(III), lutetium(III), and gadolinium(III) have all been shown to bind to transferrin, although not as strongly as do iron(III), gallium(III), and to a lesser extent indium(III) [18, 38, 70–74]. One explanation for the weaker binding of the lanthanides to transferrin is based on their large size being a poor fit into the binding sites, with the smaller charge-to-radius ratio and use of 4f orbitals (weaker interaction than 3d orbitals of iron) decreasing binding affinity [71]. Following this, it has been suggested that the main reason for the poor binding of large metal ions to transferrin is strictly a result of steric repulsion of the more crowded C-terminus binding site [72]. A convincing argument has also been made that transferrin binding strength is better related to metal ion acidity than size [75–77]. This has been demonstrated with the large but very acidic metal ion bismuth(III), which has an abnormally high binding affinity for transferrin despite its size (103 pm, log K_1 = 19.4, and log K_2 = 18.5) [75–77]. Both arguments support the prediction of a low binding affinity of Y(III) for transferrin.

The transferrin stability constants for binding one or two Lu(III) ions have been determined to be log K_1^* = 11.08 and log K_2^* = 7.93, which can be seen to be many orders of magnitude smaller than the binding affinity for Ga(III) and Fe(III) [72]. The transferrin stability constant for the rare earth metal ion Y(III) has not been reported to our knowledge; however, based on the arguments above, it can be reasonably estimated to be low, similar to that for lutetium. In light of the moderate stability binding of the lanthanides to transferrin, the stability of Y(III) and lanthanide complexes would be best evaluated by blood serum incubation or *in vivo* biodistribution studies, so that complexes would be challenged by a broad spectrum of endogenous *in vivo* competition. Yttrium and the lanthanides have especially high uptake in bone, and free ^{90}Y injected into a human is 50% associated with bone, with the next highest organ uptake being 25% in the liver [78]. There is also evidence that intact cationic lanthanide complexes can adsorb onto the surface of bone, demonstrating that even chelate-bound radiometals may accumulate in bone under certain circumstances [79]. In light of the high affinity of Y(III) complexes for bone, stability assays should be performed that take this into consideration (e.g., biodistribution, hydroxyapatite competition).

5.4.4 [86]Yttrium Radiometallation Protocols

Y(III) is a very hard metal ion, and due to its lower acidity (pKa = 7.7), it is less prone than Zr(IV), Ga(III), and In(III) to forming insoluble hydroxides at neutral pH. General radiometallation protocol follows that once a solution of [86]Y is procured (usually as an acidic solution in 0.1 M HCl or nitric acid), the desired quantity of activity is transferred to a buffered solution containing a chelator or a BFC-biovector conjugate, allowed to incubate until quantitatively radiometallated, and finally purified before administration [3, 42, 66, 80–82]. Acyclic chelators such as CHX-A"-DTPA are typically incubated for 20-60 minutes at room temperature, but are sometimes heated to improve labelling efficiency with certain biovectors. Macrocyclic chelators such as DOTA typically require temperatures in the range of 70-100 °C for 15-60 minutes. [90]Y labelling techniques are identical to those for [86]Y, with the exception that the high-energy β⁻ emissions of ^{90}Y can cause a higher degree of biomolecule radiolysis (bond cleavage leading to decomposition), depending on the amount of activity and the dilution and buffer used [83]. Another consideration is that the specific activity and concentration of radiometals may vary between different isotopes, because they are often made through different production routes. A majority of the concepts in radiometal BFC design that have been discussed for Y(III) are applicable to the other radiometals discussed in this chapter (^{68}Ga/^{64}Cu/[89]Zr).

5.4.5 ^{86}Yttrium Summary

Due to the ideal therapeutic properties of ^{90}Y, a great deal of work has focused on its incorporation into cancer-treating radiopharmaceuticals. Having no significant β^+ or γ emissions, ^{90}Y-based agents require a matched isotope pair to be used together for imaging and dosimetry. ^{86}Y is an optimal candidate as a PET imaging surrogate for ^{90}Y because their chelate complexes have identical chemical and physiological properties. Because ^{86}Y and ^{90}Y form chemically identical chelate complexes, they are considered bioequivalent as demonstrated by ^{86}Y providing more accurate dosimetry data than other non-identical surrogates such as ^{111}In. Aside from use as a dosimetry surrogate for ^{90}Y, ^{86}Y may not be an ideal candidate for mainstream imaging applications as it has a high-energy β^+ emission that results in lower resolution images (although corrections can be applied) and high-energy γ emissions that expose patients to significant absorbed doses.

5.5 ^{68}GALLIUM RADIOMETAL ION PROPERTIES

^{68}Ga has a short half-life of 68 minutes, a high positron abundance of 90%, and robust coordination chemistry with many ligands. Ga(III) is an acidic group 13 metal ion (pKa = 2.6) that is redox stable in aqueous conditions. Ga(III) hydrolyses to insoluble $Ga(OH)_3$ between pH 3-7, has a very high affinity for hydroxide ions, and tends to deligate above pH 7 to form the soluble gallate anion $[Ga(OH)_4]^-$ [16]. Ga(III) has the slowest water exchange rate of all the metal ions discussed in this chapter and prefers to form complexes with coordination numbers of 4 to 6. The most promising chelating agents for gallium are hexadentate to maximise complex stability, typically adopting a distorted octahedral geometry. The hard character of the Ga(III) ion lends preference to hard donor atoms such as carboxylate-oxygens, amine-nitrogens, phenolate-oxygens, and hydroxamate-oxygens. The short 68-minute half-life of ^{68}Ga makes fast, low temperature radiometallation with BFC agents a high priority. ^{66}Ga is also a PET isotope of Ga(III); however, its high energy β^+ emission of 4150 keV and lower positron abundance (56%) leads to poor image resolution. ^{66}Ga also has a high-energy γ emission (4000 keV) that exposes patients to very high absorbed doses. These serious limitations, combined with the lack of a portable generator system and various production/purification problems have led to clinical applications of ^{66}Ga being extremely limited [18].

One negative property of ^{68}Ga is its relatively high positron energy (~1880 keV), which results in lower resolution images when compared to lower β^+ energy emitters such as ^{64}Cu and ^{89}Zr (656 and 897 keV, respectively). However, the positron abundance (favourable branching ratio) of 90% for ^{68}Ga is much higher than that of ^{64}Cu, ^{89}Zr, and ^{86}Y (19%, 23%, and 33%, respectively), which means that lower amounts of activity and/or shorter image acquisition times can be used. Another favourable nuclear decay property of ^{68}Ga is its lack of high-energy γ emissions (unlike ^{86}Y, ^{89}Zr, and ^{124}I), which allows for easier data collection and lower patient absorbed doses. ^{68}Ga is obtained through a ^{68}Ge-based generator system using a stationary-phase substrate such as Al_2O_3, CeO_2, SnO_2, TiO_2, or ZrO_2 [84]. ^{68}Ga(III) is typically eluted with a dilute HCl solution, while the ^{68}Ge(IV) is retained on the generator column [84]. No ^{68}Ge/^{68}Ga generator system is currently approved as having pharmaceutical-grade eluent, which limits translation of ^{68}Ga-based imaging agents into the clinic [85, 86]. The ^{68}Ge/^{68}Ga generator system provides a long shelf life of ~1 year and allows easy mobilisation to hospital radiopharmacies regardless of their proximity to cyclotron or nuclear reactor production sites. The easy distribution of ^{68}Ga through a long-lived generator system is one of the reasons that ^{68}Ga is thought by many people to be the most attractive radiometal discussed in this chapter. Many ^{68}Ga-based agents such as ^{68}Ga-DOTA-TOC, ^{68}Ga-DOTA-NOC, and ^{68}Ga-DOTA-TATE have been studied and have shown excellent promise for use in diagnostic nuclear medicine, but without an approved generator system they are limited in their widespread clinical application [87–89].

5.5.1 Clinical Trials Based on ^{68}Gallium

There are currently no FDA-approved radiopharmaceuticals that utilise ^{68}Ga; however, ^{67}Ga-citrate is an FDA-approved SPECT agent for imaging of various cancers and inflammatory lesions. Once injected into a patient, ^{67}Ga is easily transmetallated from the weak citrate chelate and becomes ~99% transferrin bound. ^{68}Ga-citrate is currently in clinical trials in the European Union for use in PET imaging of inflammatory infectious diseases [53]. There are several BFC systems in clinical trials in North America and the European Union that utilise ^{68}Ga [53]. ^{68}Ga-DOTA-TOC (DOTA-D-Phe1-Tyr-octreotide) is a labelled somatostatin analogue (octreotide) that is in clinical trials as a kit for imaging and dosimetry, to be used as a theranostic agent with the therapeutic isotopes ^{90}Y/^{177}Lu (with DOTA-TOC or DOTA-TATE, Figure 5.8) [53, 87–89].

DOTA-TOC utilises the cyclic peptide octreotide, which binds to somatostatin receptors that are overexpressed on neuroendocrine tumours [88, 90]. ^{68}Ga-DOTA-TOC PET has been shown to be even more effective at identifying cancerous lesions than are CT or MRI [91]. Recent work has also shown ^{68}Ga-DOTA-TOC to be more sensitive for detecting neuroendocrine tumours than the FDA-approved SPECT agent OctreoScan™ (^{111}In-DTPA-octreotide) [92, 93]. The octreotide-based

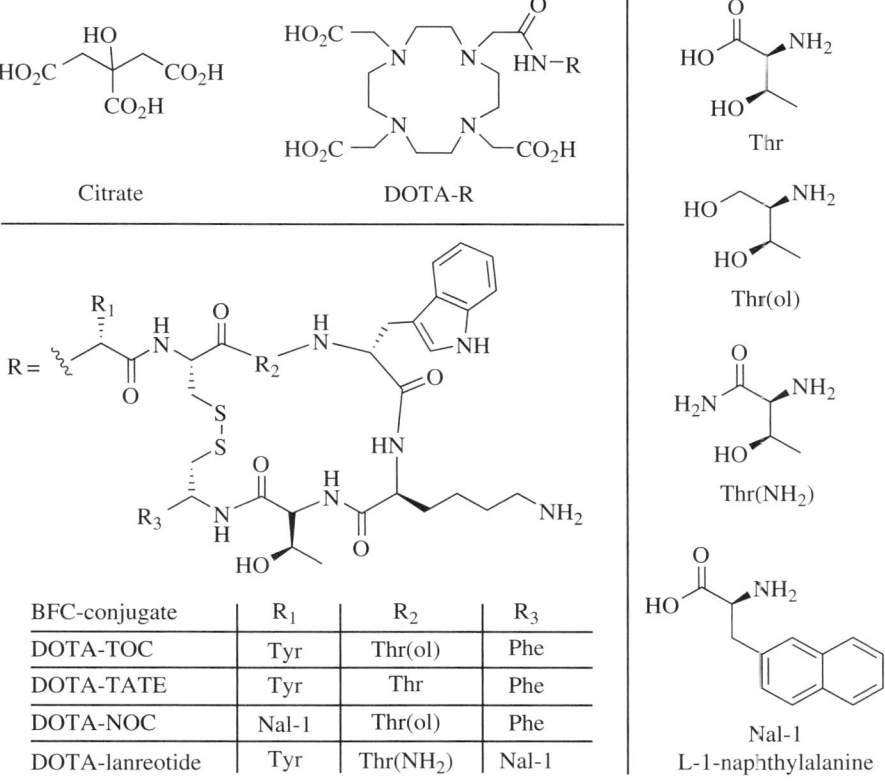

FIGURE 5.8 ^{67}Ga-citrate (FDA-approved for SPECT imaging purposes), and various DOTA-**R** octreotide-based BFC-conjugates that are currently being investigated with [67/68]Ga (**R** = octreotide derivative as defined above).

BFCs DOTA-TOC, DOTA-NOC, DOTA-TATE, and DOTA-lanreotide are examples of other ^{68}Ga-DOTA somatostatin-targeting peptide conjugates in clinical trials (Figure 5.8) [53, 88, 90, 94]. ^{68}Ga-DOTA-bombesin is another peptide BFC conjugate in clinical trials that is used for imaging prostate cancer [53, 95].

The short half-life of ^{68}Ga is not suitable for imaging with antibodies, due to their slow localisation (imaging with antibodies typically is performed 24-72 hours post injection) [26]. To overcome this limitation, small antibody fragments are used as biovectors with faster localisation times for use with short-lived isotopes. Affibodies are similar to F(ab')2 fragments (Fab = fragment antigen binding), diabodies, and minibodies, which are all fragments of antibodies, that due to their smaller size of ~6.5 kDa, are well below the ~70,000 kDa threshold for renal filtration [26]. They are able to quickly localise, bind their target receptors, and have the unbound population clear the blood through the kidneys, making imaging studies feasible with short half-life isotopes like [68]Ga [96, 97]. The ^{68}Ga-based agent ^{68}Ga-F(ab')2-trastuzumab (HERScan) is currently finishing clinical trials for breast cancer imaging [53, 98]. A similar example to HERScan is the ^{68}Ga-DOTA-$Z_{HER2:2891}$-affibody conjugate, which has also been successfully used for imaging HER2/neu positive tumours and targets the same breast cancer receptors as the full antibody Herceptin (trastuzumab), but localises in hours instead of days [96]. One universal flaw amongst all antibody fragments is exceptionally high kidney uptake [96, 99], which greatly limits the possible injected dose of activity due to radiation burden on the kidneys; however, high kidney uptake can be reduced by co-administration of the amino acid lysine, often both pre- and post-therapy [100].

5.5.2 Recent [68]Gallium Work

Although the most commonly used chelate system for ^{68}Ga radiopharmaceuticals is DOTA, its close relative NOTA and its derivatives have actually been shown to have far superior coordination properties (Figure 5.9) [101].

NOTA forms a very stable macrocyclic complex with gallium; however, contrary to DOTA, it is able to label quantitatively at room temperature and is therefore suitable for conjugation with sensitive biomolecules. DOTA has lower thermodynamic stability with Ga(III), as demonstrated by its stability constant (log K_{ML} = 21.3) [102]; however, NOTA shows exceptional thermodynamic stability with Ga(III) (log K_{ML} = 31.0) [37]. DOTA forms a large macrocyclic cavity that is not

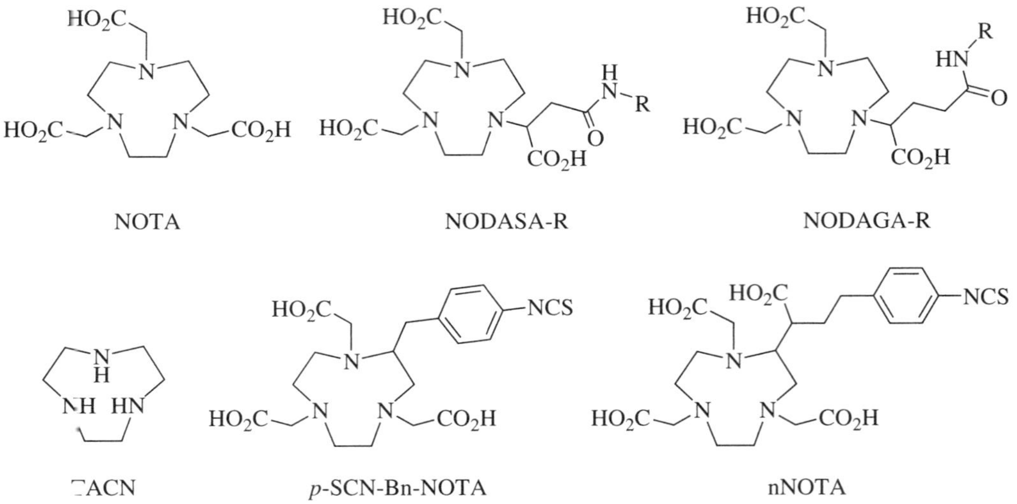

NOTA   NODASA-R   NODAGA-R

TACN   p-SCN-Bn-NOTA   nNOTA

FIGURE 5.9 The macrocyclic chelating agent TACN and the TACN derivative NOTA, which is currently the 'gold standard' for gallium complexation, along with several NOTA-based BFC derivatives (**R** = biovector).

ideal for Ga(III) and only binds in a hexadentate fashion (N_4O_2) instead of utilising the full denticity of DOTA (octadentate N_4O_4) [103]. Radiometallation of DOTA with ^{68}Ga requires heating for 30-60 minutes, which is not optimal for a short half-life isotope [103]. NOTA coordinates Ga(III) with optimal hexadentate (N_3O_3) denticity, forming a cavity size that is a near perfect fit as observed by solid-state X-ray crystallography [101]. The acid inertness of the Ga(III)-NOTA complex is superb, surviving more than 6 months in 5 M HNO_3 [101]. Many BFC derivatives of NOTA have been synthesised, both through N- and C-functionalisation [104]. The BFC derivatives NODASA [105] and NODAGA [106] retain the full hexadentate coordination sphere of NOTA and show the same radiometallation and stability properties as NOTA. Despite these details, a majority of Ga(III)-based radiopharmaceuticals that are translated to the clinic are based on DOTA. An example of this is the prostate cancer targeting ^{68}Ga-DOTA-bombesin bioconjugate, which was also studied with ^{177}Lu for therapeutic applications [95]. The widespread use of DOTA-based BFCs for isotopes such as ^{64}Cu/[86/90]Y/^{177}Lu/[89]Zr/^{111}In means that a plethora of radiochemical and clinical data and protocols are available in the literature, making isotopic translation to ^{68}Ga more streamlined than with the less investigated, although superior, chelator NOTA. In the future, this trend will hopefully shift toward chelators that provide optimal stability and coordination properties.

Largely due to the development of the attractive ^{68}Ga generator system, there has recently been a surge in development of novel BFC agents for [68]Ga. The acyclic chelator HBED shows exceptional thermodynamic stability with Ga(III), with a stability constant of log K_{ML} = 38.5 (Figure 5.10) [107]. HBED is an attractive acyclic chelator for Ga(III) because of its fast room temperature radiometallation kinetics, and the BFC derivative HBED-CC has been conjugated to various antibodies and peptides [108, 109]. The TACN-based chelator TACN-TM is a thiol-containing macrocycle and, although it is stable with Ga(III), it has not garnered much attention for BFC derivatisation [110, 111]. Several promising new ^{68}Ga chelators PrP9 (TRAP) [112], PCTA [113], CP256 [114], and H_2dedpa [115, 116] have been recently published; however, these examples have only been reported in the last two years and will require further investigation before mainstream adoption can occur (Figure 5.10). The acyclic chelators H_2dedpa and CP256 and the DOTA-based chelator PCTA have demonstrated fast room temperature radiometallation kinetics with ^{68}Ga; however, PrP9 (TRAP) requires elevated temperatures to achieve quantitative radiolabelling. A unique and novel NODAGA-like ^{68}Ga-NOTA-bisphosphonate (NOTA-BP) complex was recently synthesised as a bone-seeking agent that demonstrated high stability and high affinity for hydroxyapatite [117].

5.5.3 Stability of [68]Gallium-Based Radiopharmaceuticals

Due to the very similar size (62 pm and 65 pm, respectively [15]) and identical charge of the gallium(III) and iron(III) ions, the iron transport protein transferrin has a very high binding affinity for Ga(III) and a strong tendency to extract it from weaker chelates [38]. Transferrin is a large protein (~79 kD) with two hydrophilic binding sites, one each at the C-terminus and N-terminus [118, 119]. Each of the two binding sites coordinates metal ions with two tyrosine, one histidine, and one aspartic acid amino acid residues (carbonate binds synergistically and is a required cofactor in iron binding) [118, 119]. The problem of trans-chelation is most pressing for gallium-based radiopharmaceuticals due to the high binding affinity of

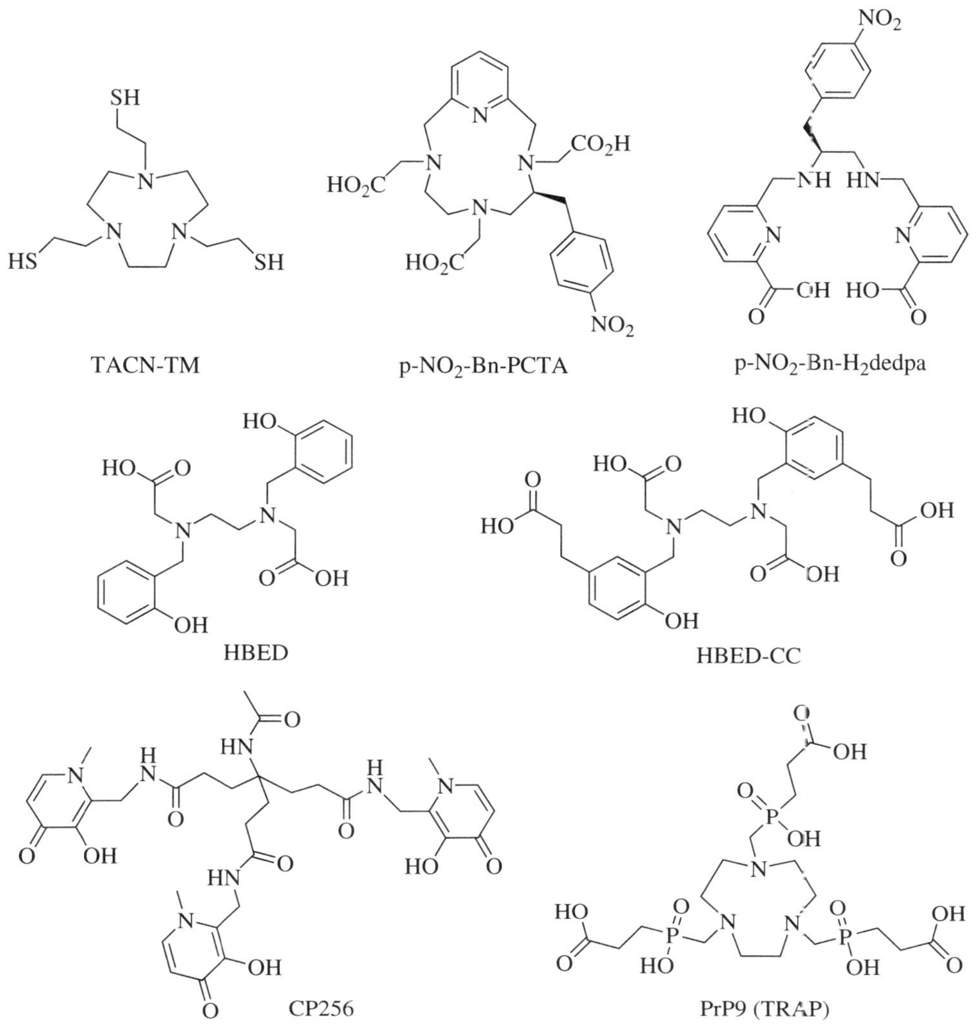

FIGURE 5.10 Examples of several promising and novel Ga(III) chelators, most of which were first published in the last few years.

Ga(III) for transferrin. Trans-chelation by transferrin can be evaluated through *in vitro* human serum and/or *apo*-transferrin competition experiments, which can be used to estimate the practical stability of metal-chelator complexes *in vivo* [113].

Because transferrin in human serum is typically only 30% saturated with iron(III) (with ~70% vacant binding sites), it can bind additional metal ions without the need for direct competition with iron(III) [118, 119]. Iron(III)-transferrin stability constants have been reported as log $K_1^* = 22.8$ and log $K_2^* = 21.5$, showing the highest stability constants of all of the metal ions discussed here (as would be expected because transferrin is an iron transport protein) [70]. The gallium-transferrin stability constants have been reported as log $K_1 = 20.3$ and log $K_2 = 19.3$ for the two metal ion binding sites found in the *C*-terminus and *N*-terminus, respectively [38]. Free ^{68}Ga can be observed to accumulate in high levels in the lungs, liver, spleen, and bone, a direct consequence of its exceptionally strong affinity for the iron transport protein transferrin [78].

5.5.4 [68]Gallium Radiometallation Protocols

As previously noted, the short half-life of [68]Ga (68 minutes) lends itself to fast radiolabelling conditions with minimal post-labelling purification. Similar to the examples given for Y(III), macrocycles such as DOTA typically require heating [120–123], and acyclic chelators such as HBED-CC and DTPA (also DTPA derivatives like mx-DTPA and CHX-A″-DTPA), and even the macrocycle NOTA use mild labelling conditions [109, 124, 125]. The peptide-conjugate ^{68}Ga-DOTA-TOC has been labelled with ^{68}Ga that was directly eluted from a generator system with no pre-labelling purification or concentration of the generator eluent (10 minutes at 100 °C, purified by C18 cartridge), which is an attractive use of the convenient generator system [87].

5.5.5 [68]Gallium Summary

[68]Ga has well-developed coordination chemistry and very attractive nuclear decay properties (high positron abundance), making it ideal for incorporation into BFC-based radiopharmaceuticals for PET imaging. The short half-life of 68 minutes makes ^{68}Ga ideal for imaging with rapidly localising and clearing biovectors, such as small peptides (i.e., RGD, octreotide, bombesin, antibody fragments), while minimising patient-absorbed doses. The commercialisation and availability of efficient ^{68}Ga generators have allowed research efforts to expand quickly and have made it one of the most promising exotic metal nuclides for PET imaging. Because Ga(III) has very similar properties to Fe(III), the iron transport protein transferrin presents a significant stability challenge *in vivo* and requires the use of very thermodynamically stable and kinetically inert chelators.

5.6 [64]COPPER RADIOMETAL ION PROPERTIES

Radiocopper is typically used in its 2+ oxidation state for radiopharmaceuticals, although oxidation states of +1/+2/+3 are accessible in aqueous conditions [18, 25]. Cu(II) is a first row transition metal with a d^9 configuration and a preference for borderline soft ligand donor atoms such as amines, imines, and thiols [18, 25]. Cu(I) (d^{10} ion) prefers softer donor atoms, and its complexes are less utilised because they are quite labile and lack the required kinetic inertness to withstand physiological conditions [18, 25]. Copper forms tetra-, penta-, and hexa-dentate complexes, typically with square planar, trigonal bipyramidal, square pyramidal, and distorted octahedral geometries (Jahn-Teller distortions) [18, 25]. By using hexadentate chelators, the coordination sphere of Cu(II) can be saturated in a distorted octahedral geometry to minimise interactions with exogenous ligands and biological chelators and maximise stability and kinetic inertness [18, 25]. Due to the fast reaction kinetics of Cu(II) (as demonstrated by its fast water exchange rate of $2 \times 10^8\,s^{-1}$, Table 5.2), complexes must be made that are very inert [16, 25].

Unlike Ga(III) and Y(III), Cu(II) can be complexed/radiometallated below pH 7 without insoluble hydroxide formation, because it has a lower affinity for hydroxide ions. The high lability of Cu(II) complexes makes their successful incorporation into radiopharmaceuticals more difficult than Ga(III), Y(III), and Zr(IV) [25]. [64]Cu is a unique metal nuclide because it has a moderate half-life of 12.7 hours and emits both β^+ and β^- particles, allowing it to be used as a multipurpose PET/therapy nuclide. In contrast to [86]Y and [89]Zr, ^{64}Cu has much lower energy γ emissions, which expose patients to lower radiation burdens. [64]Cu has a lower energy β^+ emission than does ^{68}Ga or [86]Y, providing higher resolution images. [64]Cu is typically cyclotron-produced via the nuclear reaction [64]Ni(p,n)[64]Cu and purified by anion exchange (i.e., AG1-X8) [126, 127]. A number of other isotopes of copper emit positrons: [60]Cu ($t_{1/2} = 23.7$ minutes), [61]Cu ($t_{1/2} = 3.3$ hours), and [62]Cu ($t_{1/2} = 9.7$ minutes); however, these isotopes are not as heavily investigated due to their shorter half-lives and higher positron energies. [62]Cu is a [62]Zn generator-produced copper isotope with a very short half-life of 9.7 minutes. Although [62]Cu is useful for experiments that require imaging at very short time points and for minimising radiation dose to patients, [62]Cu has not generated as much interest as the longer lived isotope [64]Cu. [62]Cu-PTSM has garnered significant interest for imaging the heart and tumour hypoxia; however, the short half-life means the generator system only lasts 1-2 days and therefore presents a significant logistical problem requiring frequent delivery and an overall high cost of operation (Figure 5.11) [128–132]. Another shortcoming of [62]Cu-PTSM is poor *in vivo* stability, resulting in high liver uptake of free [62]Cu [128].

5.6.1 Clinical Trials Based on [64]Copper

[64]Cu is not incorporated into any FDA-approved radiopharmaceuticals, and its current clinical trial offerings [53] use BFC systems similar to those used for the previously discussed metal nuclides. The DOTA antibody conjugate ^{64}Cu-DOTA-trastuzumab is currently in clinical trials for imaging and therapy of breast cancer [53, 98]. The ^{64}Cu-ATSM chelate is a more unique entry, because the chelator ATSM is not functionalised for conjugation to a biovector (Figure 5.11) [53, 128–132]. The ^{64}Cu-ATSM complex has high affinity for hypoxic tumour tissue and can be used as a way to image and measure the

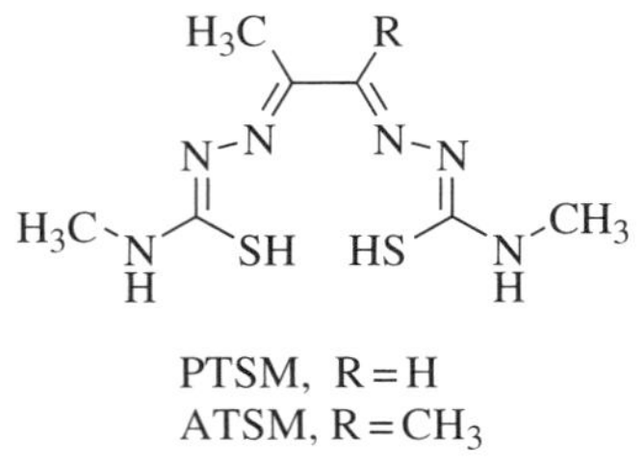

PTSM, R = H
ATSM, R = CH$_3$

FIGURE 5.11 The copper ligands PTSM and ATSM, used for imaging the heart as well as tumour hypoxia.

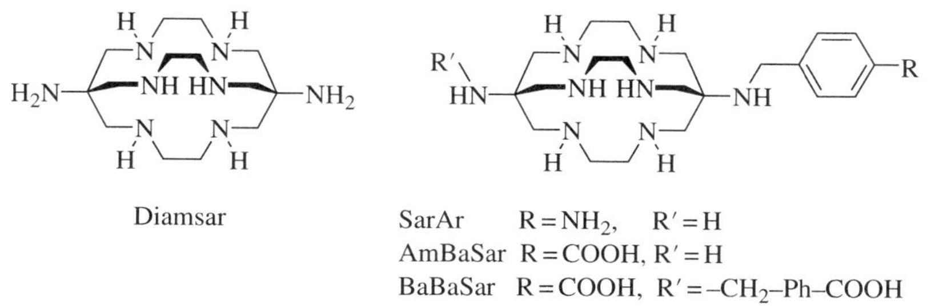

Diamsar

SarAr   R = NH$_2$,   R′ = H
AmBaSar R = COOH, R′ = H
BaBaSar R = COOH, R′ = –CH$_2$–Ph–COOH

FIGURE 5.12   The copper chelator Diamsar and three of its BFC precursors SarAr, AmBaSar, and BaBaSar.

level of oxygen depletion in tumours [53, 128–132]. The emission of β$^-$ particles from ^{64}Cu renders ^{64}Cu-ATSM a radiotherapeutic as well as an imaging agent. The combined β$^+$ and β$^-$ emission of ^{64}Cu allows for imaging/dosimetry and therapy from the same isotope, alleviating the need for a second nuclide to be used as a matched isotope pair.

The chelator Diamsar was reported to quantitatively radiolabel ^{64}Cu much faster than other macrocyclic chelates and was even able to radiolabel at room temperature (Figure 5.12) [133, 134]. The bifunctional derivative of Diamsar, SarAr, was successfully conjugated to the prostate cancer-targeting peptide bombesin, as well as the whole and fragmented antibody B72.3, whilst retaining its room temperature labelling abilities [133, 134]. A bivalent (two biovectors attached to one BFC) Diamsar-based BFC, BaBaSar, has recently been reported and conjugated to RGD [135]. BaBaSar-RGD$_2$ was quantitatively radiometalled with ^{64}Cu in 5 minutes at room temperature and demonstrated high stability as well as improved tumour targeting over monovalent AmBaSar-RGD conjugates [135]. The BFCs NODAGA, CB-TE2A, and DOTA were conjugated to the same novel somatostatin receptor antagonist p-Cl-Phecyclo(D-Cys-Tyr-D-4-amino-Phe(carbamoyl)-Lys-Thr-Cys)D-TyrNH$_2$ (LM3) and radiolabelled with ^{64}Cu and [68]Ga [136]. The results demonstrated that ^{64}Cu/^{68}Ga-NODAGA (Figure 5.9) had superior tumour-to-normal-tissue ratios and better clearance of non-receptor-bound BFC from the blood pool, when compared to the ^{64}Cu/^{68}Ga complexes of CB-TE2A and DOTA [136]. Receptor binding affinity was significantly modulated by up to 10-fold depending on the chelator and metal used, even when attached to the same targeting vector (LM3), demonstrating the strong affect that chelates can have on radiopharmaceutical pharmacokinetics [136].

5.6.2   Recent [64]Copper Work

Of all the investigated copper chelators, the cylcam/TETA-based chelator CB-TE2A appears to be one of the most promising (Figure 5.13) [35]. CB-TE2A shows high stability with Cu(II) and impressive kinetic inertness as demonstrated by acid decomplexation studies performed in 5 M HCl [35]. Animal studies performed in mice demonstrated that the mono-amide functionalised derivatives of CB-TE2A, ^{64}Cu-CB-TEAMA, and ^{64}Cu-CB-PhTEAMA had comparable stability to CB-TE2A and were suitable candidates for further elaboration as BFCs (Figure 5.13) [58]. Although CB-TE2A and its derivatives appear to be the best radiocopper chelating agents, they typically require radiolabelling conditions of ~70–95 °C, which, although suitable for peptide biovectors, is not ideal for sensitive biomolecules such as antibodies [58, 59]. The recently published CB-TE2A derivatives CB-TE2P and CB-TE1A1P contain phosphonate pendant groups. They have been shown to radiolabel ^{64}Cu at ambient temperature in high specific activities and show comparable in vivo stability to CB-TE2A [137]. Although DOTA has been used most extensively as a BFC agent for ^{64}Cu with both peptides and antibodies, it has been shown that it does not form an optimally stable complex with ^{64}Cu, as observed in vivo by high liver uptake indicating release of free [64]Cu (vida infra) [138, 139]. ^{64}Cu-NOTA-RGD (cyclic-Arg-Gly-Asp, bioactive peptide that targets integrin α$_v$β$_3$ expression) and ^{64}Cu-NOTA-BBN (bombesin, peptide that targets the gastrin-releasing peptide receptor) bioconjugates have been synthesised from p-SCN-Bn-NOTA (Figure 5.9). Although radiometallation was fast (15 minutes, 40 °C), high liver uptake was observed, suggesting poor in vivo stability [140]. A ^{64}Cu-NODAGA-c(RGDfK) conjugate has also been recently reported and radiolabelled in high specific activity that demonstrated very promising stability and tumour uptake, similar to that of ^{64}Cu-CB-TE2A-c(RGDfK) [141].

5.6.3   Stability of [64]Copper-Based Radiopharmaceuticals

Copper is very important for living organisms, and its homeostasis is tightly regulated in biological systems by a multitude of transport proteins and enzymes, which all serve as in vivo competition for BFC coordination [142]. Copper can also be redox active in vivo, and Cu(II) may be reduced to Cu(I) via ascorbic acid, which would result in a change in chelate

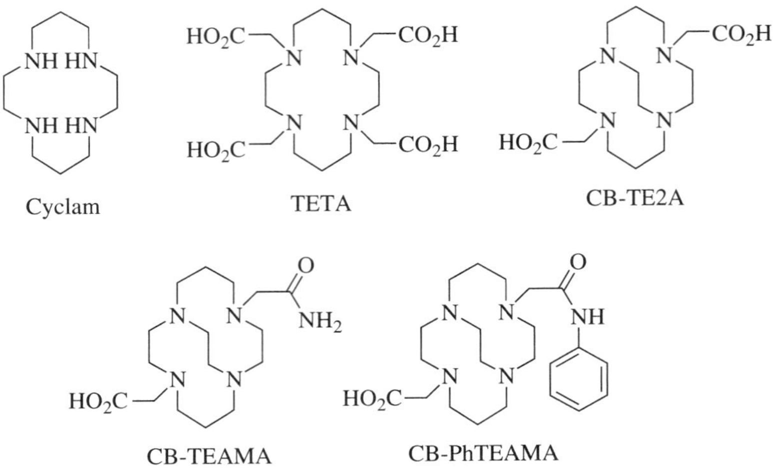

FIGURE 5.13 The macrocycle cyclam, as well as its derivatives TETA, CB-TE2A, CB-TEAMA, and CB-PhTEAMA.

coordination geometry, complex stability, and often decomplexation [143]. The acid inertness and *in vivo* stability of copper complexes has been successfully compared to their reduction potentials [35, 143]. The cyclam- and cyclen-based macrocycles have demonstrated that higher reduction potentials resulted in less stable complexes, with the most stable copper chelator CB-TE2A having the lowest reduction potential as well as a *quasi*-reversible Cu(II)/Cu(I) reduction [35, 143]. When radioactive copper complexes dissociate *in vivo*, the radiometal can be observed to accumulate in the liver, most likely associating with the proteins ceruloplasmin and/or superoxide dismutase [78, 144]. The association of free radiocopper with ceruloplasmin and superoxide dismutase in the liver has been demonstrated by the *in vivo* demetallation of an unstable [67]Cu-TETA-antibody conjugate [144].

Human serum albumin (HSA) is the most abundant serum protein in human blood, and its biological function is the nonspecific binding of various metal ions, drugs, and other small molecules [118]. HSA contains four metal ion binding sites, with site 1 being of particular interest for ^{64}Cu-based radiopharmaceuticals, because it strongly binds copper(II) and nickel(II) [145]. Although only HSA binding site 1 is known to specifically bind any of the radiometals discussed in this chapter, it can still accommodate other metal ions to varying degrees and may exhibit nonspecific binding to radiometals other than [64]Cu [145].

5.6.4   [64]Copper Radiometallation Protocols

Because ^{64}Cu has been so hotly investigated, there exists a large variety of labelling conditions with many different chelators. NOTA and DOTA conjugates are typically radiolabelled between pH 5-8 with temperatures ranging between 40-80 °C [140, 146–148], and CB-TE2A conjugates typically require high temperatures and long reaction times (60 minutes at 90 °C) [141, 149]. The chelators Diamsar and TETA can be radiolabelled under mild ambient temperatures in ~60 minutes and are therefore more compatible with antibody biovectors [149, 150]. Many ^{64}Cu antibody conjugates (targeting HER2, CEA, EGFR, PSMA, GD2, CC, etc.) have been radiolabelled, typically in 15–60 minutes and at 25–40 °C, depending on the chelator [151–153]. Many other peptides and antibodies have been conjugated to these chelators and radiolabelled with ^{64}Cu; only a few representative examples have been discussed.

5.6.5   [64]Copper Summary

[64]Cu is intrinsically a dual-modality radiometal with both β⁻ emission for therapy and β⁺ for PET imaging, negating the need for a matched isotope pair. The intermediate half-life of 12.7 hours allows it to be used with a wide variety of biovectors (peptides, antibodies, etc.); however, it may not be long enough to optimally match the slow 1- to 3-day localisation times of fully intact antibodies (~150 kDa) and image 3 to 7 days post-injection. Despite this possible limitation, many antibody conjugates have been successfully imaged 24-48 hours after using ^{64}Cu without issue. [64]Cu has ideal decay properties with a low-energy β⁺ emission that provides high-resolution PET images and low energy γ emissions that subject patients to low radiation burdens. The natural lability of Cu(II) complexes, the relatively facile redox chemistry between Cu(II)/Cu(I), and the abundance of *in vivo* competition from native copper-rich proteins means that stability problems often arise, and design of thermodynamically stable and kinetically inert chelators is a significant challenge.

5.7 [89]ZIRCONIUM RADIOMETAL ION PROPERTIES

[89]Zr has recently become very popular in the literature, and the last five years have seen a drastic increase in efforts to translate [89]Zr-based imaging agents to the clinic [22]. Zr(IV) is a highly charged and extremely hard metal ion with a relatively small ionic radius (84 and 89 pm for CN = 8 and 9, respectively [15]), and an exceptionally low pKa of 0.22 (Table 5.2). The hard nature of Zr(IV) shows in its preference for hard carboxylate and hydroxamate-oxygen anions. [89]Zr has the longest half-life (78.5 hours) of the radiometal PET nuclides discussed in this chapter and is ideally suited to theranostic applications as a matched isotope pair for long-lived therapeutic isotopes such as ^{90}Y and [177]Lu. [90]Y and ^{177}Lu have long half-lives (64.1 and 161 hours, respectively); to provide accurate imaging/dosimetry data at time points of 1 to 7 days post injection there must be an appropriate surrogate radiometal (ideally a positron emitter for PET) with established chemistry and chelate systems. Long-lived isotopes such as [89]Zr are ideally matched with antibody biovectors, because they require 2 to 3 days to fully localise and penetrate tumours (although imaging can be effectively performed after 24 hours) [26]. [89]Zr is the least investigated isotope of those presented in this chapter; however, it has been investigated as a longer half-life alternative PET isotope to ^{68}Ga for BFC-based radiopharmaceuticals. [89]Zr is thought to be superior to ^{111}In for dosimetry because it is a β+ emitter and provides more accurate and quantitative biodistribution data. The low energy β+ emission of [89]Zr (897 keV) provides high-resolution PET images; however, emission of high-energy γ rays in combination with its long half-life significantly increases the absorbed dose that patients receive. One of the major advantages of [89]Zr over other PET radiometals is that it is retained in cells after being internalised, which could provide essentially irreversible cellular delivery [154–156]. [89]Zr is typically cyclotron-produced via the nuclear reaction [89]Y(p,n)[89]Zr, and purified by anion exchange chromatography by elution from a solid-phase hydroxamate resin with 1 M oxalic acid [157].

5.7.1 Clinical Trials Based on [89]Zirconium

No FDA-approved radiopharmaceuticals currently utilise [89]Zr. Due to the long half-life of [89]Zr (78.5 hours) it is perfectly matched with the biological half-life of antibody biovectors, which can circulate *in vivo* for weeks [26]. Nearly all BFC work performed with [89]Zr relies on derivatives of the bacterial siderophore desferrioxamine (DFO) (Figure 5.14), [158], which binds [89]Zr with its three hydroxamate groups in a hexadentate fashion [159].

DTPA has been shown to form a very thermodynamically stable complex with Zr(IV) (log K_{ML} = 35.8-36.9); however, inferior *in vivo* stability has limited its use [160]. Although DFO has been the 'gold standard' for Zr(IV) chelation, its solubility is poor and causes significant synthetic challenges. [89]Zr-DFO-Zevalin was the first [89]Zr antibody conjugate imaged in humans and was shown to be a suitable PET surrogate for ^{90}Y-Zevalin dosimetry [161]. [89]Zr-DFO-U36 (anti-CD446 chimeric mAb) was in clinical trials for squamous cell carcinoma imaging and gave comparable diagnostic results as ^{18}F-FDG [53, 162]. Another similar example is [89]Zr-DFO-bevacizumab, where bevacizumab is an antibody that targets vascular endothelial growth factor (VEGF), which is a soluble ligand for the VEGF-receptor that is over-expressed in many cancers and that regulate angiogenesis [53, 163]. Antibodies such as bevacizumab that bind to VEGF provide a way to both inhibit angiogenesis in tumours and provide imaging or therapy if an appropriate radionuclide is conjugated to the antibody [53, 163]. [89]Zr-DFO-bevacizumab has demonstrated clear and significant tumour localisation after 72 hours [53, 163]. [89]Zr-DFO-trastuzumab is in clinical trials for imaging breast cancer and [89]Zr-DFO-cetuximab for imaging head, neck, and colorectal cancer [53, 164].

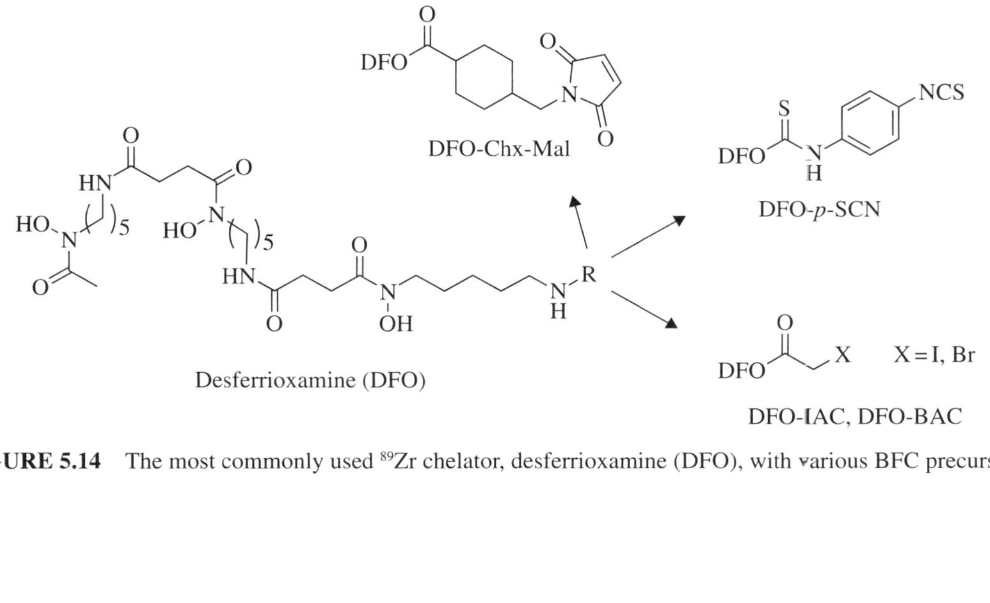

FIGURE 5.14 The most commonly used [89]Zr chelator, desferrioxamine (DFO), with various BFC precursors.

5.7.2   Recent ^{89}Zirconium Work

This emerging PET imaging nuclide is currently underutilised, and its use may be expanded through development of new chelators to provide more effective coordination, solubility, and bioconjugation chemistry. Although ^{86}Y is an ideal surrogate nuclide for use with ^{90}Y as a theranostic pair (they are chemically identical), ^{89}Zr may be a more suitable imaging surrogate for other radiometals or for imaging/dosimetry applications that require data collection beyond 72 hours. The higher charge of 4+ makes ^{89}Zr more challenging to incorporate into BFC systems while still retaining bioequivalence with other common 3+ cationic metal ions (In(III), Ga(III), Y(III), Lu(III)), because the overall chelate-complex charge is different. DFO has been functionalised with various bioconjugation handles (Figure 5.14), such as isothiocyanates [165], alkyl halides [155], and succinimidyl esters (NHS ester) [166].

^{89}Zr has an ideal half-life for use with antibody biovectors, and ^{89}Zr-DFO has been conjugated with the antibody J591, which targets prostate-specific membrane antigen (PSMA) expression and has been used for imaging prostate cancer [158]. Another example is the antibody conjugate ^{89}Zr-DFO-TRC105, which has been used for targeting CD105 expression (marker for tumour angiogenesis) [167]. Toward understanding the metabolism of simple ^{89}Zr complexes, recent work has demonstrated that ^{89}Zr-chloride accumulates in the liver with little secretion, and ^{89}Zr-oxalate accumulates mostly in bone and joints [78, 158]. This example illustrates the strong effect that the specific coordination species of the metal ion has on the biodistribution of the radiometal. In contrast to the chloride and oxalate species, non-conjugated 'bare' ^{89}Zr-DFO is observed to clear very rapidly through the kidneys with minimal nonspecific organ uptake [158]. This demonstrates a biological "clean slate" for radiopharmaceuticals based on the ^{89}Zr-DFO complex, which is a good starting point in chelator development and selection. ^{89}Zr that is lost from a BFC in vivo typically localises in bone, as demonstrated by DOTA- and DTPA-based antibody conjugates with cetuximab that show ^{89}Zr accretion in the thighbone 72 hours post injection due to ^{89}Zr-decomplexation [168]. Studies with Bi(III) have suggested that transferrin binding is strongly correlated to metal ion acidity, regardless of metal ion size, which suggests the very acidic Zr(IV) (pKa = 0.22) may bind strongly [75–77]. It has also been reported that Zr(IV) binds transferrin in an unnatural fashion, resulting in a different coordination environment than for other metal ions, which affects transferrin receptor binding and results in different biological behaviour and biodistribution [169]. Considering this, transferrin may potentially be strong competition for Zr(IV) binding in vivo, but once bound it will not have the same biodistribution pattern as other radiometal ions such as ^{68}Ga and ^{111}In [169]. ^{90}Y-Zevalin and ^{89}Zr-Zevalin have demonstrated very similar biodistribution profiles, except that ^{89}Zr-Zevalin showed significantly higher bone and liver uptake after 72 hours, suggesting ^{89}Zr-decomplexation [161]. Compromised physiological stability of ^{89}Zr with DTPA- and DOTA-based BFCs, as demonstrated by ^{89}Zr-Zevalin (based on the modified DTPA chelator tiuxetan) [161], is the main reason why DFO is most commonly chosen for BFC applications. The rapid low-temperature radiometallation kinetics of DFO is amenable toward antibody biovectors, and the long half-life of ^{89}Zr is well matched with the long biological half-lives of antibodies.

5.7.3 ^{89}Zirconium Radiometallation Protocols

A summary of radiolabelling protocols for ^{89}Zr is concise, because it is essentially restricted to DFO-based BFCs [170]. In one unique example, a ^{89}Zr-DFO-mouse serum albumin (mAlb) conjugate has been synthesised that uses the enhanced permeability and retention effect (EPR effect) and shows similar non-selective retention across various tumour phenotypes [171]. The EPR effect relies on large particles becoming trapped and retained in 'leaky' tissue that has abnormal vascular permeability, which is a hallmark of tumours and sites of inflammation and is exploited by many nanoparticle and microsphere technologies [172]. Pre-coordinating DFO with Fe(III) is a common technique that serves to protect the hydroxamate groups and modulate solubility [173, 174]. The Fe(III) bound to DFO is removed prior to ^{89}Zr-labelling with an EDTA or DTPA solution, purified by size-exclusion chromatography (PD-10 or HPLC), and then radiometallated with ^{89}Zr (typically from ^{89}Zr in oxalic acid) for ~60 minutes at ambient temperature [155, 161, 165, 174, 175]. More recently with the commercial availability of p-SCN-Bn-DFO, the Fe(III) protection is not utilised, and immunoconjugates that incorporate p-SCN-Bn-DFO can be directly radiolabelled.

5.7.4 ^{89}Zirconium Summary

This radiometal has seen a significant rise in attention over the last five years, and ^{89}Zr-based radiopharmaceuticals have shown great promise for use with antibody biovectors due to their close match in radioactive and biological half-lives. ^{89}Zr is the longest half-life PET imaging radiometal (78.5 hours) discussed in this chapter and therefore is ideal for collecting imaging and dosimetry data at time points of 3 to 7 days post injection. ^{89}Zr is ideal in its nuclear decay properties as a PET alternative to ^{111}In for use as a matched isotope pair with theranostic radiopharmaceuticals that incorporate long-lived isotopes such as ^{90}Y and ^{177}Lu. The long half-life and high-energy γ emissions of ^{89}Zr can result in patients acquiring significant absorbed doses; however, the low-energy β⁺ emission provides high-resolution PET images. In light of the challenges caused by the poor solubility of DFO and its Zr(IV) complexes, the design of novel chelators with improved solubility and chelation properties would be timely.

5.8 ^{124}IODINE NUCLIDE PROPERTIES

Iodine is unique for this chapter, because it is not incorporated into radiopharmaceuticals through coordination with a BFC as are the previously discussed metal nuclides. Isotopes of iodine must be covalently attached to molecules in order to be used in radiopharmaceuticals, much like ^{18}F (Chapter 3). This provides a different approach to molecular imaging, because one does not need to attach a bulky biovector such as an antibody or peptide to obtain site-specific delivery (although this can still be accomplished). For example, the PET imaging nuclide ^{124}I can be covalently bound to a molecule in place of a hydrogen atom, as long as the appropriate radioprecursor can be synthesised. Iodine nuclides can also be directly attached to biovectors such as antibodies (reacted with tyrosine residues) without the attachment of a metal-chelate complex, which can result in higher immunogenicity of the antibody due to lower steric hindrance, at the cost of modest instability via deiodination. Most biovectors function in a classic 'lock and key'-type receptor binding mechanism, and therefore having less 'molecular bulk' (no large BFC metal-complex) helps to retain the biovector's native binding affinity. Some of the same problems encountered in metal-based radiopharmaceutical design are faced with radioiodine-based agents, such as obtaining high specific activity products, high radiochemical yields and purity, and robust *in vivo* stability. ^{124}I is typically cyclotron produced via the nuclear reaction ^{124}Te(p,n)^{124}I [176], and purified by dry distillation [177]. Several extensive review articles have been published on general radioiodination techniques [33, 178], as well as ^{124}I production and radiosynthesis [27].

5.8.1 Clinical Trials Based on ^{124}Iodine

There are a number of FDA-approved radiopharmaceuticals that use ^{123}I (SPECT, $t_{1/2}$ = 13.3 hours), ^{125}I (SPECT, $t_{1/2}$ = 59.4 days), and ^{131}I (90% β⁻ emission for therapy, $t_{1/2}$ = 8 days); however, none are approved that use ^{124}I (PET, 23% β⁺, $t_{1/2}$ = 100.2 hours) for PET imaging. ^{122}I is another β⁺ emitting isotope that can be used for PET imaging; however, the short half-life of 3.6 minutes makes it impractical for most applications. The chemistry of radioiodination is the same for every isotope of iodine, and so the current FDA-approved offerings could easily be adapted to ^{124}I derivatives, assuming access to a supply of ^{124}I. Free radioiodine can be seen to accumulate in the thyroid and stomach, and sodium iodide (Na$^{123/ 131}$I) is routinely used to image/treat thyroid cancer, with excess radioiodide being mostly excreted in the urine [179]. Other FDA-approved agents include ^{123}I-ioflupane (DaTscan™, Parkinson's disease diagnosis), ^{123}I-iobenguane (Adreview™, MIBG, primary or metastatic neuroblastoma and pheochromacytoma brain tumours) [180], ^{125}I-iothalamate (Glofil-125, glomerular filtration evaluation), $^{125/131}$I-HSA (human serum albumin) (Jeanatope/Megatope, blood/plasma imaging), and ^{131}I-tositumomab (Bexxar®, a labelled antibody that targets CD20 antigen-expressing non-Hodgkin's lymphoma) (Figure 5.15) [51].

Although no FDA-approved ^{124}I-radiopharmaceuticals are currently available, the clinical trial offerings [53] are significant (Figure 5.16). The radiolabelled antibody ^{124}I-cG250 was recently in clinical trials to diagnose renal cancer and

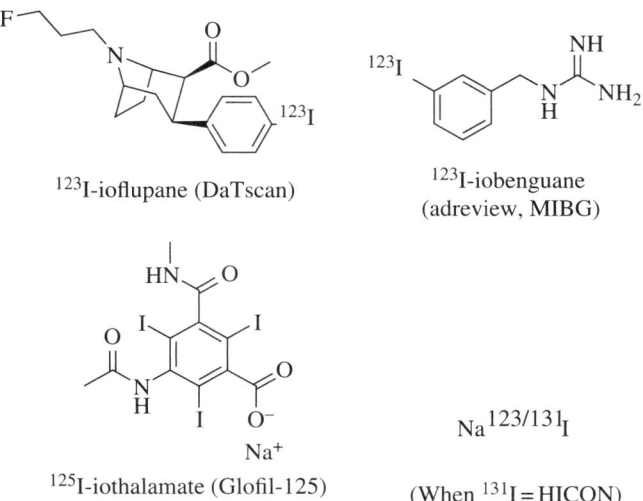

^{123}I-ioflupane (DaTscan)

^{123}I-iobenguane (adreview, MIBG)

^{125}I-iothalamate (Glofil-125)

Na$^{123/131}$I

(When ^{131}I = HICON)

FIGURE 5.15 FDA-approved radioiodine containing radiopharmaceuticals.

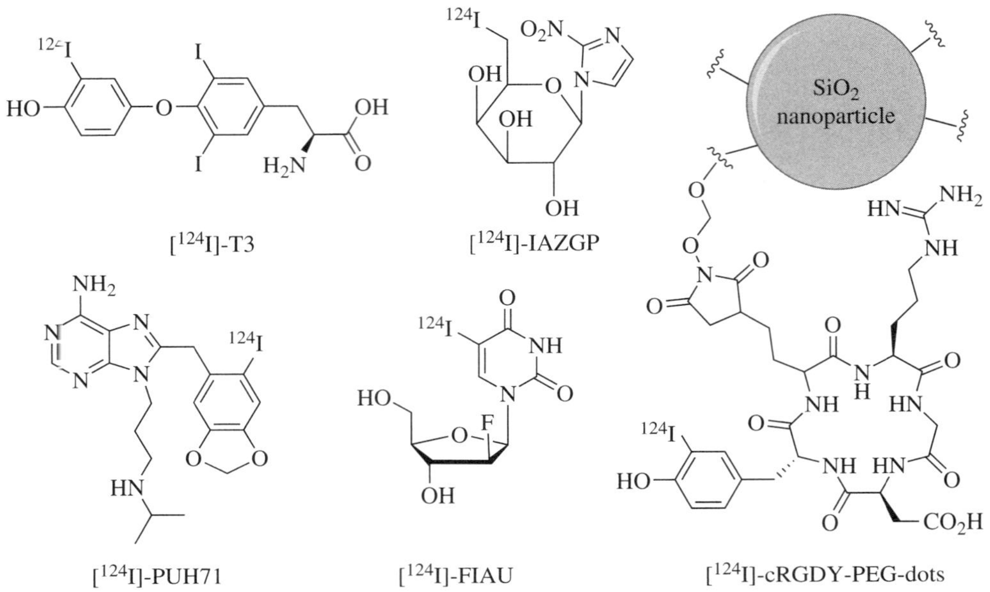

FIGURE 5.16   A selection of ^{124}I-based radiopharmaceuticals currently in clinical trials.

showed promising high specificity and selectivity [53, 181]. The non-Hodgkin's lymphoma drug PUH71 was radioiodinated with ^{124}I and used as a tool in clinical trials to evaluate the biodistribution of the drug in different compartments of the body and to measure blood clearance times [53].

2'-Fluoro-2'-deoxy-1β-D-arabinofuranosyl-5-iodouracil was labelled with ^{124}I (^{124}I-FIAU) to evaluate its biodistribution and localisation in infectious tissue. It has also been used to image herpes simplex virus type 1 thymidine kinase (HSV1-tk) gene expression [53 182, 183]. A more unique entry into recent clinical trials is ^{124}I-cRGDY-PEG-dots, which are silica nanoparticles that are conjugated through polyethyleneglycol (PEG) spacer groups to ^{124}I labelled cyclic-arginine-glycine-aspartic acid-tyrosine (cRGDY) peptides for targeting melanoma [53, 184]. Also in clinical trials is ^{124}I labelled humanised A33 (^{124}I-huA33) antibody for radioimmunodetection of colorectal cancer [53, 185]. ^{124}I-iodo-azomycin-galactopyranoside (^{124}I-IAZGP) is being evaluated in clinical trials for hypoxic tumour imaging in a variety of cancers [53, 186], and the similar compound 1-α-D-(5-deoxy-5-[^{124}I]iodo-arabinofuranosyl)-2-nitroimidazole (^{124}I-IAZA, not in clinical trials) is also being investigated for tumour hypoxia imaging [187]. The ^{124}I-labelled antibody 8H9 is being investigated for peritoneal cancer, and ^{124}I-labelled monoclonal antibody Mu11-1F4 for AL amyloidosis (primary light-chain amyloidosis) [53]. Finally, ^{124}I-labelled tri-iodothyronine (T3) is a thyroid hormone that is currently being used to study T3 metabolism [53]. These examples illustrate the large differences between radiometals and radioiodine in their methods of incorporation as radiopharmaceuticals.

5.8.2   Recent ^{124}Iodine Work

As previously discussed, matched isotope pairs for imaging/dosimetry and therapy are an integral part of nuclear medicine. Finding an isotope pair with identical chemistry is crucial in order to retain identical biological activity, as seen with the ^{86}Y/^{90}Y isotope pair. In a similar scenario, ^{124}I/^{131}I can be used as a matched isotope pair as both isotopes are chemically identical and can be radioiodinated in an identical fashion, as demonstrated with MIBG and various antibodies [188]. ^{124}I has also recently been shown to be a very effective matched isotope pair for ^{131}I metastatic differentiated thyroid cancer (DTC) treatment, in order to obtain accurate dosimetry data and minimise patient absorbed dose [189]. Currently, one of the most active areas of ^{124}I research is in antibody labelling, as observed in the current clinical trial offerings [53]. Various antibodies are being investigated such as the anti-CD20 monoclonal antibody rituximab (Mabthera®) for imaging B-cell populations in rheumatoid arthritis [190] and the anti-HER2 C6.5 diabody for obtaining predictive dosimetry data for radioimmunotherapy with antibody conjugates such as Herceptin [191]. In a very interesting recent study, the ^{124}I-labelled antibody cG250 (for renal cancer) was used to compare in vivo PET imaging dosimetry with in vitro autoradiography and γ-counting methods, and it was demonstrated that both methods yielded very similar results and that PET/CT can be reliably used for obtaining quantitative dosimetry data [192]. As with ^{86}Y, one of the major strengths of ^{124}I lies with its application in PET dosimetry studies with a matched isotope pair. A more novel application is the use of ^{124}I-labelled chitosan, which is a polysaccharide used as an ocular delivery agent [193]. Studies comparing ^{124}I-chitosan with ^{124}I-NaI showed a twofold increase in ocular retention of the labelled chitosan when compared to free radioiodide [193].

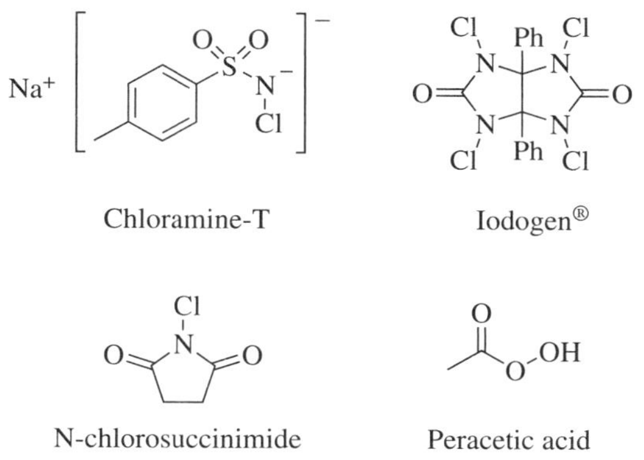

Chloramine-T   Iodogen®

N-chlorosuccinimide   Peracetic acid

FIGURE 5.17 Common oxidation reagents used in radioiodination chemistry, typically producing I^+Cl^-.

5.8.3 124Iodine Radioiodination Protocol

Unlike radiometals such as $^{86}Y/^{68}Ga/^{64}Cu/^{89}Zr$, radioiodine is not utilised through chelation, instead requiring covalent attachment to a molecule. Common radioiodination techniques include electrophilic substitution of various leaving groups via reactions such as iododeprotonation (R-H), iodododediazonisation (R-N$_2$), iododeboronation (R-B(OH)$_2$), iododestannylation (R-Sn(Me)$_3$), iododesilation (R-Si(Me)$_3$), and iodododethallation (R-Tl(OCOCF$_3$)$_2$) [178]. Iodogen® [194], chloramine-T [195], N-chlorosuccinimide, iodine monochloride (ICI) [196], peracetic acid, and various metal ions (Tl^{3+}, Ce^{4+}) are all common iodination reagents used in radioiodination reactions (Figure 5.17) [27, 178].

Many of these reagents generate HOCl, which is an oxidising agent. They generate I^+ cations *in situ* by oxidation of radioiodide anions, which can then directly radioiodinate activated aromatic rings through electrophilic substitution, such as demetallation of a good leaving group (i.e., organoboranes, Si, Ge, Sn, Hg, Tl). Generation of I^+ (usually forming I^+Cl^- when generated by *N*-Cl containing oxidising agents) allows for possible quantitative radioiodination, where I$_2$ labelling would only offer a theoretical maximum yield of 50% [33]. Of the organometallic precursors, organostannanes such as tributyltin are most common. Nucleophilic exchange reactions do not utilise these oxidising reagents because they require the native I$^-$ obtained from [124I]-NaI.

Figure 5.18 illustrates several examples of 124I-radioiodination protocols, showing examples from currently used radiopharmaceuticals. Nucleophilic exchange reactions are shown as halogen exchange reactions that utilise various metal catalysts, with activated aromatic rings providing the best reaction kinetics (level of ring activation also influences *in vivo* stability). The most typical catalysts used for nucleophilic exchange reactions are ammonium sulphate or copper(II) salts [197]. Electrophilic substitutions are the most common radioiodination reactions and are shown using oxidising agents such as Iodogen® or chloramine T (Figure 5.18). The 124I is oxidised *in situ* to I$^+$, followed by direct (radioiododeprotonation) and organometallic (radioiododemetallation) electrophilic substitution. Prosthetic groups are also used (similar to 18F chemistry) that are first radioiodinated (electrophilic substitutions shown in Figure 5.18) followed by conjugation to a biovector through standard coupling methods (e.g., amide couplings).

Examples given for 124I nucleophilic substitution are 124I-IAZGP (124I-IAZAG) [186, 198], 124I-dRFIB [199], and 124I-MIBG [180]. The radiotracer 1-(2-deoxy-β-D-ribofuranosyl)-2,4-difluoro-5-methylbenzene (124I-dRFIB) is a non-hydrogen bonding thymidine analogue used to study the importance of nucleoside base hydrogen bonding in DNA replication and cell proliferation [199]. Radiosynthesis of the compound 124I-FIAU is shown as an example of an iododeprotonation electrophilic substitution [182, 183], and *m*-[124I]-iodophenylpyrrolomorphinan (*m*-[124I]-IPPM) as an iododestannylation reaction utilising a trimethyltin 'cold' precursor [200]. *m*-[124I]-IPPM is a delta opioid (DOP) receptor agonist used for studying pain treatment [200]. Both of the examples shown in Figure 5.18 for 124I-labelled prosthetic groups utilise electrophilic radioiodination reactions. [124I]-I-HPP-VG76e is an anti-VEGF diabody conjugated to the Bolton-Hunter reagent 124I-SHPP for imaging tumour angiogenesis [201]. The 124I-SHPP prosthetic group has also been conjugated with an anti-HER2 diabody, commonly used for imaging breast cancer [202]. The 124I-SIB prosthetic group conjugate m-[124I]-IBA-Annexin-V utilises as a biovector the 36 kDa protein Annexin-V, which is used for studying cell apoptosis in tumours [203–205]. Annexin-V provides targeting access to the early stages of apoptosis, as Annexin-V has a strong affinity for phosphatidylserine (PS), which is translocated from the inner to the outer membrane at the onset of apoptosis [203–205]. An elegant class of reactions directly label tyrosine residues on proteins and antibodies (i.e., anti-CEA minibodies and diabodies, Annexin-V, A33 monoclonal antibody) without the need for prosthetic group conjugations [202]. It must be considered that the activated aromatic rings used for direct protein labelling, such as the phenol group found in tyrosine, can have stability issues and are more susceptible to *in vivo* deiodination [33].

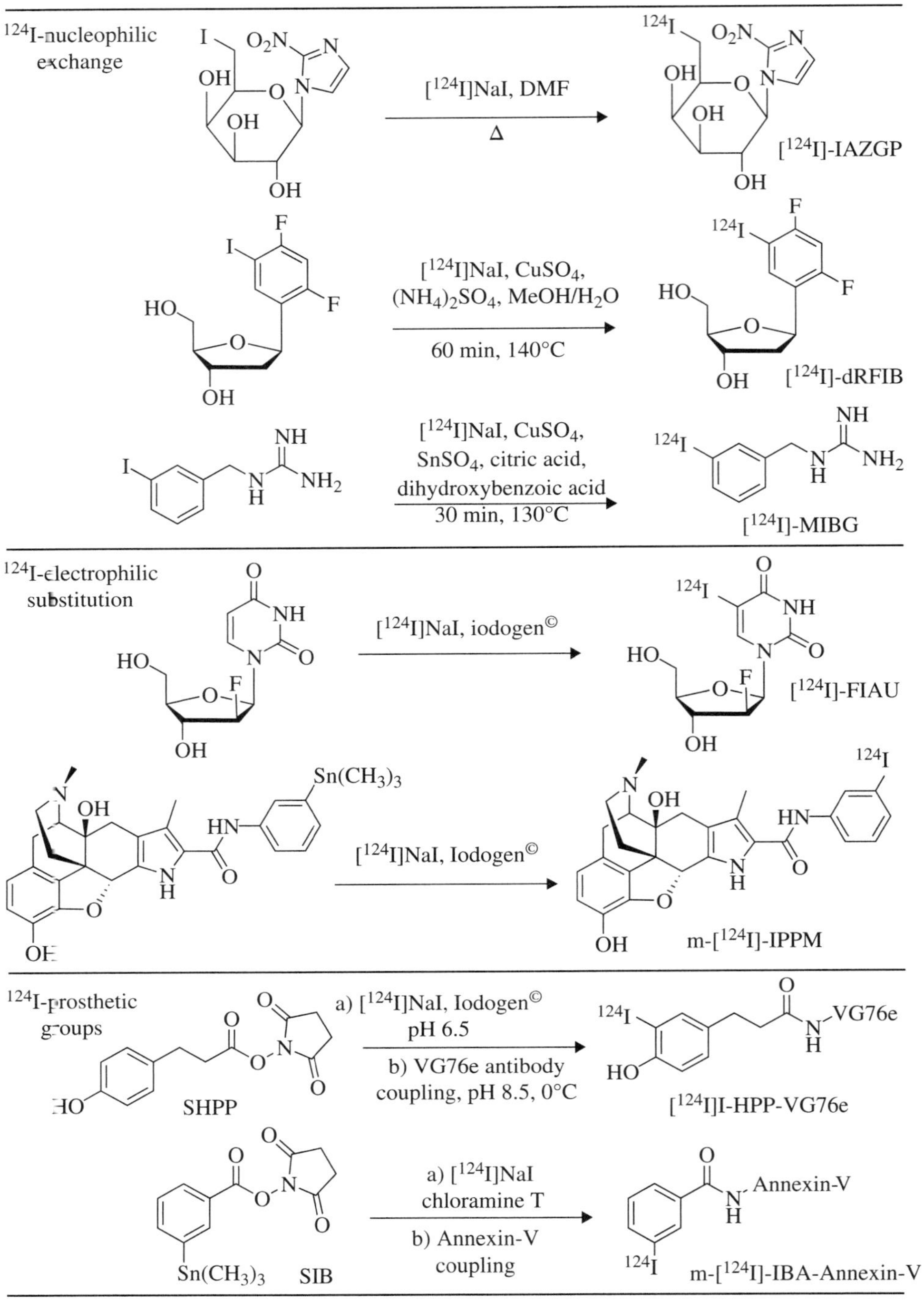

FIGURE 5.18 Examples of common ^{124}I-radioiodination protocols, grouped into nucleophilic exchange, electrophilic substitution, and prosthetic group labelling.

5.8.4 Stability of ^{124}Iodine-Based Radiopharmaceuticals

As with radiometal-containing radiopharmaceuticals, *in vivo* stability is of great concern with ^{124}I-based radiopharmaceuticals. With radioiodinated agents the issue is less straightforward, because decomposition is likely to occur through the normal biochemical metabolism of the parent drug, regardless of the stability and inertness of the carbon-radioiodine bond. Due to its long half-life and β$^+$ emission, ^{124}I is an ideal candidate for pharmacological studies in drug development and can be used to monitor the pharmacokinetics of labelled drugs *in vivo* through PET imaging over long periods of time [206]. One common method for evaluating the stability of radioiodinated compounds is to incubate them with blood serum and monitor the decomposition over time with radio-HPLC and radio-TLC methods (similar to evaluation of radiometal complex stability) [207]. Biochemical dehalogenation *in vivo* is common and occurs to varying degrees depending on the stability of the C-I bond [208]. When substituting an aromatic hydrogen for radioiodine, the resulting C-I bond is weaker than the C-H it replaces; therefore, it is more prone to decomposition and dehalogenation, with aromatic and vinyl iodinations being most stable (phenyl-H = ~460 kJ/mol, phenyl-I = 268 kJ/mol, alkyl-I = 222 kJ/mol, vinyl-I = 297 kJ/mol) [208, 209].

5.8.5 ^{124}Iodine Summary

^{124}I-radioiodination of various small molecule agents, prosthetic group conjugates, and directly labelled peptide/antibodies allow for PET imaging of a wide variety of molecular processes. Limitations on the application of ^{124}I for PET imaging is essentially to the limits of the imagination of the radiochemist, with the feasibility of non-radioactive 'cold' precursor synthesis and/or radioiodination conditions being the major hurdle. Similarly to ^{18}F radiochemistry, and in stark contrast to radiometal chemistry, ^{124}I is optimally suited to 'stealth' substitution of hydrogen atoms on pre-existing drugs with known biological activity. In this way, the native functions and pharmacology of these drugs can be explored using PET imaging in a manner that is not possible with the attachment of bulky radiometallated chelator complexes or prosthetic groups. The 100.2-hour half-life of ^{124}I allows for imaging 3+ days post-injection of biovectors with long biological half-lives such as antibodies; however, the high-energy γ emission of ^{124}I would expose patients to significant absorbed doses. One major limitation for the expansion of ^{124}I use in radiopharmaceuticals is its limited production and availability; however, new production methods and increasing availability of small biomedical cyclotrons is alleviating this problem [27]. The more lengthy and challenging radiosynthetic protocols required for producing radiohalogen-containing tracers is also in stark contrast to the simple and facile kit-preparation deployment of radiometal-containing agents. Both modalities have strengths and weaknesses, and when deployed for the right applications they provide very powerful molecular imaging tools.

5.9   CONCLUSIONS

Each of the five inorganic PET nuclides discussed in this chapter have unique characteristics that can be both positive and negative, depending on the application. One fact that should stand out is that there are no universally applicable BFC systems, because each radiometal has unique chemical and nuclear properties that require careful matching with the desired BFC-conjugate and the targeted biological process. A single chelator/BFC cannot satisfy the unique coordination chemistry of every radiometal ion discussed here, thus the modularity of BFC systems must be exploited to properly tailor the BFC to match each isotope and application. Many nuclides are used in conjunction with others as matched isotope pairs for use in theranostic radiopharmaceuticals, so decay properties such as half-lives and emission types (PET/SPECT/ therapy) must be carefully matched to optimise bioequivalence and subsequently maximise the accuracy of imaging/ dosimetry. In contrast to bulky BFC-based radiometal-biovector conjugates, radiohalogens such as radioiodine present the opportunity for 'stealth' hydrogen atom substitution (often aromatic hydrogens) in small molecule agents, which can provide pharmacological and imaging data on the small molecule's native biological functions and metabolism. Like most disciplines, there exists no 'magic-bullet' radiopharmaceutical. Many different nuclides must be evaluated so that a variety of tools are available for radiochemists and nuclear medicine practitioners to study and apply to a vast number of biochemical processes and disease states.

REFERENCES

[1] T. K. Nayak, K. Garmestani, D. E. Milenic, K. E. Baidoo and M. W. Brechbiel, *PLoS One* **6**, e18198 (2011).

[2] T. K. Nayak, K. Garmestani, K. E. Baidoo, D. E. Milenic and M. W. Brechbiel, *Int. J. Cancer* **128**, 920–926 (2010).

[3] S. Palm, R. M. Enmon, C. Matei, K. S. Kolbert, S. Xu, P. B. Zanzonico, R. L. Finn, J. A. Koutcher, S. M. Larson and G. Sgouros, *J. Nucl. Med.* **44**, 1148–1155 (2003).

[4] L. I. Gordon, T. E. Witzig, G. A. Wiseman, I. W. Flinn, S. S. Spies, D. H. Silverman, C. Emmanuolides, L. Cripe, M. Saleh, M. S. Czuczman, T. Olejnik, C. A. White and A. J. Grillo-López, *Semin. Oncol.* **29**, 87–92 (2002).

[5] M. F. Giblin, B. Veerendra and C. J. Smith, *In Vivo* **19**, 9–29 (2005).

[6] H. Herzog, F. Rösch, G. Stöcklin, C. Lueders, S. M. Qaim and L. E. Feinendegen, *J. Nucl. Med.* **34**, 2222–2226 (1993).

[7] K. S. Pentlow, M. C. Graham, R. M. Lambrecht, N.-K. V. Cheung and S. M. Larson, *Med. Phys.* **18**, 357–366 (1991).

[8] H. Williams, S. Robinson, P. Julyan, J. Zweit and D. Hastings, *Eur. J. Nucl. Med. Mol. Imaging* **32**, 1473–1480 (2005).

[9] S. Robinson, P. J. Julyan, D. L. Hastings and J. Zweit, *Phys. Med. Biol.* **49**, 5505 (2004).

[10] S. Y. F. Chu, L. P. Ekström and R. B. Firestone, *WWW table of radioactive isotopes,* database version 2/28/1999, from http://nucleardata. nuclear.lu.se/nucleardata/toi/ (1999).

[11] C. M. Lederer and V. S. Shirley, *Table of Isotopes.* 7th ed. John Wiley & Sons, New York (1978).

[12] J. P. Holland, M. J. Williamson and J. S. Lewis, *Mol. Imaging* **9**, 1–20 (2010).

[13] S. M. Qaim, *Q. J. Nucl. Med. Mol. Imaging* **52**, 111–120 (2008).

[14] K. N. Raymond, G. Müller and B. Matzanke, Complexation of Iron by Siderophores. A Review of Their Solution and Structural Chemistry and Biological Function, in *Topics in Current Chemistry,* F. L. Boschke, ed., Springer-Verlag, Berlin, Heidelberg, Vol. **123**, 50–102 (1984).

[15] R. Shannon, *Acta Crystallogr.* **A32**, 751–767 (1976).

[16] C. F. Baes, Jr. and R. E. Mesmer, *The Hydrolysis of Cations,* Wiley-Interscience: New York (1976).

[17] A. E. Martell and R. M. Smith, *Critical Stability Constants, Vol. 3: Other Organic Ligands,* Plenum Press, New York, (1977).

[18] T. J. Wadas, E. H. Wong, G. R. Weisman and C. J. Anderson, *Chem. Rev.* **110**, 2858–2902 (2010).

[19] C. J. Anderson and M. J. Welch, *Chem. Rev.* **99**, 2219–2234 (1999).

[20] F. Rösch and R. P. Baum, *Dalton Trans.* **40**, 6104–6111 (2011).

[21] M. D. Bartholomä, A. S. Louie, J. F. Valliant and J. Zubieta, *Chem. Rev.* **110**, 2903–2920 (2010).

[22] B. M. Zeglis and J. S. Lewis, *Dalton Trans.* **40**, 6168–6195 (2011).

[23] S. Liu, *Adv. Drug Delivery Rev.* **60**, 1347–1370 (2008).

[24] S. Liu, *ChemInform* **36**, 445–461 (2005).

[25] T. J. Wadas, E. H. Wong, G. R. Weisman and C. J. Anderson, *Curr. Pharm. Des.* **13**, 3–16 (2007).

[26] C. A. Boswell and M. W. Brechbiel, *Nucl. Med. Biol.* **34**, 757–778 (2007).

[27] L. Koehler, K. Gagnon, S. McQuarrie and F. Wuest, *Molecules* **15**, 2686–2718 (2010).

[28] J. S. Lewis and C. J. Anderson in G. B. Fields, ed., *Peptide Characterization and Application Protocols.* Humana Press, New York, 227–240 (2007).

[29] S. L. Rice, C. A. Foney, P. Daumar and J. S. Lewis, *Semin. Nucl. Med.* **41**, 265–282 (2011).

[30] M. W. Brechbiel, *Q. J. Nucl. Med. Mol. Imaging* **52**, 166–173 (2008).

[31] M. Pagani, S. Stone-Elander and S. A. Larsson, *Eur. J. Nucl. Med. Mol. Imaging* **24**, 1301–1327 (1997).

[32] M. Fani, J. P. André and H. R. Maecke, *Contrast Media Mol. Imaging* **3**, 53–63 (2008).

[33] M. J. Adam and D. S. Wilbur, *Chem. Soc. Rev.* **34**, 153–163 (2005).

[34] R. M. Smith and A. E. Martell, *Sci. Total Environ.* **64**, 125–147 (1987).

[35] K. S. Woodin, K. J. Heroux, C. A. Boswell, E. H. Wong, G. R. Weisman, W. Niu, S. A. Tomellini, C. J. Anderson, L. N. Zakharov and A. L. Rheingold, *Eur. J. Inorg. Chem.* **23**, 4829–4833 (2005).

[36] J.-F. Morfin and É. Tóth, *Inorg. Chem.* **50**, 10371–10378 (2011).

[37] E. T. Clarke and A. E. Martell, *Inorg. Chim. Acta* **190**, 37–46 (1991).

[38] W. R. Harris and V. L. Pecoraro, *Biochemistry* **22**, 292–299 (1983).

[39] J. Byegård, G. Skarnemark and M. Skålberg, *J. Radioanal. Nucl. Chem.* **241**, 281–290 (1999).

[40] J. B. Stimmel and F. C. Kull Jr, *Nucl. Med. Biol.* **25**, 117–125 (1998).

[41] J. B. Stimmel, M. E. Stockstill and F. C. Kull, *Bioconjug. Chem.* **6**, 219–225 (1995).

[42] L. Camera, S. Kinuya, K. Garmestani, C. Wu, M. W. Brechbiel, L. H. Pai, T. J. McMurry, O. A. Gansow, I. Pastan, C. H. Paik and J. A. Carrasquillo, *J. Nucl. Med.* **35**, 882–889 (1994).

[43] A. Harrison, C. A. Walker, D. Parker, K. J. Jankowski, J. P. L. Cox, A. S. Craig, J. M. Sansom, N. R. A. Beeley, R. A. Boyce, L. Chaplin, M. A. W. Eaton, A. P. H. Farnsworth, K. Millar, A. T. Millican, A. M. Randall, S. K. Rhind, D. S. Secher and A. Turner, *Int. J. Rad. Appl. Instrum. [B]* **18**, 469–476 (1991).

[44] R. D. Hancock, *J. Chem. Educ.* **69**, 615–621 (1992).

[45] S. Liu and D. S. Edwards, *Bioconjugate Chem.* **12**, 7–34 (2000).

[46] D. L. Kukis, S. J. DeNardo, G. L. DeNardo, R. T. O'Donnell and C. F. Meares, *J. Nucl. Med.* **39**, 2105–2110 (1998).

[47] Y. H. Jang, M. Blanco, S. Dasgupta, D. A. Keire, J. E. Shively and W. A. Goddard, *J. Am. Chem. Soc.* **121**, 6142–6151 (1999).

[48] S. Walrand, G. Flux, M. Konijnenberg, R. Valkema, E. Krenning, R. Lhommel, S. Pauwels and F. Jamar, *Eur. J. Nucl. Med. Mol. Imaging* **38**, 57–68 (2011).

[49] M. Sadeghi, M. Aboudzadeh, A. Zali, M. Mirzaii and F. Bolourinovin, *Appl. Radiat. Isot.* **67**, 7–10 (2009).

[50] G. Reischl, F. Rösch and H. J. Machulla, *Radiochim. Acta* **90**, 225–228 (2002).

[51] R. M. Sharkey, J. Burton and D. M. Goldenberg, *Expert Rev. Clin. Immunol.* **1**, 47–62 (2005).

[52] K. Hohloch, P. L. Zinzani, W. Linkesch, W. Jurczak, A. Deptala, M. Lorsbach, C. Windemuth-Kiesselbach, G. G. Wulf and L. H. Truemper, *Bone Marrow. Transplant.* **46**, 901–903 (2010).

[53] clinicaltrials.gov. Accessed Nov. 11 (2011).

[54] R. J. Lewandowski, K. G. Thurston, J. E. Goin, C.-Y. O. Wong, V. L. Gates, M. V. Buskirk, J.-F. H. Geschwind and R. Salem, *J. Vasc. Interv. Radiol.* **16**, 1641–1651 (2005).

[55] S. S. Kelkar and T. M. Reineke, *Bioconjugate Chem.* **22**, 1879–1903 (2011).

[56] F. Forrer, C. Waldherr, H. R. Maecke and J. Mueller-Brand, *Anticancer Res.* **26**, 703–707 (2006)

[57] T. Nayak, C. Regino, K. Wong, D. Milenic, K. Garmestani, K. Baidoo, L. Szajek and M. Brechbiel, *Eur. J. Nucl. Med. Mol. Imaging* **37**, 1368–1376 (2010).

[58] J. E. Sprague, Y. Peng, A. L. Fiamengo, K. S. Woodin, E. A. Southwick, G. R. Weisman, E. H. Wong, J. A. Golen, A. L. Rheingold and C. J. Anderson, *J. Med. Chem.* **50**, 2527–2535 (2007).

[59] J. C. Garrison, T. L. Rold, G. L. Sieckman, S. D. Figueroa, W. A. Volkert, S. S. Jurisson and T. J. Hoffman, *J. Nucl. Med.* **48**, 1327–1337 (2007).

[60] M. W. Brechbiel, O. A. Gansow, R. W. Atcher, J. Schlom, J. Esteban, D. Simpson and D. Colcher, *Inorg. Chem.* **25**, 2772–2781 (1986).

[61] C. Wu, H. Kobayashi, B. Sun, T. M. Yoo, C. H. Paik, O. A. Gansow, J. A. Carrasquillo, I. Pastan and M. W. Brechbiel, *Bioorg. Med. Chem.* **5**, 1925–1934 (1997).

[62] S. Liu and D. Edwards in W. Krause, ed., *Contrast Agents II: Optical, Ultrasound, X-ray and Radiopharmaceutical Imaging.* Springer-Verlag, Berlin/Heidelberg, 259–278 (2002).

[63] S. Liu, J. Pietryka, C. E. Ellars and D. S. Edwards, *Bioconjugate Chem.* **13**, 902–913 (2002).

[64] J. Fichna and A. Janecka, *Bioconjugate Chem.* **14**, 3–17 (2003).

[65] M. Regueiro-Figueroa, D. Esteban-Gómez, A. de Blas, T. Rodríguez-Blas and C. Platas-Iglesias, *Eur. J. Inorg. Chem.* 2010, 3586–3595 (**2010**).

[66] H.-s. Chong, K. Garmestani, D. Ma, D. E. Milenic, T. Overstreet and M. W. Brechbiel, *J. Med. Chem.* **45**, 3458–3464 (2002).

[67] S. Pauwels, R. Barone, S. Walrand, F. Borson-Chazot, R. Valkema, L. K. Kvols, E. P. Krenning and F. Jamar, *J. Nucl. Med.* **46**, 92S–98S (2005).

[68] J. A. Carrasquillo, J. D. White, C. H. Paik, A. Raubitschek, N. Le, M. Rotman, M. W. Brechbiel, O. A. Gansow, L. E. Top, P. Perentesis, J. C. Reynolds, D. L. Nelson and T. A. Waldmann, *J. Nucl. Med.* **40**, 268–276 (1999).

[69] G. J. Förster, M. J. Engelbach, J. J. Brockmann, H. J. Reber, H. G. Buchholz, H. R. Mäcke, F. R. Rösch, H. R. Herzog and P. R. Bartenstein, *Eur. J. Nucl. Med.* **28**, 1743–1750 (2001).

[70] W. R. Harris, *Biochemistry* **22**, 3920–3926 (1983).

[71] W. R. Harris and Y. Chen, *Inorg. Chem.* **31**, 5001–5006 (1992).

[72] W. R. Harris, B. Yang, S. Abdollahi and Y. Hamada, *J. Inorg. Biochem.* **76**, 231–242 (1999).

[73] W. R. Harris, Y. Chen and K. Wein, *Inorg. Chem.* **33**, 4991–4998 (1994).

[74] W. Harris in M. J. Clarke, ed., *Less Common Metals in Proteins and Nucleic Acid Probes.* Springer-Verlag, Berlin/Heidelberg, 121–162 (1998).

[75] H. Sun, H. Li and P. J. Sadler, *Chem. Rev.* **99**, 2817–2842 (1999).

[76] H. Sun, M. Cox, H. Li and P. Sadler in H. Hill, P. Sadler, and A. Thomson, ed., *Metal Sites in Proteins and Models.* Springer-Verlag, Berlin/Heidelberg, 71–102 (1997).

[77] H. Li, P. J. Sadler and H. Sun, *Eur. J. Biochem.* **242**, 387–393 (1996).

[78] A. Ando, I. Ando, T. Hiraki and K. Hisada, *Int. J. Rad. Appl. Instrum. [B]* **16**, 57–80 (1989).

[79] L. Wang, J. Shi, Y.-S. Kim, S. Zhai, B. Jia, H. Zhao, Z. Liu, F. Wang, X. Chen and S. Liu, *Mol. Pharm.* **6**, 231–245 (2009).

[80] T. K. Nayak, K. Garmestani, K. E. Baidoo, D. E. Milenic and M. W. Brechbiel, *J. Nucl. Med.* **51**, 942–950 (2010).

[81] D. W. Schneider, T. Heitner, B. Alicke, D. R. Light, K. McLean, N. Satozawa, G. Parry, J. Yoo, J. S. Lewis and R. Parry, *J. Nucl. Med.* **50**, 435–443 (2009).

[82] F. Rösch, H. Herzog, B. Stolz, J. Brockmann, M. Köhle, H. Mühlensiepen, P. Marbach and H.-W. Müller-Gärtner, *Eur. J. Nucl. Med. Mol. Imaging* **26**, 358–366 (1999).

[83] Q. A. Salako, R. T. O'Donnell and S. J. DeNardo, *J. Nucl. Med.* **39**, 667–670 (1998).

[84] H. R. Maecke and J. P. André in P. A. Schubiger, L. Lehmann, and M. Friebe, ed., *PET Chemistry - Ernst Schering Research Foundation Workshop Vol. 62.* Springer-Verlag, Berlin/Heidelberg, 215–242 (2007).

[85] W. Breeman and A. Verbruggen, *Eur. J. Nucl. Med. Mol. Imaging* **34**, 978–981 (2007).

[86] W. A. P. Breeman, E. de Blois, H. Sze Chan, M. Konijnenberg, D. J. Kwekkeboom and E. P. Krenning, *Semin. Nucl. Med.* **41**, 314–321 (2011).

[87] M. Asti, G. De Pietri, A. Fraternali, E. Grassi, R. Sghedoni, F. Fioroni, F. Roesch, A. Versari and D. Salvo, *Nucl. Med. Biol.* **35**, 721–724 (2008).

[88] P. Antunes, M. Cinj, H. Zhang, B. Waser, R. Baum, J. Reubi and H. Maecke, *Eur. J. Nucl. Med. Mol. Imaging* **34**, 982–993 (2007).

[89] G. Lucignani, *Eur. J. Nucl. Med. Mol. Imaging* **35**, 209–215 (2008).

[90] J. Reubi, B. Waser, J.-C. Schaer and J. Laissue, *Eur. J. Nucl. Med. Mol. Imaging* **28**, 836–846 (2001).

[91] M. Gabriel, C. Decristoforo, D. Kendler, G. Dobrozemsky, D. Heute, C. Uprimny, P. Kovacs, E. Von Guggenberg, R. Bale and I. J. Virgolini, *J. Nucl. Med.* **48**, 508–518 (2007).

[92] P. Aschoff, M. Öksüz, B. Kemke, K. Zhernosekov, M. Jennewein, F. Rösch and H. Bihl, *Nuklearmedizin* **44**, A58(V144) (2005).

[93] P. Aschoff, B. Kemke, M. Öksüz, K. Zhernosekov, M. Jennewein, F. Rösch, and H. Bihl, *Nuklearmedizin* **44**, A59(V146) (2005).

[94] J. C. Reubi, J.-C. Schär, B. Waser, S. Wenger, A. Heppeler, J. S. Schmitt and H. R. Mäcke, *Eur. J. Nucl. Med. Mol. Imaging* **27**, 273–282 (2000).

[95] H. Zhang, J. Schuhmacher, B. Waser, D. Wild, M. Eisenhut, J. Reubi and H. Maecke, *Eur. J. Nucl. Med. Mol. Imaging* **34**, 1198–1208 (2007).

[96] G. Kramer-Marek, N. Shenoy, J. Seidel, G. Griffiths, P. Choyke and J. Capala, *Eur. J. Nucl. Med. Mol. Imaging* **38**, 1967–1976 (2011).

[97] S. Demignot, M. V. Pimm and R. W. Baldwin, *Cancer Res.* **50**, 2936–2942 (1990).

[98] B. Zhao, L. H. Schwartz and S. M. Larson, *J. Nucl. Med.* **50**, 239–249 (2009).

[99] J. Löfblom, J. Feldwisch, V. Tolmachev, J. Carlsson, S. Ståhl and F. Y. Frejd, *FEBS Lett.* **584**, 2670–2680 (2010).

[100] G. Ren, J. Webster, Z. Liu, R. Zhang, Z. Miao, H. Liu, S. Gambhir, F. Syud and Z. Cheng, *Amino Acids* 1–9 (2011).

[101] C. J. Broan, J. P. L. Cox, A. S. Craig, R. Kataky, D. Parker, A. Harrison, A. M. Randall and G. Ferguson, *J. Chem. Soc., Perkin Trans.* 2 87–99 (1991).

[102] S. Chaves, R. Delgado and J. J. R. F. Da Silva, *Talanta* **39**, 249–254 (1992).

[103] A. Heppeler, S. Froidevaux, H. R. Mäcke, E. Jermann, M. Béhé, P. Powell and M. Hennig, *Chem.-Eur. J.* **5**, 1974–1981 (1999).

[104] J. P. L. Cox, A. S. Craig, I. M. Helps, K. J. Jankowski, D. Parker, M. A. W. Eaton, A. T. Millican, K. Millar, N. R. A. Beeley and B. A. Boyce, *J. Chem. Soc., Perkin Trans.* 1 2567–2576 (1990).

[105] J. P. Andre, H. R. Maecke, M. Zehnder, L. Macko and K. G. Akyel, *Chem. Commun.* 1301–1302 (1998).

[106] K.-P. Eisenwiener, M. I. M. Prata, I. Buschmann, H.-W. Zhang, A. C. Santos, S. Wenger, J. C. Reubi and H. R. Macke, *Bioconjugate Chem.* **13**, 530–541 (2002).

[107] R. Ma, R. J. Motekaitis and A. E. Martell, *Inorg. Chim. Acta* **224**, 151–155 (1994).

[108] J. Schuhmacher, S. Kaul, G. Klivényi, H. Junkermann, A. Magener, M. Henze, J. Doll, U. Haberkorn, F. Amelung and G. Bastert, *Cancer Res.* **61**, 3712–3717 (2001).

[109] M. Eder, B. Wängler, S. Knackmuss, F. LeGall, M. Little, U. Haberkorn, W. Mier and M. Eisenhut, *Eur. J. Nucl. Med. Mol. Imaging* **35**, 1878–1886 (2008).

[110] Y. Li, A. E. Martell, R. D. Hancock, J. H. Reibenspies, C. J. Anderson and M. J. Welch, *Inorg. Chem.* **35**, 404–414 (1996).

[111] R. Ma, M. J. Welch, J. Reibenspies and A. E. Martell, *Inorg. Chim. Acta* **236**, 75–82 (1995).

[112] J. Notni, P. Hermann, J. Havlíčková, J. Kotek, V. Kubíček, J. Plutnar, N. Loktionova, P. J. Riss, F. Rösch and I. Lukeš, *Chem. Eur. J.* **16**, 7174–7185 (2010).

[113] C. L. Ferreira, E. Lamsa, M. Woods, Y. Duan, P. Fernando, C. Bensimon, M. Kordos, K. Guenther, P. Jurek and G. E. Kiefer, *Bioconjugate Chem.* **21**, 531–536 (2010).

[114] D. J. Berry, Y. Ma, J. R. Ballinger, R. Tavare, A. Koers, K. Sunassee, T. Zhou, S. Nawaz, G. E. D. Mullen, R. C. Hider and P. J. Blower, *Chem. Commun.* **47**, 7068–7070 (2011).

[115] E. Boros, C. L. Ferreira, B. O. Patrick, M. J. Adam and C. Orvig, *Nucl. Med. Biol.* **38**, 1165–1174 (2011).

[116] E. Boros, C. L. Ferreira, J. F. Cawthray, E. W. Price, B. O. Patrick, D. W. Wester, M. J. Adam and C. Orvig, *J. Am. Chem. Soc.* **132**, 15726–15733 (2010).

[117] K. Suzuki, M. Satake, J. Suwada, S. Oshikiri, H. Ashino, H. Dozono, A. Hino, H. Kasahara and T. Minamizawa, *Nucl. Med. Biol.* **38**, 1011–1018 (2011).

[118] M. Van Hulle, K. De Cremer and R. Cornelis, *Fresenius. J. Anal. Chem.* **368**, 293–296 (2000).

[119] P. T. Gomme, K. B. McCann and J. Bertolini, *Drug Discov. Today* **10**, 267–273 (2005).

[120] M. H. Hofmann, H. M. Maecke, A. Börner, E. W. Weckesser, P. Schöffski, M. O. Oei, J. S. Schumacher, M. H. Henze, A. H. Heppeler, G. M. Meyer and W. K. Knapp, *Eur. J. Nucl. Med. Mol. Imaging* **28**, 1751–1757 (2001).

[121] P. M. Smith-Jones, D. Solit, F. Afroze, N. Rosen and S. M. Larson, *J. Nucl. Med.* **47**, 793–796 (2006).

[122] P. M. Smith-Jones, D. B. Solit, T. Akhurst, F. Afroze, N. Rosen and S. M. Larson, *Nat Biotech* **22**, 701–706 (2004).

[123] P. M. Smith-Jones, S. Vallabahajosula, S. J. Goldsmith, V. Navarro, C. J. Hunter, D. Bastidas and N. H. Bander, *Cancer Res.* **60**, 5237–5243 (2000).

[124] M. Eder, A. V. Krivoshein, M. Backer, J. M. Backer, U. Haberkorn and M. Eisenhut, *Nucl. Med. Biol.* **37**, 405–412 (2010).

[125] B. Koop, S. N. Reske and B. Neumaier, *Radiochim. Acta* **95**, 39–42 (2007).

[126] J. Y. Kim, H. Park, J. C. Lee, K. M. Kim, K. C. Lee, H. J. Ha, T. H. Choi, G. I. An and G. J. Cheon, *Appl. Radiat. Isot.* **67**, 1190–1194 (2009).

[127] A. Obata, S. Kasamatsu, D. W. McCarthy, M. J. Welch, H. Saji, Y. Yonekura and Y. Fujibayashi, *Nucl. Med. Biol.* **30**, 535–539 (2003).

[128] A. B. Packard, J. F. Kronauge, E. Barbarics, S. Kiani and S. T. Treves, *Nucl. Med. Biol.* **29**, 289–294 (2002).

[129] A. Obata, E. Yoshimi, A. Waki, J. Lewis, N. Oyama, M. Welch, H. Saji, Y. Yonekura and Y. Fujibayashi, *Ann. Nucl. Med.* **15**, 499–504 (2001).

[130] A. L. Vavere and J. S. Lewis, *Dalton Trans.* 4893–4902 (2007).

[131] J. P. Holland, P. J. Barnard, D. Collison, J. R. Dilworth, R. Edge, J. C. Green, J. M. Heslop, E. J. L. McInnes, C. G. Salzmann and A. L. Thompson, *Eur. J. Inorg. Chem.* **2008**, 3549–3560 (2008).

[132] T. Z. Wong, J. L. Lacy, N. A. Petry, T. C. Hawk, T. A. Sporn, M. W. Dewhirst and G. Vlahovic, *Am. J. Roentgenol.* **190**, 427–432 (2008).

[133] K. A. Lears, R. Ferdani, K. Liang, A. Zheleznyak, R. Andrews, C. D. Sherman, S. Achilefu, C. J. Anderson and B. E. Rogers, *J. Nucl. Med.* **52**, 470–477 (2011).

[134] N. Di Bartolo, A. M. Sargeson and S. V. Smith, *Org. Biomol. Chem.* **4**, 3350–3357 (2006).

[135] S. Liu, Z. Li, L.-P. Yap, C.-W. Huang, R. Park and P. S. Conti, *Chem. Eur. J.* **17**, 10222–10225 (2011).

[136] M. Fani, L. Del Pozzo, K. Abiraj, R. Mansi, M. L. Tamma, R. Cescato, B. Waser, W. A. Weber, J. C. Reubi and H. R. Maecke, *J. Nucl. Med.* **52**, 1110–1118 (2011).

[137] R. Ferdani, D. J. Stigers, A. L. Fiamengo, L. Wei, B. T. Y. Li, J. A. Golen, A. L. Rheingold, G. R. Weisman, E. H. Wong and C. . Anderson, *Dalton Trans.* **41**, 1938–1950 (2012).

[138] C. A. Boswell, X. Sun, W. Niu, G. R. Weisman, E. H. Wong, A. L. Rheingold and C. J. Anderson, *J. Med. Chem.* **47**, 1465–1474 (2004).

[139] X. Sun, M. Wuest, G. R. Weisman, E. H. Wong, D. P. Reed, C. A. Boswell, R. Motekaitis, A. E. Martell, M. J. Welch and C. J. Anderson, *J. Med. Chem.* **45**, 469–477 (2001).

[140] Z. Liu, Y. Yan, S. Liu, F. Wang and X. Chen, *Bioconjugate Chem.* **20**, 1016–1025 (2009).

[141] R. A. Dumont, F. Deininger, R. Haubner, H. R. Maecke, W. A. Weber, and M. Fani, *J. Nucl. Med.* **52**, 1276–1284 (2011).

[142] T. D. Rae, P. J. Schmidt, R. A. Pufahl, V. C. Culotta, and T. V. O'Halloran, *Science* **284**, 805–808 (1999).

[143] K. J. Heroux, K. S. Woodin, D. J. Tranchemontagne, P. C. B. Widger, E. Southwick, E. H. Wong, G. R. Weisman, S. A. Tomellini, T. J. Wadas, C. J. Anderson, S. Kassel, J. A. Golen and A. L. Rheingold, *Dalton Trans.* 2150–2162 (2007).

[144] S. V. Deshpande, S. J. DeNardo, C. F. Meares, M. J. McCall, G. P. Adams, M. K. Moi and G. L. DeNardo, *J. Nucl. Med.* **29**, 217–225 (1988).

[145] J. Lu, A. J. Stewart, P. J. Sadler, T. J. T. Pinheiro and C. A. Blindauer, *Biochem. Soc. Trans.* **36**, 1317–1321 (2008).

[146] A. F. Prasanphanich, L. Retzloff, S. R. Lane, P. K. Nanda, G. L. Sieckman, T. L. Rold, L. Ma, S. D. Figueroa, S. V. Sublett, T. J. Hoffman and C. J. Smith, *Nucl. Med. Biol.* **36**, 171–181 (2009).

[147] N. J. Baumhover, M. E. Martin, S. G. Parameswarappa, K. C. Kloepping, M. S. O'Dorisio, F. C. Pigge and M. K. Schultz, *Bioorg. Med. Chem. Lett.* **21**, 5757–5761 (2011).

[148] R. H. Kimura, Z. Cheng, S. S. Gambhir and J. R. Cochran, *Cancer Res.* **69**, 2435–2442 (2009).

[149] L. Wei, Y. Ye, T. J. Wadas, J. S. Lewis, M. J. Welch, S. Achilefu and C. J. Anderson, *Nucl. Med. Biol.* **36**, 277–285 (2009).

[150] M. Eiblmaier, R. Andrews, R. Laforest, B. E. Rogers and C. J. Anderson, *J. Nucl. Med.* **48**, 1390–1396 (2007).

[151] G. Niu, Z. Li, J. Xie, Q.-T. Le and X. Chen, *J. Nucl. Med.* **50**, 1116–1123 (2009).

[152] L. Li, J. Bading, P. J. Yazaki, A. H. Ahuja, D. Crow, D. Colcher, L. E. Williams, J. Y. C. Wong, A. Raubitschek and J. E. Shively, *Bioconjugate Chem.* **19**, 89–96 (2007).

[153] C. J. Anderson, S. W. Schwarz, J. M. Connett, P. D. Cutler, L. W. Guo, C. J. Germain, G. W. Philpott, K. R. Zinn, D. P. Greiner, C. F. Meares and M. J. Welch, *J. Nucl. Med.* **36**, 850–858 (1995).

[154] E. C. F. Dijkers, J. G. W. Kosterink, A. P. Rademaker, L. R. Perk, G. A. M. S. van Dongen, J. Bart, J. R. de Jong, E. G. E. de Vries and M. N. Lub-de Hooge, *J. Nucl. Med.* **50**, 974–981 (2009).

[155] J. N. Tinianow, H. S. Gill, A. Ogasawara, J. E. Flores, A. N. Vanderbilt, E. Luis, R. Vandlen, M. Darwish, J. R. Junutula, S. P. Williams and J. Marik, *Nucl. Med. Biol.* **37**, 289–297).

[156] G. A. M. S. van Dongen, G. W. M. Visser, M. N. Lub-de Hooge, E. G. de Vries and L. R. Perk, *The Oncologist* **12**, 1379–1389 (2007).

[157] J. P. Holland, Y. Sheh, and J. S. Lewis, *Nucl. Med. Biol.* **36**, 729–739 (2009).

[158] J. P. Holland, V. Divilov, N. H. Bander, P. M. Smith-Jones, S. M. Larson and J. S. Lewis, *J. Nucl. Med.* **51**, 1293–1300 (2010).

[159] W. E. Meijs, J. D. M. Herscheid, H. J. Haisma and H. M. Pinedo, *Int. J. Rad. Appl. Instrum. [A]* **43**, 1443–1447 (1992).

[160] A. E. Martell and R. M. Smith, *Critical Stability Constants, Vol. 1-6*, Plenum Press, New York (1974).

[161] L. Perk, O. Visser, M. Stigter-van Walsum, M. Vosjan, G. Visser, J. Zijlstra, P. Huijgens and G. van Dongen, *Eur. J. Nucl. Med. Mol. Imaging* **33**, 1337–1345 (2006).

[162] P. K. E. Börjesson, Y. W. S. Jauw, R. Boellaard, R. de Bree, E. F. I. Comans, J. C. Roos, J. A. Castelijns, M. J. W. D. Vosjan, J. A. Kummer, C. R. Leemans, A. A. Lammertsma and G. A. M. S. van Dongen, *Clin. Cancer Res.* **12**, 2133–2140 (2006).

[163] W. B. Nagengast, E. G. de Vries, G. A. Hospers, N. H. Mulder, J. R. de Jong, H. Hollema, A. H. Brouwers, G. A. van Dongen, L. R. Perk and M. N. Lub-de Hooge, *J. Nucl. Med.* **48**, 1313–1319 (2007).

[164] J. P. Holland, E. Caldas-Lopes, V. Divilov, V. A. Longo, T. Taldone, D. Zatorska, G. Chiosis and J. S. Lewis, *PLoS One* **5**, e8859 (2010).

[165] L. Perk, M. Vosjan, G. Visser, M. Budde, P. Jurek, G. Kiefer and G. van Dongen, *Eur. J. Nucl. Med. Mol. Imaging* **37**, 250–259 (2010).

[166] W. E. Meijs, H. J. Haisma, R. P. Klok, F. B. van Gog, E. Kievit, H. M. Pinedo and J. D. M. Herscheid, *J. Nucl. Med.* **38**, 112–118 (1997).

[167] H. Hong, G. Severin, Y. Yang, J. Engle, Y. Zhang, T. Barnhart, G. Liu, B. Leigh, R. Nickles and W. Cai, *Eur. J. Nucl. Med. Mol. Imaging* **39**, 138–148 (2011).

[168] L. R. Perk, G. W. M. Visser, M. J. W. D. Vosjan, M. Stigter-van Walsum, B. M. Tijink, C. R. Leemans and G. A. M. S. van Dongen, *J. Nucl. Med.* **46**, 1898–1906 (2005).

[169] W. Zhong, J. Parkinson, M. Guo and P. Sadler, *J. Biol. Inorg. Chem.* **7**, 589–599 (2002).

[170] B. M. Zeglis, P. Mohindra, G. I. Weissmann, V. Divilov, S. A. Hilderbrand, R. Weissleder and J. S. Lewis, *Bioconjugate Chem.* **22**, 2048–2059 (2011).

[171] C. Heneweer, J. F. Holland, V. Divilov, S. Carlin and J. S. Lewis, *J. Nucl. Med.* **52**, 625–633 (2011).

[172] Y. Matsumura and H. Maeda, *Cancer Res.* **46**, 6387–6392 (1986).

[173] W. E. Meijs, H. J. Haisma, R. Van Der Schors, R. Wijbrandts, K. Van Den Oever, R. P. Klok, H. M. Pinedo and J. D. M. Herscheid, *Nucl. Med. Biol.* **23**, 439–448 (1996).

[174] I. Verel, G. W. M. Visser, R. Boellaard, M. Stigter-van Walsum, G. B. Snow and G. A. M. S. van Dongen, *J. Nucl. Med.* **44**, 1271–1281 (2003).

[175] M. J. W. D. Vosjan, L. R. Perk, G. W. M. Visser, M. Budde, P. Jurek, G. E. Kiefer and G. A. M. S. van Dongen, *Nat. Protocols* **5**, 739–743 (2010).

[176] B. Scholten, Z. Kovács, F. Tárkányi and S. M. Qaim, *Appl. Radiat. Isot.* **46**, 255–259 (1995).

[177] E. J. Knust, K. Dutschka and R. Weinreich, *Appl. Radiat. Isot.* **52**, 181–184 (2000).

[178] R. H. Seevers and R. E. Counsell, *Chem. Rev.* **82**, 575–590 (1982).

[179] M. Berman, L. Braverman, E. Burke and J. Burke, *J. Nucl. Med.* **16**, 857–860 (1975).

[180] M. A. Moroz, I. Serganova, P. Zanzonico, L. Ageyeva, T. Beresten, E. Dyomina, E. Burnazi, R. D. Finn, M. Doubrovin and R. G. Blasberg, *J. Nucl. Med.* **48**, 827–836 (2007).

[181] C. R. Divgi, N. Pandit-Taskar, A. A. Jungbluth, V. E. Reuter, M. Gönen, S. Ruan, C. Pierre, A. Nagel, D. A. Pryma, J. Humm, S. M. Larson, L. J. Old and P. Russo, *The Lancet Oncology* **8**, 304–310 (2007).

[182] J. G. Tjuvajev, R. Finn, K. Watanabe, R. Joshi, T. Oku, J. Kennedy, B. Beattie, J. Koutcher, S. Larson and R. G. Blasberg, *Cancer Res.* **56**, 4087–4095 (1996).

[183] M. Doubrovin, V. Ponomarev, T. Beresten, J. Balatoni, W. Bornmann, R. Finn, J. Humm, S. Larson, M. Sadelain, R. Blasberg and J. Gelovani Tjuvajev, *Proc. Natl. Acad. Sci.* **98**, 9300–9305 (2001).

[184] M. Benezra, O. Penate-Medina, P. B. Zanzonico, D. Schaer, H. Ow, A. Burns, E. DeStanchina, V. Longo, E. Herz, S. Iyer, J. Wolchok, S. M. Larson, U. Wiesner and M. S. Bradbury, *J. Clin. Invest.* **121**, 2768–2780 (2011).

[185] J. A. Carrasquillo, N. Pandit-Taskar, J. A. O'Donoghue, J. L. Humm, P. Zanzonico, P. M. Smith-Jones, C. R. Divgi, D. A. Pryma, S. Ruan, N. E. Kemeny, Y. Fong, D. Wong, J. S. Jaggi, D. A. Scheinberg, M. Gonen, K. S. Panageas, G. Ritter, A. A. Jungbluth, L. J. Old and S. M. Larson, *J. Nucl. Med.* **52**, 1173–1180 (2011).

[186] P. Zanzonico, J. O'Donoghue, J. Chapman, R. Schneider, S. Cai, S. Larson, B. Wen, Y. Chen, R. Finn, S. Ruan, L. Gerweck, J. Humm and C. Ling, *Eur. J. Nucl. Med. Mol. Imaging* **31**, 117–128 (2004).

[187] G. Reischl, D. S. Dorow, C. Cullinane, A. Katsifis, P. Roselt, D. Binns and R. J. Hicks, *J. Pharm. Pharm. Sci* **2**, 203–211 (2007).

[188] E. Lopci, A. Chiti, M. Castellani, G. Pepe, L. Antunovic, S. Fanti and E. Bombardieri, *Eur. J. Nucl. Med. Mol. Imaging* **38**, 28–40 (2011).

[189] G. Sgouros, R. Hobbs, F. Atkins, D. Van Nostrand, P. Ladenson and R. Wahl, *Eur. J. Nucl. Med. Mol. Imaging* **38**, 41–47 (2011).

[190] L. Tran, A. D. R. Huitema, M. H. van Rijswijk, H. J. Dinant, J. W. Baars, J. H. Beijnen and W. V. Vogel, *Hum. Antib.* **20**, 29–35 (2011).

[191] S. Reddy, C. C. Shaller, M. Doss, I. Shchaveleva, J. D. Marks, J. Q. Yu and M. K. Robinson, *Clin. Cancer Res.* **17**, 1509–1520 (2011).

[192] D. A. Pryma, J. A. O'Donoghue, J. L. Humm, A. A. Jungbluth, L. J. Old, S. M. Larson and C. R. Divgi, *J. Nucl. Med.* **52**, 535–540 (2011).

[193] C. Kuntner, T. Wanek, M. Hoffer, D. Dangl, M. Hornof, H. Kvaternik and O. Langer, *Mol. Imaging Biol.* **13**, 222–226 (2011).

[194] P. J. Fraker and J. C. Speck Jr, *Biochem. Biophys. Res. Commun.* **80**, 849–857 (1978).

[195] W. M. Hunter and F. C. Greenwood, *Nature* **194**, 495–496 (1962).

[196] A. S. McFarlane, *Nature* **182**, 53–53 (1958).

[197] A. S. El-Wetery, A. A. El-Mohty, S. Ayyoub and M. Raieh, *J. Label. Compd. Radiopharm.* **39**, 631–644 (1997).

[198] R. F. Schneider, E. L. Engelhardt, C. C. Stobbe, M. C. Fenning and J. D. Chapman, *J. Label. Compd. Radiopharm.* **39**, 541–557 (1997).

[199] A. Stahlschmidt, H.-J. Machulla, G. Reischl, E. E. Knaus and L. I. Wiebe, *Appl. Radiat. Isot.* **66**, 1221–1228 (2008).

[200] E. Akgün, P. S. Portoghese, M. Sajjad and H. A. Nabi, *J. Label. Compd. Radiopharm.* **50**, 165–170 (2007).

[201] M. Glaser, V. A. Carroll, D. R. Collingridge, E. O. Aboagye, P. Price, R. Bicknell, A. L. Harris, S. K. Luthra and F. Brady, *J. Label. Compd. Radiopharm.* **45**, 1077–1090 (2002).

[202] M. K. Robinson, M. Doss, C. Shaller, D. Narayanan, J. D. Marks, L. P. Adler, D. E. González Trotter and G. P. Adams, *Cancer Res.* **65**, 1471–1478 (2005).

[203] D. R. Collingridge, M. Glaser, S. Osman, H. Barthel, O. C. Hutchinson, S. K. Luthra, F. Brady, L. Bouchier-Hayes, S. J. Martin, P. Workman, P. Price and E. O. Aboagye, *Br. J. Cancer* **89**, 1327–1333 (2003).

[204] M. Glaser, D. R. Collingridge, E. O. Aboagye, L. Bouchier-Hayes, O. C. Hutchinson, S. J. Martin, P. Price, F. Brady and S. K. Luthra, *Appl. Radiat. Isot.* **58**, 55–62 (2003).

[205] J. Koziorowski, C. Henssen and R. Weinreich, *Appl. Radiat. Isot.* **49**, 955–959 (1998).

[206] V. V. Belov, A. A. Bonab, A. J. Fischman, M. Heartlein, P. Calias, and M. I. Papisov, *Mol. Pharm.* **8**, 736–747 (2011).

[207] D. D. Rossouw and H. H. Coenen, *Nucl. Med. Biol.* **30**, 373–380 (2003).

[208] M. B. Chenoweth and L. P. McCarty, *Pharmacological Reviews* **15**, 673–707 (1963).

[209] D. F. McMillen and D. M. Golden, *Annu. Rev. Phys. Chem.* **33**, 493–532 (1982).

6

THE RADIOPHARMACEUTICAL CHEMISTRY OF TECHNETIUM AND RHENIUM

JONATHAN R. DILWORTH

Department of Chemistry, University of Oxford, Oxford, UK

SOFIA I. PASCU

Department of Chemistry, University of Bath, Bath, UK

6.1 INTRODUCTION

6.1.1 Technetium Chemistry

The position of technetium in the centre of the transition metal block enables it to exhibit a wide range of oxidation states (−1 to +7) and coordination numbers (generally 4 to 7). The accessibility to both high and low oxidation states and the availability of d orbitals of appropriate symmetry means that multiple bonding via σ and π combinations can play an important role in complex stabilisation. This ability to form multiple bonds means that even monodentate ligands can be sufficiently robust to survive in a biological environment.

Within a radiopharmaceutical context, the highly appropriate nuclear decay properties of ^{99m}Tc coupled with the availability of a generator system and facile radiolabelling has meant that nuclear medicine uses this isotope extensively. In fact, the vast majority of nuclear medical investigations worldwide are still carried out with technetium, and it has been estimated that worldwide at least 70,000 technetium SPECT scans are made daily. In the USA alone some 19 million Tc scans were performed in 2007, and approximately half of these involved cardiac imaging. Such numbers are difficult to extrapolate with any accuracy, but it is realistic to assume that around 30-40 million technetium scans are conducted annually across the world.

Although publications on technetium chemistry have remained fairly constant over the past decade, the number of papers in nuclear medicine journals has declined markedly. This indicates that while interest in the coordination chemistry of technetium chemistry and animal studies has been sustained, translation to the clinical and commercial arenas has all but disappeared. For chemistry, the regulatory difficulties of working with the long-lived ^{99}Tc radioisotope has means that there are now few centres worldwide with the capability of complete chemical characterisation of technetium complexes, and this does not augur well for the future of fundamental research. Another important factor is that the landscape of molecular imaging has changed dramatically over the past decade with the advent of PET. This has had a significant impact on the development and clinical translation of technetium-based imaging agents. Part of the driving force toward PET using isotopes such as ^{18}F and ^{11}C rather than SPECT has been the obvious advantage of being able to radiolabel a targeting molecule with minimal changes in the overall structure, which assumes particular importance in brain imaging where translocation across the BBB is crucial. However, this and the greater sensitivity and ease of quantification of PET has to be offset against the substantial investment in cyclotrons, hot cells, and automated synthesis units and the technically demanding radiolabelling procedures. It seems likely that this technology will be restricted to highly developed areas for the foreseeable future despite the pressing healthcare needs of the poorer countries [1]. The ultimate decision on whether to use PET or SPECT depends on balancing a number of issues, but it is not axiomatic

The Chemistry of Molecular Imaging, First Edition. Edited by Nicholas Long and Wing-Tak Wong.
© 2015 John Wiley & Sons, Inc. Published 2015 by John Wiley & Sons, Inc.

that PET will be the best choice in all situations. For animal imaging the modern SPECT imaging systems can achieve better resolution than PET, although the sensitivity remains some 15 times lower as shown *inter alia* by imaging of phantoms and comparative tumour imaging with ^{99m}Tc or ^{18}F labelled nanoparticles [2]. In the clinical arena it has been shown that for SPECT, at least in the case of ^{99m}Tc diphosphonate bone scans, sophisticated computer analysis techniques can be applied to permit accurate quantification [3]. Comparative studies of PET and SPECT agents in imaging patients with Parkinson's disease [4] or chest lesions [5] have shown that the SPECT and PET data provide equally useful diagnostic information.

6.1.2 Rhenium

Rhenium is located in the same transition metal group as technetium, and the two elements form many isostructural complexes. This leads to non-radioactive Re complexes frequently being used as models for technetium for structure and biological characteristics such as substrate binding. However, there are also some significant differences in their chemistries. As a third row transition element, rhenium complexes are harder to reduce and easier to oxidise than their technetium counterparts. As a consequence, the formation of perrhenate through oxidation is favoured more than for pertechnetate, and cationic rhenium complexes in high oxidation states with non-reducing ligands frequently contain perrhenate anions. Rhenium complexes are also more kinetically inert than their technetium analogues and complexation often requires more forcing conditions including heating. This can create problems in terms of labelling heat sensitive biomolecules with rhenium radioisotopes. Insertion of rhenium into a bifunctional chelator (BFC) may have to be carried out prior to conjugation to the biomolecule (preconjugation labelling). This increases the length of the labelling procedure and the risk of radiation exposure. It still remains a challenge to produce BFCs that can be labelled with a rhenium radioisotope at room temperature. A useful review of the BFCs available for the radiorhenium labelling of biomolecules is available [6].

In addition to its role as a surrogate for technetium, the major interest in rhenium lies in the potential therapeutic applications of the β-emitting ^{186}Re and ^{188}Re radioisotopes. The use of very toxic radioisotopes places stringent requirements on the stabilities of the complexes and the efficiency of targeting. As a consequence, the development of rhenium-based therapeutic agents has been much slower than for technetium imaging agents, and testing has been generally confined to animal studies. In fact, the only regular clinical use for rhenium complexes is that of the rhenium diphosphonate complex ^{188}ReHEDP for the palliation of bone pain in terminal cancer patients [7], although a number of other compounds have undergone limited clinical trials.

6.1.3 Radioisotopes of Technetium and Rhenium

As discussed above, in addition to the ideal nuclear properties, the ready availability of technetium is an important contributing factor to its popularity. However, a significant problem arose around 2000 with a potential shortage of technetium-99 m arising from the gradual decommissioning of nuclear reactors, which provided the bulk of the ^{99}Mo for the clinical generator systems as a by-product of ^{235}U fission. By 2009 the five major reactors producing ^{99}Mo were nearing the end of their productive lives [8]. However, a number of groups showed that ^{99m}Tc could be satisfactorily produced in medium energy cyclotrons in TBq quantities by the ^{100}Mo(p,2n)^{99m}Tc nuclear reaction and that the imaging characteristics of the isotope were identical with that produced from a generator [9]. It therefore appears that supplies of the ^{99m}Tc isotope can be secured into the future. The properties of the relevant isotopes of Tc and Re are summarised in Table 6.1.

The positron emitting isotope ^{94m}Tc can be cyclotron produced by the ^{94}Mo(p,n) nuclear reaction in good yields and purity [10]. There was surge of interest in the use of this radioisotope in the 1990s based on the concept that this would provide PET analogues of existing Tc agents. It was shown, for instance, that the antibody fragment 'CEA-scan' could be radiolabelled in good yield in a directly analogous manner to ^{99m}Tc [11]. The ^{94m}Tc labelled sestamibi compound was also used to image multidrug resistance in mice [12]. However, the decay characteristics of ^{94m}Tc are not ideal, and the high (2.47 MeV)

TABLE 6.1   Radioisotopes of Technetium Used in Imaging or therapy.

Isotope	Half-Life	Decay Products	Production
^{99m}Tc	6h	γ (140 KeV, 98.6%) γ (142.7 KeV, 1.4% Auger (2.1 KeV)	^{99}Mo/^{99m}Tc generator
^{94m}Tc	52min	β$^+$ (2.4 MeV, 72%)	Cyclotron from ^{94}Mo
^{186}Re	90h	β (1.07 MeV, 71%) γ (137MeV, 9%)	Reactor
^{188}Re	17h	β (2.1MeV, 100%) γ (155MeV, 15%)	^{188}W/^{188}Re generator

energy of the positron emission reduces both the resolution achievable and increases the radiation dose to the patient [13]. The current status of [94m]Tc imaging has recently been reviewed [14].

[186]Re is produced in a nuclear reactor via neutron bombardment of natural Re. The presence of both [185]Re (37.4%) and [187]Re (62.6%) in natural Re creates [188]Re, which can be removed by allowing it to decay, but this also reduces the [186]Re activity. The use of isotopically enriched [185]Re substantially increases the specific activity. The β-emission energy of [186]Re of 1.08 MeV has a range of about 5 mm in tissue and is appropriate for treatment of small tumours. In principle, the long half-life of 3.7 days would make this isotope suitable for labelling of large biomolecules such as antibodies that have slow pharmacokinetics. [188]Re can be obtained as very dilute solutions of $[^{188}ReO_4]^-$ from a [188]W/[188]Re generator analogous to that used for [99m]Tc [15]. A 0.5 Ci generator can provide enough activity to treat several hundred patients. This isotope has β emissions of 2.1 MeV with a tissue penetration of 11 mm with half-life of 17 h. This is relatively short, so targeting needs to be rapid and efficient. The Re isotopes have a potential advantage over another therapeutic radionuclide ^{90}Y in that any 'free' [90]Y produced is rapidly taken up by bone, whereas Re is excreted fairly rapidly as $[ReO_4]^-$.

6.1.4 Reviews on Technetium and Rhenium

Over the past decade there have been a considerable number of reviews on the coordination and structural chemistry of technetium and rhenium [16–31] and references to other reviews are given below in the appropriate context. A book dedicated solely to the chemical and radiopharmaceutical applications of technetium has also been published [32]. There is also a series of books edited by U. Mazzi and M. Nicolini of the proceedings of a biennial conference on technetium and rhenium that contain much relevant material. The chapters in *Comprehensive Coordination Chemistry II* on technetium (Alberto, ca. 750 references) [20] and rhenium (Abram, 1230 references) [19] are indeed comprehensive and contain a wealth of useful information. In view of the fairly extensive review literature available, this chapter does not attempt to be comprehensive but rather to give an overview of the development of the coordination chemistry of Tc and Re relevant to nuclear medicine with an emphasis on key ligand systems and imaging agents. The majority of the treatment focuses on the more important core structures, but there are also sections dealing with specific ligand systems and conjugates.

6.2 TECHNETIUM AND RHENIUM RADIOPHARMACEUTICAL CHEMISTRY

6.2.1 Monoxo Complexes of Technetium and Rhenium

In the presence of mild reductants such as stannous chloride or dithionite, pertechnetate reacts with tetradentate ligands, which are not themselves strongly reducing to give complexes containing the TcO^{3+} core Generally, rhenium direct analogues can also be made from perrhenate, although in this case stronger reductants may be needed or Re(V) precursors such as $[ReOCl_4]^-$ or $[ReOCl_3(PPh_3)_2]$ can be employed. The geometry found is almost exclusively square pyramidal with the multiply bonded oxygen occupying the apical site. The metal ligand interaction is generally regarded as being a double bond comprising a σ and a π component with a net charge of 2⁻. The strong *trans* effect exerted by the oxo group generally precludes the binding of ligands in the sixth site. For reasons not yet fully understood, some ligand systems favour equilibrium between MO and linear M_2O_3 species. The ensuing discussions of mono-oxo complexes is organised along traditional lines depending on the donor atoms on the co-ligands. Attention is focused on tetradentate ligands because these are the most commonly used for radiopharmaceutical applications because they confer greater kinetic stability than equivalent combinations of bidentate or monodentate ligands.

6.2.1.1 N₄ Donor Ligands One of the earliest Tc(V) oxo complexes to be reported was the neutral amine-oxime complex [TcO(PnAO)] **1** [33]. A series of TcPnAO complexes substituted at the central backbone carbon were shown to exist as *syn* and *anti* isomers, which interconverted in water and which were ascribed to hydration *trans* to the oxo group and proton transfer from the water, inverting the oxo-group [34]. Redistribution of the backbone methyl groups created the HMPAO ligand and the technetium complex **2** [35], which is still in clinical use an imaging agent for cerebral blood flow (CBF) under the name Ceretec (Figure 6.1). The neutral complex traverses the BBB, and washout is believed to be prevented by enzymatic conversion to a more hydrophilic species. A similar mechanism enables the use of Tc(HMPAO) for labelling lymphocytes to image infection [36]. Unusually, it has not been possible as yet to make the Re(V) analogues with these amine-oxime ligands, because under a range of conditions it appears that hydrolysis of the C=N bond occurs.

Labelling of tetraalanine with ^{99m}Tc under standard conditions gives two complexes, one of which is the expected mono-oxo complex **3** and the other which was formulated as containing a macrocyclic variant of the ligand **4** (Figure 6.2). The structure of **4** was confirmed by X-ray crystallography [37]. No rhenium complexes of this ligand system are reported.

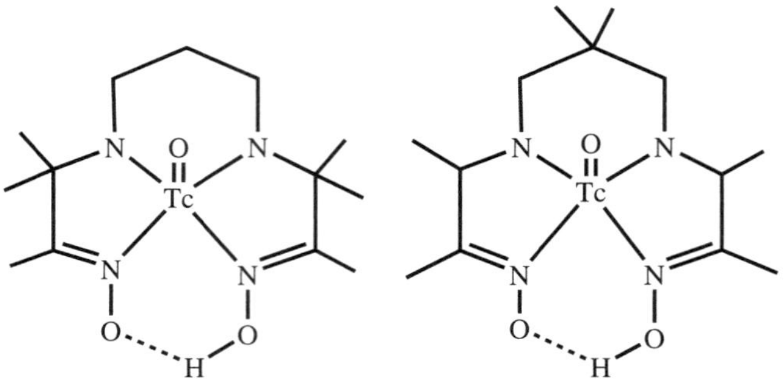

FIGURE 6.1 Amineoxime complexes.

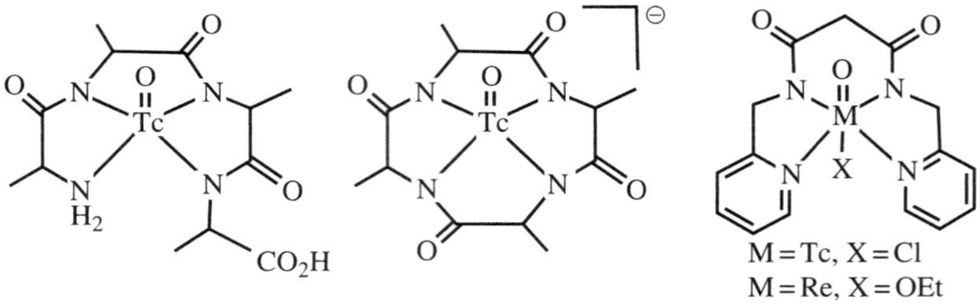

M = Tc, X = Cl
M = Re, X = OEt

FIGURE 6.2 Amineamido and amido ligands.

However, a related diammidopyridyl ligand forms both Tc and Re water-soluble mono-oxo complexes **5** and X-ray structures of both were described, but radiolabelling was not reported [38].

Porphyrins are potentially attractive ligands for technetium and rhenium because the complexes are extremely robust and also fluorescent. Generally, the insertion of the metals requires fairly forcing conditions such as dichlorobenzene under reflux [39]. However, it has been reported that a Tc(V) mono-oxo complex of octaethylporphyrin **6** can be prepared from pertechnetate by heating under reflux in glacial acetic acid, but radiolabelling and biodistribution were not reported (Figure 6.3) [40].

Haematoporphyrin **7** was radiolabelled with ^{99m}Tc, which was shown to bind at the carboxyl groups [41, 42]. Strong retention in adenocarcinomas was observed *in vivo*. This nonspecific uptake appears to be a common feature of metalloporphyrins and is exploited in photodynamic tumour therapy. The carboxyl-substituted tetraphenylporphyrin **8** has been labelled with ^{99m}Tc at the peripheral carboxylates, and the compound was shown to accumulate in a variety of murine tumours [43]. This binding to the exocyclic donors shows that for any porphyrin-bearing potential donor groups, metallation within the porphyrin ring cannot be assumed. An analogous porphyrin with pendant cyclam groups has also been labelled with ^{99m}Tc [44, 45]. It has been claimed that the sulphonated porphyrin **9** forms a rhenium(V) oxo-complex directly from perrhenate in water at 100°C with stannous tartrate as reductant. However, the structure was not verified, and some form of coordination to the sulphonate groups cannot be excluded.

The strong nonspecific binding of porphyrins is an issue when using small biologically active molecules as targeting vectors, and conjugation to larger groups such as antibody fragments would appear to be required to achieve specific targeting. The size of the targeting group will need to be selected to ensure the pharmacokinetics are not too slow for the 6 hr half-life of ^{99m}Tc. This approach does not appear to have been explored as yet for technetium but has been with other radioisotopes.

6.2.1.2 N_xO_{4-x} Donor Ligands The major types of ligand in this category that have been explored have been Schiff bases. There are reported examples of both Tc and Re mono-oxo complexes with well-known ligands such as salen and salphen, but these have not been used in a radiopharmaceutical context and will not be discussed further. Polyaminocarboxylates such as DTPA, DOTA and NOTA have been used extensively with other radionuclides as bifunctional chelators, but their behaviour with technetium and rhenium is not straightforward. Reaction of [TcOCl$_4$]$^-$ with H$_4$EDTA in anhydrous DMF gives an unusual seven coordinate Tc(V) oxo-complex [TcO(EDTA)]$^{2-}$ isolated as a barium salt [46]. The two nitrogens are coordinated approximately *trans* to the oxo-group with four carboxylates in an equatorial plane. With the high *trans* effect of the

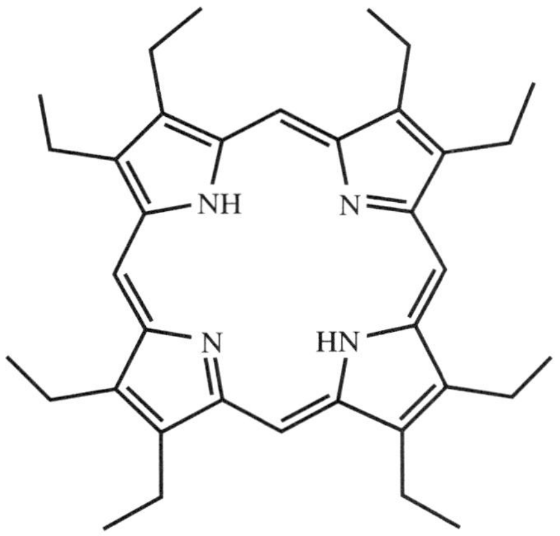

FIGURE 6.3 Porphyrin ligands and complexes.

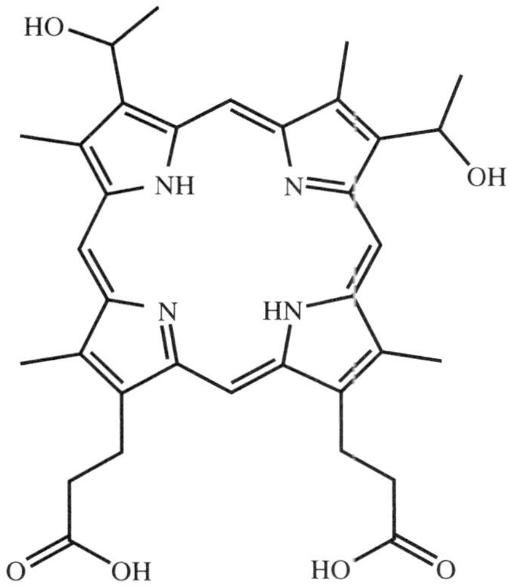

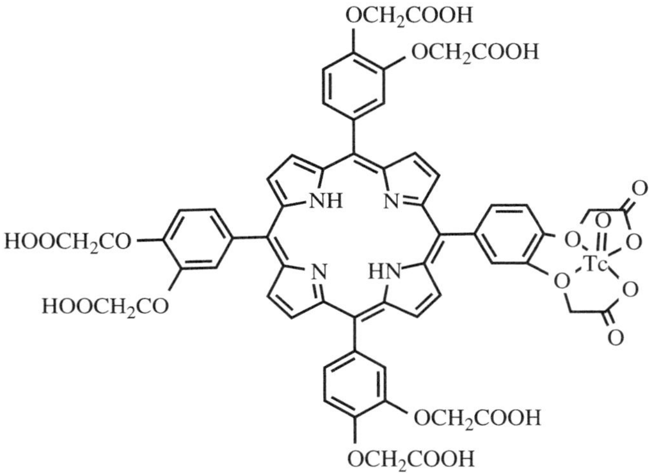

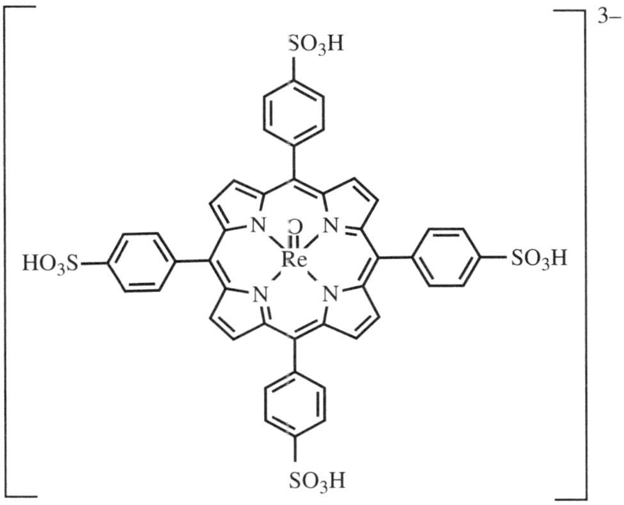

oxo group, the complexes are quite labile. Even weak donors such as ethylene glycol are able to bind, and two carboxylate oxygens are displaced from the coordination sphere [47]. The structure of the Tc complex formed with DTPA is not certain, although it has been suggested to be a mono-oxo species with the DTPA ligand bound only via carboxyl oxygens. It has been widely used for kidney imaging where it indicates glomular filtration rates. It has also been used for the labelling of biomolecules. Several reviews have appeared covering different features of the radiolabelling of biomolecules with technetium and rhenium [17, 31, 48–51].

6.2.1.3 N_xS_{4-x} Donor Ligands This classification encompasses the largest group of ligands used in conjunction with the Tc(V) and Re(V) oxo cores. Many variations have been reported and all show high serum stability. This section contains selected examples that have been tested *in vivo* or show interesting features of their coordination chemistry. One prevalent type is the diammine dithiolates variously designated as DADT or BAT ligands. A variety of backbone lengths have been explored, but the ethylene variant has been the most common choice. Figure 6.4 shows some variants on the theme. Methylation of one of the backbone nitrogens in **10** ensures that a neutral complex is formed on reaction with pertechnetate. In common with many other oxo-complexes of tetradentate ligands, the monosubstituted derivatives or conjugates lack C_2 symmmetry and *syn* or *anti* isomers with respect to the oxo group are possible. Conjugation to biomolecules can be achieved

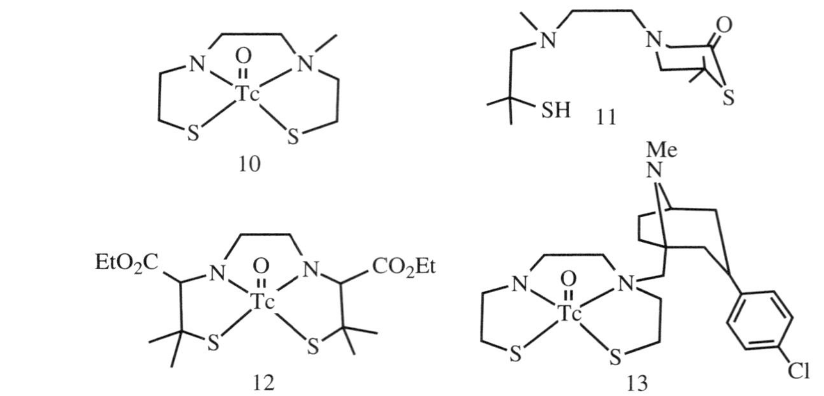

FIGURE 6.4   Diamminedithiolate (DADT or BAT) ligands and complexes.

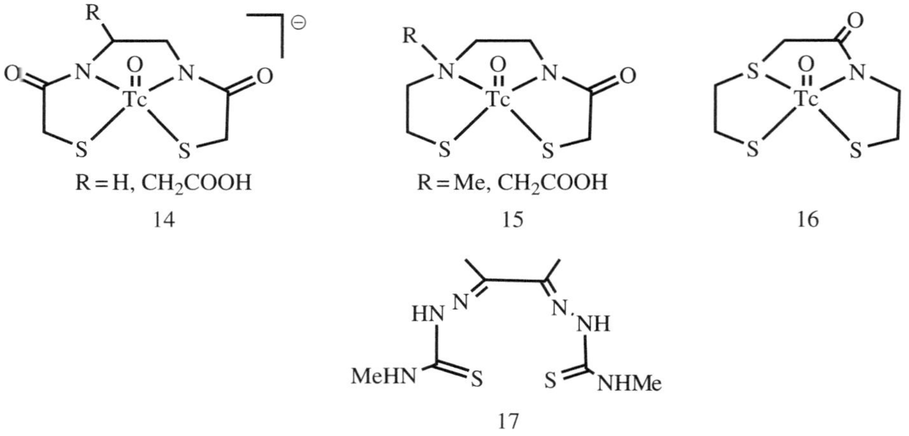

FIGURE 6.5   Diamidedithiolate (DADS) amidedithiolate and bis(thiosemicarbazone) ligands and complexes.

via attachment of a carboxylate at the methylated nitrogen and the use of the cyclic thiolester in **11** provides an elegant way to achieve this, forming an amide link and thiolate group [52]. This has been used to ^{99m}Tc-label chemotactic peptides for the imaging of inflammation [53] and bombesin for tumour imaging [54], both showing good targeting *in vivo*. A solid phase approach to the synthesis of Tc complexes of these ligands by attachment of the thiolate sulphur to a bead via an iodoacetyl group has the advantage that the large excess of ligand remains bound to the bead with only the complex in solution. This minimises the number of steps required to obtain pure complex and thereby operator dose. This method was used to Tc label chemotactic peptides in high yields [55].

The neutrality and lipophilicity of this type of complex means that they can traverse the BBB; this was exploited in complex **12,** which is in clinical use under the name Neurolite [56, 57]. This complex has chiral carbon centres. While the DD form is rapidly washed out of the brain, the LL form is enzymatically hydrolysed to a monocarboxylate, which is then retained. Interestingly, the LL form showed no brain retention in animal studies but did in man [58] – a caveat of the dangers of extrapolating from pre-clinical studies to the clinic. An example of a second-generation Tc DADT complex for imaging dopamine transporter sites in the brain is Trodat (**13**), which was shown in clinical trials to give good images despite the location of the binding site within the G-protein cavity [59].

Variants of the N_2S_2 coordination motif are provided by the introduction of amide groups in the DADS ligand systems. The CO groups can be located in a variety of positions with symmetric versions being preferred to avoid geometric isomers. Regardless of the substitution pattern, stable anionic Tc(V) and Re(V) complexes can be made in high yield directly from the tetra-oxo metallates. Complex **14** (R=H) was among the first to be investigated with Tc [60, 61], and Tc and Re complexes of the derivative with a carboxyl group on the ethylene backbone (R=CH$_2$COOH) were also described (Figure 6.5) [62].

In order to avoid the negative charge in bioconjugates the series of ligands with only one amide (**15**), designated MAMA were introduced [60] This ligand type forms very stable complexes with TcO and ReO cores, and the derivative with R=CH$_2$COOH has been used extensively to label biomolecules [62]. These are covered together with many other examples in two reviews that have appeared on the labelling of small biomolecules with ^{99m}Tc [17, 63, 64]. A further permutation of

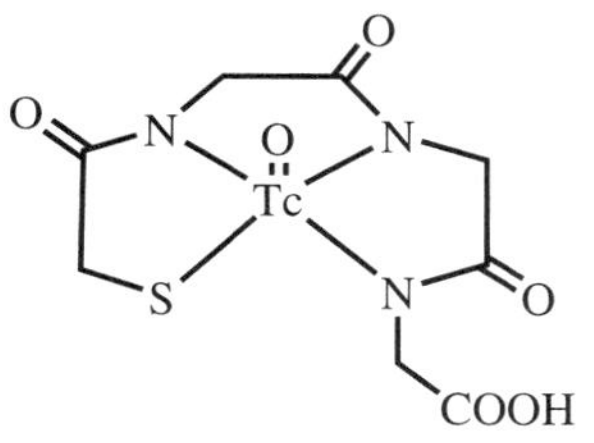

FIGURE 6.6 Aminoacid based and DMSA complexes.

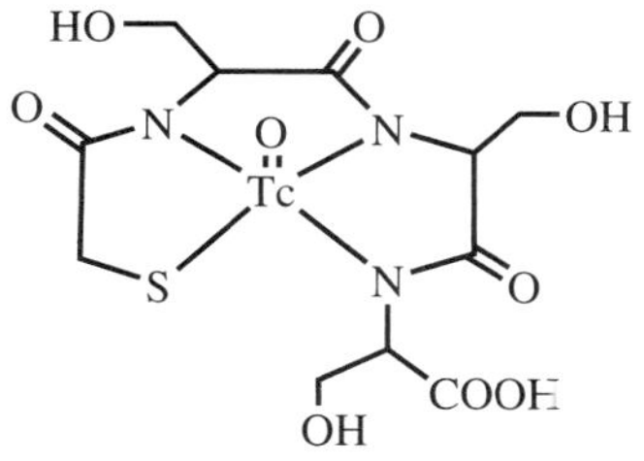

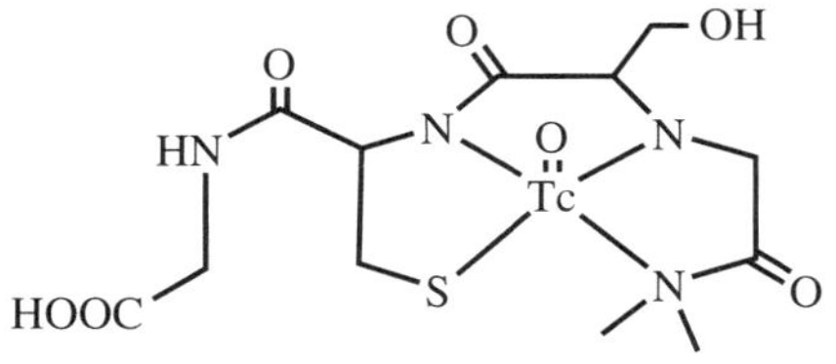

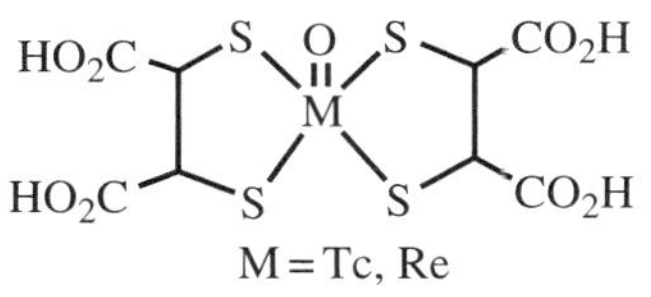

the monoamide ligand theme is the replacement of the neutral amine donor by thioether. The ^{99}Tc and Re(V) oxo complexes of a series of such ligands (**16**) have been reported, and variations of backbone lengths and substitution patterns were shown to have little impact on stability or ease of synthesis. The X-ray structures of three Re(V) oxo-complexes were reported [65].

Another N_2S_2 ligand system that has been investigated with technetium features bis(thiosemicarbazones) such as ATSM (**17**) and a variant with a single methyl group on the backbone (KTS). Reaction of $[^{99m}TcO_4]^-$ with **17** in the presence of stannous chloride gives a very stable complex that has not as yet been fully characterised. It has been suggested [66, 67] that it is a neutral Tc(IV) oxo-complex, but a Tc(V) complex with an anionic ligand such as halide *trans* to the oxo-group would seem more likely. An analogous ^{99m}Tc complex derived from a bis(thiosemicarbazone) of glucose has also been prepared with ^{99m}Tc. Interestingly, in animal studies this crosses the blood brain barrier suggesting possible applications for brain imaging [68].

Amino acids are an obvious source of amide donors, and Tc and Re complexes of both peptides and ligands containing peptide sequences have been widely pursued. The first to be described in detail by Fritzberg et al [69, 70]. was MAG_3 (mercaptoacetylcysteinetriglycine) (**18**), which has found widespread use as a kidney imaging agent (TechneScan) (Figure 6.6). This complex has also been widely used for the ^{99m}Tc labelling of biomolecules, and examples appear in the reviews quoted above. The potential issue of structural isomers has been discussed in a review covering Tc radiopharmaceuticals [71]. The X-ray structure of $ReOMAG_3$ has been reported [72], and it was shown that pre-conjugation of somastatin analogue peptides was preferable when labelling with ^{188}Re [73]. Biodistribution studies in mice showed high tumour uptake for the conjugate with values comparable to those achieved by direct labelling [74]. A similar approach has been used to radiolabel monoclonal antibodies with ^{188}Re for RAIT (radioactive immune therapy) [75]. A comparison of labelling MORFs (synthetic DNA equivalents) using ^{188}ReMAG_3 and ^{90}Y-DOTA showed some loss of both radionuclides although both were virtually intact after 48 h [76, 77]. Other examples of radiorhenium labelling of biomolecules using MAG_3 analogues appear in a review of bifunctional chelators for Re [6].

It was later reported that the Tc complex of the triserine analogue MAS_3 (**19**) was a useful alternative to MAG_3 for both renal imaging and protein labelling [78]. A comparative study of the Tc labelling of neutrophil elastase inhibitor protein (HNE-2) in primates using Tc-MAG_3, Tc$MSER_3$, Tc-DTPA, and Tc-HYNIC was carried out [79]. In terms of SPECT imaging, all of the compounds showed similar levels of accumulation in the lesion, although there were differences in the distributions in the major organs. This accentuates the problem of selecting an optimal bifunctional chelator for a given protein because, although based on very distinct Tc coordination chemistries, the actual differences in targeting ability may be relatively incremental.

An alternative to the bifunctional chelator/conjugation approach is to engineer a Tc binding site within a protein sequence. The use of the Cys-Gly-Cys, Gly-Gly-Cys, and Lys-Gly-Cys motifs for binding ^{99m}Tc in a peptide system for targeting somastatin receptors has been reported, and the binding of the Tc did not compromise receptor binding. In fact, the affinities of some of the Re(V) analogues were higher than the free proteins. The ^{99m}Tc labelled complexes showed good tumour retention in rats [80]. Similar Tc and Re binding sequences were used to label a protein to target the GPIIb/IIIa receptor as a means of thrombus imaging, and again metal binding did not interfere with biological activity [81]. The RP414 ligand system (**20**) for binding Tc or Re contains the dimethylglycine-serine-cysteine-glycine sequence and the ^{99m}Tc labelled form has been successfully conjugated to neurotensin for tumour imaging *in vivo* [82]. An interesting solid phase synthetic approach involving the initial binding of **20** to a gold surface via the cysteinyl sulphur has been reported [83]. On binding Tc, the gold-sulphur bond cleaves and the Tc complex moves into solution, leaving the excess ligand on the surface, thereby increasing the specific activity of the product.

A great advantage of utilising a peptide binding motif for the metal is that it can readily be fine-tuned to optimise *in vivo* distribution. An illustrative example is provided by the Tc Depreotide system (**22**, also known as NeoTect) for the imaging of somastatin receptors, which are overexpressed in a range of cancer types (Figure 6.7). It comprises a variant of the cyclic octreotide and a Dap-Lys-Cys (Dap = 2,3 diaminopropionic acid, a non-naturally occurring amino acid) motif to bind the TcO^{3+} core. This compound has been clinically approved and is in regular use for imaging of pulmonary masses [84, 85]. However, when it comes to using the same system for radiotherapy using ^{188}Re, the biodistribution creates significant problems. *In vivo* Tc-Depreotide shows high retention in the kidneys, which would be a significant problem for the ^{188}Re species because it would result in unacceptably high doses to a major internal organ. Also the tumour-to-muscle ratio for the retained radionuclide is somewhat on the low side. Through some elegant peptide synthesis both problems were addressed. Modification of the cyclic peptide by interchanging cysteine and phenylalanine and substituting threonine for valine improved the IC$_{50}$ value from 1.5 nM to 0.1 nM. Also, modification of the metal binding motif to that shown in **23** reduced the kidney uptake from 152% ID/g to 5% ID/g. The Re compound **23** is now in clinical trials, and the above illustrates the much more stringent measures that need to be taken when using a radionuclide for therapy.

The chemistry of Re and Tc with dithiolate ligands can be complex, but the DMSA ligand forms well defined and very stable complexes (**21**) with both elements [86]. The Tc complex has found some use for imaging kidney function and is remarkably stable *in vivo,* being excreted essentially intact [87]. The Re complex can be converted readily into a dianhydride, which was then conjugated under mild conditions to two molecules of salmon calcitonin. Acetylhydrazine can be used as a reductant for the Re; it has the advantage of giving rise to a single isomer [88].

A somewhat different strategy has been used to achieve targeting of rhenium and technetium complexes using a combination of tri- and mono-dentate ligands – the so called '3 + 1' approach (Figure 6.8). The Re(V) complex **24** is designed to target dopamine transporter sites and contains an α-tropanol derivative linked via a monodentate thiol, and the complex

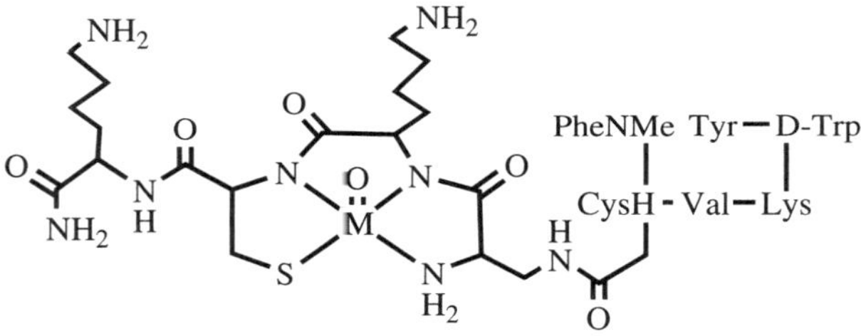

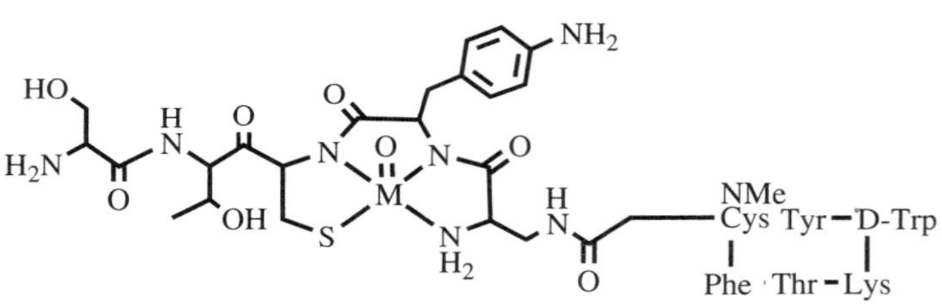

FIGURE 6.7 Peptide-based ligands for Tc and Re.

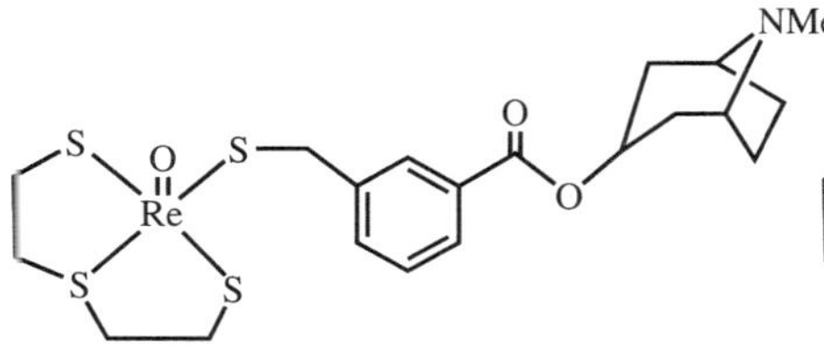

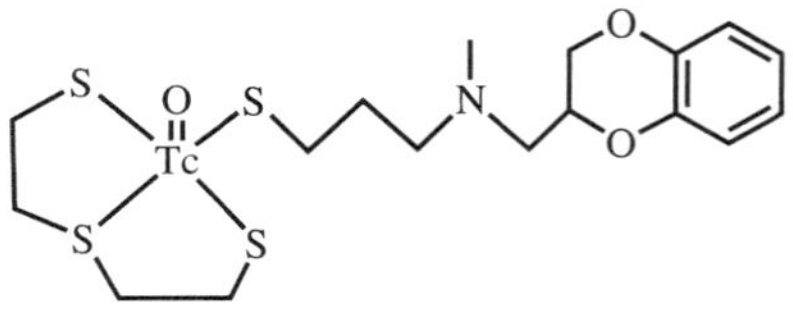

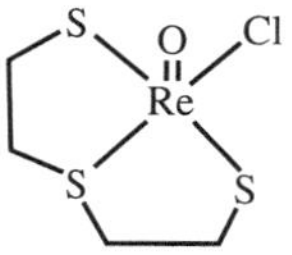

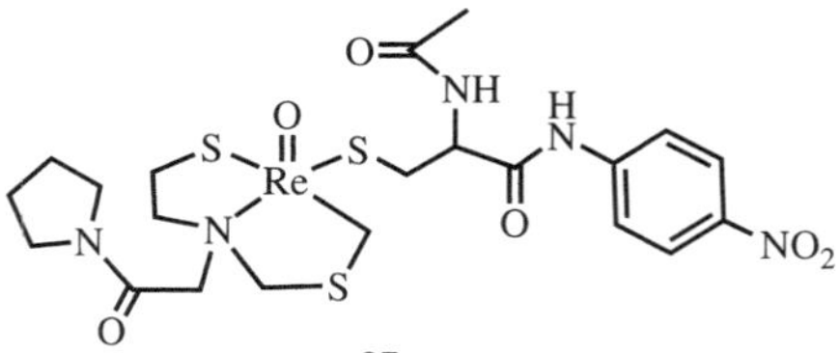

FIGURE 6.8 Examples of the '3 + 1' approach.

has an IC$_{50}$ value of 2.4µM for binding to cloned human dopamine transporter cells [89]. Analogous Tc and Re complexes containing a ketanserin analogue (**25**) to image serotonin 5-HT$_2$ receptors have also been reported [90]. Other derivatives aimed at receptor sites within the brain have been described [91, 92], but a recent paper has shown that there may be a problem with the monothiolate ligand undergoing exchange with glutathione, which raises questions as to the validity of the '3 + 1' approach [93]. Although not immediately with a radiopharmaceutical application, the reactivity of the monodentate site has been exploited in the complex **26**, which is a potent inhibitor of the endopeptidase cathepsin B binding to a cysteine at the active site [94]. This has been utilised in the construction of a combinatorial system for the synthesis of analogues of cyclophilin hCyp-18 protein, which inhibits a prolyl isomerise enzyme. One part of the binding motif is attached to the NS$_3$ ligand and the other to the thiolate. Combinations these are then assembled at the ReO core, and the most active species selected out, an example being **27**. Affinity constants of the order of 10-20 µM were achieved by this strategy [95, 96].

6.2.2 Di- and Tri-Oxo Complexes

It was shown early on that neutral N$_4$ ligands such as cyclam gave cationic M(V) dioxo-complexes with Tc and Re (**29**). Unfortunately the yield of the ^{99}Tc complex was only 7% and such complexes have been little pursued subsequently [97, 98]. However diphosphines react with pertechnetate to give *trans* dioxo-complexes and the ^{99m}Tc complex **28** is used extensively clinically for myocardial imaging under the name Myoview [99]. An interesting more recent development has been the synthesis of Re dioxo complexes (**30**) with 4-carbene ligands by reaction of an Re(V) precursor with the carbanion of the imidazolyl group [100]. A review on Tc and Re carbene complexes has also been published [101]. It remains to be seen if this type of ligand can be adapted for radiopharmaceutical applications (Figure 6.9).

The tris(pyrazolyl)borate trioxo complex [TcO$_3$(HBpz$_3$)] (pz = pyrazolyl) was reported in 1991, and the chemistry of the Re analogue was also established [102]. In general, the trioxo core is better developed for Re with examples with nitrogen heterocyclic ligands [103, 104] and the complex [ReO$_3$(oxine)], which is luminescent [105]. The last complex is included in a review of the excited state properties such as emission of high oxidation state Re complexes, which may be of interest in the context of developing dual mode imaging agents [106]. In this high oxidation state the oxo-ligands are reactive, and it was shown in 1988 that in an analogous manner to OsO$_4$ two oxo groups undergo a [3 + 2] addition reaction with olefins to give a diol [107]. This was later extended to the addition of diphenylketene to give a substituted diol (**31**) (Figure 6.10) [108]. The area of trioxo Tc and Re chemistry is now being revisited with the recognition that the [L$_3$MO$_3$] type of complex has structural similarities with the tricarbonyl system discussed below, but it is smaller and more hydrophilic and the radiopharmaceutical

FIGURE 6.9 Examples of metal dioxo complexes.

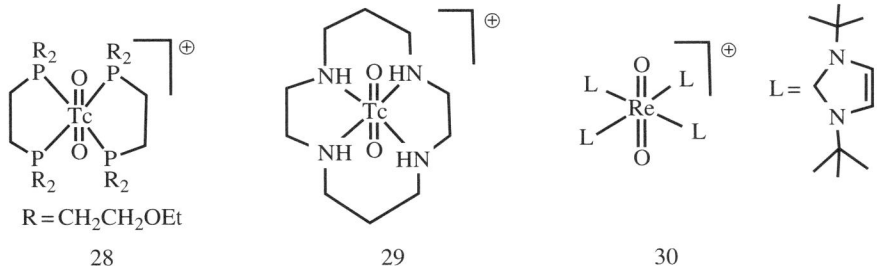

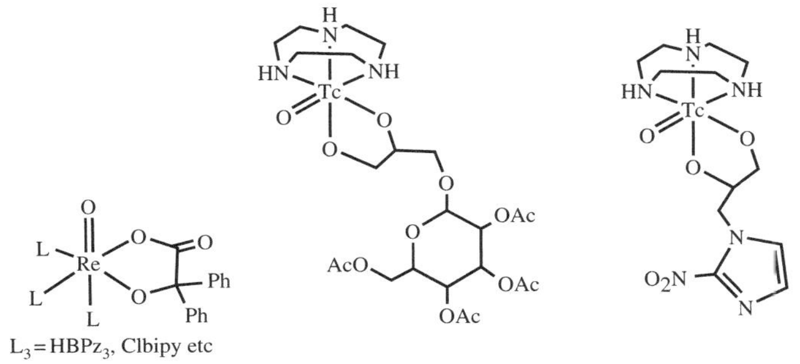

FIGURE 6.10 Examples of trioxo complexes.

possibilities are being explored. Thus the complex [^{99m}TcO$_3$(tacn)] (tacn = triazacyclonononane) has been prepared in moderate yield from pertechnetate in the presence of a phosphane attached to a solid bead. The earlier reported olefin reaction with a dioxo unit to give a diol was then exploited to attach potential targeting groups – two examples are **32** (acetylated glucose) and **33** (2-nitroimidazole) [109–111]. The biological properties of these molecules have yet to be reported.

6.2.3   Nitrido Complexes

The strong π-donating characteristics of the oxo-group makes a major contribution to the stability of high oxidation state Tc and Re oxo compounds. The trianionic nitride ligand acts in a similar manner, but the additional negative charge opens up a series of complexes with different co-ligands or overall charges to those with oxo-ligands. The first synthesis of a simple nitride of Tc or Re was carried out nearly 50 years ago by Chatt et al. who showed that [ReNCl$_2$(PPh$_3$)$_2$] and [ReNCl$_2$(PMe$_2$Ph)$_3$] could be made using hydrazine or azide as the source of nitride [112, 113]. This work was subsequently extended to ^{99}Tc and ^{99m}Tc through the work of Baldas and co-workers, who reported extensively on nitride complexes and proposed that the nitride core could be used in ^{99m}Tc-based radiopharmaceuticals [114–120]. They established that the paramagnetic Tc(VI) species [TcNCl$_4$]$^-$ could be made by the reaction of pertechnetate with azide in acid solution and investigated its substitution and redox chemistry. It was not until 1998 that the same synthetic route was successfully applied for the synthesis of [ReNCl$_4$]$^-$ [121]. The coordination chemistries and spectroscopic properties of the TcN and ReN cores were then explored further in detail by Abram and co-workers, and some representative examples of this work are shown below (**34, 35, 36**); these and others appear in the following references (Figure 6.11) [122–126].

The nitride complex baton was then taken up by Duatti and co-workers who showed that somewhat surprisingly neutral dithiocarbamato Tc nitrides showed promise as heart imaging agents. Detailed optimisation studies led to the TcNOET compound **37**, which has had clinical trials (Figure 6.12) [127–129].

The mechanism by which it is retained in myocardial tissue is still not clear. A crucial issue was to find a high yield route to the Re and Tc nitride core that did not require the use of azide. The hydrazine derivative MeSCSNMeNH$_2$ and variants were found to produce both TcN and ReN cores directly from the tetroxometallates in high yields [130, 131]. Although no intermediates could be isolated with Tc, [ReO(NHNMeCSSMe)$_2$]$^+$ can be isolated from the reaction of [ReO$_4$]$^-$ with the N-methylated hydrazine in the presence of HCl and can be subsequently be converted to [ReNCl$_2$(PPh$_3$)$_2$] by addition of PPh$_3$[132] or [ReN(dtc)$_2$] (dtc = dithiocarbamate) by adding dtc [133]. If PPh$_3$ is added to the starting [ReO$_4$]$^-$/hydrazine combination, no intermediates are accessible en route to the nitride, but removal of the hydrazinic methyl group permits the isolation of the chelated hydrazide **38**, which can subsequently be degraded to the nitride [134]. The MeS group is not a prerequisite for nitride formation and thiosemicarbazide also gives [ReNCl$_2$(PPh$_3$)$_2$], but here two consecutive intermediates are isolable: a chelated hydrazide analogous to **38** and [Re(NH)(NHNH$_2$)Cl$_2$(PPh$_3$)$_2$] [135]. This chemistry rather well illustrates both the subtleties of the product dependence on hydrazine substituent and also the greater kinetic stability of rhenium permitting the isolation of intermediates (Figure 6.13).

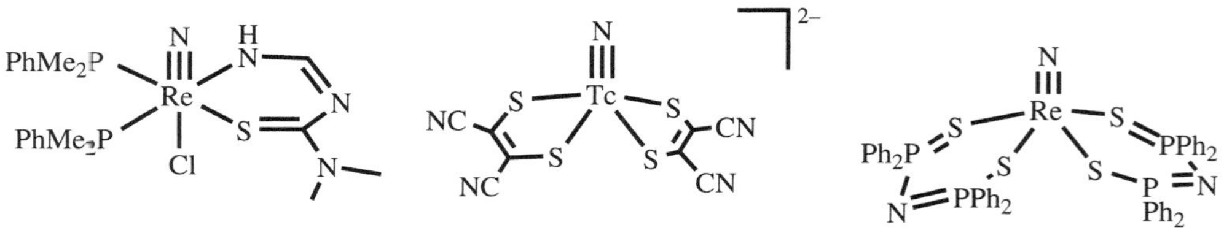

FIGURE 6.11   Technetium and rhenium(V) nitrido complexes with sulphur-containing ligands.

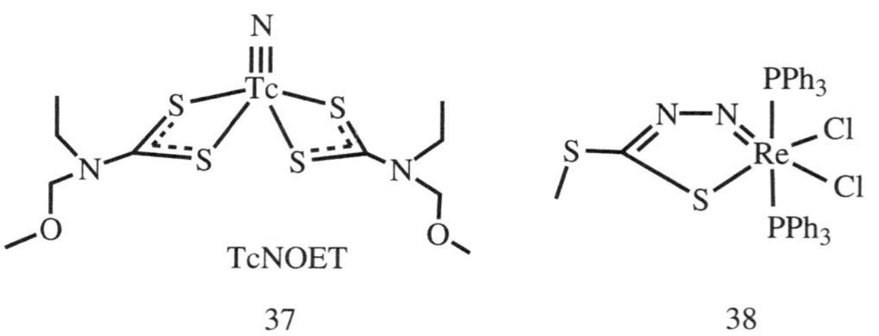

FIGURE 6.12   TcNOET and a nitride complex precursor.

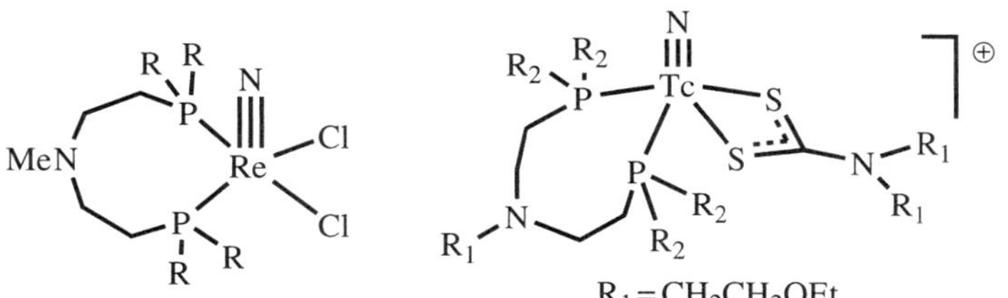

FIGURE 6.13 Nitride complexes with aminophosphine ligands.

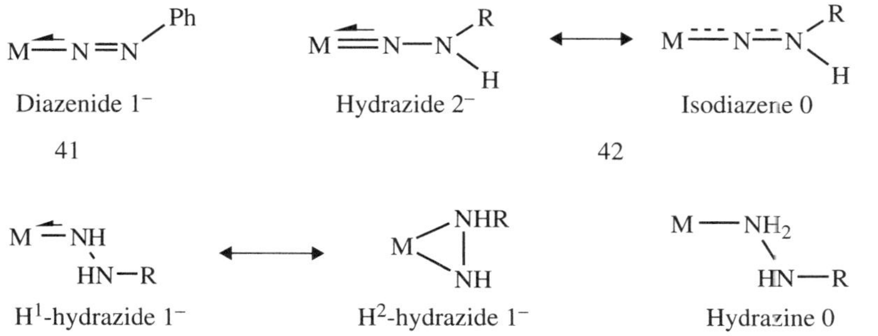

FIGURE 6.14 Structures and bonding in diazenide, isodiazene, hydrazide, and hydrazine complexes.

For maximum flexibility in deploying the nitride core, it would be desirable to synthesise complexes with a mixed ligand system analogous to the 3 + 2 systems described above for oxo complexes. Unfortunately, the same approach for the nitrides gives mixtures of compounds. However, it has been shown that the stereochemistry and stoichiometry of the Tc or Re nitride core co-ligands can be neatly controlled by the use of tridentate PNP donor ligands where the π-acceptor P donors prefer to be *trans*. This forces the halide ligands to be *cis*-oriented, and they can readily be substituted by a range of bidentate monoanionic or dianionic ligands [136]. The structure of the Re dichloride (**39**) is shown together with that of a cationic dithiocarbamate derivative (Tc-N-DBDOC, **40**), which shows promise as a cardiac imaging agent *in vivo* in both animals and humans [137]. The selection of other functionalisable bidentate ligands enables these complexes to be conjugated to a range of biomolecules [138–142].

6.2.4 Diazenide, Isodiazene, and Hydrazido Complexes

The bonding and formal charges adopted by this group of ligands are much less straightforward than for the oxo and nitride ligands considered hitherto, and a further complexity is that they can be interconverted via protic equilibria (Figure 6.14).

Structure **41** shows the most common bonding mode for the 'singly bent' diazenide ligand where it bears a monoanionic charge. Diazenide complexes can be prepared by a number of routes, but those using mono-substituted hydrazines or diazonium salts have been the most prevalent. The R group can be alkyl, aryl, or aroyl. Protonation at the diazenide nitrogen remote from the metal produces the MNNHR ligand **42,** which generally has a linear M-N-N system. Two canonical forms (hydrazide(2-) and isodiazene) can be drawn differing in the distribution of electrons in the M-N and N-N bonds. There is not a simple VB representation that covers the multicentre σ and π bonding system in the isodiazene form, thus dotted bonds have been used. X-ray crystal structures of complexes with NNR_2 ligands show an N-N distance more appropriate for a double rather than single bond; therefore, the formal nomenclature used for these ligands is isodiazene. However, it should be noted that this is solely for the purposes of providing a name and should not be interpreted that these ligands confer a neutral charge. In general, it is difficult to assign a true metal oxidation state to complexes of ligands that can undergo facile electron redistribution. If there is a donor group on the diazenide substituent (as in HYNIC, shown in Figure 6.15), then the M-N-N system is no longer linear and rehybridisation of the nitrogen adjacent to the metal from sp to sp^2 permits protonation to occur using the lone pair now available. Addition of a further proton to the nitrogen of isodiazene remote from the metal produces the hydrazido(1-) ligand **43,** which can bind end-on or side-on to the metal. A third protic addition gives the neutral hydrazine ligand **44**. Note that the positive charges that would accompany protonation have been omitted for clarity.

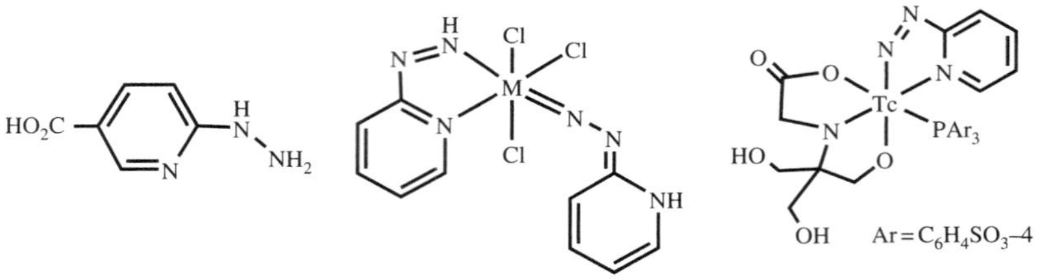

FIGURE 6.15 The HYNIC ligand and 2-pyridylhydrazine derived complexes.

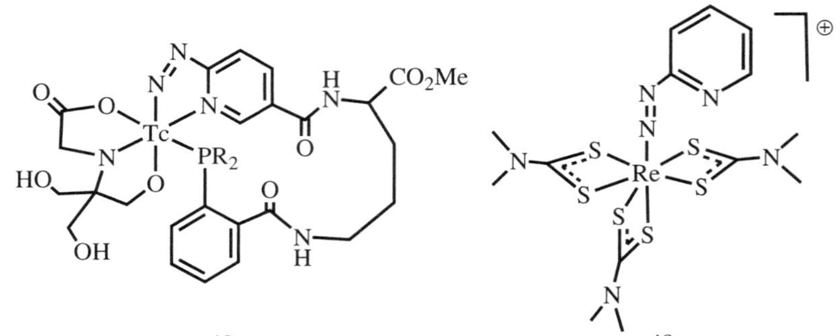

FIGURE 6.16 A complex of a tethered HYNIC type ligand and a cationic seven-coordinate Re complex.

The initial impetus for studying the Re chemistry of these ligand systems came from their involvement as intermediates in the protonation of coordinated nitrogen (i.e., R = H). The first Re diazenide complexes $[ReCl_2(NNR)(PMe_2Ph)_3]$ (R = PhCO [143], Ph [144]), prepared from the appropriate hydrazine, appeared in the early 1970s, and further papers describing mono and bis(diazenide) Re complexes with a range of co-ligands followed [145–153]. Analogues of some of these complexes with ^{99}Tc such as $[TcCl(NNPh)(Ph_2PCH_2CH_2PPh_2)_2]^+$ and $[TcCl(NNPh)_2(PPh_3)_2]$ were also described and the possibility of using the diazenide R group to attach biomolecules was discussed [154]. However, the major breakthrough in terms of radio-pharmaceutical applications came with the introduction in the early 1990s of carboxylate-substituted hydrazinopyridines by Zubieta, Babich, et al. – the so-called HYNIC system (**45**) [155, 156]. The pyridylhydrazine is conjugated to the biological targeting molecule via the carboxylate and reacted with $^{99m}TcO_4^-$, resulting in a stable metal nitrogen multiple bond. The simplicity of this process has attracted extensive use *in vivo* (over 600 references to HYNIC since 1990 in SciFinder). Some representative targeting molecules and targets are RGD peptides for tumour imaging [138, 157, 158], gastrin peptides for tumour imaging [159, 160], octreotide for somastatin receptors [161], annexin-V for apoptosis imaging [162, 163], EGF protein for tumour imaging [164], and antimicrobial peptide UBI 29-41 for infection imaging [165]. There has also been a useful recent review on the applications of HYNIC [166].

Despite its widespread use, there some issues with HYNIC that still have not been totally resolved. The number of HYNIC residues attached to the metal, their bonding mode, and state of protonation are difficult to define. Reaction of $[MO_4]^-$ with pyridylhydrazine hydrochloride gives the complexes $[MCl_3(NHNC_5H_3N)(NNC_5H_3NH)]$ (M = Tc, Re) (**46**), which were fully structurally characterised [159, 167]. Detailed mass spectroscopic studies [168, 169] of the ^{99m}Tc complexes with HYNIC and a co-ligand have confirmed that oxo and chloride ligands are absent, but there is still ambiguity as to the location of protons (Figure 6.16). An additional polydentate co-ligand such as tricine **47** or ethylenediaminediacetic acid (EDDA) is required and a range has been studied [17], but many do not confer high stability and partial replacement by donors within a conjugated peptide is possible. Also for asymmetric ligands there is the potential complication of the presence of isomers. The tethering of an additional donor group to the HYNIC molecule as in **48** has been examined, but this does not remove the need for a further co-ligand. It would be desirable to have a coordinatively saturated complex to force monodentate coordination with robust co-ligands and a positive overall charge to minimise protic equilibria. The tris(dithiocarbamate) Re complex **49**, readily prepared from **44** with excess dithiocarbamate [170], is very stable in solution and inert to protonation, and the Tc analogues might provide an alternative to the systems investigated hitherto. Although the cold precursors, such as **46** (M = Re), have been studied, there have been very few studies of the extension of the HYNIC approach to Re, probably reflecting the stability and full characterisation problems discussed above.

The discussion of the chemistry of the HYNIC system showed that it can involve both diazenide and isodiazene species in protic equilibria. The use of 1,1-disubstitued hydrazines R_2NNH_2 represents a direct route to isodiazene complexes that

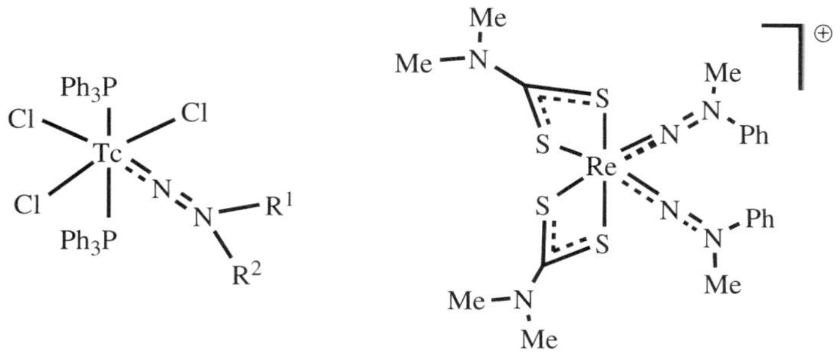

FIGURE 6.17 Isodiazene complexes of Tc and Re.

do not readily convert to diazenides; this has been applied for both Tc and Re. The first Tc hydrazide(2-) complex [TcCl$_3$(NNMePh)(PPh$_3$)$_2$] was reported in 1990 [171] from the reaction of [TcOCl$_4$]$^-$ with MePhNNH$_2$ in the presence of PPh$_3$; this was translated to ^{99m}Tc. The X-ray crystal structure (**50**) was later established [172] for the NNPh$_2$ derivative and a range of derivatives prepared via metathesis [173]. The chemistry of Re hydrazides has also been explored, and reaction of [ReOCl$_3$(PPh$_3$)$_2$] with an excess of MePhNNH$_2$ gave the cationic five coordinate species [ReCl$_2$(NNMePh)$_2$(PPh$_3$)]$^+$. Subsequent reaction with excess dithiocarbamate gave the very stable six-coordinate complex **51** (Figure 6.17) [174]. Although the hydrazide(2-) ligand is likely to be very inert toward hydrolysis, there have so far been no reports of *in vivo* investigations with this class of ligand.

6.3 TECHNETIUM AND RHENIUM(IV)

The coordination chemistry of the tetravalent oxidation state for both Tc and Re is limited compared to other oxidation states and generally has not been exploited in a radiopharmaceutical context and will not be discussed here. Comprehensive coverage of the chemistry appears in reviews [15, 16, 28].

6.4 TECHNETIUM AND RHENIUM(III)

The trivalent state for Tc and Re has the electronic configuration d^4 and complexes generally have coordination numbers from 5 through to 7. The vast majority are diamagnetic, although complexes of the type [ReCl$_3$(PMe$_2$Ph)$_3$] show temperature-independent paramagnetism and have contact-shifted NMR spectra [175]. As expected for an intermediate oxidation state, both π acceptor and σ donor ligands bind well, and the coordination chemistry with tertiary phosphines is particularly extensive. Thus reaction of [^{99}TcO$_4$]$^-$ with ditertiary phosphines in DMF gives the octahedral complexes **52**, which can be extended readily to ^{99m}Tc. This class of complex showed strong myocardial uptake in canine animal models but, sadly, none in humans, which was ascribed to reduction of the cationic Tc(III) complex to the neutral Tc(II) derivative *in vivo* [176–178]. Detailed electrochemical studies of the redox properties of both the Tc complexes and their Re analogues have been made [179]. The complexes could be reduced in two successive reversible one electron processes to Tc(II) and Tc(I) (Figure 6.18).

The complex [TcCl$_3$(MeCN)(PPh$_3$)$_2$] can be prepared by the Zn reduction of [TcCl$_4$(PPh$_3$)$_2$] or by reaction of [TcOCl$_4$]$^-$ with MeCN and PPh$_3$ [180]. This is a useful precursor for the synthesis of a range of other Tc(III) complexes such as [TcCl$_3$(py)$_3$] (py = pyridine) and [TcCl$_2$(bipy)$_2$]$^+$ (bipy = 2,2′-bipyridyl) [181]. The complex [ReCl$_3$(MeCN)(PPh$_3$)$_2$] is also known [182, 183] and was formed from the reaction of [ReOCl$_3$(PPh$_3$)$_2$] with MeCN in the presence of PPh$_3$. This presents a similarly versatile intermediate for access to a wide range of rhenium(III) derivatives.

The ability of tertiary phosphines to remove oxo groups from Tc and Re has been exploited in the synthesis of a range of Tc(III) and Re(III) Schiff base complexes by reactions of mono-oxo complexes with excess of the phosphine. This reaction is possible for a wide range of tri- and tetra-dentate ligand systems, but the most widely studied in the context of imaging applications have been complexes of acacen such as **53**. Such complexes could be made in good yield directly from [^{99m}TcO$_4$]$^-$ by sequential addition of Schiff base and phosphine and were widely studied as potential myocardial imaging agents. The furanone and phosphine substituents of **53** were used to optimise the biodistribution characteristics of the compounds [184–186]. This type of complex was later also found to be potentially useful for the imaging of multidrug resistance (MDR). Tumour cells can often become resistant to cytotoxins via overexpression of the MDR trans-membrane P-glycoprotein Pgp.

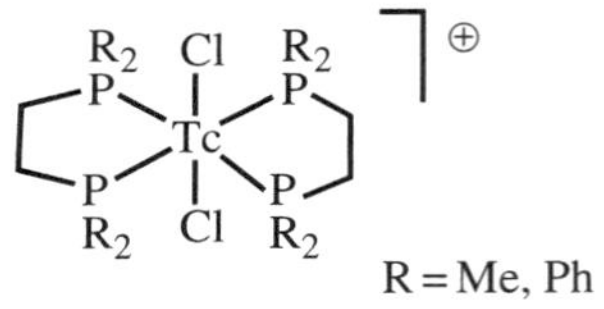

FIGURE 6.18 Tc(III) complexes with ditertiary phosphine (**52**) Schiff base (**53**) ancillary ligands.

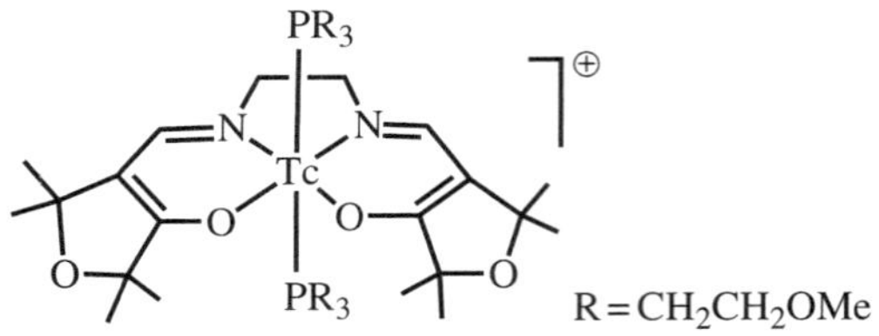

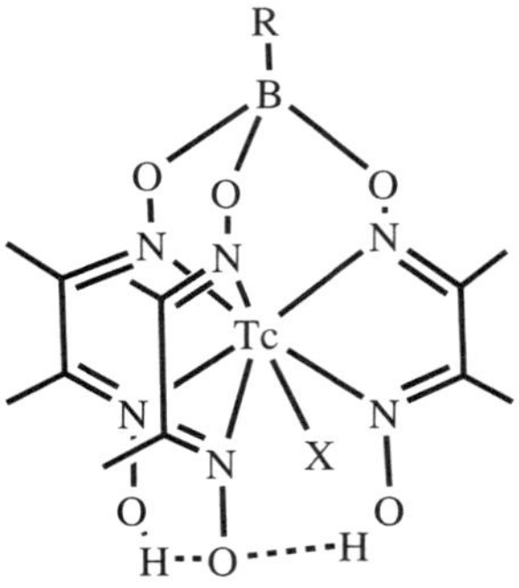

FIGURE 6.19 Tc(III) complexes with boron-capped dimethylglyoxime (54) and thiolate (55) ligands.

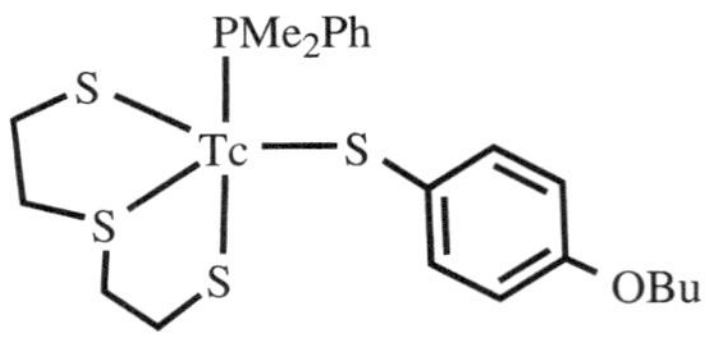

Other cationic lipophilic cations such as Myoview and Sestamibi share with **51** that uptake in tumours is decreased as Pgp expression increases, providing a route to assess MDR [187, 188]. Further examples and background appear in a useful review on MDR imaging [189].

Another well-studied class of Tc(III) complexes are those derived from dimethylglyoxime (dmg) in which three dmg units are linked together via a boron cap to give a hexadentate N_6 ligand. These are derived from dmg by reaction with boronic esters and are known as BATO ligands. The complexes formed with Tc(III) are neutral seven coordinate complexes with a halide X (**54**) completing the coordination sphere [190, 191]. The ^{99m}Tc analogues can also be made in good yield in a 'one-pot' reaction from [^{99m}TcO$_4$]$^-$ (Figure 6.19).

Despite the neutral charge, this type of complex has been used clinically for heart imaging under the name Cardiotec and can identify ischemic tissue. However, as with the TcNOET discussed above, the mechanism of uptake and retention of this neutral complex in myocardial tissue is unknown. However, it does not appear as though the halide is lost *in vivo*, and replacement of a chloride by hydroxide has little effect on the biological properties [192]. The BATO system can also be used as a bifunctional chelator; when the boron substituent is 3–isocyanatophenyl, the Tc complex can be linked to monoclonal antibodies such as B73.3, producing *in vivo* images comparable with those from a radioiodine-labelled analogue [193]. Similarly, derivatisation with a 2-nitroimidazole provided an agent capable of imaging hypoxic tissue [194] (see below for a more detailed account of hypoxia imaging). Rhenium analogues of BATO ligands have been reported but have not been radiolabelled [195].

There are many examples of Tc(III) complexes with sulphur ligands; one of the earliest to be investigated and used was [Tc(tu)$_6$]$^{3+}$, which can be used as a precursor for other Tc(III) derivatives. However, the removal of the displaced thiourea can be troublesome [196]. An X-ray crystal structure showed a pseudo-octahedral coordination with Jahn-Teller distortions [197]. Reduction of '3+1' oxo complexes discussed earlier with tertiary phosphines gives complexes such as **55**, and the same species can also be prepared directly from [^{99m}TcO$_4$]$^-$ in good yields. The geometry about Tc is now trigonal bipyramidal rather than square pyramidal, and the absence of the competing π-donating oxo-ligand means that the monodentate thiolate ligand is a better donor and exchanges more slowly with glutathione than in the oxo species [198, 199]. The trigonal bipyramidal structure motif is also found for both Tc and Re using sterically hindered thiolate ligands in complexes such as **56** [200] and also for a range of tetradentate 'umbrella' ligands with N or P capping atoms such as **57** [201]. A number of analogous Re(III) complexes to **56** have also been prepared with axial ligands such as MeCN and CO [202] and even N$_2$ [203]. Although the thiolate ligands in Tc and Re complexes of the type **56** and **57** are likely to be strongly bound, there have been no radiolabelling or biological studies reported (Figure 6.20).

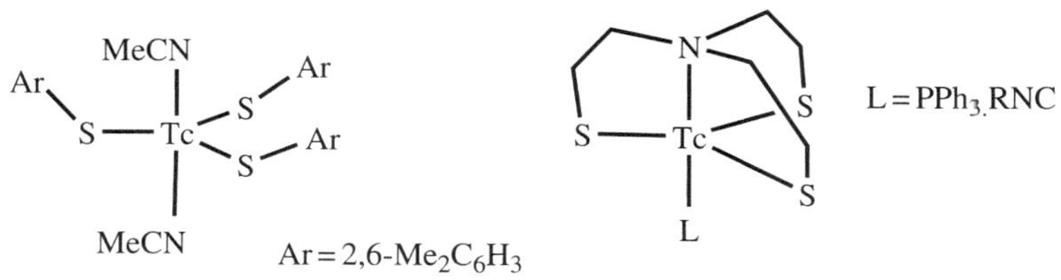

FIGURE 6.20 Thiolato Tc(III) complexes.

6.5 TECHNETIUM AND RHENIUM(I)

6.5.1 Isocyanide Complexes

In the 1960s, the early days of the development of technetium chemistry, it would have been regarded as fanciful to suggest that organometallic complexes might be used as radiopharmaceuticals. Syntheses were usually conducted in dry organic solvents in an inert atmosphere; conditions far removed from those used in a radiopharmacy. The first hints that Tc-C bonds might be viable for biological applications came with the synthesis by Davison, Jones et al. of extremely stable, water-soluble isocyanide complexes of the type $[Tc(CNBu^t)_6]^+$ in 1982 [204]. The initial synthesis was from the Tc(III) hexakis(thiourea) complex, but it was later shown that the same class of complex could be made directly from $[^{99m}TcO_4]^-$ using dithionite as reductant and a large excess of the isocyanide ligand. The same paper also showed that the complex could be used as a myocardial imaging agent and that it performed as well as ^{201}Tl, which was used clinically at that time [205, 206]. The biodistribution characteristics were later optimised by using a methoxybutyl isocyanide substituent TcMIBI, which entered clinical use under the trade name Cardiolite. The mechanism of uptake in myocardial tissue involves passive diffusion into cells followed by trapping in mitochondrial membranes due the negative potential across the membrane [207, 208]. Cancer cells have increased metabolism, resulting in a more negative potential; TcMIBI can therefore also be used for imaging tumours [209]. Analogous rhenium hexakis(isocyanide) cations can be prepared by reaction of $[ReOCl_3(PPh_3)_2]$ [210] or complexes with Re-Re multiple bonds [211] with an excess of isocyanide, but no radiolabelling or biological studies have been reported.

6.5.2 Tricarbonyl Complexes

Some purists might claim that isocyanide complexes, while containing a M-C bond, are not truly organometallic complexes. However, there is no doubt that the $[M(CO)_3Cl_3]^{2-}$ species first reported by Alberto et al. in 1995 [212] by the low pressure carbonylation of $[TcO_4]^-$ in the presence of reductant is organometallic. The imaging potential of this species and the slightly later reported $[Tc(CO)_3(H_2O)_3]^+$ [213] was quickly realised, and this class of compound has become one of the most used over the last 20 years for the development of new ^{99m}Tc-based diagnostic agents, thus representing a significant breakthrough. There have been far too many publications covering applications of this class of complex for a comprehensive account here, so we therefore present only a summary of the underlying chemistry and some selected examples. There have been several detailed reviews by Alberto et al. that give much other useful information [214–217]. The main focus of research on Re tricarbonyl complexes has been on their use a cold surrogates for the technetium analogues. Although the tri-aquo Tc cation could be made from $[TcO_4]^-$, the use of CO gas was not appropriate for radiopharmaceutical applications. A crucial step was the discovery that boranecarbonate $([H_3BCO_2H]^-)$ could be used with $[^{99m}TcO_4]^-$ in water as a simultaneous reductant and CO source, giving a high yield of the tricarbonyl derivative [218]. This enabled a kit formulation to be developed that is now available commercially under the name IsoLink™. The Re analogue can also be made from $[^{188}ReO_4]^-$ by modifying the conditions to include $H_3B.NH_3$ as well as the boranecarbonate [219, 220]. Several factors contribute to the especial suitability of the tricarbonyl core for radiopharmaceutical applications. Complexes of the type $[M(CO)_3L_3]^+$ (where L is a neutral ligand) are octahedral with a d^6 electron configuration, which confers kinetic stability of the complexes due to loss of crystal field stabilisation energy going to five or seven coordinate transition states. The CO ligands are strongly bound due to their π-acceptor properties and the low valent electron rich M(I) ion. Moreover, the balance of electron density between metal and ligand is such that the CO ligands are not open to degradation via nucleophilic or electrophilic attack at the CO carbon. The competition between the CO ligands for metal electron density leads to their obligately facial arrangement at the metal centre, which confers stereochemical rigidity on the complexes. The use of a polydentate ligand at the other three sites prevents any ligand exchange reactions. These complexes, like many classic organometallics, also conform to the 18-electron rule. The strong π-bonding creates a wide separation between the stabilised HOMO bonding and LUMO antibonding orbitals, thus oxidation and reduction are both difficult. Finally, the C_{3v} symmetry of the tricarbonyl motif also has the advantage of avoiding stereoisomers even when asymmetric ligands occupy the other three sites. The presence of such isomers was a persistent problem when dealing with the M(V) oxo-cores.

6.5.2.1 Technetium Essential Agents

This class of compound is targeted by virtue of its intrinsic physical properties and does not require conjugation to a biologically active molecule. The three water molecules in $[Tc(CO)_3(H_2O)_3]^+$ can be replaced by a very wide range of tridentate (or bidentate plus monodentate) ligands. One of the first explored was histidine, which readily formed neutral complex **58** in high yield in water (Figure 6.21) [214]. The so-called technetium essential imaging agents depend solely on the physical characteristics of the complex to achieve targeting exemplified by the use of cationic complexes for heart imaging such as Myoview™ and Cardiolite™ discussed earlier. The triaquotricarbonyl cations are too hydrophilic to be used, but a number of more lipophilic variants have been prepared, such as **59**, which has shown promising *in vivo* behaviour [221]. Alternatives to TcMAG$_3$ with the tricarbonyl core have been explored, including complex **60** with lanthionine; in clinical studies the ^{99m}Tc derivative shows comparable imaging performance to ^{131}I-hippuric acid [222]. Examples of other carbonyl complex-based Tc essential agents appear in the reviews given above.

6.5.2.2 Bioconjugates

Functionalisation of the co-ligands for the tricarbonyl core enables the addition of a very wide range of biologically active ligands for site-specific imaging. The histidine binding motif in **58** has been used as the basis of a route to attach simple amino acids as shown in **61** (Figure 6.22). The Re complex is taken up by the LAT-1 amino acid transporter as measured by the displacement of tritiated phenylalanine. The transportation is very sensitive to small changes in structure such as replacement of a CH_2 group of the linking chain by sulphur, which greatly reduces the activity [223, 224]. A variant on this theme is complex **62** with a diquinolyl-lysine ligand where the availability of pendant amino and carboxylate groups enables the orthogonal coupling of a wide range of peptide and other functional groups. This has been described as a single amino acid chelate (SAAC) system, and the chemistry and imaging applications have been detailed in a recent review [51]. The high stability of the coordination about the metal means that solid phase peptide synthesis can be used. In addition, the Re analogues are fluorescent, and the behaviour of the complexes at the cellular level can be followed using fluorescence microscopy. Some other recent examples of the labelling of biomolecules with the tricarbonyl fragment are ciproflaxin and nitrofurylthiosemicarbazone for imaging of infection [225, 226], Herceptin for imaging breast and other cancers [227], cRGD peptide (both ^{99m}Tc for angiogenesis imaging and ^{188}Re for therapy) [228, 229], histidine-tagged HER-2 targeting

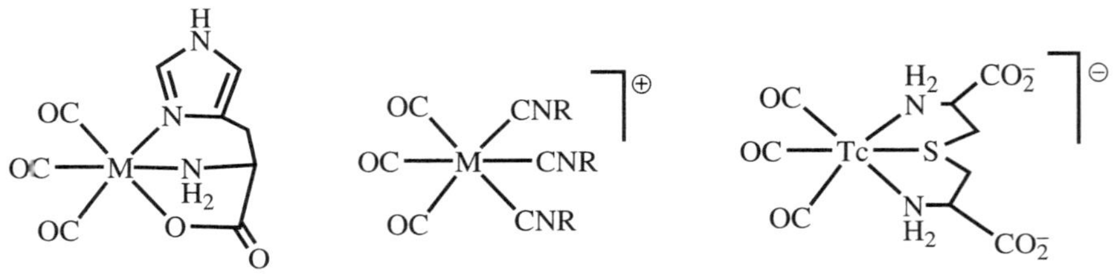

FIGURE 6.21   Tc(I) tricarbonyl complexes.

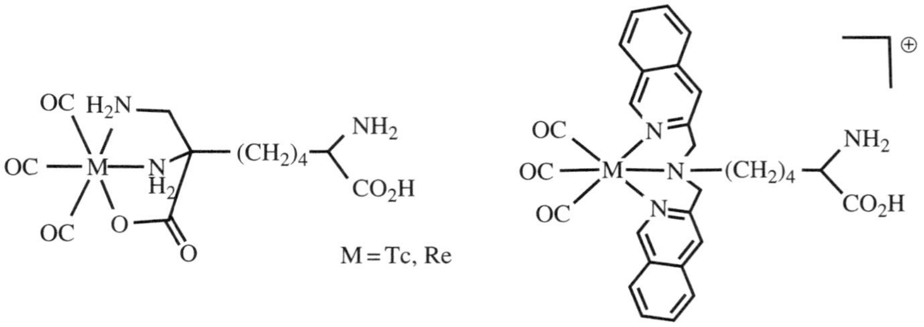

FIGURE 6.22   Tricarbonyl complexes with bifunctional chelators.

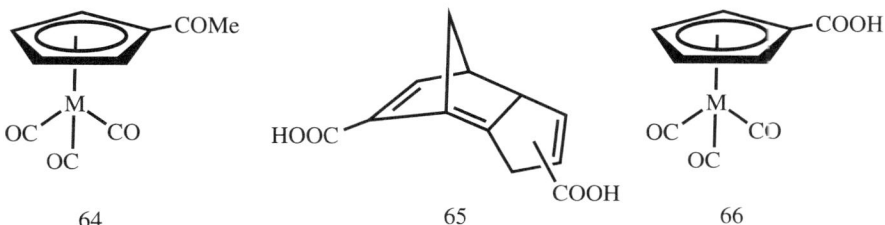

FIGURE 6.23   The 'click to chelate' approach with the Tc(I) tricarbonyl core.

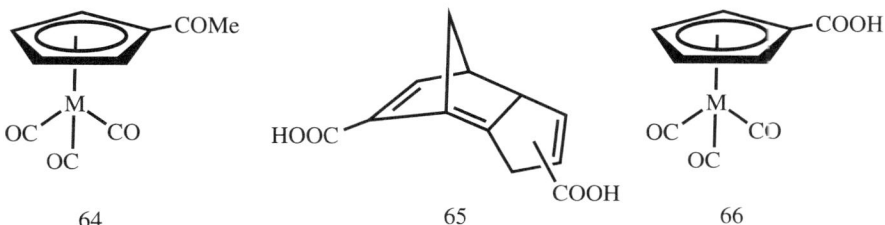

FIGURE 6.24   M(I) cyclopentadienyl complexes and precursor.

affibodies [230–232], folate for cancer imaging [233, 234], bombesin analogues for cancer imaging [234, 235], cyclophenil for imaging estrogen receptor-positive breast cancer [236], recombinant Annexin(V) for apoptosis imaging [237], nitro-L-arginine for targeting and imaging inducible nitric oxide synthetase [238], and vitamin B_{12} for cancer imaging [239, 240].

The majority of the bioconjugates discussed above have been made via some form of peptide coupling reaction between the bifunctional chelator and bioactive molecule. An interesting alternative approach has been developed by Schibli et al., the so called 'click to chelate' strategy. This exploits Cu-catalysed click chemistry to couple an azide to an acetylene group to create a link between complex and biomolecule; the triazole group created also functions as donor to the tricarbonyl unit. An example of a model system using benzyl azide is shown in **63** (Figure 6.23). Thence derivatisation of thymidine with azide generates a conjugate potentially capable of imaging proliferation in cancer by acting as a substrate for hTK1 (human cytosolic thymidine kinase). The radiolabelling with ^{99m}Tc proceeded in high yield to give a single species. Specific activity for hTK1 was achieved by identification of up to 20% of the phosphorylated thymidine conjugate [241, 242]. The click to chelate approach was subsequently modified to produce bistriazole-based ligands to bind to the $M(CO)_3$ core [243].

6.5.3   Cyclopentadienyl and Carborane Complexes

The high stability of metal cyclopentadienyl complexes even in aqueous media suggested their possible role in Tc-based imaging. The first approach was by Wenzel et al., who prepared complexes of the type $[RCp^{99m}Tc(CO)_3]$ ($Cp=C_5H_4$, R=COMe, CO_2Me, COPh, etc.) by a ligand transfer reaction involving $[^{99m}TcO_4]^-$ and $(RCp)_2Fe$ and $MnCl(CO)_5$ as CO provider. The reagents were heated in a sealed tube in MeOH at 120-150°C to give acceptable yields of the technetium complex. The same procedure was later adapted to the labelling of octreotide with ^{99m}Tc, and the observed uptake in pancreas and adrenals was shown to be receptor specific by blocking with excess octreotide [244]. Cyclofenilcyclopentadienyltricarbonyltechnetium-94m and rhenium complexes were also prepared and their estrogen receptor binding properties studied [245, 246].

Although the synthetic routes above give the required ^{99m}Tc cyclopentadienyl complexes, they are not appropriate for routine radiopharmaceutical use. However, Alberto et al. have shown that acetylcyclopentadiene reacts rapidly with $[^{99m}Tc(CO)_3(H_2O)_3]^+$ in water under mild conditions to give **64** (Figure 6.24). Also Thiele's acid **65** is cleaved by reaction with the tricarbonyl aquo ion to give the carboxylcyclopentadienyl derivative **66**. Clearly, this new synthetic approach means that cyclopentadienyl systems could form the basis of another method for labelling biomolecules for imaging, but it remains to be seen whether these offer any significant advantages over the many alternatives. Carborane clusters of the type $R^1R^2C_2B_9H_9^{2-}$ (L; $R^1=CH_2CH_2COOH$, $R^2=H$) are isoelectronic and isolobal equivalents of the cyclopentadienyl anion. Valliant et al. have shown that these can be used to prepare complexes of the type $[LM(CO)_3]^-$ (M=Re, ^{99m}Tc) by reaction of the carborane anion with the IsoLink kit with fluoride as catalyst [247–249]. Derivatives with $R=C_6H_4OH$-4 have been evaluated for binding to estrogen receptors [250]. Borane clusters have been investigated for boron neutron capture therapy (BNCT), and the incorporation of a SPECT probe will enable their exact location *in vivo* to be identified.

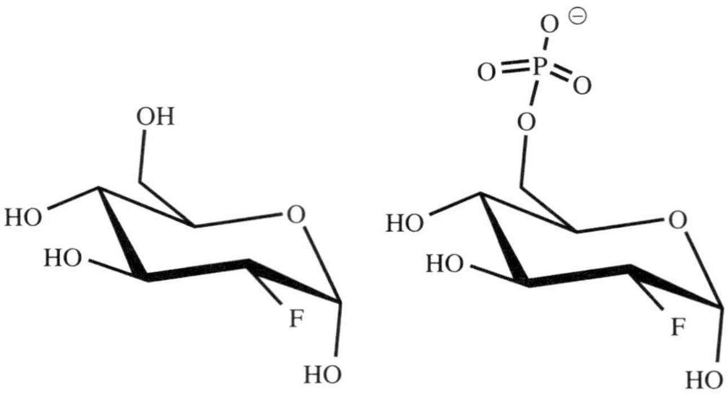

FIGURE 6.25   FDG before and after phosphorylation.

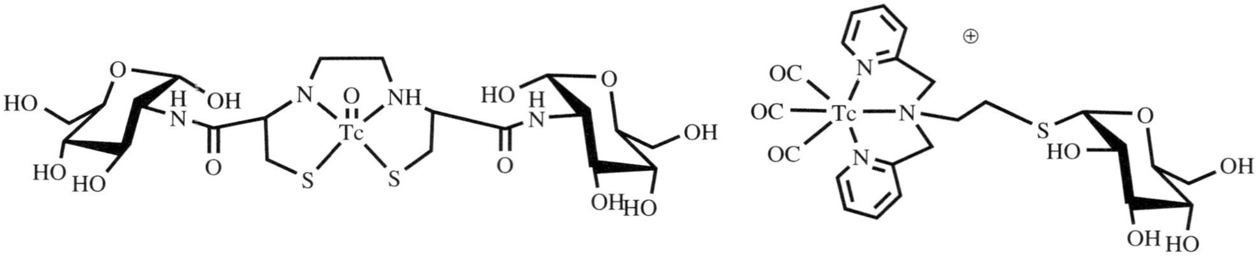

FIGURE 6.26   Glucose-derived conjugates of technetium complexes.

6.5.4   Carbohydrate Conjugates

Cancer cells have less efficient metabolism than normal cells, and there is increased glucose uptake to compensate, using glycolysis as an energy source. This has led to the extensive use of radiolabelled glucose derivatives for cancer diagnosis. Glucose is taken into cells via the GLUT-1 uptake pathway and then undergoes glycolysis, the first step of which is phosphorylation via hexokinase (HK). The most widely used derivative is ^{18}F-labelled glucose FDG **67,** which after cell uptake is phosphorylated at the 6 position to give **68** (Figure 6.25). The next step in the glycolytic pathway is dehydration by 6-phosphate dehydrogenase, but this is blocked by the fluoride, and phosphorylated FDG accumulates in the cell by virtue of its negative charge.

A ^{99m}Tc-based equivalent of ^{18}FDG would be a highly attractive alternative due to the low cost and wide availability of ^{99m}Tc. However, the challenge is to make a technetium complex that is both taken up by the GLUT-1 pathway and can act as a substrate for HK. This area was very comprehensively reviewed by Orvig et al. in 2008 [251], thus this account is restricted to a few selected examples. Many compounds containing a form of glucose linked to ^{99m}Tc or Re have been prepared over the past 20 years. Although radiolabelling and *in vivo* studies have been conducted with many of these, they have generally been restricted to measuring tumour uptake without determining GLUT-1 or HK activities; the tumour uptake could therefore be non-specific.

Glucosamine is a convenient functionalised glucose molecule; two of these have been attached to ECD as shown in **69** (Figure 6.26). *In vitro* cellular uptake was suppressed by D- but not L-glucose, suggesting the use of the GLUT-1 pathway. The ^{99m}Tc derivative was proposed to also act as a substrate for HK, but this was based on NADH detection rather than identification of the phosphorylated product [252]. Overall, the tumour uptake of the ^{99m}Tc complex **69** was comparable to ^{18}FDG. As with many other labelling targets, the ^{99m}Tc tricarbonyl core has been used to link to glucose in compounds such as **70**. Here the thioether group has been introduced to minimise the possibility of loss of the glucose molecule by enzymatic cleavage. The radiolabelling yield and stabilities were high but there was no evidence for HK activity or use of the GLUT-1 uptake route [253]. Clear evidence for glucose conjugates with a ^{99m}Tc core that truly mimic FDG is therefore sparse, but this is a rapidly evolving field and there is enough evidence to suggest that a SPECT alternative to FDG is a realistic possibility. Although glucose surrogates bearing ^{99m}Tc have been elusive, sugar groups can also be extremely useful to modify the biological behaviour of complexes and to act as linker groups to other targeting molecules.

6.6 IMAGING OF HYPOXIA WITH ^{99m}Tc

The rapid growth of some tumours is associated with reduced blood supply, particularly toward the centre and the creation of hypoxic zones. A high proportion of hypoxic tissue generally indicates poor prognosis for the patient partly due to the problems of delivering effective radiotherapy at low oxygen concentrations [254] and also to the release by hypoxic cells of activators that can *inter alia* stimulate blood vessel and tumour growth. Oxygen concentrations can be determined in tumours by use of oxygen electrodes, but this method is invasive and can be unreliable [255], particularly with smaller tumours. Hypoxia is also important in other diseases such as diabetes, arthritis, and heart ischemia. The molecular imaging of hypoxia [256] therefore has vital role both in determining the optimal course of treatment for a cancer patient [257] and generally distinguishing poorly perfused tissue from healthy or necrotic tissue. 2-Nitroimidazoles (nims) accumulate in hypoxic tissue via a redox trapping mechanism, and ^{18}F-labelled derivatives such as F-MISO **71** are used in the clinic for assessing hypoxia (Figure 6.27). The copper bis(thiosemicarbazone) complex ^{64}CuATSM **72** is also in use but operates via a very different mechanism from F-MISO. None of the agents in current use have optimal imaging characteristics and involve the use of expensive PET radioisotopes, thus alternatives using technetium-99m are a worthwhile goal. The confirmation of the hypoxic selectivity for a new agent is not straightforward. The extent of hypoxia and the time dependence of specific hypoxia uptake *in vivo* are highly dependent on the cancer type. It is therefore necessary to show that there is hypoxic selectivity in tumour cells *in vitro*, then demonstrate that any tumour uptake *in vivo* is indeed in hypoxic tissue and not merely non-specific by carrying out immunohistochemical (IHC) staining of tumour sections or at least correlation with oxygen electrode measurements.

The first approaches to technetium agents involved the conjugation of nims to well established chelators such as the BATO system via the capping boron (**73**). However, enzymatic reduction experiments suggested that rate of reduction was too low for imaging use [194]. Several nim conjugated derivatives of propyleneamineoxime ^{99m}Tc complexes have been explored as potential hypoxic selective agents that differ in the location of the nim group. Complex **74** designated BMS181321 showed hypoxic selective uptake in cancer cells [258] but was later shown to decompose in saline and have non-optimal biodistribution characteristics (Figure 6.28) [259]. The modification of the propylene backbone and attachment of nim in **75** (designated BRU59-21) improved the biodistribution with rapid blood clearance and improved tumour uptake, which was shown in phase 1 clinical trials to correlate with pimonidazole hypoxia staining [260].

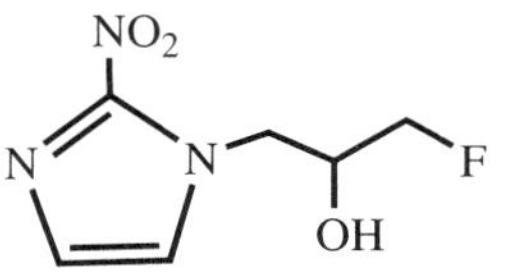

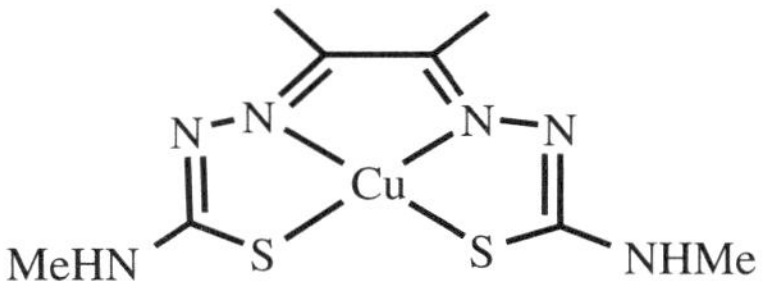

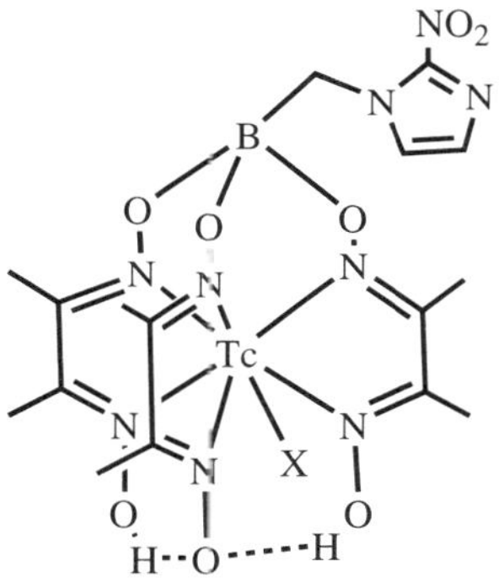

FIGURE 6.27 FMISO, CuATSM, and a nitroimidazole conjugate.

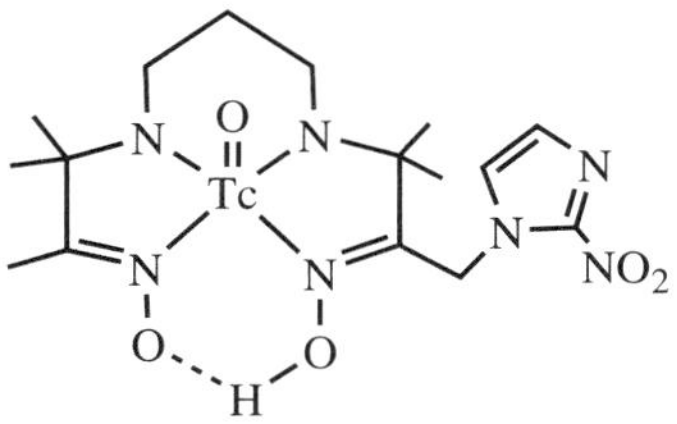

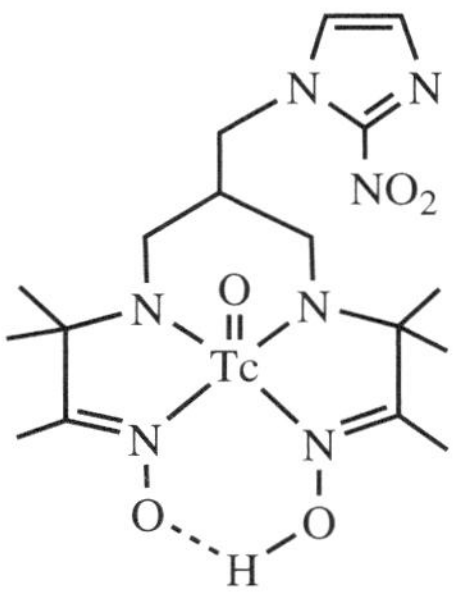

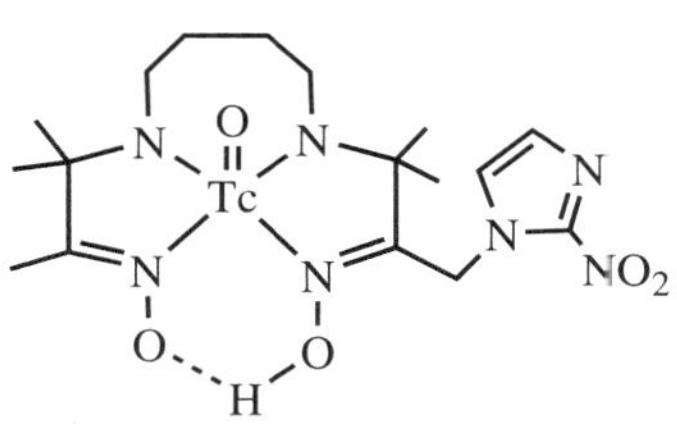

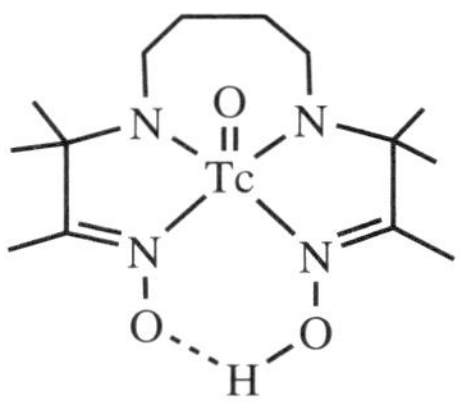

FIGURE 6.28 Nitroimidazole conjugates of Tc amineoxime complexes.

A variation on the amineoxime ligand theme was the synthesis of complex **76** with the butylene backbone, named TcHL-91M. This showed promising hypoxia selective uptake but remarkably this was lower than complex **77** (Figure 6.28), TcHL-91, or Prognox™, which was originally selected as a non-hypoxic selective control. When the reaction of the amine-oxime with [$^{99m}TcO_4$]$^-$ is carried out in non-aqueous solvents, a mono-oxo complex is formed, but this is rapidly converted in aqueous media to the *trans* dioxo complex. Cellular uptake studies showed that Tc HL-91 retention was significantly higher under hypoxic conditions and that it also has higher uptake in a live isolated ischemic rat heart. In mice CaNT tumour retention was comparable with that of a ^{123}I-labelled nitroimidazole IAZA, and the uptake was normalised to that of tritiated MISO [261, 262]. Thus, the hypoxic selectivity of HL-91 is unambiguous and has been confirmed by subsequent pre-clinical and clinical investigations [263–269], including correlation with FDG [270], but the mechanism of selectivity remains unknown. Autoradiography studies show that it accumulates in the cell membrane under hypoxic conditions, but it is not clear if any bioreductive process is occurring or if this is a response to hif (hypoxia inducible factor) initiated changes in cell biochemistry. 2-, 4-, and 5-Nitroimidazoles [271], misonidazole [272], and nitrophenyl derivatives [273] have been coupled to the ^{99m}Tc tricarbonyl core (**78** and**79**), and preliminary biodistributions in tumour-bearing mice carried out. Although tumour retention was observed and found to be highest for the 2-nitroimidazole derivatives, this has not yet been correlated with hypoxia by IHC.

There have also been investigations of the effect of coupling more than one nim molecule to a ^{99m}Tc chelator; examples appear in **80** and **81** (Figure 6.29). For the propylene amine complex **80**, the introduction of a second 2-nim group raised the retention of ^{99m}Tc in murine sarcoma cells under hypoxic conditions from approximately 25% for BMS 181321 (discussed above) to approximately 60%, suggesting that hypoxic selectivity is increased by adding a second 2-nim group [274]. However, *in vivo* studies have not been carried out. An alternative strategy involved the coupling of four nitroimidazole-based groups to a ^{99m}Tc cyclam complex as shown in **81**. Presumably this was the *trans* dioxo derivative, although this was not explicitly confirmed. For the ^{99m}Tc derivative of cyclam with misonidazole groups attached, the uptake under hypoxia in DU-145 prostate tumour cells was only marginally higher than under air, and the *in vivo* tumour uptake was similarly low [275]. A cyclam ^{99m}Tc derivative analogous to **81** but bearing four molecules of the hypoxic selective nitrotriazole **82** (AK 2123) has also been reported (Figure 6.30) [276]. Here uptake in rat mammary tumours was correlated with both oxygen electrode measurements of pO$_2$ and uptake of a ^{125}I-labelled nitroimidazole derivative, suggesting that this compound was promising for the clinical imaging of hypoxia, although other cancer types need to be studied. There has also been an

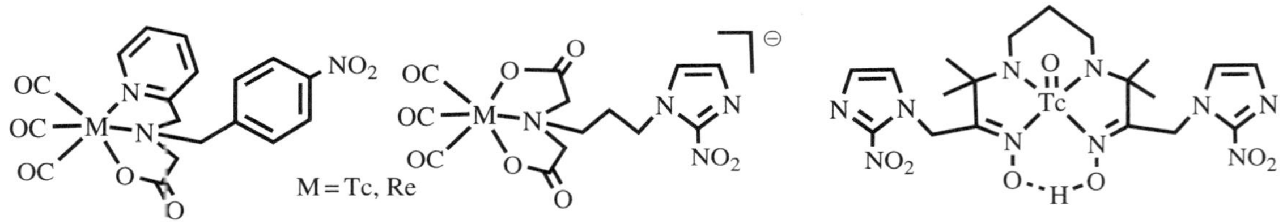

FIGURE 6.29 Further nitroimidazole and nitroaromatic conjugates.

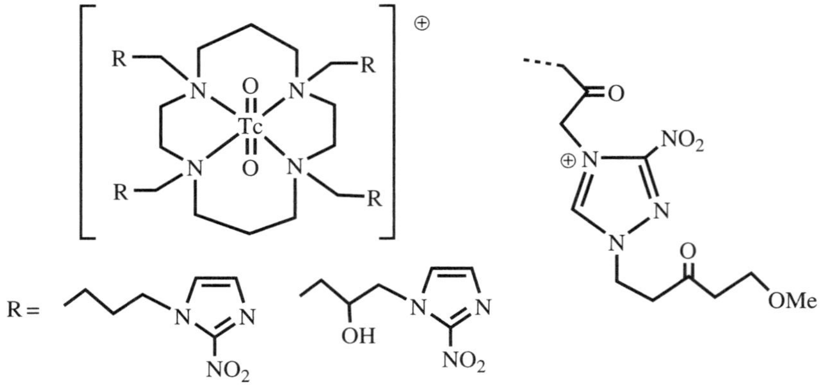

FIGURE 6.30 Nitroimidazole and nitrotriazole conjugates with Tc cyclam complexes.

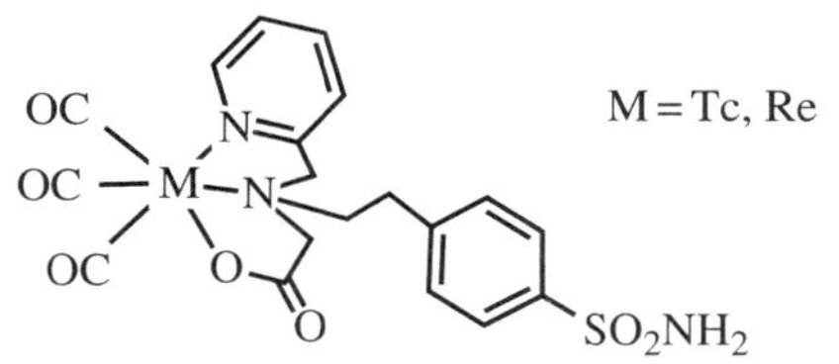

FIGURE 6.31 Conjugates of metal complexes with aromatic sulphonamides for CAIX targeting.

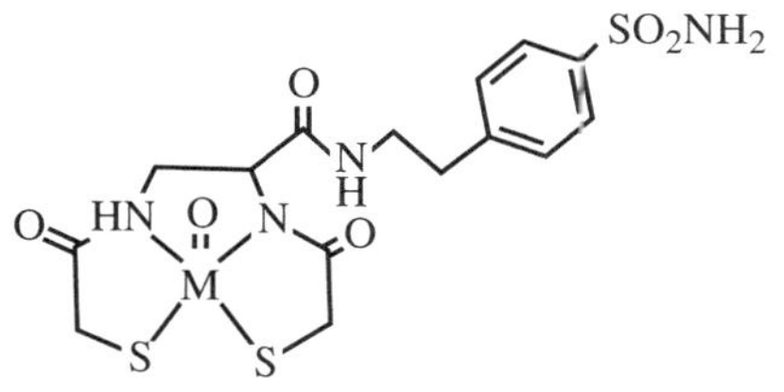

interesting report that a ^{99m}Tc-labelled bis(thiosemicarbazone) complex (Figure 6.5) also shows selective uptake in the myocardium in an isolated rat heart ischemic model [277]. Uptake studies in HeLa cancer cells also showed selective uptake under hypoxic conditions, but *in vivo* studies in mice with CaNT hypoxic tumours revealed only small retention of ^{99m}Tc in the tumour [278].

There is now data to suggest that it is possible to have a ^{99m}Tc hypoxia agent based on conjugation to a nim molecule, but further biological evaluation will be required. The complex TcHL-91 is definitely hypoxic selective and of perhaps of more interest than the nim derivatives because it appears to operate by a very different mechanism. This raises the possibility that it is providing information on hypoxia that is complementary to that from a nim-based agent such as F-MISO.

An alternative approach to imaging hypoxia is to utilise the hif controlled increase of expression on the external surface of cancer cells of carbonic anhydrase IX (CAIX) [279], which may play a role in the control of pH within the cancer cell [280]. CAIX therefore acts as an indirect marker for hypoxia [281], and ^{18}F-labelled aromatic sulphonamides (inhibitors of CAIX) or radioiodine-labelled antibodies to the CAIX enzyme have been used to image CAIX expression *in vivo*. Aromatic sulphonamides have been coupled to ^{99m}Tc or Re complexes as shown in **83** and **84** (Figure 6.31). The Re analogue of **83** was an effective inhibitor of the CAIX enzyme *in vitro,* but the ^{99m}Tc derivative showed little uptake in mice with xenografted CAIX-expressing tumours [282]. Similarly, the Re complex **84** was an effective enzyme inhibitor *in vitro* but showed only marginal uptake in cells with CAIX over expression [276]. Enzyme inhibition studies are evidently not a reliable guide to the behaviour in cells or animals. It appears that the sulphonamide class of CAIX inhibitors may be rather sensitive to the bulk and structure of the appended radionuclide and that an approach using labelled antibodies or fragments may be more productive, although the comparatively short half-life of ^{99m}Tc may be an issue and a pre-targeting approach might be necessary.

6.7   TECHNETIUM AND RHENIUM DIPHOSPHONATE COMPLEXES

It has been known for a long time that ^{99m}Tc polyphosphate compounds can be used to detect calcification in bone [283], and the ^{99m}Tc complex of hydroxyethyldiphosphonate HEDP (**85**) is sold commercially as Medronate™ for the imaging of bone metastases in cancer patients. The coordination chemistry of this type of ligand with both Tc and Re is complex, and a mixture of oligomers of unknown structures and metal oxidation states are formed from reaction of diphosphonate with $[TcO_4]^-$ with stannous chloride as reductant. Only one X-ray structure has been reported of a polymeric product from $[^{99}TcBr_6]^{2-}$ and methylene diphosphonate; a portion of the structure is shown in **86** (Figure 6.32) [284].

The bone-targeting ability of these complexes is believed to be due to the uncoordinated phosphate oxygen atoms binding to exposed calcium ions at the bone surface. EXAFS studies on the ReHEDP species show Re-O bonding and also some Re-Re interactions [285]. This Re complex is in clinical use for the palliation of pain accompanying bone metastases. The complex $[^{188}ReO(DMSA)]$ that was discussed earlier in the context of therapy of medullary carcinoma is also taken up in bone metastases and can also be used for pain relief [86, 286].

6.8   THE FUTURE FOR TECHNETIUM AND RHENIUM RADIOPHARMACEUTICALS

The chemistry described above illustrates that the coordination chemistry for radiopharmaceuticals of these two elements is very much alive and flourishing with significant advances having been made over the past two decades. At first sight it is perhaps surprising that so few of the many new compounds have been widely adopted for clinical use. This in part reflects the advent of PET imaging, which is perceived to offer advantages over SPECT. This issue was discussed in the introduction to this section. However, it has to be said that many of the new bifunctional chelators provide alternative ways of targeting the same classes of biomolecule with improvements in imaging performance that are sometimes relatively small. This does

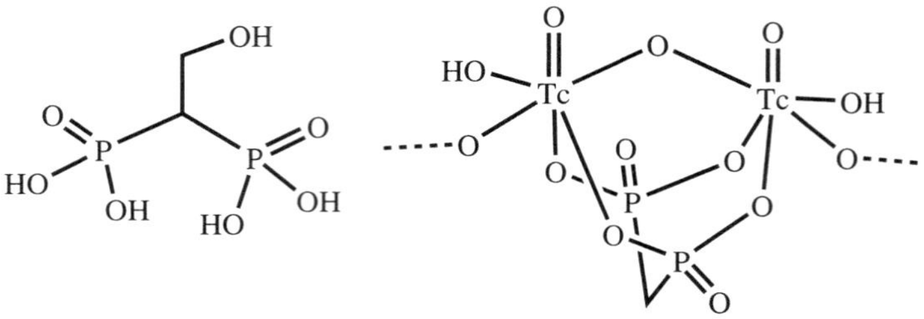

85   86

FIGURE 6.32   HEDP and the structure of Tc methylenediphosphonate complex.

not provide enough incentive for clinicians to change the technetium compounds that are in use or indeed to switch from PET to SPECT imaging. The way forward might be to exploit the aspects of technetium chemistry that are not available for PET radionuclides such as ^{18}F. These might include utilising the ability of Tc in intermediate oxidation states such as (III) and (IV) to undergo redox reactions at biologically accessible potentials. Hitherto the approach has been to make Tc complexes that would not oxidise or reduce *in vivo,* but embracing the redox capabilities could create new types of biological behaviour. The imaging of hypoxia is one obvious example. Technetium-99m also has Auger electron emissions that can be deployed therapeutically, particularly if targeted to the nucleus of cells where maximal DNA damage can occur. There have been recent developments of Tc complexes that have both nuclear penetrating peptides and a fluorophore attached to a stable core that are localised in the nucleus [287, 288]. As the targeting of complexes is refined to the cellular level, the need for multimodal agents that can be observed by fluorescence microscopy will increase and there is certainly scope here for novel chemistry. Much remains to be discovered about the chemistry of these elements in the context of molecular imaging, and the coordination chemistry of these elements will surely continue to flourish for the foreseeable future.

REFERENCES

[1] G. Mariani, L. Bruselli and A. Duatti, *Eur.J. Nucl. Med. Mol.Im.* **35**, 1560–1565 (2008).

[2] D. Cheng, Y. Wang, X. Liu, P. Pretorius, M. Liang, M. Rusckowski and D. Hnatowich, *Bioconj. Chem.* **21**, 1565–1570 (2010).

[3] J. Zeintl, A. Vija, A. Yahil, J. Hornegger and T. Kuwert, *J Nucl. Med.* **51**, 921–928 (2010).

[4] S. Eshuis, P. Jager, R. Maguire, S. Jonkman, R. Dierckx and K. Leenders, *Eur.J. Nucl. Med. Mol. Imag.* **36**, 454–462 (2009).

[5] M. Santini, A. Fiorelli, G. Vicidomini, P. Laperuta, L. Busiello, P. Rambaldi, L. Mansi and A. Rotondo, *Respiration* **80**, 524–533 (2010).

[6] G. Liu and D. Hnatowich, *Anticancer Agents Med. Chem.* **7**, 367–377 (2007).

[7] H. Biersack, H. Palmedo, A. Andris, S. Rogenhofer, F. Knapp, S. Guhlke, S. Ezziddin, J. Bucerius and D. von Mallek, *J. Nucl. Med.* **52**, 1721–1726 (2011).

[8] P. Gould, *Nature* **460**, 312–313 (2009).

[9] B. Guerin, S. Tremblay, S. Rodrigue, J. Rousseau, V. Dumulon-Perreault, R. Lecomte, J. van Lier, A. Zyuzin and E. van Lier, *J. Nucl. Med.* **51**, 13N–16N (2010).

[10] S. Qaim, *Nucl. Med. Biol.* **27**, 323–328 (2000).

[11] G. Griffiths, D. Goldenburg, F. Roesch and H. Hansen, *Clin. Canc. Res.* **5**, 3001–3003 (1999).

[12] H. Bigott, J. Prior D. Piwinica-Worms and M. Welch, *Mol. Imag.* **4**, 30–39 (2005).

[13] E. Watson, M. Stabinand J. Stubbs, *J. Nucl. Med.* **35**, 923–924 (1994).

[14] K. Gagnon, S. McQuarrie, D. Abrams, A. McEwan and F. Wuest, *Curr. Radiopharm.* **4**, 90–101 (2011).

[15] F. Knapp, A. Beets and S. Mirzadeh, *Abstracts of Papers of the American Chemical Society* **215**, U934–U934 (1998).

[16] A. J. West, *Ann. Rep. Prog. Chem., Sect. A Inorg. Chem.* **105**, 211–220 (2009).

[17] S. R. Banerjee, K. P. Maresca, L. Francesconi, J. Valliant, J. W. Babich and J. Zubieta, *Nucl. Med. Biol.* **32**, 1–20 (2005).

[18] U. Abram and R. Alberto, *J. Braz. Chem. Soc.* **17**, 1486–1500 (2006).

[19] U. Abram, *Comp. Coord. Chem. II* **5**, 271–402 (2004).

[20] R. Alberto, *Comp Coord. Chem. II* **5**, 127–270 (2004).

[21] P. Thornton, *Ann. Rep. Prog. Chem., Sect. A Inorg. Chem.* **100**, 217–227 (2004).

[22] M. Kohlickova, V. Jedinakova-Krizova and F. Melichar, *Chem. Listy* **94**, 151–158 (2000).

[23] P. J. Blower and S. Prakash, *Perspect. Bioinorg. Chem.* **4**, 91–143 (1999).

[24] A. M. W. Cargill Thompson, *Ann. Rep. Prog. Chem., Sect. A Inorg. Chem.* **95**, 153–164 (1999).

[25] J. C. Vites and M. M. Lynam, *Coord. Chem. Rev.* **172**, 357–388 (1998).

[26] J. C. Vites and M. M. Lynam, *Coord. Chem. Rev.* **169**, 201–235 (1998).

[27] J. R. Dilworth and S. J. Parrott, *Developments in Nuclear Medicine*, S. Mather, ed., **30**, 1–29 (1996).

[28] J. Dilworth and P. Donnelly, in *Metallotherapeutics*, M. Gielen and E. Tiekink, eds. Wiley, New York, 2005, 463–486.

[29] J. R. Dilworth and S. J. Parrott, *Chem. Soc. Rev.* **27**, 43–55 (1998).

[30] R. Garcia, A. Paulo and I. Santos, *Inorg. Chim. Acta* **362**, 4315–4327 (2009).

[31] S. Jurisson and J. Lydon, *Chem. Rev.* **99**, 2205–2218 (1999).

[32] K. Schwochau, *Technetium*, Wiley, Weinheim, 2000.

[33] D. Troutner, W. Volkert, T. Hoffman and R. Holmes, *Appl. Radiat. Isot.* **35**, 467–470 (1984).

[34] J. Cyr, D. Nowotnik, Y. Pan, J. Gougoutas, M. Malley, J. Di Marco, A. Nunn and K. Linder, *Inorg Chem.* **40**, 3555–3561 (2001).

[35] R. Neirinckx, L. Canning, I. Piper, D. Nowotnik, R. Pickett, R. Holmes, W. Volkert, A. Forster, P. Weisner, J. Marriott and S. Chaplin, *J. Nucl. Med.* **28**, 191–202 (1987).

[36] M. Roca, J. Martin-Comin, W. Becker, M. Bernardo, B. Gutfilen, A. Moisan, M. Peters, E. Prats, M. Rodrigues, C. Sampson, A. Signore, H. Sinzinger and M. Thakur, *Eur. J. Nucl. Med.* **25**, 797–799 (1998).

[37] G. Bormans, O. Peeters, H. Vanbilloen, N. Blaton and A. Verbruggen, *Inorg. Chem.* **35**, 6240–6244 (1996).

[38] L. Kurti, D. Papagiannopoulou, M. Papadopoulos, I. Pirmettis, C. P. Raptopoulou, A. Terzis, E. Chiotellis, M. Harmata, R. R. Kuntz and R. S. Pandurangi, *Inorg. Chem.* **42**, 2960–2967 (2003).

[39] J. W. Buchler and S. B. Kruppa, *Z. Naturforsch., B Chem. Sci.* **45**, 518–530 (1990).

[40] A. J. Lawrence, J. R. Thornback, G. D. Zanelli and A. Lawson, *Inorg. Chim. Acta* **141**, 165–166 (1988).

[41] D. W. Wong, A. Mandal, I. C. Reese, J. Brown and R. Siegler, *Int. J. Nucl. Med. Biol.* **10**, 211–218 (1983).

[42] A. K. Babbar, A. K. Singh, H. C. Goel, U. P. Chauhan and R. K. Sharma, *Nucl. Med. Biol.* **27**, 537–592 (2000).

[43] A. K. Babbar, A. K. Singh, H. C. Goel, U. P. Chauhan and R. K. Sharma, *Nucl. Med. Biol.* **27**, 419–426 (2000).

[44] S. Murugesan, S. J. Shetty, O. P. D. Noronha, A. M. Samuel, T. S. Srivastava, C. K. K. Nair and L. Kothari, *Appl. Radiat. Isot.* **54**, 81–88 (2000).

[45] S. Murugesa, S. J. Shetty, T. S. Srivastava, O. P. Noronha and A. M. Samuel, *Appl. Radiat. Isot.* **55**, 641–646 (2001).

[46] G. Bandoli, U. Mazzi and E. Roncari, *Coord. Chem. Rev.* **44**, 191–227 (1982).

[47] K. Linder, A. Davison and A. Jones, *Technetium in Chemistry and Nuclear Medicine*, M. Nicolini, G. Bandoli, U. Mazzi, eds., Raven, New York, 1986, 43–45.

[48] R. Alberto, H. Braband and H. W. P. N'Dongo, *Curr. Radiopharm.* **2**, 254–267 (2009).

[49] A. Stephenson, R. Karin, S. Banerjee, O. Sangeeta, S. Oyebola, M. K. Levadala, N. McFarlane, R. D. Boreham, K. P. Maresca, J. W. Babich, J. Zubieta and J. F. Valliant, *Bioconjug. Chem.* **16**, 1189–1195 (2005).

[50] R. W. Riddoch, P. Schaffer and J. F. Valliant, *Bioconjug. Chem.* **17**, 226–235 (2006).

[51] K. P. Maresca, S. M. Hillier, F. J. Femia, C. N. Zimmerman, M. K. Levadala, S. R. Banerjee, J. Hicks, C. Sundararajan, J. Valliant, J. Zubieta, W. C. Eckelman, J. L. Joyal and J. W. Babich, *Bioconjug. Chem.* **20**, 1625–1633 (2009).

[52] K. Baidoo, U. Scheffel and S. Lever, *Cancer Res.* **50**, S799–S803 (1990).

[53] K. Baidoo, U. Scheffel, M. Stathis, P. Finley, S. Lever, Y. Zhan and H. Wagner, *Bioconj. Chem.* **9**, 208–217 (1998).

[54] K. Baidoo, K. Lin, Y. Zhan, P. Finley, U. Scheffel and H. Wagner, *Bioconj. Chem.* **9**, 218–225 (1998).

[55] C. Kao, K. Baidoo and S. Lever, *Nucl. Med. Biol.* **24**, 499–505 (1997).

[56] H. Kung, Y. Guo, C. Yu, J. Billings, V. Subramanyam and J. Calabrese, *J. Med. Chem.* **32**, 433–437 (1989).

[57] E. Cheesman, M. Blanchette, M. Ganey, L. Maheu, S. Miller, R. Morgan, R. Walovitch, A. Watson and S. Williams, *Eur. J. Nucl. Med.* **14**, 304–304 (1988).

[58] R. Walovitch, T. Hill, S. Garrity, E. Cheesman, B. Burgess, D. Oleary, A. Watson, M. Ganey, R. Morgan and S. Williams, *J. Nucl. Med.* **30**, 1892–1901 (1989).

[59] S. Meegalla, K. Plossl, M. Kung, D. Svenson, L. Liablesands, A. Rheingold and H. Kung, *J. Am. Chem. Soc.* **117**, 11037–11038 (1995).

[60] A. Davison, A. Jones, C. Orvig and M. Sohn, *Inorg. Chem.* **20**, 1629–1632 (1981).

[61] A. Davison, A. Jones, C. Orvig, M. Sohn, M. Lategola and G. Freeman, *J. Nucl. Med.* **22**, P57–P58 (1981).

[62] T. Rao, D. Adhikesavalu, A. Camerman and A. Fritzberg, *J. Am. Chem. Soc.* **112**, 5798–5804 (1990).

[63] S. Liu, D. Edwards and J. Barrett, *Bioconj. Chem.* **8**, 621–636 (1997).

[64] S. Liu and D. Edwards, *Chem. Rev.* **99**, 2235–2268 (1999).

[65] C. Archer, J. Dilworth, D. Griffiths, M. AlJeboori, J. Kelly, C. Lu, M. Rosser and Y. Zheng, *Dalton Trans.* 1403–1410 (1997).

[66] T. Hosotani, A. Yokoyama, Y. Arano, K. Horiuchi, H. Wasaki, H. Saji and K. Torizuka, *Int. J. Nucl. Med. Biol.* **12**, 431–437 (1986).

[67] A. Yokoyama, Y. Terauchi, K. Horiuchi, H. Tanaka, T. Odori, R. Morita, T. Mori and K. Torizuka, *J. Nucl. Med.* **17**, 816–819 (1976).

[68] A. Yokoyama, A. Yamada, Y. Arano, K. Horiuchi, K. Yamamoto and K. Torizuka, *J. Nucl. Med. Allied. Sci.* **26**, 159–160 (1982).

[69] A. Fritzberg, S. Kasina, D. Eshima and D. Johnson, *J. Nucl. Med.* **27**, 111–116 (1986).

[70] D. Nosco, R. Manning and A. Fritzberg, *J. Nucl. Med.* **27**, 939 (1986).

[71] L. Hansen, L. Marzilli and A. Taylor, *J. Nucl. Med.* **42**, 280–293 (1998).

[72] L. Hansen, R. Cini, A. Taylor and L. Marzilli, *Inorg. Chem.* **31**, 2801–2808 (1992).

[73] S. Guhlke, A. Schaffland, P. Zamora, J. Sartor, D. Diekmann, H. Bender, F. Knapp and H. Biersack, *Nucl. Med. Biol.* **25**, 621–631 (1998).

[74] M. Hosono, M. Hosono, T. Haberberger, P. Zamora, S. Guhlke, H. Bender, F. Knapp and H. Biersack, *Jap. J. Cancer Res.* **87**, 995–1000 (1996).

[75] G. Visser, M. Gerretsen, J. Herscheid, G. Snow and G. Vandongen, *J. Nucl. Med.* **34**, 1953–1963 (1993).

[76] C. Liu, G. Liu, N. Liu, Y. Zhang, J. He, M. Rusckowski and D. Hnatowich, *Nucl. Med. Biol.* **30**, 207–214 (2003).

[77] G. Liu, J. He, S. Zhang, C. Liu, M. Rusckowski and D. Hnatowich, *Antisense & Nucleic Acid Drug Develop.* **12**, 393–398 (2002).

[78] F. Chang, T. Qu, M. Rusckowski and D. Hnatowich, *Appl. Radiat. Isot.* **50**, 723–732 (1999).

[79] M. Rusckowski, T. Qu, S. Gupta and D. Hnatowich, *J. Nucl. Med.* **42**, 1870–1877 (2001).

[80] D. Pearson, J. ListerJames, W. McBride, D. Wilson, L. Martel, E. Civitello, J. Taylor, B. Moyer and R. Dean, *J. Med. Chem.* **39**, 1361–1371 (1996).

[81] D. Pearson, J. ListerJames, W. McBride, D. Wilson, L. Martel, E. Civitello and R. Dean, *J. Med. Chem.* **39**, 1372–1382 (1996).

[82] K. Chavatte, E. Wong, T. K. Fauconnier, L. Lu, T. Nguyen, D. Roe, A. Pollak, D. Eshima, D. Terriere, J. Mertens, K. Iterbeke, D. Tourwe, J. Thornback and A. Bossuyt, *J. Label. Compd. Radiopharm.* **42**, 415–421 (1999).

[83] A. Pollak, D. G. Roe, C. M. Pollock, L. F. L. Lu and J. R. Thornback, *J. Am. Chem. Soc.* **121**, 11593–11594 (1999).

[84] J. Cyr, D. Pearson, C. Nelson, B. Lyons, Y. Zheng, J. Bartis, J. He, M. Cantorias, R. Flowell and L. Francesconi, *J. Med. Chem.* **50**, 4295–4303 (2007).

[85] M. Cantorias, R. Howell, L. Todaro, J. Cyr, D. Berndorff, R. Rogers and L. Francesconi, *Inorg. Chem.* **46**, 7326–7340 (2007).

[86] P. J. Blower, A. S K. Lam, M. J. O'Doherty, A. G. Kettle, A. J. Coakley and F. F. Knapp, Jr., *Eur. J. Nucl. Med.* **25**, 613–621 (1998).

[87] M. Delange, D. Fiers, J. Kosterink, W. Vanluijk, S. Meijer, D. Dezeeuw and G. Vanderhem, *J. Nucl. Med.* **30**, 1219–1223 (1989).

[88] U. Choudhry, W. E. P. Greenland, W. A. Goddard, T. A. J. MacLennan, S. J. Teat and P. J. Blower, *Dalton Trans.* 311–317 (2003).

[89] A. Hoepping, P. Brust, R. Berger, P. Leibnitz, H. Spies, S. Machill, D. Scheller and B. Johannsen, *Bioorg. Med. Chem.* **6**, 1663–1672 (1998).

[90] B. Johannsen, R. Berger, P. Brust, H. Pietzsch, M. Scheunemann, S. Seifert, H. Spies and R. Syhre, *Eur. J. Nucl. Med.* **24**, 316–319 (1997).

[91] M. Kretzschmar, P. Brust, S. Elz, H. H. Pertz, H. J. Pietzsch, M. Scheunemann, S. Seifert, J. Zessin and B. Johannsen, *J. Label. Comp. Radiopharm.* **42**, S345–S347 (1999).

[92] A. Hoepping, M. Reisgys, P. Brust, S. Seifert, H. Spies, R. Alberto and B. Johannsen, *J. Label. Comp. Radiopharm.* **42**, S48–S50 (1999).

[93] R. Syhre, S. Seifert, H. Spies, A. Gupta and B. Johannsen, *Eur. J. Nucl. Med.* **25**, 793–796 (1998).

[94] R. Mosi, I. R. Baird, J. Cox, V. Anastassov, B. Cameron, R. T. Skerlj and S. P. Fricker, *J. Med. Chem.* **49**, 5262–5272 (2006).

[95] C. Clavaud, M. Heckenroth, C. Stricane, M.-A. Lelait, A. Menez and C. Dugave, *ChemBioChem.* **7**, 1352–1355 (2006).

[96] C. Clavaud, J. Le Gal, R. Thai, M. Moutiez and C. Dugave, *ChemBioChem.* **9**, 1823–1829 (2008).

[97] J. Simon, S. Zuckman, D. Troutner, W. Volkert and R. Holmes, *J. Label. Comp. Radiopharm.* **18**, 151–152 (1981).

[98] S. Zuckman, G. Freeman, D. Troutner, W. Volkert, R. Holmes, D. Vanderveer and E. Barefield, *Inorg. Chem.* **20**, 2386–2389 (1981).

[99] J. Kelly, A. Forster, B. Higley, C. Archer, F. Booker, L. Canning, K. Chiu, B. Edwards, H. Gill, M. Mcpartlin, K. Nagle, I. Latham, R. Pickett, A. Storey and P. Webbon, *J. Nucl. Med.* **34**, 222–227 (1993).

[100] H. Braband, T. I. Zahn and U. Abram, *Inorg. Chem.* **42**, 6160–6162 (2003).

[101] H. Braband, T. I. Kueckmann and U. Abram, *J. Organomet. Chem.* **690**, 5421–5429 (2005).

[102] A. Davison and J. Thomas, *Inorg. Chim. Acta* **190**, 231–235 (1991).

[103] T. Lis, *Acta Cryst. C-Crystal Struct. Comm.* **43**, 1710–1711 (1987).

[104] F. Kuhn, J. Haider, E. Herdtweck, W. Herrmann, A. Lopes, M. Pillinger and C. Romao, *Inorg. Chim. Acta* **279**, 44–50 (1998).

[105] H. Kunkely and A. Vogler, *J Photochem. Photobiol.* **136**, 175–177 (2000).

[106] A. Vogler and H. Kunkely, *Coord. Chem. Rev.* **200**, 991–1008 (2000).

[107] R. Pearlstein and A. Davison, *Polyhedron* **7**, 1981–1989 (1988).

[108] M. Middleditch, J. Anderson, A. Blake and C. Wilson, *Inorg. Chem.* **46**, 2797–2804 (2007).

[109] Y. Tooyama, H. Braband, B. Spingler, U. Abram and R. Alberto, *Inorg. Chem.* **47**, 257–264 (2008).

[110] Y. Tooyama, H. Braband, T. Fox and R. Alberto, *Chem. Eur. J.* **15**, 633–638 (2009).

[111] H. Braband, Y. Tooyama, T. Fox, R. Simms, J. Forbes, J. Valliant and R. Alberto, *Chem. Eur. J.* **17**, 12967–12974 (2011).

[112] J. Chatt, G. Rowe, J. Garforth and N. Johnson, *J. Chem. Soc.* 1012–1014 (1964).

[113] J. Chatt, C. Falk, G. Leigh and R. Paske, *J. Chem. Soc.* 2288–2291 (1969).

[114] J. Baldas and J. Bonnyman, *J. Nucl. Med.* **29**, 187 (1985).

[115] J. Kanellos, G. Pietersz, I. Mckenzie, J. Bonnyman and J. Baldas, *J. Nat. Cancer Inst.* **77**, 431–439 (1986).

[116] G. Williams, J. Bonnyman and J. Baldas, *Austr. J. Chem.* **40**, 27–33 (1987).

[117] G. Williams and J. Baldas, *Austr. J. Chem.* **42**, 875–884 (1989).

[118] J. Baldas, *Pure App. Chem.* **62**, 1079–1080 (1990).

[119] J. Baldas, J. Boas, J. Bonnyman, S. Colmanet and G. Williams, *Inorg. Chim. Acta* **179**, 151–15ₓ (1991).

[120] J. Baldas, *Topics Curr. Chem. Springer-Verlag*, **176**, 37–76 (1996).

[121] U. Abram, M. Braun, S. Abram, R. Kirmse and A. Voigt, *J. Chem. Soc., Dalton Trans.* 231–238 (1998).

[122] U. Abram, S. Abram, H. Spies, R. Kirmse, J. Stach and K. Kohler, *Z. Anorg. Allg. Chem.* **544**, 167–180 (1987).

[123] U. Abram, B. Lorenz, L. Kaden and D. Scheller, *Polyhedron* **7**, 285–289 (1988).

[124] J. R. Dilworth, R. Huebener and U. Abram, *Z. Anorg. Allg. Chem.* **623**, 880–882 (1997).

[125] C. Abram, M. Braun, S. Abram, R. Kirmse and A. Voigt, *Dalton Trans.* 231–238 (1998).

[126] U. Abram, E. S. Lang, S. Abram, J. Wegmann, J. R. Dilworth, R. Kirmse and J. D. Woollins, *J. Chem. Soc., Dalton Trans.* 623–630 (1997).

[127] A. Boschi, A. Duatti and L. Uccelli, *Topics Curr. Chem.* **252**, 85–115 (2005).

[128] R. Pasqualini and A. Duatti, *J. Chem. Soc. Chem. Commun.* 1354–1355 (1992).

[129] R. Pasqualini, E. Bellande, V. Comazzi and A. Duatti, *Eur. J. Nucl. Med.* **19**, 593 (1992).

[130] A. Boschi, A. Massi, L. Uccelli, M. Pasquali and A. Duatti, *Nucl. Med. Biol.* **237**, 927–934 (2010).

[131] A. Duatti, A. Marchi and R. Pasqualini, *Dalton Trans.* 3729–3733 (1990).

[132] A. Marchi, L. Uccelli, L. Marvelli, R. Rossi, M. Giganti, V. Bertolasi and V. Ferretti, *Dalton Trans* 3105–3109 (1996).

[133] F. Demaimay, A. Roucoux, N. Noiret and H. Patin, *J. Organomet. Chem.* **575**, 145–148 (1999).

[134] F. Mevellec, N. Lepareur, A. Roucoux, N. Noiret, H. Patin, G. Bandoli, M. Porchia and F. Tisato, *Inorg. Chem.* **41**, 1591–1597 (2002).

[135] J. R. Dilworth, J. S. Lewis, J. R. Miller and Y. Zheng, *J. Chem. Soc., Dalton Trans.* 1357–1361 (1995).

[136] C. Bolzati, A. Boschi, A. Duatti, S. Prakash, L. Uccelli, F. Refosco, F. Tisato and G. Bandoli, *J. Am. Chem. Soc.* **122**, 4510–4511 (2000).

[137] C. Cittanti, L. Uccelli, M. Pasquali, A. Boschi, C. Flammia, E. Bagatin, M. Casali, M. Stabin, L. Feggi, M. Giganti and A. Duatti, *J. Nucl. Med.* **49**, 1299–1304 (2008).

[138] C. Decristoforo, I. Santos, H. J. Pietzsch, J. U. Kuenstler, A. Duatti, C. J. Smith, A. Rey, R. Alberto, E. Von Guggenberg and R. Haubner, *J. Nucl. Med. Mol. Imaging* **51**, 33–41 (2007).

[139] C. Bolzati, A. Mahmood, E. Malago, L. Uccelli, A. Boschi, A. G. Jones, F. Refosco, A. Duatti and F. Tisato, *Bioconjug. Chem.* **14**, 1231–1242 (2003).

[140] A. Boschi, L. Uccelli, A. Duatti, C. Bolzati, F. Refosco, F. Tisato, R. Romagnoli, G. Baraldi Pier, K. Varani and A. Borea Pier, *Bioconjug. Chem.* **14**, 1279–1288 (2003).

[141] A. Boschi, C. Bolzati, E. Benini, E. Malago, L. Uccelli, A. Duatti, A. Piffanelli, F. Refosco and F. Tisato, *Bioconjug. Chem.* **12**, 1035–1042 (2001).

[142] F. Refosco, C. Bolzati, A. Duatti, F. Tisato and L. Uccelli, *Recent Res. Dev. Inorg. Chem.* **2**, 89–98 (2000).

[143] J. Chatt, J. R. Dilworth, G. J. Leigh and V. D. Gupta, *J. Chem. Soc. A* 2631–2639 (1971).

[144] V. Duckworth, P. Douglas, R. Mason and B. Shaw, *J. Chem. Soc. Chem. Comm.* 1083–1086 (1970).

[145] J. R. Dilworth, S. A. Harrison, D. R. M. Walton and E. Schweda, *Inorg. Chem.* **24**, 2594–2595 (1985).

[146] C. M. Archer, J. R. Dilworth, P. Jobanputra, M. E. Harman, M. B. Hursthouse and A. Karaulov, *Polyhedron* **10**, 1539–1543 (1991).

[147] B. Coutinho, J. R. Dilworth, P. Jobanputra, R. M. Thompson, S. Schmid, J. Straehle and C. M. Archer, *J. Chem. Soc., Dalton Trans.* 1663–1669 (1995).

[148] A. R. Cowley, J. R. Dilworth, P. S. Donnelly and S. J. Ross, *Dalton Trans.* 73–82 (2007).

[149] A. R. Cowley, J. R. Dilworth and P. S. Donnelly, *Inorg. Chem.* **42**, 929–931 (2003).

[150] A. Moehlenkamp and R. Mattes, *Z. Naturforsch., B Chem. Sci.* **47**, 969–977 (1992).

[151] T. Nicholson, P. Lombardi and J. Zubieta, *Polyhedron* **6**, 1577–1585 (1987).

[152] T. Nicholson and J. Zubieta, *Inorg. Chem.* **26**, 2094–2101 (1987).

[153] M. Teresa, A. R. S. da Costa, M. F. C. Guedes da Silva, J. J. R. Frausto da Silva and A. J. L. Pombeiro, *Collect. Czech. Chem. Commun.* **72**, 599–608 (2007).

[154] C. M. Archer, J. R. Dilworth, P. Jobanputra, R. M. Thompson, M. McPartlin and W. Hiller, *J. Chem. Soc., Dalton Trans.* 897–904 (1993).

[155] M. Abrams, M. Juweid, C. Tenkate, D. Schwartz, M. Hauser, F. Gaul, A. Fuccello, R. Rubin, H. Strauss and A. Fischman, *J. Nucl. Med.* **31**, 2022–2028 (1990).

[156] D. Schwartz, M. Abrams, M. Hauser, F. Gaul, S. Larsen, D. Rauh and J. Zubieta, *Bioconj. Chem.* **2**, 333–336 (1991).

[157] Y. Zhou, S. Chakraborty and S. Liu, *Theranostics* **1**, 58–82 (2011).

[158] Z.-F. Su, G. Liu, S. Gupta, Z. Zhu, M. Rusckowski and D. J. Hnatowich, *Bioconj. Chem.* **13**, 561–570 (2002).

[159] R. King, M. B.-U. Surfraz, C. Finucane, S. C. G. Biagini, P. J. Blower and S. J. Mather, *J. Nucl. Med.* **50**, 591–598 (2009).

[160] E. Von Guggenberg, M. Behe, T. M. Behr, M. Saurer, T. Seppi and C. Decristoforo, *Bioconj. Chem.* **15**, 864–871 (2004).

[161] M. Gabriel, C. Decristoforo, E. Donnemiller, H. Ulmer, C. W. Rychlinski, S. J. Mather and R. Moncayo, *J. Nucl. Med.* **44**, 708–716 (2003).

[162] F. Blankenberg, J. Vanderheyden, H. Strauss and J. Tait, *Nat. Prot.* **1**, 108–110 (2006).

[163] F. Blankenberg, C. Kalinyak, L. Liu, M. Koike, D. Cheng, M. Goris, A. Green, J. Vanderheyden, D. Tong and M. Yenari, *Eur. J. Nucl. Med. Mol. Imaging* **33**, 566–574 (2006).

[164] Z. Levashova, M. V. Backer, G. Horng, D. Felsher, J. M. Backer and F. G. Blankenberg, *Bioconj. Chem.* **20**, 742–749 (2009).

[165] M. M. Welling, R. Visentin, H. I. J. Feitsma, A. Lupetti, E. K. J. Pauwels and P. H. Nibbering, *Nucl. Med. Biol.* **31**, 503–509 (2004).

[166] L. K. Meszaros, A. Dose, S. C. G. Biagini and P. J. Blower, *Inorg. Chim. Acta* **363**, 1059–1069 (2010).

[167] D. J. Rose, K. E. Maresca, T. Nicholson, A. Davison, A. G. Jones, J. Babich, A. Fischman, W. Graham, J. R. D. DeBord and J. Zubieta, *Inorg. Chem.* **37**, 2701–2716 (1998).

[168] L. K. Meszaros, A. Dose, S. C. G. Biagini and P. J. Blower, *Dalton Trans.* **40**, 6260–6267 (2011).

[169] R. C. King, M. B.-U. Surfraz, S. C. G. Biagini, P. J. Blower and S. J. Mather, *Dalton Trans.* 4998–5007 (2007).

[170] A. Hinni, *D. Phil. Thesis*, University of Oxford (2010).

[171] C. M. Archer, J. R. Dilworth, P. Jobanputra, R. M. Thompson, M. McPartlin, D. C. Povey, G. W. Smith and J. D. Kelly, *Polyhedron* **9**, 1497–1502 (1990).

[172] T. Nicholson, M. Hirsch-Kuchma, A. Davison and A. Jones, *Inorg. Chim. Acta* **271**, 191–194 (1998).

[173] T. Nicholson, D. J. Kramer, A. Davison and A. G. Jones, *Inorg. Chim. Acta* **353**, 269–275 (2003).

[174] J. R. Dilworth, P. Jobanputra, S. J. Parrott, R. M. Thompson, D. C. Povey and J. A. Zubieta, *Polyhedron* **11**, 147–155 (1992).

[175] J. Chatt, G. Leigh, D. Mingos, E. Randall and D. Shaw, *Chem. Comm.* 419–421 (1968).

[176] H. Nishiyama, R. Adolph, E. Deutsch, V. Sodd, K. Libson, M. Gerson, E. Saenger, S. Lukes, M. Gabel, J. Vanderheyden and D. Fortman, *J. Nucl. Med.* **23**, 1102–1110 (1982).

[177] H. Nishiyama, E. Deutsch, R. Adolph, V. Sodd, K. Libson, E. Saenger, M. Gerson, M. Gabel, S. Lukes, J. Vanderheyden, D. Fortman, K. Scholz, L. Grossman and C. Williams, *J. Nucl. Med.* **23**, 1093–1101 (1982).

[178] H. Nishiyama, V. Sodd, E. Deutsch, R. Adolph, K. Libson, M. Gerson, J. Vanderheyden, M. Gabel, S. Lukes and E. Saenger, *J. Nucl. Med.* **23**, P12–P12 (1982).

[179] J. Kirchhoff, W. Heineman and E. Deutsch, *Inorg. Chem.* **27**, 3608–3614 (1988).

[180] C. M. Archer, J. R. Dilworth, R. M. Thompson, M. McPartlin, D. C. Povey and J. D. Kelly, *J. Chem. Soc., Dalton Trans.* 461–466 (1993).

[181] J. Barrera, A. Burrell and J. Bryan, *Inorg. Chem.* **35**, 335–341 (1996).

[182] G. Rouschias and G. Wilkinson, *J. Chem. Soc.* 993–995 (1967).

[183] G. Rouschias and G. Wilkinson, *J. Chem. Soc.* 489–490 (1968).

[184] S. Jurisson, K. Dancey, M. Mcpartlin, P. Tasker and E. Deutsch, *Inorg. Chem.* **23**, 4743–4749 (1984).

[185] S. Jurisson, L. Lindoy, K. Dancey, M. Mcpartlin, P. Tasker, D. Uppal and E. Deutsch, *Inorg. Chem.* **23**, 227–231 (1984).

[186] M. Marmion, S. Woulfe, W. Neumann, D. Nosco and E. Deutsch, *Nucl. Med. Biol.* **26**, 755–770 (1999).

[187] D. Piwnicaworms, M. Chiu, M. Budding, J. Kronauge, R. Kramer and J. Croop, *Cancer Res.* **53**, 977–984 (1993).

[188] V. Rao, M. Chiu, J. Kronauge and D. Piwnicaworms, *J. Nucl. Med.* **35**, 510–515 (1994).

[189] F. Mendes, A. Paulo and I. Santos, *Dalton Trans.* **40**, 5377–5393 (2011).

[190] K. Linder, M. Malley, J. Gougoutas, S. Unger and A. Nunn, *Inorg. Chem.* **29**, 2428–2434 (1990).

[191] K. Linder, D. Nowotnik, M. Malley, J. Gougoutas and A. Nunn, *Inorg. Chim. Acta* **190**, 249–255 (1991).

[192] S. Jurisson, W. Hirth, K. Linder, R. Dirocco, R. Narra, D. Nowotnik and A. Nunn, *Nucl. Med. Biol.* **18**, 735–744 (1991).

[193] K. Linder, M. Wen, D. Nowotnik, M. Malley, J. Gougoutas, A. Nunn and W. Eckelman, *Bioconj. Chem.* **2**, 160–170 (1991).

[194] K. E. Linder, Y. W. Chan, J. E. Cyr, D. P. Nowotnik, W. C. Eckelman and A. D. Nunn, *Bioconj. Chem.* **4**, 326–333 (1993).

[195] S. Jurisson, L. Francesconi, K. Linder, E. Treher, M. Malley, J. Gougoutas and A. Nunn, *Inorg. Chem.* **30**, 1820–1827 (1991).

[196] M. Abrams, D. Brenner, A. Davison and A. Jones, *J. Label. Comp. Radiopharm.* **21**, 1061–1062 (1984).

[197] M. Abrams, A. Davison, R. Faggiani, A. Jones and C. Lock, *Inorg. Chem.* **23**, 3284–3288 (1984).

[198] H. Pietzsch, H. Spies and S. Hoffmann, *Inorg. Chim. Acta* **168**, 7–9 (1990).

[199] S. Seifert, A. Drews, A. Gupta, H. Pietzsch, H. Spies and B. Johannsen, *Appl. Radiat. Isot.* **53**, 431–438 (2000).

[200] N. Devries, J. Dewan, A. Jones and A. Davison, *Inorg. Chem.* **27**, 1574–1580 (1988).

[201] H. Spies, M. Glaser, H. Pietzsch, F. Hahn, O. Kintzel and T. Lugger, *Angew. Chem.* **33**, 1354–1356 (1994).

[202] P. J. Blower and J. R. Dilworth, *J. Chem. Soc., Dalton Trans.* 2305–2309 (1985).

[203] J. R. Dilworth, J. Hu, R. M. Thompson and D. L. Hughes, *J. Chem. Soc., Chem. Commun.* 551–553 (1992).

[204] A. Jones, M. Abrams and A. Davison, *J. Nucl. Med.* **26**, 149–150 (1982).

[205] B. Holman, A. Jones, J. ListerJames, A. Davison, M. Abrams, S. Tumeh and R. Nesto, *Circulation* **70**, 124 (1984).

[206] A. Jones, M. Abrams, A. Davison, J. Brodack, A. Toothaker, S. Adelstein and A. Kassis, *J. Nucl. Med. Biol.* **11**, 225–234 (1984).

[207] D. Piwnicaworms, J. Kronauge and M. Chiu, *Circulation* **82**, 1826–1838 (1990).

[208] M. Chiu, J. Kronauge and D. Piwnicaworms, *J. Nucl. Med.* **31**, 1646-1653 (1990).

[209] L. Maffioli, J. Steens, E. Pauwels and E. Bombardieri, *Tumori* **82**, 12–21 (1996).

[210] M. Abrams, A. Davison, A. Jones, C. Costello and H. Pang, *Inorg. Chem.* **22**, 2798–2800 (1983).

[211] C. Cameron, S. Tetrick and R. Walton, *Organometallics* **3**, 240–247 (1984).

[212] R. Alberto, R. Schibli, A. Egli, P. Schubiger, W. Herrmann, G. Artus, U. Abram and T. Kaden, *J. Organomet. Chem.* **493**, 119–127 (1995).

[213] R. Alberto, R. Schibli, A. Egli, A. P. Schubiger, U. Abram and T. A. Kaden, *J. Am. Chem. Soc.* **120**, 7987–7988 (1998).

[214] R. Alberto, R. Schibli, U. Abram, A. Egli, F. F. Knapp and P. A. Schubiger, *Radiochim. Acta* **79**, 99–103 (1997).

[215] R. Alberto, *Top. Organomet. Chem.* **32**, 219–246 (2010).

[216] R. Alberto, *Eur. J. Inorg. Chem.* 21– 31 (2009).

[217] R. Alberto, R. Schibli, R. Waibel, U. Abram and A. P. Schubiger, *Coord. Chem. Rev.* **190–192**, 901–919 (1999).

[218] R. Alberto, K. Ortner, N. Wheatley, R. Schibli and A. P. Schubiger, *J. Am. Chem. Soc.* **123**, 3135–3136 (2001).

[219] R. Schibli, R. Schwarzbach, R. Alberto, K. Ortner, H. Schmalle, C. Dumas, A. Egli and P. Schubiger, *Bioconj. Chem.* **13**, 750–756 (2002).

[220] H. He, M. Lipowska, X. Xu, A. T. Taylor, M. Carlone and L. G. Marzilli, *Inorg. Chem.* **44**, 5437–5446 (2005).

[221] G. Hao, J. Zang, L. Zhu, Y. Guo and B. Liu, *J. Label. Comp. Radiopharm.* **47**, 513–521 (2004).

[222] H. He, M. Lipowska, X. Xu, A. Taylor and L. Marzilli, *Inorg. Chem.* **46**, 3385–3394 (2007).

[223] Y. Liu, J. Pak, P. Schmutz, M. Bauwens, J. Mertens, H. Knight and R. Alberto, *J. Am. Chem. Soc.* **128**, 15996–15997 (2006).

[224] Y. Liu, B. Oliveira, J. Correia, I. Santos, I. Santos, B. Springler and R. Alberto, *Org.Biomol.Chem.* **8**, 2829–2839 (2010).

[225] K. Halder, D. Nayak, R. Baishya, B. Sarkar, S. Sinha, S. Ganguly and M. Debnath, *Metallomics* **3**, 1041–1048 (2011).

[226] P. Kyprianidou, C. Tsoukalas, A. Chiotellis, D. Papagiannopoulou, C. Raptopoulou, A. Terzis, M. Pelecanou, M. Papadopoulos and I. Pirmettis, *Inorg. Chim. Acta* **370**, 236–242 (2011).

[227] W. Chen, Y. Chao-Liang, S. Lo, K. Chen and J. Lo, *Allied Radiation and Isotopes* **66**, 340–345 (2008).

[228] B. Lee, J. Kim, J. Jung, B. Moon, N. Im, H. Lee, J. Son, S. Kwak, D. Oh and S. Kim, *J. Label. Comp. Radiopharm.* **54**, S185–S185 (2011).

[229] B. Lee, J. Jung, B. Moon, J. Kim, N. Im, H. Lee, J. Son, S. Kwak and S. Kim, *J. Label. Comp. Radiopharm.* **54**, S186–S186 (2011).

[230] A. Orlova, C. Hofstrom, J. Malmberg, S. Ahlgren, T. Graslund and V. Tolmachev, *Eur. J. Nucl. Med. Mol. Imag.* **37**, S264–S265 (2010).

[231] V. Tolmachev, C. Hofstrom, J. Malmberg, S. Ahlgren, S. Hosseinimehr, M. Sandstrom, L. Abrahmsen, A. Orlova and T. Graslund, *Bioconj. Chem.* **21**, 2013–2022 (2010).

[232] V. Tolmachev, C. Hofstrom, J. Malmberg, S. Ahlgren, A. Orlova and T. Graslund, *Nucl. Med. Biol.* **37**, 698 (2010).

[233] C. Muller, C. Dumas, U. Hoffmann, P. Schubiger and R. Schibli, *J. Organomet. Chem.* **689**, 4712–4721 (2004).

[234] C. Schweinsberg, V. Maes, L. Brans, P. Blaeuenstein, D. Tourwe, P. Schubiger, R. Schibli and E. Garayoa, *Bioconj. Chem.* **19**, 2432–2439 (2008).

[235] L. Brans, V. Maes, E. Garcia-Garayoa, C. Schweinsberg, S. Daepp, P. Blauenstein, P. Schubiger, R. Schibli and D. Tourwe, *Chem. Biol.Drug Des.* **72**, 496–506 (2008).

[236] H. Zhu, L. Huang, Y. Zhang, X. Xu, Y. Sun and Y. Shen, *J. Biol. Inorg. Chem.* **15**, 591–599 (2010).

[237] M. Teran, E. Martinez, A. Reyes, A. Paolino, M. Vital, P. Esperon, J. Pacheco and E. Savio, *Nucl. Med. Biol.* **38**, 279–285 (2011).

[238] B. Oliveira, P. Raposinho, F. Mendes, I. Santos, I. Santos, A. Ferreira, C. Cordeiro, A. Freire and J. Correia, *J. Organomet. Chem.* **696**, 1057–1065 (2011).

[239] S. Kunze, F. Zobi, P. Kurz, B. Springler and R. Alberto, *Angew. Chem. Int. Ed.* **43**, 5025–5029 (2004).

[240] R. Alberto, C. Knight, H. Hector and S Mundweiler, *WO Pat.*, 2005–EP168, 2005068483 (2005).

[241] H. Struthers, B. Springler, T. Mindt and R. Schibli, *Chem.Eur. J.* **14**, 6173–6183 (2008).

[242] H. Struthers, A. Hagenbach, U. Abram and R. Schibli, *Inorg. Chem.* **48**, 5154–5163 (2009).

[243] T. Mindt, H. Struthers, B. Spingler, L. Brans, D. Tourwe, E. Garcia-Garayoa and R. Schibli, *Chemmedchem* **5**, 2026–2038 (2010).

[244] T. Spradau, W. Edwards, C. Anderson, M. Welch and J. Katzenellenbogen, *Nucl. Med. Biol.* **26**, 1–7 (1999).

[245] H. M. Bigott, E. Parent, L. G. Luyt, J. A. Katzenellenbogen and M. J. Welch, *Bioconj. Chem.* **16**, 255–264 (2005).

[246] L. G. Luyt, H. M. Bigott, M. J. Welch and J. A. Katzenellenbogen, *Bioorg. Med. Chem.* **11**, 4977–4989 (2003).

[247] A. Louie, O. Sogbein, P. Schaffer and J. Valliant, *J. Label. Comp. Radiopharm.* **48**, S45–S45 (2005).

[248] O. Sogbein, A. Green and J. Valliant, *Inorg. Chem.* **44**, 9585–9591 (2005).

[249] O. Sogbein, A. Green, P. Schaffer, R. Chankalal, E. Lee, B. Healy, P. Morel and J. Valliant, *Inorg. Chem.* **44**, 9574–9584 (2005).

[250] P. W. Causey, T. R. Besanger and J. F. Valliant, *J. Med. Chem.* **51**, 2833–2844 (2008).

[251] M. Bowen and C. Orvig, *Chem. Comm.* 5077–5091 (2008).

[252] D. Yang, C. Kim N. Schechter, A. Azhdarinia, D. Yu, C. Oh, J. Bryant, J. Won, E. Kim and D. Podoloff, *Radiology* **226**, 465–473 (2003).

[253] C. L. Ferreira, F. L. N. Marques, M. R. Y. Okamoto, A. H. Otake, Y. Sugai, Y. Mikata, T. Storr, M. Bowen, S. Yano, M. J. Adam, R. Chammas and C. Orvig, *Appl. Radiat. Isot.* **68**, 1087–1093 (2010).

[254] J. Brown, H. Sies and B. Brune, *Oxygen Biology and Hypoxia* **435**, 297–299 (2007).

[255] R. Iyer, P. Haynes, R. Schneider, B. Movsas and J. Chapman, *J. Nucl. Med.* **42**, 337–344 (2001).

[256] L. Wiebe, H. Machulla and H. Machulla, *Imaging of Hypoxia*, **33**, 1–18 (1999).

[257] K. Krohn, J. Link and R. Mason, *J. Nucl. Med.* **49**, 129S–148S (2008).

[258] K. Linder, Y. Chan, J. Cyr, M. Malley, D. Nowotnik and A. Nunn, *J. Med. Chem.* **37**, 9–17 (1994).

[259] T. Melo, J. Duncan, J. R. Ballinger and A. M. Rauth, *J. Nucl. Med.* **41**, 169–176 (2000).

[260] F. Hoebers, H. Janssen, R. Olmos, D. Sprong, A. Nunn, A. Balm, C. Hoefnagel, A. Begg and K. Haustermans, *Eur. J. Nucl. Med. Mol. Imaging* **29**, 1206–1211 (2002).

[261] C. Archer, B. Edwards, J. Kelly, A. KIng and A. Riley, *Technetium and Rhenium in Chemistry and Nuclear Medicine*, M. Nicolini, G.Bandoli, U.Mazzi, eds. SGEditorali, Padua, **4**, 535–539 (1995).

[262] R. D. Okada, G. Johnson, III, K. N. Nguyen, B. Edwards, C. M. Archer and J. D. Kelly, *Circulation*, **95**, 1892–1899 (1997).

[263] Y. Diao, Y. Li, L. Zhou, D. Li and J. Gao, *Zhonghua Heyixue Zazhi* **25**, 156–158 (2005).

[264] D. J. Honess, S. A. Hill, D. R. Collingridge, B. Edwards, G. Brauers, N. A. Powell and D. J. Chaplin, *Int. J. Radiat. Oncol. Biol. Phys.* **42**, 731–735 (1998).

[265] B.-H. Huang, G.-D. Li, X.-Z. Chen, R.-Z. Wu and Z.-W. Zhen, *Zhonghua Heyixue Zazhi* **27**, 97–99 (2007).

[266] K. Imahashi, K. Morishita, H. Kusuoka, Y. Yamamichi, S. Hasegawa, K. Hashimoto, Y. Shirakami, M. Kato-Azuma and T. Nishimura, *J. Nucl. Med.* **41**, 1102–1107 (2000).

[267] X. Jiang, P. Dai, D. Song, J. Wu and S. Li, *Xiandai Zhongliu Yixue* **19**, 43–46 (2011).

[268] M. Tatsumi, K. Yutani, H. Kusuoka and T. Nishimura, *Eur. J. Nucl. Med.* **26**, 91–94 (1999).

[269] A. M. Abrantes, M. E. S. Serra, A. C. Goncalves, J. Rio, B. Oliveiros, M. Laranjo, A. M. Rocha-Gonsalves, A. B. Sarmento-Ribeiro and M. F. Botelho, *Nucl. Med. Biol.* **37**, 125–132 (2010).

[270] G. Cook, S. Houston, S. Barrington and I. Fogelman, *J. Nucl. Med.* **39**, 99–103 (1998).

[271] M. Mallia, S. Subramanian, A. Mathur, H. Sarma, M. Venkatesh and S. Banerjee, *J. Label. Comp. Radiopharm.* **53**, 535–542 (2010).

[272] M. Mallia, A. Mathur, S. Banerjee, H. Sarma and M. Venkatesh, *Nucl. Med. Biol.* **37**, 682–683 (2010).

[273] J. Giglio, G. Patsis, I. Pirmettis, M. Papadopoulos, C. Raptopoulou, M. Pelecanou, E. Leon, M. Gonzalez, H. Cerecetto and A. Rey, *Eur. J. Med. Chem.* **43**, 741–748 (2008).

[274] H. Huang, H. Zhou, Z. Li, X. Wang and T. Chu, *Bioorg. Med. Chem. Lett.* **22**, 172–177 (2012).

[275] E. Engelhardt, R. Schneider, S. Seeholzer, C. Stobbe and J. Chapman, *J. Nucl. Med.* **43**, 837–850 (2002).

[276] S. Murugesan, S. Shetty, O. Noronha, A. Samuel, T. Srivastava, C. Nair and L. Kothari, *Appl. Radiat. Isot.* **54**, 81–88 (2001).

[277] K. Horiuchi, T. Tsukamoto, M. Saito, M. Nakayama, Y. Fujibayashi and H. Saji, *Nucl. Med. Biol.* **27**, 391–399 (2000).

[278] X. Sun, *D. Phil. Thesis*, University of Oxford (2009).

[279] C. Supuran, *Nat. Rev. Drug Discovery* **7**, 168–181 (2008).

[280] E. Svastova, A. Hulikova, M. Rafajova, M. Zat'ovicova, A. Gibadulinova, A. Casini, A. Cecchi, A. Scozzafava, C. Supuran, J. Pastorek and S. Pastorekova, *Febs. Lett.* **577**, 439–445 (2004).

[281] P. Ebbesen, E. O. Pettersen, T. A. Gorr, G. Jobst, K. Williams, J. Kieninger, R. H. Wenger, S. Pastorekova, L. Dubois, P. Lambin, B. G. Wouters, T. Van Den Beucken, C. T. Supuran, L. Poellinger, P. Ratcliffe, A. Kanopka, A. Gorlach, M. Gasmann, A. L. Harris, P. Maxwell and A. Scozzafava, *J. Enzyme Inhib. Med. Chem.* **24**, 1–39 (2009).

[282] V. Akurathi, L. Dubois, N. Lieuwes, S. Chitneni, B. Cleynhens, D. Vullo, C. Supuran, A. Verbruggen, P. Lambin and G. Bormans, *Nucl. Med. Biol.* **37**, 557–564 (2010).

[283] A. Richards, *J. Nucl. Med.* **15**, 1057–1060 (1974).

[284] K. Libson, E. Deutsch and B. Barnett, *J. Am.Chem. Soc.* **102**, 2476–2478 (1980).

[285] R. Elder, J. Yuan, B. Helmer, D. Pipes, K. Deutsch and E. Deutsch, *Inorg. Chem.* **36**, 3055–3063 (1997).

[286] P. J. Blower, A. G. Kettle, M. J. O'Doherty, A. J. Coakley and F. F. Knapp, Jr., *Eur. J. Nucl. Med.* **27**, 1405–1409 (2000).

[287] K. Zelenka, L. Borsig and R. Alberto, *Bioconj. Chem.* **22**, 958–967 (2011).

[288] K. Zelenka, L. Borsig and R. Alberto, *Org. Biomol. Chem.* **9**, 1071–1077 (2011).

7

THE RADIOPHARMACEUTICAL CHEMISTRY OF GALLIUM(III) AND INDIUM(III) FOR SPECT IMAGING

JONATHAN R. DILWORTH

Department of Chemistry, University of Oxford, Oxford, UK

SOFIA I. PASCU

Department of Chemistry, University of Bath, Bath, UK

7.1 INTRODUCTION TO GALLIUM AND INDIUM CHEMISTRY

The radiochemistry-relevant coordination chemistry of gallium and indium provides an interesting contrast to that of technetium and rhenium highlighted in Chapter 6. The similarities and differences have been highlighted in a fairly recent review by Zubieta et al. on molecular imaging [1]. Another review compared the coordination chemistry of technetium and indium [2], and the uses of ^{99m}Tc and ^{111}In for the detection of inflammation/infection were compared in a review of the molecular imaging of these diseases [3]. The radiopharmaceutical chemistry of Ga and In was included in a recent, comprehensive review that covered both coordination chemistry and applications in diagnostic SPECT imaging [4].

Only the trivalent oxidation state is realistically accessible for Ga and In in aqueous, biocompatible, media. There is generally little π-bonding to stabilise Ga or In ligand bonds, and compounds with monodentate ligands tend to undergo rapid exchange reactions. Stabilities of the level needed for radiopharmaceuticals can only be achieved through polydentate ligands, preferably with substituents that provide steric shielding and additional kinetic stabilisation. Although predominantly classified as hard metals with a preference for N and O, both also show good stabilities with anionic S donors. The octahedral ionic radii of the metals differ considerably (Ga = 62 pm, In = 92 pm); this impacts the stability of macrocyclic ligands where matching of the ionic radius of the metal ion to the size of the cavity within the ligand is important in the design of radiotracers for SPECT imaging. The ionic radius of Ga(III) is very similar to that of Fe(III), and the biological systems designed to sequester Fe(III) also bind Ga(III) very effectively. The smaller ionic radius of Ga relative to In also creates a stronger bond to water molecules, and water exchange rates for Ga are noticeably slower. However, this difference does not impact significantly on the relative rates of complexation of Ga and In to polydentate ligands, and similar radiolabelling conditions can be used for both. Coordination numbers for Ga complexes used in a radiopharmaceutical context vary between 4 and 6, whereas the larger In can accommodate 7 or 8 donors.

This chapter focuses on the coordination chemistry of the polydentate ligands most used for radiopharmaceutical applications and gives selected examples of the applications of the radiolabelled complexes in imaging and therapy.

7.1.1 The Radioisotopes

Table 7.1 summarises the nuclear properties of the isotopes discussed in this chapter. Both Ga and In have a wide range of accessible radioisotopes but only ^{67}Ga, ^{68}Ga, and ^{111}In have been used extensively in SPECT imaging. The applications of ^{68}Ga for PET chemistry are discussed elsewhere in this volume (Chapter 5). ^{67}Ga is produced commercially by the proton irradiation

The Chemistry of Molecular Imaging, First Edition. Edited by Nicholas Long and Wing-Tak Wong.
© 2015 John Wiley & Sons, Inc. Published 2015 by John Wiley & Sons, Inc.

TABLE 7.1   Details of Ga^{67} and In^{111} Radioisotopes.

Isotope	Half-Life	Decay Emissions	Production
^{67}Ga	78 h	γ-emissions at : 93, 185, 300, 393 KeV Auger 7.2–9.7 KeV Decay 100% by EC	Cyclotron ^{68}Zn (p,2n)^{67}Ga
^{111}In	68 h	γ-emissions at 172 and 245 KeV Auger 19–23 KeV Decay 100% by EC	Cyclotron ^{111}Cd (p,n)^{111}In

of a ^{68}Zn target. Its relatively long half-life means that it can be readily transported to radiopharmacies distant from the cyclotron. It is also relatively inexpensive at approximately \$20/mCi. It decays entirely by electron capture with 10 γ emissions, the four most intense being shown in Table 7.1. There is accompanying Auger electron emission that in energy terms is comparable to that of ^{99m}Tc. It can be supplied in HCl solution or as the citrate following addition of citrate and neutralisation.

While ^{110}In, ^{110m}In, and ^{114m}In have been investigated for radiopharmaceutical applications, ^{111}In has been by far the most widely used. It is made from ^{111}Cd by proton bombardment in a medium energy cyclotron. Decay occurs exclusively via electron capture to give excited states of ^{111}Cd and then photons of 175 and 245 KeV are emitted en route to the ground state. There is also some internal conversion generating Auger electron emissions in the range 19-23 KeV, which are suitable for therapeutic applications.

7.2   GALLIUM AND INDIUM COMPLEXES AND RELATED BIOCONJUGATES

In recent times, the use of ^{67}Ga has decreased substantially with a much increased emphasis on the positron-emitting ^{68}Ga. We have here focused on ligands that have been used with ^{67}Ga or ^{111}In but have included some recent examples involving ^{68}Ga to illustrate what may also be possible with ^{67}Ga. The following section follows the format used for technetium and rhenium (Chapter 6) with ligands organised according to the donor atoms available.

7.2.1   O$_6$ Donor Ligands

^{67}Gallium citrate has been used for many years as an imaging agent [5, 6] sold under the name Neoscan™ by GE Healthcare. It targets metastatic tumours and focal sites of infection but cannot distinguish reliably between these. The gallium is readily *trans*-chelated from the citrate complex **1** to iron transport systems, so high concentrations of ^{67}Ga tend to reflect areas of high iron turnover. This is an interesting case where imaging capabilities depend on the complex being labile to provide ^{67}Ga for biological Fe sequestering agents. The X-ray crystal structure is shown in Figure 7.1 (**1**) and reveals a distorted octahedral coordination about gallium [7]. The similarities of Ga and Fe coordination chemistry means that biological ligands for iron such as deferrioxamine (DFO) form extremely stable complexes with both metals. The X-ray structure of a Ga complex (**2**) (Figure 7.1) with a model hydroxamate has been determined and shows octahedral O$_6$ coordination. In contrast to Zirconium-89, there have been few, if any, examples of the use of bioconjugates based on DFO with radioactive gallium or indium. Interestingly, a ^{67}Ga transferrin complex has found use as a tumour imaging agent [8] and is incorporated in cells by clathrin-mediated endocytosis.

Two groups have independently synthesised the capped tris(hydroxypyridone) ligands **3, 4** (Figure 7.1). The C-capped ligand **3** radiolabels in extremely high yield at room temperature, pH 6.5 with ^{68}Ga and can readily be derivatised at the capping carbon for attachment of biomolecules. A variant with a pendant maleimide group has been used to label C2Ac protein for the potential imaging of apoptosis [9]. The N-capped ligand **4** does not appear to be as well pre-organised for coordination to gallium, but nevertheless it has been labelled with ^{67}Ga and showed promising biodistribution *in vivo* with no binding to blood protein and rapid renal clearance [10].

7.2.2   N$_2$O$_x$S$_{4-x}$ Donor Ligands

The classic hexadentate ligand EDTA (x = 4) forms very stable (log K = 24.9) octahedral complexes with Ga and the 7-coordinate complexes formed with In are even more stable. Replacement of two carboxyl substituents in EDTA by thiolato groups, for example, **5** (x = 2), causes an impressive increase in stability of the Ga complex, increasing log K to 41 (Figure 7.2). This ligand has been labelled with ^{67}Ga in good yield, and *in vivo* studies showed high stability with excretion occurring

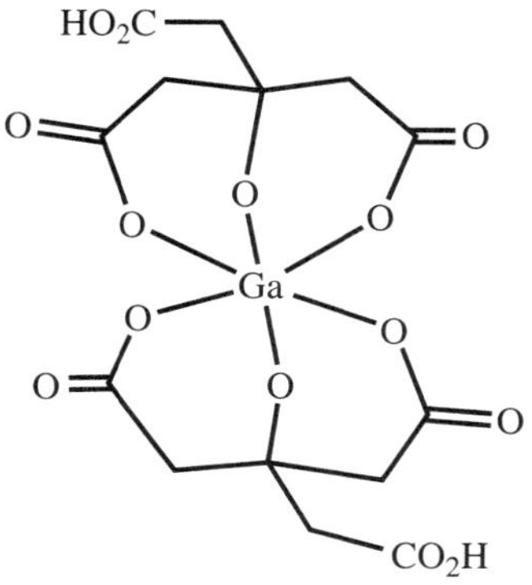

FIGURE 7.1 Ligands that provide an O_6 donor set and Ga complexes.

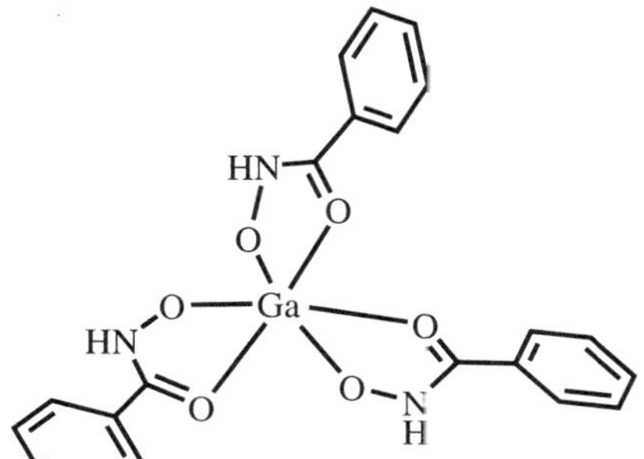

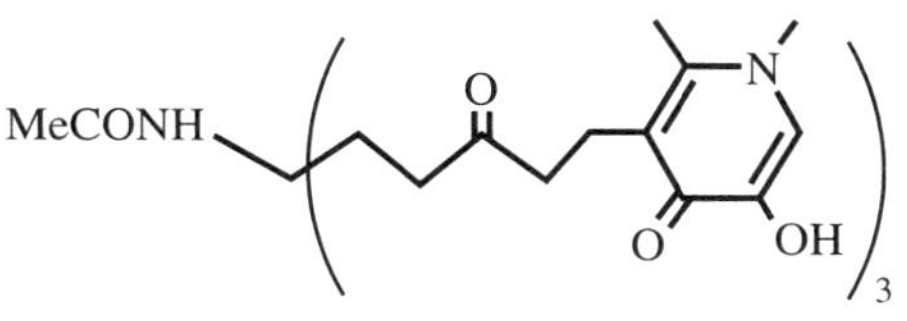

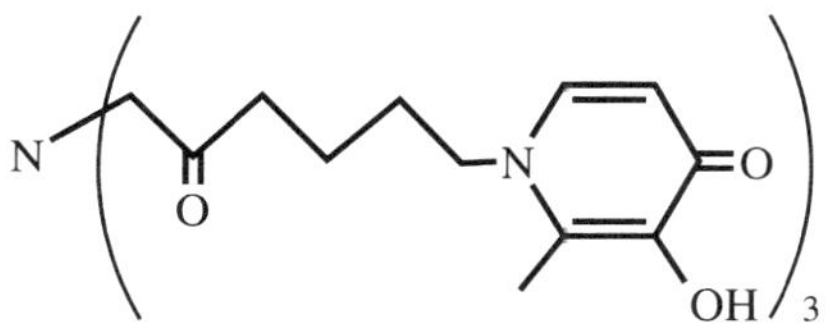

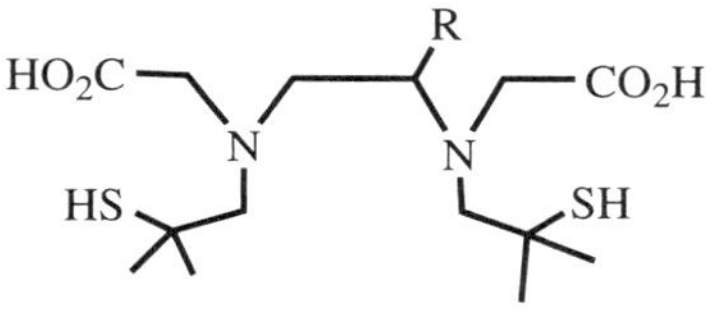

FIGURE 7.2 Some $N_2O_2S_2$ donor ligands.

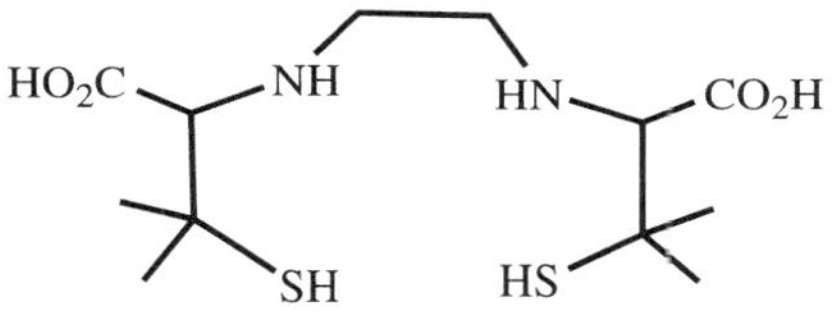

largely via the liver [11]. Attachment of a carboxymethoxybenzyl group to the ethylene backbone provides a potential route to bioconjugation. Ligand **6** (x = 2), has been encountered in Chapter 6 as the diester where it was used with ^{99m}Tc as a myocardial imaging agent. Here, with the free carboxylates, a very stable octahedral Ga complex is formed (log K = 31.5). The In complex is slightly more stable and also has a distorted octahedral geometry. There are no reports as yet of biological studies of ^{67}Ga or ^{111}In complexes with this ligand system.

7.2.3 N_3O_3 Donor Ligands

The neutral [^{111}In(oxine)$_3$] (oxine = 8-hydroxyquinoline) complex has been used for many years for the labelling of platelets and white blood cells [12]. Blood samples withdrawn from the patient led to separated white blood cells that were labelled with [^{111}In(oxine)$_3$] and injected back into the patient, thereby providing SPECT images of sites of infection. Within cells, the ^{111}In is *trans*-chelated, probably to biological iron chelators, leading to irreversible trapping of ^{111}In [13, 14]. The analogous ^{67}Ga oxine derivative has not been investigated as a potential radiopharmaceutical. The X-ray crystal structure of [In(oxine)$_3$] has been determined [15]. The indium is, as expected, pseudo octahedral and the 3 N donors are in a *mer* configuration. There was no evidence for other structural isomers.

Gallium forms an extremely stable octahedral complex with NOTA (**7**) (Figure 7.3) [16]. The kinetics of the reaction of Ga citrate with NOTA have been studied, and the rate determining step is proton reorganisation in a Ga citrate-NOTA intermediate [17]. The indium NOTA complex is reported to be less robust and has a seven-coordinate structure where the In retains one chloride and one carboxylate is protonated to give overall neutral charge [16, 18]. Two isomers of NOTA and related complexes are possible depending on the helicity of the pendant carboxyl groups when coordinated to the Ga. In the Ga structure discussed above, a single isomer has been formed. However, the potentially hexadentate ligand **8** forms a

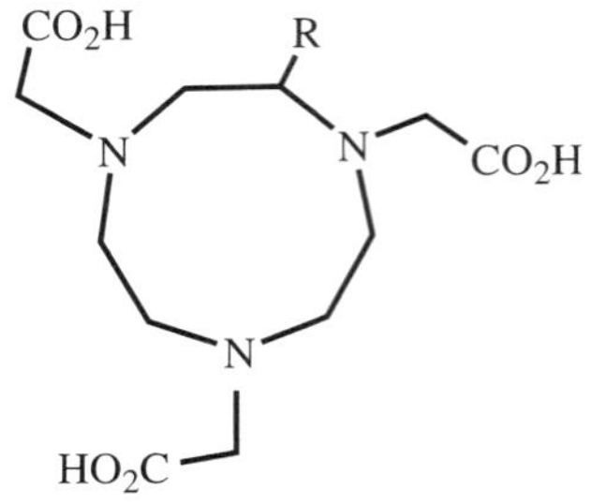

R = H, CH₂C₆H₄NCS, CH₂CH₂CO₂H

R = H, $CH_2C_6H_4NCS$, $CH_2CH_2CO_2H$

7

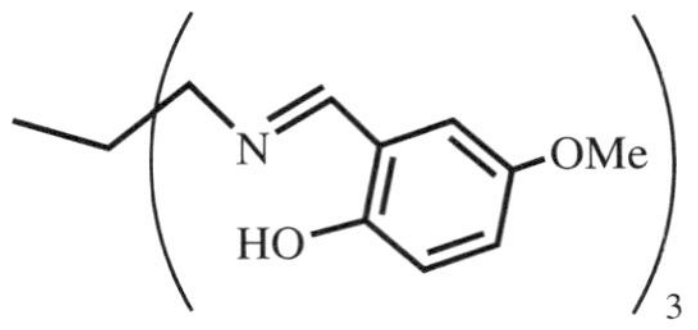

8

FIGURE 7.3 Hexadentate N_3O_3 donor NOTA and Schiff base ligands.

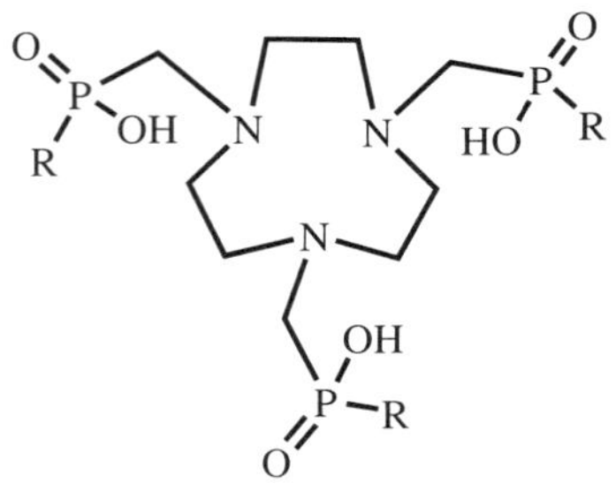

9

FIGURE 7.4 Structure of TRAP ligand system, where R = H, Ph, OH.

cationic complex with Ga that has been investigated as a myocardial agent. Two stereoisomers are possible depending on the disposition of the chelate ring systems. The (+) and (-) forms of the ^{67}Ga complex have been separated on a chiral HPLC column and biodistributions studied in wild type and MDRIa knockout mice.

The biodistributions of the two isomers were significantly different, particularly with respect to liver uptake [19]. Thus the stereoisomerism of octahedral Ga complexes, although different in nature to those of the technetium and rhenium oxo complexes discussed earlier, should be taken into account when considering their biological behaviour. The issue of isomerism in gallium complexes has been well summarised in a review covering their X-ray crystal structures [20].

To avoid compromising the stable N_3O_3 coordination core, conjugation of NOTA to biomolecules is achieved by addition of a functional group at a backbone carbon. The derivative of NOTA with $R=CH_2CH_2COOH$ is designated NODA and has been used to label (Tyr³)-octreatide, TOC, with ^{67}Ga, ^{68}Ga, and ^{111}In for the imaging of somastatin-receptor-expressing tumours. Both the ^{67}Ga and ^{111}In complexes showed good stability with IC_{50} values in the low nM range for binding to the receptor. The biodistribution of the ^{67}Ga complex is very favourable with rapid clearance from all non-target organs [21]. NOTA derivatised with an isothiocyanatobenzyl group has been used to conjugate [68]GaNOTA to cyclic RGD peptides for angiogenesis imaging [22]. This RGD peptide labelling approach appears not to have been used for [67]Ga. The same isothiocyanato NOTA derivative labelled with ^{67}Ga has been conjugated to human epidermal growth factor hEGF and showed good stability and specific binding following injection of an anti EGFR affibody [23]. Conjugation of [68]GaNOTA to a nitroimidazole molecule has been investigated as a possible route to PET imaging of hypoxia. *In vivo* studies of the conjugates showed increasing retention of ^{68}Ga in the tumour with time. However, this has not been correlated with immuno-histochemical staining of hypoxia, and the integrity of the complex *in vivo* has not been explored [24]. A comparison of NOTA and DOTA derivatives with ^{68}Ga has been made that revealed that the NOTA complexes were formed in higher radiochemical yield and showed superior stability. Moreover, the NOTA derivatives *in vivo* were more rapidly cleared from the blood and muscle, permitting better contrast in images [25].

More recently, gallium(III) complexes with TRAP (1,4,7-triazacyclononane phosphinic acid) l and 1,4,7-triazacyclonon-ane-1,4,7-triacetic acid were reported (Figure 7.4) [26]. Complexes of several phosphinic acid 1,4,7-triazacyclononane derivatives bearing methylphosphinic (TRAP-H), methyl(phenyl)phosphinic (TRAP-Ph), or methyl(hydroxymethyl)phos-phinic acid (TRAP-OH) pendant arms were investigated by potentiometry, multinuclear NMR, and DFT calculations, and a new family of efficient Ga^{3+} chelators was established. TRAP ligands were shown to exhibit high thermodynamic selectivity for Ga^{3+} over the other metal ions ($\log K_{GaL} - \log K_{ML} = 7$–9).

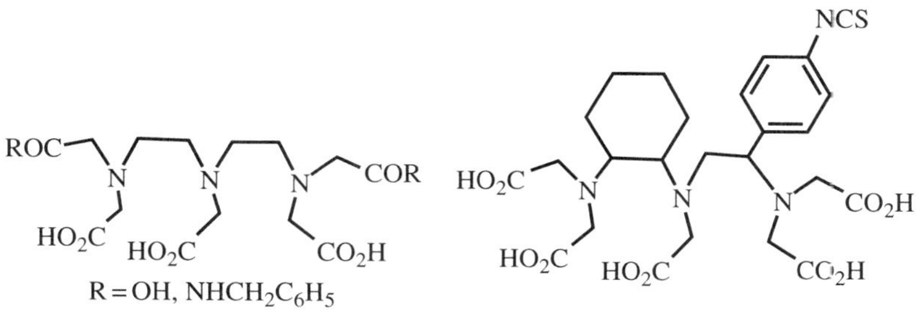

FIGURE 7.5 DTPA and CHX-A″-DTPA derivatives.

Stabilities of the Ga^{3+} complexes are dependent on the basicity of the donor atoms: [Ga(NOTA)] (log $K_{GaL} = 29.6$) > [Ga(TRAP-OH)] (log $K_{GaL} = 23.3$) > [Ga(TRAP-H)] (log $K_{GaL} = 21.9$). The [Ga(TRAP-OH)] complex exhibits unusual reversible rearrangement of the "in-cage" N_3O_3 complex to the "out-of-cage" O_6 complex. These forms are pH-dependent; the in-cage complex is present in acidic solutions, and at neutral pH, Ga^{3+} ion binds hydroxide anion, which induces deprotonation and coordination of the *P*-hydroxymethyl group(s), and the Ga moves out of the macrocyclic cavity. Complex formation studies in acidic solutions indicate that Ga^{3+} complexes of the phosphinate ligands are formed quickly (minutes) and quantitatively even at pH < 2. Compared to common Ga^{3+} chelators (e.g. 1,4,7,10-tetraazacyclododecane-1,4,7,10-tetraacetic acid (DOTA) derivatives), these novel ligands show fast complexation of Ga^{3+} over a broad pH range. The TRAP ligands have therefore been proposed as suitable alternatives for the development of new radiopharmaceuticals but await radiochemistry and *in vitro/in vivo* investigations.

7.2.4 N_3O_5 Donor Ligands

The DTPA-based ligands **10-11** have been widely used for coordinating and targeting radioisotopes (Figure 7.5). The structure of the Ga complex has not been determined but the structure of a bis(benzylamide) variant (R = H_2NCH_2Ph) with In has been shown to have an eight-coordinate structure with an N_3O_5 donor set with a water molecule also bound. This exists as at least three isomers in solution [27].

[111]In labelled DTPA complexes have been used to image EGFR for breast cancer [28] by coupling to a non-coordinated carboxyl group, and the use of such compounds for Auger electron therapy is discussed below. [111]In labelled DTPA has also been used to radiolabel peptidic nucleic acids that are antisense to m-RNA with the objective of imaging brain tumours [29]. Somastatin-expressing tumours have been imaged using ^{111}In-DTPA conjugated to octreotide, which was sold commercially under the name Octreoscan™ and was in regular clinical use for many years [30]. However, the binding to the somostatin receptor types was relatively poor, and this agent has been superceded by other radiolabelled analogues of octreotide (see under DOTA ligands below). The Fab satumomab targets a glycoprotein overexpressed in ovarian and colorectal cancer. It has been conjugated to ^{111}In-DTPA and was one of the first antibody-based imaging agents to be FDA approved in the mid-1990s, marketed under the name Oncoscint™. However, this was withdrawn from the market in 2002 due to the advent of equivalent PET agents [31].

Li et al. designed a dual modality probe for tumour imaging, ^{111}In-DTPA-Lys(IRDye800)-cyclic(KRGDf), which was found to bind efficiently to integrin $\alpha_v\beta^3$ present in cancerous melanoma cells [32]. The tumours were visualised using NIR, optical, and SPECT imaging [32]. Several versatile probes for PET, SPECT, NIR fluorescence, or MRI imaging based on a single folic acid-based precursor and using ^{67}Ga, ^{111}In, or ^{99m}Tc DTPA for SPECT, the fluorescent dye Cy 5.5 for optical imaging, and ^{18}F for PET have been described [33]. The ^{111}In-DTPA folate complex has also been reported to have the capacity to quantify macrophage activation [34].

The analogue of DTPA **10** (CHX-A″-DTPA) was designed both to improve the binding characteristics of the ligand by using the stereochemically rigid cyclohexyl backbone and to provide an isothiocyanato group for conjugation to biomolecules. This ligand system, like DTPA, is perhaps ideally set up for tetravalent metal ions, but it has been used for the ^{111}In labelling of trastumuzab (Herceptin), and the binding of the conjugate to tumour cells is comparable to that of native Herceptin [35]. Furthermore, an interesting approach to ^{111}In SPECT imaging of apoptosis has been to conjugate In-DTPA to a small phenylarsonous acid which binds to heat shock protein Hsp90, which is one of the most prevalent in the cytosol of cells. *In vivo,* this conjugate provided good images of apoptosis and could prove useful to monitor the impact of chemotherapy regimens [36].

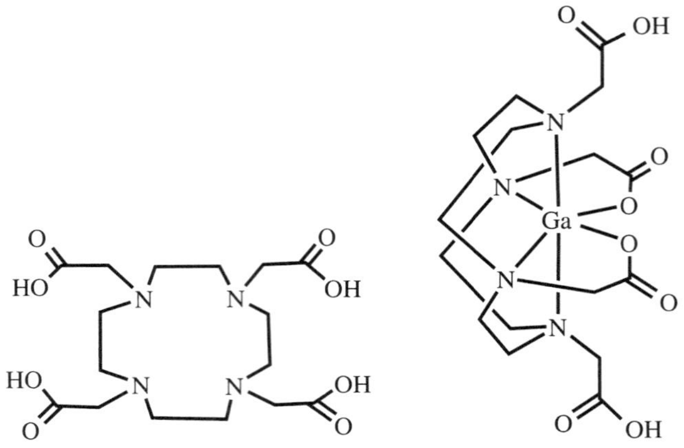

FIGURE 7.6 DOTA ligand (12) and the structure of its Ga complex (13).

7.2.5 N_4O_4 Donor Ligands and Related Species

Gallium and indium radioisotopes have also been used extensively with bifunctional chelators incorporating N_4 macrocycles with pendant carboxylic acid groups [37] such as DOTA (1,4,7,10-tetraazacyclododecane-1,4,7,10-tetraacetic acid) (Figure 7.6) DOTA has been shown by an X-ray structure determination to be bound to Ga(III) via the four macrocyclic nitrogens and two *cis* carboxylate oxygens (Figure 7.6) [38]. One of the two pendant carboxylates is protonated, and the negative charge on the other gives the complex an overall neutral charge. For the larger In(III) ion, the coordination in a complex of an amine derivatised DOTA ligand was shown to be square anti-prismatic with four nitrogens and four carboxylate oxygens bound to the metal [39].

DOTA has been very widely used for the targeting of ^{67}Ga and ^{111}In by conjugation to biomolecules via a pendant carboxylate group. Much interest has been focused on the radiolabelling of DOTA conjugates with both imaging (e.g., ^{67}Ga, ^{111}In) and therapeutic (e.g., ^{90}Y) radionuclides to provide the opportunity of using SPECT imaging for assessment both before and after targeted radiotherapy. The most widely used peptides in clinical imaging have been those that the target somatostatin receptor sites, which are overexpressed by a range of human tumour types. There are five subtypes of receptor sites (ssts 1-5); however, the sst2 is the most prevalent [40]. The majority of these are variants of the cyclic peptide octreotide: for example, NOC, TOC, and TATE. Conjugation to the radionuclide chelator is achieved via the phenylalanine amino group.

^{111}In has now largely been replaced by PET-emitting ^{68}Ga, and $^{68}GaDOTA$ TOC [41, 42] and ^{68}Ga DOTA NOC [43–45] have been widely investigated in clinical trials and been shown to be highly effective in imaging neuroendocrine tumours. These three PET imaging agents all show subtle differences in receptor binding and specificity with ^{68}Ga DOTA TOC having been the most widely used. A recent review discusses these and other examples of sst imaging using octreatide receptors with DOTA derivatives and also other peptidic targeting agents [46] including some recent examples which showed enhanced melanoma uptake and reduced renal uptake [46b].

These DOTA conjugates with octreotide analogues also bind the therapeutic radionuclides ^{90}Y and ^{177}Lu, which have been explored clinically for therapy with good responses to neuroendocrine tumours reported [46]. The proto-oncogene C-kit is overexpressed in gastrointestinal stromal tumours (GIST) and small cell lung cancer (SCLC) and is a target molecule for cancer diagnostics and therapeutics.

Recently, an ^{111}In labelled C-kit (incorporating the ligands DTPA and DOTA for radiometal chelation) was synthesised and tested by *in vitro* binding and cellular internalisation assays. The analogous ^{64}Cu system was evaluated *in vivo* by PET and enabled clear tumour visualisation in a mouse model, indicating its possible use as a tool to enable an informed decision to be made prior to targeted therapy [47].

Human serum albumin (HSA) was recently modified by DOTA mono-N-hydroxysuccinimide ester (DOTA-NHS ester) as well as with the bifunctional cross-linker sulfosuccinimidyl 4-[N-maleimidomethyl]cyclohexane-1-carboxylate (Sulfo-SMCC). Subsequently, a HER2 affibody analogue, Ac-Cys-Z(HER2:342), was covalently conjugated with HSA, and the resulting bioconjugate DOTA-HSA-Z (HER2:342) was radiolabelled with ^{111}In and evaluated *in vitro* and *in vivo* by SPECT. The results compared to the corresponding ^{64}Cu PET imaging. Radiolabelled DOTA-HSA-Z (HER2:342) conjugates displayed a significant and specific cell uptake into SKOV3 cell cultures. Both SPECT and PET *in vivo* imaging in mice models indicated a high tumour and liver uptake [48].

Furthermore, a bivalent single-chain antibody dimer fragment (called a diabody, which is a noncovalent dimer of a single-chain antibody fragment) that retained the avidity of intact IgG but has more favourable blood clearance than intact IgG, was attached to a DOTA-PEG ligand. This was then labelled with ^{125}I, ^{111}In, and ^{64}Cu, and the biodistributions investigated by ^{64}Cu PET in athymic mice with xenografted colon tumours showed good tumour and low kidney uptakes [49].

A DOTA-GlyGlu-Cyc lactam bridged cyclised α-melanocyte stimulating hormone peptide was radiolabelled with ^{67}Ga, which allowed visualisation of primary and metastatic melanoma, showing potential as a theranostic agent if an analogous DOTA conjugate was also used to coordinate a therapeutic radionuclide such as [90]Y [50]. The first report of effectively synthesised lactam bridge-cyclised alpha-MSH peptides involved In[111]. This displayed high *in vivo* and *in vitro* stability, and melanoma metastases were clearly visualised using SPECT. This procedure confirmed the effectiveness of lactam bridge-cyclised alpha-MSH peptides for diagnostic imaging of melanoma. Whole-body SPECT imaging revealed that ^{67}Ga-DOTA-GlyGlu-CycMSH exhibited a rapid tumour uptake, reaching its peak of 12.93% +/- 1.63 two hours after injection. Body clearance was rapid with 82% of the injected radioactivity being cleared through the urinary system. Normal organ uptakes were low, an exception being the kidneys, which still retained high amounts of ^{67}Ga after 2 hours (23.94% +/- 7.15). Furthermore, a DOTA-PEG4-BN bombesin analogue that was labelled with [67/68]Ga and ^{177}Lu showed good tumour uptake in PC-3 xenografted nude mice, where [67/68]Ga can act as a diagnostic radionuclide and ^{177}Lu as a therapeutic [51].

Yoshimoto et al. radiolabelled DOTA-c(RGDfK) with ^{111}In and ^{90}Y, which showed high tumour uptake due to specificity for $\alpha_v\beta_3$ integrin and therefore promise as a theranostic pair [52]. DOTA-neurotensin (DOTA-NT) analogues were designed by Gruaz-Guyon et al. for targeted radiotherapy with ^{90}Y or ^{177}Lu and PET or SPECT imaging labelled with ^{68}Ga or [111]In [53]. The ^{111}In DOTA-NT conjugate showed higher tumour and renal uptake (by SPECT) than the ^{68}Ga DOTA-NT conjugate imaged by PET, with very low background in tissues with the exception of the kidney. Yttrium displayed greater affinity than indium for DOTA-NT, confirming the potential for tumour targeting as a radiotherapeutic. Recently, cyclic peptides have also been applied for combined near infrared fluorescence and SPECT imaging of tumours, allowing unambiguous visualisation of the tumour by both modalities [54].

Several DOTA-conjugated gonadotropin-releasing hormone Receptor-Targeting (GnRH) peptides have been designed and synthesised. The DOTA group was conjugated to the epsilon or *alpha*-amino group of D-lysine or the epsilon amino group of L-lysine via an aminohexanoic acid (Ahx) linker to generate the corresponding DOTA–Ahx–(D-Lys[6]-GnRH) conjugates. The radiopharmaceutical ^{111}In-DOTA-Ahx-(d-Lys[6]-GnRH) (whereby the radiometal was attached at the epsilon amino group of D-lysine) was synthesised (95% yield) and characterised as a possible agent for imaging prostate cancer. In order to ensure maximum binding affinities, both the N and C termini of the peptide chain need to be preserved [55]. The introduction of Ahx was used to increase the lipophilicity of the attached d-Lys[6]-GnRH peptide to favour receptor binding [55]. The chelator DOTA was used because of its known thermodynamic stability and that its conjugation to the epsilon amino group of D-lysine best preserved the nanomolar GnRH receptor binding affinity. Further studies are required to confirm the existence of these compounds, but a possible way to minimise renal uptake is the co-injection of lysine, which in the case in In[111] labelled DOTA-*alpha*-melanocyte stimulating hormone peptide caused reductions of up to 70% [55].

7.2.6 N₂S₂ and N₂S₄ Donor Ligands

Gallium and indium complexes containing Ga/In-S bonds are comparatively unusual; the coordination of these metal ions being dominated by their 'hard' acceptor nature with bonding most commonly found to electronegative non-polarizable donors such as oxygen and nitrogen. However, Ga(III) and In(III) complexes of the type shown in Figure 7.7 have been prepared and shown to be stable in aqueous solution, where they are cationic with loss of the apical chloride. These ligands were labelled with ^{67}Ga, ^{68}Ga, and unusually, with [113m]In in good yield. All the labelled complexes exhibited high myocardial uptake *in vivo* in rats but with some washout over time.

Copper *bis*(thiosemicarbazonate) complexes such as Cu(ATSM) have been primarily studied for their hypoxia selectivity, but this ligand type has recently been used to generate Group 13 complexes of potential interest as a bifunctional chelators for small peptides and biomolecules. Several straightforward peptide coupling routes are now available with the advent of *bis*(thiosemicarbazone) ligands with pendant carboxylic acid groups (Figure 7.8). However, the ^{64}Cu-labelled complexes of

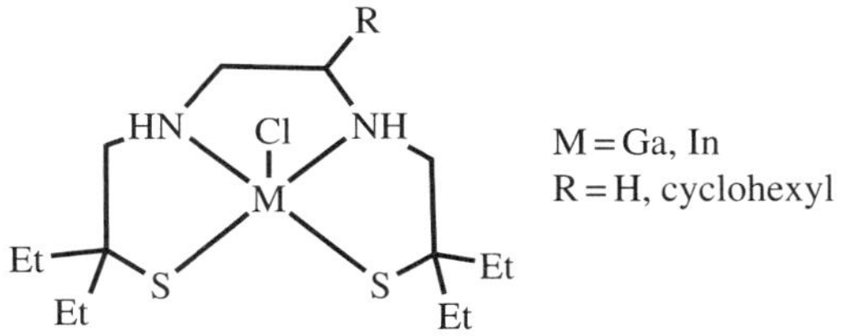

M = Ga, In
R = H, cyclohexyl

14

FIGURE 7.7 Ga and In complexes of aminothiolate ligands.

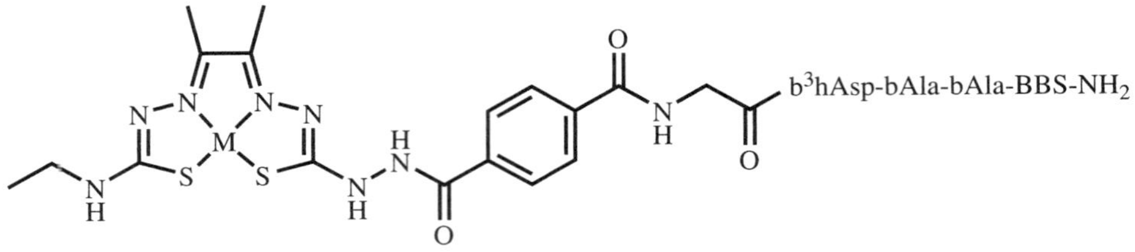

FIGURE 7.8 Structure of a metallic ATSM-Bombesin derivative, M = Cu(II) [57].

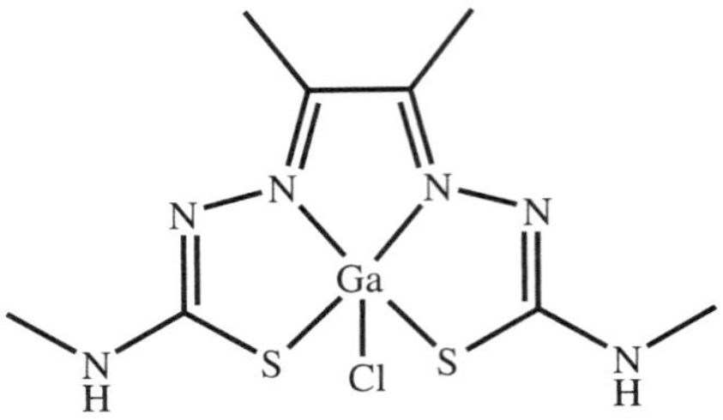

FIGURE 7.9 Ga complexes of bis(thiosemicarbazones).

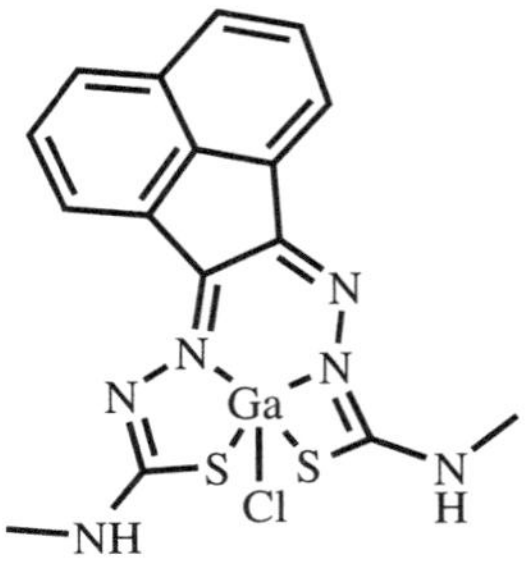

these generally suffer from high liver accumulation *in vivo* [56]. Introduction of hydrophilic groups on the ligand exocyclic structure have lowered the uptake of ^{64}Cu in the liver somewhat but it still remains high, probably due to the Cu metabolism liberated from the ligand [56]. In light of this, interest has focused on the synthesis of [67]Ga- and ^{111}In-radiolabelled *bis*(thiosemicarbazone) ligands to establish if the biodistribution pathways of such gallium or indium chelators was significantly different to that of their copper analogues [57].

Whilst it is not expected that the gallium or indium derivatives will be hypoxic selective because they lack metal-based redox chemistry, the authors' recent work has shown that hypoxic selective groups such as nitroimidazoles can be coupled to the *bis*(thiosemicarbazone) ligand *via* a hydrazinic linker [58, 59]. Through this strategy hypoxia selectivity may be conferred, albeit indirectly, to the gallium/indium *bis*(thiosemicarbazone) unit, and studies of such complexes will make an interesting comparison with the known copper analogues.

We recently reported several gallium *bis*(thiosemicarbazonate) complexes (Figure 7.9) [56, 60–62]. and X-ray diffraction studies showed both symmetric and asymmetric coordination of the ligand, depending on the ligand backbone. An analogous zinc compound that bound DABCO in the fifth coordination site of the zinc ion has recently been published and shows a similar asymmetric binding mode (Figure 7.9). Distorted square pyramidal geometries were observed at both gallium(III) centres in compounds **16** and **17**.

All metal complexes of the *bis*(thiosemicarbazonate) ligand incorporating an aromatic naphthalene backbone (Figure 7.9, e.g., compound **17**) studied so far (for M=Zn(II), Cu(II), Ga(III) and In(III)) have been shown to display increased kinetic stability with respect to their M(ATSM) analogues by virtue of the enforced rigidity of the ligand system. These compounds are intrinsically fluorescent and show cytotoxicities in the micromolar region in a range of human cancer cell lines [63]. Synthesis of the gallium and indium complexes is readily achieved by direct *trans*-metallation of the zinc derivative. A similar method was employed for ^{68}Ga and ^{111}In radiolabelling, and this proceeded rapidly in almost quantitative radiochemical yield [60–62, 64].

The cell uptake of the fluorescent indium bis(thiosemicarbazonato) isostructural analogues has been investigated using confocal fluorescence microscopy [61]. This showed localisation in mitochondria, lysosomes, and additionally for indium complexes in the nucleus, therefore opening up the possibility of Auger electron emission therapy via [111]In. A recent study led to the development of a new gallium bis(thiosemicarbazonate) complex for tumour imaging. A ^{67}Ga-acetylacetonate bis(thiosemicarbazonate) complex [67]Ga(AATS) was prepared by reaction of ^{67}Ga acetate with acetylacetonate bis(thiosemicarbazone) (AATS) for 30 minutes at 90 °C. The radiolabelled Ga complex was prepared with high radiochemical purity (>97%, HPLC) and shown to be stable in serum. The biodistribution of the labelled compound in wild-type and fibrosarcoma-bearing rodents was determined and revealed significant tumour accumulation of the tracer at two hours. This

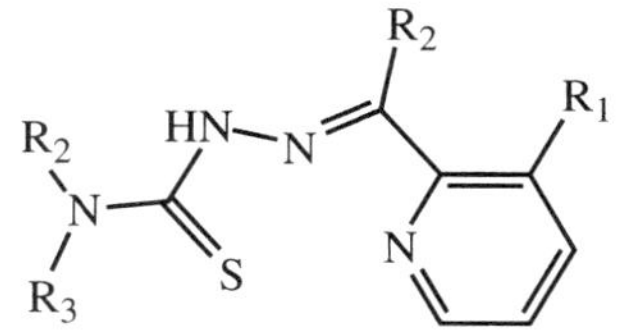

$R_1, R_2, R_3 = H, Me, Et$
HL

FIGURE 7.10 Triapine-related ligands of type HL.

was far higher than that of simple ^{67}Ga cations, and the compound wash-out was significantly faster. The pharmacokinetic properties suggested that this may be an interesting radio-gallium complex of relevance to both SPECT and PET (with ^{68}Ga) imaging of fibrosarcoma tumours and other malignancies [65, 66].

Metal complexes of 3-aminopyridine-2-carboxaldehyde thiosemicarbazone (Triapine) were first synthesised by the Keppler group [67, 68], and the Fe(III) and Ga(III) complexes were studied by X-ray crystallography. Ga(III) and Fe(III) complexes $[M(L)_2]^+$ were also prepared with related ligands HL (where HL = 2-formylpyridinethiosemicarbazone, 2-acetylpyridinethiosemicarbazone, or 2-pyridineformamidethiosemicarbazone, Figure 7.10) and exhibited moderate fluorescence in aqueous solutions. All ligands and complexes were tested for their *in vitro* anti-proliferative activity in two human cancer cell lines (41 M and SK-BR-3). Selected compounds were studied for the capacity of inhibiting ribonucleotide reductase measured by incorporation of 3H-cytidine into DNA. Formation of high stability bis-ligand complexes was found in all cases, and these are predominant at physiological pH with Fe(III)/Fe(II), but only at the acidic pH range with Ga(III). It appears that the N-terminal dimethylation does not affect the redox potential, but has the highest impact on the stability of the complexes. Structure-activity relations established that in general, coordination to Ga(III) increased the cytotoxicity, while the Fe(III) complexes show reduced cytotoxic activity compared to the metal-free thiosemicarbazones, especially in the absence of a NH$_2$ group on the pyridine [67, 68].

In parallel work, gallium(III) and iron(III) complexes of alpha-N-heterocyclic thiosemicarbazones **HL** (2-acetylpyridineN,N-dimethylthiosemicarbazone, 2-acetylpyridineN-pyrrolidinylthiosemicarbazone, acetylpyrazineN,N-dimethylthiosemicarbazone, acetylpyrazineN-pyrrolidinylthiosemicarbazone and acetylpyrazine N-piperidinylthiosemicarbazone) with the general formula $[GaLCl_2]$ and $[ML_2][Y]$ (M = Ga(III), Y = PF$_6$) have also been developed for biomedical imaging and therapeutic applications. Their syntheses, characterization, cytotoxicity, and interaction with ribonucleotide reductase were recently reported. The *in vitro* antitumour potency was studied in two human cancer cell lines (41 M and SK-BR-3). As was the case with the earlier work, gallium(III) enhances, whereas iron(III) reduces, the cytotoxicity of the ligands [69].

A parallel investigation of the synthesis and structural studies of gallium(III) and indium(III) complexes of 2-acetylpyridine thiosemicarbazones has also been reported by another group. Several new 2-acetylpyridine 4N-alkyl/phenyl thiosemicarbazones and their Ga(III) and In(III) complexes were prepared and fully characterised. A comparison of the crystal structures demonstrated the preference for $[ML_2]^+$ type complexes with gallium and $[MLX_3]$ species with indium. Stability studies indicated that complexes of the type $[InLCl_2MeOH]$ are stable for prolonged periods in human serum, suggesting they have potential for radiolabelling and SPECT imaging [70].

A related ^{67}Ga complex of 2-acetylpyridine N^4-*ortho* fluorophenylthiosemicarbazone (PhoF) was developed as a radiotracer for brain imaging *in vivo*. The labelling of the free ligand with ^{67}GaCl$_3$ was performed in methanol with high radiochemical yield, as demonstrated by radioHPLC (97.5±0.6% of radiochemical purity and high specific activity, 1.0 TBq / mmol). The biodistribution and SPECT imaging of this new ^{67}Ga-based SPECT imaging agent were evaluated in Swiss mice and nude mice bearing glioblastoma multiforme tumours (U87-MG).

In biodistribution studies, ^{67}Ga- PhoF displayed not only significant tumour uptake, but also rapid blood clearance and low accumulations in nontarget tissues, resulting in high target-to-nontarget ratios. Scintigraphic images of ^{67}Ga-PhoF in nude mice bearing U87-MG tumours showed a significant activity in tumour (~7% of total activity), and tumour-to-normal tissue ratio was more than 10-fold higher depending on the organ. It is believed that this compound fulfils the characteristics required for a radiotracer for brain tumour diagnosis [71]. Further structural investigations are needed to establish the precise structure and physico-chemical characteristics of this gallium species, in particular to establish whether the complex involved in binding to the brain tumour site is neutral or cationic.

Related Ga(III) octahedral complexes of the general formula $[GaL_2]NO_3$ (L = 2-pyridineformamidethiosemicarbazone) were also prepared. These thiosemicarbazones were cytotoxic against malignant RT2 glioblastoma cells (expressing the p53 protein) with IC$_{50}$ values ranging from 7.3 to 360 mM, and against malignant T98 glioblastoma cells (expressing mutant p53 protein) with IC$_{50}$ values ranging from 3.6 to 143 mM. Coordination to Ga strongly increased the cytotoxicity of the N(4)-alkylated

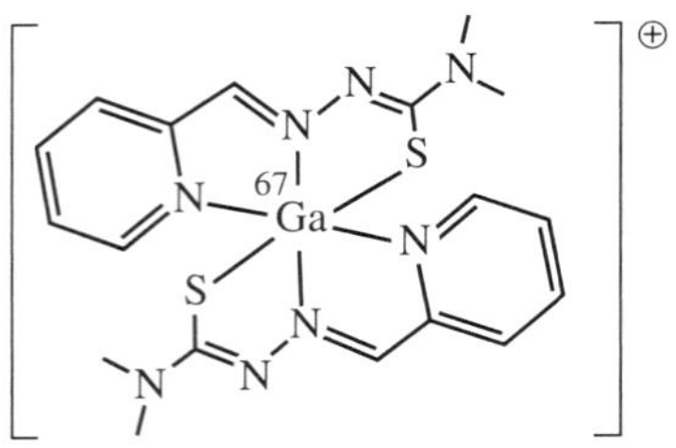

18

FIGURE 7.11 A monocationic octahedral gallium-67 complex of the $APTSM_2$ ligand studied *in vivo*.

complexes (which showed IC_{50} values ranging from 0.81 to 9.57 mM range against RT2 cells and from 3.6 to 11.30 mM against T98 cells). These compounds were found to be 20-fold more potent than *cis*-platin and to induce cell death by apoptosis [72].

Due to the interesting anti-proliferative properties of gallium-thiosemicarbazone complexes of this family, [^{67}Ga]-labelled 2-acetylpyridine 4,4-di-Me thiosemicarbazone, $APTSM_2$ (Figure 7.11) was also investigated for SPECT imaging applications. Radiolabelling was performed by treating fresh [^{67}Ga]GaCl$_3$ with the free ligand for 60 minutes at 90° to yield [^{67}Ga(APTSM$_2$)$_2$]$^+$ with an almost quantitative radiochemical yield and a specific activity of approximately. 370-740 MBq/mmol (10-20 Ci/mmol). The high stability in human serum (at 37 °C) led to SPECT and biodistribution studies of the [^{67}Ga] 2-acetylpyridine 4,4-dimethyl thiosemicarbazonate complex in normal mice. These were higher than the uptake of free Ga^{3+} cation, and it was suggested that this compound represents a promising agent for the detection of malignancies [65, 73].

7.2.7 N_4 Donor Porphyrin Ligands

Porphyrins form extremely stable metal complexes that are highly resistant to demetallation *in vivo* and have therefore been of interest as radiolabelling hosts. The majority of the research efforts involving medical applications of porphyrins have been directed toward specific targeting for Photodynamic Therapy (PDT) through conjugation with biomolecules. However, ^{67}Ga and ^{111}In radiolabelled porphyrins have also been studied and conjugated to biologically active molecules such as antibodies, antibody fragments, peptides, and proteins to achieve specific targeting.

The reaction between InCl$_3$ or GaCl$_3$ with tetraphenylporphyrins gives five-coordinate complexes of the type [MCl(porphyrin)] (M = Ga, In) where the metal ion sits above the plane of the ligand [74]. The complex [^{111}InCl(TTP)] (TTP = tetraphenylporphyrin) has been prepared and was shown to be completely stable in the presence of serum proteins [75]. The metallation process occurs under much milder conditions (30 minutes at 25 °C) than those for Tc and Re discussed previously. One advantage of the porphyrin complexes of Ga and In is that they are strongly fluorescent, which permits cellular distributions to be determined using confocal fluorescence spectroscopy.

The ^{67}Ga and ^{111}In radiolabelling of tetraphenylporphyrin derivatives has also been achieved with further functionalities such as sulphonate, pyridinium, and anilinium located at the periphery of the porphyrin phenyl groups to modulate solubility and lipophilicity. Following *in vivo* imaging with these functionalised labelled porphyrins, the accumulation of ^{111}In and ^{67}Ga in the liver and kidneys was observed in all cases, but the blood clearance was notably slower with the sulphonate derivatives.

The InTTAP complex (Figure 7.12) has been shown to give a highly favourable lymph node to muscle uptake ratio of 85:1 [76]. Similarly, InTMPyP (Figure 7.12) was initially shown to delineate malignant melanoma tumours in hamsters [77, 78], but subsequent studies indicated strong liver and kidney uptake in human carcinoma models [79].

The radiolabelling of TPP with ^{67}Ga has recently been revisited. Labelling was achieved in 30 min. at room temperature in an acetate buffer, but HPLC revealed the presence of two ^{67}Ga species. The structures of these were not discussed but may involve different ligands such as acetate or hydroxide on the Ga. The radiolabelled complexes were very stable in serum, and biodistribution studies in wild-type rats showed excretion predominantly via the liver [80]. Replacement of one of the pyridyl groups of tetrapyridylporphyrin with a carboxyphenyl group enabled conjugation of the ^{111}In porphyrin to an anti-CEA antibody and this was shown to have retained immunoreactivity *in vitro* [81].

In addition, other porphyrin cores based on deuteroporphyrin (Figure 7.12) have been functionalised on the periphery using conventional chelating groups for isotope labelling. The DTPA ester of Ga4-[1-(2-hydroxy-ethyloxy)ethyl]-2-vinyldeuteroporphyrin-IX has been radiolabelled with ^{111}In in greater than 95% radiochemical purity, although is not clear why the ^{111}In radiolabel was not simply inserted into the porphyrin. However, the ^{111}In-DTPA conjugate was subjected to *in vivo* testing in a Lewis lung cancer mouse model and showed strong tumour uptake with a tumour/blood ratio of 16.4 +/- 6.6 [82].

Hematoporphyrins (Figure 7.12) have also been labelled with ^{111}In and examined *in vivo* in mice with induced breast tumours. Malignant tumours showed tumour-to-blood ratios of up to 50 [83]. As for the ^{99m}Tc complexes (discussed in Chapter 6) the mechanism of tumour uptake is not known.

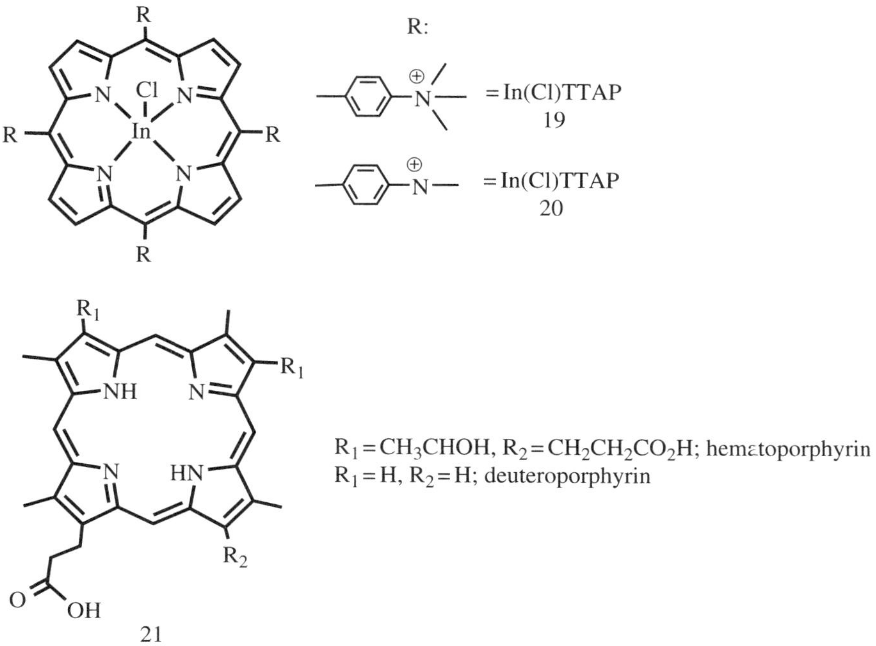

$R_1 = CH_3CHOH$, $R_2 = CH_2CH_2CO_2H$; hematoporphyrin
$R_1 = H$, $R_2 = H$; deuteroporphyrin

FIGURE 7.12 Gallium and indium porphyrins, deutero- and haemato-porphyrins.

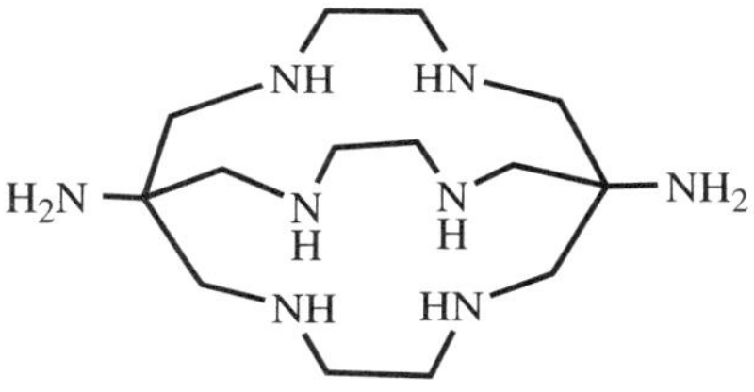

FIGURE 7.13 A cryptand ligand used to bind Ga(III) giving highly kinetically stable complexes.

7.2.8 N$_6$ Donor Ligands

Cryptand ligands (Figure 7.13) have been known for many years and form extremely stable complexes with a range of metal ions [84, 85]. The ligand system has now been shown to form very robust tricationic complexes with Ga(III), and an X-ray crystal structure shows distorted N$_6$ octahedral coordination about the metal. The conjugate with two molecules of a cyclic RGD peptide attached via an extended linker was radiolabelled in high yield with ^{68}Ga at 80 °C for 35 min. PET studies *in vivo* on a xenografted mouse model showed good specific accumulation of ^{68}Ga in tumours with high integrin expression [86].

7.3 AUGER ELECTRON THERAPY WITH ^{111}INDIUM

The most common form of treatment of tumours involves external beam X-ray therapy (XRT) and is used in approximately 50% of patients with proven efficacy. It is, however, not as effective when dealing with metastases or situations where the cancer has spread. The alternative of using ^{186}Re or ^{188}Re bioconjugates for targeted radionuclide therapy has been discussed above. The Auger electrons emitted by the decay of ^{111}In offer an additional way to achieve DNA damage and eventual cell death. Their penetration range is only 1 to 25 nM and to have maximum impact on DNA, the ^{111}In needs ideally to be localised in the nucleus. A recent review provides an excellent summary of the field of Auger electron radiotherapy [87]. Earlier reviews are also available [88–90]. Our account is not comprehensive but presents some examples of the strategies employed in ^{111}In-based therapeutic applications.

Somatostatin analogues bind to sstrs on the external surface of cells and are thence internalised and locate in the nucleus. Nuclear uptake can however, be increased markedly by using a nuclear localising peptide (NLS) [91]. The conjugate

^{111}In-NLS-DOTA-TOC (TOC = octreotide analogue) has 45 times more nuclear uptake than ^{111}In-DOTA-TOC [92]. An alternative route to target the nucleus has been to use ^{111}In-EDTA conjugated to an antisense nucleotide sequence that was encapsulated in a vesicle. This has been used as a potential therapy for neuroblastoma. The conjugate was shown to inhibit tumour cell proliferation [93].

Antibodies have also been explored as the targeting vectors. ^{111}In-DTPA has been conjugated to the HuM195 antibody, which binds to the transmembrane CD33 receptor. Addition of NLS to the antibody exterior significantly increased the nuclear localisation and the cytotoxicity toward clonogenic tumour cells [94]. The antibody Trastuzumab (Herceptin) binds to the HER2 protein and has also been decorated with NLS peptide and labelled with ^{111}In with the radioconjugate dramatically decreasing clonogenic tumour cell survival [95, 96]. Two biological targeting molecules have proved more effective than one in the radiolabelled conjugate ^{111}In-DTPA-RGD-octreotide, which shows enhanced tumour cell death relative to ^{111}In-DTPA-octreotide [30], [97]. Several of the conjugates described have been subjected to clinical trials, and the prospects for Auger electron therapy are very positive. Doubtless there will be further advances in this area in the future, inclusively alongside advances of nanomedicine and nanotechnology for medical imaging applications: indeed some recent work (2013) led to the radiolabeling of nanographene oxide (NGO) with indium-111 and the simultaneous incorporation of trastuzumab. Such nano-constructs (NGO-trastuzumab) when radiolabeled showed enhanced pharmacokinetics in vivo with respect to indium-trastuzumab alone (i.e. in the absence of NGO) coupled with a rapid clearance from circulation and the promising targeting of Her2 receptors [97].

7.4 PROSPECTS FOR ^{67}Ga AND ^{111}In RADIOCHEMISTRY

The relatively few papers covering SPECT imaging with ^{67}Ga and ^{111}In that have appeared in the last 10 years partly reflect the pronounced shift toward PET imaging and apart perhaps for Ga citrate, ^{67}Ga has been superseded by [68]Ga. Without the convenience of a generator system, it is difficult to envisage either ^{67}Ga or ^{111}In competing effectively with ^{99m}Tc, let alone PET radionuclides. The fundamental coordination chemistry of Ga and In lacks the diversity of Tc and Re, and the current range of chelators appears to meet all the current radiopharmaceutical requirements. There is, therefore, not the driving force of new coordination chemistry to stimulate the development of new SPECT agents. However, ^{111}In also offers the possibility of highly effective targeted Auger electron therapy, and this would appear to be the area most likely to expand in the future inclusively alongside recent developments in nanomedicine for imaging applications.

REFERENCES

[1] M. D. Bartholomä, A. S. Louie, J. F. Valliant and J. Zubieta, *Chem. Rev.* **110**, 2903–2920 (2010).

[2] S. Chakraborty and S. Liu, *Curr. Top. Med. Chem.* **10**, 1113–1134 (2010).

[3] A. Signore, S. J. Mather, G. Piaggio, G. Malviya and R. A. Dierckx, *Chem. Rev.* **110**, 3112–3145 (2010).

[4] T. J. Wadas, E. H. Wong, G. R. Weisman and C. J. Anderson, *Chem. Rev.* **110**, 2858–2902 (2010).

[5] C. Edwards and R. Hayes, *J. Nucl. Med.* **10**, 103–104 (1969)

[6] C. Edwards and R. Hayes, *J. Nucl. Med.* **10**, 332–334 (1969).

[7] P. O'Brien, H. Salacinski and M. Motevalli, *J. Am. Chem. Soc.* **119**, 12695–12696 (1997).

[8] C. L. Edwards and R. L. Hayes, *J. Nucl. Med.* **10**, 103–105 (1969).

[9] D. J. Berry, Y. Ma, J. R. Ballinger, R. Tavare, A. Koers, K. Sunassee, T. Zhou, S. Nawaz, G. E. D. Mullen, R. C. Hider and P. J. Blower, *Chem. Commun.* **47**, 7068–7070 (2011).

[10] S. Chaves, A. C. Mendonca, S. M. Marques, M. I. Prata, A. C. Santos, A. F. Martins, C. F. G. C. Geraldes and M. A. Santos, *J. Inorg. Biochem.* **105**, 31–38 (2011).

[11] Y. Sun, C. Anderson, T. Pajeau, D. Reichert, R. Hancock, R. Motekaitis, A. Martell and M. Welch, *J. Med. Chem.* **39**, 458–470 (1996).

[12] M. Loken, M. Clay, R. Carpenter, R. Boudreau and J. McCullough, *Clin. Nucl. Med.* **10**, 902–911 (1985).

[13] W. Heaton, H. Davis, M. Welch, C. Mathais J. Joist, L. Sherman and B. Siegel, *British Journal of Haematology* **42**, 613–622 (1979).

[14] C. Mathais and M. Welch, *J. Nucl. Med.* **20**, 659–659 (1979).

[15] M. Green and J. Huffman, *J. Nucl. Med.* **29**, 417–420 (1988).

[16] C. Broan, J. Cox, A. Craig, R. Kataky, D. Parker, A. Harrison, A. Randall and G. Ferguson, *J. Chem. Soc.-Perkin Trans. 2* 87–99 (1991).

[17] J.-F. Morfin and E. Toth, *Inorg. Chem.* **50**, 10371–10378 (2011).

[18] A. Craig, I. Helps, D. Parker, H. Adams, N. Bailey, M. Williams, J. Smith and G. Ferguson, *Polyhedron* **8**, 2481–2484 (1989).

[19] M. Green, M. Welch and J. Huffman, *J. Am. Chem. Soc.* **106**, 3689–3691 (1984).

[20] G. Bandoli, A. Dolmella, F. Tisato, M. Porchia and F. Refosco, *Coord. Chem. Rev.* **253**, 56–77 (2009).

[21] K. Eisenweiner, M. Prata, I. Buschmann, H. Zhang, A. Santos, S. Wenger, J. Reubi and H. Maecke, *Bioconj. Chem.* **13**, 530–541 (2002).

[22] J. Jeong, M. Hong, Y. Chang, Y. Lee, Y. Kim, G. Cheon, D. Lee, J. Chung and M. Lee, *J. Nucl. Med.* **49**, 830–836 (2008).

[23] K. Sandstrom, A. Haylock, I. Velikyan, D. Spiegelberg, H. Kareem, V. Tolmachev, H. Lundqvist and M. Nestor, *Canc. Biother. Radiopharm.* **26**, 593–601 (2011).

[24] L. Hoigebazar, J. M. Jeong, M. K. Hong, Y. J. Kim, J. Y. Lee, D. Shetty, Y.-S. Lee, D. S. Lee, J.-K. Chung and M. C. Lee, *Bioorg. Med. Chem.* **19**, 2176–2181 (2011).

[25] C. L. Ferreira, E. Lamsa, M. Woods, Y. Duan, P. Fernando, C. Bensimon, M. Kordos, K. Guenther, P. Jurek and G. E. Kiefer, *Bioconj. Chem.* **21**, 531–536 (2010).

[26] J. Šimeček, M. Schulz, J. Notni, J. Plutnar, V. Kubíček, J. Havlíčková and P. Hermann, *Inorg. Chem.* **51** 577–590 (2012).

[27] W. Hsieh and S. Liu, *Inorg. Chem.* **43**, 6004–6006 (2004).

[28] R. M. Reilly, R. Kiarash, J. Sandhu, Y. W. Lee, R. G. Cameron, A. Hendler, K. Vallis and J. Gariepy, *J. Nucl. Med.* **41**, 903–911 (2000).

[29] T. Suzuki, D. Wu, F. Schlachetzki, J. Li, R. Boado and W. Pardridge, *J. Nucl. Med.* **45**, 1766–1775 (2004).

[30] M. De Jong, W. Breeman, D. Kwekkeboom, R. Valkema and E. Krenning, *Acc. Chem. Res.* **42**, 873–880 (2009).

[31] P. Bohdiewicz, *J. Nucl. Med. Technol.* **26**, 155–163 (1998).

[32] C. Li, W. Wang, Q. P. Wu, K. Shi, J. Houston, E. Sevick-Muraca, L. Dong, D. Chow, C. Charnsangavej and J. G. Gelovani, *Nucl. Med. Biol.* **33**, 349–358 (2006).

[33] T. L. Mindt, C. Mueller, F. Stuker, J.-F. Salazar, A. Hohn, T. Mueggler, M. Rudin and R. Schibli, *Bioconj. Chem.* **20**, 1940–1949 (2009).

[34] T. M. Piscaer, C. Mueller, T. L. Mindt, E. Lubberts, J. A. N. Verhaar, E. P. Krenning, R. Schibli, M. De Jong and H. Weinans, *Arthritis and Rheumatism*, **63**, 1898–1907 (2011).

[35] H. Xu, K. Baidoo, A. Gunn, C. Boswell, D. Milenic, P. Choyke and M. Brechbiel, *J. Med. Chem.* **50**, 4759–4765 (2007).

[36] D. Park, A. Don, T. Massamiri, A. Karwa, B. Warner, J. MacDonald, C. Hemenway, A. Naik, K. Kuan, P. Dilda, J. Wong, K. Camphausen, L. Chinen, M. Dyszlewski and P. Hogg, *J. Am. Chem. Soc.* **133**, 2832–2835 (2011).

[37] S. Liu, *Adv. Drug Delivery Rev.* **60**, 1347–1370 (2008).

[38] N. Viola, R. Rarig, W. Ouellette and R. Doyle, *Polyhedron* **25**, 3457–3462 (2006).

[39] S. Liu, Z. He, W. Hsieh and P. Fanwick, *Inorg. Chem.* **42**, 8831–8837 (2003).

[40] J. Reubi, J. Schar, B. Waser, S. Wenger, A. Heppeler, J. Schmitt and H. Macke, *Eur. J. Nucl. Med.* **27**, 273–282 (2000).

[41] M. Hofmann, A. Boerner, H. Maecke, D. Otto, H. Kalbacher and W. Knapp, *J. Nucl. Med.* **43**, 311P–311P (2002).

[42] M. Henze, J. Schuhmacher, P. Hipp, A. Dimitrakopoulou-Strauss, H. Maecke, L. Strauss and U. Haberkorn, *J. Nucl. Med.* **44**, 117P–117P (2003).

[43] C. Pettinato, A. Sarnelli, M. Di Donna, S. Civollani, C. Nanni, G. Montini, D. Di Pierro, M. Ferrari, M. Marengo and C. Bergamini, *Eur. J. Nucl. Med. Mol. Imag.* **35**, 72–79 (2008).

[44] V. Ambrosini, P. Tomassetti, P. Castellucci, D. Campana, G. Montini, D. Rubello, C. Nanni, A. Rizzello, R. Franchi and S. Fanti, *Eur. J. Nucl. Med. Mol. Imag.* **35**, 1431–1438 (2008).

[45] S. Fanti, V. Ambrosini, P. Tomassetti, P. Castellucci, G. Montini, V. Allegri, G. Grassetto, D. Rubello, C. Nanni and R. Franchi, *Biomed. Pharmaco.* **62**, 667–671 (2008).

[46] (a) V. Ambrosini, M. Fani, S. Fanti, F. Forrer and H. Maecke, *J. Nucl. Med.* **52**, 42S–55S (2011); (b) H. Guo, F. Gallazzi, Y. Miao, *Bioconj. Chem.* **23**(6), 1341–1348 (2012).

[47] C. Yoshida, A. B. Tsuji, H. Sudo, A. Sugyo, C. Sogawa, M. Inubushi, T. Uehara, T. Fukumura, M. Koizumi, Y. Arano and T. Saga, *Nucl. Med. Biol.* **38**, 331–337 (2011).

[48] S. Hoppmann, Z. Miao, S. L. Liu, H. G. Liu, G. Ren, A. D. Bao and Z. Cheng, *Bioconj. Chem.* 2011, **22**, 413–421.

[49] L. Li, F. Turatti, D. Crow, J. R. Bading, A. L. Anderson, E. Poku, P. J. Yazaki, L. E. Williams, D. Tamvakis, P. Sanders, D. Leong, A. Raubitschek, P. J. Hudsony, D. Colcher and J. E. Shively, *J. Nucl. Med.* **51**, 1139–1146 (2010).

[50] H. X. Guo, J. Q. Yang, N. Shenoy and Y. B. Miao, *Bioconj. Chem.* **20**, 2356–2363 (2009).

[51] H. W. Zhang, J. Schuhmacher, B. Waser, D. Wild, M. Eisenhut, J. C. Reubi and H. R. Maecke, *Eur. J. Nucl. Med. Mol. Imag.* **34**, 1198–1208 (2007).

[52] M. Yoshimoto, K. Ogawa, K. Washiyama, N. Shikan, H. Mori, R. Amano and K. Kawai, *Int. J. Canc.* **123**, 709–715 (2008).

[53] F. Alshoukr, A. Prignon, L. Brans, A. Jallane, S. Mendes, J. N. Talbot, D. Tourwe, J. Barbet and A. Gruaz-Guyon, *Bioconj. Chem.* **22**, 1374–1385 (2011).

[54] W. Wang, S. Ke, S. Kwon, S. Yallampalli, A. G. Cameron, K. E. Adams, M. E. Mawad and E. M. Sevick-Muraca, *Bioconj. Chem.* **18**, 397–402 (2007).

[55] H. Guo, F. Gallazzi, L. A. Sklar and Y. Miap, *Bioorg. Med. Chem. Lett.* **21**, 5184–5187 (2011).

[56] R. L. Arrowsmith, B. M. Zeglis, N. Viola-Villegas, V. Divilov, M. Jones, P. A. Waghorn, F. L. Phillips, T. L. Mindt, I. M. Eggleston, S. W. Botchway, J. R. Dilworth, F. I. Aigbirhio, J. S. Lewis and S. I. Pascu, *J. Lab. Comp. Radiopharm.* **54** (Supplement 1), S59–S63 (2011).

[57] R. Hueting, M. Christlieb, J. R. Dilworth, E. G. Garayoa, V. Gouverneur, M. W. Jones, V. Maes, R. Schibli, X. and T. D. A., *Dalton Trans.*, 3620–3632 (2010).

[58] P. D. Bonnitcha, A. L. Vavere, J. S. Lewis and J. R. Dilworth, *J. Med. Chem.* **51**, 2985–2991 (2008).

[59] J. P. Holland, F. I. Aigbirhio, H. M. Betts, P. D. Bonnitcha, P. Burke, M. Christlieb, G. C. Churchill, A. R. Cowley, J. R. Dilworth, P. S. Donnelly, J. C. Green, J. M. Peach, S. R. Vasudevan and J. E. Warren, *Inorg. Chem.* **46**, 465–485 (2007).

[60] R. L. Arrowsmith, S. I. Pascu and H. Smugowski, *Royal Society of Chemistry, Specialist Periodical Reports Organometallic Chemistry*, **38** (2012).

[61] R. L. Arrowsmith, P. A. Waghorn, M. W. Jones, A. Bauman, S. K. Brayshaw, Z. Hu, G. Kociok-Koehn, T. L. Mindt, R. M. Tyrrell, S. W. Botchway, J R. Dilworth and S. I. Pascu, *Dalton Trans.* **40**, 6238–6252 (2011).

[62] R. L. Arrowsmith, J. Zhong, G. Kociok-Kohn, P. A. Waghorn, P. Burgos, J. R. Dilworth, S. W. Botchway, F. I. Aigbirhio, R. M. Tyrrell and S. I. Pascu, 240th ACS National Meeting, August 22–26, 2010 Boston, MA, United States (2010).

[63] S. I. Pascu, P. A. Waghorn, T. D. Conry, H. M. Betts, J. R. Dilworth, G. C. Churchill, T. Pokrovska, M. Christlieb, F. I. Aigbirhio and J. E. Warren, *Dalton Trans.* 4988–4997 (2007).

[64] S. I. Pascu, P. A. Waghorn, T. Conry, B. Lin, C. James and J. M. Zayed, *Adv. Inorg. Chem.* **61**,131–178 (2009).

[65] A. R. Jalilian, P. Mehdipour, M. Akhlaghi, H. Yousefnia and K. Shafaii, *Sci. Pharm.* **77**, 343–354 (2009).

[66] A. R. Jalilian, H. Yousefnia, J. Garousi, A. Novinrouz, A. A. Rajamand and K. Shafaee, *Nucl. Med. Rev. Cent. East. Eur.* **12**, 65–71 (2009).

[67] C. R. Kowol, R. Trondl, P. Heffeter, V. B. Arion, M. A. Jakupec, A. Roller, M. Galanski, W. Berger and B. K. Keppler, *J. Med. Chem.* **52**, 5032–5043 (2009).

[68] E. A. Enyedy, M. F. Primik, C. R. Kowol, V. B. Arion, T. Kiss and B. K. Keppler, *Dalton Trans.* **40**, 5895–5905 (2011).

[69] C. R. Kowol, R. Berger, R. Eichinger, A. Roller, M. A. Jakupec, P. P. Schmidt, V. B. Arion and B. K. Keppler, *J. Med. Chem.* **50**, 1254–1265 (2007)

[70] J. Chan, A. L. Thompson, M. W. Jones and J. M. Peach, *Inorg. Chim. Acta* **363**, 1140–1149 (2010).

[71] M. A. Soares, P. B. Pujatti, J. A. Lessa, H. Beraldo, E. B. de Araújo, J. L. Pesquero and R. G. dos Santos, International Nuclear Atlantic Conference, INAC 2011, Brazil (2011).

[72] I. C. Mendes, M. A. Soares, S. R. G. dos, C. Pinheiro and H. Beraldo, *Eur. J. Med. Chem.* **44**, 1870–1877 (2009).

[73] F. Haghighi Moghadam, A. R. Jalilian, A. Nemati and M. Abedini, *J. Radioanal. Nucl. Chem.* **272**, 115–121 (2007).

[74] C. Bedelcloutour, L. Mauclaire, M. Pereye, S. Adams and M. Drager, *Polyhedron*, **9**, 1297–1303 (1990).

[75] C. Cloutour, D. Ducassou, J. Pommier and L. Vuillemin, *Int. J. Appl. Rad. Isot.* **33**, 1311ff (1982).

[76] G. Robinson, A. Alavi, R. Vaum and M. Staum, *J. Nucl. Med.* **27**, 239–242 (1986).

[77] N. Foster, D. Woo, F. Kaltovich, J. Emrich and C. Ljungquist, *J. Nucl. Med.* **26**, 756–760 (1985).

[78] N. Foster, D. Woo, F. Kaltovich, J. Emrich, C. Ljungquist and S. Hemperly, *J. Nucl. Med.* **26**, 829–830 (1985).

[79] N. Maric, S. Chan, P. Hoffer and P. Duray, *Nucl. Med. Biol.* **15**, 543ff (1988).

[80] Y. Fazaeli, A. Jalilian, M. Amini, A. Rahiminejad-kisomi, S. Rajabifar, F. Bolourinovin and S. Moradkhani, *J. Radioanal. Nucl. Chem.* **288**, 17–24 (2011).

[81] C. BedelCloutour, L. Mauclaire, A. Saux and M. Pereyre, *Bioconj. Chem.* **7**, 617–627 (1996).

[82] S. Nakajima, H. Hayashi, K. Oshima, K. Yamazaki, Y. Kubo, N. Samejima, Y. Kakiuchi, Y. Shindoh, H. Koshimizu, I.Sakata and H. Yamauchi *Photochem. Photobiol.* **46**, 783–788 (1987).

[83] D. Wong, A. Mandal, J. Brown, I. Reese, R. Siegler and S. Hyman, *Nucl. Med. Biol.* **16**, 269–281 (1989).

[84] R. Geue, T. Hambley, J. Harrowfield, A. Sargeson and M. Snow, *J. Am. Chem. Soc.* **106**, 5478–5488 (1984).

[85] A. Sargeson, *Pure Appl. Chem.* **56**, 1603–1619 (1984).

[86] T. Ma Michelle, C. Neels Oliver, D. Denoyer, P. Roselt, A. Karas John, B. Scanlon Denis, M. White Jonathan, J. Hicks Rodney and S. Donnelly Paul, *Bioconjug. Chem.* **22**, 2093–2103 (2011).

[87] B. Cornelissen and K. A. Vallis, *Curr. Drug Disc. Technol.* **7**, 263–279 (2010).

[88] G. Mariani, L. Bodei, S. Adelstein and A. Kassis, *J. Nucl. Med.* **41**, 1519–1521 (2000).

[89] W. Bloomer, W. McLaughlin, S. Adelstein and A. Wolf, *Strahlentherapie*, **160**, 755–757 (1984).

[90] J. O'Donoghue and T. Wheldon, *Phys. Med. Biol.* **41**, 1973–1992 (1996).

[91] V. Kersemans, K. Kersemans and B. Cornelissen, *Curr. Pharm. Design* **14**, 2415–2427 (2008).

[92] P. Chen, J. Wang, K. Hope, L. Jin, J. Dick, R. Camron, J. Brandwein, M. Minden and R. Reilly, *J. Nucl. Med.* **47**, 827–836 (2006).

[93] N. Watanabe, H. Sawai, I. Ogihara-Umeda, S. Tanada, E. Kim, Y. Yonekura and Y. Sasaki, *J. Nucl. Med.* **47**, 1670–1677 (2006).

[94] V. Kersemans, B. Cornelissen, M. Minden, J. Brandwein and R. Reilly, *J. Nucl. Med.* **49**, 1546–1554 (2008).

[95] D. L. Costantini, C. Chan, Z. Cai, K. A. Vallis and R. M. Reilly, *J. Nucl. Med.* **48**, 1357–1368 (2007).

[96] B. Bernard, A. Capello, M. van Hagen, W. Breeman, A. Srinivasan, M. Schmidt, J. Erion, A. van Gameren, E. Krenning and M. de Jong, *Canc. Biother. Radiopharm.* **19**, 173–180 (2004).

[97] B. Cornelissen, S. Able, V. Kersemans, P.A. Waghorn, S. Myhra, K. Jurkshat, A. Crossley, K. A. Vallis, **34**(4), 1146–1154 (2013).

8

THE CHEMISTRY OF LANTHANIDE MRI CONTRAST AGENTS

STEPHEN FAULKNER AND OCTAVIA A. BLACKBURN

Chemistry Research Laboratory, University of Oxford, Oxford, UK

8.1 INTRODUCTION

Over the last quarter century, the use of magnetic resonance imaging has revolutionised diagnostic medicine and soft tissue imaging. Contrast agents containing gadolinium are widely used to assist in the acquisition and interpretation of MRI images [1]. Normal NMR spectroscopy relies on the presence of a uniform magnetic field (B_0) through a sample. When a magnetic field is applied to the sample, the nuclear spin populations are perturbed by the magnetic field. A second field (B_1) is then applied at right angles to the first, giving rise to a realignment of the nuclear spins. Magnetic resonance imaging is slightly different. In a typical biological sample, the vast majority of the NMR active nuclei belong to protons in water. The NMR spectrum is therefore likely to be rather uninformative. However, while the spin alignment is hard to separate, the relaxation of the water molecules from the aligned state is highly dependent on their environment, that is, some protons will take longer to get back to normal than others. There are two kinds of relaxation time:

(a) *The spin-lattice, or longitudinal, relaxation time, T_1* : a measure of how long the nucleus takes to return to its equilibrium state after the pulse has been applied. Essentially T_1 determines how often an experiment can be repeated.

(b) *The spin-spin relaxation time, T_2*: the rate at which nuclear spins get out of phase with one another. T_2 determines the rate at which the signal dies away.

Whereas spectroscopy can be achieved with a uniform field, that in magnetic resonance imaging must be varied across the sample, using what is called a gradient field to provide spatial resolution (usually mm). Following the transmitter pulse, the nuclear spins induce small voltages in a receiver coil. These constitute the free induction decay and are amplified before Fourier transform. This procedure is repeated at a range of field gradients, and the data are then processed to generate an image. Images can be obtained either by taking spectra from a series of thin slices, or by imaging the whole area and then deconvoluting the signal. Relaxation-sensitive pulse sequences are often used to provide contrast. Even in the absence of probe molecules, the relaxation properties of water will vary in the body. Well-resolved NMR signals come from water molecules in free solution, making MRI a very good method to probe the environment inside a patient. A number of pulse sequences can be used to distinguish between different environments – inversion recovery, saturation recovery, spin echo methods, and proton density imaging methods are all used to achieve this goal.

Up to this point, we have not mentioned the role of the contrast agent in imaging by MRI. Non-endogenous paramagnetic species can provide enhanced image contrast and shorter image acquisition times through changing the local relaxation rates of bulk water protons [2]. Both T_1 and T_2 can be influenced in this way, but for the purposes of this chapter we will only consider the role of lanthanide complexes as T_1 contrast agents, which produce T_1 weighted images when suitable imaging pulse sequences are applied and deconvoluted.

The Chemistry of Molecular Imaging, First Edition. Edited by Nicholas Long and Wing-Tak Wong.
© 2015 John Wiley & Sons, Inc. Published 2015 by John Wiley & Sons, Inc.

8.2   GADOLINIUM COMPLEXES AS MRI CONTRAST AGENTS

In the search for paramagnetic species that can increase the relaxation rates of water protons, gadolinium, with its seven unpaired electrons, presents itself as a natural choice. Also importantly, its isotropic electronic distribution (symmetric ⁸S ground state) confers relatively slow electronic relaxation rates and gives rise to no paramagnetically induced chemical shift. The dipolar shift induced by the presence of a lanthanide cation is relatively small, and rapid exchange of bound solvent with bulk gives rise to a single peak for which the weighted average chemical shift is indistinguishable from bulk water under the conditions of an MRI experiment. These properties, combined with the chemical stability of gadolinium(III), make gadolinium contrast agents ideal for the purposes of imaging – provided that issues of toxicity of the free metal can be dealt with by encapsulation into stable complexes (a topic that we will discuss in detail later in this chapter).

Europium(II) complexes would also be expected to give rise to similar MRI properties. Indeed, they combine more rapid water exchange at the metal centre (as a consequence of the reduced charge on the lanthanide) with longer electronic relaxation times than their gadolinium analogues – meaning that they should be good candidates for use in generating MRI contrast [3]. However, despite the fact that some europium(II) species can be isolated in water and despite considerable effort from a section of the MRI contrast community [4], no chelates have been identified in which europium(II) is sufficiently stable for clinical application.

For use *in vivo*, the toxicity of the contrast agent is a vital consideration [5, 6]. Toxicity in the free gadolinium(III) ion results, for example, from having similar size to calcium(II) ions, disrupting Ca^{2+}-mediated signalling processes within the body, and also through the effects of colloidal precipitate formation and accretion in membranes [7]. Thus, Gd(III) must be packaged in such a way as to enhance MRI images but not present a toxic hazard. This is achieved through complexation with appropriate polydentate ligands. Gd(III) complexes need to have both high thermodynamic stability and kinetic inertness whereby they remain intact for the duration of their residence within the patient. Because bonding to lanthanide ions is primarily ionic in nature, due to the lack of participation of the contracted 4f orbitals in bonding, sufficiently stable complexes form only with charged polydentate ligands or with macrocyclic ligands bearing pendant donor groups.

The eight-coordinate ligands DTPA and DOTA and their derivatives have become almost ubiquitous in this field, with several in clinical use (Figure 8.1), including [Gd.DOTA]⁻ (Dotarem®), [Gd.DTPA]²⁻ (Magnevist®), [Gd.DTPA-BMA]

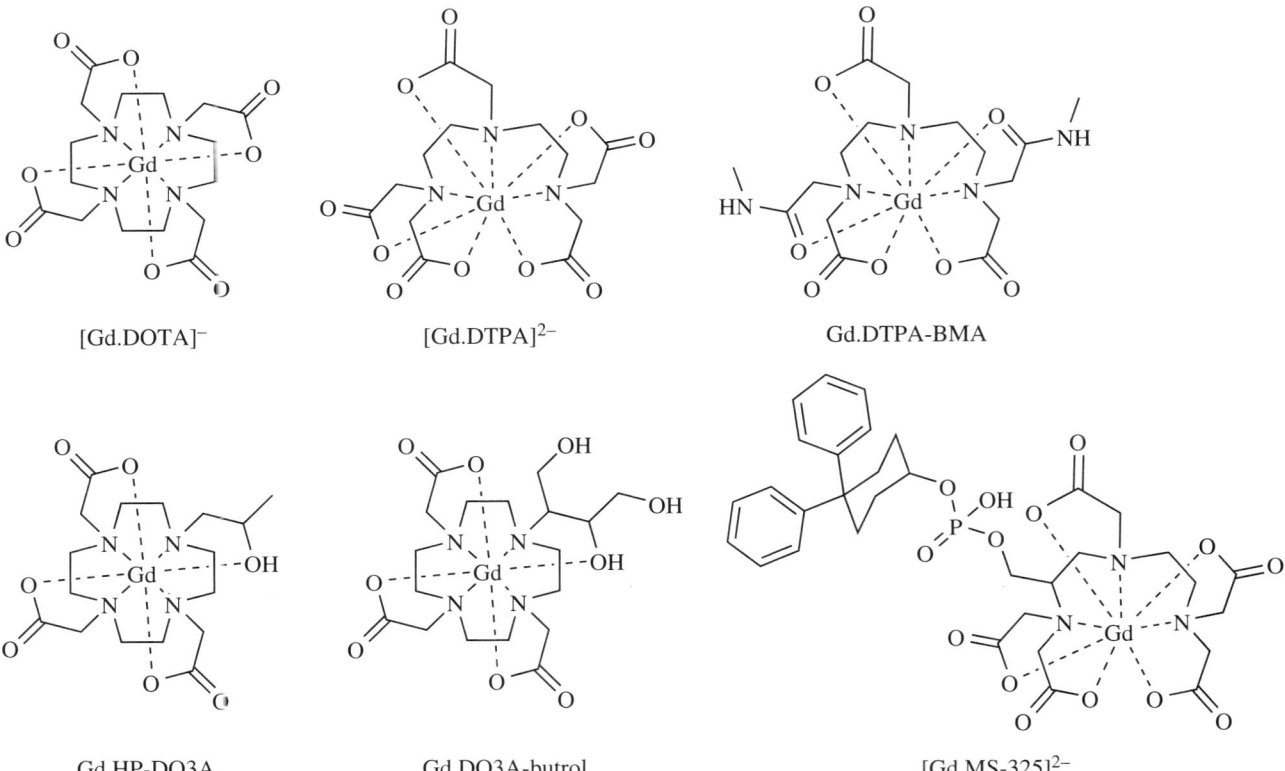

[Gd.DOTA]⁻	[Gd.DTPA]²⁻	Gd.DTPA-BMA

Gd.HP-DO3A	Gd.DO3A-butrol	[Gd.MS-325]²⁻

FIGURE 8.1   Widely studied MRI contrast agents. Gadolinium complexes of DOTA, DTPA, DTPA-BMA, HP-DO3A, DO3A-butrol, and MS-325 are all in clinical use [8–13].

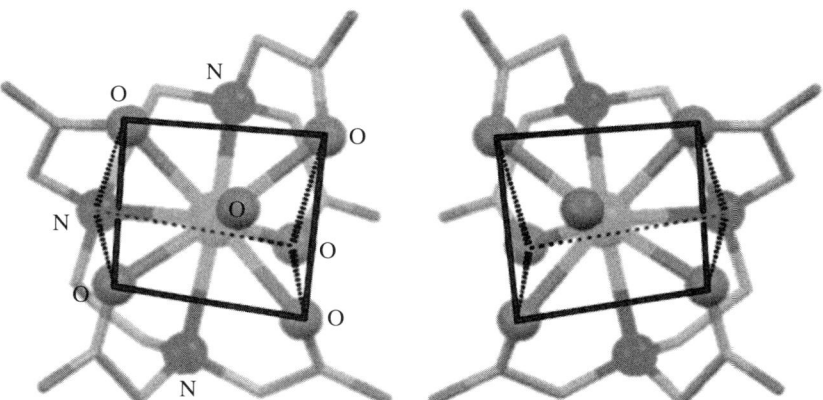

FIGURE 8.2 The two enantiomers of Dy.DTPA found in the X-ray structure [14] viewed approximately along the H_2O–Dy bond, with donor atoms and the central metal shown as spheres and trigonal prisms superimposed on the atoms.

(Omniscan®), [Gd.HP-DO3A] (Prohance®), [Gd.DO3A-butrol] (Gadovist) and [Gd.MS-325]$^{2-}$ (Angiomark®) [8–13]. The ninth coordination site in these complexes is occupied by a water molecule (see below). DTPA and DOTA bind to trivalent lanthanide ions through three or four amine nitrogen atoms and five or four carboxylate oxygen atoms respectively. In the body, these complexes behave similarly, presenting a hydrophilic surface where the ligand surrounds the toxic Gd(III) ion, but allows access to a single water molecule.

Closer examination of the solid-state structures reveals differences. The most favourable geometries for nine-coordinate lanthanide complexes are the tricapped trigonal prism (TTP) and the monocapped square anti-prism (SAP). In the solid state, DTPA and its derivatives and analogues tend to adopt a distorted TTP geometry [14]. There are two possible ways in which the acyclic DTPA can encase the lanthanide ion, giving two enantiomeric 'wrapping' isomers and making the central nitrogen chiral. This can be seen in Figure 8.2, where both enantiomers of the complex are shown with trigonal prisms superimposed upon the atoms. It is also clear from this figure how close such geometries come to being rationalised by a distorted square anti-prism geometry (although this is not shown explicitly, it is clear that the upper plane of oxygen atoms resembles a square, while the N_3O donor set is distorted from square geometry). Clearly, all possible isomers are broadly comparable, meaning that interconversion between them should be facile (and that ligand geometry is the ultimate arbiter of the observed structure). In solution, interconversion between the two enantiomers is achieved by a rearrangement of the acetate arms and a simultaneous flip between staggered conformations of the diethylenetriamine backbone. Proton NMR studies of the Yb(III), Eu(III), and Pr(III) complexes reveal rapid exchange of isomers at high temperatures, with 18 resonances coalescing to nine with increasing temperature [15]. The activation barriers to exchange between these isomers are 49.4, 55.4, and 56.5 kJ mol^{-1} for Yb(III), Eu(III), and Pr(III) complexes, respectively, and reflect the generally flexible nature of the structure. Flexibility in this system also accounts for the limited kinetic stability of the lanthanide complexes.

In derivatives of DTPA, substitution of acetate arms for other donor groups leads to increased numbers of stereoisomers. When the diethylenetriamine backbone of DTPA is substituted, for example in MS-325, the chiral centre introduced in addition to the chirality at the central nitrogen and the wrapping isomers generate eight possible stereoisomers, but also act as a chiral auxiliary – favouring the formation of some isomers over others and increasing the rigidity of the backbone. By contrast, in bisamide derivatives of DTPA, such as DTPA-BMA, all three nitrogens are chiral, and there are four possible diastereomers (*cis, trans, syn,* and *anti*), which can each exist as two wrapping isomers, giving a total of eight isomers. In such systems, the reduction in charge that results from replacing two carboxylate donors with uncharged carboxamide donors results in a reduction in both the thermodynamic and kinetic stability of such complexes. We will deal with this issue in greater detail later in this chapter.

In the solid state, lanthanide DOTA complexes and derivatives display common features [16]. The four nitrogen donors form the basal plane of a square anti-prism with four coplanar oxygen donors making up the opposite and parallel face of the prism. The twist angle between N_4 and O_4 planes defines two possible geometries: a square anti-prism (SAP) with a twist angle of about 40° or a twisted SAP (TSAP) with a twist angle of about 30° (Figure 8.3). The metal ion is shifted from the centre of the prism toward the O_4 plane [17, 18], probably due to differing donor atom affinities. In the solid-state, [Gd. DOTA]$^-$ and isostructural Eu(III), Lu(III), Y(III), and Ho(III) complexes [19–22], have the expected N_4O_4 ligand coordination with a capping water molecule. The twist angle in these complexes is 39°, indicative of a SAP geometry. By contrast, the La(III) complex of DOTA forms a carboxylate bridged chain structure with TSAP geometry with a twist angle of 22°. Gd.HP-DO3A and the isostructural Y(III) complex crystallise with two independent molecules in the unit cell, one with SAP

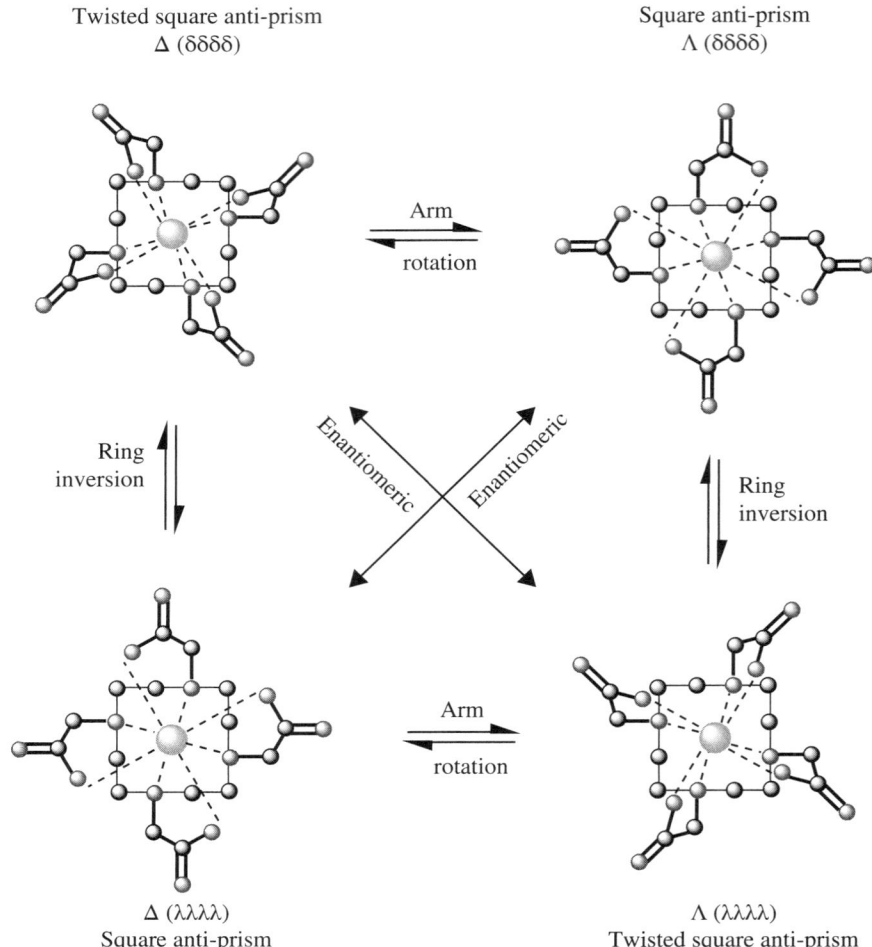

Twisted square anti-prism
Δ ($\delta\delta\delta\delta$)

Square anti-prism
Λ ($\delta\delta\delta\delta$)

Arm
rotation

Ring
inversion

Ring
inversion

Enantiomeric

Enantiomeric

Arm
rotation

Δ ($\lambda\lambda\lambda\lambda$)
Square anti-prism

Λ ($\lambda\lambda\lambda\lambda$)
Twisted square anti-prism

FIGURE 3.3   Schematic diagram showing the four possible stereoisomers of lanthanide DOTA complexes.

and one with TSAP geometry [23, 24]. Gd.DO3A-butrol crystallises as a carboxylate dimer with TSAP geometry [25]. Study of Figures 8.2 and 8.3 reveals considerable similarity between DTPA and DOTA derivatives when viewed from above the carboxylate plane: The key to ensuring stability in both kinds of complex lies in maximising the barrier to interconversion between closely related structures.

As we will see as this chapter develops, complexes with macrocyclic ligands generally offer higher degrees of kinetic stability in comparison with multidentate ligands, particularly for octadentate ligand systems, owing to the increased pre-organisation of the host molecules and the greater barriers to conformational change conferred by the cyclic nature of their backbones. Unlike DTPA derivatives, backbone functionalisation can profoundly reduce the stability of lanthanide complexes. This is particularly true if the backbone is rigidified to the point where the macrocycle becomes sufficiently inflexible to prevent formation of an appropriate conformation in the complex. Several examples of such systems have been noted where aryl substituents are appended to a cyclen backbone (e.g., Ph-DOTA, Figure 8.4): In these systems, kinetically unstable lanthanide complexes form that have no practical use as contrast agents [26].

Lanthanide complexes with DOTA and its derivatives exist as four possible diastereomers (two pairs of enantiomers) resulting from two independent elements of chirality associated with the macrocyclic ring conformation ($\lambda\lambda\lambda\lambda$ and $\delta\delta\delta\delta$) and the spatial arrangement of the acetate arms (Λ and Δ), as illustrated in Figure 8.3. In solution, isomers can interconvert by ring inversion or acetate arm rotation, with a single process converting SAP and TSAP geometries and two successive processes exchanging enantiomers. The activation energy barrier to such interconversion is relatively high, and such species are commonly in slow exchange for octadentate cyclen derivatives. By contrast, heptadentate ligands such as DO3A have a much lower energy barrier to arm rotation, and exchange between the SAP and TSAP forms can be fast on the NMR timescale [27].

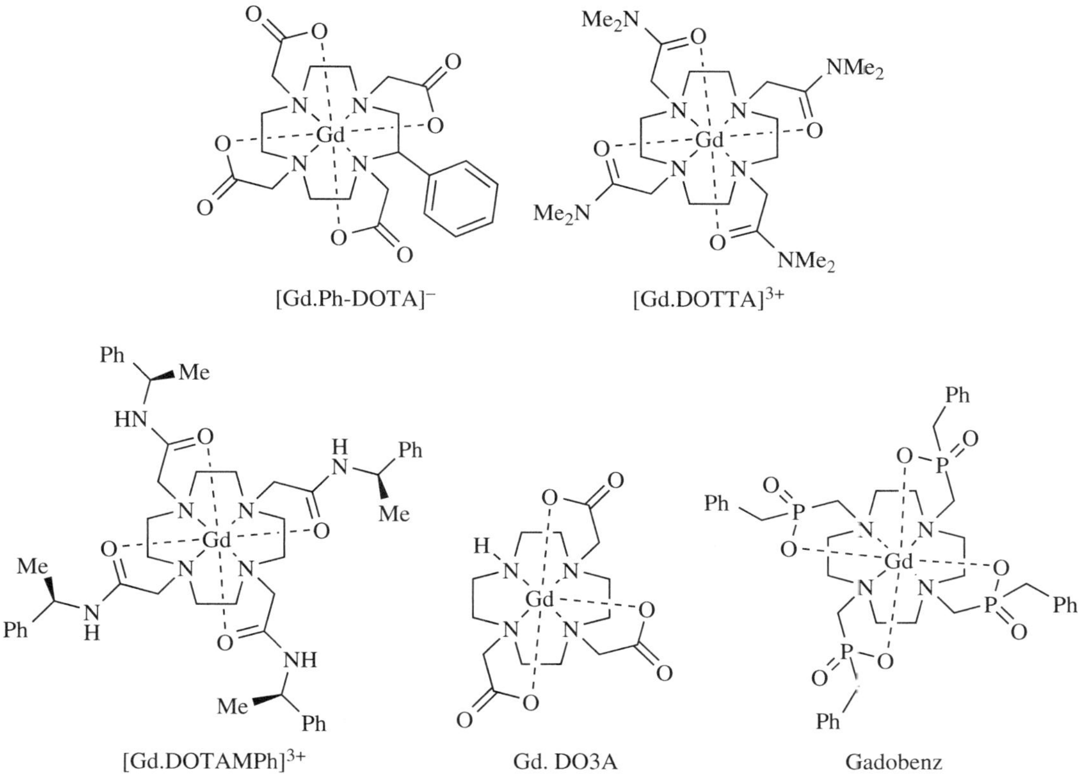

FIGURE 8.4 The structures of [Gd.Ph-DOTA]⁻, [Gd.DOTTA]³⁺, [Gd.DOTAMPh]³⁺, [Gd. DO3A], and Gadobenz.

Changes in structure can influence the balance between these two diastereoisomers. With virtually all octadentate cyclen-derived ligands, the SAP isomers predominate. Accordingly, they have also been referred to as M isomers (where M stands for major), while the TSAP isomers are often referred to by m (where m stands for minor) [28]. Occasionally, the situation is reversed, which can make the M/m descriptors confusing. For instance, DOTTA (Figure 8.4) predominantly forms TSAP isomers with lanthanide ions (and we will discuss the consequences of this in detail later) [29]. It has also proved possible to simplify the picture with regard to isomerism to the point where a single isomer forms in solution. For instance, DOTAMPh (Figure 8.4) [30, 31] exists solely as a single SAP isomer in solution – the chiral amide pendant arms effectively favour just one form of the complex.

NMR spectroscopy can be utilised as a tool for determining the structure of most lanthanide complexes in solution. In the presence of paramagnetic lanthanides, the chemical shifts of nearby nuclei are shifted with respect to their usual positions; this is termed the lanthanide induced shift (LIS, $\Delta\omega$) and has contact ($\Delta\omega_{contact}$) and pseudo-contact ($\Delta\omega_{pseudocontact}$) contributions [32]:

$$\Delta\omega = \Delta\omega_{contact} + \Delta\omega_{pseudocontact} \tag{8.1}$$

This can also be expressed as:

$$\Delta\omega = S_Z F + C_D G \tag{8.2}$$

where S_Z and C_D are lanthanide specific terms, and F and G are ligand-specific parameters for contact and pseudo-contact terms respectively. Rearrangement of these equations gives:

$$\left(\frac{\Delta\omega}{S_Z}\right) = \left(\frac{C_D}{S_Z}\right)G + F \tag{8.3}$$

or

$$\left(\frac{\Delta\omega}{C_D}\right) = \left(\frac{S_Z}{C_D}\right)F + G \tag{8.4}$$

Values of S_Z and C_D are known, and therefore plots of $\Delta\omega/S_Z$ versus C_D/S_Z or $\Delta\omega/C_D$ versus S_Z/C_D should give straight lines for iso-structural complexes within a series of lanthanide complexes of the same ligand [33]. Structural variations across a series of lanthanide complexes are usually ascribed to the gradual decrease in ionic radii owing to the lanthanide contraction. It is necessary to establish trends in structure across a series because lanthanide ions are often substituted for one another to allow for useful information to be obtained in the solution state. Gd(III) is therefore often substituted for its neighbours in the f-series, Eu(III) or Tb(III) ions, because they possess useful luminescence characteristics and NMR shifts in a usable range.

For non-symmetric ligands, the polar coordinates ϕ, θ, and r relative to the lanthanide centre can be used to define the pseudo-contact term:

$$G = C\left(\frac{(3cos^2\theta - 1) + (sin^2\phi\,cos^2 2\theta)}{r^3}\right)$$ (8.5)

for axially symmetric systems, the second term vanishes [32], so

$$G = C\left(\frac{(3cos^2\theta - 1)}{r^3}\right)$$ (8.6)

In both cases, C varies with temperature and the nature of the complex (particularly with B_0^2, the second order crystal field parameter, and the nature of the J in the lanthanide state):

$$C = \frac{-2C_J\beta^2 B_0^2}{kT^2}$$ (8.7)

and

$$C_J = g^2 J(J+1)(2J-1)(2J+3)\langle J|a|J'\rangle$$ (8.8)

where J is $L - S$ for the early lanthanides and $L + S$ for the later lanthanides in accordance with the Russell-Saunders coupling scheme. Such parameters may be used to understand and assess the structure of related complexes. In general, the most effective approach is to ensure that similar structures are adopted by the complexes with (smaller) terbium and (larger) europium ions. The ^{1}H NMR spectra of terbium and europium complexes appear very different, because C takes the opposite sign for each ion (this in itself is a simplification, because C in europium complexes arises from thermal population of excited J states in the ^7F_J manifold - because $C_J = 0$ for J = 0). In the proton NMR spectra of cyclen-derived complexes such as DOTA, the most shifted proton environment will correspond to one of the axial ring protons (H_{ax}) [16]. When we consider the axial ring protons with reference to known crystal structures [16, 29, 34, 35], a greater magnitude of pseudo-contact shift is always to be expected for H_{ax}(SAP) than for H_{ax}(TSAP) because the former subtend a smaller angle (θ) at the molecular axis. In both cases, H_{ax}(SAP) predominates (as can be seen from the marked resonances on the spectra). Thus it may be assumed with confidence that similar behaviour is exhibited by the gadolinium complexes.

8.3 MINIMISING THE TOXICITY OF GADOLINIUM CONTRAST AGENTS

Clearly, it is a prerequisite that complexes containing toxic non-endogenous ions that are intended for use *in vivo* must have high thermodynamic stability if toxicity is to be minimised [5, 6, 18]. All commercial agents fulfil this criterion, and we have already outlined aspects of their design that can be applied to other systems. It is perhaps less obvious that kinetic stability is equally important: Unless a system is under kinetic control, even equilibria that involve very strong binding can be perturbed by kinetic traps such as membrane transport or precipitation. Kinetic stability can be assessed by considering the rate constants for dissociation or competitive binding. Brücher and co-workers showed that the rate of exchange of lanthanide ions could be described by an observed rate constant that incorporates pH dependent terms and pH independent terms [36, 37]:

$$k_{exch} = k_{ind} + k_{dep}[H^+]$$ (8.9)

TABLE 8.1 Rate Constants for Gadolinium Exchange in a Range of Complexes.

Complex	k_{ind}/s^{-1}	$k_{dep}/\text{M}^{-1}\,\text{s}^{-1}$	k_{obs} (pH = 1)	k_{obs} (pH = 7.4)	ref
[Gd.DOTA]$^-$	$<5 \times 10^{-8}$	5×10^{-6}	3–5×10^{-7}	$<5 \times 10^{-8}$	[36, 97]
Gd.HP-DO3A	$-7(\pm 20) \times 10^{-10}$	2.6×10^{-4}	2.6×10^{-5}	$< 1.3 \times 10^{-9}$	[36]
Gd.DO3A-butrol	2×10^{-10}	2.8×10^{-5}	2.8×10^{-6}	2×10^{-10}	[36]
Gd.DTPA-BMA	–	–	$>2 \times 10^{-2}$	?	[6]
Gd.DO3A	–	–	2.3×10^{-3}	?	[5]
[Gd.DTPA]$^{2-}$	–	–	1.2×10^{-3}	?	[6]

For all clinically approved complexes, slow exchange of gadolinium is observed at physiological pH (where the pH independent term is likely to dominate). Because it is desirable that complexes be kinetically stable, it will come as no surprise that studies on the most effective complexes can be time-consuming. Many authors have aimed to speed up the process by measuring k_{exch} at low pH, where (again unsurprisingly) the pH dependent term is dominant [5, 6].

Release of metal ions from polydentate ligands requires some displacement of ligand donor atoms from the metal coordination sphere. This can either occur spontaneously or through protonation of a donor atom (giving rise to the pH-dependent component). Studies on DTPA-bisamide derivatives show multiple diastereoisomers that interconvert through dissociation of up to three ligand donor atoms from the metal centre [38, 39]. Such an extreme process is not observed in DTPA itself, as a result of the improved donor ability of the extra carboxylate donors in DTPA. It also goes a long way to explaining how DTPA-bisamides are relatively unstable. In normal circumstances, this may appear an academic consideration. However, renally compromised patients have exhibited a debilitating disorder known as nephrogenic systemic fibrosis (NSF) following contrast imaging [40, 41]. This has been found to correlate very strongly with the use of bisamide DTPA derivatives and gadolinium dissociation from the chelate, suggesting that kinetic stability is of great importance when considering suitable contrast agents. This is one area in which there is likely to be a need for more data to be obtained: The majority of kinetic measurements have been made at low pH. However, study of equation 8.9 and the rate constants in Table 8.1 clearly reveal that, for all the complexes where data on both k_{ind} and k_{dep} have been obtained, the pH independent pathway for complex dissociation will be almost completely dominant in the pH range available to viable organisms. For the pH dependent pathway to be important at physiological pH, k_{dep} must be very much greater than k_{ind}. In the absence of a more complete body of data, it would be unwise to draw more detailed conclusions at this time. However, even with the data available in the table, it is clear that the pH dependent pathway will only dominate for DOTA when the pH is less than 2, while in the case of HP-DO3A and gadobutrol, pH-mediated lanthanide release will become important at pHs as high as 6. Thus it is clear that structure and ligand pK_a will play a key role.

8.4 RATIONALISING THE BEHAVIOUR OF MRI CONTRAST AGENTS

Relaxivity is a useful measure of the effectiveness of a complex as a contrast agent; the best contrast agents will have large relaxivities per mmol of gadolinium, effectively meaning that good images can be obtained with lower concentrations of contrast agent. The relaxivity r_i is defined by the observed concentration dependence of $(1/T_i)_{obs}$ according to the equation

$$(1/T_i)_{obs} = (1/T)_{dia} + r_i[\text{Gd}] \qquad i = 1, 2 \tag{8.10}$$

As such, the relaxivity is readily determined from a number of T_i measurements over a range of contrast agent concentrations as the slope of a plot of $1/T_i$ against concentration of gadolinium. It is worth noting that the relaxivity itself will have units of mmol(Gd)$^{-1}$ s^{-1}. In the later part of this chapter, where we address multimetallic complexes, it should be remembered that r_i is established per unit of gadolinium, rather than per unit of complex.[1] In a paramagnetic system, T_1 and T_2 are both influenced by both diamagnetic and paramagnetic contributions. In general,

$$(1/T_i)_{obs} = (1/T_i)_{dia} + (1/T_i)_{para} \qquad i = 1, 2 \tag{8.11}$$

where $(1/T_i)_{obs}$ is the reciprocal of the observed T_i, and $(1/T_i)_{dia}$ and $(1/T_i)_{para}$ are the reciprocals of the diamagnetic and paramagnetic contributions respectively.

---

[1] In other words, the effectiveness of a multimetallic complex might be thought of as being given by r_i. n where n is the number of gadolinium ions in the system.

Gadolinium complexes tend to be T_1 relaxation agents (i.e., they give rise to large changes in T_1 and relatively small changes in T_2, and T_1 weighted images produce the most effective results), and for all subsequent discussion of the factors affecting inner sphere contrast, we will refer to T_1 for simplicity.[2]

To obtain a more detailed understanding of contrast agent behaviour, paramagnetic relaxation can be divided up in other ways, that is

$$(1/T_1)_{para} = (1/T_1^{IS}) + (1/T_1^{OS})$$ (8.12)

where T_1^{IS} and T_1^{OS} refer to contributions to T_1 from inner and outer sphere solvent molecules respectively. Even complexes with q=0 give rise to a significant increase in contrast as a consequence of solvation of the molecule as a whole: Outer sphere solvent molecules are more numerous than inner sphere solvent molecules and, by their nature, tend to exchange more rapidly with bulk solvent [42, 43]. To a first approximation, and given the similarity in structure of complexes in current use, the outer sphere contribution is likely to be similar for all complexes. Inner sphere solvents (i.e., water molecules bound directly to the lanthanide) will have a much greater effect, provided that they are in rapid exchange on the NMR time-scale. The relationship between T_1^{IS} and q, the number of inner sphere water molecules, is given by:

$$\frac{1}{T_1^{IS}} = \frac{qP_m}{T_{1m} + \tau_m}$$ (8.13)

where P_m is the mole fraction of bound solvent molecules relative to bulk (and thus related to the concentration of the complex under study), τ_m is the reciprocal of the rate of exchange of bound solvent molecules (i.e., the residence lifetime of a bound solvent molecule), and T_{1m} is the relaxation rate enhancement for a coordinated solvent molecule [44, 45]. There is thus a clear relationship between the number of inner sphere solvent molecules bound to an individual gadolinium complex and the magnitude of the relaxivity. Furthermore, where $T_{1m} \gg \tau_m$, T_{1m} will dominate the observed relaxivity.

In practice, this equation represents a slight simplification, because a given complex may not always have a single value for q; indeed, non-integer values are often reported in the literature as a consequence of population weighted averaging for different inner sphere hydration states of a complex. It might be more strictly correct to think of $1/T_1^{IS}$ being represented by a sum of terms that reflect the populations of all possible q values for a given complex. However, the equation above stands up well in practice, and a more complicated treatment would usually be over-parameterised. The value of q can be determined in the solid state using crystallographic techniques. In solution, time-resolved luminescence measurements have been used to determine the luminescence lifetimes of the europium and terbium complexes, which are then used to determine q using the relations:

$$q_{Eu} = 1.2(\tau_{H2O} - \tau_{D2O} - 0.25 - 0.075x) \text{ and}$$ (8.14)

$$q_{Tb} = 5(\tau_{H2O} - \tau_{D2O} - 0.06)$$ (8.15)

where τ_{H2O} and τ_{D2O} are the luminescence lifetimes of the complex under study in ms in H_2O and D_2O respectively, and x is the number of close diffusing amide N-H oscillators [46]. The lanthanide contraction means that the value of q for a gadolinium complex is likely to be between those of its europium and terbium analogues. Alternatively, q can be determined by measurements of the induced shifts in the ^{17}O NMR spectrum for dysprosium complexes; for complexes in which the bound solvent is in fast exchange, it has been shown that there is a linear relationship between the dysprosium induced shift to the bulk water resonance and q [47].

This simple treatment allows us to rationalise the behaviour of contrast agents in a general sense and illustrates that the inner sphere solvation at the metal centre is clearly very important in determining the effectiveness of a contrast agent. Values of r_1 at 20 MHz are shown in Table 8.2. It is clear from these data that, while hydration is clearly important, other factors also play a role in determining the relaxivity.

The Solomon-Bloembergen-Morgan (SBM) equations have been used to provide some rationalisation of the properties of lanthanide complexes [18, 48]. These divide the contributions to $1/T_{im}$ into contributions from dipole-dipole (DD) and contact, or scalar (SC) pathways: Since rates are additive.

$$\frac{1}{T_{im}} = \frac{1}{T_i^{DD}} + \frac{1}{T_i^{SC}}$$ (8.16)

---

[2] For those who wish to understand T_2 in more detail, refer to [18] P. Caravan, J.J. Ellison, T.J. McMurry, R.B. Lauffer, *Chem. Rev.* **99**, 2293–2352 (1999).

TABLE 8.2 Relaxivities and Inner Sphere Hydration Numbers (q) for a Range of Gadolinium Complexes.

Complex	r_1/mmol(Gd)$^{-1}$ s^{-1}	q	ref
Gadobenz	1.8	0	[43]
[Gd.DTPA]$^{2-}$	3.8–4.3	1	[98, 99]
[Gd.DOTA]$^-$	3.5–4.8	1	[98–100]
Gd.DTPA-BMA	3.9–4.6	1	[101]
Gd.HP-DO3A	3.6–3.7	1	[100, 102, 103]
[Gd.MS-325]$^{2-}$	6.6	1	[13]
Gd.DO3A	4.8	2	[103]
[Gd.AAZTA]$^-$	7.1	2	[65]
Gd.tren-1,2-HOPO	10.5	2	[61]

All values refer to r_1 at 20 MHz. All r_1 values are dependent upon observed field and temperature (*vide infra*).

The SBM equations describe the relationship between the rate of relaxation and the magnetic field. Because the Larmor frequency ω is related to the applied field B_0 by the equation:

$$\omega = \gamma . B_0 \tag{8.17}$$

where γ is the gyromagnetic ratio, it is possible to use the SBM equations to rationalise both variations in relaxation rates and variations in observed relaxivity with magnetic field. If we are considering contributions to T_1, then:

$$\frac{1}{T_1^{DD}} = \frac{2}{15} \frac{\gamma_I^2 g^2 \mu_B^2 (S(S+1))}{r^6} \left[\frac{3\tau_{c1}}{1+\omega_s^2 \tau_{c1}^2} + \frac{7\tau_{c2}}{1+\omega_s^2 \tau_{c2}^2} \right] \tag{8.18}$$

$$\frac{1}{T_1^{SC}} = \frac{2}{3} S(S+1) \left(\frac{A}{\hbar} \right)^2 \left[\frac{\tau_{e2}}{1+\omega_s^2 \tau_{e2}^2} \right] \tag{8.19}$$

where γ_I is the nuclear gyromagnetic ratio, g the electronic g factor (which can be obtained from EPR measurements), μ_B is the Bohr magneton, r is the apparent separation of electronic and nuclear spin centres, ω_I is the nuclear Larmor precession frequency, ω_s the electron Larmor precession frequency, $(A/\hbar)$ is the electronic nuclear hyperfine coupling constant, S is the total spin (for uncoupled gadolinium complexes $S = 7/2$), τ_{Ci} are the correlation times for dipole-dipole terms, and τ_{ei} are the correlation times for contact terms.

These equations look complicated when taken all of a piece, but reveal some important information. For instance, when considering $1/T_1^{DD}$, it is important to note that the equation given is the sum of two terms. The first is a function of the nuclear precession frequency, while the second is a function of the electronic precession frequency. Because the magnetogyric ratio of an electron is much greater than that of a proton, the electronic term may dominate at low magnetic fields, while the nuclear term dominates at higher applied fields. When the nuclear term dominates, $1/T_1^{DD}$ approaches a maximum as the ratio of ω_I to τ_{Ci} approaches unity. Furthermore, these equations show that the contact contribution to relaxation in gadolinium complexes is likely to be small. Not only are the f-orbitals on gadolinium effectively core-like, but additionally the two-bond (Gd-O and O-H) coupling required to give rise to a contact interaction between the water proton and the lanthanide centre will mean that $A/\hbar$ will be very small and that contact contributions will be very minor at high fields.

Having thus established that the dipole-dipole mechanism is likely to be dominant under the conditions of the MRI experiment, it also becomes clear that the separation between the electron and the proton nucleus will have a profound influence on the observed relaxation rate. However, it is difficult to engineer this separation except by accident, because the distance of close approach of water will be determined by both ligand structure and the residual charge on the lanthanide centre in the complex.[3] $1/\tau_{ci}$ and $1/\tau_{ei}$ can themselves be expressed as a sum of terms as shown in equations 8.20 and 8.21, where τ_m is the residence lifetime of a bound water molecule, τ_R is the rotational correlation time of the complex as a whole, and T_{ie} (i = 1,2) is the electronic relaxation time.

$$\frac{1}{\tau_{ci}} = \frac{1}{T_{ie}} + \frac{1}{\tau_m} + \frac{1}{\tau_R} \tag{8.20}$$

---

[3] However, see the section on optimisation below.

$$\frac{1}{\tau_{ei}} = \frac{1}{T_{ie}} + \frac{1}{\tau_m} \tag{8.21}$$

Influencing T_{1e}, τ_m and τ_r raises the possibility of optimising r_1 relaxivities. As we will see in the subsequent sections, τ_m can be varied by controlling both ligand structure and hydrogen bonding to bound solvent molecules, while τ_r can be optimised by making large rigid molecules that tumble slowly in solution.[4] The electronic relaxation time also has potential to be varied between structures, but is hard to predict - not least because it is itself dependent upon the applied field [49].

8.5 STRATEGIES FOR INCREASING RELAXIVITY

We have already seen that the observed relaxivity for a contrast agent can be influenced by changes to the number of inner sphere solvent molecules (q); by the residence time of solvent (τ_m); and by the variety of factors that influence T_{1m}, namely the separation between the gadolinium ion and a proton on the coordinated water (r), the electronic relaxation time (T_{1e}), and the rotational correlation time (τ_r). Of these, the electronic relaxation time is the most difficult to predict, measure, and control - indeed, related gadolinium complexes can exhibit very different zero field splittings and electronic relaxation times [50], and a much larger body of data is required before a rational design approach can be applied to optimise T_{1e}. Until that time, the variability in this parameter represents a potential pitfall for those engaged in contrast agent design.

Variation in r is also difficult to predict (and measure), but changes in this separation induced by changes in the equilibrium position between SAP and TSAP isomers can have a profound influence upon both r and τ_m. In the TSAP isomer, the upper plane of donor atoms imposes a smaller space for access by an axial ninth donor. Accordingly, the separation of the gadolinium and the bound water protons is increased, while the greater Gd-O separation increases the relative rate of water exchange. Indeed, in the TSAP isomer, the rate of water exchange has been shown to be up to 50 times faster than in the SAP isomer of the same complex [29, 51, 52]. Thus, the two effects compete in that an increase in r would be expected to lead to a decrease in relaxivity, while the decrease in τ_M should increase relaxivity. In fact, the changes in separation are relatively small, and the large change in residence time dominates the observed relaxivity. Indeed, in complexes where both SAP and TSAP forms are present, the TSAP form can provide the dominant contribution to the relaxivity even in cases where the SAP form is the major isomer; effectively, a weighted average is observed in which the dramatic difference in exchange rates ensures that TSAP dominates exchange with bulk water.

At high fields, the usefulness of lanthanide complexes will be determined chiefly by q, τ_M and τ_R [18]. Accordingly, we will consider these in greater detail. τ_M can also vary dramatically between complexes. The residual charge upon the metal centre determines its affinity for inner sphere water. Thus solvent exchange rates in [Gd.DOTA]$^-$ (k_{ex}(SAP) $= 4.1 \times 10^6$ s^{-1}; k_{ex}(TSAP) $= 24 \times 10^6$ s^{-1}) are dramatically faster than in tetraamide complexes such as [Gd.DOTAM]$^{3+}$ (k_{ex}(SAP) $= 0.05 \times 10^6$ s^{-1}), while the fastest rates of all are observed with highly anionic complexes such as [Gd.$RRRR$-gDOTA]$^{5-}$ (k_{ex}(SAP) $= 15.4 \times 10^6$ s^{-1}; k_{ex}(TSAP) $= 22 \times 10^6$ s^{-1}) [29, 53–55]. There are exceptions to this simple trend. For instance, [Gd.DOTAGly]$^-$ (Figure 8.5) exhibits slow exchange of water on the NMR timescale despite the overall negative charge on the complex - to the point where other lanthanide analogues of this complex have been used in saturation transfer imaging [56]. This is a consequence of the position of the glycine pendant residues in this complex, because hydrogen bonding with these residues stabilises the bound solvent molecules and slows exchange.

At first sight, the benefits of increasing q seem straightforward. Relaxivity should increase more or less in proportion with the increase in number of bound water molecules at the gadolinium centre (provided the residence time and relaxation rate of the bound water remain approximately the same). However, all clinical agents in current use have q = 1. There are a variety of reasons for this. First, the stability of lanthanide complexes decreases as the denticity of the ligand donor set is reduced from eight to seven. Indeed, while DO3A forms stable complexes with lanthanide ions, we have already seen that the thermodynamic and kinetic stability of such complexes is significantly reduced compared to DOTA complexes. Furthermore, while uncharged donor ligands (such as DOTA-tetraamides) form highly stable complexes with lanthanides, DO3A-triamide complexes tend to be more labile in solution. While a number of stable complexes with seven coordinate ligands exist, the story does not end here. It is also necessary to consider displacement of coordinated water by endogenous bidentate anions. It is widely known that phosphate, hydrogencarbonate, and other anions can displace water from the inner coordination sphere in a range of DO3A derivatives [57]. In the case of gadolinium complexes, this displacement clearly negates the advantage of having q = 2 in the first place, and the anion bound forms effectively behave like outer sphere contrast media. Parker and

---

[4] It should also be noted that the rotational correlation time will vary between tissue types, because viscosity can vary significantly between intra- and extracellular fluids and between biological domains.

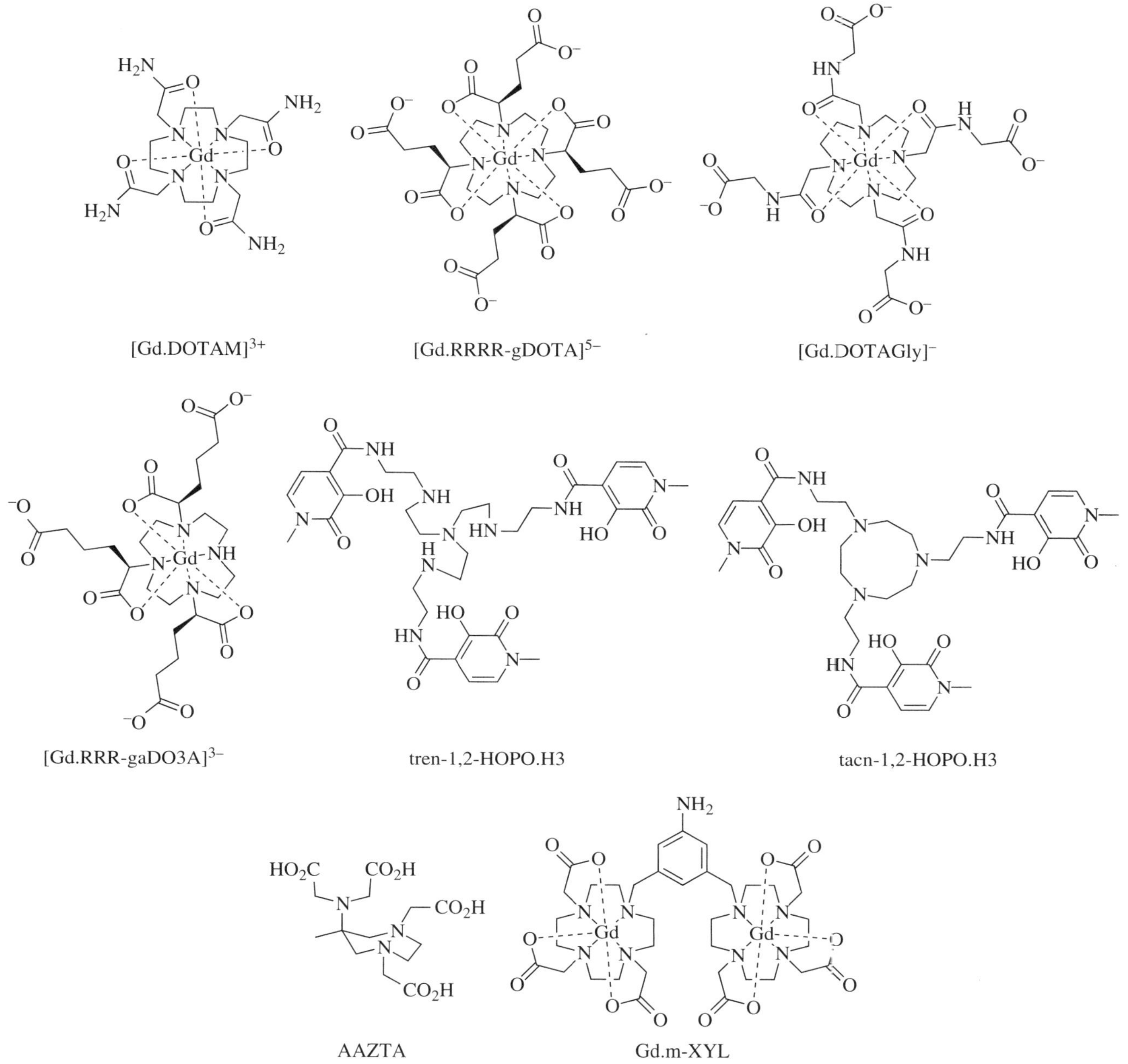

FIGURE 8.5 The structures of [Gd.DOTAM]$^{3+}$, Gd.m-XYL, and complexes/ligands investigated with respect to increasing relaxivity by increasing τ_M or q.

co-workers have established that it is possible to inhibit anion and protein binding in such systems: [Gd.*RRR*-gaDO3A]$^{3-}$ (Figure 8.5) exploits a large negative charge to minimise the interactions with ions [58]. In this system, the high relaxivity observed in water (12.3 mMGd^{-1}s^{-1} at 20 MHz) is maintained in serum. The bulky chiral pendant arms in this system also impart additional rigidity to the complex, inhibiting exchange between stereoisomers and giving rise to high kinetic stability. The authors used the method of Muller and co-workers [59] to screen the kinetic stability of [Gd.*RRR*-gaDO3A]$^{3-}$ by challenging with endogenous ions and reported that the kinetic stability was ten times greater than that of [Gd.DTPA]$^{2-}$. This system has much appeal because it also offers rapid water exchange (k_{ex} = 33 × 10^{6}s^{-1}), as a consequence of the more open water binding site and the increased negative charge on the complex (for comparison in Gd.DO3A, k_{ex} = 6.25 × 10^{6}s^{-1}) [58].

Other q = 2 systems have also been widely investigated. Raymond and co-workers have used an alternative approach, focusing on the preference of lanthanide ions for hard donor ligands, and using all oxygen donors [60]. Gd.tren-1,2-HOPO

(Figure 8.5) uses an entirely different ligand framework to achieve a diaqua complex, exploiting an acyclic backbone and all oxygen donor set to achieve q = 2 in an eight-coordinate complex in which there is sufficient space for associative (and therefore rapid) water exchange [61–63]. This approach has been extended to q = 3 systems in Gd.tacn-1,2-HOPO [64]. [Gd.AAZTA]$^-$ (Figure 8.5) offers another approach to achieving q = 2 complexes; these are highly stable with unchanged relaxivities over a broad range of pH (7.1 mMGd^{-1}s^{-1} at 20 MHz in the range pH = 2-11) and show no anion or protein binding [65]. Furthermore, they offer high kinetic stability when challenged with endogenous ions and can be synthesised without recourse to the use of a macrocyclic backbone. This not only reduces the cost of synthesis, but also allows for facile variation of structure and bioconjugation [66]. As with the other systems we have discussed, there are two isomeric forms in solution with greatly differing solvent exchange rates [67]. At the time of writing, AAZTA-derived systems appear to have great promise.

We have yet to mention control of rotational correlation time. From the discussion above, it is clear that extending τ_R should enhance the observed relaxivity. τ_R for a spherical molecule is directly related to the size of the molecule. For a sphere of radius a in solution in a medium with viscosity η,

$$\tau_R = 4\pi\eta a^3 / 3kT. \tag{8.22}$$

Because τ_R will vary with the cube of the molecular radius, large effects are to be expected from changing the size of the molecule. However, the disadvantage of this purely theoretical approach is obvious in that solvated molecules are not spherical. What's more, they tend to become less spherical as molecular size increases. Furthermore, a molecule must be rigid if it is to attain maximum theoretical relaxivity. If a small(ish) molecule like a DOTA-monoamide is tethered to a larger edifice by a single tether, rotational motion around the tether is often still possible, meaning that τ_R is often much shorter than would be estimated from a simple equation. τ_R can be measured in number of ways, most notably by making measurements on analogous vanadyl(IV) complexes (although this method has an obvious limitation in that the coordination chemistry of vanadyl is very different to that of Gd(III)) [68], through ^{17}O NMR measurements on the Gd complex, or through fluorescence polarisation measurement (although this obviously requires the presence of a fluorescent chromophore within the molecule) [69].

Parker and co-workers have controlled the relaxivity by placing the lanthanide binding motif at the centre of a dendritic architecture derived from [Gd.gDOTA]$^{5-}$ [70]. These systems essentially place the lanthanide at the hub of a wheel in which the rest of the dendrimer provides the spokes and hub, ensuring that all motion relating to the lanthanide centre is coupled to the molecular tumbling and removing any potential alternative routes that would decrease the rotational correlation time. For the best of these systems [Gd.DENgDOTA]$^-$ (Figure 8.6), the observed relaxivity of 23.5 mmolGd^{-1}s^{-1} is very high indeed, and the rotational correlation time of 390 ns, which the authors estimated by fitting the field-dependence of the observed T$_1$ data, is very long compared to that observed in simple complexes (for Gd.DOTA, τ_R = 56 ps) [71].

Alternatively, it is possible to exploit protein binding to change the rotational correlation time. Many lanthanide complexes bind to serum albumin and other proteins in blood. This phenomenon has long been recognised and leads to an enhancement of the rotational correlation time as a consequence of the increased bulk of the assembly [45]. Binding to human serum albumin is favoured by lipophilic side chains appended to the binding site, including aryl groups and steroids. Both DOTA and DTPA analogues have been shown to demonstrate these effects [72–75]. To deal with one example in detail, MS-325 (Figure 8.1) not only demonstrates strong affinity for serum albumin proteins [76], but also gives rise to very different variations in relaxivity with applied field between the protein bound and protein unbound forms of the complex. Because the protein-bound form displays strong field dependence of the relaxivity around 1.5 T, while the unbound form displays very little variation with field in this region, this can be exploited to provide direct images of albumin concentration. Use of dreMR (a variable field approach to MRI imaging) allowed Caravan and co-workers to obtain images of protein concentration following inflammation in the human hand (Figure 8.7) [77]. This quantitative approach to MRI (where the same probe is used to provide both a signal and a reference channel depending upon the applied field) represents a landmark and one that is sure to be developed further once whole body dreMR scanners become available.

Perhaps the most obvious trick of all is to incorporate more than one gadolinium centre into a molecular complex, thereby increasing the local concentration of gadolinium. Provided such systems remain soluble, this approach has much to recommend it, because there will often be complementary enhancements to the relaxivity arising from increases in the rotational correlation time. For instance, in the m-xylyl bridged DO3A derivative (m-XYL, Figure 8.5), the observed relaxivity at 400 MHz in water is 13.8 mmolGd^{-1}s^{-1} (cf. 5.5 mmolGd^{-1}s^{-1} for Gd.DO3A) [27]. In other words, each gadolinium contributes more than twice the relaxivity of an analogous monomeric system as a consequence of the increase in molecular diameter. On a larger scale, Meade and co-workers have incorporated seven lanthanide complexes onto a β-cyclodextrin scaffold, giving a system (HEPTAGADODEXTRIN, Figure 8.6) with q = 1 on each of the seven sites. The resulting array has r$_1$ = 12.2 mmol^{-1}Gds^{-1} (cf. 3.2 for a monomeric analogue) [78]. This approach gives rise to relatively small molecules with

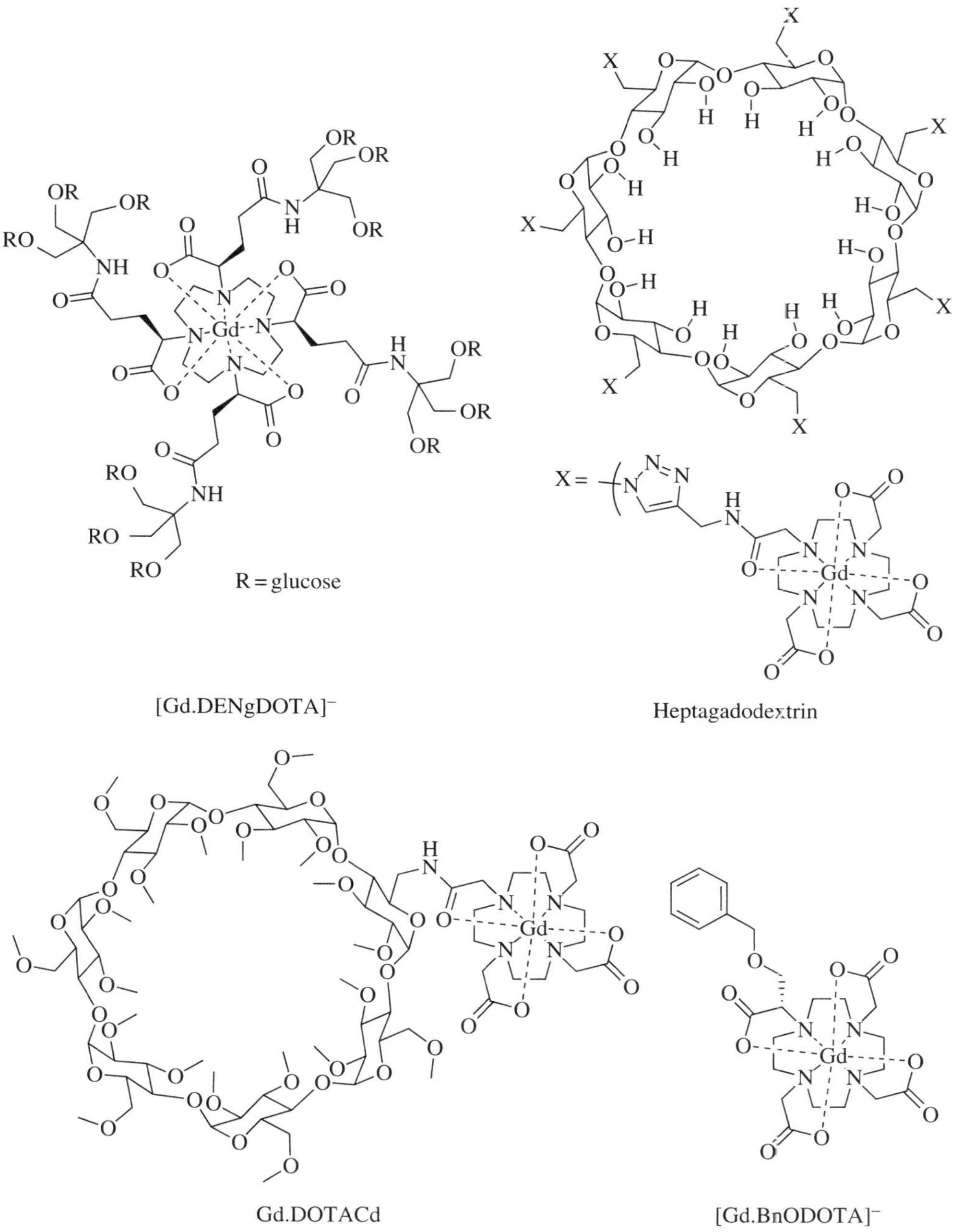

FIGURE 8.6 Structures of assemblies investigated with respect to increasing relaxivity through increasing τ_R or the number of Gd per complex.

large relaxivities: It is worth noting that this system has a molecular weight in the region of 5500, and a 'relaxivity per mole of complex' of around $85\,mmol^{-1}\,s^{-1}$, compared with a molecular weight of around 3200 for the mono-metallic Gd. DENgaDOTA complex discussed above. In applications where molecular size and the consequent slow diffusion through biological media is not an issue, there is clearly much to be said in favour of the approach of using many metal centres.

Parker and co-workers have exploited cyclodextrins in a different way, assembling a multimetallic complex through the interaction between a DOTA-monoamide derivative (Gd.DOTACd) appended with a permethylated cyclodextrin, and Gd.BnODOTA (Figure 8.6) [79]. They showed that the resulting assembly formed with an association constant at 298 K of around $200\,M^{-1}$; this means that a mixture of species will exist in solution at the mM concentrations associated with MRI measurements, and in this case only very small variations were observed between sum of the relaxivities of the individual components and the total relaxivity of the mixture. By contrast, the same cyclodextrin interacts strongly with MS-325,

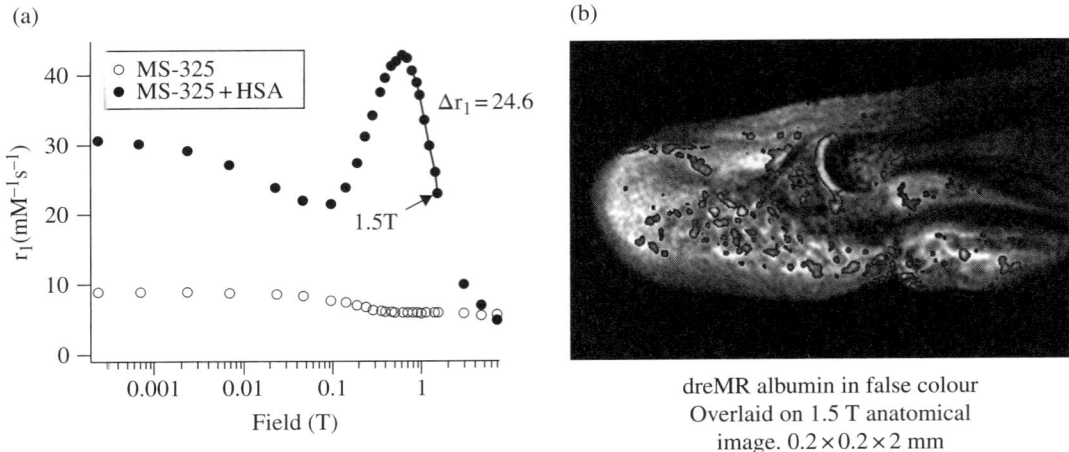

(a)

(b)

dreMR albumin in false colour
Overlaid on 1.5 T anatomical
image. $0.2 \times 0.2 \times 2$ mm

FIGURE 8.7 (a) The variation in r_1 for MS-325 in albumin free and albumin bound forms with magnetic field. (b) dreMR albumin false colour image overlaid on 1.5 T anatomical image of a finger fracture. Figure drawn from data kindly provided by Prof. Peter Caravan. (*See insert for colour representation of the figure.*)

giving rise to a much greater enhancement of r_1 as a consequence of greater affinity of MS-325 for the cyclodextrin host. This approach has been extended further by Aime and co-workers, who used poly-β-cyclodextrin and an adamantly appended gadolinium complex with high affinity for cyclodextrin: Combining these with functionalised dextran derivatives yields high values of r_1 [80]. This has been elaborated to a system that can be used to assess pH through combined $^1H/^{19}F$ MRI [81]. In this system, a gadolinium component contributes to the proton relaxivity, while a fluorine-appended adamantyl derivative contributes toward the fluorine MRI signal. The pH responsive behaviour in this case derives from pH dependent changes in q, which will be discussed in greater detail in the next section.

8.6 RESPONSIVE MRI

In an effort to produce ever more clinically useful MRI agents, considerable research has been directed to developing 'smart' contrast agents capable of sensing their biological surroundings. Upon administration of such a compound, the relaxivity of water protons should be dependent on some biochemical variable - for example, pH - ion concentrations, oxygen concentration, or temperature. One attractive concept is that of targeting biological entities that are specific to a given disease, for example, a receptor or enzyme. The challenge encountered in this strategy is usually that the target molecule is present in insufficient concentration to be able to deliver enough gadolinium to provide sufficient contrast for imaging, given the relatively low sensitivity and resolution of the MR technique. Either the relaxivity of the agent must be increased or multi-Gd agents must be employed (see above).

A recent example of receptor targeting includes the activation of immunologically important macrophage cells using the tetrapeptide, tuftsin (TKPR), and closely related pentapeptide (TKPPR) as targeting units, conjugated with Gd.DOTA-monoamide complexes to provide MR contrast. The terbium(III) analogues were also investigated by virtue of their optical properties. Fortunately, N-terminal substituents do not affect the biological activity of tuftsin, and the peptide conjugate is internalised by the macrophage, a process that is mediated by the tuftsin receptor. The Gd-tuftsin conjugate was found to have similar relaxivity in comparison with other Gd.DOTA-monoamide complexes. Both T_1 measurements of macrophage cells incubated with Gd-tuftsin and luminescence of Tb-tuftsin incubated cells confirmed the desired targeting ability of these compounds. The analogous pentapeptide conjugates were found to have a stronger affinity for the macrophage cells than tuftsin [82, 83].

Considerable effort has been directed toward the targeted imaging of integrins [84]. Integrins are transmembrane proteins that mediate interactions between cells, and the $\alpha_v\beta_3$ integrin has been shown to regulate angiogenesis, the formation of new blood vessels key to the development of malignant tumours. The overexpression of the $\alpha_v\beta_3$ integrin in diseased tissue can be targeted by exploiting its recognition of short amino acid sequences, such as RGD. Shukla et al. conjugated a G-5 PAMAM (polyamidoamine) dendrimer with a doubly cyclised RGD peptide and an Alexa Fluor 488 fluorescent probe. The conjugate was shown to accumulate in cells expressing $\alpha_v\beta_3$ integrin [85]. Later, Boswell and co-workers extended this strategy to include MR contrast by using a G-3 PAMAM dendrimer as a platform for conjugation of approximately two RGD-cyclopeptides and one Alexa Fluor 594 dye, along with 27 modified Gd.DTPA chelates using orthogonal coupling methodologies [86]. The RGD-conjugated dendrimer prior to Gd(III) complexation was selectively taken up by M21 cells, known to express $\alpha_v\beta_3$ integrin, compared with the control RAD-conjugated assembly. However, *in vivo* studies employing ^{111}In in empty DTPA sites on the dendrimer revealed little accumulation in tumour cells ($1.25 \pm 0.51\%$ after 2 h).

Contrast agents whose relaxivities are dependent on pH are highly sought after, owing to the observed difference in extra-cellular pH found within tumours (ca. 6.8 – 6.9) with respect to normal tissue (ca. 7.4) due to excess lactic acid production in tumour cells. A critical MRI parameter, for example, q, τ_r, or τ_m, must be substantially affected by pH in the proposed system. The image intensity will depend upon both the relaxivity and the concentration of the contrast agent and therefore non-uniform dispersion of the agent creates a challenge. This can be surmounted, for example, by taking the difference of images obtained using two contrast agents with differing pH-relaxivity functions but similar biodistributions administered in quick succession.

Parker and co-workers investigated DO3A-based systems incorporating sulfonamide pendant arms with various *p*-aryl groups and β-carboxyalkyl substituents α to the ring nitrogens (to disfavour the binding of endogenous anions, as shown for [Gd.*RRR*-gaDO3A]³⁻ above). The hydration number in these systems is approximately zero in alkaline media and greater than one in acidic media, corresponding to ligation to Gd and protonation of the sulfonamide nitrogen respectively, leading to relaxivity enhancement with decreasing pH [87, 88]. The change in relaxivity of the Gd(III) complexes in the biologically relevant pH range 6.8 – 7.4 reached a maximum of 48% (for Gd.Sulfon-gaDO3A, Figure 8.8) in a human serum solution, with protein binding enhancing the effect with respect to a simulated extracellular anion background. The pH response of the complexes depends heavily on structure, with a closely related structure showing a change in serum solution of 15%, thus providing a means of circumventing the concentration problem. The pH-sensitive sulfonamide unit was utilised in ratio-metric measurement through the application of ¹⁹F/¹H MRI methods to self-assembled systems based on cyclodextrins (see above) [81]. A good example of the effect of pH on τ_r is found with the macromolecular [Gd(DO3ASQ)]$_{30}$-Orn$_{114}$ complex studied by Aime et al. (Figure 8.8) [89]. Thirty Gd.DO3A units are linked to a polyornithine chain via squaric acid moieties, and a linear increase in relaxivity from 23 to 32 mM⁻¹ s⁻¹ ongoing from pH 4 to 8 is measured. This response is explained by considering the structural changes occurring with pH. At acidic pH, protonated amino groups repel one another to give a highly flexible structure with mobile Gd.DO3A units. As the pH rises and amino groups become deprotonated, hydrogen bonding within the peptide backbone causes a rigidification of the structure associated with an increase in τ_r, confirmed by assessment of the NMRD profile. These macromolecular complexes were used again in an investigation into a ratiometric approach to the concentration problem. In the case where one of T_1 or T_2 is dependent on the biological variable of choice, but not the other, the ratio T_1/T_2 becomes independent of local concentration of contrast agent. This concept was successfully

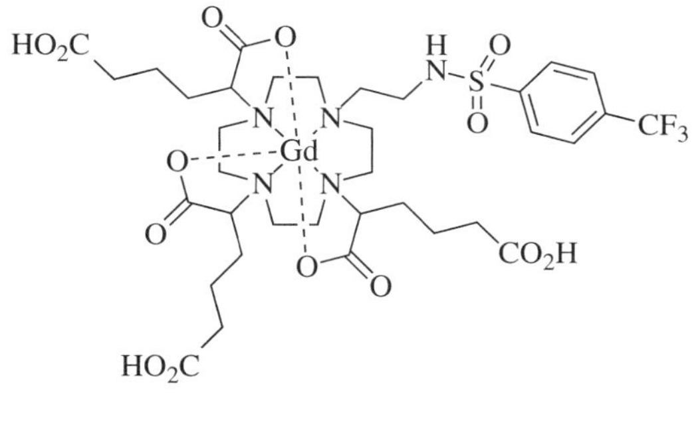

Gd.sulfon-gaDO3A 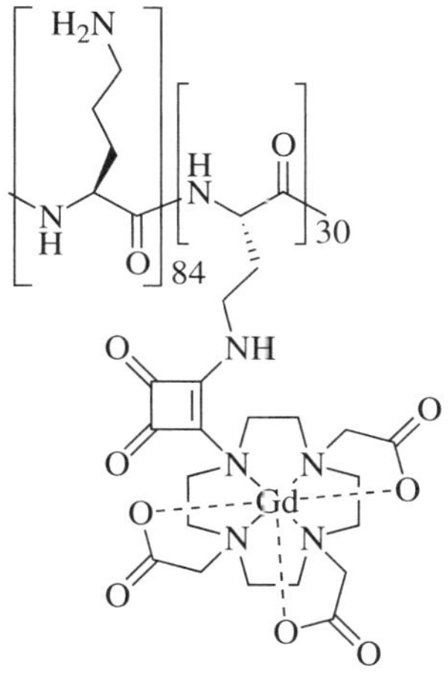   [Gd(DO3ASQ)]$_{30}$-Orn$_{114}$

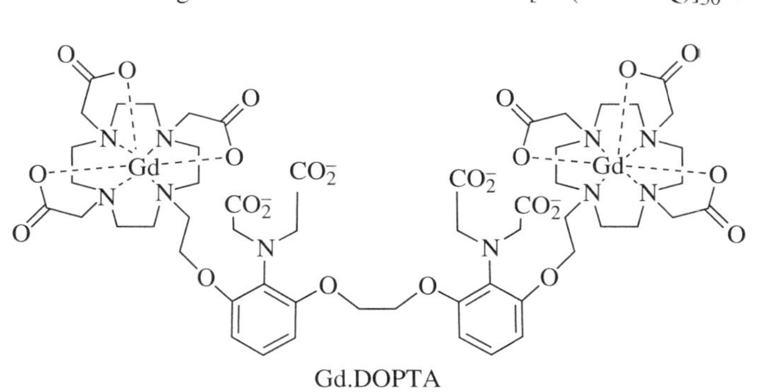

Gd.DOPTA

FIGURE 8.8 Structures of pH and pCa responsive contrast agents.

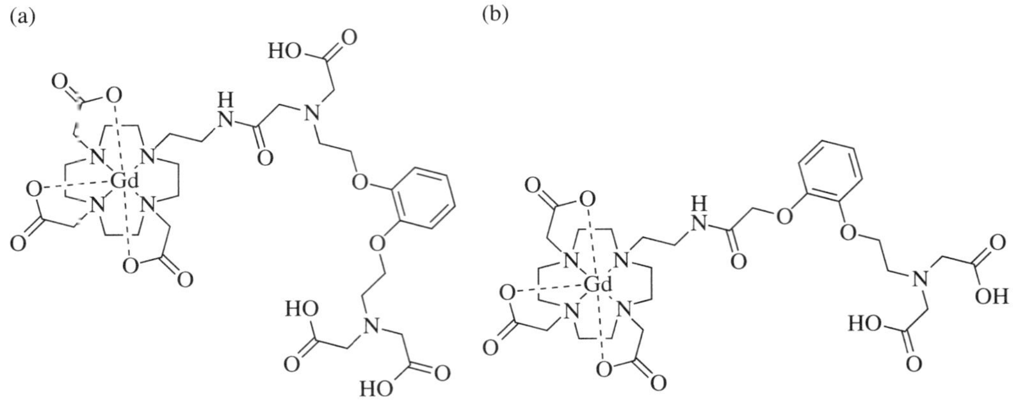

(a)   (b)

FIGURE 8.9 Variations in structure between calcium and zinc responsive contrast agents (a and b respectively).

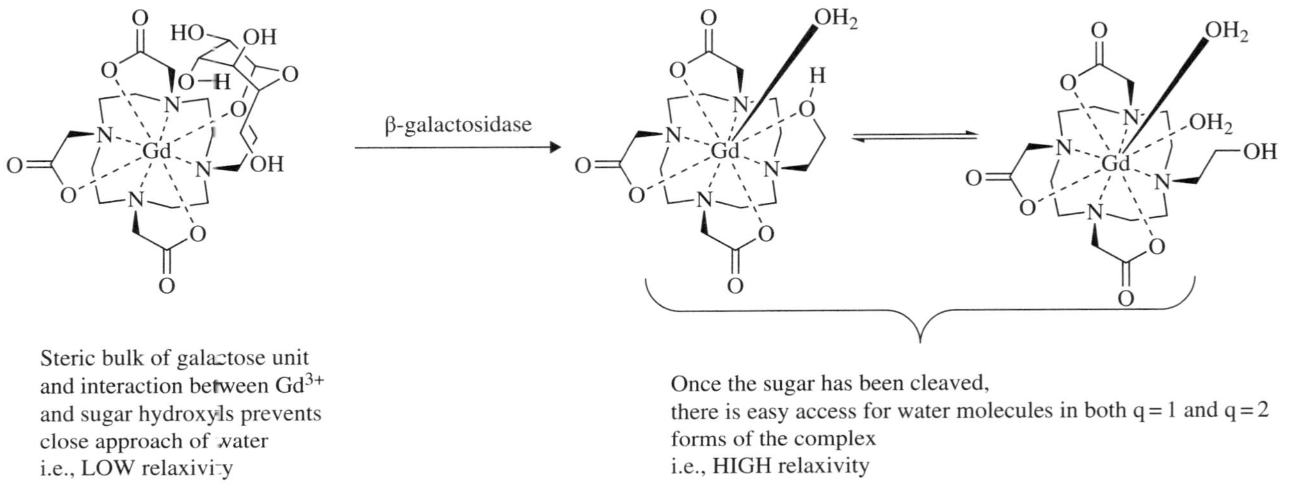

Steric bulk of galactose unit
and interaction between Gd^{3+}
and sugar hydroxyls prevents
close approach of water
i.e., LOW relaxivity

Once the sugar has been cleaved,
there is easy access for water molecules in both $q=1$ and $q=2$
forms of the complex
i.e., HIGH relaxivity

FIGURE 8.10 The mechanism of action of a contrast agent responsive to the enzyme galactosidase. We have represented the cyclen backbone structure to enable comparison with the representation of DOTA structures in Figure 8.3.

demonstrated for a $[Gd(DO3ASQ)]_{33}$-Orn_{205} complex, whose T_1 and T_2 show differing pH dependencies, especially at higher fields [90].

Another highly desirable target is a response to biologically relevant metal ions [91]. For example, calcium is important as a signalling ion in biology, especially in the brain, where changes in Ca^{2+} concentration can be indicative of neural function. The first Gd-based contrast agent to display relaxivity modulation selective for Ca^{2+} was presented by Meade and co-workers and is known as Gd.DOPTA (Figure 8.8) [92]. The design combines two Gd.DO3A units with a BAPTA core via propyl linkers, the latter being a well-known selective binder of calcium (over, e.g., H^+ and Mg^{2+}). In the absence of calcium ions, the acetate groups of the BAPTA core interact with the Gd centres and inhibit close approach of water. Addition of calcium causes a rearrangement of the ligand with the BAPTA now binding calcium, thus allowing access of water to the Gd sites and resulting in an increase in relaxivity of approximately 80% (in the $[Ca^{2+}]$ range 0.1 to 10 μM). Further studies into the mechanism of the response involved luminescence measurements of the analogous terbium complex. Assessment of the luminescence lifetimes in H_2O and D_2O in the absence and presence of Ca^{2+} revealed an increase in the number of bound water molecules from 0.47 to 1.05, consistent with the proposed mechanism [93].

Mishra et al. have investigated Ca^{2+} and Zn^{2+}-binding gadolinium complexes of related structures (A and B respectively in Figure 8.9), with eight and six coordinate binding environments for the guest metal respectively [94]. As with Meade's complex, the mechanism of increased relaxivity upon Ca^{2+}/Zn^{2+} binding is also related to the hydration state of the Gd(III) ion. Luminescence studies of the corresponding europium complexes show an increase in q from about 0.4 to 1 upon ion binding, with a change in coordination environment at Eu(III) indicated by a change in the ratio of the hypersensitive $\Delta J=2$ transition to the $\Delta J=1$ transition. The reversibility of the response was demonstrated by competitive binding of the divalent metal ions to EDTA to give restoration of the initial relaxivity values. Studies in mouse serum reveal approximately 30–40%

increases in relaxivity, with the binding of bicarbonate to the Ca.Gd adduct perturbing the equilibrium. Given the speed of modern instrumentation, these responsive agents have potential to track changes in ion concentrations within the brain.

Meade and co-workers have established the use of enzyme dependent contrast agents such as that shown in Figure 8.10 [95, 96]. In this case, a galactose unit is attached to a DO3A derivative via an acetal link. The bulk of this substituent and its capacity as an O-donor ligand restricts access by water to the gadolinium site, resulting in a low relaxivity. However, galactosidase selectively cleaves the acetal, giving rise to an increase in q and a concomitant increase in relaxivity. This allows such a probe to be used to assess galactosidase concentration (provided the concentration of the probe can be established by other means).

8.7 CONCLUSIONS AND PROSPECTS

Magnetic resonance imaging has already made a dramatic contribution to human well-being, even though clinical agents are currently devoted to imaging blood pool. Over the last decade, significant contributions to the field have raised the prospect of extending the use of such systems to other forms of diagnosis and analysis. The examples given in the text above are just that: Many groups are working toward extending the effectiveness of gadolinium-containing contrast media. In theory, the maximum contrast achievable could be as high as 350 mmolGd^{-1}s^{-1}; such high values would raise the prospect of effective receptor targeting, combined with low dose for more conventional contrast measurements.

We have seen how it is beginning to be possible to control the relaxivity through design that influences molecular size and tumbling, close approach of water and the rate of water exchange, and through electronic relaxation properties of coordinated gadolinium. This last point represents an area in which immediate progress toward understanding and quantifying the electronic relaxation time is both possible and desirable. Studies on self-assembled and multimeric systems have revealed alternative strategies for maximising relaxivity: These also need further development before clinical application.

Perhaps the key to progress is the development of new generation techniques for quantitative imaging that combine the best that synthetic chemistry and data analysis have to offer with new approaches to developing instrumentation. With these in place, the future for gadolinium contrast agents looks rosy. Alternative strategies to MRI imaging are discussed elsewhere in this volume.

REFERENCES

[1] D. Parker, Imaging and targeting [in supramolecular technology], in *Comprehensive Supramolecular Chemistry*. Pergamon, Oxford, 487–536 (1996).

[2] A.E. Merbach and E. Toth, *The Chemistry of Contrast Agents in Medical Magnetic Resonance Imaging*. Wiley, Chichester (2001).

[3] S. Seibig, E. Toth and A.E. Merbach, *J. Am. Chem. Soc.* **122**, 5822–5830 (2000).

[4] L. Burai, E. Toth, S. Seibig, R. Scopelliti and A.E. Merbach, *Chem.Eur. J.* **6**, 3761–3770 (2000)

[5] J.-M. Idée, M. Port, I. Raynal, M. Schaefer, S. Le Greneur and C. Corot, *Fundam. Clin. Pharmccol.* **20**, 563–576 (2006).

[6] A.D. Sherry, P. Caravan and R.E. Lenkinski, *J. Magn. Reson. Imaging* **30**, 1240–1248 (2009).

[7] A. Spencer, S. Wilson and E. Harpur, *Hum. Exp. Toxicol.* **17**, 633–637 (1998).

[8] J.F. Desreux, *Inorg. Chem.* **19**, 1319–1324 (1980).

[9] J.R. Alger and J.A. Frank, *Annu. Rev. Physiol.* **54**, 827–846 (1992).

[10] D. Pubanz, G. Gonzalez, D.H. Powell and A.E. Merbach, *Inorg. Chem.* **34**, 4447–4453 (1995).

[11] M.F. Tweedle, *Eur. Radiol.* **7**, S225–S230 (1997).

[12] H. Vogler, J. Platzek, G. Schuhmann, T. Giampieri, T. Frenzel, H.J. Weinmann, B. Raduchel and W.R. Press, *Eur. J. Radiol.* **21**, 1–10 (1995).

[13] D.J. Parmelee, R.C. Walovitch, H.S. Ouellet and R.B. Lauffer, *Invest. Radiol.* **32**, 741–747 (1997).

[14] N. Sakagami, J.-I. Homma, T. Konno and K.-I. Okamoto, *Acta. Cryst. C* **C53**, 1376–1378 (1997).

[15] B.G. Jenkins and R.B. Lauffer, *Inorg. Chem.* **27**, 4730–4738 (1988).

[16] S. Aime, M. Botta and G. Ermondi, *Inorg. Chem.* **31**, 4291–4299 (1992).

[17] D. Parker, R.S. Dickins, H. Puschmann, C. Crossland, J.A.K. Howard, *Chem. Rev.* **102**, 1977–2010 (2002).

[18] P. Caravan, J.J. Ellison, T.J. McMurry and R.B. Lauffer, *Chem. Rev.* **99**, 2293–2352 (1999).

[19] C.A. Chang, L.C. Francesconi, M.F. Malley, K. Kumar, J.Z. Gougoutas, M.F. Tweedle, D.W. Lee and L.J. Wilson, *Inorg. Chem.* **32**, 3501–3508 (1993).

[20] M.R. Spirlet, J. Rebizant, J.F. Desreux and M.F. Loncin, *Inorg. Chem.* **23**, 359–363 (1984).

[21] S. Aime, A. Barge, M. Botta, M. Fasano, J.D. Ayala and G. Bombieri, *Inorg. Chim. Acta* **246**, 423–429 (1996).

[22] F. Benetollo, G. Bombieri, S. Aime and M. Botta, *Acta. Cryst. C* **C55**, 353–356 (1999).

[23] S. Aime, A. Barge, F. Benetollo, G. Bombieri, M. Botta and F. Uggeri, *Inorg. Chem.* **36**, 4827–4829 (1997).

[24] K. Kumar, C.A. Chang, L.C. Francesconi, D.D. Dischino, M.F. Malley, J.Z. Gougoutas and M.F. Tweedle, *Inorg. Chem.* **33**, 3567–3575 (1994).

[25] J. Platzek, P. Blaszkiewicz, H. Gries, P. Luger, G. Michl, A. Muller-Fahrnow, B. Raduchel and D. Sulzle, *Inorg. Chem.* **36**, 6086–6093 (1997).

[26] C. Edlin, S. Faulkner, D. Parker, M. Wilkinson, M. Woods, J. Lin, E. Lasri, F. Neth and M. Port, *New J. Chem.* **22**, 1359–1364 (1998).

[27] M.P. Placidi, L.S. Natrajan, D. Sykes, A.M. Kenwright and S. Faulkner, *Helv. Chim. Acta* **92**, 2427–2438 (2009).

[28] S. Aime, M. Botta, M. Fasano, M.P.M. Marques, C. Geraldes, D. Pubanz and A.E. Merbach, *Inorg. Chem.* **36**, 2059–2068 (1997).

[29] S. Aime, A. Barge, J.I. Bruce, M. Botta, J.A.K. Howard, J.M. Moloney, D. Parker, A.S. de Sousa and M. Woods, *J. Am. Chem. Soc.* **121**, 5762–5771 (1999).

[30] R.S. Dickins, D. Parker, J.I. Bruce and D.J. Tozer, *Dalton Trans.* 1264–1271 (2003).

[31] R.S. Dickins, J.A.K. Howard, C.L. Maupin, J.M. Moloney, D. Parker, J.P. Riehl, G. Siligardi and J.A.G. Williams, *Chem.-Eur. J.* **5**, 1095–1105 (1999).

[32] B. Bleaney, *J. Magn. Reson.* **8**, 91–100 (1972).

[33] J.A. Peters, J. Huskens and D.J. Raber, *Prog. Nucl. Magn. Reson. Spectrosc.* **28**, 283–350 (1996).

[34] S. Aime, M. Botta, M. Fasano, P.M. Marques, C.F.G.C. Geraldes, D. Pubanz and A.E. Merbach, *Inorg. Chem.* **36**, 2059–2068 (1997).

[35] L.S. Natrajan, N.M. Khoabane, B.L. Dadds, C.A. Muryn, R.G. Pritchard, S.L. Heath, A.M. Kenwright, I. Kuprov and S. Faulkner, *Inorg. Chem.* **49**, 7700–7709 (2010).

[36] E. Toth, R. Kiraly, J. Platzek, B. Radüchel and E. Brucher, *Inorg. Chim. Acta* **249**, 191–199 (1996).

[37] L. Sarka, L. Burai and E. Brucher, *Chem. Eur. J.* **6**, 719–724 (2000).

[38] C.F.G.C. Geraldes A.M. Urbano, M.C. Alpoim, M.A. Hoefnagel and J.A. Peters, *J. Chem. Soc. Chem. Commun.* 656–658 (1991).

[39] C.F.G.C. Geraldes A.M. Urbano, M.A. Hoefnagel and J.A. Peters, *Inorg. Chem.* **32**, (1993).

[40] Food and Drug Adminstration, in http://www.fda.gov/NewsEvents/Newsroom/PressAnnouncements/2010/ucm225286.htm

[41] P. Stratta, C. Canavese and S. Aime, *Curr. Med. Chem.* **15**, 1229–1235 (2008).

[42] S. Aime, A.S. Batsanov, M. Botta, R.S. Dickins, S. Faulkner, C.E. Foster, A. Harrison, J.A.K. Howard, J.M. Moloney, T.J. Norman, D. Parker, L. Royle and J.A.G. Williams, *J. Chem. Soc. Dalton Trans.* 3623–3636 (1997).

[43] S. Aime, A.S. Batsanov, M. Botta, J.A.K. Howard, D. Parker, K. Senanayake and G. Williams, *Inorg. Chem.* **33**, 4696–4706 (1994).

[44] Z. Luz and S. Meiboom, *J. Chem. Phys.* **40**, 2686–2688 (1964).

[45] S. Aime, M. Botta, M. Fasano and E. Terreno, *Chem. Soc. Rev.* **27**, 19–29 (1998).

[46] A. Beeby, I.M. Clarkson, R.S. Dickins, S. Faulkner, D. Parker, L. Royle, A.S. de Sousa, J.A.G. Williams and M. Woods, *J. Chem. Soc. Perkin Trans.* **2** 493–503 (1999).

[47] M.C. Alpoim, A.M. Urbano, C. Geraldes and J.A. Peters, *J. Chem. Soc. Dalton Trans.* 463–467 (1992).

[48] M.L. Wood and P.A. Hardy, *JMRI-J. Magn. Reson. Imaging* **3**, 149–156 (1993).

[49] A. Borel, L. Helm, A.E. Merbach, V.A. Atsarkin, V.V. Demidov, D.M. Odintsov, R.L. Belford and R.B. Clarkson, *J. Phys. Chem. A* **106**, 6229–6231 (2002).

[50] M. Benmeloukaa, J. Van Tolb, A. Borel, S. Nellutla, M. Port, L. Helm, L.-C. Brunel and A.E. Merbach, *Helv. Chim. Acta* **92**, 2173–2185 (2009).

[51] F.A. Dunand, R.S. Dickins, D. Parker and A.E. Merbach, *Chem. Eur. J.* **7**, 5160–5167 (2001).

[52] M. Woods, A. Pasha, P. Zhao, G. Tircso, S. Chowdhury, G. Kiefer, D.E. Woessner and A.D. Sherry, *Dalton Trans.* **40**, 6759–6764 (2011).

[53] M. Woods, S. Aime, M. Botta, J.A.K. Howard, J.M. Moloney, M. Navet, D. Parker, M. Port and O. Rousseaux, *J. Am. Chem. Soc.* **122**, 9781–9792 (2000).

[54] J.P. Dubost, J.M. Leger, M.H. Langlois, D. Meyer and M. Schaefer, *Acad. Sci. Paris Ser. II* **312**, 349 (1991).

[55] S. Aime, M. Botta, A. Barge, L. Frullano, U. Merlo, K.I. Hardcastle, *J. Chem. Soc., Dalton Trans.* 3435–3441 (2000).

[56] A.D. Sherry and M. Woods, *Annu Rev Biomed Eng* **10**, 391–411 (2008).

[57] D. Parker, *Coord. Chem. Rev.* **205**, 109–130 (2000).

[58] D. Messeri, M.P. Lowe, D. Parker and M. Botta, *Chem Commun.* 2742–2743 (2001).

[59] S. Laurent, L. Van der Elst, F. Copoix and R.N. Muller, *Invest. Radiol.* **36**, 115–118 (2001).

[60] A. Datta and K.N. Raymond, *Acc. Chem. Res.* **42**, 938–947 (2009).

[61] J. Xu, S.J. Franklin, D.W. Whisenhunt, Jr. and K.N. Raymond, *J. Am. Chem. Soc.* **117**, 7245–7246 (1995).

[62] S. Hajela, M. Botta, S. Giraudo, J. Xu, K.N. Raymond and S. Aime, *J. Am. Chem. Soc.* **122**, 11228–11229 (2000).

[63] D.M.J. Doble, M. Botta, J. Wang, S. Aime, A. Barge and K.N. Raymond, *J. Am. Chem. Soc.* **123**, 10758–10759 (2001).

[64] E.J. Werner, S. Avedano, M. Botta, B.P. Hay, E.G. Moore, S. Aime and K.N. Raymond, *J. Am. Chem. Soc.* **129**, 1870–1871 (2007).

[65] S. Aime, L. Calabi, C. Cavallotti, E. Gianolio, G.B. Giovenzana, P. Losi, A. Maiocchi, G. Palmisano and M. Sisti, *Inorg. Chem.* **43**, 7588–7590 (2004).

[66] G. Gugliotta, M. Botta and L. Tei, *Org. Biomol. Chem.* **8**, 4569–4574 (2010).

[67] E.M. Elemento, D. Parker, S. Aime, E. Gianolio and L. Lattuada, *Org. Biomol. Chem.* **7**, 1120–1131 (2009).

[68] J.W. Chen, F.P. Auteri, D.E. Budil, R.L. Belford and R.B. Clarkson, *J. Phys. Chem. A* **98**, 13452–13459 (1994).

[69] A. Visser, T. Ykema, A. van Hoek, D. O'Kane and J. Lee, *Biochem.* **24**, 1489–1496 (1985).

[70] D.A. Fulton, E.M. Elemento, S. Aime, L. Chaabane, M. Botta and D. Parker, *Chem Commun.* 1064–1066 (2006).

[71] C.F.G.C. Geraldes, A.D. Sherry, I. Lazar, A. Miseta, P. Bogner, E. Berenyi, B. Sumegi, G.E. Kiefer, K. McMillan, F. Maton and R.N. Muller, *Magn. Reson. Med.* **30**, 696–670 (1993).

[72] P. Caravan, M.T. Greenfield, X. Li and A.D. Sherry, *Inorg. Chem.* **40**, 6580–6587 (2001).

[73] S. Aime, S.G. Crich, E. Gianolio, G.B. Giovenzana, L. Tei and E. Terreno, *Coord. Chem. Rev.* **250**, 1562–1579 (2006).

[74] C.P. Montgomery, E.J. New, D. Parker and R.D. Peacock, *Chem Commun.* 4261–4263 (2008).

[75] D.M. Dias, J.M.C. Teixeira, I. Kuprov, E.J. New, D. Parker and C.F.G.C. Geraldes, *Org. Biomol. Chem.* **9**, 5047–5050 (2011).

[76] P. Caravan, N.J. Cloutier, M.T. Greenfield, S.A. McDermid, S.U. Dunham, J.W.M. Bulte, J.C. Amedio, Jr., R.J. Looby, R.M. Supkowski, W.D. Horrocks, Jr., T.J. McMurry and R.B. Lauffer, *J. Am. Chem. Soc.* **124**, 3152–3162 (2002).

[77] J.K. Alford, A.G. Sorensen, T. Benner, B.A. Chronik, W.B. Handler, T.J. Scholl, G. Madan and P. Caravan, *Proc. Intl. Soc. Mag. Reson. Med.* **19**, 452–456 (2011).

[78] Y. Song, E.K. Kohlmeir and T.J. Meade, *J. Am. Chem. Soc.* **130**, 6662–6668 (2008).

[79] P.J. Skinner, A. Beeby, R.S. Dickins, D. Parker, S. Aime and M. Botta, *J. Chem. Soc. Perkin Trans.* 2 1329–1338 (2000).

[80] E. Battistini, E. Gianolio, R. Gref, P. Couvreur, S. Fuzerova, M. Othman, S. Aime, B. Badet and P. Durand, *Chem.Eur. J.* **114**, 4551–4561 (2008).

[81] E. Gianolio, R. Napolitano, F. Fedeli, F. Arena and S. Aime, *Chem. Commun.* 6044–6046 (2009).

[82] R.J. Aarons, J.K. Notta, M.M. Meloni, J.H. Feng, R. Vidyasagar, J. Narvainen, S. Allan, N. Spencer, R.A. Kauppinen, J.S. Snaith and S. Faulkner, *Chem. Commun.* 909–911 (2006).

[83] J.H. Feng, M.M. Meloni, S.M. Allan, S. Faulkner, J. Narvainen, R. Vidyasagar and R. Kauppinen, *Contrast Media Mol. I* **5**, 223–230 (2010).

[84] M. Tan and Z.-R. Lu, *Theranostics* **1,** 83–101 (2011).

[85] R. Shukla, T.P. Thomas, J.A. Peters, A. Kotlyar, A. Myc and J.R.J. Baker, *Chem. Commun.* 5739–5741 (2005).

[86] C.A. Boswell, P.K. Eck, C.A.S. Regino, M. Bernardo, K.J. Wong, D.E. Milenic, P.L. Choyke and M.W. Brechbiel, *Mol. Pharmaceut.* **5**, 527–539 (2008).

[87] M.P. Lowe and D. Parker, *Chem. Commun.* 707–708 (2000).

[88] M.P. Lowe, D. Parker, O. Reany, S. Aime, M. Botta, G. Castellano, E. Gianolio and R. Pagliarin, *J. Am. Chem. Soc.* **123**, 7601–7609 (2001).

[89] S. Aime, S.G. Crich, M. Botta, G. Giovenzana, G. Palmisano and M. Sisti, *Chem Commun.* 1577–1578 (1999).

[90] S. Aime, F. Fedeli, A. Sanino and E. Terreno, *J. Am. Chem. Soc.* **128**, 11326–11327 (2006).

[91] J.L. Major and T.J. Meade, *Acc. Chem. Res.* **42**, 893–903 (2009).

[92] W.H. Li, S.E. Fraser and T.J. Meade, *J. Am. Chem. Soc.* **121**, 1413–1414 (1999).

[93] W.-H. Li, G. Parigi, M. Fragai, C. Luchinat and T.J. Meade, *Inorg. Chem.* **41**, 4018–4024 (2002).

[94] A. Mishra, N.K. Logothetis and D. Parker, *Chem. Eur. J.* **17**, 1529–1537 (2011).

[95] R.A. Moats, S.E. Fraser and T.J. Meade, *Angew. Chem. Int. Ed.* **36**, 726–728 (1997).

[96] L.L. M., M. Urbanczyk-Pearson and T.J. Meade, *Nature Protocols* **3**, 341–350 (2008).

[97] E. Toth, E. Brucher, I. Lazar and I. Toth, *Inorg. Chem.* **33**, 4070–4076 (1994).

[98] R.B. Shukla, K. Kumar, R. Weber, X. Zhang and M. Tweedle, *Acta Radiol.* **38**, 121–123 (1997).

[99] S. Aime, M. Botta, M. Panero, M. Grandi and F. Uggeri, *Magn. Reson. Chem.* **29**, 923–927 (1991).

[100] R. Shukla, M. Fernandez, R.K. Pillai, R. Ranganathan, P.C. Ratsep, X. Zhang and M.F. Tweedle, *Magn. Reson. Med.* **35**, 928–931 (1996).

[101] D.H. Powell, O.M. Dhubhghaill, D. Pubanz, L. Helm, Y.S. Lebedev, W. Schlaepfer and A.E. Merbach, *J. Am. Chem. Soc.* **118**, 9333–9346 (1996).

[102] X. Zhang, C.A. Chang, H.G. Brittain, J.M. Garrison, J. Telser and M.F. Tweedle, *Inorg. Chem.* **31**, 5597–5600 (1992).

[103] S.I. Kang, R.S. Ranganathan, J.E. Emswiler, K. Kumar, J.Z. Gougoutas, M.F. Malley and M.F. Tweedle, *Inorg. Chem.* **32**, 2912–2918 (1993).

9

NANOPARTICULATE MRI CONTRAST AGENTS

JUAN GALLO AND NICHOLAS J. LONG

Department of Chemistry, Imperial College London, London, UK

9.1 INTRODUCTION

Nanotechnology is destined to play a major role in future medicine. One of the most advanced examples of nanotechnology in biomedicine is the use of magnetic nanoparticles. Commercially available formulations of iron oxide nanoparticles are currently approved for their application in hospitals to diagnose liver carcinomas by magnetic resonance imaging (MRI). This detection is based on the unspecific accumulation of the contrast agent in the liver as a 'detoxifying' organ, but it cannot easily be extended to the detection or treatment of other tissues. However, due to the excellent performance of the nanoparticles in this specific example, it has encouraged further research in this general field of nanotechnology [1–4]. Most of the studies are based on the same kind of iron oxide nanoparticles or on their derivatives (ferrites), because they have proven themselves to be a powerful starting point due to their lack of toxicity, stability, biodegradability, and acceptable contrast enhancement [5]. Other magnetic materials, such as FeCo or FePt [6], have also been tested as possible contrast agents, although they present different issues, as will be discussed later in the chapter.

All the materials mentioned so far act as T_2 contrast agents. This means that they can be detected by MRI because they create their own magnetic field that disturbs the field from the instrument, negating the signal from their surroundings (the final effect is a darkening of the image in the areas where the T_2 contrast agent is present). However, these contrast agents present several inherent problems [7]. First, their effect on tissues results in labelled areas appearing as hypointense regions, sometimes making it difficult to distinguish labelled areas from pathogenic conditions, for example, bleeding. In addition, the high susceptibility of these kinds of agents perturbs the magnetic field of the neighbouring unlabelled tissues in what is called the *susceptibility artefact* or *blooming effect*, making it difficult to identify the exact state of the lesion [8]. These facts are the reason why researchers have also started to examine T_1 nanoparticle-based contrast agents [7, 9]. T_1 contrast agents are already in clinical use but mainly in the form of Gd chelates. As nanoparticles, the main two T_1-based materials tested to date have been gadolinium(III) oxide and manganese oxide.

In general, the use of nanoparticles in medicine presents several advantages over the use of traditional chemical structures. Nanoparticles can be used as beacons over which different molecules with different roles can be attached. In this way a nanoparticle can carry, on one hand, a targeting molecule toward a diseased area, and on the other a drug to treat this disease, while the metallic core of the nanoparticle can provide the means for detection. Also, nanoparticle structures are capable of carrying hundreds to thousands of ligand molecules that can result in an enhanced local concentration of the drug once it reaches the target. In the same way, multivalent presentation of ligands on the surface of the nanoparticles allows for the use of molecules with low affinity toward their target (i.e., carbohydrates), widening the range of options to achieve specific labelling.

The Chemistry of Molecular Imaging, First Edition. Edited by Nicholas Long and Wing-Tak Wong.
© 2015 John Wiley & Sons, Inc. Published 2015 by John Wiley & Sons, Inc.

9.2 T_2 CONTRAST AGENTS

Magnetic nanoparticles, especially those featuring iron oxide, are the leading type of nanoparticle structures in biomedicine, especially for diagnosis. Since the first *in vivo* MRI images were obtained back in 1974 [10], the technique has evolved rapidly with many companies and research groups becoming involved in the field. Magnetic nanoparticles were developed at the end of the 1970s—in 1978 Ohgushi and co-workers [11] used dextran-coated iron oxide nanoparticles in *in vitro* studies for the first time, whilst in 1986 the first *in vivo* studies with magnetic nanoparticles were carried out by Lauterbur in dog models [12].

Generally, there are two different approaches for the preparation of a nanoparticle: the top-down physical protocols and the bottom-up chemical methods. The main advantage of the physical methodologies is that they can produce large batches of the product, but they have been hardly used in this field because it is difficult to control the size and size distribution of the final products. On the other hand, a number of different chemical protocols have been developed that achieve, with varying degrees of success, the goals desired for preparation of nanoparticles for this application—high crystallinity, controllable size and chemical nature, and narrow size distribution [6].

9.2.1 T_2 Nanoparticle Preparation

A variety of different chemical approaches for the preparation of magnetic nanoparticles have been explored to date, but with regard to MRI applications, only two of them account for more than 90% of the published works. These two methodologies, co-precipitation of iron(II) and iron(III) salts and thermal decomposition of organometallic compounds, are the 'gold standard' for the preparation of magnetic nanoparticles.

9.2.1.1   *Co-precipitation of Fe^{2+} and Fe^{3+} Salts*

Co-precipitation protocols are widely used for the preparation of magnetite (Fe_3O_4) and maghemite (γ-Fe_2O_3) nanoparticles due to their simplicity, good-enough quality of the resulting nanoparticles, and the large amounts of product that can be obtained in a single batch. However, particle size distribution and crystallinity are the weak points of these protocols because only kinetic factors control the growth of the crystals [13].

$$Fe^{2+} + 2\,Fe^{3+} + 8\,OH^- \rightarrow Fe_3O_4 + 4\,H_2O \tag{9.1}$$

In this methodology, a basic solution is added to an aqueous 2:1 mixture of ferric and ferrous salts (equation 9.1) to obtain magnetite nanoparticles [14]. Magnetite (Fe_3O_4) is the preferred chemical nature of iron oxide for MRI applications due to its superior performance when compared to other iron oxides such as maghemite (γ-Fe_2O_3) or hematite (α-Fe_2O_3) [15]. According to the thermodynamics of this reaction, if a stoichiometry of Fe^{3+}:Fe^{2+} (2:1) is maintained and the reaction is carried out under non-oxidising and oxygen-free conditions, a complete precipitation is expected at a pH between 8 and 14 [16]. The size and, to some extent, the shape of the resulting nanoparticles can be tuned by exerting a close control of the pH, the reaction temperature, the ionic strength of the reaction media, the chemical nature of the iron salts (usually chlorides, but perchlorates, nitrates and sulphates have also been used), and the ratio of Fe^{3+} and Fe^{2+}. To prevent magnetite from further oxidation and to avoid particle aggregation, whilst at the same time controlling the size of the final product, organic coating molecules (e.g., polymers) are usually introduced into the reaction media during the precipitation process. The most widely used of these polymeric materials is dextran, a branched glucose polymer (for a review on dextran-coated iron oxide nanoparticles see [17]) that due to its biocompatibility, biodegradability, and good performance has become the most popular option even for commercial T_2 contrast agents. Other popular polymeric choices include chitosan; synthetic polymers such as PEG, poly(vinyl alcohol) (PVA), poly(acrylic acid) (PAA), poly(methylacrylic acid) (PMAA), poly(lactic acid), polyvinylpyrrolidone (PVP), polyethyleneimine (PEI); and AB- and ABC-type block copolymers containing the above polymers as segments [5]. The main drawback of using polymeric materials is that while the size of the magnetic core remains in the low nanometre range, the overall hydrodynamic radii is far larger (i.e., Feridex: magnetic core diameter=4.96 nm; hydrodynamic diameter=160 nm). This feature makes it easier for the immune system to detect and clear away these contrast agents and reduces their half-life in the body. Small molecules, usually containing carboxylic acid moieties, have also been used *in situ* as coating ligands, the most common example being citric acid [18]. These molecules act as chelating agents, absorbing onto the magnetic surface and helping to control the size of the final particles.

9.2.1.2   *Thermal Decomposition and/or Reduction Routes*

Thermal decomposition protocols appeared in the late 1990s to early 2000s. They were designed to overcome the two main disadvantages of co-precipitation routes: poor crystallinity and wide size distribution. As the name suggests, they are based on the decomposition of organometallic compounds at high temperatures, usually in high boiling point non-polar solvents. In 1999, Alivisatos et al. developed the thermal

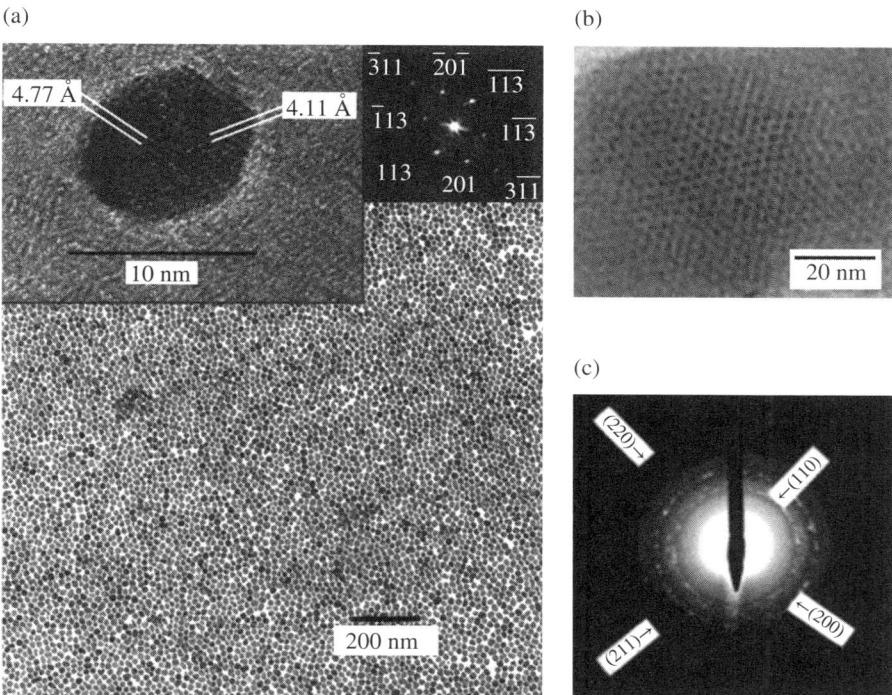

FIGURE 9.1 (a) Low-resolution TEM image of a monolayer of individual γ-Fe$_2$O$_3$ nanocrystals (10.0 ± 1.5 nm). Top left: High-resolution TEM image of one of these nanocrystals. The indicated lattice plane distances correspond to the (113) and (201) lattice planes of tetragonal γ-Fe$_2$O$_3$ with ordered superlattice of the cation vacancies. Top right: FFT of the high-resolution TEM image looking down the [512] zone-axis. Adapted with permission from Ref. [19]. (b) Transmission electron micrographs (TEM) of spherical iron nanoparticles with diameters of 2 nm. (c) Electron microdiffraction pattern of rod-shaped iron oxide nanoparticles. (b) and (c) reproduced with permission from Ref. [20].

decomposition method to prepare γFe$_2$O$_3$, Mn$_3$O$_4$, and Cu$_2$O nanoparticles using metal cupferron complexes as starting materials (Me^{x}Cup$_x$, Me = Fe^{3+}, Mn^{2+}, or Cu^{2+}, Cup = N-nitrosophenylhydroxylamine) (Figure 9.1a) [19].

Hyeon's group developed a similar method for the preparation of magnetic nanoparticles using iron pentacarbonyl as the iron source and TOPO (trioctylphosphine oxide) as coordinating solvent (Figure 9.1b, c) [20]. In the following years, this protocol was further developed to obtain high crystallinity and closer control over the final size of the nanoparticles [21, 22]. The initial methods (hot injection methods) required the swift injection of the metal precursors over preheated solvent solutions, which presented several disadvantages, mainly the difficulty to scale up the processes (due to temperature inhomogeneities) and the risks associated with the injection over solvents at high temperatures and subsequent vigorous reactions. Both problems were solved in 2004 with the introduction of new organometallic compounds as starting materials: oleates and acetylacetonates. Park et al. described a very general route for the large-scale preparation of monodisperse metal oxide nanoparticles (Figure 9.2) from oleates that were easily produced from salts of the desired metal and sodium oleate [23].

In the same year, Sun and co-workers published the preparation of magnetic nanoparticles from commercial acetylacetonates with high crystallinity and narrow size distribution [24]. A further advantage of thermal decomposition protocols is their flexibility in terms of chemical nature of the nanoparticles. The methods have been used to prepare not only iron oxide nanoparticles, but also ferrites of different nature (MnFe$_2$O$_4$ and CoFe$_2$O$_4$ [24], NiFe$_2$O$_4$ [25], (Zn$_x$Mn$_{1-x}$)Fe$_2$O$_4$, and (Zn$_x$Fe$_{1-x}$)Fe$_2$O$_4$ [26]) and other types of magnetic (FePt [27–29]) and non-magnetic (Mn$_x$O$_y$ [30, 31]) nanoparticles for MRI imaging applications.

These protocols produce nanoparticles with optimal properties for use as MRI contrast agents. The weak point of this methodology is that an intermediate extra step has to be introduced to produce water-soluble nanoparticles. To date, there are only a few examples in the literature where polar high boiling point solvents have been used to prepare iron oxide nanocrystals following thermal decomposition protocols. This advantageous modification gives a product that is directly soluble in water, avoiding the ligand exchange step. For example, Li et al. used 2-pyrrolidone as the solvent to obtain Fe$_3$O$_4$ nanoparticles. In this case, the solvent also acts as a coating molecule, rendering nanoparticles dispersible in both acidic and alkaline aqueous media, but not at neutral pH [32]. This methodology was further improved by using derivatised PEG molecules as ligands to improve water solubility and at the same time provide functional groups for further modifications [33].

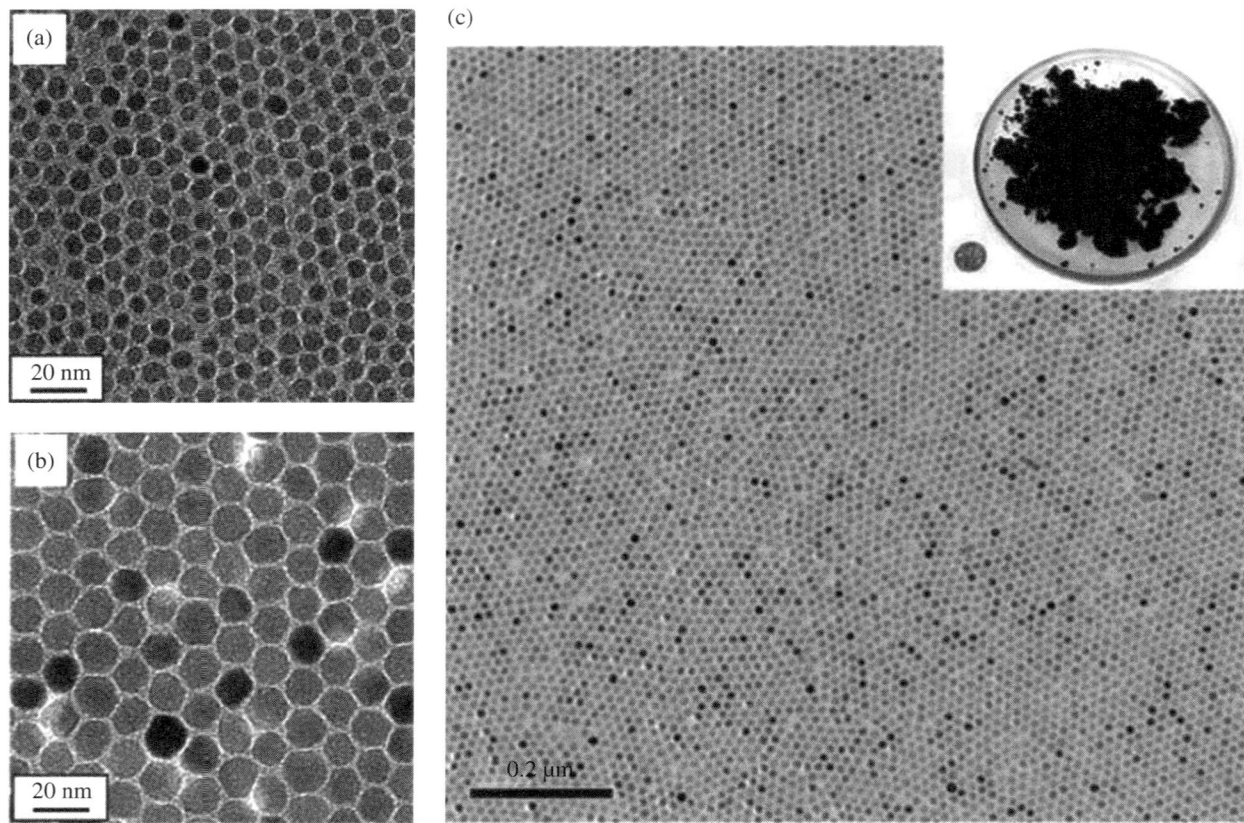

FIGURE 9.2 TEM bright field images of (a) 6 nm and (b) 12 nm Fe_3O_4 nanoparticles deposited from their hexane dispersion on an amorphous carbon-coated copper grid and dried at room temperature. Reproduced with permission from [24]. (c) TEM overview micrograph of 12 nm Fe_3O_4 nanoparticles. Inset, photograph showing a Petri dish containing 40 g of the monodisperse magnetite nanocrystals and a U.S. one-cent coin for comparison. Reproduced with permission from Ref. [23].

9.2.1.3 Other Methods of Synthesis

The methodologies described so far are the most popular ones for the preparation of magnetic nanoparticles for imaging applications, but a number of other chemical approaches have also been described.

- Microemulsion protocols: This kind of protocol can be considered to be an evolution of the classic co-precipitation methodology. The weak point of the iron salts co-precipitation method is the broad size distribution of the obtained particles, thus numerous modifications of this synthesis have been described in which the reaction takes place in constrained environments. Surfactant molecules spontaneously form nanodroplets of different sizes when in the correct solvents and concentrations. Inside these droplets iron salts can be encapsulated in water solutions isolated from outer organic non-polar solvents. In the case of these reverse microemulsions, the size of the droplets or micelles can be tuned (by varying the molar ratio of water to surfactant) to be usually between 1 and 50 nm [3]. Two such microemulsions can be mixed together; upon stirring, the micelles will collide, coalesce, and break up again, thus mixing their contents and starting the chemical reaction under constrained conditions, thus working as a nano-reactor [34]. Finally, in the case of nanoparticles, the final size and homogeneity of the population of nanocrystals will depend on the size and quality of the starting microemulsion. The reactions that can be carried out inside these systems are not restricted to co-precipitation of iron salts. Indeed, the first example of magnetic nanoparticles produced using microemulsion approaches was based on the oxidation of Fe^{2+} salts into Fe_2O_3 and Fe_3O_4 [35]. Depending on the chemical nature of the reagents needed for the reaction, different types of surfactants can be used, both ionic and non-ionic ones. Some common examples are sodium bis(2-ethylhexyl sulfosuccinate) (AOT) and cetyltrimethylammonium bromide (CTAB) [36] as ionic ones, whilst Triton X-100, Igepal CO-520, and Brij-97 [37, 38] are some common examples of non-ionic surfactants.

A further advantage of the microemulsion method is its versatility in terms of the chemical nature of the products that can be prepared. For example, Carpenter et al. prepared cobalt, cobalt/platinum, and gold-coated cobalt/platinum nanoparticles using CTAB in water/octane with 1-butanol as co-surfactant [39]. Liu and co-workers prepared $MnFe_2O_4$ nanoparticles with sizes between 4 and 15 nm in water/toluene systems with sodium dodecylbenzenesulfonate (NaDBS) as surfactant [40].

To summarise, this methodology allows a closer control over the size distribution of the prepared particles when compared to co-precipitation protocols. However, these are not ideal protocols due to the complex purifying procedures required (large volume of solvents), the low yield of nanoparticles obtained when compared to co-precipitation methodologies, and difficulties in scale-up.

- Solvothermal/hydrothermal synthesis: In these less well-studied strategies, nanoparticles are obtained in aqueous solutions at high temperatures and under high pressures inside an autoclave. As with the other methodologies, the addition of stabilising molecules can help control the final size and narrow the size distribution. Li et al. published [41] a general hydrothermal method for the preparation of nanoparticles. In this report they describe the preparation of nanoparticles of several chemical natures, from noble metal nanocrystals to quantum dots, magnetic nanoparticles, and rare earth nanoparticles at the interfaces between liquid, solid, and solution. Baruwati and co-workers used nitrate or chloride ions and oleic acid to obtain different ferrite (Ni, Co, and Mn) and maghemite nanoparticles under both conventional and microwave hydrothermal conditions [42]. In addition, Ge et al. prepared Fe_3O_4 nanoparticles with tunable sizes from iron(II) chloride and ammonia. In this example, the concentration of the iron salt was used to control the final size of the nanocrystal (higher Fe^{2+} concentration, smaller particles) [43].

- Miscellaneous: Other methodologies that have been used to some extent to prepare magnetic nanoparticles include an electrochemical approach in which the current density through an iron electrode controlled the final size of maghemite nanoparticles [44]. Laser pyrolysis was used by Alexandrescu et al. to prepare iron oxide nanoparticles of a different nature. Here, the laser hits a gaseous mixture of the iron precursor and carrier gas to produce small and non-aggregated nanoparticles usually with a size below 10 nm [45]. Abu Mukh-qasem et al. used a sonochemical approach to prepare 9 nm Fe_3O_4 nanocrystals from iron pentacarbonyl ($Fe(CO)_5$) in an aqueous solution and in the presence of sodium dodecylsulfate (SDS), although in this case the resulting nanoparticles are amorphous [46].

9.3 T_1 CONTRAST AGENTS

The use of nanoparticulate systems as T_1 contrast agents has been far less explored than their T_2 counterparts. To date, the only T_1 contrast agents used in clinics are based on organic chelates of paramagnetic ions that possess a large number of unpaired electrons capable of promoting a very effective T_1 relaxation. The Gd^{3+} ion has been heavily utilised because it has seven unpaired electrons combined with a large magnetic moment; however, gadolinium is highly toxic so it has to be used in the form of kinetically and thermodynamically stable chelates. Another ion that has been used in a clinical setting is Mn^{2+}. Manganese participates in a large number of biochemical processes, but in high concentrations it can cause hepatic failures and cardiac toxicity. Mn^{2+} has been used as a contrast agent even in the form of the chloride salt ($MnCl_2$) due to its prominent contrast effect but is currently limited to animal studies.

Even though iron oxide and related nanoparticles are being shown to be promising tools for medical diagnosis, the development of a new generation of T_1 nanocrystals would present some advantages. T_1 contrast agents produce a positive signal (brightening of the images), which is less likely to be confused with pathological conditions (like bleeding) as can occur with T_2 agents. Also, magnetic nanoparticles disrupt the magnetic homogeneity of their surroundings disturbing the anatomic background; the use of paramagnetic T_1 nanoparticles would minimise this kind of artefact. Alongside these advantages, there are benefits over current chelates: (i) Most organic chelates in use have poor cell penetration abilities, and (ii) nanoparticles can bring more versatility to the field because several different moieties can be bound to their surface at the same time, allowing for simultaneous specific imaging and drug delivery.

9.3.1 Gadolinium-based nanoparticles

As stated previously, current clinical T_1 contrast agents are mostly based on gadolinium chelates, so gadolinium nanoparticles offer an attractive alternative. To date, three different types of gadolinium(III) crystals have been prepared for their application as contrast agents: gadolinium oxide (Gd_2O_3), gadolinium fluoride (GdF_3), and gadolinium phosphate ($GdPO_4$). Gd_2O_3 nanoparticles of around 30 nm were first prepared and studied for MRI imaging purposes in 2006 using a reduction and precipitation method in the presence of dextran as stabilising agent. The reaction was carried out in aqueous media using gadolinium chloride ($GdCl_3$) as starting material [47]. Polyol protocols have also been used for the preparation of gadolinium nanoparticles [48, 49]. The resulting gadolinium oxide nanoparticles present a narrow size distribution centred around 5 nm that can be tuned by the initial concentration of the metal ion and by the water/DEG ratio in the reaction. More recently, Park and co-workers used a thermal decomposition methodology to obtain ultrasmall gadolinium oxide nanoparticles with excellent relaxivity properties [50, 51]. Using three different Gd ion precursors (chloride, acetate, and acetylacetonate) in tripropylene glycol as the high boiling point solvent and three different compounds as ligands (PEG, D-glucuronic acid, and lactobionic acid), they prepared 1 nm Gd_2O_3 nanocrystals by taking the solution to reflux for 24 h while bubbling air through the solvent.

There are only several other examples of gadolinium-based nanoparticles. Evanics and co-workers prepared both GdF_3 and GdF_3/LaF_3 directly in aqueous solutions [52]. GdF_3 nanoparticles were prepared by dropwise addition of a water solution of $Gd(NO_3)_3$ into a preheated water solution of NaF and citric acid as the coating molecule at pH 7. GdF_3/LaF_3 nanoparticles were prepared likewise but using 2-aminoethyl phosphate instead of citric acid as the stabilising molecule. The main problem in their protocol is the broad size distribution of the nanoparticles obtained, between 10 and 150 nm. Hifumi et al. [53] report the preparation of $GdPO_4$ nanoparticles using a hydrothermal protocol. Water solutions of gadolinium(III) nitrate $(Gd(NO_3)_3)$ as Gd source, ammonium hydrogen-phosphate $((NH_4)_2HPO_4)$, dextran as stabiliser, and NaOH to take the pH to 12.5 are sealed in glass pressure tubes and heated to 200°C for several hours. The inclusion of dextran in the reaction media helps to control both the size and the shape of the nanoparticles (rod-shaped nanocrystals with an average hydrodynamic diameter of 23.2 nm).

9.3.2 Manganese-Based Nanoparticles

Recently, MnO nanoparticles have captured researchers' attention due to their good T_1 relaxation properties and reduced toxicity when compared to gadolinium. The preparation of MnO nanoparticles for medical applications follows, in most of the cases, thermal decomposition protocols similar to those described previously for the preparation of iron oxide nanoparticles. For example, in 2003, Yin et al. described the preparation of MnO nanoparticles from manganese acetate $(Mn(CO_2CH_3)_2)$ in trioctylamine with oleic acid as ligand [30]. All these reagents were mixed together under N_2 and the reaction mixture was then heated rapidly to 320°C and maintained at that temperature for 1 hour. This protocol provided monodisperse nanoparticles of 7 nm diameter. The size of the resulting nanoparticles could be further increased to be between 12 and 20 nm by including a final heating step at 100°C. In 2004, Park and co-workers described the preparation of MnO nanoparticles following a thermal decomposition route [23]. The product of this reaction is a monodisperse population of MnO nanoparticles with a size of 25 nm. Larger nanoparticles (35 nm) can be obtained by increasing the heating time to 2 hours, and smaller populations (from 7 to 20 nm) can be obtained using 1-hexadecene as the solvent combined with a slightly different temperature (280°C) and heating times of up to 1 hour.

Different methodologies have also been described for the preparation of manganese oxide nanoparticles. Recently Baek et al. reported the preparation of MnO nanoparticles following a polyol route [54]. In 2010, Huang and co-workers published a hydrothermal preparation of Mn_3O_4 nanoparticles for MRI applications [55]. In their synthesis, manganese chloride is used to first prepare manganese stearate $(Mn(SA)_2)$, which is then dissolved in toluene with dodecylamine and mixed with an aqueous solution of *tert*-butylamine to form a two-phase solution. This two-phase mixture was sealed in an autoclave and submitted to different temperature protocols to prepare either Mn_3O_4 nanocubes, nanoplates, or nanospheres.

A recent development in the field of MnO T_1 nanoparticles features the preparation of hollow manganese oxide (Mn_xO_y) nanoparticles. Protocols have been developed to etch the nanoparticles' core after the standard synthesis. The erosion of the core allows an increased surface exposure to the solvent increasing the r_1 relaxivity of the nanoparticles. An et al. [56] prepared MnO nanoparticles following one of the thermal decomposition routes mentioned before [23], and the nanoparticles were dissolved in trioctylphosphine oxide (TOPO) and heated at 300°C for 2 hours (Figure 9.3).

A different methodology was used by Shin et al. [57] to obtain the same result. In this case, the same protocol was also used for the preparation of the starting MnO nanoparticles, which were then transferred to water through an encapsulation inside a PEG-phospholipid (1,2-distearoyl-sn-glycero-3-phosphoethanolamine-N-[methoxy(polyethylene glycol)-2000] shell. The etching of the cores was achieved by the dispersion of the water soluble nanoparticles in phthalate buffer pH 4.6 for 12 hours.

Finally, there is an example worth mentioning that combines both Gd^{3+} and Mn^{2+} ions within the same nanoparticle. Choi et al. were able to prepare nanoparticles comprising a gadolinium oxide core (Gd_2O_3) and a manganese oxide coating (MnO). The resulting product showed improved relaxivity values compared to those of pure Gd_2O_3 or MnO nanocrystals. For the preparation of the nanoparticles, they followed a three-step, one-pot strategy. Using a polyol methodology, they prepared Gd_2O_3 nanoparticles by heating $GdCl_3$ in triethylene glycol to 250°C for 24 hours. After cooling the reaction to 100°C, $MnCl_2$ was injected into the reaction medium, which was heated at 250°C for 24 more hours. Finally, after cooling again to 150°C, lactobionic acid was injected, and the reaction was kept at that temperature for another 24 hours. The resulting product was a monodisperse population of ultra-small $Gd_2O_3@MnO$ nanoparticles with a diameter around 1.5 nm.

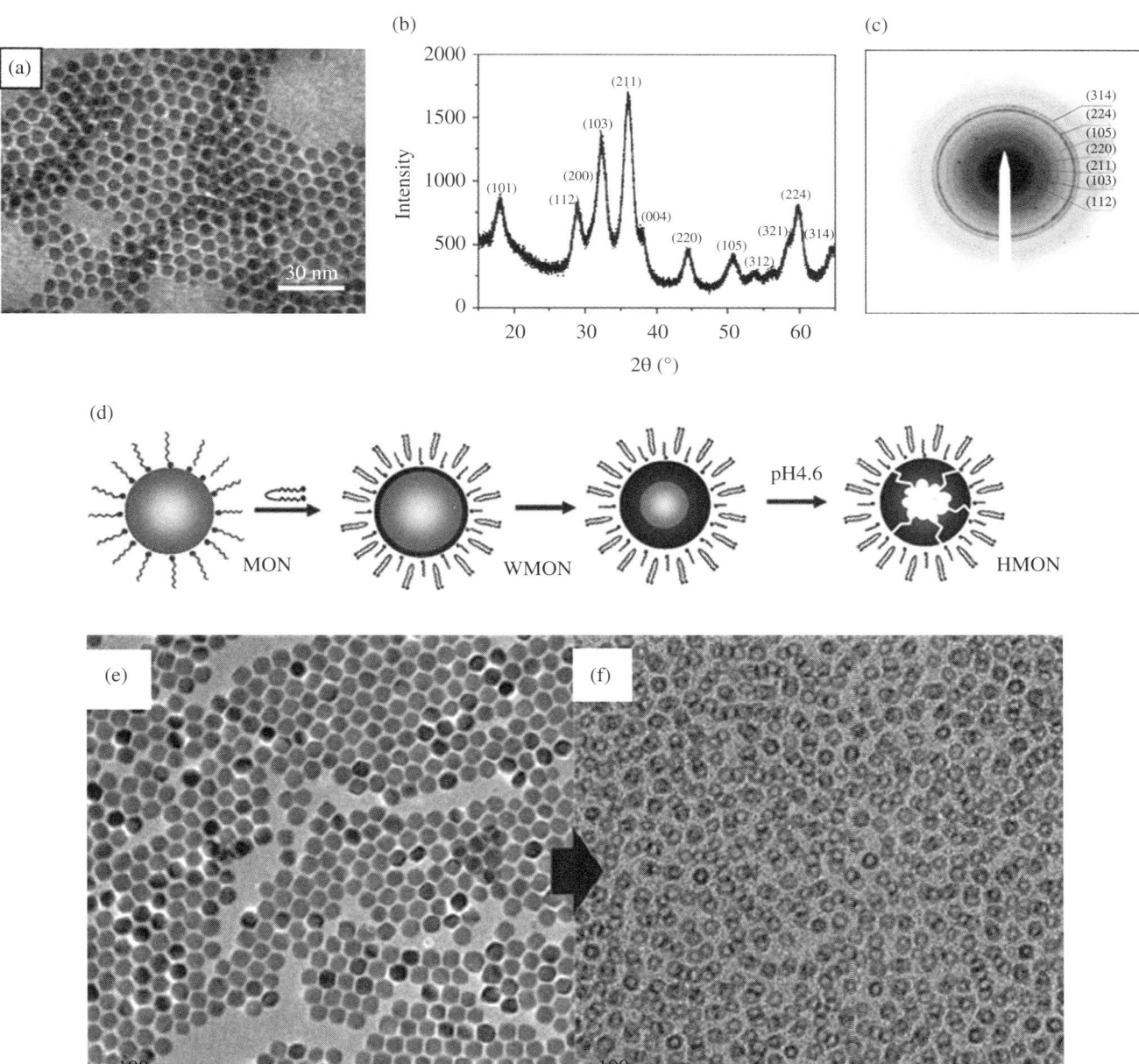

FIGURE 9.3 (a) Transmission electron micrograph. (b) X-ray powder diffraction pattern. (c) Indexed selected area electron diffraction pattern (negative shown for clarity) of 7 nm diameter Mn$_3$O$_4$ nanocrystals. Reproduced with permission from Ref. [30]. (d) Scheme of the formation of Mn$_3$O$_4$ hollow nanoparticles. Reproduced with permission from Ref. [168] **e** and **f**: TEM images showing the changes from solid nanocrystals of 18 nm-sized MnO (e) to the corresponding hollow oxide nanoparticles (f) through the etching process. Adapted with permission from Ref. [56].

9.3.3 Paramagnetic Additions to Nanoparticles

A different approach toward the production of nanoparticulate T$_1$ agents is an evolution of the classic chelates strategy. Basically, DOTA (1,4,7,10-tetraazacyclododecane tetraacetic acid) or related chelates are functionalised with a linker featuring a functional group that allows its conjugation to a nanoparticle. In this case, the nanoparticle acts only as a platform to bind the paramagnetic agents. In this way some solubility issues can be overcome, the local concentration of the contrast agent is increased, and more importantly, the chelates can be combined with other ligands to provide other properties (as for example, cell-penetrating abilities). This strategy has been useful for the design of multifunctional nanoparticles, because the nanoparticles themselves can be active in another imaging field (e.g., quantum dots for fluorescence imaging). Here, we will only give a few examples of purely T_1 MRI agents; multifunctional nanomaterials will be discussed in Chapter 16.

Silica and gold nanoparticles have been used as platforms onto which the different chelates can be coupled. Silica nanoparticles present the advantage that they can be synthesised as mesoporous materials comprising channels with diameters tunable between 2 and 10 nm. This feature allows their use either as platforms onto which molecules can be attached or as encapsulating materials containing active species in their channels. The work by Weili and Lin [58, 59] demonstrated that gadolinium chelates can be bound both to the surface or the channels of silica nanoparticles. In reference [58] 37 nm silica (SiO_2) nanoparticles incorporating gadolinium chelates in their surface were prepared following a two-step one-pot reverse microemulsion protocol. Alternatively, in reference [59], mesoporous SiO_2 nanoparticles were prepared following a surfactant template, base-catalysed condensation procedure. Cetyltrimethylammonium bromide (CTAB) was used as the surfactant in water, and TEOS was used as SiO_2 precursor (Figure 9.4a, b). The polymerisation was catalysed by NaOH and carried out for 2 hours at 80°C. According to their data, the Gd chelate is mainly attached to the channels in the structure of the nanoparticles. This channel location improves the performance as T_1 contrast agents because it seems to present highly efficient accessibility of the magnetic centres to available water molecules.

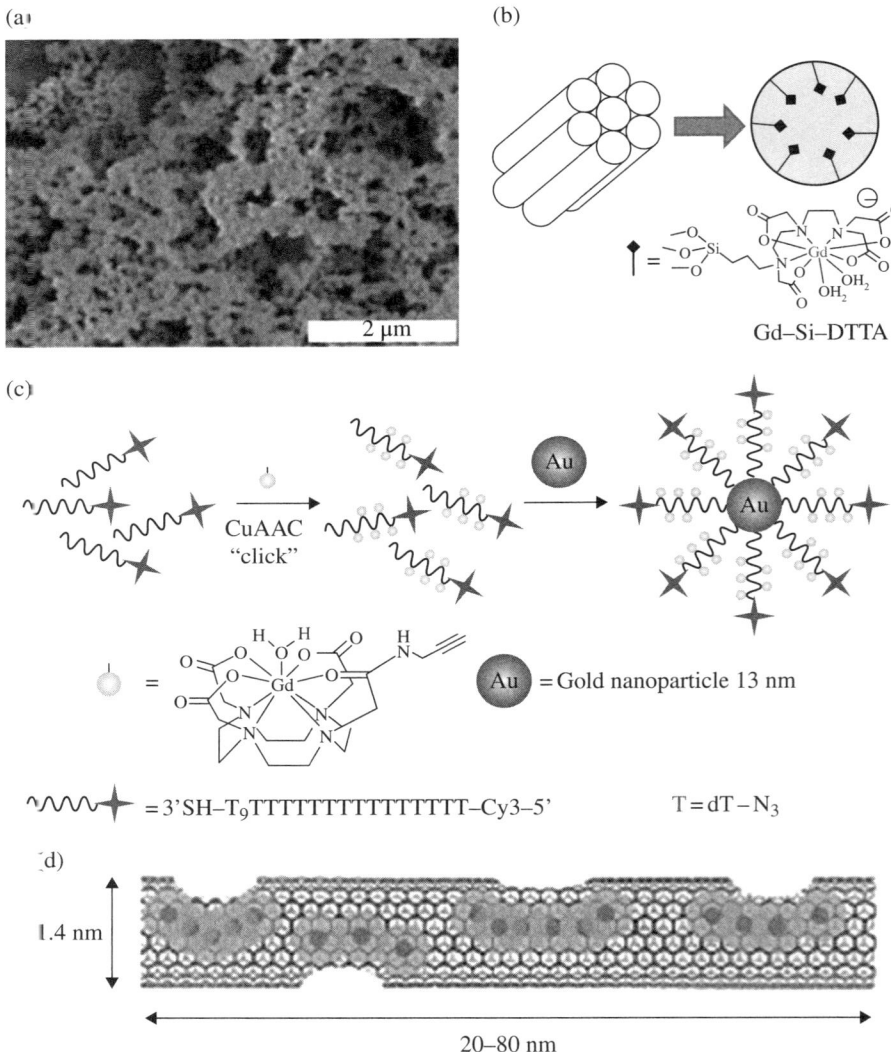

FIGURE 9.4   (a) SEM image of mesoporous silica nanospheres showing the formation of monodisperse, water-dispersible nanoparticles. (b) Scheme showing the Gd-Si-DTTA complexes residing in hexagonally ordered nanochannels of 2.4 nm in diameter. Reproduced with permission from Ref. [59]. (c) Scheme of the DNA–GdIII@AuNP conjugates prepared in Ref. [63]. (d) Representative illustration of a gadonanotube (GNT). Clusters of internally loaded Gd^{3+} ions are located at defect sites along the nanocapsule sidewalls. Reproduced with permission from Ref. [65].

The use of gold nanoparticles has also been widely explored as a 'beacon' to which chelates can be attached. Gold nanoparticles are easily produced, and well-known thiol-gold chemistry can be explored for the stable functionalisation of the nanoparticles in a versatile and efficient way. Alric et al. published the *in situ* preparation of 2.4 nm gold nanoparticles protected with a chelate, DTDTPA, derived from diethylenetriaminepentacetic acid bis-anhydride [60]. Their synthesis was carried out in a mixture of water, ethanol, and acetic acid using HAuCl$_4$ as the gold source and sodium borohydride (NaBH$_4$) as the reducing agent. In a second step, the nanoparticles plus the chelate were incubated in the presence of gadolinium chloride to force the paramagnetic ion into the complex. A very similar strategy was followed by Mayer and co-workers to obtain T$_1$ gold nanoparticles [61]. They prepared 3.5 nm gold nanoparticles *in situ* with a thiol derivative of DTTA chelate, and in a second step incubated them with GdCl$_3$. A similar strategy was followed by Marradi and co-workers [62]. In this case they also used an *in situ* reaction to produce gold nanoparticles functionalised via thiol chemistry with gadolinium chelates (DO3A type) and different carbohydrates. They showed that it is possible to alter the performance of the nanoparticles as T$_1$ contrast agents by not only modifying the number and type of chelates, but also by changing the nature of the surrounding molecules. In this case, the organisation of water molecules around carbohydrates is configuration dependent; this can either bring water protons closer to the paramagnetic metal centre or keep water molecules apart. Using a different strategy, Song et al. prepared 13 and 30 nm gold nanoparticles with gadolinium chelates [63]. First, gold nanoparticles were prepared, and in a second step a ligand exchange reaction was carried out. The new ligands were thiol-ended modified DNA chains bearing gadolinium chelates (Gd595) (Figure 9.4c). The ligand exchange with the DNA derivatives was performed over 24 hours, gradually increasing the concentration of NaCl in the reaction medium (10 mM phosphate buffer saline) up to 0.1 M. In this case the modified DNA ligands provide not only the T$_1$ contrast enhancement, but also increase cell permeation and can serve as cell tracking ligands.

Finally, carbon nanotubes (CNTs) have been used as 'inert' nanoparticulate platforms to couple T$_1$ chelates. Richard et al. functionalised the outer surface of multi-walled carbon nanotubes (MWNT) with modified DTPA chelates [64]. They designed the modified DTPA chelate in such a way that it presents an aliphatic chain capable of surface adsorption on the nanotubes via van der Waals interactions, while the rest of the molecule is a polar head group that confers water solubility and at the same time is capable of chelating Gd^{3+} ions. To functionalise MWNTs with the amphiphilic ligands, solutions of the DTPA derivative at a concentration above the critical micelle concentration were sonicated with powders of the nanotubes. A completely different approach was chosen by Tran et al. In their design, paramagnetic ions are located only inside the nanotubes clustered around defect sites along the sidewalls (Figure 9.5d) [65]. They used single-walled nanotubes produced by an electric-arc discharge method that were then cut into ultra-short SWNTs (20-80 nm in length) by fluorination and pyrolysis at 1000°C under inert atmosphere. The nanotubes were reduced with sodium to produce individualised or small bunches of nanotubes and then sonicated in a water solution of gadolinium chloride. Finally, to solubilise the product in aqueous media, the nanotubes were sonicated with a surfactant (Pluronic F-108).

(a)   (b)

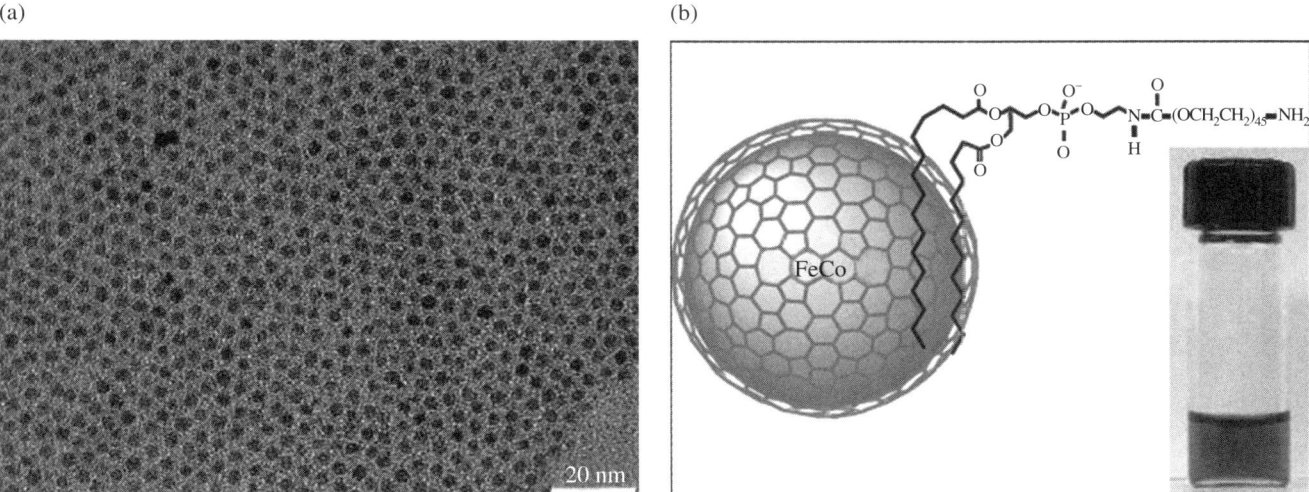

FIGURE 9.5 (a) TEM micrograph of 4 nm Fe$_3$O$_4$ nanoparticles. From Ref. [67]. (b) Schematic diagram of a FeCo/GC nanocrystal, structure of the phospholipid molecule used for functionalisation and a photograph of a PBS suspension of functionalised FeCo nanocrystals taken after heating to 80°C for 1 hour. Reproduced with permission from Ref. [69].

9.3.4   Classical T_2 Agents Used for T_1 Measurements

In this last group of T_1 nanoparticles, some examples of 'classical' T_2 nanoparticles that have been used for T_1 measurements are discussed. A very important parameter for the application of a contrast agent in T_1 measurements is the r_2/r_1 ratio. This value has to be as small as possible; the ideal T_1 contrast agent should have a large paramagnetic property (large r_1) but negligible magnetic anisotropy (small r_2) [66]. Regarding this requirement, certain T_2 contrast agents with an unusually low r_2 can be used for T_1 imaging, as shown by Tromsdorf and co-workers [67] and Kim and co-workers [68]. Tromsdorf et al. prepared 4 nm iron oxide nanoparticles by the well-known thermal decomposition method and then modified them with PEG molecules of different lengths. It was demonstrated that increasing the length of the PEG molecule can reduce the r_2 of the nanoparticles, leaving the r_1 nearly unaffected. This effect seems to reach a limit with PEG molecules of around 1000 Da as using PEG molecules of 2000 Da gave a rise in the r_2.

In the other example, Kim et al. used a modified thermal decomposition methodology to obtain highly monodispersed maghemite (Fe_2O_3) nanoparticles with a small size (below 4 nm). They used iron oleate as the starting material in a mixture with oleic acid and oleyl alcohol in diphenyl ether as the high boiling point solvent. The mixture was heated to 250°C for 30 minutes under an inert atmosphere to form the final nanoparticles. Varying the stoichiometry of the iron precursor, oleic acid, and oleyl alcohol allowed the authors to obtain iron oxide nanoparticles of different sizes, between 1.5 and 3.7 nm. The addition of oleyl alcohol allowed a reduction in reaction temperature by reducing the iron oleate complex, resulting in the production of small nanoparticles.

A different example features the use of FeCo nanoparticles as T_1 contrast agents (Figure 9.5) [69]. Seo et al. reported the synthesis and application of these magnetic nanoparticles as T_1 contrast agents. FeCo nanoparticles stabilised with a layer of carbon were reported to present both high r_1 and r_2 values, with the r_1 value being 15 times higher than that of Magnevist (commercial T_1 contrast agent used in clinics), while the r_2 was six-fold higher than that of Ferridex (commercial T_2 contrast agent also used in clinics). The main drawback of this work is the complex methodology used for the preparation of the starting nanoparticles. Fumed silica was impregnated with iron(III) nitrate ($Fe(NO_3)_3$) and cobalt(II) nitrate ($Co(NO_3)_3$) methanolic solutions and sonicated for an hour. It was then dried at 80°C and ground to a powder before subjecting it to methane chemical vapour deposition (CVD) at 800°C under H_2 flow and followed by a methane flow. After cooling, the sample was etched with HF to dissolve the silica and collect the nanoparticles.

9.4 T_1-T_2 DUAL MRI CONTRAST AGENTS

T_1 and T_2 MRI modalities have traditionally been separate techniques sharing the same instrumentation, and contrast agents have been specifically designed either for one or the other modality. This trend began to change several years ago with the application of classical T_2 contrast agents as T_1 agents (see previous section), and even more recently, preparation of dual T_1-T_2 contrast agents has been reported. The justification of this development concerns the extra accuracy that the combination of two imaging techniques can supply without the inconvenience of complex image reconstruction needed in the case of other dual applications like MRI-PET, -SPECT, or -CT. To date, all the examples of dual T_1-T_2 probes feature a central magnetic nanoparticle. Around this magnetic core a T_1 contrast agent is built, either in the form of an inorganic paramagnetic shell, a doped inorganic shell, or an organic shell formed by paramagnetic chelates.

In 2010, Choi et al. reported the preparation of core@shell nanoparticles comprising a manganese ferrite core, a separating SiO_2 layer, and an outer paramagnetic $Gd_2O(CO_3)_2$ shell [70]. First, $MnFe_2O_4$ nanoparticles were prepared following a thermal decomposition methodology from iron(III) acetylacetonate and manganese(II) chloride at 300°C for 1 hour. Then a silica layer was built around the magnetic core using a reverse microemulsion system, and to prepare the final $MnFe_2O_4@SiO_2@Gd_2O(CO_3)_2$ nanoparticles, the purified product from the previous reaction was dispersed in diethylene glycol and heated to 80°C in the presence of $Gd(NO_3)_3$ and $(NH_2)_2CO$ (Figure 9.6). The authors show that by controlling the thickness of the intermediate separating SiO_2 layer, both r_1 and r_2 can be tuned. The nanoparticles' r_1 values increase with the thickness of the silica layer, reaching a maximum at around 16 nm of silica shell. On the other hand, the r_2 value behaves in an opposite way, that is, decreasing with increasing thickness. Therefore, a compromise has to be reached to allow dual T_1-T_2 imaging.

A different strategy was followed by Yang and co-workers [71]. They prepared magnetite nanoparticles following the thermal decomposition protocol described by Sun et al. [24] (see previous sections). Then, using the same protocol as in the previous example, they coated the nanoparticles with a silica shell. APTES ((3-aminopropyl)triethoxysilane) was then added to functionalise the surface of the resulting particles with amine groups. These amine groups were used to couple diethylenetriamine pertaacetic acid (DTPA) moieties, which were incubated with $Gd(NO_3)_3$ in aqueous solution to incorporate the paramagnetic ion. Here, the relaxivity properties of the nanoparticles were optimised: controlling the ratio Fe/

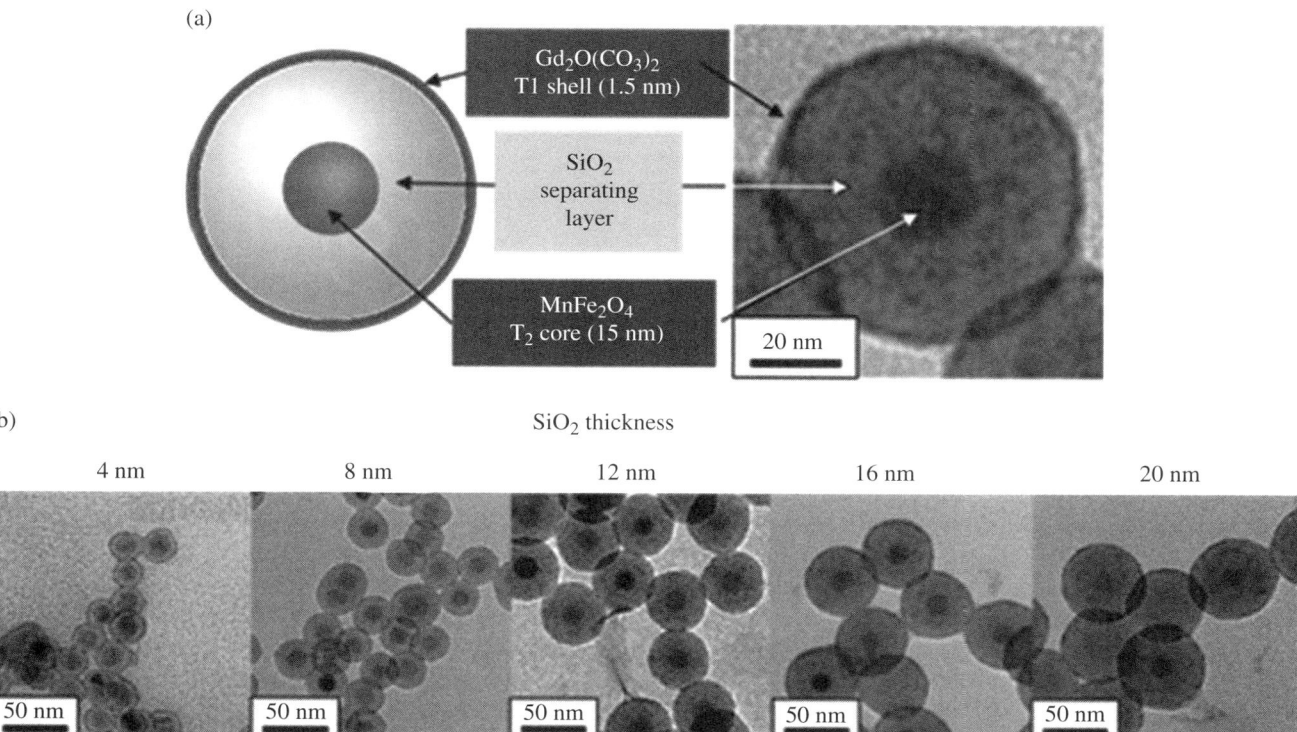

FIGURE 9.6 (a) Schematic and transmission electron microscope (TEM) image of core-shell type $MnFe_2O_4@SiO_2@Gd_2O(CO_3)_2$ nanoparticles. (b) TEM images of the same nanoparticles with variable separating layer thickness (4, 8, 12, 16, and 20 nm), having a fixed $MnFe_2O_4$ core (15 nm in diameter) and a $Gd_2O(CO_3)_2$ shell (1.5 nm). Reproduced with permission from Ref. [70].

Gd by means of the number of DTPA molecules attached per nanoparticle. In a final example, Huang and co-workers followed a similar strategy except that gadolinium chelates were coupled to the channels of a mesoporous silica shell [72]. 22 nm Fe_3O_4 nanoparticles were prepared from $Fe(acac)_3$, oleic acid, and trioctylamine, following a thermal decomposition route. The hydrophobic ligands of these nanoparticles were exchanged for more hydrophilic ones (CTAB), and the nanoparticles were encapsulated inside mesoporous silica shells in basic water solution using TEOS as the silicon source. After purification, the particles were heated to reflux in ethanol, and APTES was added to yield amine functionalised nanoparticles. DOTA-NHS was subsequently coupled to the amine groups, and the resulting product was incubated with $GdCl_3$ to yield the final dual MRI probe. In this case, the effect of the T_1 probe over the T_2 one or vice versa was not further investigated.

9.5 WATER SOLUBILISATION

There are now a number of reliable and well-characterised protocols for the preparation of magnetic nanoparticles to suit their application or design. The real challenge comes in the functionalisation of the nanoparticles to achieve water soluble, biocompatible, and specific probes. For years the *in vivo* performance of nanoparticulate MRI contrast agents relied on an intrinsic difference in uptake between different organs or tissues. This is the case for iron oxide nanoparticle-based contrast agents currently approved in the clinic. The nanoparticles are cleared away from the blood stream by the liver, thus they are used for the diagnosis of hepatic diseases. This trend has been moving progressively toward a specific labelling design using targeting molecules [73], although the strategy has not, as yet, provided good enough results to be applied in the clinic.

As mentioned previously, some of the preparation protocols, for example, co-precipitation ones, are flexible enough to allow the *in situ* introduction of different ligands. In these cases, water-soluble nanoparticles ready for further functionalisation can be obtained in one reaction. However, most preparations do not give the option of ligand selection because molecules with highly specific properties are required. In these other cases, the first mandatory step toward the specific functionalisation of the nanoparticles is the replacement or modification of the ligands (Figure 9.7). There have been a number of different strategies proposed [34], with the most popular one being ligand exchange.

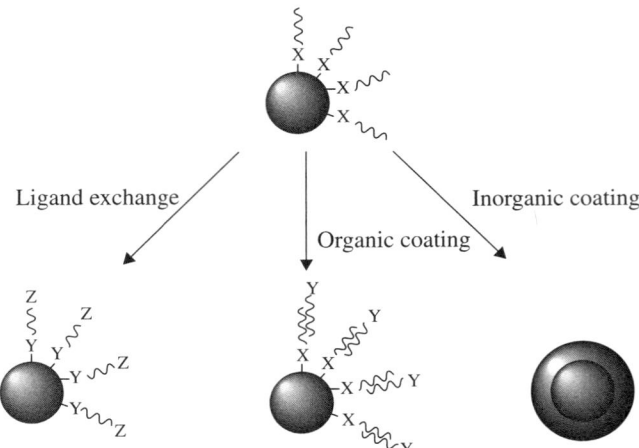

FIGURE 9.7 Schematic representation of the different strategies available to form water soluble and functionalisable nanoparticles. The final ligands on the nanoparticles can be either individual molecules or polymers.

9.5.1 Ligand Exchange

In this process, the ligands on the nanoparticles are replaced by more suitable ones, usually bifunctional species. This can be performed in one step, just mixing the nanoparticles with a large excess of the new ligands, or in two steps, stripping first the nanoparticles and then introducing the new ligand. With metal oxide nanoparticles, molecules bearing carboxylic acids are the preferred ligands [25, 50, 54], but alternatives include diols (usually dopamine-based molecules) [74–77], molecules with phosphate/bisphosphonates groups [78, 79], or hydroxamic acids [80]. Phosphonate groups have been introduced more recently than carboxylic acids, and their bonds to oxide nanoparticles are more stable even at high temperatures [81]. The way in which the ligands interact with the metallic atoms on the surface of the nanoparticles depends on (i) the chemical nature of the ligand itself, (ii) their ability to pack and form self-assembled monolayers (SAMs) on the surface of the nanoparticles, (iii) the nature of the nanoparticle, and (iv) the curvature of the nanoparticle's surface. For the case of carboxylic ligands, three different interaction modes have been proposed [82]. In the monodentate situation, only one oxygen atom binds to the surface. When both oxygen atoms are involved in the bonding, they can either be attached to the same metallic ion (chelating situation) or to different atoms (bridged situation). A similar situation has been proposed for phosphates; one or two oxygen atoms from the phosphate group can interact with Fe^{3+} ions on the surface of the nanoparticle [83]. The case of FePt nanoparticles is different because the presence of platinum atoms on the surface of the nanoparticle allows the use of thiol-noble metal chemistry for the ligand exchange [28, 29]. This chemistry has been widely studied for the functionalisation of gold nanoparticles [84] and the sulphur-metal bond is more stable than the ones described above, making the ligand exchange process easier. In general, this strategy can provide individual and monodisperse nanoparticles without a dramatic increase in the size of the particles, but due to the nature of the bond between the ligands and the inorganic core, some stability issues can arise.

A similar approach toward the preparation of water soluble nanoparticles is the coating of the nanoparticles with a polymer. In this case, the stability of the resulting nanoparticles is generally higher, but the hydrodynamic size of the nanoparticles is drastically increased, and samples in which several nanoparticles are encapsulated inside the same polymer chain are often found. Bifunctional or natural biocompatible and biodegradable polymers can be used; among the natural polymers (usually carbohydrate-based materials) dextran [17] is the most preferred. Dextran is a non-charged polymer of D-glucose that can be easily modified (even once on the nanoparticles [85]) to present carboxylic or amino groups on its surface. As mentioned in previous sections, dextran can be introduced directly into the nanoparticle formation reaction, or it can be used after their formation to provide water solubility and biocompatibility. When considering synthetic polymers, many different options have been explored (Figure 9.8), from polystyrene sulfonate (PSS) [55], to the biodegradable poly(lactic-*co*-glycolic acid) (PLGA) [86], to polyethyleneimine (PEI) [87], to more complex approaches such as layer-by-layer protocols [88]. However, by some way the most used synthetic polymer is polyethylene glycol (PEG) due to its outstanding properties for biological applications. PEG is a linear, biocompatible polymer whose length can be tailored to suit particular applications. It has been approved for human applications in pharmacology [89] and has been shown to minimise unspecific interactions, helping these nanoparticles to evade the mononuclear phagocyte system (MPS) [90]. It is not possible to use simple PEG molecules for a ligand exchange reaction because the bond between the hydroxyl terminal group of the PEG and the nanoparticle is not strong enough, but PEG molecules can be chemically functionalised to present different

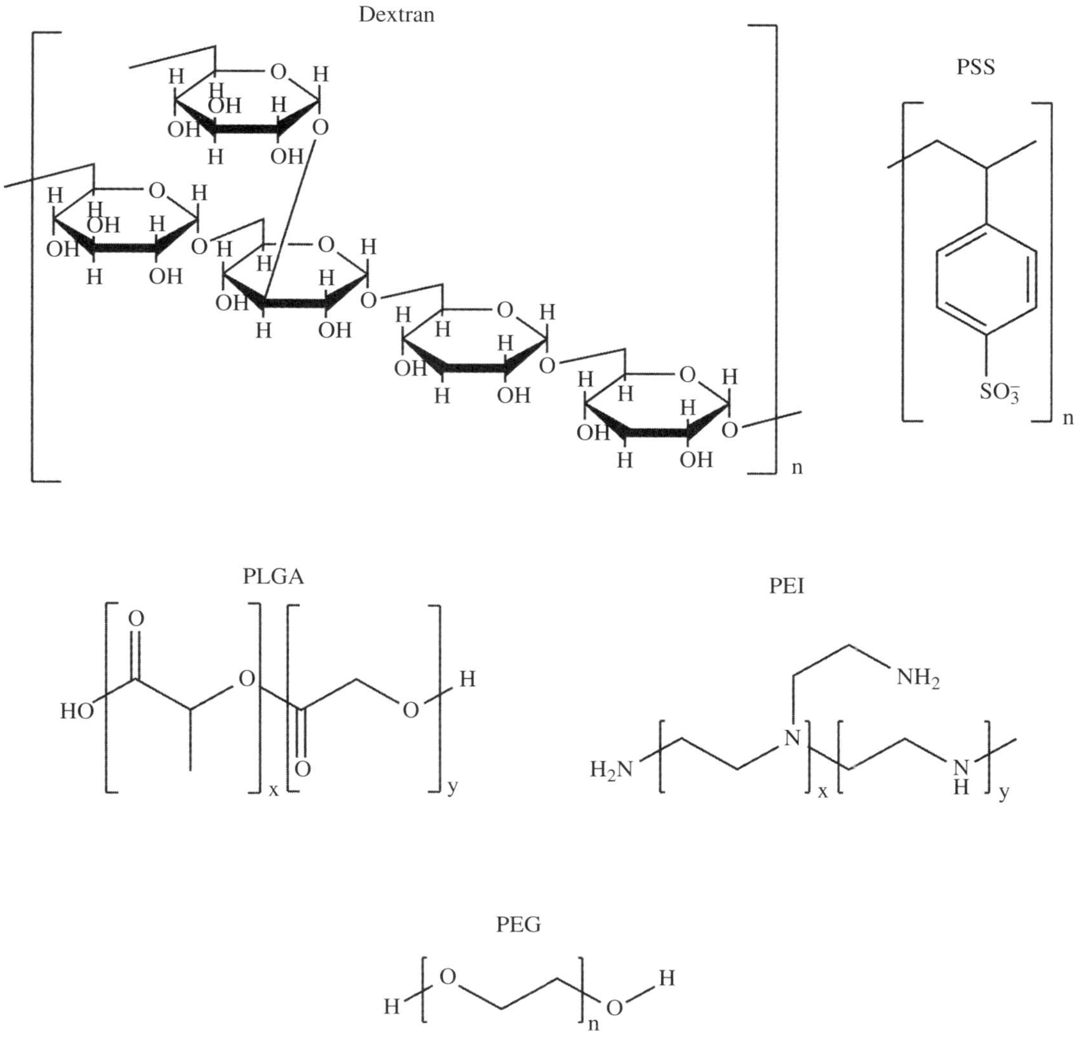

FIGURE 9.8 Structure of some of the most common polymers used for the stabilisation of nanoparticulate MRI contrast agents in aqueous media. PSS: polystyrene sulfonate, PLGA: poly(lactic-*co*-glycolic acid), PEI: polyethyleneimine, PEG: polyethylene glycol.

terminal groups. They can also, after terminal modification, be coupled to other molecules such as polymers. There are now examples in the literature of water-soluble nanoparticles stabilised by phosphate-PEG molecules [67], silane-PEGs [48], polyethyleneimine-PEGs [91], phosphine oxide-PEG [68], poly(α, βaspartic acid)-PEG [92], and gallol-PEG [93].

9.5.2 Organic Coatings: Hydrophobic Interactions

A different way of stabilising the nanoparticles in water is to maintain the original organic layer on the surface and use it to attach a second organic layer by means of hydrophobic interactions. A similar strategy to this is the encapsulation of the nanoparticles inside liposomes or micelles, which are stabilised by the same kind of hydrophobic forces. In this case the interaction between the outer organic layer and the nanoparticle is weaker and highly dependent on the environment. Results obtained with this kind of nanoparticle have to be carefully analysed, and control experiments should be thoroughly designed because some reports have demonstrated the low stability of this approach *in vivo* [94]. However, this approach presents the advantage that the coating can be used to simultaneously solubilise the nanoparticles and to encapsulate pharmaceuticals for possible drug delivery applications. As demonstrated in the previous section, individual molecules [91, 95–98] or polymers [91, 99–101] can be used. All of them are amphiphilic molecules with a hydrophobic part (usually an aliphatic chain) plus a more hydrophilic head that facilitates water solubility as well as a starting point for further functionalisation. This approach involving hydrophobic interactions has not only been used with metal oxides nanomaterials. For example, Seo et al. used phospholipids to solubilise FeCo nanoparticles coated with a layer of graphite [69]. The hydrophobic fatty acid chains of the phospholipids interact with the carbon layer on the nanoparticles while the polar charged head provides solubility in aqueous media and can be modified, for example, with PEG molecules to reduce nonspecific interactions. In a different example, Sitharaman and co-workers used surfactant molecules to solubilise gadolinium loaded carbon nanotubes (CNTs) [102]. Either sodium dodecylsulfate (SDS) or the more biocompatible surfactant pluronic F98 is used to keep their CNTs in aqueous media for their measurements.

9.5.3 Inorganic Coatings

The coating of prepared nanoparticles with a shell of a different inorganic material can also facilitate water solubilisation. This method presents several advantages: First, the coating material is chosen to facilitate the chemical reactions needed for the functionalisation of the particles, but at the same time it can protect the active core from oxidation or degradation (further oxidation in the case of metal oxide nanoparticles). If the shell is porous, drugs can be encapsulated to facilitate the nanoparticles' use as drug delivery agents as well as imaging probes. The addition of an outer layer of inorganic material increases the total mass of the nanoparticles with redundant (from an imaging standpoint) material, but the overall size increase facilitates the clearance of the particles *in vivo* by the reticulo-endothelial system.

Currently, there are three main options for the inorganic coating of MRI nanoprobes: gold, silica, or carbon. When considering *in vivo* applications, silica is by far the most commonly used of the three. Although gold presents unbeatable properties of stability and chemical versatility, its molecular weight is very high, so it dramatically increases the average molecular weight of the nanoparticles. In contrast, carbon layers add a negligible amount of mass to the final nanoparticles but feature complex preparation protocols.

(i) SiO_2 shells: The coating of magnetic nanoparticles with silica shells is a popular option for a number of reasons: Silica is chemically inert and exceptionally stable, (especially in aqueous media), there are several well-established and simple protocols for its preparation, and its porosity can be controlled [3]. From the range of synthetic methodologies available, the so-called Stöber method [103] and the microemulsion approach [104, 105] are the most popular ones. Both methods are based on the polymerisation of tetraethyl orthosilicate (TEOS) under basic conditions. While the Stöber approach uses polar solvents (alcohols or water-based solutions), microemulsion involves a water-in-oil reverse emulsion. Thus, in general the second method is more useful because the starting nanoparticles are usually not soluble in polar solvents. The addition of a silica shell generally comprises two steps: the polymerisation of TEOS, followed by addition of an outer thin shell of a different silane bearing a functional group for further functionalisation. The most common example of these silanes is 3-aminopropyl triethoxysilane (APTES) that, after application, renders nanoparticles with terminal amine groups. There has been an increasing number of examples reported in recent years, not only with iron oxide nanoparticles [70–72, 106–108], (Figure 9.9a) but also with other materials, such as gadolinium oxide [49, 109].

(ii) Au shells: The application of gold-coated nanoparticles in the biomedical field is limited because of the high atomic weight of gold atoms. Apart from that, the properties of gold make it ideal for *in vivo* applications. It is chemically inert, extremely stable, and most importantly, easy to functionalise. Thiol-gold chemistry has been extensively studied for the functionalisation of gold surfaces through self-assembled monolayers (SAMs) and can be applied in this case. Several synthetic protocols are available for the nanoparticle coating with gold. The thermal decomposition of gold acetate over iron oxide nanoparticles immediately after their preparation and without further purification [110] is a convenient approach that has been further modified to also be successful with manganese and cobalt ferrites nanoparticles [111]. The thickness of the gold coating can be tuned by controlling the amount of the starting gold precursor. A different strategy is the one described by Xu and co-workers [112] in which they coat Fe_3O_4 nanoparticles with a gold shell in chloroform at room temperature using hydrogen tetrachloroaurate hydrate ($HAuCl_4$) as the gold source and oleylamine as a mild reducing agent as well as a surfactant. In contrast, Lyon et al. [113] brought their magnetic nanoparticles to water using citric acid and then coated them with a gold shell by the reduction of Au^{3+} ions via iterative hydroxylamine seeding. A more complex strategy was followed by Bouchard and co-workers to prepare what they called gold-coated cobalt nanowontons (Figure 9.9b) [114]. Over silicon pillars they deposited sequentially four metallic layers: 5 nm chromium, 10 nm gold, 10 nm cobalt, and finally 10 nm gold again. The silicon and the chromium layers were etched away by immersion in 10% KOH solution at 80°C for 10 min. The result was a batch of gold-coated nanoparticles with a size of around 60 nm featuring promising T_2 contrast properties.

(iii) Carbon shells: The addition of a carbon layer around the nanoparticle core is the less well explored option to date due to the complexity of its preparation. However, graphitic carbon coating presents some outstanding properties, such as increased chemical stability against acids and bases and reduced mass. So far, there is only one report on its

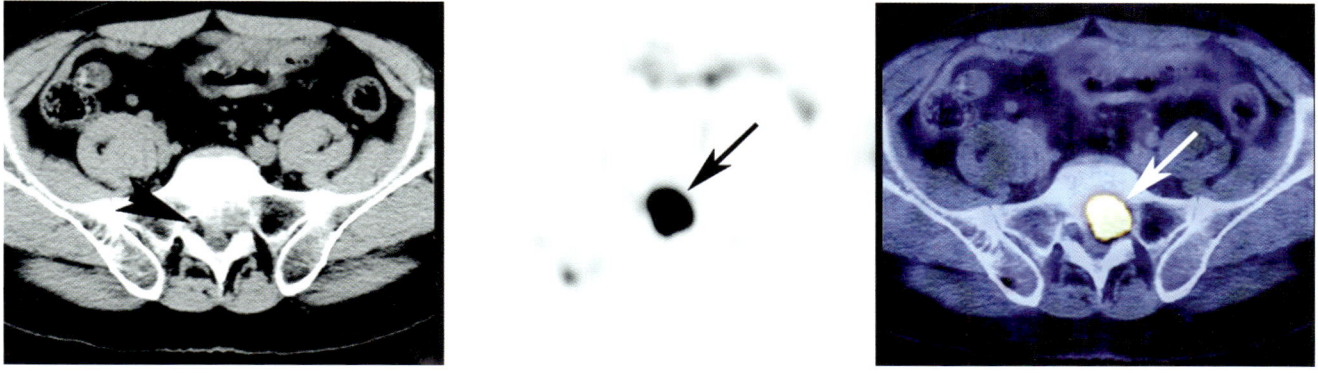

FIGURE 1.13   Left: Image from a CT scan; Middle: Image from a PET scan; Right: Image from a CT-PET scan [23].

(a)   (b)

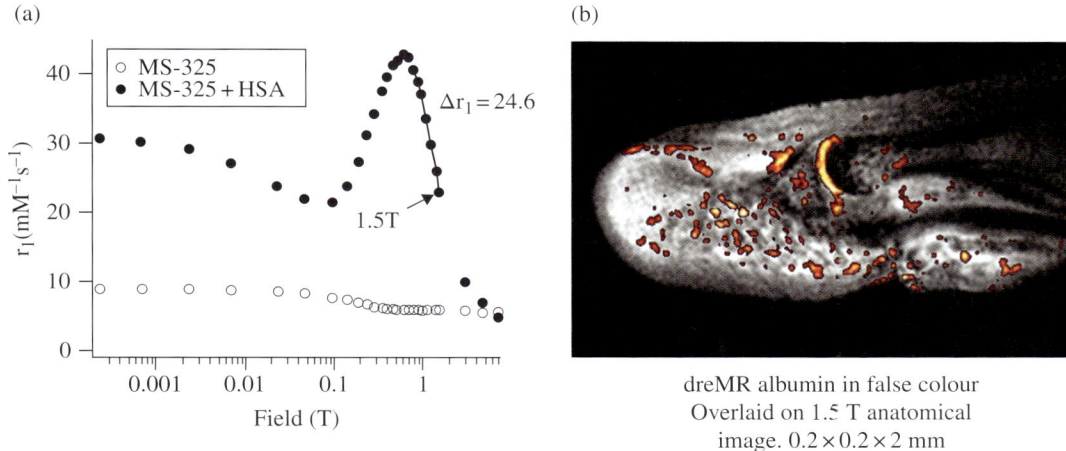

dreMR albumin in false colour
Overlaid on 1.5 T anatomical
image. $0.2 \times 0.2 \times 2$ mm

FIGURE 8.7   (a) The variation in r_1 for MS-325 in albumin free and albumin bound forms with magnetic field. (b) dreMR albumin false colour image overlaid on 1.5 T anatomical image of a finger fracture. Figure drawn from data kindly provided by Prof. Peter Caravan.

(a)   (b)

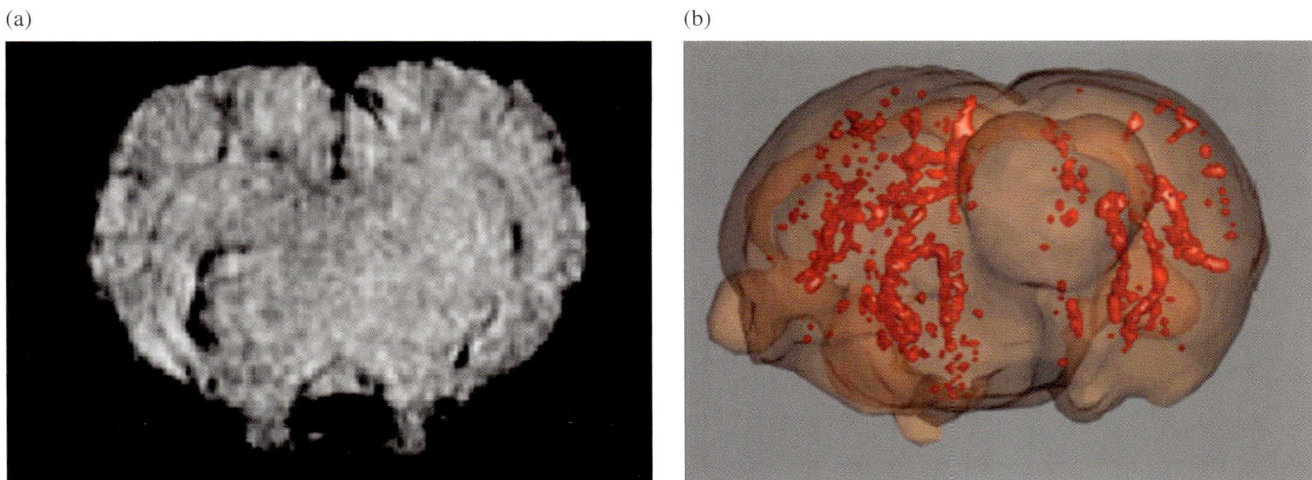

FIGURE 9.12   (a) Selected image taken from T_2^* weighted 3D dataset of rat brain. (b) 3D reconstruction of the accumulation of contrast agent. The images reveal that magnetic nanoparticles functionalised with sLe^x enable detection of lesions in models of stroke. Reproduced with permission from Ref. [154].

The Chemistry of Molecular Imaging, First Edition. Edited by Nicholas Long and Wing-Tak Wong.
© 2015 John Wiley & Sons, Inc. Published 2015 by John Wiley & Sons, Inc.

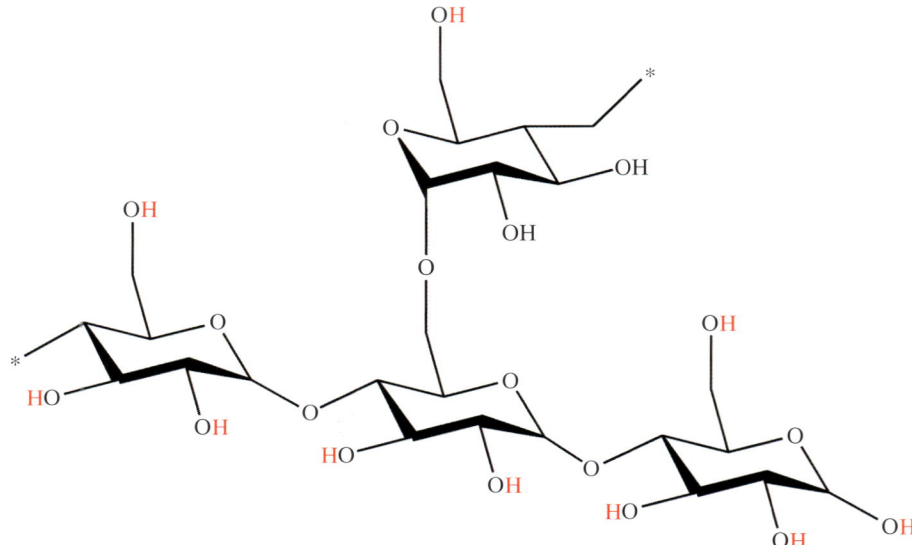

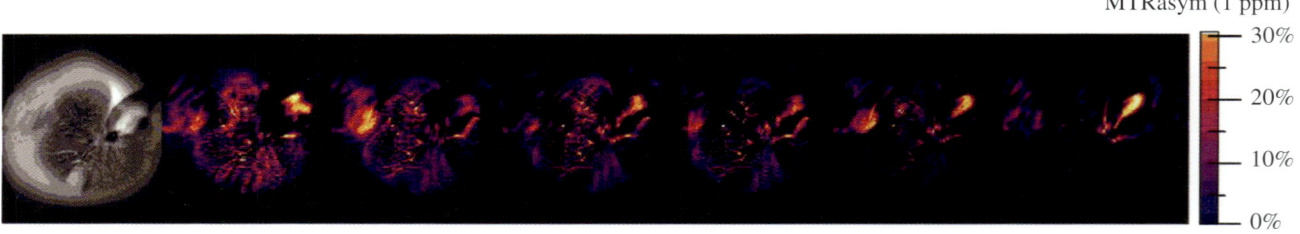

FIGURE 10.4   Top: Chemical structure of a fragment of glycogen showing exchangeable –OH protons. Bottom: GlycoCEST imaging of a perfused fed-mouse liver at 4.7 T and 37 °C. The colourised images show the gradual reduction in the CEST signal originating from glycogen as a result of glucagon stimulation of glycogenolysis. Reproduced with permission from Ref. [26].

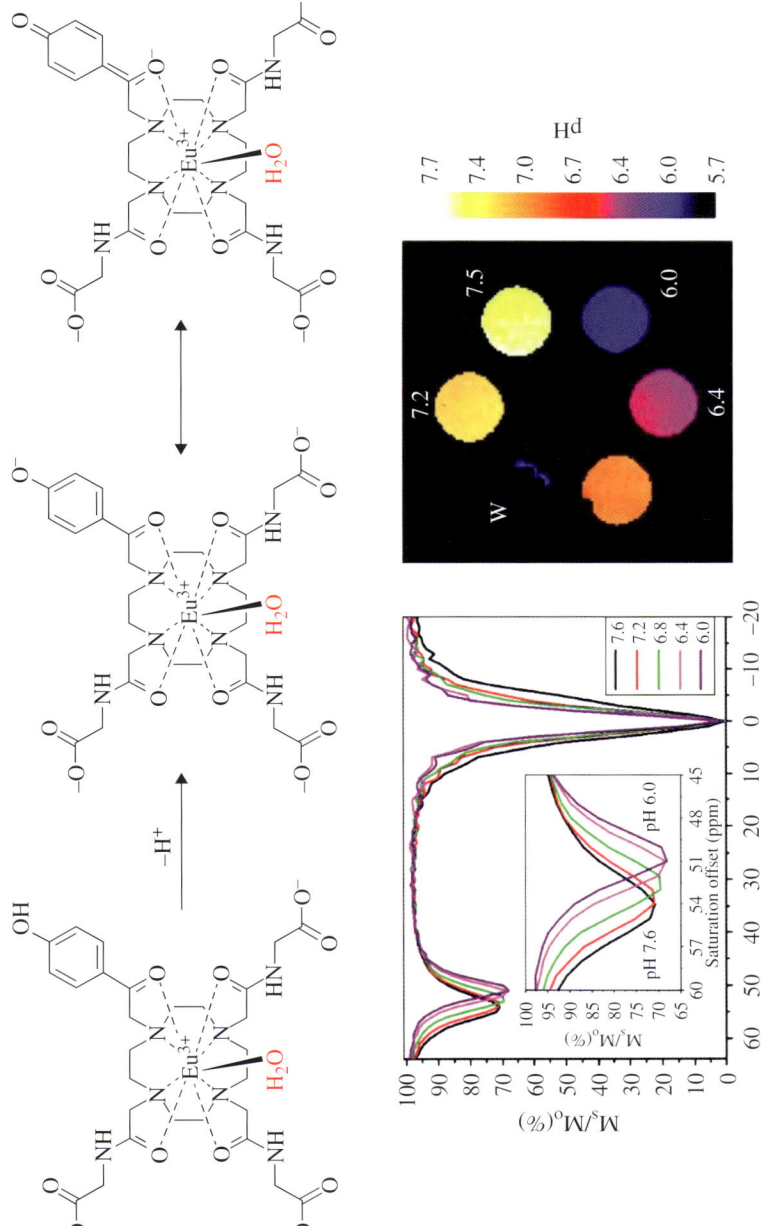

FIGURE 10.8 Top: Change in chemical structure of pH-responsive PARACEST agent as pH is increased. Bottom: pH dependence of CEST spectra showing changes in chemical shift (left), and CEST pH images of a phantom containing water (w) and PARACEST agent after activation at 54 ppm and 47 ppm (right). Adapted with permission from Ref. [41].

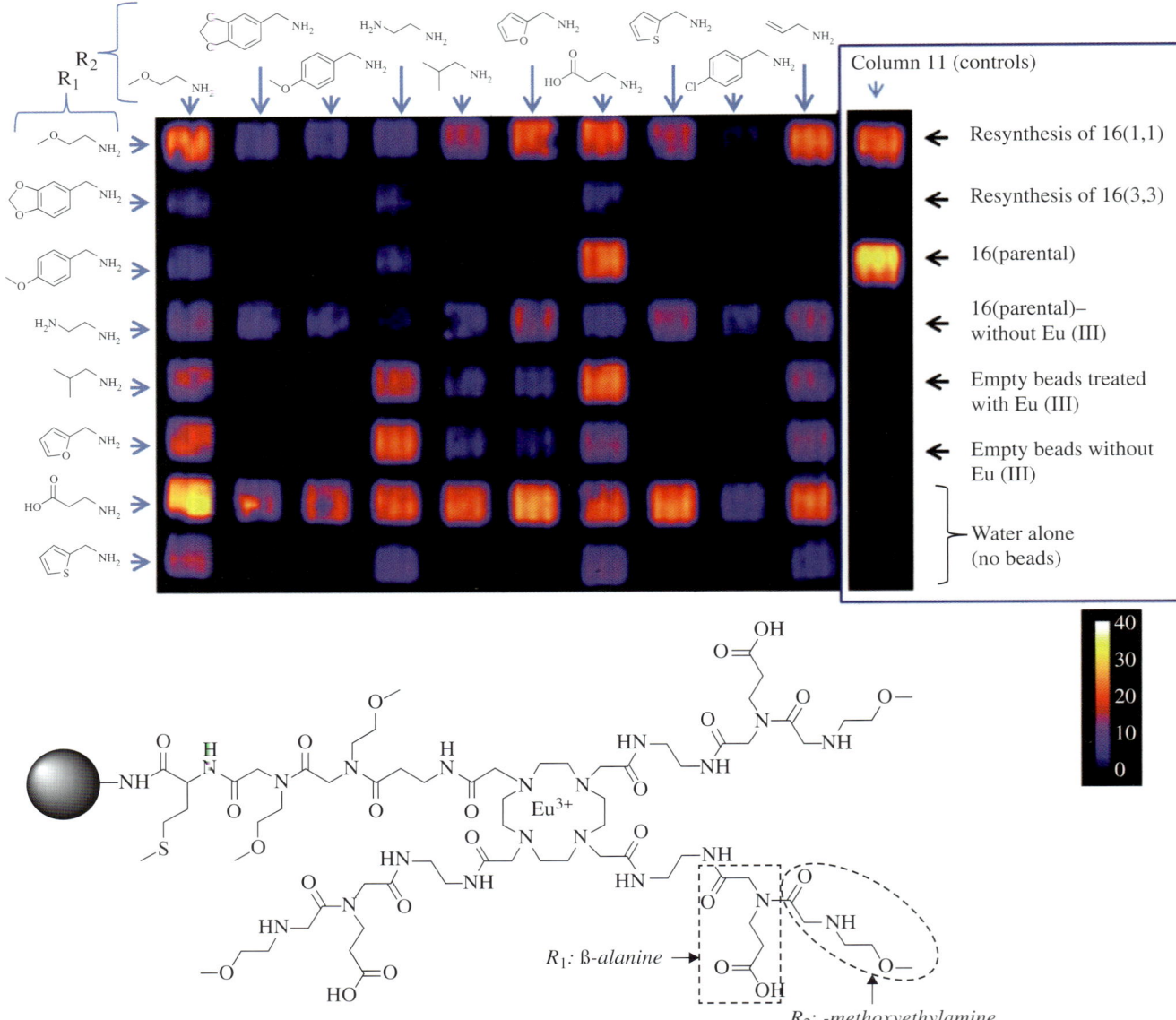

FIGURE 10.10 Top: CEST image of an 80-compound library of Eu(III)DOTA-tetraamide-peptoid derivatives attached to beads. The chemical structures of each amine used to build the library are shown along each horizontal row (R_1) and each vertical column (R_2). Bottom: Structure of Eu(III)DOTA-tetraamide-peptoid in well (7, 1) that showed the highest CEST intensity. Reproduced with permission from Ref. [35].

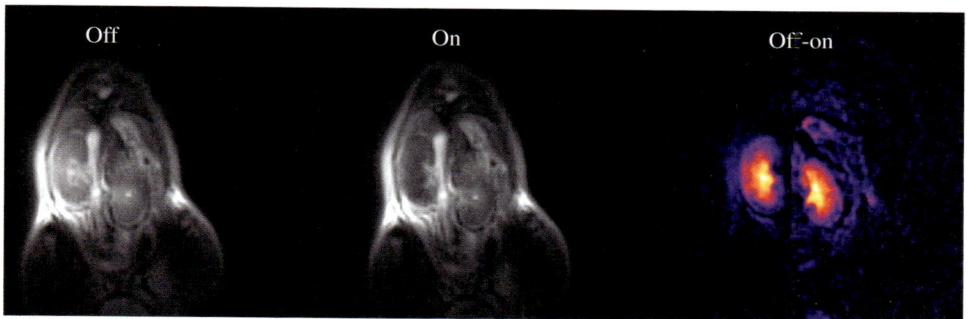

FIGURE 10.12 *In vivo* CEST images of mouse kidney after intravenous injection of 1 mmol/kg EuDOTA-(gly)$_4^-$ using the SWIFT pulse sequence. Adapted with permission from Ref. [64].

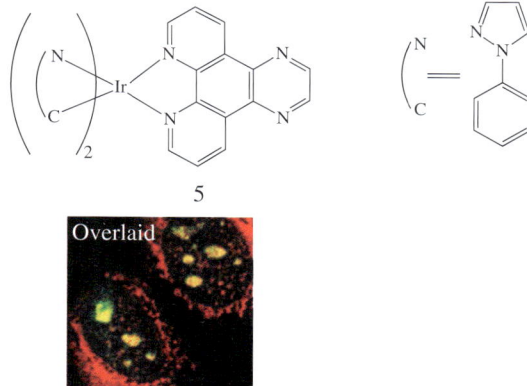

FIGURE 12.5 Nucleolar-targeting iridium complex **5** and cell image showing nucleolar localisation (Reproduced with permission from Ref. [10]).

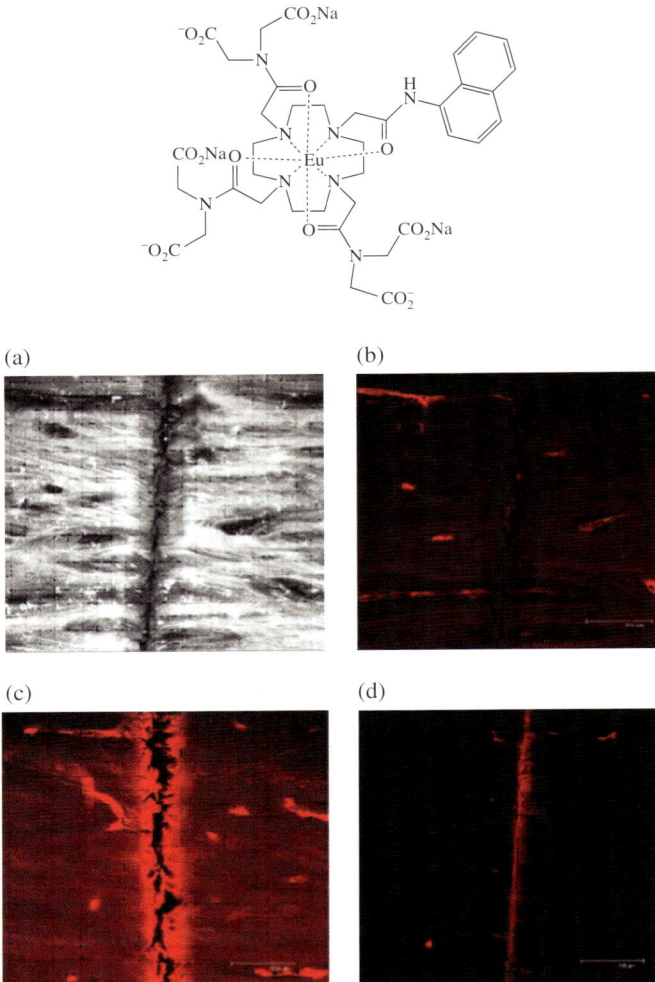

FIGURE 12.16 A polycarboxylate terminated Eu^{III} complex (left) and microscopy images of bone sample immersed in 10^{-3} M solution of the complex. (a) Reflected light image: 0 h (b) control (c) 4 h (d) 24 h.

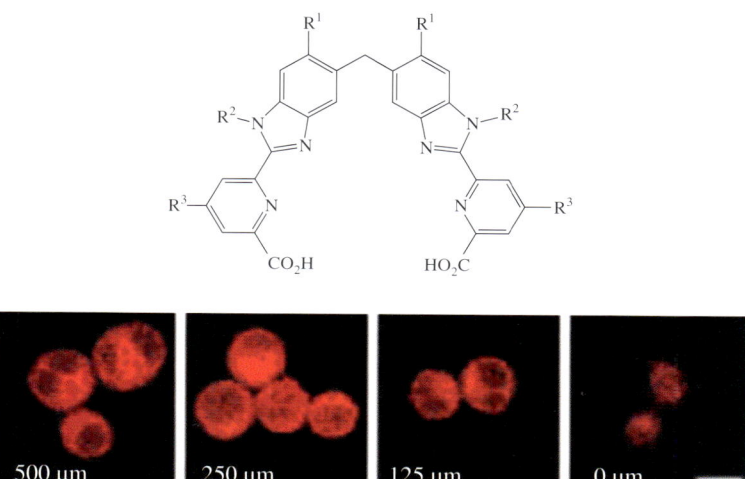

FIGURE 12.26 Ligand utilised for the bimetallic helicate [Eu$_2$L$_3$] (R^1=H, R^2=Me, R^3=PEG chain). Cells were incubated in the presence of different concentrations of the helicate in RPMI-1640 for 24 h. The images were taken using a Zeiss LSM 500 META confocal microscope (λ_{ex} 405 nm).

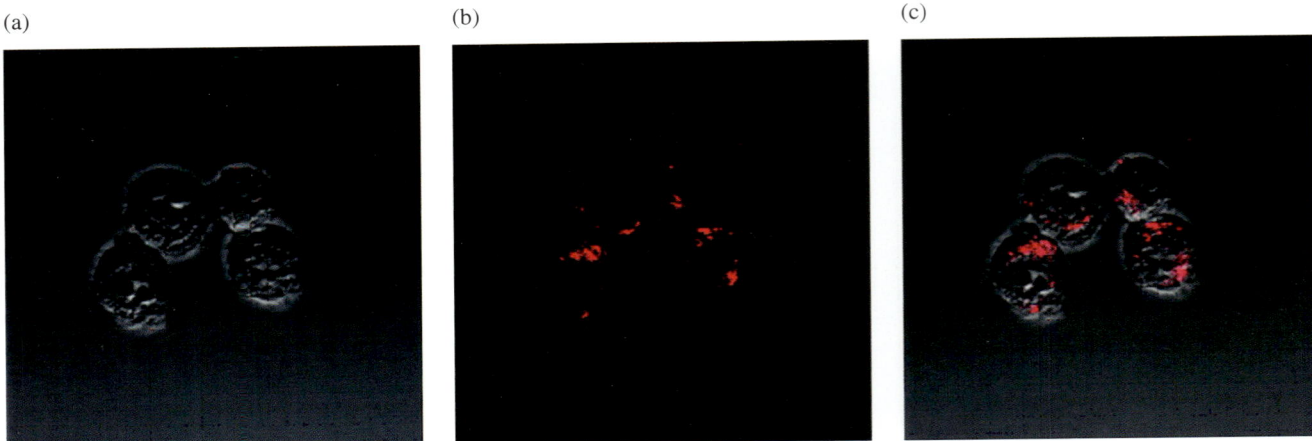

FIGURE 12.27 Two-photon microscopy images of HeLa cells incubated with $200\,\mu M$ of a bimetallic Eu^{III} helicate $[Eu_2(L)_3]$ in RPMI-1640 culture medium for 12 h at 37 °C, 5% CO_2: (a) bright field image; (b) luminescence ($\lambda_{ex} = 750\,nm$, $\lambda_{em} = 570 - 650\,nm$); (c) merged image.

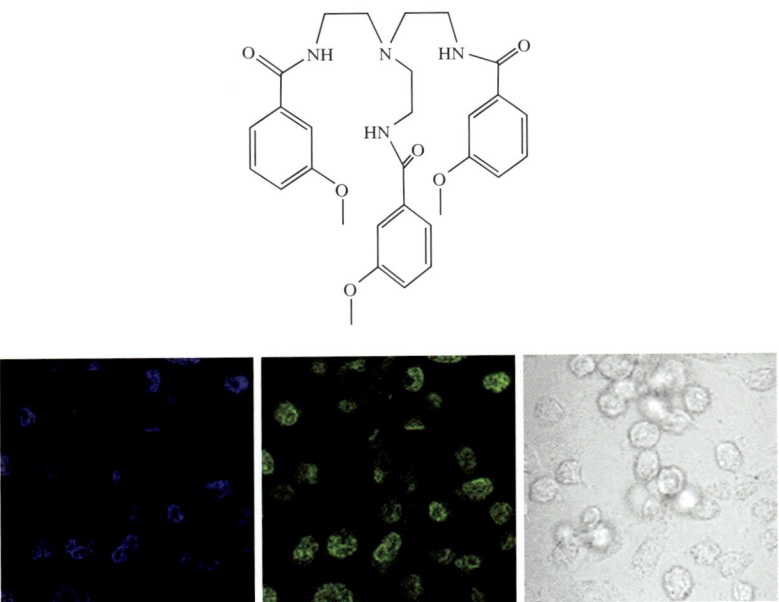

FIGURE 12.28 Top left: Free ligand. Bottom: Multi-photon confocal microscopy images of the terbium complexes. Hoechst 33342-labelled nuclei (blue) in the HONE1 cells with excitation at 800 nm, Filter BP = 400 − 480 nm (left); (middle) terbium complexes loaded into the cell for 1 hour (Figure 2c) with nuclei labelled, filter BP = 400 − 615 nm; (right) bright field image of the HONE1 cells.

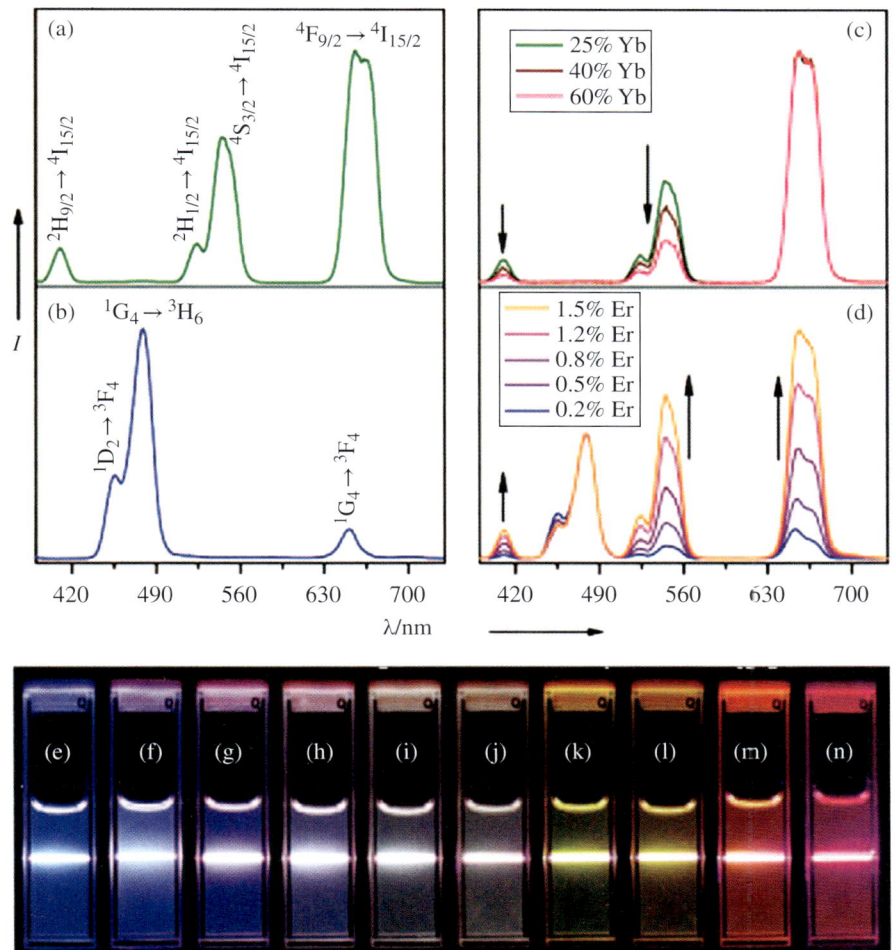

FIGURE 13.1 UCL emission spectra of (a) NaYF$_4$:Yb,Er (18/2 mol%), (b) NaYF$_4$:Yb,Tm (20/0.2 mol%), (c) NaYF$_4$:Yb,Er (25–60/2 mol%), and (d) NaYF$_4$:Yb,Tm/Er (20/0.2/0.2–1.5 mol%) particles in ethanol. (e)–(n) are compiled luminescent photos showing corresponding colloidal solutions of the samples shown in (a)–(c). Reprinted with permission from Ref. [12]. Copyright 2008 American Chemical Society.

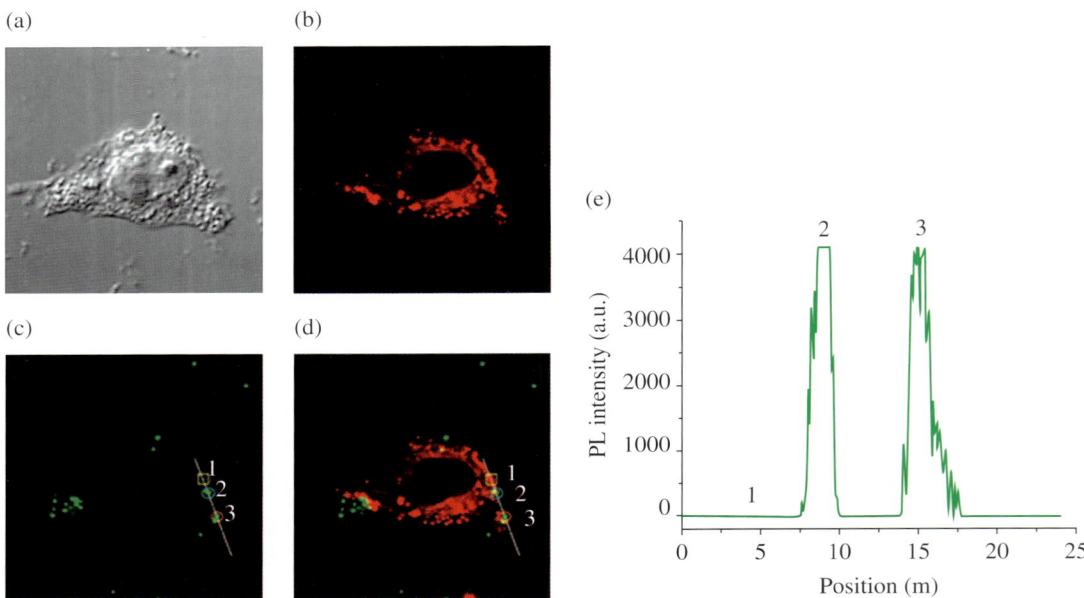

FIGURE 13.4 Bright field (a) and luminescent images (b–d) of a live HeLa cell dual-labelled with DiI and UCNPs. (b) Image of DiI when emission was collected at 560–600 nm (λ_{ex} = 543 nm). (c) UCL image of UCNPs when emission was collected at 500–600 nm (λ_{ex} = 980 nm). (d) The image of DiI and UCNPs when overlaid. (e) The detected upconversion signals along the line in (c) and (d). Reprinted with permission from Ref. [2]. Copyright 2009 American Chemical Society.

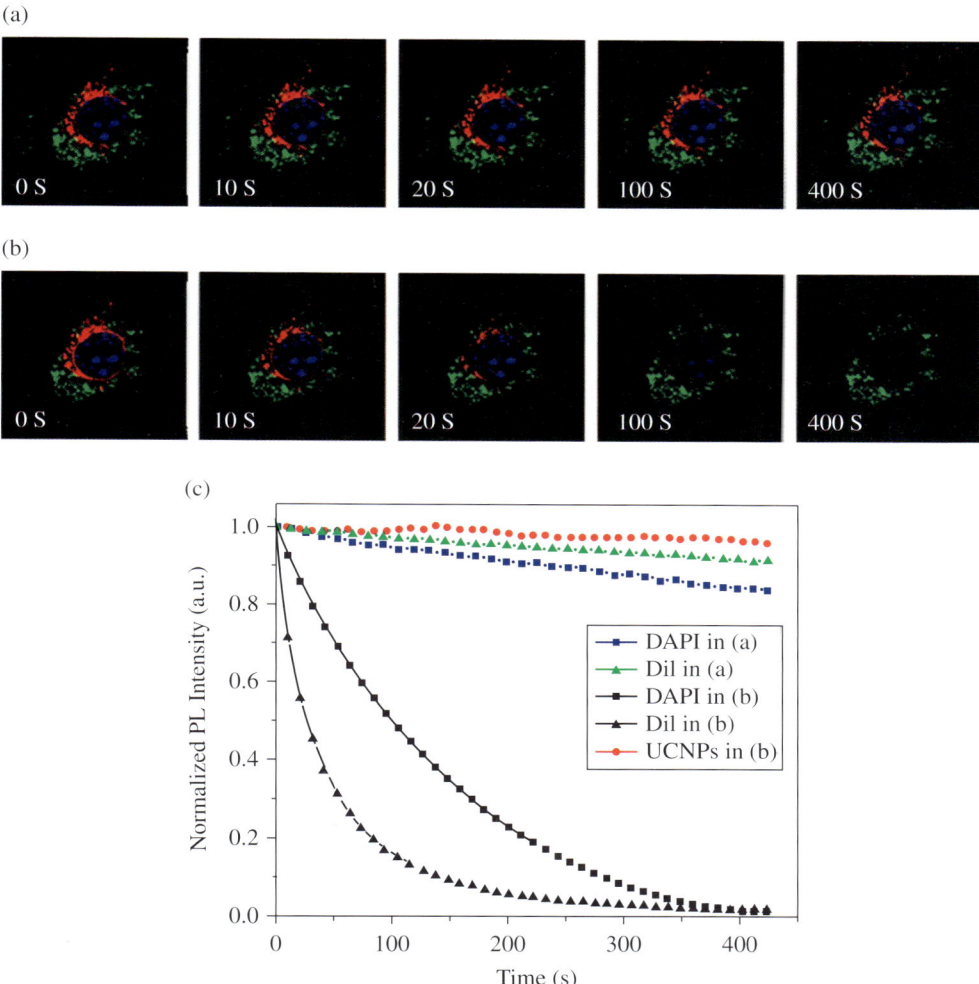

FIGURE 13.5 (a–b) Comparison of photobleaching of DAPI, DiI, and Ln-UCNPs in LSUCLM and conversional confocal microscopy imaging. The luminescence signals of DAPI, DiI, and Ln-UCNPs are shown in blue, red, and green, respectively. Simultaneous excitation was provided by CW lasers at 405, 543, and 980 nm with powers of approximately 19 mW, 15 μW, and 0.8 μW in (a) and 1.6, 0.13, and 19 mW in (b) in the focal plane, respectively. (c) Quantitative analysis of the signals in (a) and (b). Reprinted with permission from Ref. [2]. Copyright 2009 American Chemical Society.

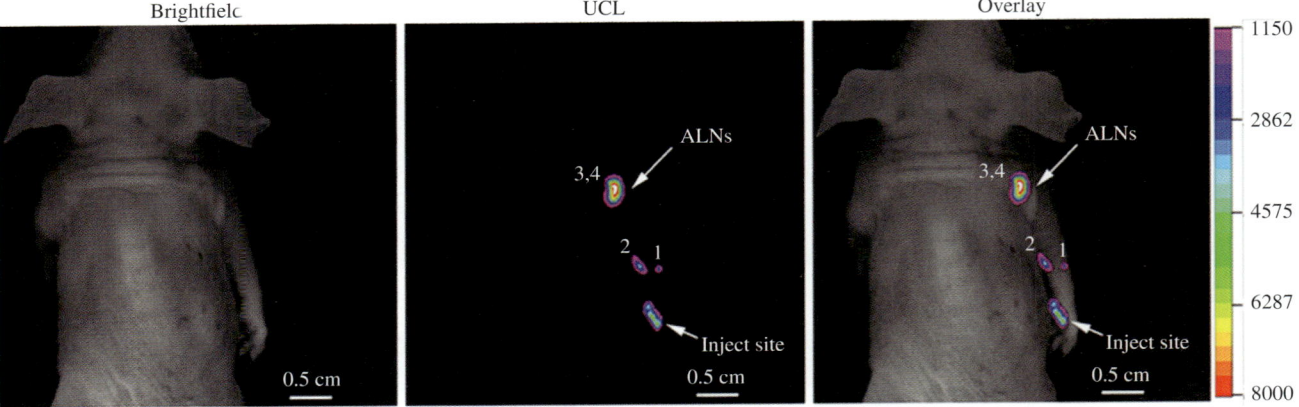

Brightfield	UCL	Overlay

FIGURE 13.7 *In vivo* lymphatic drainage upconversion imaging at 800 nm to show four different draining lymph basins (1, 2, 3, 4) along the right antebrachium of the nude mouse. Reprinted with permission from Ref. [77]. Copyright 2011 Elsevier Ltd.

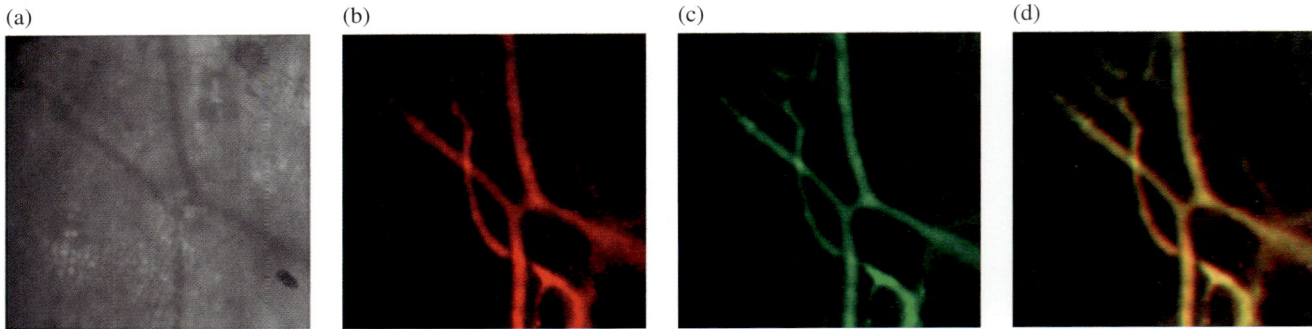

FIGURE 13.8 Optical imaging of blood vessels in the mouse ear following tail vein injection of the nanoparticles (10 mg); (a) blood vessels imaged with a blue light filter, (b) upconversion image with excitation at 980 nm, (c) fluorescence image of the carbocyanine dye with excitation at 737 nm, (d) merged image of the upconversion and fluorescence signals. Reprinted with permission from Ref. [78]. Copyright 2009 The Royal Society of Chemistry.

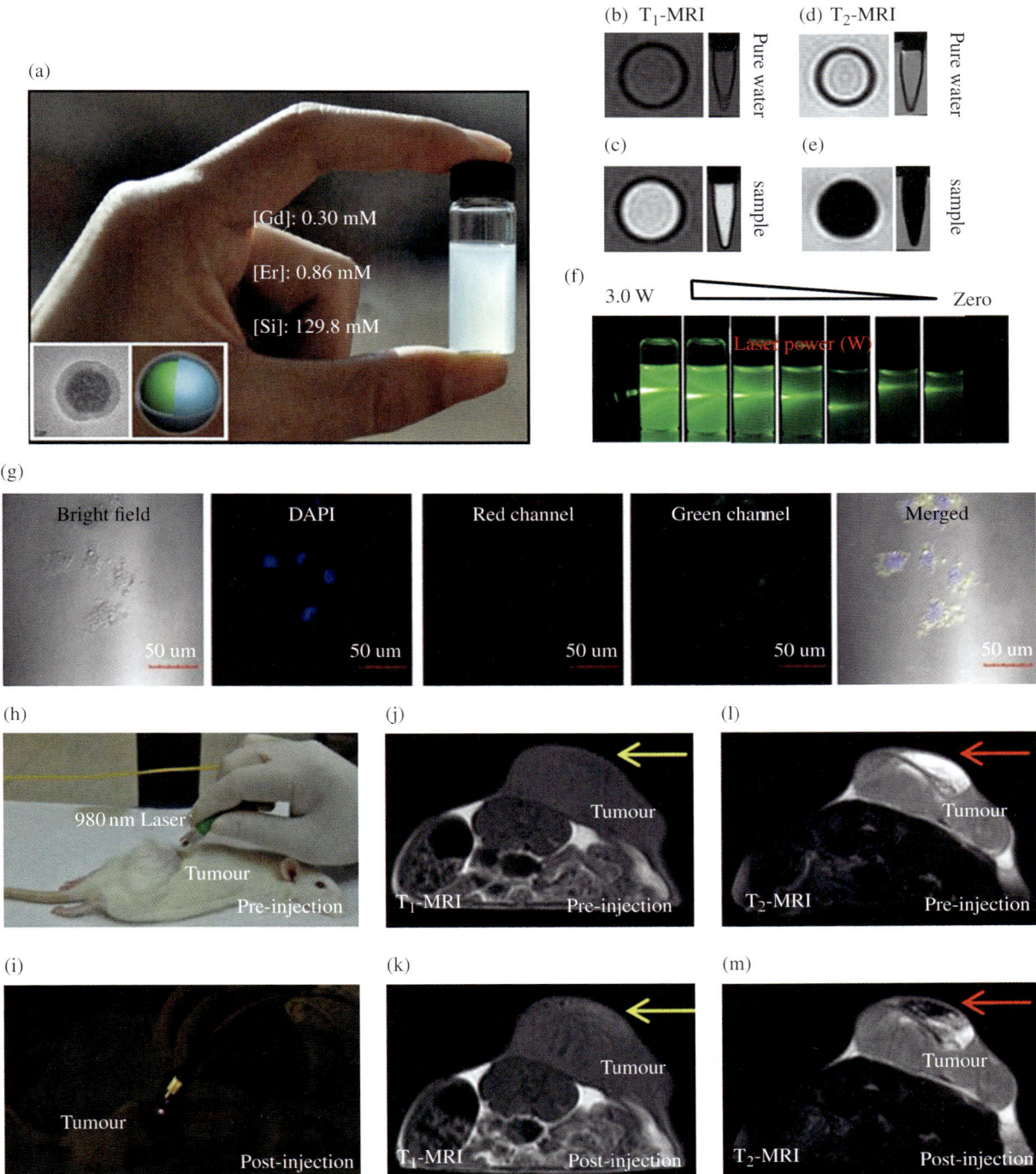

FIGURE 13.10 (a) Digital photograph of water soluble silica-protected NaGdF$_4$-based UCNPs. T$_1$-MR images of (b) pure water (r$_1$ = 0.35 1/s) and (c) UCNPs (0.2 nm) (r$_1$ = 2.03 1/s). T$_2$-MR images of (d) pure water (r$_2$ = 6.05 1/s) and (e) UCNPs (r$_2$ = 50 1/s). (f) Digital photos of samples under varied NIR laser powers. (g) Confocal images of MCF-7 cells incubated with the probe. *In vivo* optical imaging in tumour: (h) pre-injection, (i) post-injection. *In vivo* T$_1$-MR images of tumour: (j) pre-injection, (k) post-injection. *In vivo* T$_2$-MR images of tumour: (l) pre-injection, (m) post-injection. Tumour sites were marked with yellow and red arrows. Reprinted with permission from Ref. [91]. Copyright 2011 Wiley-VCH Verlag GmbH & Co. KgaA, Weinheim.

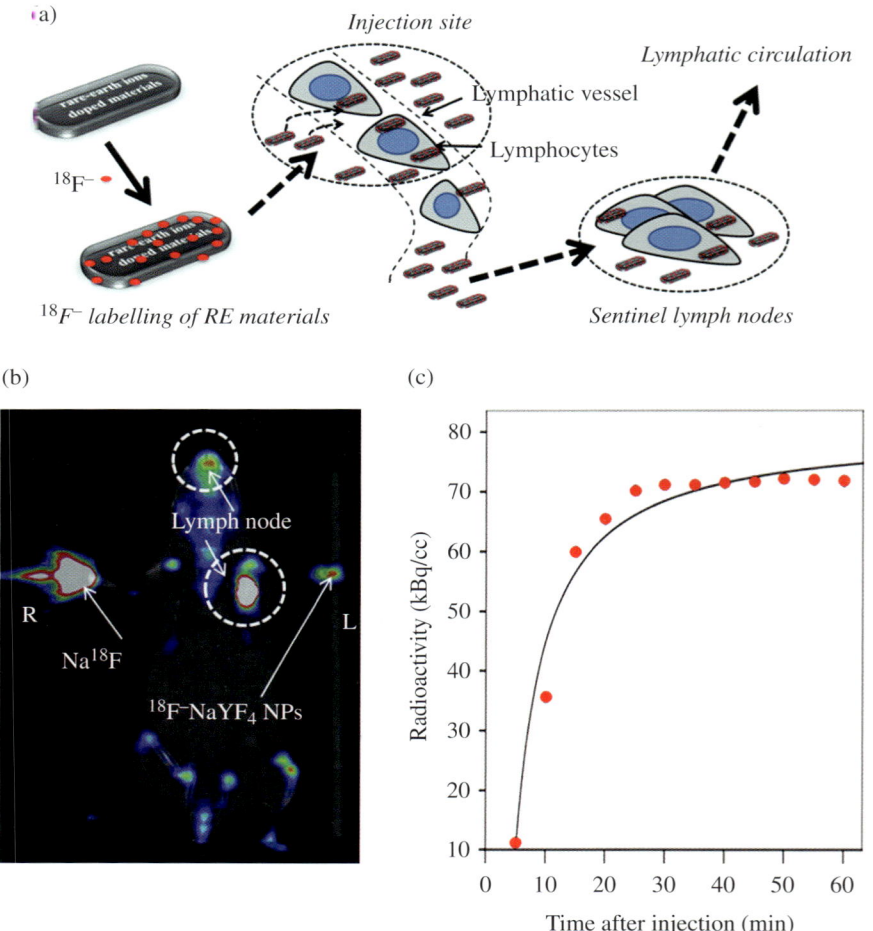

FIGURE 13.11 (a) Schematic representation of the preparation of ^{18}F-labelled UCNPs and the lymph node imaging mechanism. (b) PET/CT imaging of lymph node 30 min. after subcutaneous injection of 740 kBq/0.05 mL ^{18}F UCNPs into the left paw footpad. Na[18]F was injected into the right paw as a control experiment. (c) Curvilinear regression of real-time PET quantification detection of lymph node signals. Reprinted with permission from Ref. [102]. Copyright 2011 Elsevier Ltd.

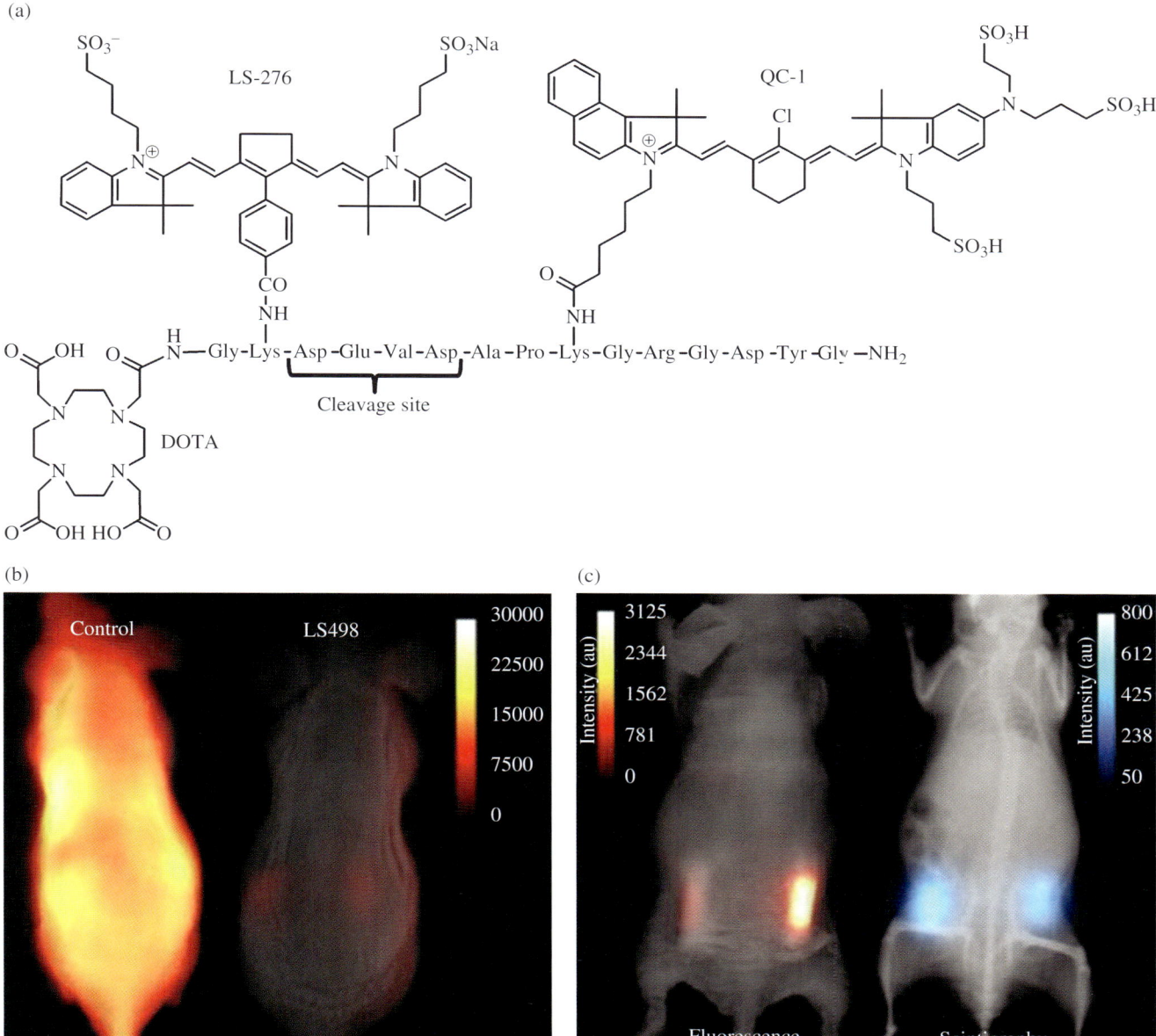

FIGURE 15.4 Noninvasive imaging of Caspase-3 activity with an activatable probe. (a) Chemical structure of the probe. (b) Activated (left) versus quenched (right) fluorescence signal of the probe in mice. (c) Fluorescence (left) and scintigraphy (right) images of subcutaneously implanted capsules containing ⁶⁴Cu-labelled activatable probe (i.e., LS498). Tubes were implanted subcutaneously below the surface of the skin. Comparable signal intensity was observed in scintigraphy ('always on'). However, the fluorescence signal was much stronger in the presence of Caspase-3. Adapted with permission from Ref. [49].

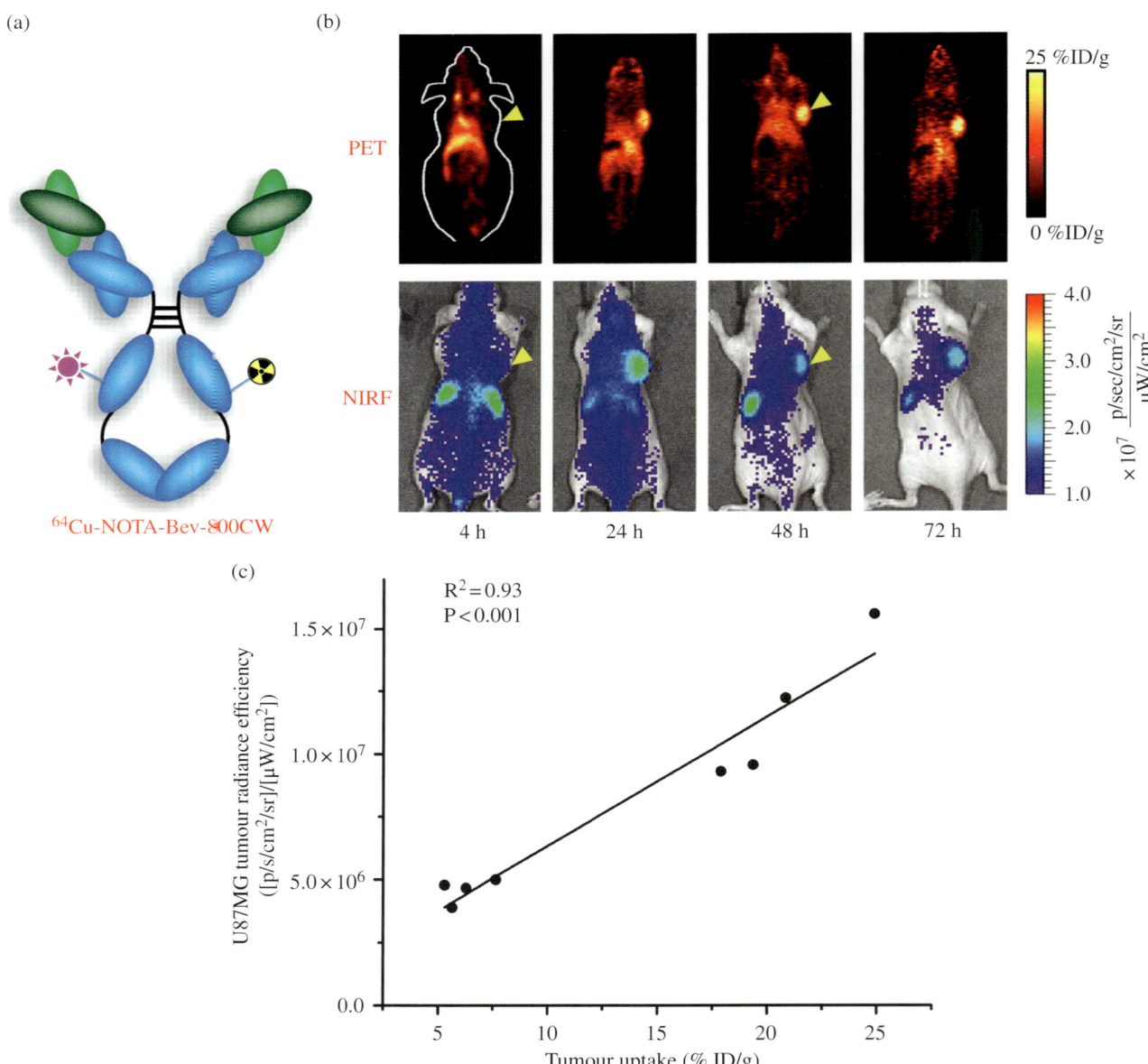

(a)

^{64}Cu-NOTA-Bev-800CW

(b)

PET

25 %ID/g

0 %ID/g

NIRF

4.0

3.0

2.0

1.0

$\times 10^{7}$ $\dfrac{\text{p/sec/cm}^2/\text{sr}}{\mu\text{W/cm}^2}$

4 h   24 h   48 h   72 h

(c)

R^2 = 0.93
P < 0.001

U87MG tumour radiance efficiency ([p/s/cm^2/sr]/[μW/cm^2])

Tumour uptake (% ID/g)

FIGURE 15.5   PET/NIRF imaging of VEGF. (a) Schematic representation of the dual-modality probe ^{64}Cu-NOTA-Bev-800CW. (b) Serial *in vivo* PET/NIRF imaging of U87MG tumour-bearing mice at 4, 24, 48 and 72 h post-injection of ^{64}Cu-NOTA-Bev-800CW. Arrowheads indicate the U87MG tumours. (c) Linear correlation of the *ex vivo* NIRF signal intensity with the percentage injected dose per gram of tissue (%ID/g) values based on PET in all U87MG tumour-bearing mice at 72 h post-injection [54].

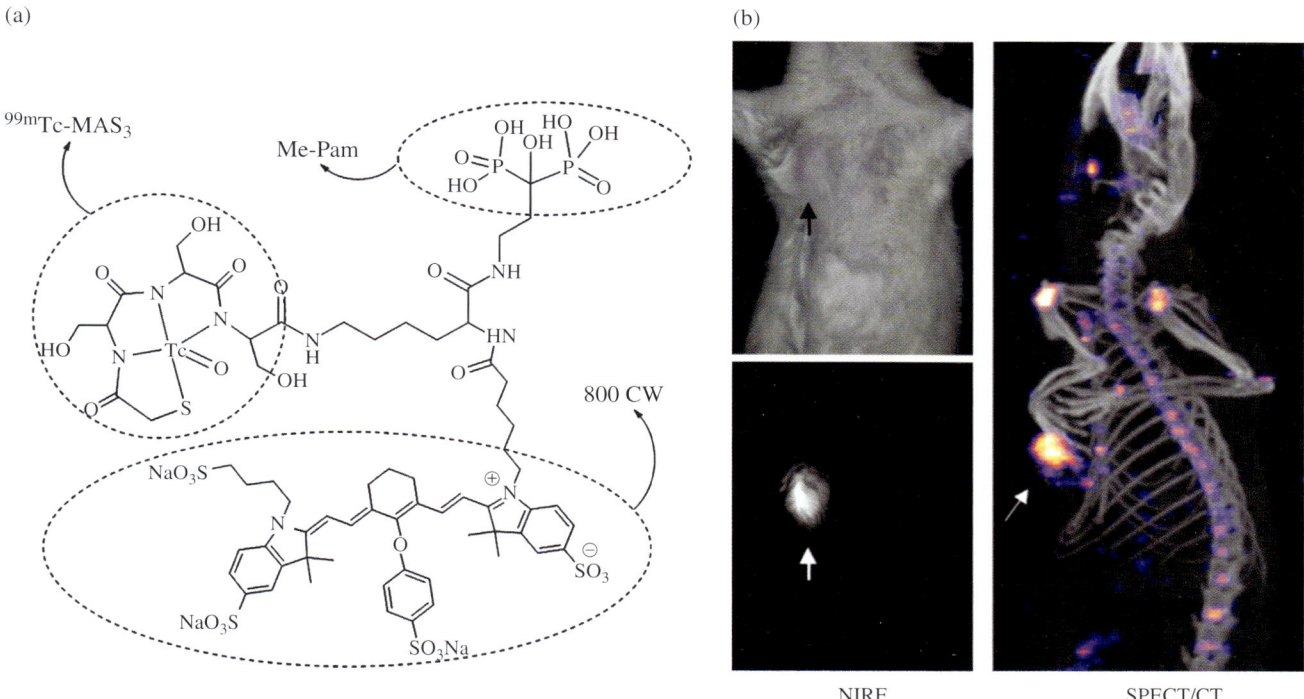

FIGURE 15.6 *In vivo* SPECT/NIRF imaging of breast cancer microcalcification. (a) Chemical structure of the Pam-Tc-800 probe. (b) NIRF and SPECT/CT images after administration of the probe in a breast cancer rat model. Arrows indicate the tumour. Adapted with permission from Ref. [74].

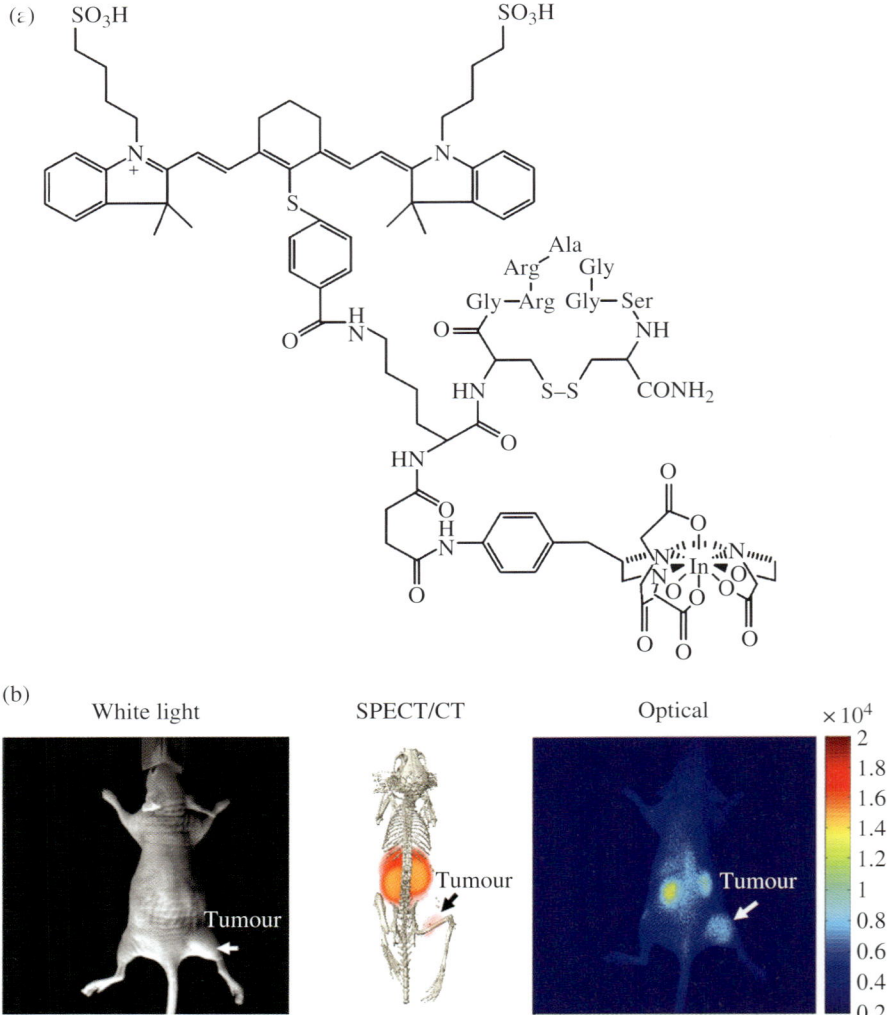

FIGURE 15.8 Dual-modality SPECT/optical imaging of IL-11Rα. (a) Chemical structure of the dual-modality probe. (b) SPECT/CT and optical images of a tumour-bearing mouse at 24 h after probe injection. Adapted with permission from Ref. [95].

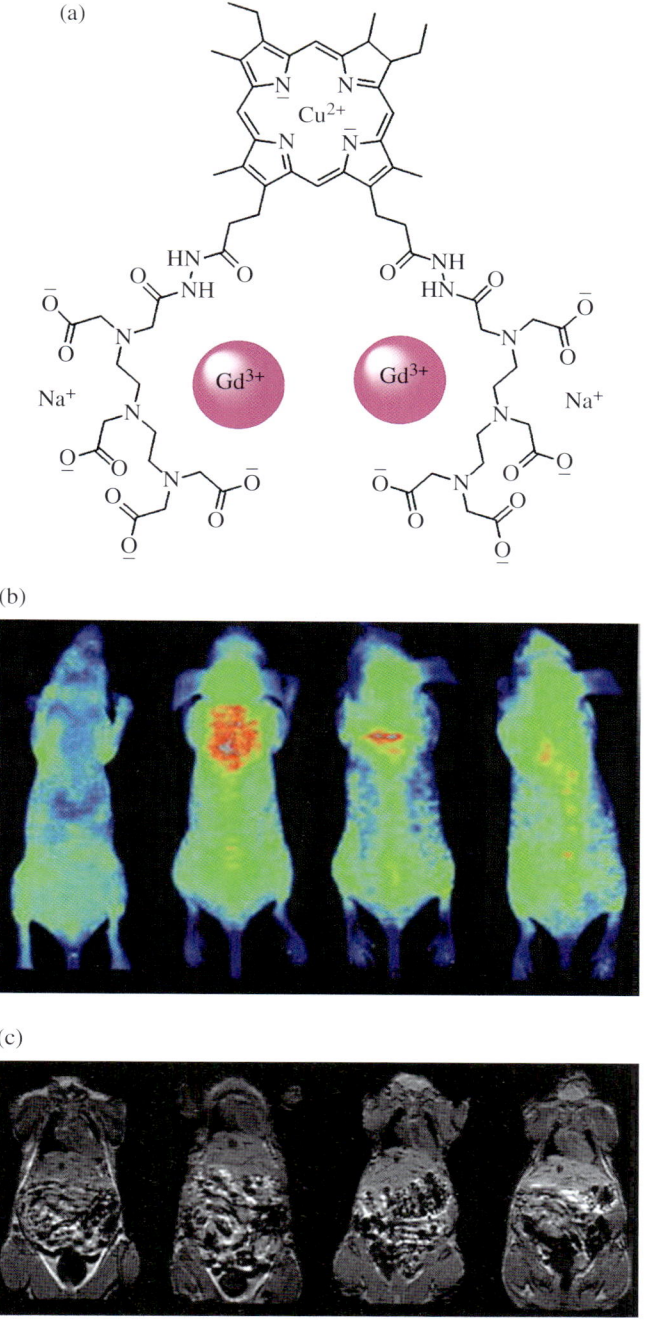

| 0 h | 1 h | 4 h | 24 h |

FIGURE 15.9 Dual-modality MRI/optical imaging with gadophrin-2. (a) Chemical structure of gadophrin-2. (b) Optical images and MRI images (c) of a mouse before and after injection of gadophrin-2-labelled hematopoietic cells, which showed a progressive cell accumulation in various organs (e.g., the lung). Adapted with permission from Ref. [121].

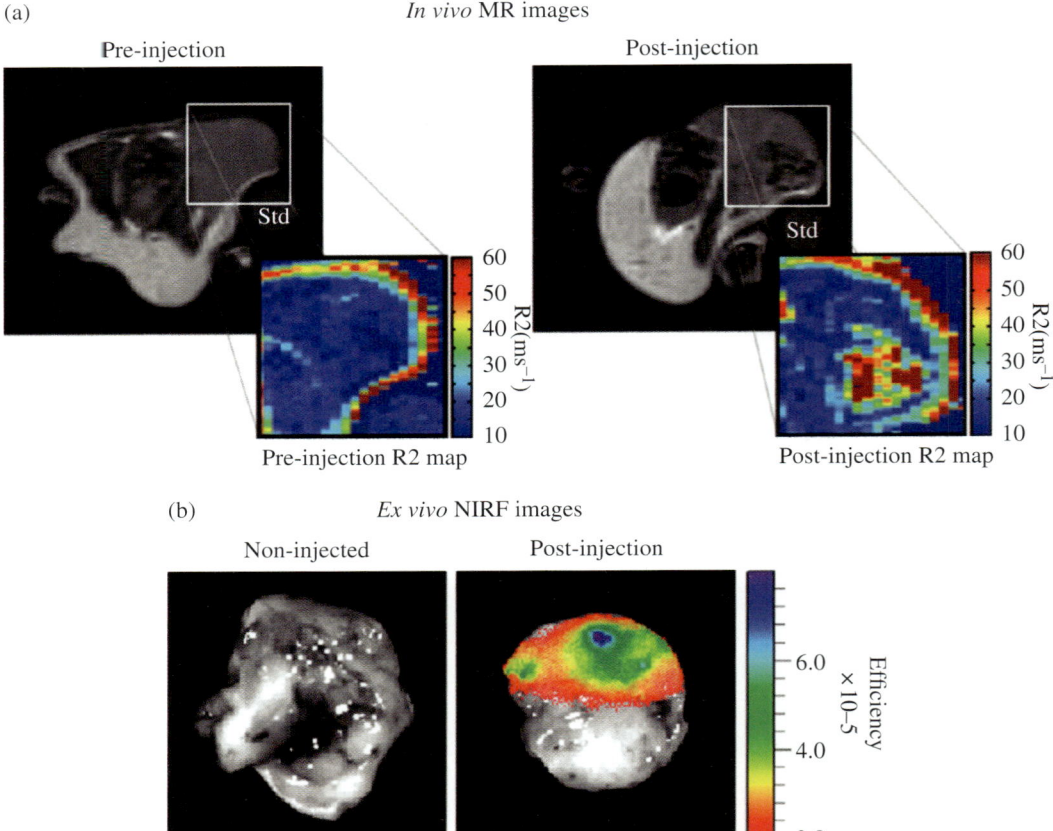

FIGURE 16.12   (a) MRI of the mouse bearing a 9 L tumour pre- and post-injection (48 h) of the CTX-Cy5.5 nanoparticles. Heterogeneous accumulation of the nanoparticles is observed by negative contrast enhancement and changes in the R_2 measurements in the region of interest. (b) *Ex vivo* NIRF image of xenograft tumours acquired from a control mouse (left) and the mouse injected with CTX-Cy5.5 nanoparticles (right). Reprinted with permission from Ref. [47]. Copyright 2010 American Chemical Society.

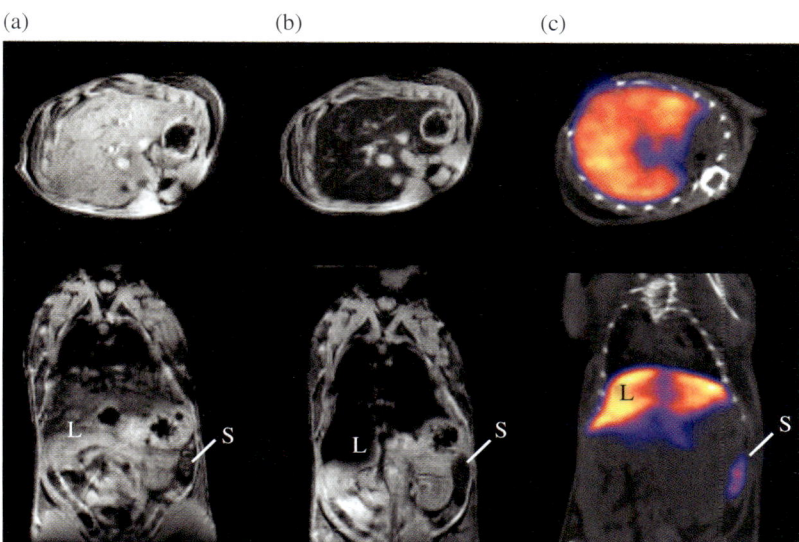

(a)   (b)   (c)

FIGURE 16.25   Short-axis view (top) and coronal view (bottom) images: (a) T_2*-weighted MR images before injection of [99m]TcDPA-ale-Endorem, (b) T_2*-weighted MR image 15 min. post injection, and (c) nanoSPECT-CT image of the same animal in a similar view 45 min. post injection. Contrast in the liver (L) and spleen (S) changes after injection due to accumulation of [99m]TcDPA-ale-Endorem, in agreement with the nanoSPECT-CT image, which shows almost exclusively liver and spleen accumulation of radioactivity. Reproduced with permission from Ref. [87].

(a)   (b)

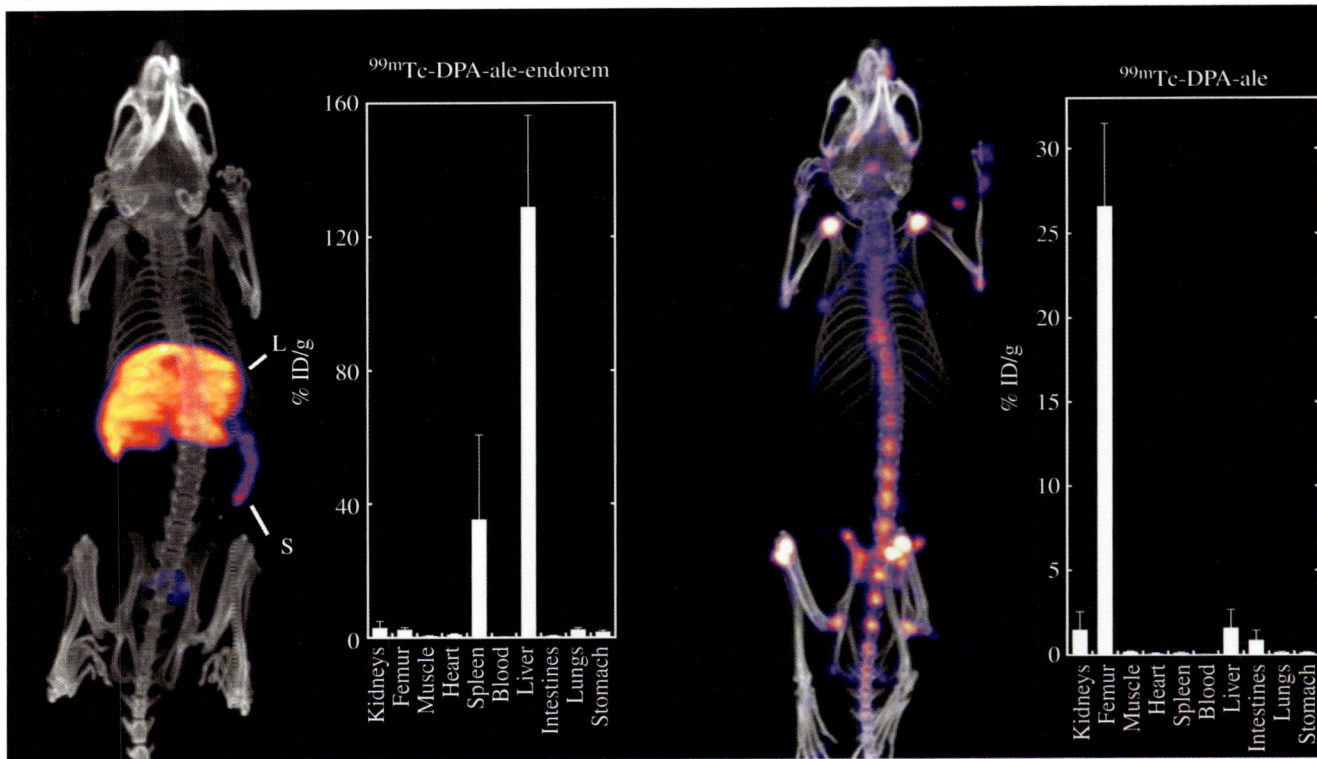

FIGURE 16.26 Whole-body SPECT-CT maximum intensity projection (left) and biodistribution studies (right) of ^{99m}TcDPA-ale-Endorem (a) and ^{99m}TcDPA-ale (b). Reprinted with permission from Ref. [87]. Copyright 2011 American Chemical Society.

(a)   (b)   (c)   (d)

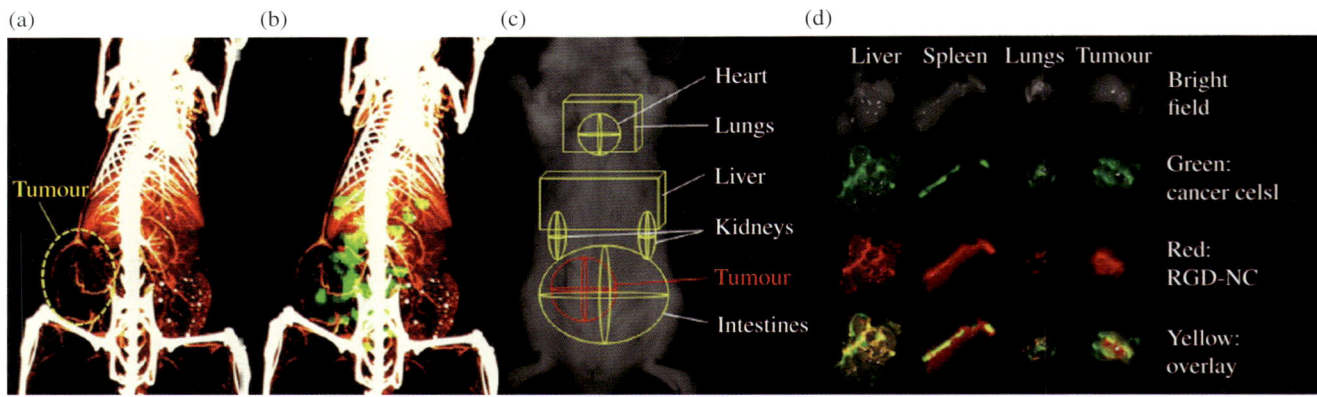

FIGURE 16.27 CT/fluorescence imaging of metastases. (a) Micromorphological imaging of normal and tumour vasculature at 99 μm resolution of a metastatic 4T1 tumour (week 5) using a liposome-based iodinated contrast agent. (b) Co-registration of the micro-CT image with the fluorescence image of the same animal injected with the RGD-NC nanoparticles. (c) Regions of interest indicate the location of the tumour and organs. (d) *Ex vivo* imaging of organs indicates the co-localisation of RGD-NC particles and 4T1 metastatic cells expressing GFP. Reprinted with permission from Ref. [92]. Copyright 2012 American Chemical Society.

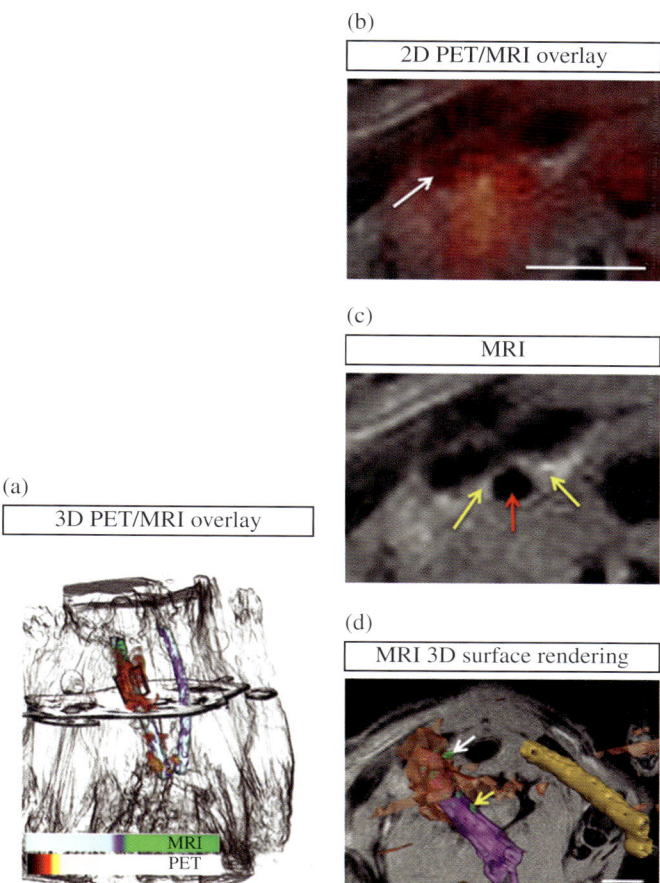

FIGURE 16.30 Multimodal macromolecular probes localise to the injured vessel in the rat copper cuff model. (a) Overlay of PET with MRI 24 h post injection (head at top, out of field of view). The PET and MR signal from vessels was rendered in color using the assignments shown in the scale bar. The plane through the image indicates the position of the image shown in (b). This co-registered MRI/PET image shows a diffuse cloud of PET signal in the region around the injured vessel and region of higher MR intensity on the right side of the vessel. The clavicle is the dark region indicated by the white arrow. Scale bar = 2.5 mm. (c) The MR image from the same plane clearly shows elevated MR contrast in the walls of the vessel. (d) 3D reconstruction of MRI and PET data in an oblique orientation shows the discrete accumulations of macrophages on the vessel wall. Scale bar = 5 mm. This view is zoomed out from the FOV in panels (b) and (c) to include both vessels. Injured carotid artery is purple, increased MR signal intensity relative to vessel background is green, PET signal is orange, and contralateral vessel is gold. Reproduced with permission from Ref. [8].

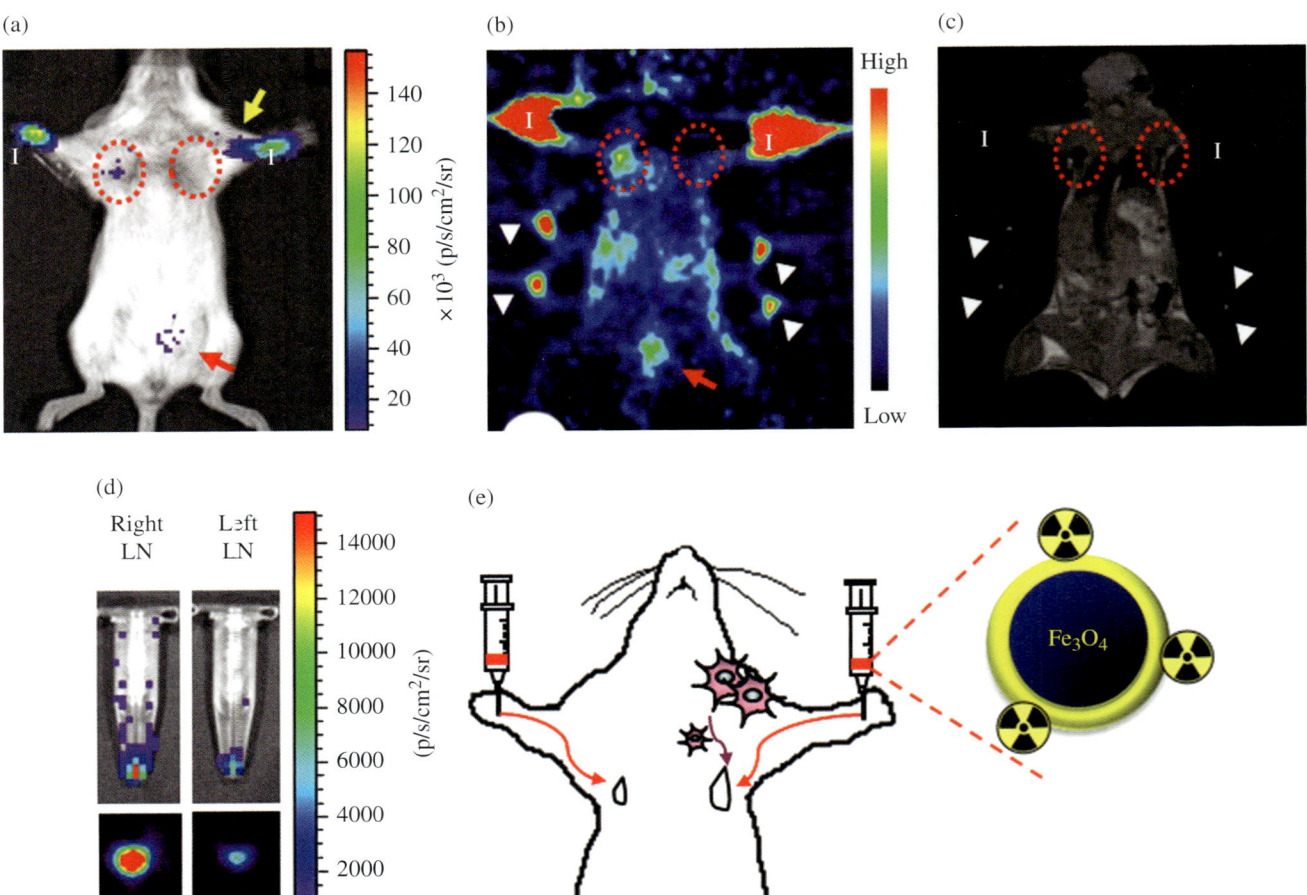

FIGURE 16.31 Triple-modality imaging of radiolabelled nanoparticles: (a) optical, (b) microPET, and (c) MRI of [124]I-labelled SPIONs injected into the front paws of a BALB/c mouse bearing a 4T1 tumour implanted on its shoulder. Tumour: arrow; sentinel lymph node: dotted circle; injection site: "I"; bladder: arrow; fiduciary markers: white arrow head. (d) *Ex vivo* luminescence (top) and microPET (bottom) images of the dissected lymph nodes. (e) Schematic diagram of tumour metastasis model and injection route of radiolabelled nanoparticles. Reprinted with permission from Ref. [94].

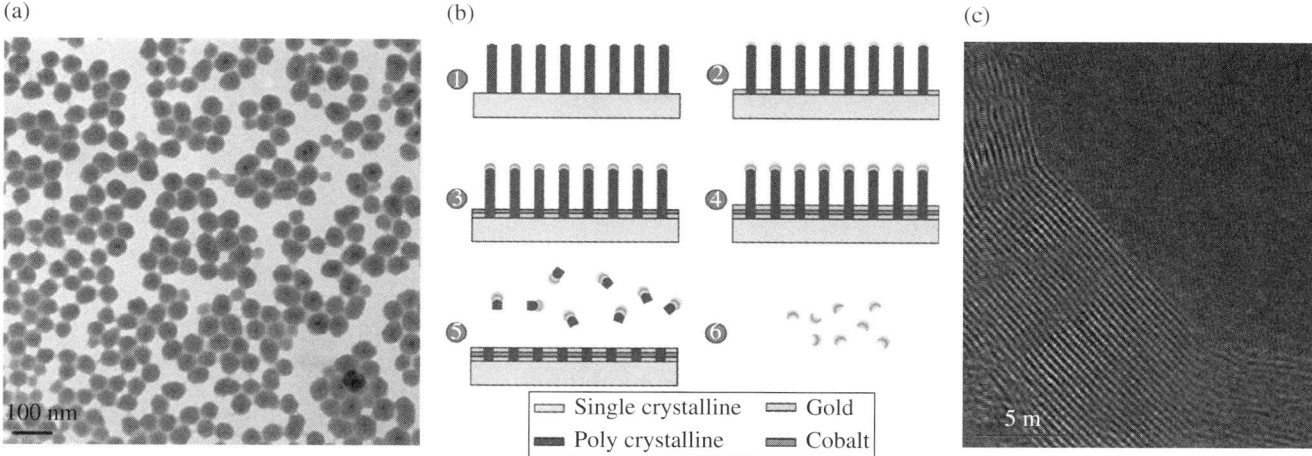

FIGURE 9.9 (a) TEM micrograph of SiO$_2$ coated iron oxide nanoparticles. Reproduced with permission from Ref. [106]. (b) Fabrication procedure of the nanowontons including six steps: (1) Etching polysilicon nanopillars on the surface of single crystalline silicon wafer; (2) deposition of a 10 nm gold thin film; (3) deposition of a 10 nm cobalt thin film; (4) deposition of a 10 nm gold thin film; (5) etching polysilicon nanopillars in KOH batch solution; and (6) removal of polysilicon by basic etching. Adapted with permission from Ref. [114]. (c) High-resolution transmission electron microscopy (HRTEM) image of graphite encapsulated iron nanoparticle. Reproduced with permission from Ref. [117].

application within MRI probes [69]. Seo and co-workers prepared FeCo nanoparticles and then coated them with a few layers of graphitic carbon by chemical vapour deposition at 800°C using methane as carbon provider. Finally, to stabilise these nanoparticles in aqueous conditions, they used modified phospholipids exploiting the interaction between the carbon layers and the fatty acid chains of the phospholipids. There are a few papers where this approach has been used to stabilise other types of nanoparticles, most of them metal-only nanoparticles that are highly sensitive to air oxidation (Figure 9.9c) [115–118]. To date, though, none of these materials have reached MRI testing.

9.6 FUNCTIONALISATION AND SURFACE MODIFICATION

The current state of the art in MRI contrast agent design requires functionalisation of the contrast agent that gives it specificity. Therefore, the last steps in the preparation of these probes is often the coupling of the targeting moiety. Many molecules have been used as targeting motifs, with special attention given to natural biomolecules. For the coupling of these biomolecules to the nanoparticles, it is highly desirable to use rapid, high-yielding reactions carried out under mild conditions in, as far as possible, aqueous media. An extensive description of conjugation techniques is featured in Chapter 2, thus only the most common functionalisation methodologies followed in nanoparticle work will be mentioned here (Figure 9.10). The simplest strategy is to just incubate a solution of the nanoparticles with a solution of the targeting ligand, where the high surface to volume ratio of the nanoparticles helps in the adsorption of the biomolecules. The biomolecule-functionalised nanoparticle is then stabilised either by electrostatic forces or, in most cases by weaker van der Waals interactions or hydrogen bonds [119–121]. For example, the amount of proteins that can be bound in this way to the nanoparticles is not negligible [119]. Both DNA and RNA molecules are biopolymers that present a global negative charge under physiological conditions, thus they can be electrostatically attached to positively charged nanoparticles (usually functionalised with terminal amine groups). Here there are issues with the stability of the resulting nanoparticles, that is, pH, ionic strength, or concentration variations can desorb the bound biomolecules. This is the main reason why more complex covalent approaches are generally preferred nowadays.

- Peptidic bonds: The coupling of an amine group to a carboxylic acid to form an amide is a versatile reaction that can be carried out in aqueous buffers under mild conditions. Most peptide sequences and proteins contain both primary amines and terminal carboxylic acids. Other common nanoparticle ligands also present these kind of groups (e.g., citric acid, APTES) or they can be chemically modified to present them (e.g., dextran can be easily transformed into carboxy-dextran or amino-dextran). An advantage for *in vivo* work is that the new bond is already present in the organism thus no adverse effects are expected. The coupling reaction requires activation of the carboxylic group and EDC, 1-ethyl-3-(3-dimethylaminopropyl)carbodiimide hydrochloride), sometimes combined with NHS or sulfo-NHS (N-hydroxysuccinimide,

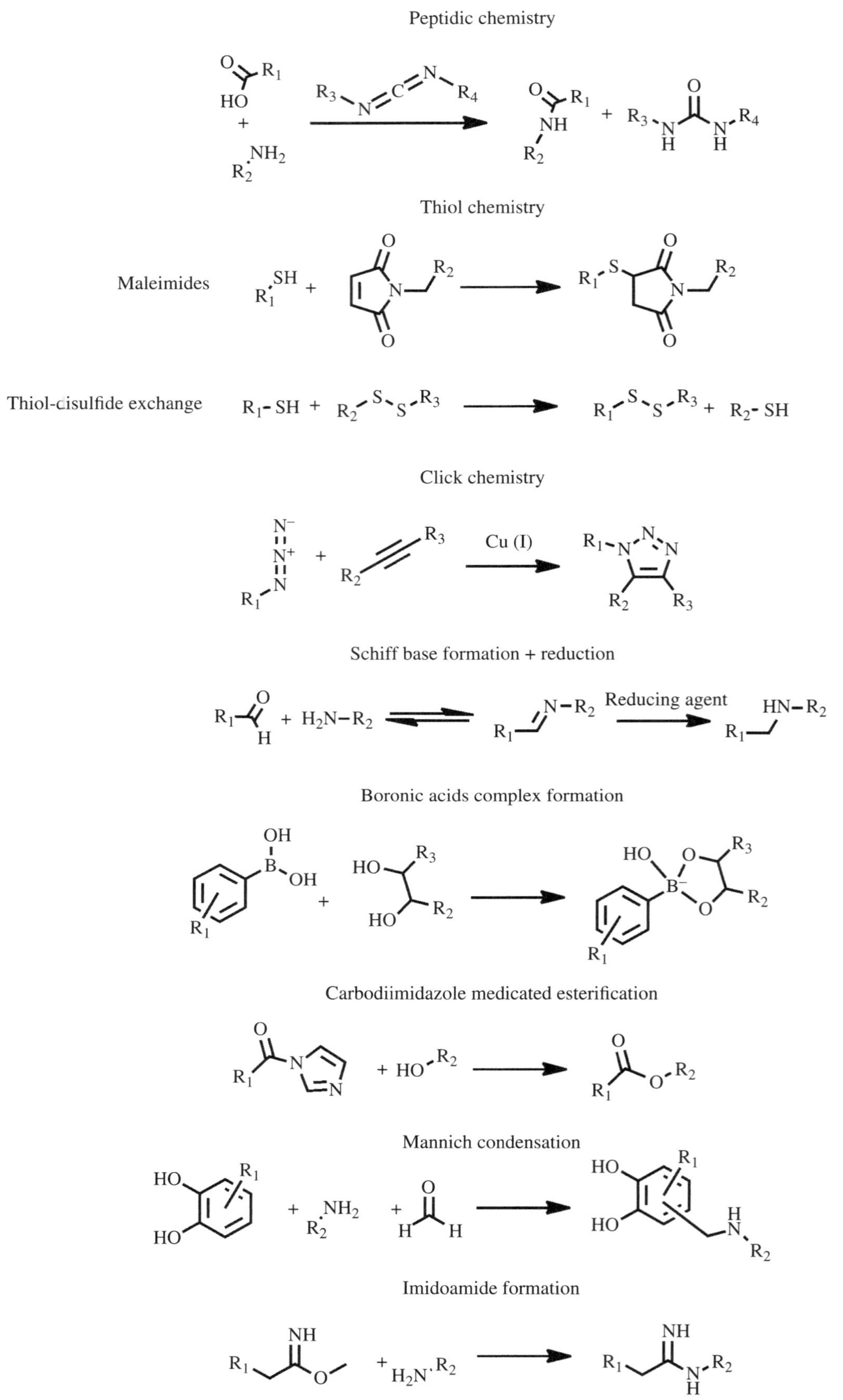

FIGURE 9.10 Some of the most common chemical reactions used for the functionalisation of nanoparticles.

sulfo-N-hydroxysuccinimide), is often used to stabilise the activated form of the carboxylic acid. This chemical approach has been widely used not only for the conjugation of targeting molecules (small molecules [88, 122], aptamers [123], antibodies [29, 124–126], peptides [71, 127, 128], proteins [129], carbohydrates [130]), but also of intermediate linkers [49, 64, 131–135], reporter moieties [72, 77, 131], and dyes [109].

- Thiol chemistry: Thiol groups are present in many proteins and peptides (coming from the side chain of cystein) either as a single group or in the form of disulfide bonds. Generally, there are two possibilities for involving thiol groups: (i) reaction between a thiol group and a maleimide group and (ii) thiol-disulfide exchange. Maleimides are prepared by the reaction of maleic anhydride and an amine derivative and react with high specificity with thiol groups in the pH range of 6.5-7.5. This methodology has been recently applied to the preparation of nanoparticulate MRI probes with antibodies [31, 119, 136, 137], peptides [71], or small molecules [138] as targeting molecules.

    Thiol-disulfide exchange is the other main possibility concerning sulfhydryl groups. Basically a thiol-ended molecule attacks a disulfide bond, the bond is then broken, and the new mixed disulfide is formed. The main concern with this thiol exchange reaction is that the new bond is as labile as the initial one, thus the same process can take place later on in vivo (there can be high local concentrations of thiol-containing molecules such as proteins or peptides, i.e., glutathione) displacing the targeting ligand. The strategy has been successfully followed to obtain nanoprobes functionalised with peptides and proteins [132, 139–141]. The use of heterobifunctional linkers is of interest in the functionalisation of nanoparticles, because the linkers can be used either as spacers, for example, to separate the targeting molecule from the nanoparticle core, or just to alter the chemical functionality of the nanoparticle. An example is SPDP (N-succinimidyl-3-(2-pyridyldithio)propionate). On one end, it presents a succinimidyl moiety to react with amines, and on the other it has a disulfide bond able to react with a thiol molecule, thus providing a convenient way of linking peptide chemistry to thiol chemistry [142].

- Click chemistry: This now ubiquitous reaction within many areas of chemistry features an azide group reacting with an alkyne to form a [3 + 2] cycloaddition product. Generally, the reaction requires high temperatures; this was the main reason why it was not used until recently for the coupling of biomolecules. However, Tornoe et al. [143] and Rostovtsev et al. [144] demonstrated that the reaction could take place in aqueous conditions and at room temperature, in the presence of Cu (I) as catalyst. The reaction can even take place in vivo without copper catalysis if strained alkynes are utilised [145]. One of the main advantages of the click reaction is its bioorthogonality: None of the two reactant groups required for the cycloaddition can react with any functional groups found in biological systems. This provides great selectivity for the conjugation, even in complex biological media, and in recent years an increasing number of examples have been published exploring this methodology with various nanoparticles [63, 79, 146, 147].

- Other chemical reactions: Another useful reaction is that between an aldehyde group and an amine group to form a Schiff base. The product of this reaction is unstable, it stays in equilibrium with the free form of the molecules, but it can be stabilised by reduction—for example, the addition of a mild reducing agent will form a C-N covalent secondary amine bond. This strategy has been followed to functionalise magnetic nanoparticles with proteins and antibodies [148, 149].

    Boronic acid derivatives are capable of forming ring structures with molecules, such as 1,2- or 1,3-diols, 1,2- or 1,3-hydroxyacids, 1,2- or 1,3-hydroxylamines, and 1,2- or 1,3-hydroxyoximes. The products of these reactions are sometimes reversible with a change in pH, thus making them an ideal option for the functionalisation of nanoparticles with carbohydrates [150–152]. An analogous reaction is CDI (1,1′-carbonyldiimidazole)-mediated esterification, where carbonyldiimidazole will react with hydroxyl groups in organic solvents in the complete absence of water [138]. Another reaction useful for the modification of catechol-like protected nanoparticles is the Mannich reaction: the condensation of an amine-containing molecule with an active hydrogen-containing compound (catechol derivative) in the presence of formaldehyde. The product is formed as a mixture of ortho- and para-isomers, and the condensation has been utilised for the functionalisation of magnetic nanoparticles with peptides [153].

    Another strategy toward the coupling of biomolecules to amine groups (or vice versa) is the use of imidoesters. There are specific acylating agents available for the modification of primary amine groups, because they present minimal affinity toward other nucleophilic groups. The imidoamide bond formed is stable even at low pH, but it can be hydrolysed at high pH, and the methodology has been applied to the functionalisation of iron oxide nanoparticles with carbohydrates [154]. In theory, any chemical reaction can be used to couple targeting molecules to nanoparticles as long as both the nanoparticle and the ligand can withstand the reaction conditions.

- Biological interactions/interactions involving biomolecules: As stated before, most of the targeting molecules in nanoparticle functionalisation are biomolecules, thus biological interactions have also been explored as driving forces to couple such molecules to the nanoparticle core. The strategy incorporates any of the chemical coupling

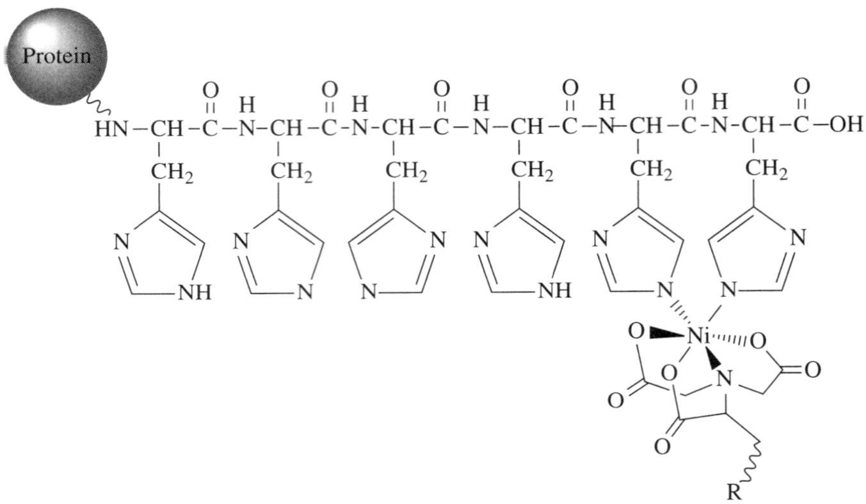

FIGURE 9.11   Schematic representation of a histidine tag-nickel nitrilotriacetic acid (NTA) interaction.

strategies detailed in Figure 9.10 to bind a biomolecules to the nanoparticle and then exploit its affinity and specificity toward the targeting molecule. The most commonly used of these interactions is the one between biotin and avidin/streptavidin. Biotin can be easily coupled to any other linker/spacer, protein/peptide, or chemical through peptide chemistry. It is small enough not to substantially modify the properties of proteins when it is attached to them, and its partner for the interaction is avidin, neutravidin, or streptavidin. All of these proteins present high affinity and specificity toward biotin, with avidin having a slightly higher affinity ($K_d \approx 10^{-15}$ M). These proteins present several binding sites in the same molecule to attach biotin (avidin can bind up to four biotin molecules), so they can be used sandwich-like, first to couple to biotin in the nanoparticle and then to bind a biotin-modified targeting molecule to the final complex. Examples using this methodology include the coupling of antibodies and dyes, for example [93, 131, 155, 156].

A pair of proteins that have been tested in the functionalisation of MRI nanoprobes are A/G-IgG antibodies. A and G are small proteins originally isolated from the cell wall of bacteria (they are now available from recombinant sources) that bind with high affinity and specificity to IgG antibodies (protein A also binds with lower affinity to IgE, M, and A). As before, the strategy details the use any of the chemical approaches described above, usually peptidic chemistry, to couple either of the proteins to the nanoparticle and then incubate these nanoparticles with a buffer solution of the desired antibody[129, 141, 157, 158]. A further strategy involves the use of genetically modified proteins bearing what is known as the histidine tag (Figure 9.11). This is a sequence of between 5 and 9 histidine amino acids fused usually at either the N- or the C-terminus of the protein. This motif allows the protein to bind to metal chelates, such as nitrilotriacetic acid (NTA) or iminodiacetic acid (IDA).

An advantage of this approach is that there are commercially available antibodies targeted to the histidine tag sequences that are very convenient to detect. However, the preparation of recombinant proteins with these sequences is not easy, and this limits the applicability of this approach, but the strategy has been followed to prepare MRI probes conjugated to peptides and proteins [141, 158, 159].

9.7   APPLICATIONS

This section discusses some selected applications to illustrate the different fields in which nanoparticles and MRI have been applied within biomedicine, but does not aim to provide an exhaustive list. Much work with nanoparticles has been carried out toward the application of MRI not only to diagnose but also to follow the evolution of lesions along the therapeutic process. The scientific community is currently working hard on the design and preparation of targeted contrast agents, but the examples of *in vivo* application of these kinds of probes is still scarce. El-Boubbou et al. reported the preparation of magnetic nanoparticles as MRI probes but not to use within the MRI imaging technique but to obtain information from the T_2 values [108]. They prepared a series of magnetic nanoparticles functionalised with different carbohydrate moieties to ascertain the differences between healthy and tumoural cells. Carbohydrate display on cells is only now beginning to be considered as a possible key feature for the diagnosis of diseases. To date, carbohydrate biological interactions have been much less explored within biology due to the fact that

(a)   (b)

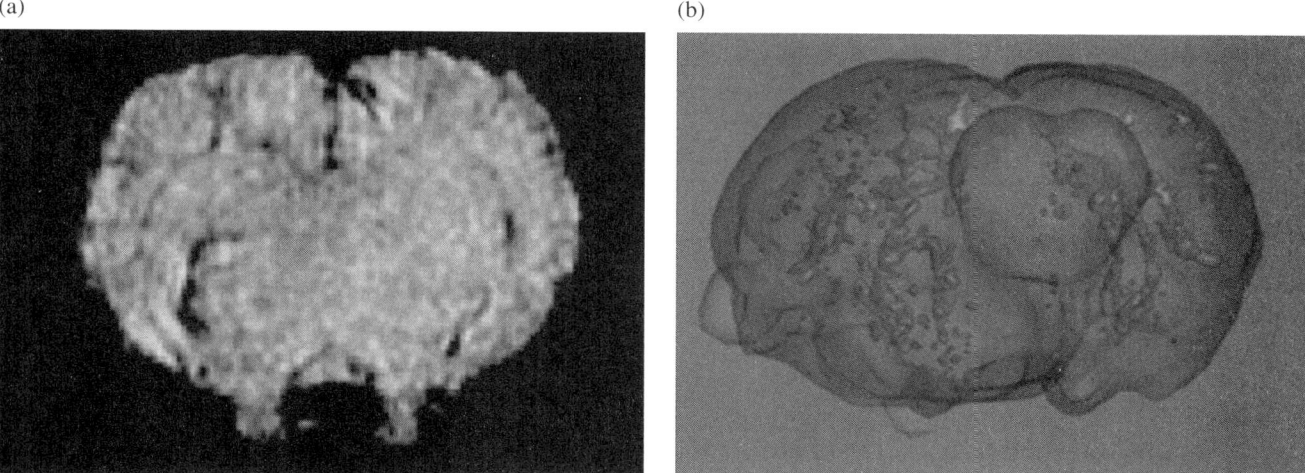

FIGURE 9.12 (a) Selected image taken from T_2^* weighted 3D dataset of rat brain. (b) 3D reconstruction of the accumulation of contrast agent. The images reveal that magnetic nanoparticles functionalised with sLe^x enable detection of lesions in models of stroke. Reproduced with permission from Ref. [154]. (*See insert for colour representation of the figure.*)

they are weaker than other biological interactions so a multivalent effect is usually needed to measure or observe them. Nanoparticles are thus a perfect platform to display carbohydrate motifs with the multivalent presentation needed to study these interactions. In this report, iron oxide nanoparticles (Fe_3O_4) were prepared following a co-precipitation protocol starting from iron(III) chloride and iron(II) sulphate salts and without the presence of any ligand. Naked Fe_3O_4 nanoparticles were subsequently coated with a polyvinylpyrrolidone (PVP) polymer and a stable silica shell (using TEOS as the SiO_2 precursor). After some manipulation, simple carbohydrates were coupled to the nanoparticles using standard peptidic coupling conditions (mannose, galactose, fucose, or sialic acid) with the APTES-functionalised nanoparticles, or click chemistry was used with the azide-functionalised ones (N-acetylglucosamine). The group acquired T_2-weighed MRI images and calculated the variation in T_2 for each nanoparticle. Their results showed that each kind of cells has a different carbohydrate-binding characteristic that could be used for differentiation, and more importantly, that the carbohydrate-binding pattern changes from healthy to tumoural cells.

Van Kasteren et al. presented research using carbohydrates as targeting molecules [154]. In this case, they wanted to demonstrate that magnetic nanoparticles functionalised with complex carbohydrate structures could be used to image activated brain endothelium. This activation is associated with the inflammation process that takes place in diseases such as multiple sclerosis, ischemic stroke, and HIV-related dementia, and cannot be visualised by conventional imaging techniques. To image this process, E-/P-selectins (CD62E/CD62P) were chosen that are endothelial markers in acute inflammation. These molecules are carbohydrate-binding transmembrane proteins whose known ligand is sialyl Lewis^x, so a chemical strategy was designed to form iron oxide nanoparticles with these ligands on their surface. The coupling of the ligand to the nanoparticle was achieved via the reaction of an imidoester in the carbohydrate to an amine group on the nanoparticle. The final results obtained by MRI (Figure 9.12) using these probes *in vivo* with different animal models of brain diseases combined with hystochemical corroboration, confirmed that these nanoparticulate probes were able to target the activated endothelia when the contrast agent was administrated via the tail vein.

A different and simpler strategy was followed by Nieman and co-workers to study the migration of stem cells in brain [160]. Commercial iron oxide-based nanoparticles encapsulated inside a polymer microbead were purchased and used without further modification. In this report, instead of trying to design a complex strategy to target certain cells, the probes were delivered directly to their target through a stereotactic injection. The intention was to follow *in vivo* the migration of stem cells from their niche in the subventricular zone to the olfactory bulb in a physiological process to maintain the homeostasis of olfactory neurons using MRI (Figure 9.13). They demonstrated that upon labelling of a certain population of cells (neural stem cells in this case), these can then be tracked non-invasively *in vivo* along several days with a precision that allows the calculation of speeds of migration and allows the anatomic placement of the cells in their destination.

Hadjipanayis et al. have prepared and functionalised magnetic nanoparticles to be used as MRI probes for the detection and treatment of glioblastoma multiforme [161]. The preparation of the magnetic nanoparticle probes was from commercial iron oxide nanoparticles coated with an amphiphilic triblock copolymer. This copolymer included PEG moieties to minimise non-specific interactions and carboxylic groups that allow them to continue the functionalisation of the nanoparticles. Using amide-coupling conditions (EDC/sulfo-NHS), a specific antibody anti-epidermal growth factor receptor (EGFR) deletion

Parasagittal   RMS path

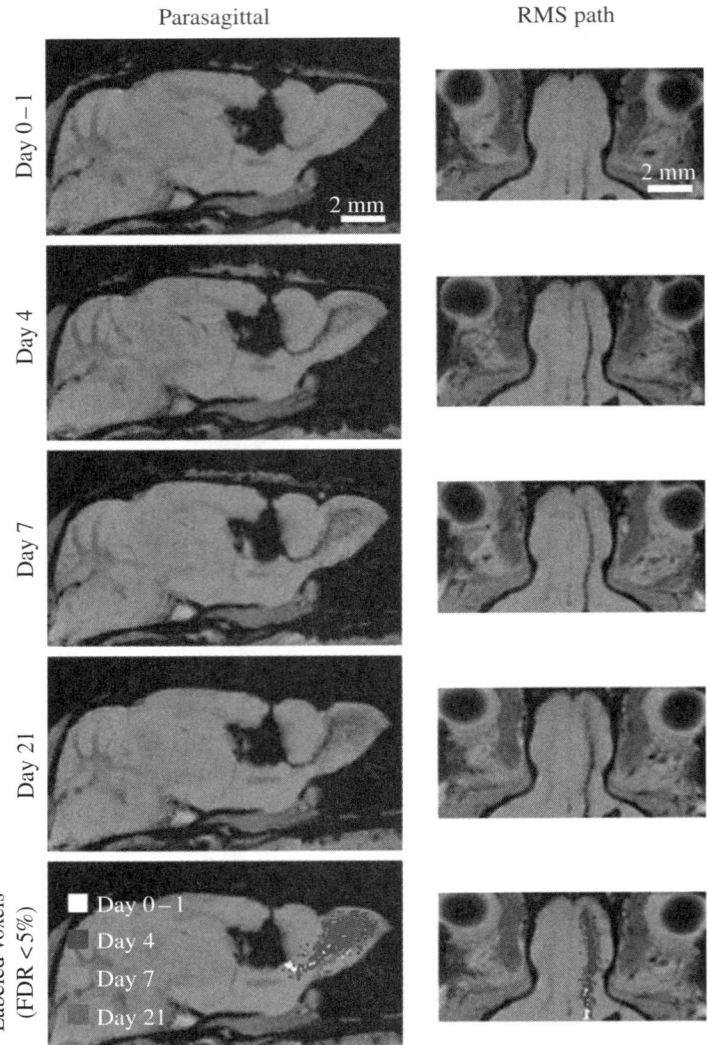

FIGURE 9.13 Distribution of magnetic nanoparticles in the olfactory bulb (OB). Parasagittal and RMS reformatted images are shown at Days 0–1 (combining images from 3 to 24 h post-injection), Day 4, Day 7, and Day 21. Each image is representative of a composite of three or more individual animals. Distribution through the bulb occurs primarily in the parasagittal plane (left column) with relatively little motion laterally (as seen in the right column). Images obtained with permission from Ref. [160].

mutant (EGFRvIII) present on human glioblastoma multiforme was coupled to the nanoparticles. In this strategy, the antibody is coupled to the nanoparticles in a random way through a ε amino group of the side chain of a lysine. Through this random coupling, part of the antibodies will probably lose their bioactivity because their recognition areas will be blocked or hindered, but the overall population of nanoparticles will gain specificity. Apart from the preparation of nanoparticles, the other main problem encountered was the delivery of the probes to the tumour inside the brain. The blood-brain barrier (BBB) is a closely packed layer of cells that prevents the delivery of most products inside the brain. To overcome this barrier they used a procedure called convection-enhanced delivery (CED) which is a minimally invasive surgical procedure that provides fluid convection in the brain by a pressure gradient that bypasses the BBB. The group demonstrated that glioblastomas could be labelled and visualised by MRI *in vivo* in mice with functionalised magnetic nanoparticles and that the use of the antibody anti EGFR in the nanoparticles also produces a significant increase in the animal survival.

Sosnovik et al. [162] reported the application of iron oxide-based T_2 contrast agents to image cardiomyocyte apoptosis and necrosis *in vivo* in an ischemia model. To do so [139], an MRI probe was prepared consisting of iron oxide nanoparticles functionalised with a protein called annexin V. Annexin V is a relatively small protein that binds strongly and with high specificity to phosphatidylserine (PS). The group prepared first amine-functionalised cross-linked iron oxide nanoparticles (CLIO) following co-precipitation protocols and using dextran as the ligand. These were then reacted with SPDP

(a)   (b)

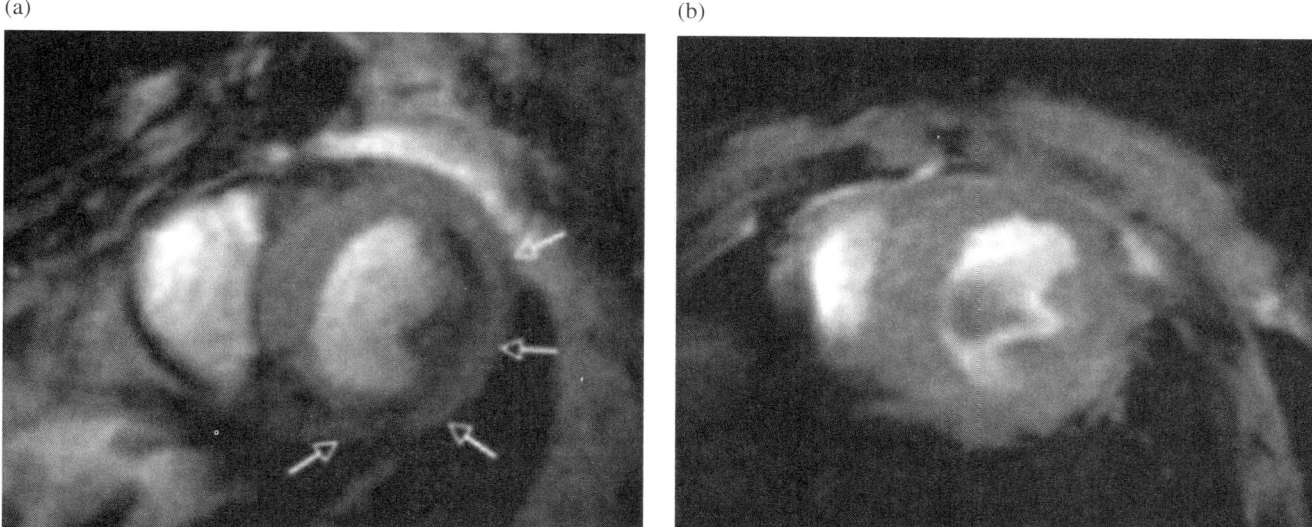

FIGURE 9.14 T_2^* weighted molecular MRI of cardiomyocytes apoptosis in myocardium exposed to moderate injury. (a) Mouse injected with the specific Annexin V CLIO contrast agent. (b) mouse injected with non-functionalised CLIO contrast agent. Reproduced with permission from Ref. [162].

(N-succinimidyl-3-(2-pyridyldithio)propionate) molecules to change the functionality for thiol chemistry, and the targeting molecule was coupled to the nanoparticles. Using these protein functionalised nanoparticles together with a T_1 probe (Gd-DTPA-NBD) specific for the imaging of necrosis, the group was able to demonstrate that cardiomyocytes apoptosis can be imaged *in vivo* with high specificity, high spatial resolution, and within the first 4 to 6 hours of ischemia, during which apoptosis is prevalent to necrosis (Figure 9.14).

Cho and co-workers have designed a magnetic nanoplatform for the imaging of angiogenesis in orthotopic bladder tumours [163]. $MnFe_2O_4$ nanoparticles of 12 nm were first prepared following a thermal decomposition methodology starting from acetylacetonates of both Fe(III) and Mn(II); lauric acid and laurylamine were used as the ligand molecules. To bring the nanoparticles into water, they used a carboxylated form of Polysorbate 80. The hydrophobic chain of these molecules intercalates the original hydrophobic coating of the nanoparticles, rendering water soluble nanoparticles functionalised with carboxylic acid groups. The final step was coupling the targeting molecule, which was a fusion protein, $VEGF_{121}/rGEL$, selected to deliver the nanoparticles to the tumour vasculature. VEGF (vascular endothelial growth factor) and its receptors have been established as primary proangiogenic factors [164]. The group showed that the use of this contrast agent in animal models could provide detailed anatomic information on intratumoural neoangiogenesis vessels that, according to the authors, could be used to decide on treatment strategies and therapeutic monitoring.

A completely different approach toward the specific imaging of cancer was adopted by Crayton and co-workers [165]. In this case, instead of preparing nanoparticles relying on a passive targeting or receptor-ligand interactions, a metabolic approach was followed. Their work is based on the different conditions of the tumour microenvironment when compared to healthy tissues. Primary and secondary events in tumourigenesis alter the metabolic profile of developing and metastatic cancers. One of these alterations is the reduced pH found in the extracellular matrix in tumours [166]. To take advantage of this, the authors designed a pH-responsible iron oxide nanoparticle. They prepared lanthanide-doped iron oxide nanoparticles following a co-precipitation protocol in the presence of dextran. These particles were then grafted with glycol chitosan using epichlorohydrin as crosslinking agent. The final coating of these nanoparticles (glycol chitosan) was a polymer with pH-dependent charge. The amine groups of this polymer can be protonated when the pH is slightly lower than physiological ($pk_a \sim 6.5$), rendering the nanoparticles with a positive charge that can electrostatically interact with negative charges in the tumour microenvironment. Using this strategy, it was shown that the percentage of injected dose accumulated per gram of tumour went from around 1.9% for the comparable non-responsible control nanoparticles, to around 3% with the pH-responsible analogues.

Finally, Kievit et al. [167] have reported iron oxide nanoparticles coated by a copolymer of chitosan and polyethylene glycol and further modified with a fluorescent dye and functionalised with an antibody anti Neu receptor (overexpressed in 30% of breast cancers). Chitosan-coated Fe_3O_4 nanoparticles were prepared *in situ* following co-precipitation techniques. The nanocrystals were further functionalised using peptidic chemistry and thiol-maleimide linkages to bear the antibodies in their surface. The nanoparticles were studied *in vivo* for the labelling of orthotropic breast tumours in mice. It was found that the targeted nanoparticles were able to decrease the intensity of the MRI signal up to threefold compared to similar nanoparticles functionalised with a nonspecific antibody, and up to tenfold compared to control non-injected animals. More important is the fact that biodistribution studies showed that there was a statistical difference between the uptake of the contrast agent by the tumour and the uptake by the liver. Furthermore, these targeted nanoparticles were able to recognise and tag spontaneous micrometastases in the lungs, livers, and bone marrow of these mice indicating the potential for MRI detection of micrometastases. Control NPs showed no labelling of metastatic cells, highlighting the importance of targeting for delivery to metastatic disease.

New examples of the application of nanoparticles as MRI contrast agents are published every week – though many now feature as multimodal or theranostic nanoparticles and not solely as contrast agents. Here are a few key contrast agent examples that have appeared recently. Bianchi et al. published the preparation of ultrasmall silica nanoparticles functionalised with gadolinium chelates for the detection of non-small-cell lung cancer [169]. The main novelties of this work are the top-down approach used for the preparation of the nanoparticles (final hydrodynamic size <5nm), and the combination of orotracheal administration of the particles with ultrashort echo time (UTE) free-breathing MRI acquisitions to identify and segment the tumours. Also in the field of T_1 contrast agents, Xing and co-workers published the preparation and application of ultrasmall $NaGdF_4$ nanodots for MR angiography and atherosclerotic plaque imaging [170]. In this work the authors show that the performance of Gd containing inorganic fluoride nanoparticles can be superior to commercially available T_1 contrast agents (Magnevist), while their safety can be preserved by a rapid renal excretion (due to the small size) and by the inclusion of chelating molecules (DTPA) on the surface of the nanoparticles. Moving now to T_2 contrast agents, Liu et al. prepared iron oxide nanoparticles following thermal decomposition protocols, and after bringing them to water by their encapsulation into phospholipidic shells, applied them to quantitatively test reticuloendothelial system function by MRI [171]. In this piece of work the authors took advantage of one of the most common drawbacks of the application of nanoparticles *in vivo*, their rapid clearance from circulation by the immune system. In another T_2 example, Gallo et al. prepared magnetite nanoparticles functionalised with a complex ligand system designed to give an MRI signal enhancement only in the tumour environment [172]. Using the CXCR4 receptor as a target and matrix metalloproteinase (MMP) enzymes as triggers, the nanoparticles self-assemble only in the tumour to give an enhanced signal.

9.8   CONCLUSIONS AND OUTLOOK

The examples summarised in the previous section show how the application of nanotechnology to human health, although still in its initial stages, is most promising. One of the main advantages of nanoparticle formulations is the possibility of using the motif as a functional platform onto which a number of different ligands can be assembled. This allows the simultaneous delivery of different drugs, or the combination of different imaging probes (treated in Chapter 16), together with targeting molecules to gain specificity. It also allows the multivalent presentation of ligands (giving even more versatility to the system as weaker targeting molecules can be used). All these properties can give rise to the appearance of synergistic effects between drugs, and their targeted delivery to the diseased site has the potential to reduce side effects from the aggressive treatment of diseases like cancer. The targeted delivery of contrast agents will also help in the early diagnosis of diseases because the specific delivery of the imaging probes will allow the imaging of smaller features by a reduction in the background signal.

Despite the good results obtained so far in *in vivo* tests mainly in mice, there are still a number of problems to be solved before the widespread application of targeted MRI contrast agents in humans. The accumulation of nanoparticles in general in the liver/spleen represents a problem for their application in this field, and issues such as optimised blood half-lives of these systems are still not completely solved. The toxicity of nanoparticles is an area onto which great efforts are being concentrated upon, and long-term toxicity issues with nanomaterials are only really starting to be fully investigated. More studies are also needed to fully understand the relationship between the size, shape, and charge and the behaviour of the probes *in vivo*.

Finally, the intrinsic low sensitivity of MRI is still an important, and sometimes limiting, factor in any healthcare application. A good deal of work is being carried out in this respect; the development of new materials with higher relaxivity properties, the design of more complex ligands to enhance the performance of magnetic materials, and also developmental work on MRI technology, are all expected to help increase the sensitivity of MRI and thus help translate the magnetic nanoparticles from the laboratory to the clinic.

REFERENCES

[1] A. H. Lu, E. L. Salabas and F. Schuth, *Angew. Chem. Int. Ed. Engl.* **46**, 1222–1244 (2007).

[2] D. E. Sosnovik, M. Nahrendorf and R. Weissleder, *Basic Res. Cardiol.* **103**, 122–130 (2008).

[3] T. D. Schladt, K. Schneider, H. Schild and W. Tremel, *Dalton Trans.* **40**, 6315–6343 (2011).

[4] J. Gao, H. Gu and B. Xu, *Accounts Chem. Res.* **42**, 1097–1107 (2009).

[5] R. Qiao, C. Yang and M. Gao, *J. Mater. Chem.* **19**, 6274–6293 (2009).

[6] Y. W. Jun, J. H. Lee and J. Cheon, *Angew. Chem. Int. Ed. Engl.* **47**, 5122–5135 (2008).

[7] T. Hyeon and H. B. Na, *J. Mater. Chem.* **19**, 6267–6273 (2009).

[8] J. W. Bulte and D. L. Kraitchman, *NMR Biomed.* **17**, 484–499 (2004).

[9] W. Y. Huang and J. J. Davis, *Dalton Trans.* **40**, 6087–6103 (2011).

[10] P. C. Lauterbur, *Pure Appl. Chem.* **40**, 149–157 (1974).

[11] M. Ohgushi, K. Nagayama and A. Wada, *J. Magn. Reson.* **29**, 599–601 (1978).

[12] M. H. M. Dias and P. C. Lauterbur, *Magnet. Reson. Med.* **3**, 328–330 (1986).

[13] S. Laurent, D. Forge, M. Port, A. Roch, C. Robic, L. Vander Elst and R. N. Muller, *Chem. Rev.* **108**, 2064–2110 (2008).

[14] R. Massart, *Ieee T. Magn.* **17**, 1247–1248 (1981).

[15] O. Veiseh, J. W. Gunn and M. Q. Zhang, *Adv. Drug Deliver. Rev.* **62**, 284–304 (2010).

[16] J. P. Jolivet, C. Chaneac and E. Tronc, *Chem. Commun.* 481–487 (2004).

[17] C. Tassa, S. Y. Shaw and R. Weissleder, *Accounts Chem. Res.* **44**, 842–852 (2011).

[18] A. Bee, R. Massart and S. Neveu, *J. Magn. Magn. Mater.* **149**, 6–9 (1995).

[19] A. P. Alivisatos, J. Rockenberger and E. C. Scher, *J. Am. Chem. Soc.* **121**, 11595–11596 (1999).

[20] T. Hyeon, S. J. Park, S. Kim, S. Lee, Z. G. Khim and K. Char, *J. Am. Chem. Soc.* **122**, 8581–8582 (2000).

[21] T. Hyeon, S. S. Lee, J. Park, Y. Chung and H. Bin Na, *J. Am. Chem. Soc.* **123**, 12798–12801 (2001).

[22] T. Hyeon, J. Park, E. Lee, N. M. Hwang, M. S. Kang, S. C. Kim, Y. Hwang, J. G. Park, H. J. Noh, J. Y. Kini and J. H. Park, *Angew. Chem. Int. Ed.* **44**, 2872–2877 (2005).

[23] J. Park, K. An, Y. Hwang, J. G. Park, H. J. Noh, J. Y. Kim, J. H. Park, N. M. Hwang and T. Hyeon, *Nat. Mater.* **3**, 891–895 (2004).

[24] S. Sun, H. Zeng, D. B. Robinson, S. Raoux, P. M. Rice, S. X. Wang and G. Li, *J. Am. Chem. Soc.* **126**, 273–279 (2004).

[25] J. H. Lee, Y. M. Huh, Y. W. Jun, J. W. Seo, J. T. Jang, H. T. Song, S. Kim, E. J. Cho, H. G. Yoon, J. S. Suh and J. Cheon, *Nat. Med.* **13**, 95–99 (2007).

[26] J. T. Jang, H. Nah, J. H. Lee, S. H. Moon, M. G. Kim and J. Cheon, *Angew. Chem. Int. Ed. Engl.* **48**, 1234–1238 (2009).

[27] S. Sun, C. B. Murray, D. Weller, L. Folks and A. Moser, *Science* **287**, 1989–1992 (2000).

[28] S. Chen, L. Wang, S. L. Duce, S. Brown, S. Lee, A. Melzer, A. Cuschieri and P. Andre, *J. Am. Chem. Soc.* **132**, 15022–15029 (2010).

[29] S. W. Chou, Y. H. Shau, P. C. Wu, Y. S. Yang, D. B. Shieh and C. C. Chen, *J. Am. Chem. Soc.* **132**, 13270–13278 (2010).

[30] M. Yin and S. O'Brien, *J. Am. Chem. Soc.* **125**, 10180–10181 (2003).

[31] H. B. Na, J. H. Lee, K. An, Y. I. Park, M. Park, I. S. Lee, D. H. Nam, S. T. Kim, S. H. Kim, S. W. Kim, K. H. Lim, K. S. Kim, S. O. Kim and T. Hyeon, *Angew. Chem. Int. Ed. Engl.* **46**, 5397–5401 (2007).

[32] M. Y. Gao, Z. Li, H. Chen and H. B. Bao, *Chem. Mater.* **16**, 1391–1393 (2004).

[33] F. Hu, Z. Li, C. Tu and M. Gao, *J. Colloid. Interface Sci.* **311**, 469–474 (2007).

[34] A. K. Gupta and M. Gupta, *Biomaterials.* **26**, 3995–4021 (2005).

[35] K. Inouye, R. Endo, Y. Otsuka, K. Miyashiro, K. Kaneko and T. Ishikawa, *J. Phys. Chem.* **86**, 1465–1469 (1982).

[36] B. Chu, P. A. Dresco, V. S. Zaitsev and R. J. Gambino, *Langmuir* **15**, 1945–1951 (1999).

[37] G. T. Dimitrova, T. F. Tadros, P. F. Luckham and M. R. Kipps, *Langmuir* **12**, 315–318 (1996).

[38] J. A. L. Perez, M. A. L. Quintela, J. Mira, J. Rivas and S. W. Charles, *J. Phys. Chem. B.* **101**, 8045–8047 (1997).

[39] E. E. Carpenter, C. T. Seip and C. J. O'Connor, *J. Appl. Phys.* **85**, 5184–5186 (1999).

[40] C. Liu, A. J. Rondinone, R. M. Anderson and Z. J. Zhang, *Abstr. Pap. Am. Chem. Soc.* **219**, U777 (2000).

[41] X. Wang, J. Zhuang, Q. Peng and Y. Li, *Nature.* **437**, 121–124 (2005).

[42] R. S. Varma, B. Baruwati and M. N. Nadagouda, *J. Phys. Chem. C.* **112**, 18399–18404 (2008).

[43] S. Ge, X. Shi, K. Sun, C. Li, J. R. Baker, M. M. Banaszak Holl and B. G. Orr, *J. Phys. Chem. C. Nanomater. Interfaces* **113**, 13593–13599 (2009).

[44] F. Favier, C. Pascal, J. L. Pascal, M. L. E. Moubtassim and C. Payen, *Chem. Mater.* **11**, 141–147 (1999).

[45] R. Alexandrescu, I. Morjan, I. Voicu, F. Dumitrache, L. Albu, I. Soare and G. Prodan, *Appl. Surf. Sci.* **248**, 138–146 (2005).

[46] R. Abu Mukh-Qasem and A. Gedanken, *J. Colloid. Interface Sci.* **284**, 489–494 (2005).

[47] M. A. McDonald and K. L. Watkin, *Acad. Radiol.* **13**, 421–427 (2006).

[48] M. A. Fortin, R. M. Petoral, F. Soderlind, A. Klasson, M. Engstrom, T. Veres, P. O. Kall and K. Uvdal, *Nanotechnology* **18**, 395501, (2007).

[49] J. L. Bridot, A. C. Faure, S. Laurent, C. Riviere, C. Billotey, B. Hiba, M. Janier, V. Josserand, J. L. Coll, L. V. Elst, R. Muller, S. Roux, P. Perriat and O. Tillement, *J. Am. Chem. Soc.* **129**, 5076–5084 (2007).

[50] J. Y. Park, M. J. Baek, E. S. Choi, S. Woo, J. H. Kim, T. J. Kim, J. C. Jung, K. S. Chae, Y. Chang and G. H. Lee, *ACS Nano.* **3**, 3663–3669 (2009).

[51] G. H. Lee, J. Y. Park, E. S. Choi, M. J. Baek, S. Woo and Y. Chang, *Eur. J. Inorg. Chem.* 2477–2481 (2009).

[52] F. C. J. M. van Veggel, F. Evanics, P. R. Diamente, G. J. Stanisz and R. S. Prosser, *Chem. Mater.* **18**, 2499–2505 (2006).

[53] H. Hifumi, S. Yamaoka, A. Tanimoto, D. Citterio and K. Suzuki, *J. Am. Chem. Soc.* **128**, 15090–15091 (2006).

[54] M. J. Baek, J. Y. Park, W. Xu, K. Kattel, H. G. Kim, E. J. Lee, A. K. Patel, J. J. Lee, Y. Chang, T. J. Kim, J. E. Bae, K. S. Chae and G. H. Lee, *ACS Appl. Mater. Interfaces* **2**, 2949–2955 (2010).

[55] C. C. Huang, N. H. Khu and C. S. Yeh, *Biomaterials* **31**, 4073–4078 (2010).

[56] K. An, S. G. Kwon, M. Park, H. B. Na, S. I. Baik, J. H. Yu, D. Kim, J. S. Son, Y. W. Kim, I. C. Song, W. K. Moon, H. M. Park and T. Hyeon, *Nano Lett.* **8**, 4252–4258 (2008).

[57] T. L. Ha, H. J. Kim, J. Shin, G. H. Im, J. W. Lee, H. Heo, J. Yang, C. M. Kang, Y. S. Choe, J. H. Lee and I. S. Lee, *Chem. Commun.* **47**, 9176–9178 (2011).

[58] W. J. Rieter, J. S. Kim, K. M. Taylor, H. An, W. Lin and T. Tarrant, *Angew. Chem. Int. Ed. Engl.* **46**, 3680–3682 (2007).

[59] W. B. Lin, K. M. L. Taylor, J. S. Kim, W. J. Rieter, H. An and W. L. Lin, *J. Am. Chem. Soc.* **130**, 2154ff (2008).

[60] S. Roux, C. Alric, J. Taleb, G. Le Duc, C. Mandon, C. Billotey, A. Le Meur-Herland, T. Brochard, F. Vocanson, M. Janier, P. Perriat and O. Tillement, *J. Am. Chem. Soc.* **130**, 5908–5915 (2008).

[61] C. R. Mayer, L. Moriggi, C. Cannizzo, E. Dumas, A. Ulianov and L. Helm, *J. Am. Chem. Soc.* **131**, 10828ff (2009).

[62] M. Marradi, D. Alcantara, J. M. de la Fuente, M. L. Garcia-Martin, S. Cerdan and S. Penades, *Chem. Commun.* 3922–3924 (2009).

[63] Y. Song, X. Xu, K. W. MacRenaris, X. Q. Zhang, C. A. Mirkin and T. J. Meade, *Angew. Chem. Int. Ed. Engl.* **48**, 9143–9147 (2009).

[64] C. Richard, B. T. Doan, J. C. Beloeil, M. Bessodes, E. Toth and D. Scherman, *Nano Lett.* **8**, 232–236 (2008).

[65] L. A. Tran, R. Krishnamurthy, R. Muthupillai, G. Cabreira-Hansen Mda, J. T. Willerson, E. C. Perin and L. J. Wilson, *Biomaterials* **31**, 9482–9491 (2010).

[66] T. Hyeon, H. B. Na and I. C. Song, *Adv. Mater.* **21**, 2133–2148 (2009).

[67] U. I. Tromsdorf, O. T. Bruns, S. C. Salmen, U. Beisiegel and H. Weller, *Nano Lett.* **9**, 4434–4440 (2009).

[68] S. H. Choi, B. H. Kim, N. Lee, H. Kim, K. An, Y. Il Park, Y. Choi, K. Shin, Y. Lee, S. G. Kwon, H. B. Na, J. G. Park, T. Y. Ahn, Y. W. Kim, W. K. Moon and T. Hyeon, *J. Am. Chem. Soc.* **133**, 12624–12631 (2011).

[69] W. S. Seo, J. H. Lee, X. Sun, Y. Suzuki, D. Mann, Z. Liu, M. Terashima, P. C. Yang, M. V. McConnell, D. G. Nishimura and H. Dai, *Nat. Mater.* **5**, 971–976 (2006).

[70] J. S. Choi, J. H. Lee, T. H. Shin, H. T. Song, E. Y. Kim and J. Cheon, *J. Am. Chem. Soc.* **132**, 11015–11017 (2010).

[71] H. Yang, Y. Zhuang, Y. Sun, A. Dai, X. Shi, D. Wu, F. Li, H. Hu and S. Yang, *Biomaterials* **32**, 4584–4593 (2011).

[72] C. C. Huang, C. Y. Tsai, H. S. Sheu, K. Y. Chuang, C. H. Su, U. S. Jeng, F. Y. Cheng, H. Y. Lei and C. S. Yeh, *ACS Nano.* **5**, 3905–3916 (2011).

[73] C. C. Berry, *J. Phys. D Appl. Phys.* **42**, 224003 (2009).

[74] S. Peng, C. Wang, J. Xie and S. Sun, *J. Am. Chem. Soc.* **128**, 10676–10677 (2006).

[75] R. Hong, N. O. Fischer, T. Emrick and V. M. Rotello, *Chem. Mater.* **17**, 4617–4621 (2005).

[76] J. Gao, G. Liang, J. S. Cheung, Y. Pan, Y. Kuang, F. Zhao, B. Zhang, X. Zhang, E. X. Wu and B. Xu, *J. Am. Chem. Soc.* **130**, 11828–11833 (2008).

[77] D. Patel, A. Kell, B. Simard, B. Xiang, H. Y. Lin and G. Tian, *Biomaterials* **32**, 1167–1176 (2011).

[78] Y. Lalatonne, C. Paris, J. M. Serfaty, P. Weinmann, M. Lecouvey and L. Motte, *Chem. Commun.* 2553–2555 (2008).

[79] M. A. White, J. A. Johnson, J. T. Koberstein and N. J. Turro, *J. Am. Chem. Soc.* **128**, 11356–11357 (2006).

[80] A. Hofmann, S. Thierbach, A. Semisch, A. Hartwig, M. Taupitz, E. Ruhl and C. Graf, *J. Mater. Chem.* **20**, 7842–7853 (2010).

[81] L. L. Zhou, J. Y. Yuan and Y. Wei, *J. Mater. Chem.* **21**, 2823–2840 (2011).

[82] A. Weddemann, I. Ennen, A. Regtmeier, C. Albon, A. Wolff, K. Eckstadt, N. Mill, M. K. Peter, J. Mattay, C. Plattner, N. Sewald and A. Hutten, *Beilstein J. Nanotechnol.* **1**, 75–93 (2010).

[83] C. Yee, G. Kataby, A. Ulman, T. Prozorov, H. White, A. King, M. Rafailovich, J. Sokolov and A. Gedanken, *Langmuir* **15**, 7111–7115 (1999).

[84] M. C. Daniel and D. Astruc, *Chem. Rev.* **104**, 293–346 (2004).

[85] J. H. Park, G. von Maltzahn, L. Zhang, M. P. Schwartz, E. Ruoslahti, S. N. Bhatia and M. J. Sailor, *Adv. Mater.* **20**, 1630–1635 (2008).

[86] M. F. Bennewitz, T. L. Lobo, M. K. Nkansah, G. Ulas, G. W. Brudvig and E. M. Shapiro, *ACS Nano.* **5**, 3438–3446 (2011).

[87] I. K. Park, C. P. Ng, J. Wang, B. Chu, C. Yuan, S. Zhang and S. H. Pun, *Biomaterials* **29**, 724–732 (2008).

[88] X. Y. Shi, S. H. Wang, S. D. Swanson, S. Ge, Z. Y. Cao, M. E. Van Antwerp, K. J. Landmark and J. R. Baker, *Adv. Mater.* **20**, 1671–1677 (2008).

[89] F. Fuertges and A. Abuchowski, *J. Control. Release* **11**, 139–148 (1990).

[90] J. M. Harris and R. B. Chess, *Nat. Rev. Drug Discov.* **2**, 214–221 (2003).

[91] U. I. Tromsdorf, N. C. Bigall, M. G. Kaul, O. T. Bruns, M. S. Nikolic, B. Mollwitz, R. A. Sperling, R. Reimer, H. Hohenberg, W. J. Parak, S. Forster, U. Beisiegel, G. Adam and H. Weller, *Nano Lett.* **7**, 2422–2427 (2007).

[92] M. Kumagai, M. R. Kano, Y. Morishita, M. Ota, Y. Imai, N. Nishiyama, M. Sekino, S. Ueno, K. Miyazono and K. Kataoka, *J. Control. Release* **140**, 306–311 (2009).

[93] E. Amstad, S. Zurcher, A. Mashaghi, J. Y. Wong, M. Textor and E. Reimhult, *Small* **5**, 1334–1342 (2009).

[94] K. Chen, J. Xie, H. Xu, D. Behera, M. H. Michalski, S. Biswal, A. Wang and X. Chen, *Biomaterials* **30**, 6912–6919 (2009).

[95] D. B. Robinson, H. H. Persson, H. Zeng, G. Li, N. Pourmand, S. Sun and S. X. Wang, *Langmuir* **21**, 3096–3103 (2005).

[96] M. S. Martina, J. P. Fortin, C. Menager, O. Clement, G. Barratt, C. Grabielle-Madelmont, F. Gazeau, V. Cabuil and S. Lesieur, *J. Am. Chem. Soc.* **127**, 10676–10685 (2005).

[97] H. Yang, H. Zhou, C. X. Zhang, X. J. Li, H. Hu, H. X. Wu and S. P. Yang, *Dalton Trans.* **40**, 3616–3621 (2011).

[98] N. Lee, Y. Choi, Y. Lee, M. Park, W. K. Moon, S. H. Choi and T. Hyeon, *Nano Lett.* **12**, 3127–3131 (2012).

[99] W. W. Yu, E. Chang, C. M. Sayes, R. Drezek and V. L. Colvin, *Nanotechnology* **17**, 4483–4487 (2006).

[100] N. Nasongkla, E. Bey, J. M. Ren, H. Ai, C. Khemtong, J. S. Guthi, S. F. Chin, A. D. Sherry, D. A. Boothman and J. M. Gao, *Nano Lett.* **6**, 2427–2430 (2006).

[101] J. Qin, S. Laurent, Y. S. Jo, A. Roch, M. Mikhaylova, Z. M. Bhujwalla, R. N. Muller and M. Muhammed, *Adv. Mater.* **19**, 1874–1878 (2007).

[102] B. Sitharaman, K. R. Kissell, K. B. Hartman, L. A. Tran, A. Baikalov, I. Rusakova, Y. Sun, H. A. Khant, S. J. Ludtke, W. Chiu, S. Laus, E. Toth, L. Helm, A. E. Merbach and L. J. Wilson, *Chem. Commun.* 3915–3917 (2005).

[103] W. Stöber, A. Fink and E. Bohn, *J. Colloid. Interf. Sci.* **26**, 62–69 (1968).

[104] S. Y. Chang, L. Liu and S. A. Asher, *J. Am. Chem. Soc.* **116**, 6739–6744 (1994).

[105] A. P. Philipse, M. P. B. Vanbruggen and C. Pathmamanoharan, *Langmuir* **10**, 92–99 (1994).

[106] K. S. Park, J. Tae, B. Choi, Y. S. Kim, C. Moon, S. H. Kim, H. S. Lee, J. Kim, J. Park, J. H. Lee, J. E. Lee, J. W. Joh and S. Kim, *Nanomedicine* **6**, 263–276 (2010).

[107] C. W. Lu, Y. Hung, J. K. Hsiao, M. Yao, T. H. Chung, Y. S. Lin, S. H. Wu, S. C. Hsu, H. M. Liu, C. Y. Mou, C. S. Yang, D. M. Huang and Y. C. Chen, *Nano Lett.* **7**, 149–154 (2007).

[108] K. El-Boubbou, D. C. Zhu, C. Vasileiou, B. Borhan, D. Prosperi, W. Li and X. Huang, *J. Am. Chem. Soc.* **132**, 4490–4499 (2010).

[109] D. Kryza, J. Taleb, M. Janier, L. Marmuse, I. Miladi, P. Bonazza, C. Louis, P. Perriat, S. Roux, O. Tillement and C. Billotey, *Bioconjug. Chem.* **22**, 1145–1152 (2011).

[110] L. Y. Wang, J. Luo, M. M. Maye, Q. Fan, R. D. Qiang, M. H. Engelhard, C. M. Wang, Y. H. Lin and C. J. Zhong, *J. Mater. Chem.* **15**, 1821–1832 (2005).

[111] J. Gallo, I. Garcia, D. Padro, B. Arnaiz and S. Penades, *J. Mater. Chem.* **20**, 10010–10020 (2010).

[112] Z. Xu, Y. Hou and S. Sun, *J. Am. Chem. Soc.* **129**, 8698–8699 (2007).

[113] J. L. Lyon, D. A. Fleming, M. B. Stone, P. Schiffer and M. E. Williams, *Nano Lett.* **4**, 719–723 (2004).

[114] L. S. Bouchard, M. S. Anwar, G. L. Liu, B. Hann, Z. H. Xie, J. W. Gray, X. D. Wang, A. Pines and F. F. Chen, *Proc. Natl. Acad. Sci. USA.* **106**, 4085–4089 (2009).

[115] A. H. Lu, W. C. Li, E. L. Salabas, B. Spliethoff and F. Schuth, *Chem. Mater.* **18**, 2086–2094 (2006).

[116] A. H. Lu, W. C. Li, N. Matoussevitch, B. Spliethoff, H. Bonnemann and F. Schuth, Chem. Commun. 98–100 (2005).

[117] J. F. Geng, D. A. Jefferson and B. F. G. Johnson, *Chem. Commun.* 2442–2443 (2004).

[118] V. V. Baranauskas, M. A. Zalich, M. Saunders, T. G. St Pierre and J. S. Riffle, *Chem. Mater.* **17**, 5246–5254 (2005).

[119] K. H. Bae, K. Lee, C. Kim and T. G. Park, *Biomaterials* **32**, 176–184 (2011).

[120] J. Huang, J. Xie, K. Chen, L. Bu, S. Lee, Z. Cheng, X. Li and X. Chen, *Chem. Commun.* **46**, 6684–6686 (2010).

[121] E. Schellenberger, F. Rudloff, C. Warmuth, M. Taupitz, B. Hamm and J. Schnorr, *Bioconjugate Chem.* **19**, 2440–2445 (2008).

[122] E. N. M. Cheung, R. D. A. Alvares, W. Oakden, R. Chaudhary, M. L. Hill, J. Pichaandi, G. C. H. Mo, C. Yip, P. M. Macdonald, G. J. Stanisz, F. C. J. M. van Veggel and R. S. Prosser, *Chem. Mater.* **22**, 4728–4739 (2010).

[123] A. Z. Wang, V. Bagalkot, C. C. Vasilliou, F. Gu, F. Alexis, L. Zhang, M. Shaikh, K. Yuet, M. J. Cima, R. Langer, P. W. Kantoff, N. H. Bander, S. Jon and O. C. Farokhzad, *Chem. Med. Chem.* **3**, 1311–1315 (2008).

[124] F. Q. Hu, L. Wei, Z. Zhou, Y. L. Ran, Z. Li and M. Y. Gao, *Adv. Mater.* **18**, 2553–2556 (2006).

[125] P. Zou, Y. K. Yu, Y. A. Wang, Y. Q. Zhong, A. Welton, C. Galban, S. M. Wang and D. X. Sun, *Mol. Pharmaceut.* **7**, 1974–1984 (2010).

[126] L. L. Yang, H. Mao, Y. A. Wang, Z. H. Cao, X. H. Peng, X. X. Wang, H. W. Duan, C. C. Ni, Q. G. Yuan, G. Adams, M. Q. Smith, W. C. Wood, X. H. Gao and S. M. Nie, *Small* **5**, 235–243 (2009).

[127] C. J. Xu, Z. L. Yuan, N. Kohler, J. M. Kim, M. A. Chung and S. H. Sun, *J. Am. Chem. Soc.* **131**, 15346–15351 (2009).

[128] E. J. Cha, E. S. Jang, I. C. Sun, I. J. Lee, J. H. Ko, Y. I. Kim, I. C. Kwon, K. Kim and C. H. Ahn, *J. Control. Release* **155**, 152–158 (2011).

[129] J. Gallo, I. Garcia, N. Genicio, D. Padro and S. Penades, *Biomaterials* **32**, 9818–9825 (2011).

[130] R. Weissleder, K. Kelly, E. Y. Sun, T. Shtatland and L. Josephson, *Nat. Biotechnol.* **23**, 1418–1423 (2005).

[131] Y. Liu, Z. Chen, C. Liu, D. Yu, Z. Lu and N. Zhang, *Biomaterials* **32**, 5167–5176 (2011).

[132] Z. Medarova, W. Pham, C. Farrar, V. Petkova and A. Moore, *Nat. Medicine.* **13**, 372–377 (2007).

[133] Y. M. Huh, Y. W. Jun, H. T. Song, S. Kim, J. S. Choi, J. H. Lee, S. Yoon, K. S. Kim, J. S. Shin, J. S. Suh and J. Cheon, *J. Am. Chem. Soc.* **127**, 12387–12391 (2005).

[134] C. Sun, O. Veiseh J. Gunn, C. Fang, S. Hansen, D. Lee, R. Sze, R. G. Ellenbogen, J. Olson and M. Zhang, *Small* **4**, 372–379 (2008).

[135] C. R. Sun, K. Du, C. Fang, N. Bhattarai, O. Veiseh, F. Kievit, Z. Stephen, D. H. Lee, R. G. Ellenbogen, B. Ratner and M. Q. Zhang, *ACS Nano.* **4**, 2402–2410 (2010).

[136] Y. W. Jun, Y. M. Huh, J. S. Choi, J. H. Lee, H. T. Song, S. Kim, S. Yoon, K. S. Kim, J. S. Shin, J. S. Suh and J. Cheon, *J. Am. Chem. Soc.* **127**, 5732–5733 (2005).

[137] Y. S. Cho, T. J. Yoon, E. S. Jang, K. S. Hong, S. Y. Lee, O. R. Kim, C. Park, Y. J. Kim, G. C. Yi and K. Chang, *Cancer Lett.* **299**, 63–71 (2010).

[138] L. A. Wang, K. G. Neoh, E. T. Kang and B. Shuter, *Biomaterials.* **32**, 2166–2173 (2011).

[139] E. A. Schellenberger, D. Sosnovik, R. Weissleder and L. Josephson, *Bioconj. Chem.* **15**, 1062–1067 (2004).

[140] T. J. Harris, G. von Maltzahn, A. M. Derfus, E. Ruoslahti and S. N. Bhatia, *Angew. Chem. Int. Edit.* **45**, 3161–3165 (2006).

[141] S. Mazzucchelli, M. Colombo, C. De Palma, A. Salvade, P. Verderio, M. D. Coghi, E. Clementi, P. Tortora, F. Corsi and D. Prosperi, *ACS Nano.* **4**, 5693–5702 (2010).

[142] F. Corsi, L. Fiandra, C. De Palma, M. Colombo, S. Mazzucchelli, P. Verderio, R. Allevi, A. Tosoni, M. Nebuloni, E. Clementi and D. Prosperi, *ACS Nano.* **5**, 6383–6393 (2011).

[143] C. W. Tornoe, C. Christensen and M. Meldal, *J. Org. Chem.* **67**, 3057–3064 (2002).

[144] V. V. Rostovtsev, L. G. Green, V. V. Fokin and K. B. Sharpless, *Angew. Chem. Int. Edit.* **41**, 2596ff (2002).

[145] N. J. Agard, J. A. Prescher and C. R. Bertozzi, *J. Am. Chem. Soc.* **127**, 11196 (2005).

[146] K. El-Boubbou, D. C. Zhu, C. Vasileiou, B. Borhan, D. Prosperi, W. Li and X. F. Huang, *J. Am. Chem. Soc.* **132**, 4490–4499 (2010).

[147] G. von Maltzahn, Y. Ren, J. H. Park, D. H. Min, V. R. Kotamraju, J. Jayakumar, V. Fogal, M. J. Sailor, E. Ruoslahti and S. N. Bhatia, *Bioconjugate Chem.* **19**, 1570–1578 (2008).

[148] A. Toma, E. Otsuji, Y. Kuriu, K. Okamoto, D. Ichikawa, A. Hagiwara, H. Ito, T. Nishimura and H. Yamagishi, *Brit. J. Cancer.* **93**, 131–136 (2005).

[149] M. Zhao, D. A. Beauregard, L. Loizou, B. Davletov and K. M. Brindle, *Nat. Medicine.* **7**, 1241–1244 (2001).

[150] P. C. Lin, S. H. Chen, K. Y. Wang, M. L. Chen, A. K. Adak, J. R. R. Hwu, Y. J. Chen and C. C. Lin, *Anal. Chem.* **81**, 8774–8782 (2009).

[151] L. Zhang, Y. Xu, H. Yao, L. Xie, J. Yao, H. Lu and P. Yang, *Chemistry.* **15**, 10158–10166 (2009).

[152] J. Tang, Y. Liu, P. Yin, G. P. Yao, G. Q. Yan, C. H. Deng and X. M. Zhang, *Proteomics.* **10**, 2000–2014 (2010).

[153] J. Xie, K. Chen, H. Y. Lee, C. J. Xu, A. R. Hsu, S. Peng, X. Y. Chen and S. H. Sun, *J. Am. Chem. Soc.* **130**, 7542–7543 (2008).

[154] S. I. van Kasteren, S. J. Campbell, S. Serres, D. C. Anthony, N. R. Sibson and B. G. Davis, *Proc. Natl. Acad. Sci. USA.* **106**, 18–23 (2009).

[155] G. von Maltzahn, T. J. Harris, J. H. Park, D. H. Min, A. J. Schmidt, M. J. Sailor and S. N. Bhatia, *J. Am. Chem. Soc.* **129**, 6064–6065 (2007).

[156] R. Bardhan, W. X. Chen, M. Bartels, C. Perez-Torres, M. F. Botero, R. W. McAninch, A. Contreras, R. Schiff, R. G. Pautler, N. J. Halas and A. Joshi, *Nano Lett.* **10**, 4920–4928 (2010).

[157] C. Kaittanis, S. A. Naser and J. M. Perez, *Nano Lett.* **7**, 380–383 (2007).

[158] Y. T. Lim, M. Y. Cho, J. M. Lee, S. J. Chung and B. H. Chung, *Biomaterials* **30**, 1197–1204 (2009).

[159] P. C. Wu, C. H. Su, F. Y. Cheng, J. C. Weng, J. H. Chen, T. L. Tsai, C. S. Yeh, W. C. Su, J. R. Hwu, Y. Tzeng and D. B. Shieh, *Bioconj. Chem.* **19**, 1972–1979 (2008).

[160] B. J. Nieman, J. Y. Shyu, J. J. Rodriguez, A. D. Garcia, A. L. Joyner and D. H. Turnbull, *Neuroimage* **50**, 456–464 (2010).

[161] C. G. Hadjipanayis, R. Machaidze, M. Kaluzova, L. Wang, A. J. Schuette, H. Chen, X. Wu and H. Mao, *Cancer Res.* **70**, 6303–6312 (2010).

[162] D. E. Sosnovik, E. Garanger, E. Aikawa, M. Nahrendorf, J. L. Figuiredo, G. P. Dai, F. Reynolds, A. Rosenzweig, R. Weissleder and L. Josephson, *Circ-Cardiovasc. Imag.* **2**, 460–467 (2009).

[163] E. J. Cho, J. Yang, K. A. Mohamedali, E. K. Lim, E. J. Kim, C. J. Farhangfar, J. S. Suh, S. Haam, M. G. Rosenblum and Y. M. Huh, *Invest. Radiol.* **46**, 441–449 (2011).

[164] N. Ferrara, *Endocr. Rev.* **25**, 581–611 (2004).

[165] S. H. Crayton and A. Tsourkas, *ACS Nano.* **5**, 9592–9601 (2011).

[166] L. M. R. Ferreira, *Exp. Mol. Pathol.* **89**, 372–380 (2010).

[167] F. M. Kievit, Z. R. Stephen, O. Veiseh, H. Arami, T. Z. Wang, V. P. Lai, J. O. Park, R. G. Ellenbogen, M. L. Disis and M. Q. Zhang, *ACS Nano.* **6**, 2591–2601 (2012).

[168] J. M. Shin, R. M. Anisur, M. K. Ko, G. H. Im, J. H. Lee and I. S. Lee, *Angew. Chem. Int. Edit.* **48**, 321–324 (2009).

[169] A. Bianchi, S. Dufort, F. Lux, P. Y. Fortin, N. Tassali, O. Tillement, J. L. Coll and Y. Cremillieux, *Proc. Natl. Acad. Sci. U. S. A.* **111**, 9247–9252 (2014)

[170] H. Y. Xing, S. J. Zhang, W. B. Bu, X. P. Zheng, L. J. Wang, Q. F. Xiao, D. L. Ni, J. M. Zhang, L. P. Zhou, W. J. Peng, K. L. Zhao, Y. Q. Hua and J. L. Shi, *Adv. Mater.* **26**, 3867–3872 (2014).

[171] T. Liu, H. Choi, R. Zhou and I. W. Chen, *Plos One* **9**, 723–730 (2014).

[172] J. Gallo, N. Kamaly, I. Lavdas, E. Stevens, Q.-D. Nguyen, M. Wylezinska-Arridge, E. O. Aboagye and N. J. Long, *Angew. Chem. Int. Ed.* **53**, 9550–9554 (2014).

10

CEST AND PARACEST AGENTS FOR MOLECULAR IMAGING

OSASERE M. EVBUOMWAN

Department of Chemistry, University of Texas at Dallas, Richardson, TX, USA

ENZO TERRENO AND SILVIO AIME

Departmento di Chimica I.F.M. and Centro di Imaging Moleculare, Universita di Torino, Torino, Italy

A. DEAN SHERRY

Advanced Imaging Research Center, UT Southwestern Medical Center, Dallas, TX, USA; Department of Chemistry, University of Texas at Dallas, Richardson, TX, USA

10.1 INTRODUCTION

Magnetic resonance imaging (MRI) is a diagnostic technique that is well known for its ability to produce high spatial resolution of soft tissue in clinical imaging [1]. The quality of anatomical information obtained in an MR image can be further augmented by taking advantage of a paramagnetic contrast agent. Current clinically approved MRI contrast agents provide little functional information about tissue other than information gained by dynamic contrast enhancement (DCE) measurements [1]. To better understand specific physiological, biochemical, or metabolic abnormalities associated with disease, it will be important to obtain information at the cellular and molecular level. This goal may be best accomplished by use of a new class of contrast media referred to as chemical exchange saturation transfer (CEST) agents.

CEST agents offer some advantages over typical Gd^{3+}-based MRI contrast agents because of the unique contrast mechanism by which they operate. Typically, these agents possess one or more labile protons that have NMR frequencies chemically distinct from bulk water protons. Application of a radiofrequency (RF) pulse at this unique frequency causes selective saturation of the CEST agent protons, and those saturated spins subsequently appear in the bulk water proton pool as a result of chemical exchange (Figure 10.1). This leads to a partial reduction in solvent proton signal intensity, and the difference in water intensity or contrast can be displayed as a dark signal, a bright signal (negative image intensity), or a colour map in an MR image [2, 3]. CEST principles offer a fundamental advantage for MRI because contrast can be introduced only when a frequency-selective RF pulse is applied at the resonance frequency of the labile protons. Hence, image contrast can be turned on and off again at will, thus eliminating the need for pre- and post-contrast image acquisition.

The CEST signal is typically reported as the ratio of the bulk water magnetisation after application of a long pre-saturation pulse (M_s) relative to the water intensity without saturation (M_o). A plot of M_s/M_o *versus* pre-saturation frequency is now commonly referred to as a CEST spectrum or a Z-spectrum [4, 5]. The magnitude of the CEST contrast is assessed by subtracting the magnetisation ratio at a specific frequency of interest (on) from the effect at an equal frequency on the opposite side of the water frequency (off).

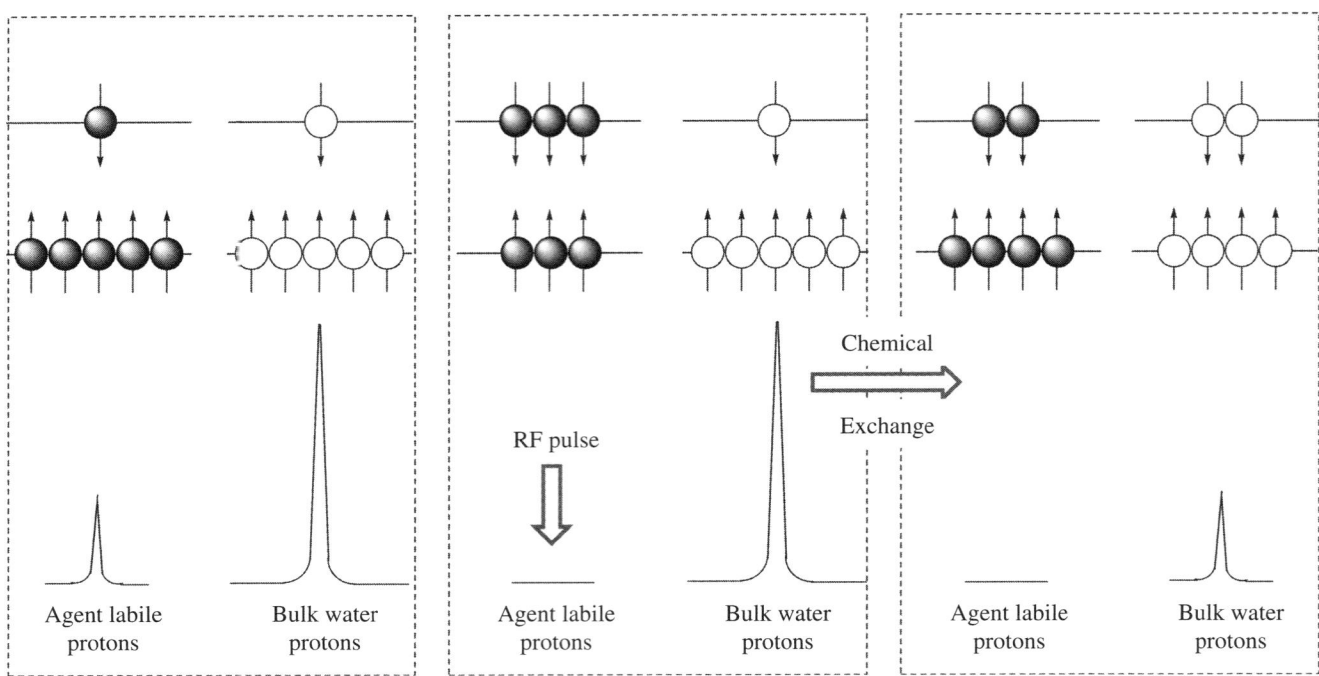

FIGURE 10.1 Illustration of the CEST mechanism showing the Boltzmann distribution of proton spins and the simulated NMR spectra for two chemically distinct pools of protons. Application of a frequency selective RF pulse causes saturation of the NMR signal, which is then transferred to the bulk water proton pool by chemical exchange.

Any molecule with labile protons (–NH, –OH) can potentially be used as a CEST agent, provided that the exchange of these protons with the bulk water protons occurs at a rate (k_{ex}) that is slower than the frequency difference ($\Delta\omega$) between the exchanging pools (Eq. 10.1).

$$k_{ex} \ll \Delta\omega \tag{10.1}$$

Many types of molecules fit this requirement, including hundreds of low-molecular weight compounds containing –NH or –OH protons [4–7], macromolecules [8–11], and nanoparticles of various types [8–15]. In this chapter, small-molecule CEST agents will be divided into classifications of diamagnetic (DIACEST) and paramagnetic (PARACEST) agents. Given that DIACEST agents typically have proton chemical shifts less than 5 ppm away from water, proton exchange must be quite slow to meet the CEST requirement, whereas the chemical shift of protons in PARACEST agents can vary widely, often lying more than 100 ppm from bulk water. This means that chemical exchange can be quite fast in PARACEST agents and still meet the CEST requirement defined by Eq. 10.1.

10.2 DIAMAGNETIC CEST AGENTS

Agents of this class are characterised by exchangeable protons from a variety of functional groups such as hydroxyl, amide, and amine groups that reside within 10 ppm of bulk water protons [16]. Although the small $\Delta\omega$ of these agents could prove disadvantageous due to an increased probability of directly saturating the bulk water protons, if the exchange rates of the labile protons are slow enough on the NMR timescale, appreciable CEST contrast can be generated using DIACEST agents.

While a number of other functional groups are capable of producing CEST contrast, amide protons represent the most attractive means of generating CEST because they exchange with water protons more slowly than hydroxyl and amine protons and the chemical shifts of amide protons are typically further downfield of water than most other types of protons. The abundance of amide protons in endogenous proteins and peptides has also facilitated the *in vivo* translation of DIACEST MRI through an imaging technique now referred to as amide proton transfer (APT) [17, 18]. Note that APT and DIACEST are identical processes but the acronym APT should be used only when the experiment reflects exchange of amide protons specifically.

Amide proton exchange is highly pH dependent. Below ~ pH 5, amide proton exchange is typically too slow for CEST; above pH 5, base catalysis of amide protons accelerates proton exchange almost linearly with increasing pH [2, 18]. This

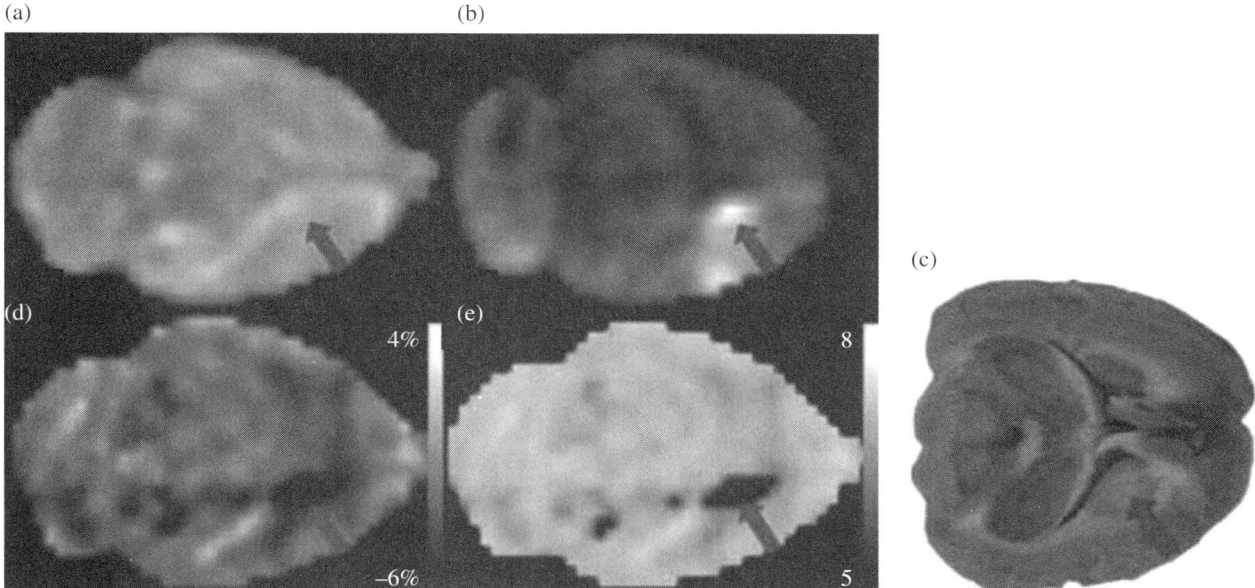

FIGURE 10.2 APT effects during focal ischemia in rat brain. No effect of ischemia was visible on the T_2-weighted image (a), but the pH-sensitive images (d and e) show the ischemic region. As confirmed by the diffusion weighted image (b), and histology (c). Adapted with permission from Ref. [18].

results in an interesting phenomenon of generating little to no CEST contrast at low pH, maximal CEST at pH values between 6 and 8 depending on the compound, then a loss in CEST signal again at high pH values [19]. Given that the magnitude of CEST contrast is heavily influenced by k_{ex}, the CEST signal can be used to monitor changes in tissue pH. This strong pH dependence of the APT signal has been used to detect ischemic regions of rat brain tissue [18, 20]. In this case, ischemic tissue resulting from lack of sufficient oxygen delivery also results in a decrease in tissue pH and a corresponding reduction in APT contrast (Figure 10.2). APT could prove useful for detecting early ischemic events such as those occurring in stroke where conventional MRI methods are not particularly sensitive.

APT imaging has also been used to detect brain tumours in both humans and animals by taking advantage of their higher protein/peptide content compared to normal brain tissue. When applied to rats implanted with 9L gliosarcoma tumours, the APT signal outlined a well-defined brain tumour region and was able to distinguish between edema and tumour, whereas T_1, T_2, and diffusion-weighted MRI methods yielded a more diffuse tumour pattern [17]. The above technique was extended into imaging human brain tumour, and the results obtained were very consistent with the animal studies [21]. More recently, the potential of utilising APT imaging for tumour grading and tumour zone delineation was noninvasively demonstrated in human brain tumours [22, 23]. APT intensities were found to be significantly higher in the core of high-grade tumours than in normal-appearing white matter (NAWM) and peri-tumoural edema in patients with high-grade tumours, whereas no such signal differences were observed in low-grade tumours. The utility of APT MRI in the differentiation of viable glioma from radiation necrosis in rats has also been reported [24].

Exogenous agents with exchangeable amide protons can also be used for CEST imaging. Iopamidol, a clinically approved X-ray contrast agent having two non-equivalent amide protons that display different exchange rates, has been exploited for pH mapping *in vivo* [25]. Upon injection into healthy mice, the distribution of Iopamidol in the kidney was imaged using CEST methods, and the corresponding pH map of the kidneys was obtained using ratiometric imaging methods (Figure 10.3). This study further reinforced the advantage of CEST over conventional MRI methods in providing functional information.

The –OH protons of endogenous molecules have also been used for CEST contrast in tissue. Although –OH protons are characterised by small chemical shifts and fast exchange, the CEST signal due to the exchange of these protons becomes more defined at higher magnetic field strengths due to an increase in $\Delta\omega$ that favours slow-to-intermediate exchange. For example, glycogen, the primary storage form of glucose in mammalian tissue, constitutes a large number of exchangeable –OH protons that can be saturated simultaneously by use of a frequency-selective RF pulse. Given the high abundance of glycogen in liver, the possibility of monitoring glycogen metabolism by CEST in perfused mouse livers at 4.7T was elegantly demonstrated by van Zijl et al. [26]. Upon addition of glucagon to induce glycogenolysis, a reduction in the CEST signal resulting from glycogen –OH exchange was observed. The reduction in CEST contrast correlated nicely with depletion of glycogen and export of glucose into the medium (Figure 10.4). Because glycogen content may be abnormal in

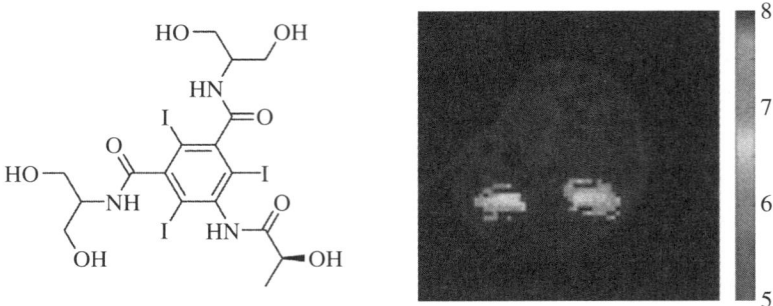

FIGURE 10.3 Left: Chemical structure of Iopamidol showing exchangeable amide protons. Right: Corresponding pH map of a mouse kidney obtained after applying ratiometric analysis to the saturation transfer effects from the amide proton pools at 4.2 ppm and 5.5 ppm. Adapted with permission from Ref. [25].

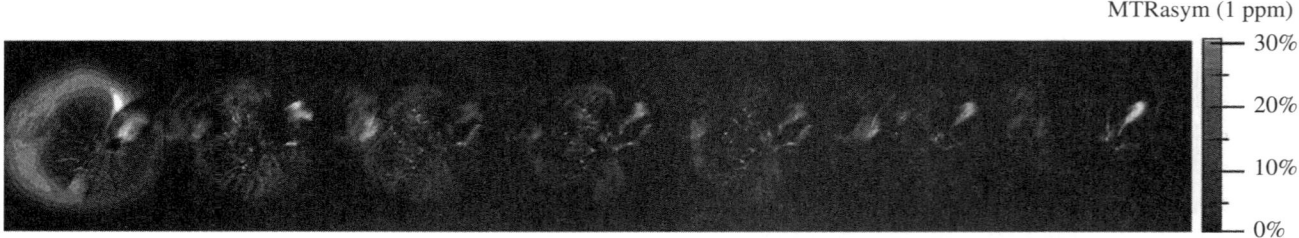

FIGURE 10.4 Top: Chemical structure of a fragment of glycogen showing exchangeable –OH protons. Bottom: GlycoCEST imaging of a perfused fed-mouse liver at 4.7 T and 37°C. The colourised images show the gradual reduction in the CEST signal originating from glycogen as a result of glucagon stimulation of glycogenolysis. Reproduced with permission from Ref. [26]. (*See insert for colour representation of the figure.*)

pre-diabetic individuals and obese subjects, noninvasive CEST imaging of liver and skeletal glycogen could be useful in developing a better understanding of the pathophysiology of these diseases.

Another exciting example of *in vivo* CEST imaging of endogenous –OH protons was presented by Ling et al. [27], who obtained a map of glycosaminoglycan (GAG) distribution in the patella of a human knee using CEST imaging. CEST images showed a clear delineation of a cartilage lesion on the medial facet of the knee in addition to a decrease in GAG concentration on the medial side of the patellofemoral knee joint (Figure 10.5). Considering that GAG loss is associated with a number of diseases such as osteoarthritis, the ability to directly image GAG in tissues could prove useful in diagnosis.

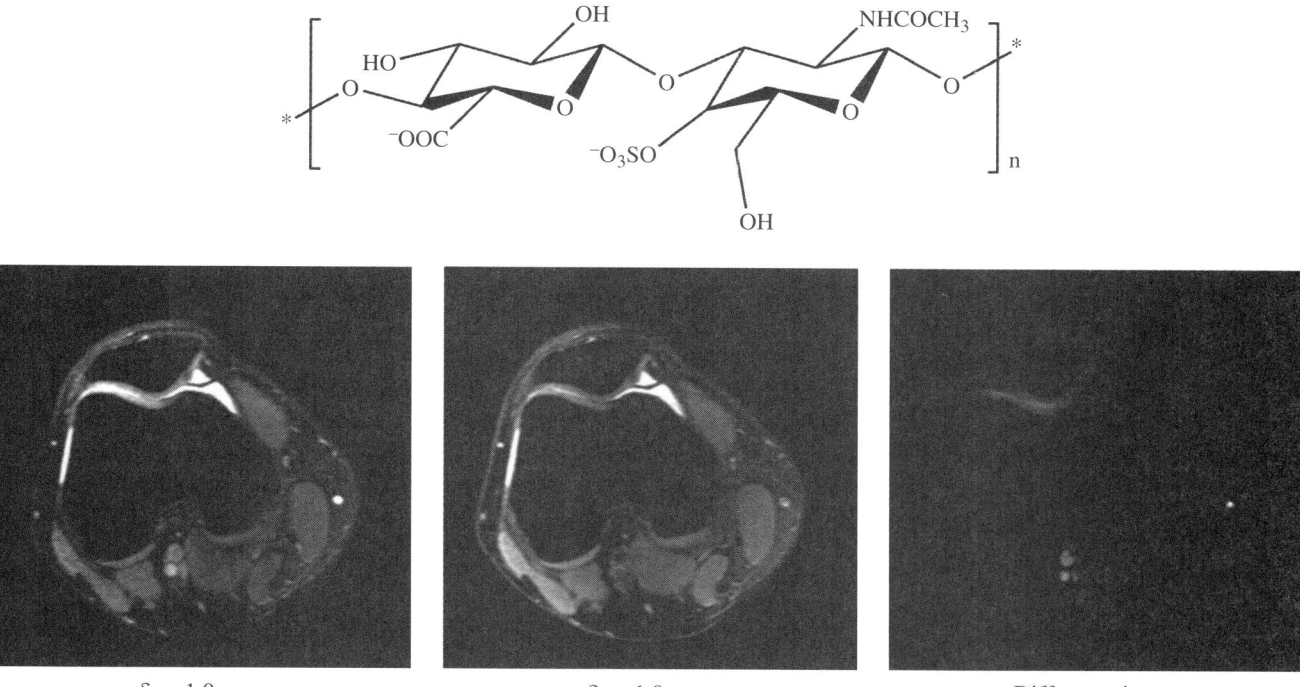

FIGURE 10.5 Top: Structure of chondroitin sulphate, a basic unit of GAG. Bottom: CEST images of a human patella *in vivo* with irradiation at −1.0 ppm (left), and at 1.0 ppm (centre). The difference image outlined a clear demarcation of a cartilage lesion on the medial facet (right). Reproduced with permission from Ref. [27].

10.3 PARAMAGNETIC CHEMICAL EXCHANGE SATURATION TRANSFER (PARACEST) AGENTS

Paramagnetic CEST agents represent another class of chemical exchange systems that offer some potential advantages and disadvantages over DIACEST agents. The most common of these agents are derived from paramagnetic lanthanide ion complexes with either exchangeable –NH protons or lanthanide ion-bound water molecules. The major advantage paramagnetic agents have over their diamagnetic counterparts is a much larger frequency difference ($\Delta\omega$) between the exchanging proton (or water molecule) with respect to the bulk water frequency. This large frequency difference allows much easier RF saturation of the paramagnetically shifted proton without indirect partial saturation of bulk water protons. Thus, the saturation selectivity improves and there is less need to perform an asymmetry analysis (difference between saturation at±offset frequencies on each side of the water peak). The larger frequency difference also permits the use of faster exchanging species without approaching the fast exchange limit. This means that unlike a typical DIACEST compound containing a –NH proton exchanging with water at a rate of 500–1000 s^{-1}, one can envision using systems that undergo exchange much more frequently, perhaps as fast as 5000–10,000 s^{-1}. Unfortunately, one major disadvantage of such faster exchanging systems is that they require more RF power for saturation, which could potentially result in tissue heating, depending upon coil efficiency and other factors. Thus, paramagnetic systems that combine large hyperfine shifts in chemical sites with moderate-to-slow proton exchange groups (–NH, –OH, or bound water molecules) offer the advantages of both DIACEST and PARACEST. Slow proton or water molecule exchanging systems will likely be a requirement of those systems that are ultimately moved forward toward clinical use. This requirement has increased interest in the chemical community for identifying lanthanide complexes that have water exchange kinetics two to three orders of magnitude slower than their DOTA and DTPA counterparts [28–30].

The slower water exchange properties of lanthanide complexes formed with DOTA-tetraamide ligands originates from the relatively poor electron-donor ability of the amide oxygen donors compared to carboxylate oxygen donors. This condition renders the lanthanide ion more electron-deficient and increases its affinity for any directly coordinated water molecule. The observation that the exchange rate of water in various Eu^{3+} complexes is the slowest amongst all other lanthanide complexes of the same ligand [31] has led to the conclusion that the EuDOTA-tetraamide complexes have the best, although not optimal, water exchange kinetics for *in vivo* use. Hence, the majority of water exchange PARACEST agents reported to date have been modifications of the basic Eu(III) DOTA-tetraamide structure. These symmetric complexes typically have Eu^{3+}-bound water resonances near 50 ppm, depending upon temperature [32].

MRI contrast produced by PARACEST agents can be generated via exchange of highly shifted lanthanide-bound water molecules (H_2O) [10, 11, 15, 32–42], amide protons (–NH) [43–48], or hydroxyl protons (–OH) [49–52], depending upon which lanthanide ion is used. Because one can chemically modify one or more ligand side chains and alter the water exchange kinetics of a system rather easily, the development of responsive or 'smart' PARACEST agents has become a focus of many research laboratories involved in the design of novel MRI contrast agents.

10.4   RESPONSIVE PARACEST AGENTS

Since k_{ex} and $\Delta\omega$ are both quite temperature sensitive (with k_{ex} typically increasing and $\Delta\omega$ decreasing with temperature), the magnitude of the CEST signal also happens to be temperature sensitive, sometimes in unexpected ways. This makes PARACEST agents particularly interesting for monitoring temperature in biological systems. Because the hyperfine shift induced by a paramagnetic lanthanide ion is always inversely proportional to temperature (can have both T_1 and T_2 components), one can quite easily and predictably obtain temperature maps [33, 53] by simply measuring $\Delta\omega$ and using a calibration curve.

The rate of proton exchange (k_{ex}) for –NH systems is strongly pH dependent. Above pH ~5, base catalysis of –NH exchange becomes dominant, resulting in an acceleration in k_{ex}. Because the magnitude of the CEST effect is strongly influenced by k_{ex}, changes in the CEST spectrum can be used to assess differences in pH. This concept was elegantly demonstrated by Aime et al. where a mixture of Eu^{3+} and Yb^{3+}DOTA-(gly)$_4^-$ and a ratiometric imaging approach was used to estimate pH [43, 54]. More recently, Opina et al. used a series of different Yb^{3+}DOTA-tetraamide complexes to show that the range of pH sensitivity could be further adjusted by altering the electrostatic properties of the amide substituents (Figure 10.6) [19].

The rate of proton exchange from –OH groups is also pH dependent. For DIACEST compounds, this dependence is not terribly useful for measuring pH in the physiological range, but for PARACEST systems involving an –OH group directly bound to a lanthanide ion, the pH sensitivity of –OH proton exchange once again becomes interesting. In a recent report, the CEST spectrum of Yb(III)HPDO3A, an analogue of GdHP-DO3A(Prohance), shows two highly shifted –OH exchange peaks (at 99 ppm and 71 ppm) reflecting the presence of either two different stereoisomers or coordination isomers in solution [52]. Interestingly, the exchange rates for these two –OH protons differ quite substantially, and this feature allows construction of a ratiometric calibration curve to evaluate solution pH (Figure 10.7).

Furthermore, it has been demonstrated that the chemical shift between the lanthanide-bound water and bulk water ($\Delta\omega$) in various LnDOTA-tetraamide complexes can be altered by varying the electron density either on one [36] or all four [40] of the amide substituents. This observation led to the development of a pH sensor [41] that responds to changes in pH by a change in chemical shift of the water exchange peak position rather than a change in CEST intensity. This feature is highly advantageous because the sensor readout (in this case, chemical shift) is independent of concentration. To accomplish this, one of the amide side chains of a DOTA-tetraglycinate was replaced with a ketone moiety having an oxygen ligand donor atom in conjugation with a phenolic group. At pH values below the pK_a of the phenol group (in this case, about 6.7), the ligand donor is essentially an isolated ketone oxygen atom with relatively poor electron donating capabilities. The result of

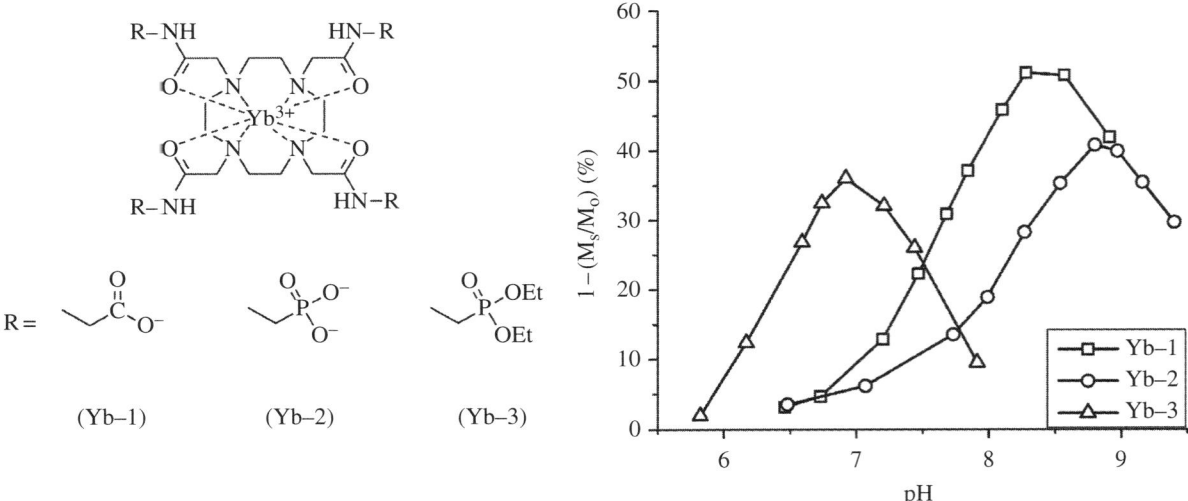

FIGURE 10.6   Left: Chemical structures of Yb^{3+}DOTA-tetraamide complexes with differing amide substituents. Right: Plots of CEST magnitude *versus* pH for each complex. Adapted with permission from Ref. [19].

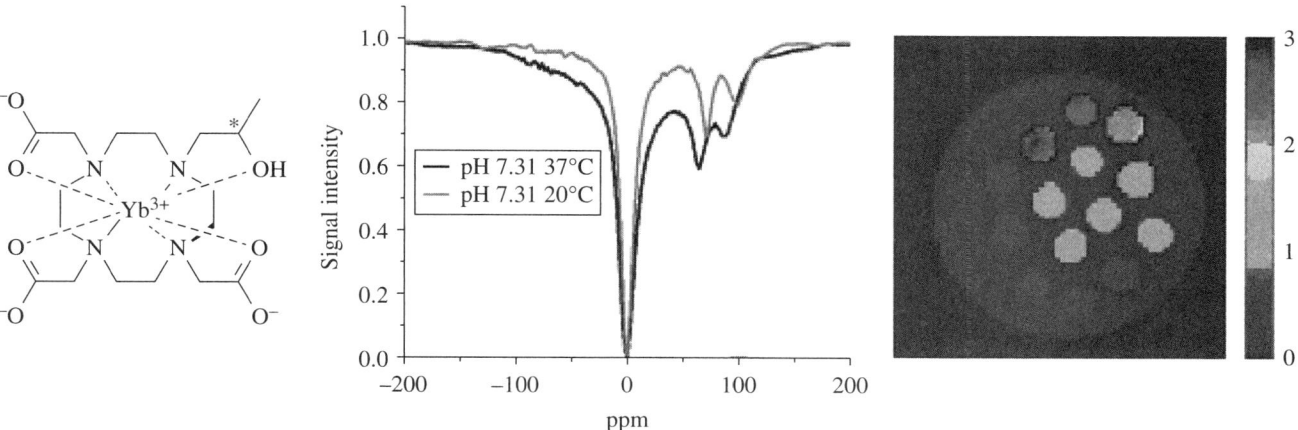

FIGURE 10.7 Left: Chemical structure of Yb(III)-HPDO3A showing exchangeable –OH proton. Centre: CEST spectra of Yb(III)-HPDO3A at 20°C and 37°C showing the –OH exchange peaks of both isomers in solution. Right: Ratiometric CEST images of a phantom containing 14 capillaries of Yb(III) HPDO3A. Images were acquired after irradiation at 71 ppm and 99 ppm and at pH values ranging from 5.2 to 8.8. Adapted with permission from Ref. [52].

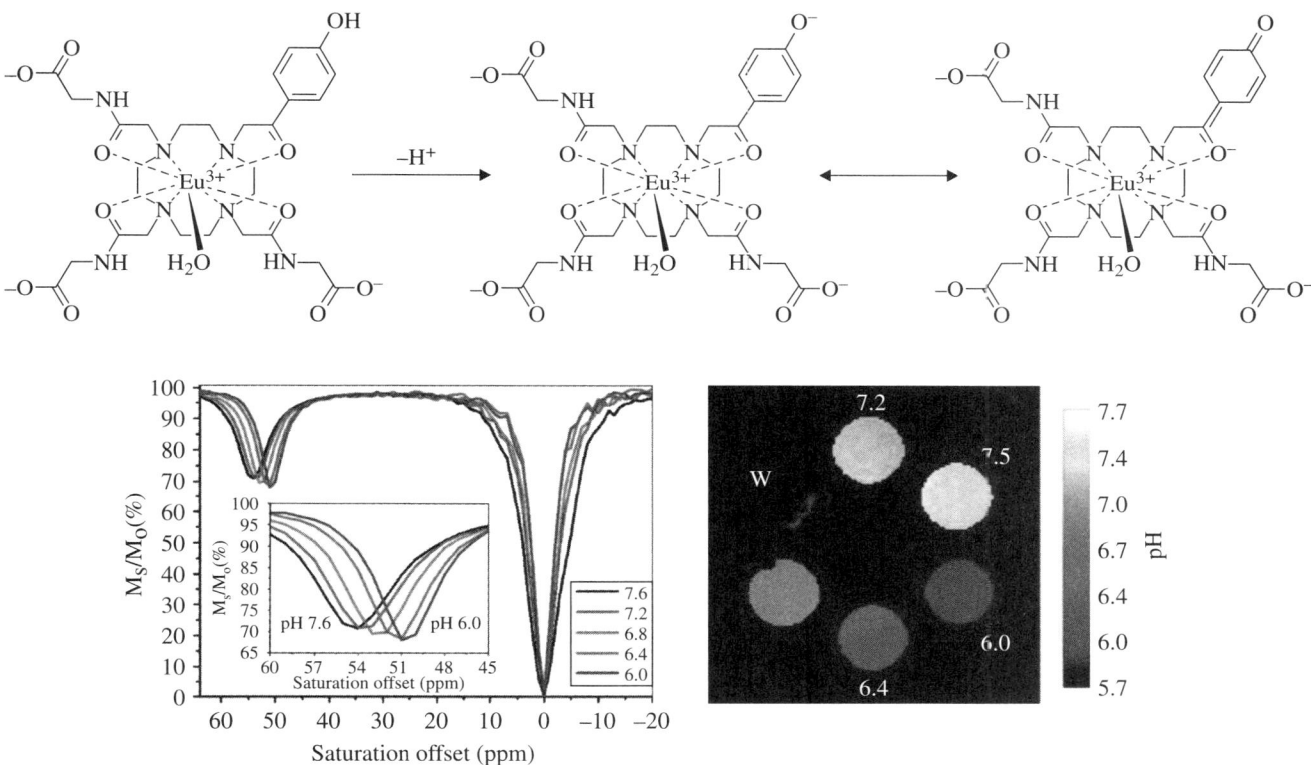

FIGURE 10.8 Top: Change in chemical structure of pH-responsive PARACEST agent as pH is increased. Bottom: pH dependence of CEST spectra showing changes in chemical shift (left), and CEST pH images of a phantom containing water (w) and PARACEST agent after activation at 54 ppm and 47 ppm (right). Adapted with permission from Ref. [41]. (*See insert for colour representation of the figure.*)

this is an overall weaker ligand field around the Eu^{3+} ion and therefore, a smaller $\Delta\omega$. As the pH is increased, the phenolic group loses a proton and the negative charge on the phenolate anion is delocalised through the ligand π-system onto the ketone oxygen donor atom. This results in greater electron density on the ketone oxygen, an overall stronger ligand field for the Eu^{3+} ion, and an increase in $\Delta\omega$ for the exchanging water molecule (Figure 10.8, top). The difference in $\Delta\omega$ between the high and low pH extremes was surprisingly large, ~7 ppm (Figure 10.8, bottom left), making it feasible once again to use ratiometric imaging to obtain a direct readout of pH. The pH of five aqueous phantom samples as determined by imaging (Figure 10.8, bottom right) were in perfect agreement with those measured directly by use of a pH electrode.

The accessibility of bulk water molecules to the coordination sphere of a lanthanide cation in PARACEST complexes is another factor that affects k_{ex}. This can be exploited in the design of responsive agents. Using this approach, sensors for glucose [38, 42], Zn(II) [37], and Ca(II) [55], have been designed and evaluated. In the case of the glucose sensor, water exchange was somewhat too fast to produce a CEST signal from the exchanging water molecule, but CEST was turned on upon binding of glucose to the two phenylboronate moieties situated directly above the exchanging water molecule (Figure 10.9, top). This binding event served to slow water exchange by hindering access of bulk water molecules to the water coordination site, thus resulting in an increase in the CEST intensity. A similar strategy was used in the design of a Zn^{2+} sensor, but, in this case, binding of a single Zn^{2+} ion directly above the water coordination site had the opposite effect, where k_{ex} was found to increase upon Zn^{2+} binding due to catalysis of proton exchange by a proximate Zn^{2+}-coordinated OH group (Figure 10.9, bottom). This resulted in switching the CEST signal 'off' when the Eu^{3+} complex was fully bound with Zn^{2+}. The Ca^{2+} sensor also registered a reduction in CEST intensity upon Ca^{2+} binding but this probe was not selective for Ca^{2+} as it was able to elicit a similar response in the presence of Mg^{2+}. Additionally, PARACEST agents that detect lactate [44] and a number of anions [49, 50, 56] have also been reported.

Chemical modification of a PARACEST agent by the action of an enzyme or a specific reaction with another substance can result in a new entity with different CEST properties. This approach has been used for the detection of caspase-3 [47] and esterase [45] enzyme activity by following a decrease in the CEST signal from a slowly exchanging amide proton in the initial sensor as well as the appearance of a new CEST signal from the more rapidly exchanging amine protons. A similar approach was also reported for the detection of nitric oxide [46].

One of the major challenges in the development of PARACEST agents is that conventional solution-phase synthesis of these agents is time consuming and generally limits studies to one agent at a time. In an effort to overcome this limitation, Napolitano et al. [35] recently reported a rapid, convenient on-bead synthesis of a library of 80 different PARACEST agents that differed only in the chemical identity of three amide side-chain groups. Without removing the complexes from beads, CEST imaging was performed to quickly identify those agents (and side-chains) that exhibited the most favourable water exchange kinetics for CEST (Figure 10.10).

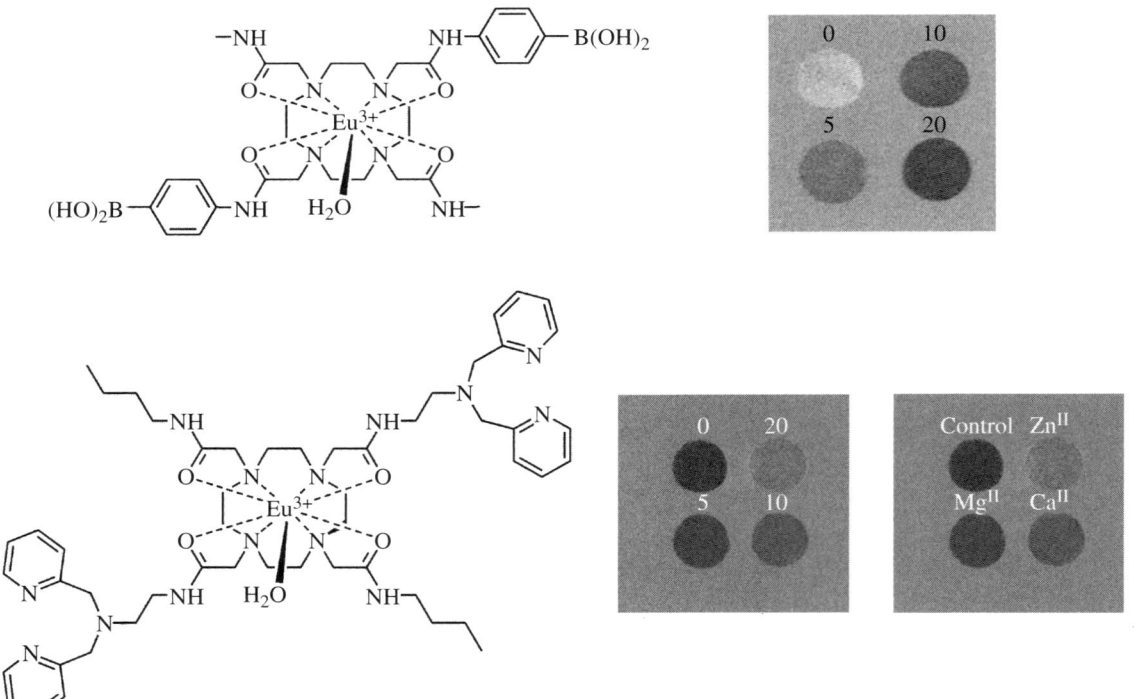

FIGURE 10.9 Top: Structure of glucose-responsive PARACEST agent (left), and CEST images of phantoms containing PARACEST agent and different concentrations of glucose (right). Reproduced with permission from [42]. Bottom: Structure of zinc-responsive PARACEST agent (left), CEST images of phantoms containing PARACEST agent and different concentrations of Zn^{2+} (centre), and PARACEST agent with different metal ions (right). Reproduced with permission from Ref. [37].

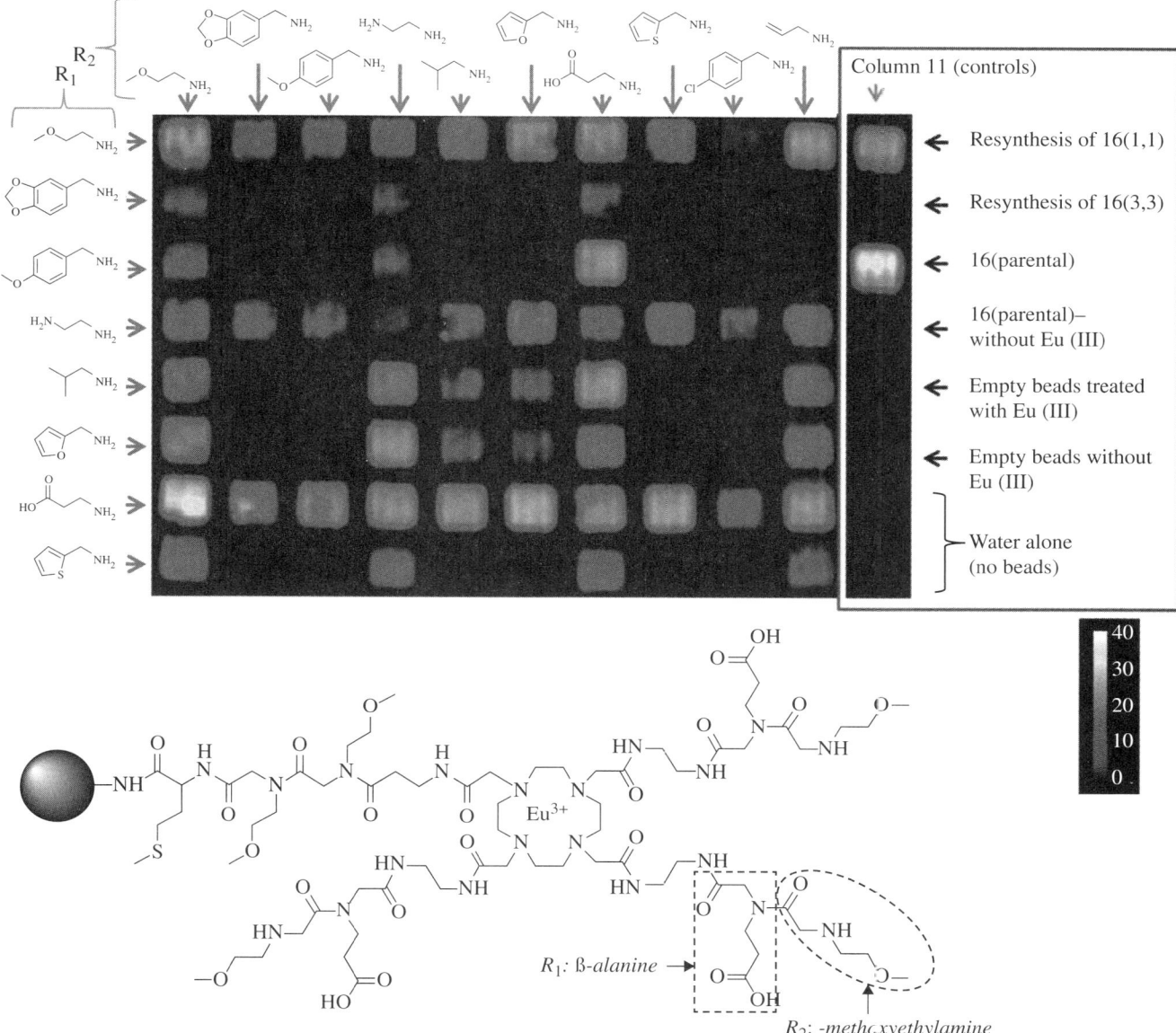

FIGURE 10.10 Top: CEST image of an 80-compound library of Eu(III)DOTA-tetraamide-peptoic derivatives attached to beads. The chemical structures of each amine used to build the library are shown along each horizontal row (R_1) and each vertical column (R_2). Bottom: Structure of Eu(III)DOTA-tetraamide-peptoid in well (7, 1) that showed the highest CEST intensity. Reproduced with permission from Ref. [35]. (*See insert for colour representation of the figure.*)

10.5 *IN VIVO* DETECTION OF PARACEST AGENTS

Although PARACEST agents show considerable promise as responsive MRI agents *in vitro*, translation of these sensors to *in vivo* measurements has so far been limited, especially for those PARACEST systems that operate through a water exchange mechanism. One problem is that the water exchange rates and CEST spectra of most paramagnetic complexes designed as CEST agents are typically measured on samples held at or near room temperature only. However, at body temperature typical of small animals (38–39 °C), water exchange in these systems is, not surprisingly, considerably faster, adding an additional complexity. Faster water exchange results in broader CEST peaks not only for the lanthanide-shifted bound water exchange peak but also for the bulk water resonance. This added line broadening is not so noticeable in aqueous samples at concentrations typically used in an *in vitro* CEST experiment (10–30 mM) even at 38 °C but can become quite evident once one of these agents is injected into an animal and the agent concentrates in the kidneys as it is being filtered. This additional T_2 exchange (T_{2exch}) contribution to the water linewidth can result in as much as a 50% loss in water signal intensity in kidney images after a normal intravenous injection of a Eu^{3+}-based PARACEST agent [57]. The T_{2exch} effect is obviously detrimental to CEST imaging because the signal loss is present even before a frequency-selective RF pulse can be applied to activate the PARACEST agent. Thus, the desired CEST signal contributed by the sensor is masked by the less desirable T_{2exch} contribution to the image. Although the above effect has slowed translation of these water exchange-based PARACEST agents into animal imaging, this new mechanism could pave the way for a new class of activatable T_{2exch} agents.

Nevertheless, there have been a few reports of successful detection of PARACEST agents *in vivo* and in isolated, perfused tissues (*ex vivo*). The first report demonstrated the use of a train of 360°WALTZ-16* preparation pulses applied directly on bulk water (the OPARACHEE method) to provide a direct measure of water signal loss caused by the paramagnetic agent. This method essentially presents an indirect way to detect the agent without the need of applying a frequency-selective pre-saturation pulse to activate the agent. Given that a Tm^{3+}-based PARACEST agent was found to be optimal for OPARACHEE and that this Tm^{3+}-based agent also induces a much larger hyperfine shift in its bound water molecule compared to the more common Eu^{3+}-based agents, this effort effectively represented the first report of the detection of a very efficient agent that can now be referred to as a T_{2exch} agent. Nonetheless, it was the first imaging method used to successfully monitor the accumulation and clearance of a PARACEST agent in mouse kidney [58]. More recently, OPARACHEE was also successfully used to detect a different PARACEST agent in mouse kidney [59] and in a mouse glioblastoma multiforme tumour [60]. Additionally, the contrast from PARACEST agents conjugated to second-generation and fifth-generation polyamidoamine (PAMAM) dendrimers has also been used to track the relative pharmacokinetics of these nanoparticles *in vivo* [61]. Following the injection of the dendrimer conjugates into a mouse model of mammary carcinoma, a gradual increase in image contrast resulting from accumulation of the PARACEST conjugates in the tumour was observed.

The same Eu^{3+}-based glucose sensor described in Figure 10.9 has been used to image the tissue distribution of glucose in perfused mouse livers [62]. This was accomplished by perfusing two different livers to steady state, one with the agent plus 10 mM glucose and the second liver with the agent alone. Comparisons were then made of the water image intensities in the two livers upon activation of the PARACEST agent by a long pre-saturation pulse at 42 ppm (the chemical shift of the Eu^{3+}-bound water exchange peak). Interestingly, the liver perfused with the agent plus glucose showed a 17% decrease in water intensity compared to the liver with the same amount of agent but no glucose (Figure 10.11). This important experiment

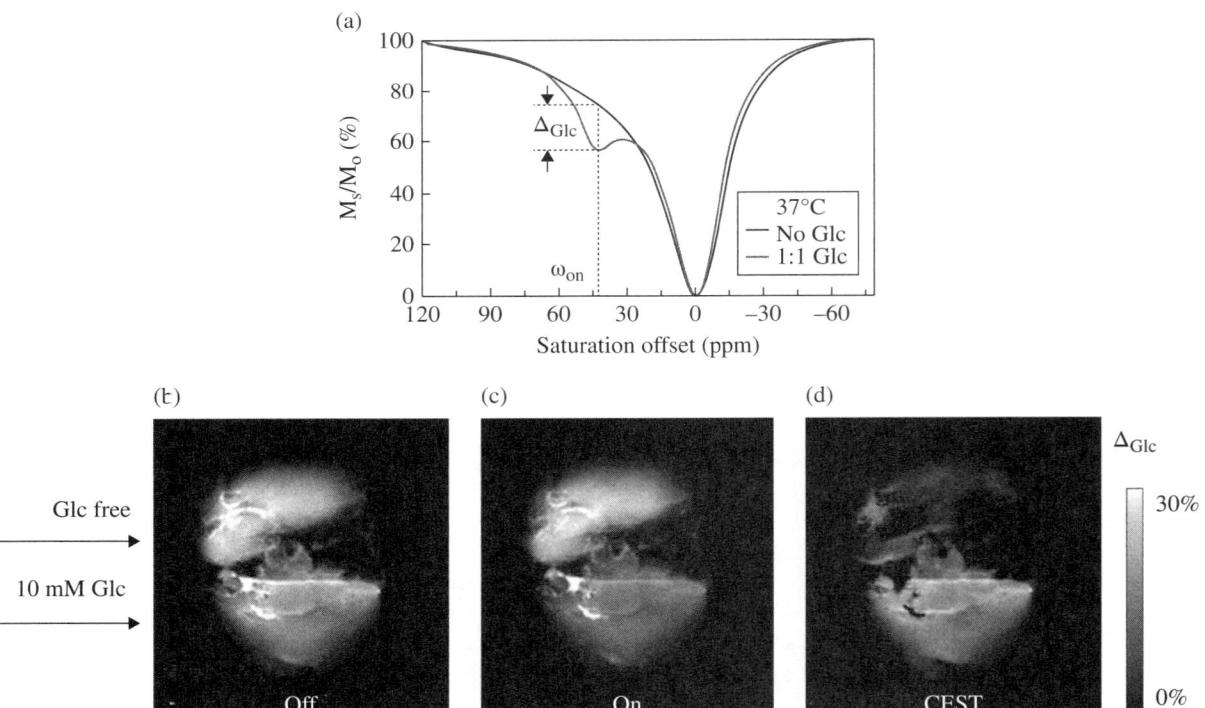

FIGURE 10.11   (a) CEST spectra of fresh effluent from a liver isolated from a 24-hr fasted mouse and perfused at 37°C showing a glucose-induced CEST peak at 42 ppm. Both perfusates contained 10 mM agent at pH = 7.4. (b) single slice selected images of mouse livers collected without a pre-saturation pulse ('off') and (c) with a pre-saturation pulse ('on'). The two livers were perfused using a physiological buffer without glucose (top) or with 10 mM glucose (bottom). Both perfusates contained 10 mM agent at 37°C. (d) The difference CEST image showing the glucose induced CEST contrast between the fed and fasted mouse livers. Reproduced with permission from Ref. [62].

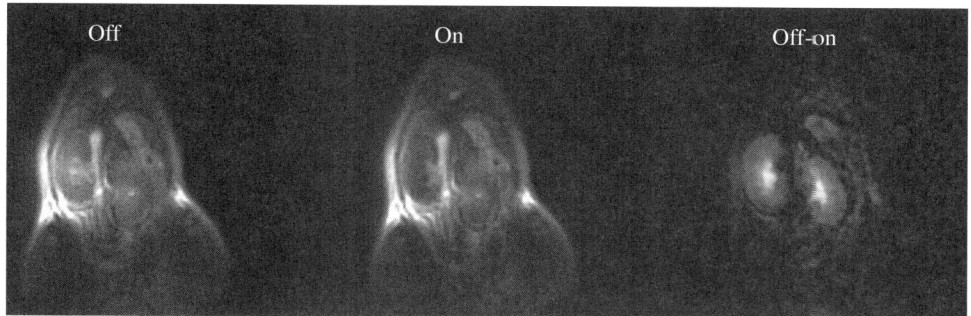

FIGURE 10.12 *In vivo* CEST images of mouse kidney after intravenous injection of 1 mmol/kg EuDOTA-(gly)$_4^-$ using the SWIFT pulse sequence. Adapted with permission from Ref. [64]. (*See insert for colour representation of the figure.*)

demonstrated that carbohydrate moieties from endogenous structures such as glycoproteins and glycolipids present in liver did not give a sufficient background signal to interfere with glucose sensing by the PARACEST agent. In separate experiments, the agent has also been shown to be capable of detecting glucose exported from hepatocytes after hormonal stimulation of glycogenolysis. It is important to note that the amount of agent used in this perfused tissue experiment was not high enough to cause a dramatic change in water linewidth due to the T_{2exch} contribution described above. This bodes well for eventual translation of PARACEST-base sensors into the *in vivo* setting as long as they can be used at moderate concentrations.

To take full advantage of the on/off RF activation properties of PARACEST agents *in vivo*, imaging sequences that recapture the T_2 shortening effects induced by the PARACEST agents are required. A potential solution to this problem was highlighted by Soesbe et al. [63], who implemented the SWIFT imaging sequence characterised by ultra-short TE (<10 µs) in an attempt to reclaim the loss in signal due to T_2 exchange. This experiment demonstrated for the first time that the shortened T_{2exch} component can be recaptured and used to detect a water exchange-based PARACEST agent *in vivo* using traditional on/off CEST imaging (Figure 10.12).

The future of PARACEST agents as biological sensors remains bright because significant progress is being made in the development of imaging sequences that will allow their detection *in vivo*. The ultimate sensitivity or detection limit of these agents *in vivo* will also need to be optimised; however, this can be done by fine-tuning the water exchange kinetics of these agents to make them optimal at 37 °C. One common approach to improve sensitivity is to prepare low molecular weight polymers of the agents [10, 11] or add multiple copies to a dendrimer [65] or a nanoparticle [12, 15]. Such strategies make it feasible to lower the detection limits of these novel sensors into the µM range.

10.6 SUPRAMOLECULAR CEST AGENTS

As discussed above, the sensitivity issue of CEST agents can be tackled either by exploiting larger k_{ex} values or by designing systems containing a high number of mobile protons. In an attempt to address the latter parameter, van Zijl and co-workers explored the use of several diamagnetic macromolecular systems (polyaminoacids, dendrimers, RNA-like polymers) and found that a CEST effect of approximately 50% could be generated for a 5 µM sample of polyuridylic acid (poly(rU)) [66]. Such a high sensitivity is the result of a high number of irradiated mobile protons (ca. 2000 per polymer). Although these findings represent an important step in the search for more sensitive CEST agents, it is evident that a much higher improvement could have been realised if such exchangeable proton pools displayed much larger $\Delta\omega$ values. On this basis, it was deemed of interest to explore the use of supramolecular adducts between species containing a high number of exchanging protons and a paramagnetic shift reagent. The major advantage of this system resides in the fact that the resonance frequency of the mobile protons, upon interacting with the paramagnetic agent, is shifted away from its diamagnetic position, thus the observed CEST effect is not affected by the contribution of protons belonging to endogenous molecules. Furthermore, the marked increase in $\Delta\omega$ would also allow the exploitation of faster proton exchanging systems (larger k_{ex}).

As a model system to prove the efficacy of this approach, the supramolecular adduct formed by the cationic polypeptide poly-L-arginine (53.5 kDa) and the negatively charged lanthanide complex Tm(HDOTP)$^{4-}$, whose ability to act as NMR shift reagent for cationic species is well documented, was investigated [8, 67, 68]. In the absence of the shift reagent, the exchange rate of the mobile guanidine protons of the polypeptide at pH 7.4 and 312 K was so fast that no CEST effect could be observed (Figure 10.13, open squares). However, upon the addition of Tm(HDOTP)$^{4+}$ to the poly-L-arginine solution, a remarkable transfer of saturated magnetisation was observed around 20 to 30 ppm downfield of the water resonance (Figure 10.13, filled squares), which was an unambiguous indication of the formation of a tightly associated ion pair. It is

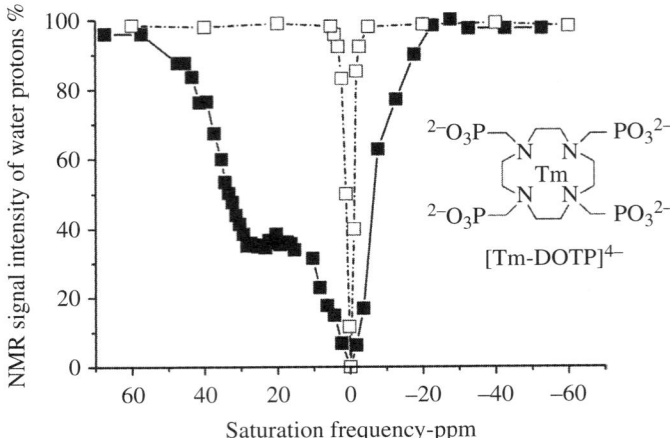

FIGURE 10.13 CEST spectra of a solution of 0.11 mM poly-L-arginine alone (open squares) and 0.11 mM poly-L-arginine in the presence of 2 mM of Tm(HDOTP)$^{4-}$ (filled squares) acquired at 7.05 T, pH 7.4, and 312 K. Irradiation conditions: square pulse, duration: 2 s, power: 25.0 μT. Reproduced with permission from Ref. [8].

very likely that the Tm(HDOTP)$^{4-}$ chelate sweeps along the polypeptide surface, thus causing an overall shift of all guanidine protons that, upon saturation, resulted in a strong decrease in the water signal intensity. An 18:1 ratio between the paramagnetic chelate and the polypeptide was found to provide an optimal CEST effect, whereas higher ratios yielded extensive precipitation of the adduct owing to the decrease of the effective residual electric charge.

Interestingly, a 5% CEST effect was observed at very low concentrations of polymer (1.7 μM) and metal complex (30 μM), thus demonstrating the great sensitivity enhancement that can be attained by this approach. Although the observed sensitivity enhancement significantly decreased the detection limit of an MRI-CEST experiment to values that approached conventional Gd(III)-based contrast agents, a further step toward high sensitivity CEST probes has been attained by the introduction of lipoCEST agents.

10.7 LipoCEST AGENTS

A dramatic increase in sensitivity has been obtained by considering the ensemble of water molecules contained in the inner cavity of liposomes as a pool of exchangeable protons. Depending on the size of these liposomes (50–300 nm in diameter), the number of exchangeable protons can be as high as 10^6 to 10^8 per liposome. In order to distinguish between the internal and external water protons, a paramagnetic shift reagent (SR) must be entrapped in the inner aqueous cavity. Ln(III)DOTA-like complexes, in which the ninth coordination site around the lanthanide ion is occupied by a water molecule in fast exchange with the bulk molecules, have been shown to work very well in this respect. The coordinated water molecule lies along the molecular axis, and this structural arrangement endows the water protons with a large hyperfine shift. In Figure 10.14, the ^{1}H-NMR spectrum of such a suspension containing a lipoCEST agent (Tm-DOTMA as SR) is shown [69, 70].

LipoCEST agents display exceptional sensitivity in the picomolar concentration range (on a per liposome basis) without the need for high saturation powers. However, the major disadvantage of first-generation lipoCEST agents lies in the small chemical shift difference between the inner water resonance (±4 ppm depending on the Bleaney constant of the lanthanide ion of choice) and the bulk water signal. Hence, the *in vivo* application of these lipoCEST agents is limited by the fact that endogenous molecules may exhibit exchangeable protons in the same frequency region. To overcome this drawback, it is necessary to develop lipoCEST agents with much larger separation between the inner water and the bulk water resonances ($\Delta^{intralipo}$), a feature that can be achieved by either changing the nature and concentration of the SR unit or by altering the shape of the liposome. Basically, $\Delta^{intralipo}$ is the sum of two contributions: (i) the chemical interaction arising from the labile coordination of a water molecule to the metal centre of the SR (pseudocontact shift, Δ^{pseudo}), and (ii) the changes in bulk magnetic susceptibility (BMS shift, Δ^{BMS}) due to compartmentalisation of the paramagnetic SR inside the vesicle (Eq. 10.2).

$$\Delta^{int\,ralipo} = \Delta^{pseudo} + \Delta^{BMS}$$

(10.2)

The latter term is null for spherical compartments but can dominate for asymmetric systems (Figure 10.15) [71]. It is worth noting that both the Δ^{pseudo} and Δ^{BMS} contributions are directly dependent on the intraliposomal concentration of the SR,

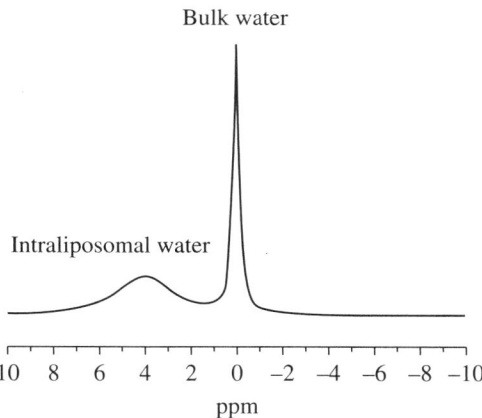

FIGURE 10.14 ¹H NMR spectrum (14.1 T and 39°C) of a suspension of a lipoCEST agent entrapping Tm-DOTMA (inset). The signal at 3.1 ppm downfield from the bulk water corresponds to the water protons entrapped within the liposomes that contain 0.1M of SR.

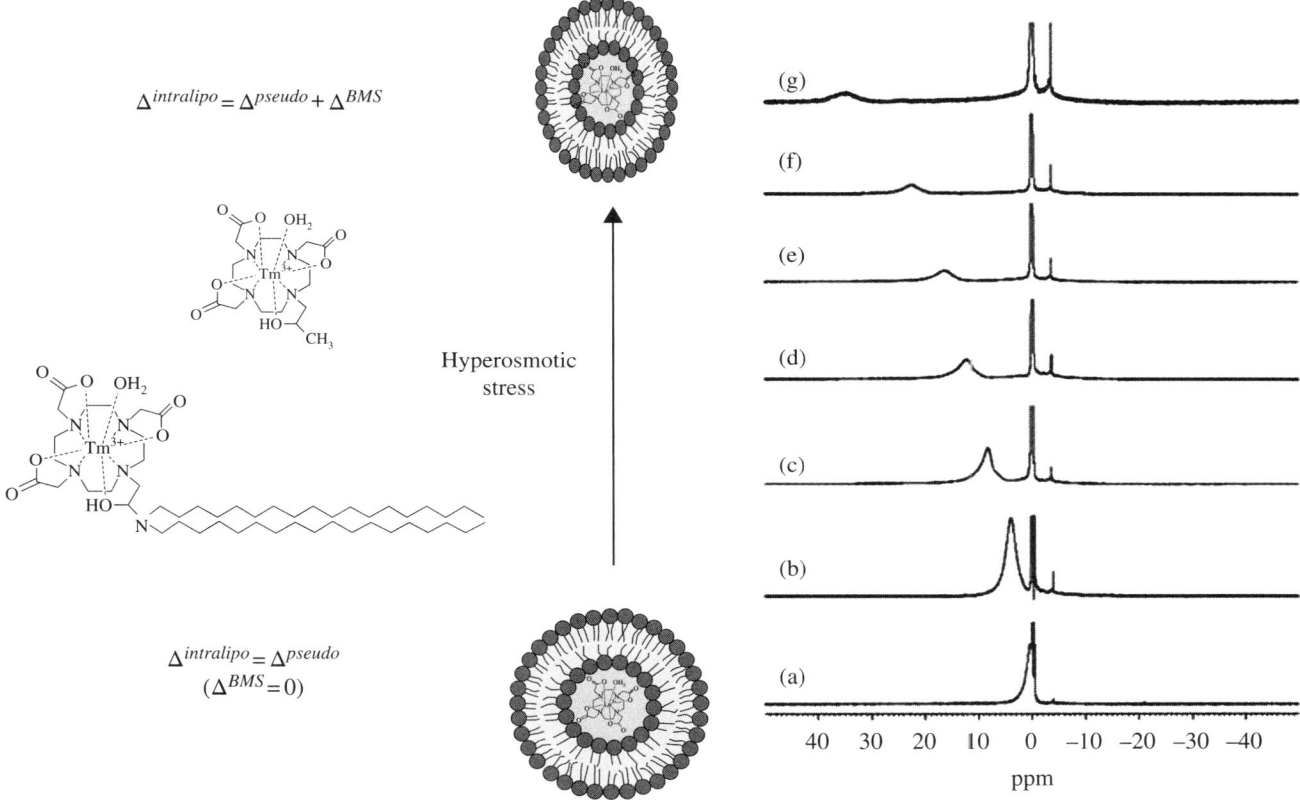

FIGURE 10.15 ¹H-NMR spectra (14 T, 25°C) of a suspension of liposomes made of DPPC/Tm-1/DSPE-PEG2000 (65/30/5 mol%) encapsulating 40 mM Tm-HPDO3A (shown on the top left) and suspended in a buffered medium (pH 7.4) with increasing osmolarity: (a) 40 mOsm. (b) 80 mOsm. (c) 110 mOsm. (d) 160 mOsm. (e) 230 mOsm. (f) 300 mOsm (isotonic). (g) 600 mOsm. The structure of the Tm-1 complex is shown on the bottom left. Reproduced with permission from Ref. [71].

and this concentration is influenced by osmotic effects for both spherical and non-spherical lipoCEST probes. In essence, the amount of encapsulated SR will depend on the concentration of the paramagnetic agent in the solution that is used for hydrating the thin lipid film. Successively, the intraliposomal SR concentration can then increase or decrease according to the relative osmolarity of this solution with respect to the isotonic buffer used in the dialysis purification procedure.

On the other hand, non-spherical liposomes can be prepared by shrinking spherical ones through osmotic stress, and the orientation of the resulting liposomes in the magnetic field (which determines the sign of $\Delta^{intralipo}$) can be modulated by

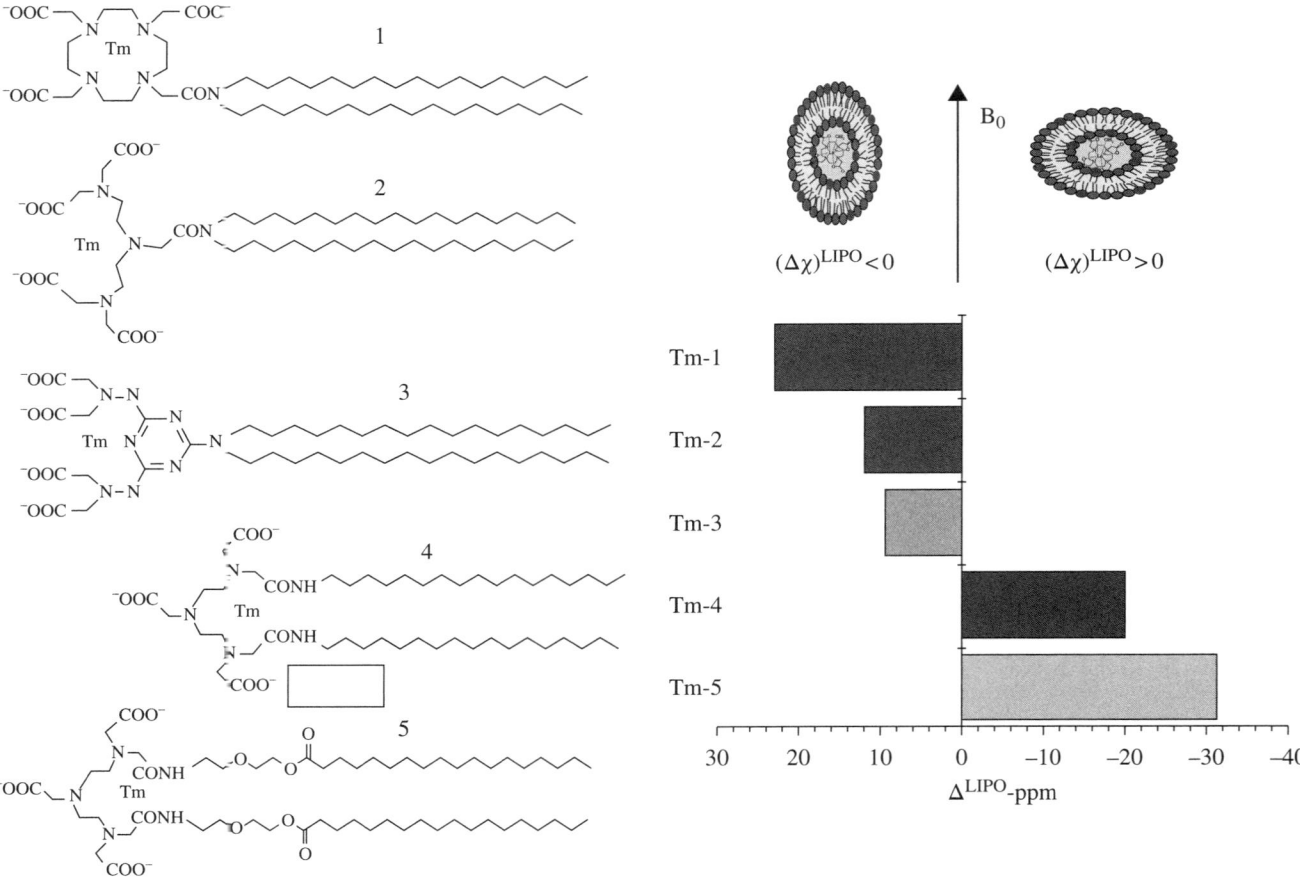

FIGURE 10.16 $\Delta^{intralip}$ values (25°C) for a series of nonspherical lipoCEST agents encapsulating Tm-HPDO3A and incorporating the amphiphilic Tm-complexes reported on the left.

incorporating amphiphilic paramagnetic complexes, endowed with the appropriate magnetic susceptibility anisotropy, in the liposomal membrane. To demonstrate this concept, a number of nonspherical lipoCEST agents with intracavity water proton frequencies ranging from +30 to −45 ppm have been investigated (Figure 10.16).

This significant separation of the intracavity liposomal water from the resonance of bulk water could drastically reduce the artefacts in the MRI-CEST images generated by the asymmetry of the bulk water signal and/or the inhomogeneity of the imaging coil. Furthermore, the extension of the irradiation frequency values could essentially facilitate the setup of imaging protocols for the visualisation of multiple lipoCEST probes.

A very elegant demonstration showing the relationship between the magnetic field orientation and the chemical shift of the intravesicular water protons for nonspherical lipoCEST agents has been reported by Burdinski et al. [72] A 100 μm (inner diameter) capillary, coated with a monolayer of cyclodextrins and capable of hosting liposomal surface-exposed adamantane moieties, was utilised. Amphiphilic Dy(III) and Tm(III) complexes were incorporated into the liposomal membrane while aqueous solutions of paramagnetic SRs were encapsulated in the inner cavity. The liposomes were tightly bound to the capillary surface and upon parallel or perpendicular alignment of the capillary with respect to B_0, their magnetic alignment-dependent CEST properties were observed. These results showed that the orientation induced by the binding of nonspherical liposomes to a target surface could be determined by using routine CEST methods. As a corollary, these findings offer unique opportunities in molecular MRI applications as bound and unbound lipoCEST agents could be easily distinguished based on the differences in their CEST resonance frequencies. Further enhancement of the magnitude of $\Delta^{intralipo}$ has also been achieved by encapsulating neutral multimers in the intraliposomal cavity [73].

One of the major advantages of CEST methodology is the possibility of visualising multiple probes in the same MR imaging voxel. For *in vivo* applications, it is imperative that the CEST agent displays excellent sensitivity and their mobile protons exhibit significantly different resonance frequencies. Current lipoCEST agents fulfil both requirements; this has facilitated their translation *in vivo*. The first *ex-vivo* co-localisation of two lipoCEST agents was carried out on a bovine muscle used as a tissue surrogate [74]. The individual responses of the two agents did not appear to interfere with each other, thus allowing the separate visualisation of nanomolar concentrations of both agents in the same image voxels. Subsequently, two lipoCEST agents were injected under the skin of a mouse in different body regions and were successfully detected independently upon irradiation at the respective absorption frequencies of their exchangeable proton pools [75].

More recently, preliminary results on *in vivo* targeting of $\alpha_v\beta_3$ integrin receptors (which are known to be overexpressed during angiogenesis in many tumour vessels) have been reported by Flament et al. [76] In this study, an RGD-functionalised lipoCEST agent was prepared and injected intravenously in a mouse model bearing a U87 glioma brain tumour. An enhancement in CEST contrast in the tumour and in other brain regions was observed. This was attributed to nonspecific binding and/or distribution of the RGD-lipoCEST. Evidently, these studies highlight the feasibility of the *in vivo* detection of targeted lipoCEST agents, even though the issue of specificity needs to be addressed further.

Opina et al. have tackled the task of endowing lipoCEST agents with pH responsiveness by encapsulating lanthanide complexes of the DOTA-tetraglycinate ligand (endowed with four magnetically equivalent amide protons that exchange with protons of bulk water) in liposomes [77]. The base-catalysed amide proton exchange of the respective complexes followed the order Yb > Tb > Dy. Even though the Dy(III) complex showed the largest hyperfine shift, the combination of favourable chemical shift and amide proton CEST linewidth in the Tm(III) complex was deemed more amenable for future *in vivo* applications where tissue magnetisation effects can interfere.

TmDOTA-(gly)$_4^-$ at various concentrations was encapsulated in the core interior of liposomes to yield lipoCEST particles for molecular imaging. The resulting nanoparticles showed less than 1% leakage of the agent from the interior over a range of temperatures and pH values. A plot of the magnitude of the CEST effect from the amide proton exchange as a function of pH differed for the free versus encapsulated agents over the acidic pH regions, consistent with a lower proton permeability across the liposomal bilayer for the encapsulated agent. Nevertheless, the resulting lipoCEST nanoparticles amplified the CEST sensitivity by a factor of approximately 10^4 compared to the free, unencapsulated agent. Such pH sensitive nanoprobes could prove useful for pH mapping of liposomes targeted to tumours.

A very interesting temperature-responsive agent that could potentially be applied in the field of imaging-guided drug delivery has been reported by Langereis et al. [78]. Localised delivery of anticancer drugs based on liposomal nanocarriers promises a large therapeutic window with reduced side effects of the treatment. If the drug is released from the liposomal nanocarrier locally by external stimulation (e.g., causing a temperature increase), therapy is expected to be particularly effective. In response to a mild hyperthermic treatment (39–42 °C), temperature-sensitive liposomes are known to release the contents of their inner cavity near the melting temperature (T_m) of their lipid membrane. Based on this idea, Langereis et al. designed a new type of temperature-responsive liposome useful for both ^1H CEST and ^{19}F imaging. These liposomes contained both a chemical shift agent (for lipoCEST detection) as well as a fluorine compound (NH_4PF_6, for ^{19}F detection) in their lumen. Inside the liposome, the ^{19}F spectral lines are strongly broadened and not detectable due to fast relaxation induced by the paramagnetic SR. At the T_m of the liposomal membrane, the chemical shift agent and the fluorinated compound are both released. This results in the disappearance of the lipoCEST contrast and a simultaneous appearance of the ^{19}F MR signal because it is no longer influenced by the SR (Figure 10.17). Hence, the ^{19}F signal could be used to quantify the amount of released drug payload, while the CEST signal could measure the local nanocarrier concentration before the release.

In principle, Gd(III) complexes are not considered for the design of paramagnetic CEST agents for two reasons: (i) They cause a marked relaxation enhancement of water protons (both T_1 and T_2) that is detrimental to CEST contrast detection, and

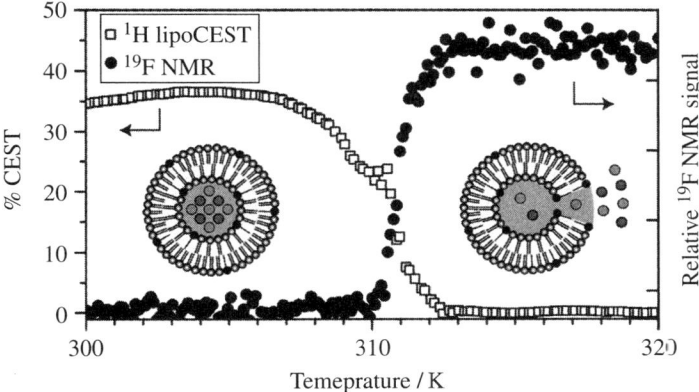

FIGURE 10.17 ^1H CEST effect and ^{19}F NMR signal intensity of liposomes containing Tm-HPDO3A and NH_4PF_6 acquired as a function of temperature, with irradiation field intensity of 4.5 µT and at 7 T. Reproduced with permission from Ref. [78].

(ii) the isotropic distribution of the seven unpaired electrons in the f-orbitals prevents these complexes from acting as SRs. However, when the Gd(III) complexes are entrapped in nonspherical liposomes, they yield a lipoCEST agent analogous to the above described systems based on other paramagnetic lanthanide ions. Actually, it has been shown that Gd-HPDO3A (a commercial relaxation agent widely used in the clinical practice) entrapped in a liposome yields a system that works both as a T_1/T_2 contrast agent and a CEST agent [79]. As the osmolarity of the suspension was increased by adding NaCl, the liposomes shrunk, thus releasing part of the intraliposomal water to attain the same osmolarity as the outside medium. This phenomenon was accompanied by a loss of the spherical shape of the vesicles, a change that was clearly detected as a progressive downfield shift in the resonance of the entrapped water molecules. When the osmolarity of the suspension was in the range of biological fluids, the shift of the internal water was approximately 7 ppm from bulk water. This increase in the shift was accompanied by an increase in linewidth that probably accounted for both an enhanced intraliposomal Gd complex concentration and the consequent increase in R_2 of the encapsulated water protons.

Delli Castelli et al. [80] recently attempted to gain more insight into the understanding of the *in vivo* fate of liposomes and their payload by comparing contrast changes induced by the presence of a classical relaxation agent with the effect induced by a CEST agent. Liposomes were loaded with the paramagnetic complexes Gd-HPDO3A and Tm-DOTMA in order to endow the nanovesicles with the characteristic properties of T_1/T_2 and CEST/T_2 MRI agents, respectively. The paramagnetically loaded liposomes were injected directly into the tumour (B16 melanoma grafted in mice) where they generated T_1, T_2, and CEST contrast in MR images that was quantitatively monitored over time (0–48 h) (Figure 10.18, left). A kinetic model was devised to fit the experimental multi-contrast data in order to extract the relevant information about the cellular uptake of the liposomes and the release of their payload (Figure 10.18, right). Upon comparing conventional stealth liposomes with pH-sensitive liposomes, it was shown that the latter type differed substantially in the step associated with release of the drug, which most likely occurred in the endosomal acidic vesicles.

Finally, analogues of lipoCEST agents can be obtained by using di-block copolymer vesicles loaded with paramagnetic SRs. Block copolymer vesicles have some key advantages over liposomes, such as their lower critical aggregation concentration and their tunable membrane properties that are controlled by the nature and molecular weight of the hydrophobic block [81]. Such vesicles having biocompatible hydrophilic poly(ethylene glycol) (PEG) blocks are well-known for their long blood circulation time due to reduced opsonisation as well as their water diffusion across the polymer membrane.

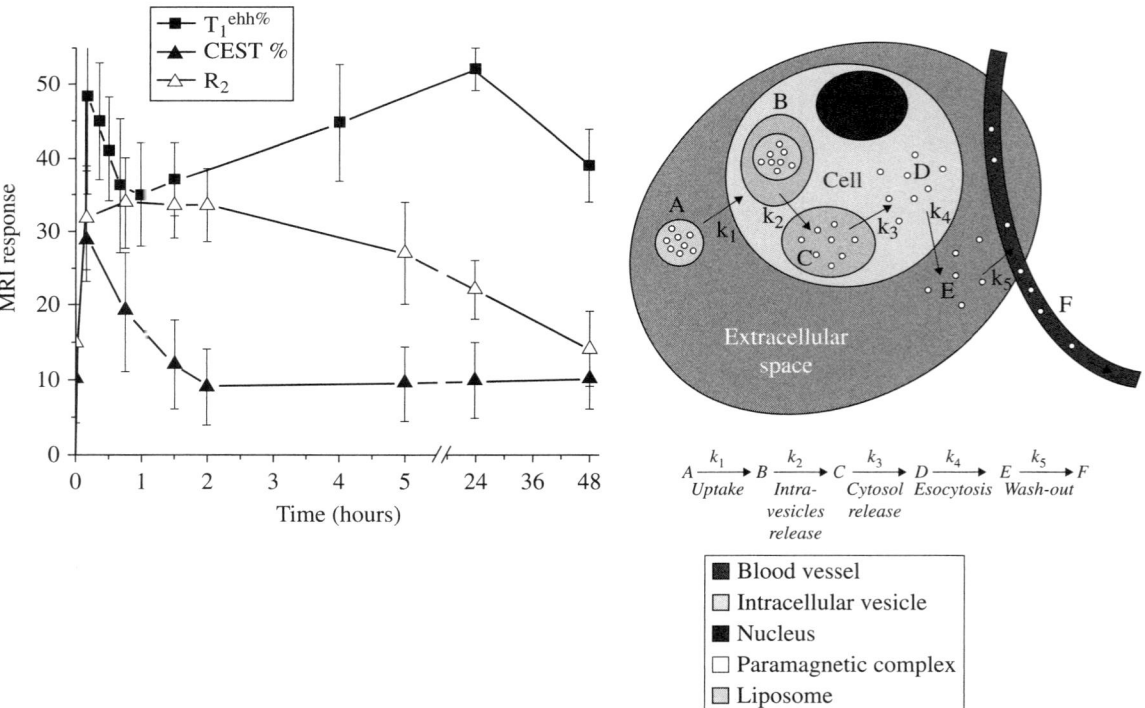

FIGURE 10.18   Left: Temporal evolution of the three contrast modes (means and standard deviations of 6 mice) after intratumoural injection of paramagnetic stealth liposomes. The reported R_2 values refer to data obtained from mice treated with liposomes loaded with Tm-DOTMA. Right: Schematic representation of the kinetic model used for the analysis of the temporal evolution of the MRI responses after the administration of paramagnetic liposomes acting as multi-contrast agents. Reproduced with permission from Ref. [80].

These properties can easily be fine-tuned by varying the molecular weight and microstructure of the block copolymer. Polymersomes entrapping high amounts of TmDOTMA have also been prepared and their ability to generate CEST contrast has been assessed *in vitro*. Interestingly, polymersomes behaved like liposomes in terms of chemical shift of intravesicular protons, contrast efficiency, and sensitivity of the nanovesicles to osmotic changes [82].

In the recent years other nanosised particles such as micelles [83], dendrimers [84], silica- [85,86], polymer- [87], and perfluorocarbon nanoparticles [88], have been considered as carriers of PARACEST agents. Additional innovative applications of such systems have been further explored (e.g. see Ref. [87]).

10.8 CONCLUSIONS

CEST agents have origins from the well-established magnetisation transfer (MT) phenomenon widely exploited both in high resolution NMR spectroscopy and MRI. Compared to MT effects, CEST effects are characterised by more defined NMR signals, which allows selective RF irradiation. The availability of these frequency-encoding agents allows the setup of experiments in which the contrast in the MR image is generated 'at will' only if the appropriate frequency corresponding to the labile protons of the exogenous agent is irradiated. This approach offers the possibility of detecting more than one agent in the same region, thus allowing for multiplex detection that is not permitted with classical relaxation enhancers based on Gd(III) complexes or iron oxide particles. Furthermore, the acquisition of a pre-contrast image may no longer be necessary because the CEST contrast results from the acquisition of two post-contrast experiments simply differing in the on/off switch of the irradiation RF field.

Another important advantage that CEST agents have over conventional relaxation agents is the possibility of designing responsive agents whose MRI signal can be made independent of the probe concentration. Although the poor sensitivity of small-molecule CEST agents may still present a problem for molecular imaging applications, the design of supramolecular and nanosized systems with much lower detection limits could potentially resolve this issue. The increasing number of scientific reports on the CEST topic supports the view that this new family of MR contrast agents can certainly be considered as one of the most interesting innovations that has come out in the field of MRI contrast media development in the recent years.

REFERENCES

[1] A. E. Merbach and É. Tóth, *The Chemistry of Contrast Agents in Medical Magnetic Resonance Imaging*, Wiley, Chichester, u.a., p. xii, 471 S (2001).

[2] A. D. Sherry and M. Woods, *Ann. Rev. Biomed. Eng.* **10**, 12.11–12.21 (2008).

[3] M. Woods, E. W. C. Donald and A. D. Sherry, *Chem. Soc. Rev.* **35**, 500–511 (2006).

[4] K. M. Ward, A. H. Aletras and R. S. Balaban, *J. Magn. Reson.* **143**, 79–87 (2000).

[5] J. Y. Zhou and P. C. M. van Zijl, *Prog. Nucl. Magn. Reson. Spectrosc.* **48**, 109–136 (2006).

[6] K. M. Ward and R. S. Balaban, *Magn. Reson. Med.* **44**, 799–802 (2000).

[7] V. Guivel-Scharen, T. Sinnwell, S. D. Wolff and R. S. Balaban, *J. Magn. Reson.* **133**, 36–45 (1998).

[8] S. Aime, D. Delli Castelli and E. Terreno, *Angew. Chem. Int. Ed.* **42**, 4527–4529 (2003).

[9] N. Goffeney, J. W. M. Bulte, J. Duyn, L. H. Bryant and P. C. M. van Zijl, *J. Am. Chem. Soc.* **123**, 8628–8629 (2001).

[10] Y. K. Wu, P. Y. Zhao, G. E. Kiefer and A. D. Sherry, *Macromolecules.* **43**, 6616–6624 (2010).

[11] Y. K. Wu, Y. F. Zhou, O. Ouari, M. Woods, P. Y. Zhao, T. C. Soesbe, G. E. Kiefer and A. D. Sherry, *J. Am. Chem. Soc.* **130**, 13854–13855 (2008).

[12] K. Cai, G. E. Kiefer, S. D. Caruthers, S. A. Wickline, G. M. Lanza and P. M. Winter, *NMR Biomed.* **25**, 279–285 (2011).

[13] O. Vasalatiy, R. D. Gerard, P. Zhao, X. K. Sun and A. D. Sherry, *Bioconjugate Chem.* **19**, 598–606 (2008).

[14] O. Vasalatiy, P. Zhao, S. Zhang, S. Aime and A. D. Sherry, *Contrast Media Mol. Imaging* **1**, 10–14 (2006).

[15] P. M. Winter, K. J. Cai, J. Chen, C. R. Adair, G. E. Kiefer, P. S. Athey, P. J. Gaffney, C. E. Buff, J. D. Robertson, S. D. Caruthers, S. A. Wickline and G. M. Lanza, *Magn. Reson. Med.* **56**, 1384–1388 (2006).

[16] P. C. van Zijl and N. N. Yadav, *Magn. Reson. Med.* **65**, 927–948 (2011).

[17] J. Y. Zhou, B. Lal, D. A. Wilson, J. Laterra and P. C. M. van Zijl, *Magn. Reson. Med.* **50**, 1120–1126 (2003).

[18] J. Y. Zhou, J. F. Payen, D. A. Wilson, R. J. Traystman and P. C. M. van Zijl, *Nat. Med.* **9**, 1085–1090 (2003).

[19] A. C. L. Opina, Y. K. Wu, P. Y. Zhao, G. Kiefer and A. D. Sherry, *Contrast Media Mol. Imaging* **6**, 459–464 (2011).

[20] P. Z. Sun, J. Y. Zhou, W. Y. Sun, J. Huang and P. C. M. van Zijl, *J. Cereb. Blood Flow Metab.* **27**, 1129–1136 (2007).

[21] C. K. Jones, M. J. Schlosser, P. C. M. van Zijl, M. G. Pomper, X. Golay and J. Y. Zhou, *Magn. Reson. Med.* **56**, 585–592 (2006).

[22] J. Y. Zhou, J. O. Blakeley, J. Hua, M. Kim, J. Laterra, M. G. Pomper and P. C. M. van Zijl, *Magn. Reson. Med.* **60**, 842–849 (2008).

[23] Z. B. Wen, S. G. Fu, F. H. Huang, X. L. Wang, L. L. Guo, X. Y. Quan, S. L. Wang and J. Y. Zhou, *NeuroImage* **51**, 616–622 (2010).

[24] J. Zhou, E. Tryggestad, Z. Wen, B. Lal, T. Zhou, R. Grossman, S. Wang, K. Yan, D. X. Fu, E. Ford, B. Tyler, J. Blakeley, J. Laterra and P. C. van Zijl, *Nat. Med.* **17**, 130–134 (2011).

[25] D. L. Longo, W. Dastru, G. Digilio, J. Keupp, S. Langereis, S. Lanzardo, S. Prestigio, O. Steinbach, E. Terreno, F. Uggeri and S. Aime, *Magn. Reson. Med.* **65**, 202–211 (2011).

[26] P. C. M. van Zijl, C. K. Jones, J. Ren, C. R. Malloy and A. D. Sherry, *Proc. Natl. Acad. Sci. U. S. A.* **104**, 4359–4364 (2007).

[27] W. Ling, R. R. Regatte, G. Navon and A. Jerschow, *Proc. Natl. Acad. Sci. U. S. A.* **105**, 2266–2270 (2008).

[28] M. Woods, M. Botta, S. Avedano, J. Wang and A. D. Sherry, *Dalton Trans.* **24**, 3829–3837 (2005).

[29] S. R. Zhang, X. Y. Jiang and A. D. Sherry, *Helv. Chim. Acta* **88**, 923–935 (2005).

[30] S. R. Zhang, M. Merritt, D. E. Woessner, R. E. Lenkinski and A. D. Sherry, *Acc. Chem. Res.* **36**, 783–790 (2003).

[31] S. Zhang, K. Wu and A. D. Sherry, *J. Am. Chem. Soc.* **124**, 4226—4227 (2002).

[32] S. Zhang, P. Winter, K. Wu and A. D. Sherry, *J. Am. Chem. Soc.* **123**, 1517–1518 (2001).

[33] A. X. Li, F. Wojciechowski, M. Suchy, C. K. Jones, R. H. Hudson, R. S. Menon and R. Bartha, *Magn. Reson. Med.* **59**, 374–381 (2008).

[34] T. Mani, G. Tircso, O. Togao, P. Zhao, T. C. Soesbe, M. Takahashi and A. D. Sherry, *Contrast Media Mol. Imaging* **4**, 183–191 (2009).

[35] R. Napolitano, T. C. Soesbe, L. M. De Leon-Rodriguez, A. D. Sherry and D. G. Udugamasooriyam, *J. Am. Chem. Soc.* **133**, 13023–13030 (2011).

[36] S. J. Ratnakar, M. Woods, A. J. M. Lubag, Z. Kovacs and A. D. Sherry, *J. Am. Chem. Soc.* **130**, 6–7 (2008).

[37] R. Trokowski, J. M. Ren, F. K. Kalman and A. D. Sherry, *Angew. Chem. Int. Ed.* **44**, 6920–6923 (2005).

[38] R. Trokowski, S. R. Zhang and A. D. Sherry, *Bioconjugate Chem.* **15**, 1431–1440 (2004).

[39] O. Vasalatiy, P. Zhao, M. Woods, A. Marconescu, A. Castillo-Muzquiz, P. Thorpe, G. E. Kiefer and A. Dean Sherry, *Bioorg. Med. Chem.* **19**, 1106–1114 (2011).

[40] S. Viswanathan, S. J. Ratnakar, K. N. Green, Z. Kovacs, L. M. De Leon-Rodriguez and A. D. Sherry, *Angew. Chem. Int. Ed.* **48**, 9330–9333 (2009).

[41] Y. Wu, T. C. Soesbe, G. E. Kiefer, P. Zhao and A. D. Sherry, *J. Am. Chem. Soc.* **132**, 14002–14003 (2010).

[42] S. R. Zhang, R. Trokowski and A. D. Sherry, *J. Am. Chem. Soc.* **125**, 15288–15289 (2003).

[43] S. Aime, A. Barge, D. D. Castelli, F. Fedeli, A. Mortillaro, F. U. Nielsen and E. Terreno, *Magn. Reson. Med.* **47**, 639–648 (2002).

[44] S. Aime, D. Delli Castelli, F. Fedeli and E. Terreno, *J. Am. Chem. Soc.* **124**, 9364–9365 (2002).

[45] Y. Li, V. R. Sheth, G. Liu and M. D. Pagel, *Contrast Media Mol. Imaging* **6**, 219–228 (2011).

[46] G. Liu, Y. Li and M. D. Pagel, *Magn. Reson. Med.* **58**, 1249–1256 (2007).

[47] B. Yoo and M. D. Pagel, *J. Am. Chem. Soc.* **128**, 14032–14033 (2006).

[48] S. R. Zhang, L. Michaudet, S. Burgess and A. D. Sherry, *Angew. Chem. Int. Ed.* **41**, 1919–1921 (2002).

[49] C.-H. Huang, J. Hammell, S. J. Ratnakar, A. D. Sherry and J. R. Morrow, *Inorg. Chem.* **49**, 5963–5970 (2010).

[50] C.-H. Huang and J. R. Morrow, *J. Am. Chem. Soc.* **131**, 4206–4207 (2009).

[51] M. Woods, D. E. Woessner, P. Zhao, A. Pasha, M. Yang, C. Huang, O. Vasalitiy, J. Morrow and D. Sherry, *J. Am. Chem. Soc.* **128**, 10155–10162 (2006).

[52] D. D. Castelli, E. Terreno and S. Aime, *Angew. Chem. Int. Ed.* **50**, 1798–1800 (2011).

[53] S. R. Zhang, C. R. Malloy and A. D. Sherry, *J. Am. Chem. Soc.* **127**, 17572–17573 (2005).

[54] S. Aime, D. Delli Castelli and E. Terreno, *Angew. Chem. Int. Ed.* **41**, 4334–4336 (2002).

[55] G. Angelovski, T. Chauvin, R. Pohmann, N. K. Logothetis and E. Toth, *Bioorg. Med. Chem.* **19**, 1097–1105 (2011).

[56] J. Hammell, L. Buttarazzi, C.-H. Huang and J. R. Morrow, *Inorg. Chem.* **50**, 4857–4867 (2011).

[57] T. C. Soesbe, M. E. Merritt, K. N. Green, F. A. Rojas-Quijano and A. D. Sherry, *Magn. Reson. Med.* **66**, 1697–1703 (2011).

[58] E. Vinogradov, H. He, A. Lubag, J. A. Balschi, A. D. Sherry and R. E. Lenkinski, *Magn. Reson. Med.* **58**, 650–655 (2007).

[59] C. K. Jones, A. X. Li, M. Suchy, R. H. E. Hudson, R. S. Menon and R. Bartha, *Magn. Reson. Med.* **63**, 1184–1192 (2010).

[60] A. X. Li, M. Suchy, C. Li, J. S. Gati, S. Meakin, R. H. E. Hudson, R. S. Menon and R. Bartha, *Magn. Reson. Med.* **66**, 67–72 (2011).

[61] M. M. Ali, B. Yoo and M. D. Pagel, *Mol. Pharmaceutics* **6**, 1409–1416 (2009).

[62] J. Ren, R. Trokowski, S. Zhang, C. R. Malloy and A. D. Sherry, *Magn. Reson. Med.* **60**, 1047–1055 (2008).

[63] T. C. Soesbe, O. Togao, M. Takahashi and A. D. Sherry, *Proc. Intl. Soc. Magn. Reson. Med.* **19**, 1668–1672 (2011).

[64] T. C. Soesbe, O. Togao, M. Takahashi and A. D. Sherry, *Magn. Reson. Med.* **68**, 816–821 (2012).

[65] J. A. Pikkemaat, R. T. Wegh, R. Lamerichs, R. A. van de Molengraaf, S. Langereis, D. Burdinski, A. Y. F. Raymond, H. M. Janssen, B. F. M. de Waal, N. P. Willard, E. W. Meijer and H. Grull, *Contrast Media Mol. Imaging* **2**, 229–239 (2007).

[66] K. Snoussi, J. W. M. Bulte, M. Gueron and P. C. M. van Zijl, *Magn. Reson, Med.* **49**, 998–1005 (2003).

[67] S. Aime, M. Botta, S. G. Crich, E. Terreno, P. L. Anelli and F. Uggeri, *Chem. Eur. J.* **5**, 1261–1266 (1999).

[68] A. D. Sherry, J. Ren, J. Huskens, E. Brucher, E. Toth, C. F. C. G. Geraldes, M. M. C. A. Castro and W. P. Cacheris, *Inorg. Chem.* **35**, 4604–4612 (1996).

[69] S. Aime, D. D. Castelli and E. Terreno, *Angew. Chem. Int. Ed.* **44**, 5513–5515 (2005).

[70] S. Aime, S. G. Crich, E. Gianolio, G. B. Giovenzana, L. Tei and E. Terreno, *Coord. Chem. Rev.* **250**, 1562–1579 (2006).

[71] E. Terreno, D. D. Castelli, E. Violante, H. M. H. F. Sanders, N. A. J. M. Sommerdijk and S. Aime, *Chem. Eur. J.* **15**, 1440–1448 (2009).

[72] D. Burdinski, J. A. Pikkemaat, M. Emrullahoglu, F. Costantini, W. Verboom, S. Langereis, H. Grull and J. Huskens, *Angew. Chem. Int. Ed.* **49**, 2227–2229 (2010).

[73] E. Terreno, A. Barge, L. Beltrami, G. Cravotto, D. D. Castelli, F. Fedeli, B. Jebasingh and S. Aime, *Chem. Commun.* **5**, 600–602 (2008).

[74] E. Terreno, D. D. Castelli, L. Milone, S. Rollet, J. Stancanello, E. Violante and S. Aime, *Contrast Media Mol. Imaging* **3**, 38–43 (2008).

[75] E. Terreno, D. D. Castelli, C. Cabella, W. Dastru, A. Sanino, J. Stancanello, L. Tei and S. Aime, *Chem. Biodiversity* **5**, 1901–1912 (2008).

[76] J. Flament, B. Marty, C. Giraudeau, S. Meriaux, J. Valette, C. Medina, C. Robic, M. Port, F. Lethimonnier, G. Bloch, D. Le Bihan and F. Boumezbeur, *Proc. Int. Soc. Magn. Reson. Med.* **19**, 4497 (2011).

[77] A. C. L. Opina, K. B. Ghagada, P. Zhao, G. Kiefer, A. Annapragada and A. D. Sherry, *PLoS One* **6**, e27370 (2011).

[78] S. Langereis, J. Keupp, J. L. J. van Velthoven, I. H. C. de Roos, D. Burdinski, J. A. Pikkemaat and H. Grull, *J. Am. Chem. Soc.* **131**, 1380–1381 (2009).

[79] S. Aime, D. D. Castelli, D. Lawson and E. Terreno, *J. Am. Chem. Soc.* **129**, 2430–2431 (2007).

[80] D. D. Castelli, W. Dastru, E. Terreno, E. Cittadino, F. Mainini, E. Torres, M. Spadaro and S. Aime, *J. Controlled Release* **144**, 271–279 (2010).

[81] B. M. Discher, Y. Y. Won, D. S. Ege, J. C. M. Lee, F. S. Bates, D. E. Discher and D. A. Hammer, *Science* **284**, 1143–1146 (1999).

[82] H. Grull, S. Langereis, L. Messager, D. Delli Castelli, A. Sanino, E. Torres, E. Terreno, and S. Aime, *Soft Matter* **6**, 4847–4850 (2010).

[83] O. M. Evbuomwan, G. Kiefer, and A. D. Sherry, *Eur. J. Inorg. Chem.* **12**, 2126–2134 (2012).

[84] M. M. Ali, M. P. I. Bhuiyan, B. Janic, N. R. S. Varma, T. Mikkelsen, J. R. Ewing, R. A. Knight, M. D. Pagel, and A. S. Arbab, *Nanomedicine* **7**, 1827–1837 (2012).

[85] O. M. Evbuomwan, M. E. Merritt, G. E. Kiefer, and A. D. Sherry, *Contrast Media Mol. Imaging* **7**, 19–25 (2012).

[86] G. Ferrauto, F. Carniato, L. Tei, H. Hu, S. Aime, and M. Botta, *Nanoscale* **21**, 9604–9607 (2014).

[87] Y. Wu, C. E. Carney, M. Denton, E. Hart, P. Zhao, D. N. Streblow, A. D. Sherry, and M. Woods, *Org. Biomol. Chem.* **8**, 5333–5338 (2010).

[88] K. Cai, G. E. Kiefer, S. D. Caruthers, S. A. Wickline, G. M. Lanza, and P. M. Winter, *NMR Biomed.* **25**, 279–285 (2012).

11

ORGANIC MOLECULES FOR OPTICAL IMAGING

Michael Hon-Wah Lam

Department of Biology and Chemistry, City University of Hong Kong, Kowloon, Hong Kong SAR, China

Ga-Lai Law

Department of Applied Biology and Chemical Technology, Hong Kong Polytechnic University, Hung Hom, Kowloon, Hong Kong SAR, China

Chi-Sing Lee

Laboratory of Chemical Genomics, School of Chemical Biology and Biotechnology, Peking University Shenzhen Graduate School, Shenzhen University Town, Xili, Shenzhen, China

Ka-Leung Wong

Department of Chemistry, Ho Sin Hang Campus, Hong Kong Baptist University, Kowloon Tong, Kowloon, Hong Kong SAR, China

11.1 INTRODUCTION

What is the meaning of life? This is not merely a philosophical or religious question, but also one of the ultimate scientific mysteries that humans, as the conscious species, seek to comprehend. The cell can be regarded as the fundamental functioning unit of living organisms. It is generally believed that by understanding various biochemical processes that sustain a living cell, we might be able to get a glimpse of the wonder of life itself. In fact, since the establishment of the cell theory in 1839 [1, 2], we have already gained a considerable amount of knowledge about the basic functioning of living cells—enough for us to realise their complexity. In a sense, a cell resembles a city, where different activities take place at different locations to maintain its overall operation. There are ports at the border for the selective importation and exportation of materials and products. There are defence mechanisms to counteract potentially dangerous intruders. There are power plants to generate energy from burning of fuels. There is a control centre that issues commands to regulate the entire system. All of these activities are interconnected through the exchange and trafficking of materials, products, signals, and messengers, to and from different activity units, within and outside the system. Of course, a cell is more complicated than a city. It undergoes cycles of growth and reproduction. Genetic materials within its nucleus can make exact copies of themselves upon cell division. All these syntheses and fine manipulations of biomolecules and subcellular structures imply the existence of very delicate cytokinetic mechanisms, intra- and intercellular communications, and multiple loops of control and regulations. These life processes are what we would like to understand.

Even after four centuries since its invention, optical microscopy remains one of the major tools for biologists to study cellular structures, and histo- and cytochemical processes [3, 4]. Visual observations and optical/spectral measurements allow convenient recording and qualitative and quantitative determination of the spatial and temporal changes of selected subcellular features throughout the cell cycle upon specific treatments and stimulations. Some endogenous biomolecules possess specific spectroscopic properties that permit their direct detection using a dedicated optical setup. For example, reduced pyridine nucleotides NADH and NADPH are able to produce two-photon induced (with excitation at ca. 800 nm) fluorescence at 400–450 nm [5–9], This enables their fluorescent imaging against the high single-photon autofluorescent cellular

The Chemistry of Molecular Imaging, First Edition. Edited by Nicholas Long and Wing-Tak Wong.
© 2015 John Wiley & Sons, Inc. Published 2015 by John Wiley & Sons, Inc.

background. Expression of fluorescent proteins in genetically modified cells is another frequently adopted approach for the visualisation of specific protein transcription processes in cells [10–14]. Nevertheless, the majority of biomolecules and intracellular components do not possess any special spectroscopic or luminescent properties for their easy observation and study. Such an approach cannot reveal any post-transcriptional information. Thus, the use of exogenous chromophores and luminophores for direct staining of intracellular components, in live and proliferated cells, and fluoro-tagging/labelling and tracing of biomolecules *in vitro*, is still the most common microscopic imaging practice in biomedical research and cell biology.

From the dawn of optical microscopy in the 17th century to the present, state-of-the-art near-infrared and multi-photon confocal laser scanning microscopy, staining/labelling agents, and probes ranging from extracts of natural plants and insects to simple inorganic salts, synthetic organic dyes, coordination and organometallic complexes, and nanoparticles and nanocomposite materials have been developed for many purposes. The scope of this chapter is to introduce readers to a series of specially designed organic-based dyes and probes that are useful in bioimaging and *in vitro/in vivo* chemosensing to reveal the locales and conditions of specific subcellular structures, the activity of selected cytochemical processes, and the *in vitro* syntheses, trafficking, interactions, and degradations of specific biomolecules.

11.2 DESIGNING MOLECULAR PROBES FOR BIO-IMAGING

Before the turn of the 20th century, colorimetric staining agents were the only tools available for cell biologists. The invention of fluorescent microscopy in 1904, its subsequent popularity in the life science community in the late 1970s and early 1980s, and the eventual emergence of the laser scanning confocal microscopy in 1986 revolutionised research in cell biology. The discoveries enabled the development of a suite of novel techniques to divulge cellular architectures and intracellular parameters with high optical resolution that had never been revealed before. Designated subcellular features can now be made to glow at preselected wavelengths with the corresponding fluoro-tagged antibodies. Intra- and intercellular transportation of metal ions and biomolecules can now be followed in real time with the use of specific luminescent chemosensors. Histochemical processes taking place deep within thick tissues can now be observed and analysed by optical sectioning with laser scanning confocal microscopy and deep tissue-penetrating near-infrared and multi-photon imaging techniques.

Biomolecules that bear natural fluorophores are ideal subjects for bioimaging because they can be directly observed at the appropriate excitation and observation wavelengths, with minimal perturbations to the biological system under investigation. Common natural fluorophores present in proteins are aromatic amino acids, including tryptophan, tyrosine, and phenylalanine (Figure 11.1). Among these three, tryptophan is the most highly fluorescent. However, its quantum yield in different proteins can vary from <0.01 to 0.35 with lifetimes from <0.1 to 7 ns. Such luminescent efficiency is much poorer than those of most other conventional fluorescent dyes. Another serious limitation of native fluorescence detection of aromatic amino acids is their low photostability under single- and two-photon excitation. All these have rendered native fluorophores not suitable for highly sensitive applications, such as single-molecule fluorescence detection.

With these limitations of natural fluorophores, the use of exogenous synthetic organic fluorophores (fluorescence dyes) remains the mostly adopted approach in bioimaging, biomedical, and bioanalytical studies. There are two general tactics in the application of synthetic organic fluorophores: (a) the *covalent association approach*—the fluorophores are covalently tagged onto the targeted biomolecules so that their transportation and transformation can be traced, and (b) the *non-covalent association approach*—the fluorophores are not covalently bound to any biomolecule prior to *in vitro/in vivo* applications and are free to engage in any interaction with various biochemical species and bio-transformation processes.

As shown in Figure 11.2, tagging of a fluorescence dye covalently to an analyte biomolecule can be achieved using (a) amine-reactive functional moieties such as isothiocyanates, chlorotriazinyl derivatives, and hydroxysuccinimido active esters; (b) sulfhydryl-reactive moieties such as iodoacetamido and maleimido functional groups; and (c) via click

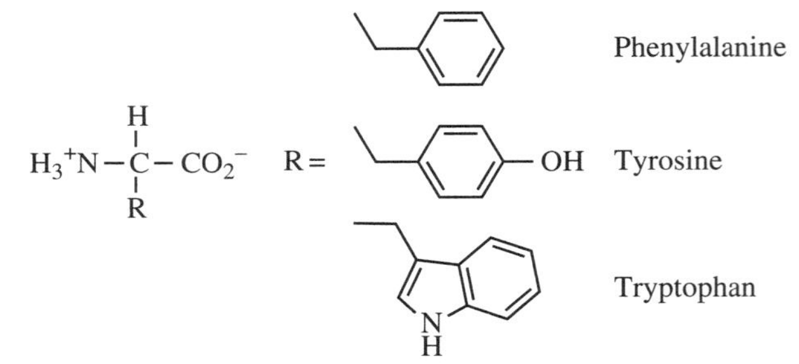

FIGURE 11.1 Structures of naturally fluorescent aromatic amino acids.

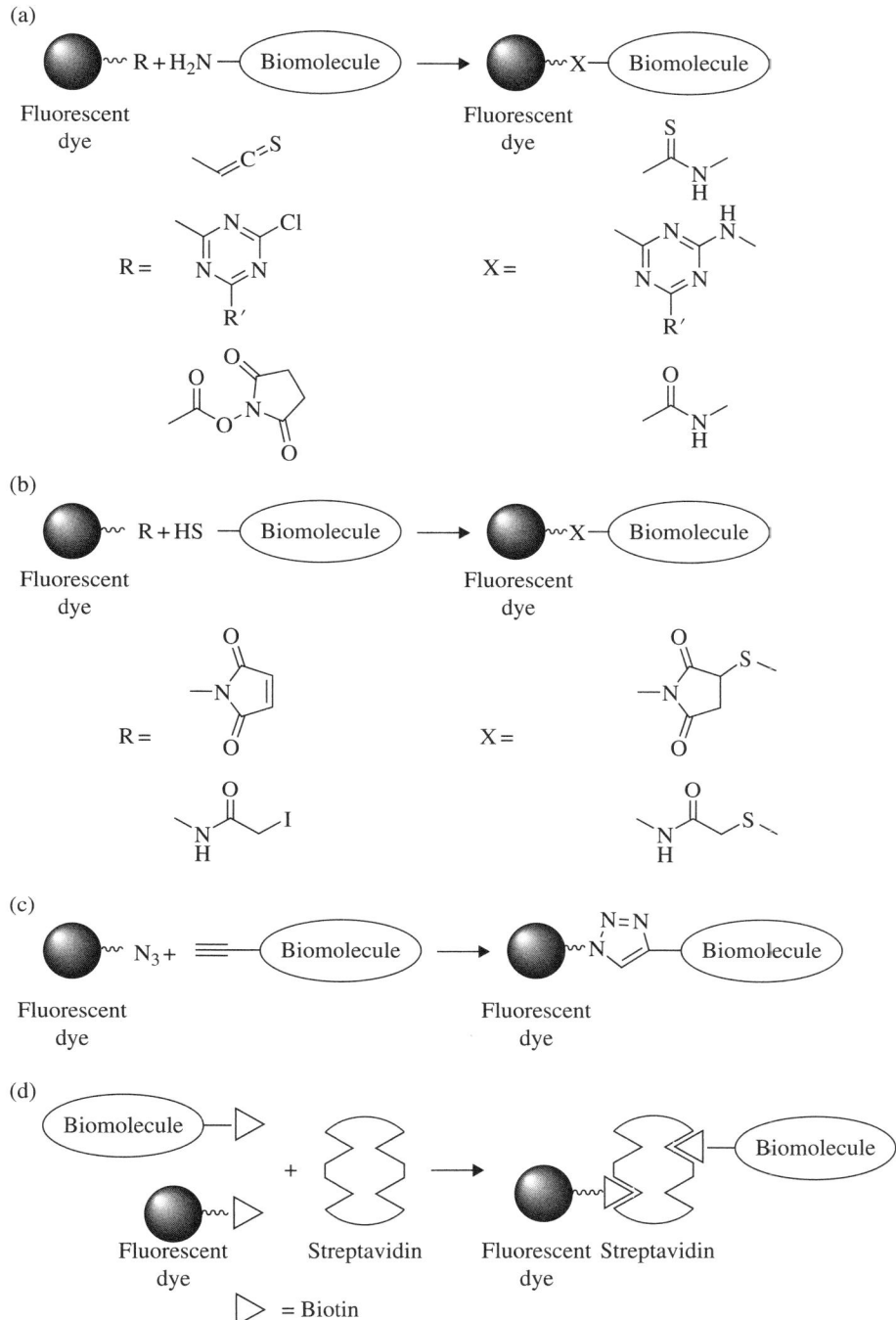

FIGURE 11.2 Fluorescence labelling via (a) amino groups, (b) sulfhydryl groups, (c) click chemistry, and (d) streptavidin-biotin bridges.

chemistry—the [3+2] cycloaddition between azides and alkynes. Alternatively, both the fluorescent dye and the biomolecule can be labelled with biotin and then coupled together via the formation of the streptavidin-biotin bridges.

Despite the rapid growth in the field of bioconjugation, tagging of biomolecules with specific fluorescent probes is still a very challenging task, and hence, a lot of the bioimaging studies nowadays are carried out with 'non-covalently associating' probes. Without prior anchorage to any biomolecules, the histological and cellular localisation profiles of these non-covalently associated probes are directed by their chemical nature, as well as the establishment of specific interactions within the system under investigation. Hydrophilicity/hydrophobicity and acid-base characteristics of these fluorescent probes are two important properties governing their *in vitro* and *in vivo* behaviour. For example, a series of commercially available fluorescent probes bearing basic amine groups have been developed for the staining of lysosomes (Figure 11.3). Protonation of the amine functionalities within the acidic microenvironment of lysosomes switches on their fluorescence due to the suppression of the photo-induced electron transfer (PET) processes (the PET mechanism will be introduced in a later section). This also converts the probes into cationic species and assists their retention in the lysosomes.

Non-covalent associating fluorescent probes that give signals upon binding with their analytes (molecular recognition) are referred to as fluorescent responsive probes (or chemosensors). This type of fluorescent probe is generally composed of two major components, a receptor for molecular recognition and a fluorophore as the signalling source, which are linked together with an appropriate spacer, forming the well-known fluorophore-spacer-receptor motif. Alternatively, the receptor and the fluorophore can be integrated without a spacer (Figure 11.4). When the analyte is bound to the receptor, the physiochemical properties of the fluorophore, such as fluorescence intensity, emission wavelength, and fluorescence lifetime, will be changed via different photophysical mechanisms. These changes provide a signal that indicates the recognition event. A variety of fluorescent biosensors have been developed for specific binding with biologically important cations (such as Mg^{2+}, Zn^{2+}, Cu^{2+}, Ca^{2+}, K^+, and Na^+), anions (such as phosphate, citrate, and carbonate) and neutral molecules.

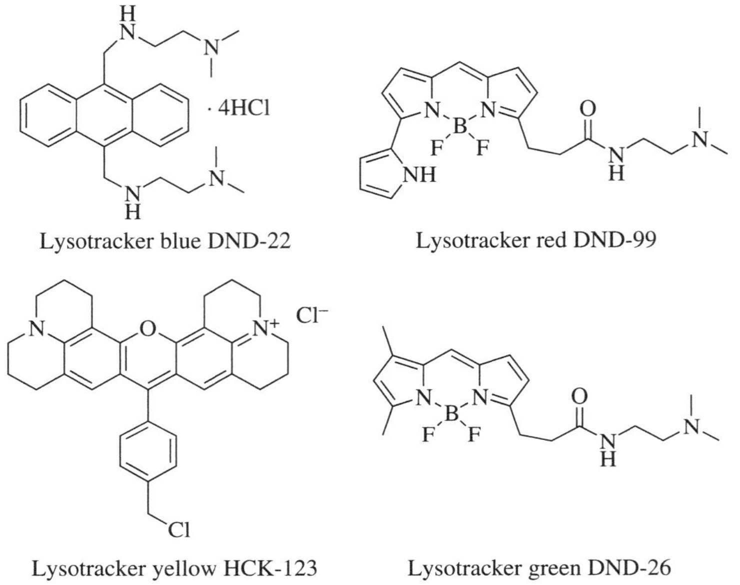

Lysotracker blue DND-22
Lysotracker red DND-99
Lysotracker yellow HCK-123
Lysotracker green DND-26

FIGURE 11.3 Structures of selected fluorescent probes for lysosomal staining.

(a)
Spacer
Fluorophore
Receptor
(b)
Fluorophore
Receptor
(c)
Spacer
Biosensor
= Analyte
Spacer
Change in physical properties

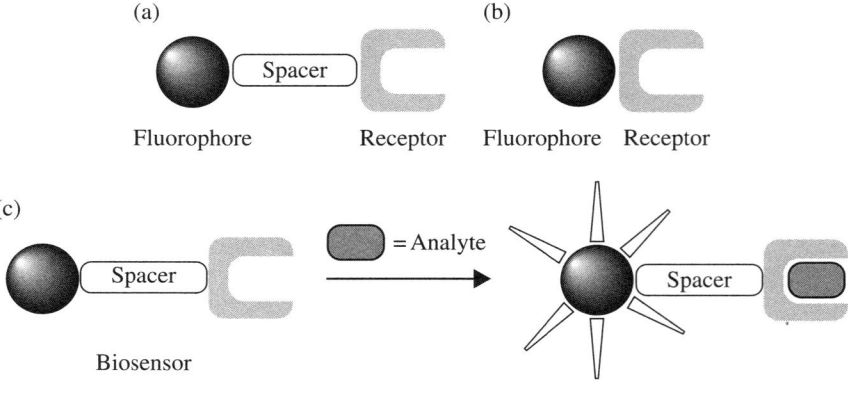

FIGURE 11.4 (a) The spaced and (b) integrated model for fluorescent sensors. (c) The schematic representation of fluorescent sensing.

11.3 DIFFERENT TYPES OF ORGANIC-BASED CHROMOPHORES AND FLUOROPHORES FOR BIOIMAGING

Before we discuss various advanced molecular imaging and sensing tools for the *in vitro* and *in vivo* studies of life processes, it is perhaps necessary for us to take a look at the basic building blocks of these tools: the numerous chromophores and fluorophores that have been developed over the last few centuries.

11.3.1 Natural Dyes as Staining for Early Optical Microscopy

Early microscopists had already recognised the necessity to increase the contrast of transparent tissue specimens in order to aid the observation and visual discrimination of different tissue and cellular edifices [15–18]. Colorimetric staining of specimens or specific regions of a specimen have rapidly become the mainstream approach. Natural dyes from plants and insects, such as extracts of saffron, hematoxylins from logwood, and carminic acids from cochineal, were amongst the first batch of staining agents for microscopy (Figure 11.5; hereafter, the absorption maxima, λ_{abs}, and emission maxima, λ_{em}, reported are measured in aqueous solution at neural pH unless otherwise stated).

They were then followed by a rapid exploration of a wide range of natural and synthetic organic dyes of better abilities to reveal more detailed cellular and subcellular features in all kinds of tissues, micro- and macro-flora, and fauna. In essence, the history of microscopic staining is the history of the development of synthetic organic dyestuffs [19].

11.3.1.1 *Azine Dyes*
Manuveine, invented in 1856, was the first synthetic chemical dye belonging to the azine dye family (more precisely, the diaminophenazine family, Figure 11.6) and was first used as a staining agent in optical microscopy in 1862. Other common staining agents in the family include Neutral Red and Safranine O. Both produce red fluorescence and are useful staining agents for cell nuclei and as counterstains for Gram-negative bacteria [20–24]. The related thiazine family of the dyes has been utilised, in combination with xanthene dyes, for the identification of blood cells and hematopoietic tissues since 1882 [25–27]. Such a thiazine-eosinate neutral staining technique is able to produce characteristic purple-colour staining due to the formation of neutral ion-pairs between the blue-colour cationic thiazine dyes and the red-colour anionic xanthene dyes, the nuclear chromatin of cells, and the cytoplasmic granules of polymorphonuclear leukocytes that were proliferated by acute bacterial infection. The mostly used thiazine dye for this purpose is Methylene Blue, which is transformed into Azure B upon prolonged boiling of its aqueous solution for ion-pairing with the xanthene dye Eosin-Y.

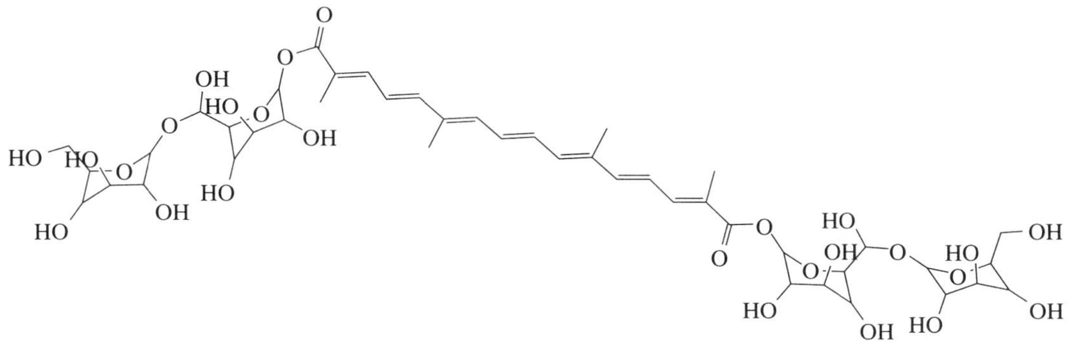

β - Gentobiose crocetin of saffron (λ_{abs} = 427, 452 nm; λ_{em} = 543 nm, ethanol)

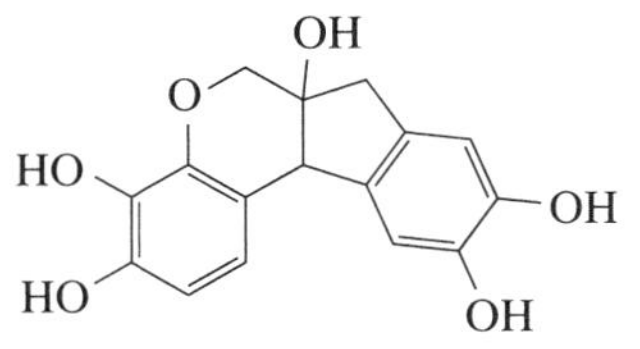

Hamatoxylin from the heartwood
of the logwood tree
(λ_{abs} = 445 nm, pH < 7.0; 560 nm, pH > 7.0)

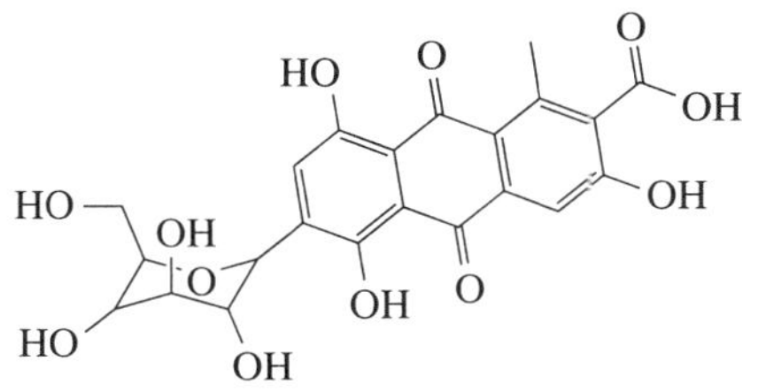

Carminic acid from cochineal
(λ_{abs} = 490 nm; 590 nm, pH = 1.7)

FIGURE 11.5 Natural dyes from plants and insects applied as staining agents in the early stage of development of microscopy.

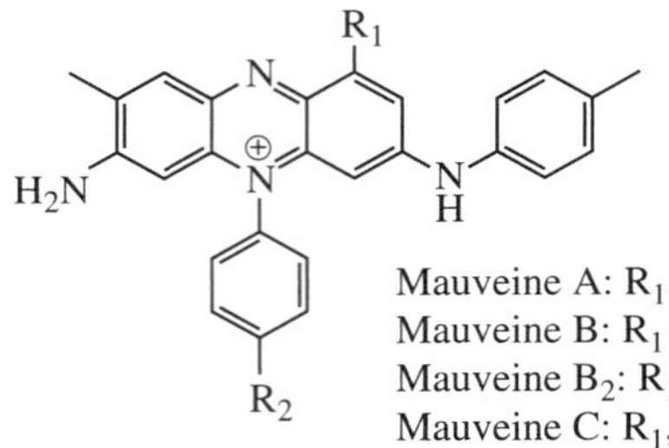

Mauveine A: R_1, R_2 = H
Mauveine B: R_1 = Me; R_2 = H
Mauveine B_2: R_1 = H; R_2 = Me
Mauveine C: R_1, R_2 = Me

Mauveine dyes

Azine & thiazine dyes	λ_{abs}(nm)	λ_{em}(nm)	Remarks
Mauveine	~548		
Neutral red	454 (pH 8.1)	640	50% aqueous
	529 (pH 5.8)		EtOH
Safranine O	530	587	50% aqueous EtOH
Methylene blue	656 (H2O)		
	661 (MeOH)		
Azure B	639		MeOH

FIGURE 11.6 The azine and thiazine dye families.

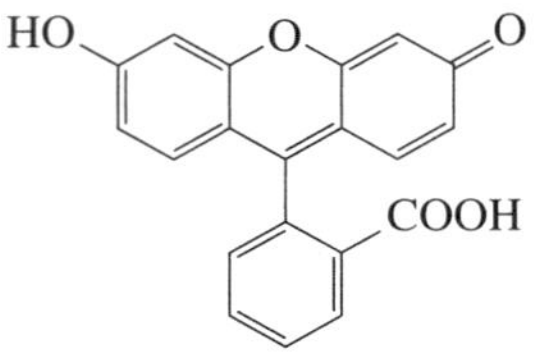

Fluorescein

Hydroxyxanthene dyes	λ_{abs}(nm)	λ_{em}(nm)	Remarks
Fluorescein	494	521	/
Eosin-Y	516	538	/
Erythrosin B	525	555	pH > 7.0
Rose bengal B	548	567	pH > 7.0
Calcium-green^{TM}-1	506	531	When bound with Ca^{2+}

FIGURE 11.7 The hydroxyxanthene dyes.

11.3.1.2 Xanthene Dyes The first anionic hydroxyxanthene dye, fluorescein (Figure 11.7), was synthesised in 1871. Soon afterward, a series of structurally related organic dyes appeared and were found to be extremely important in microscopy staining, biolabelling, and bioimaging applications because of their outstanding luminescent properties and biocompatibility. As mentioned in the previous section, the brominated hydroxyxanthene, Eosin-Y, is used, coupled with Methylene Blue, in the thiazine-eosinate staining. The iodinated hydroxyxanthene Erythrosin, and the mixed chlorinated iodinated hydroxyxanthene Rose Bengal are also important chromophores and luminophores in microscopy staining, bioimaging, and chemosensing. Besides these hydroxyxanthenes, another eminent subfamily of xanthene is the aminoxanthene dyes—the rhodamines. These dyes will be discussed in the later sections concerning fluorescent bioimaging.

11.3.1.3 Triarylmethines Another class of organic dyes that were developed and applied as microscopy staining agents around the same period of time was the triarylmethines (Figure 11.8) [28]. Representatives of this dye family include Fuchsine, Aniline Blue, Methyl Violet, and Malachite Green. Basic Fuchsine is routinely being used as a nuclear and mucin stain in histology [29, 30]. It is also used as the visualising reagent for enzyme labelling in immunostaining. Aniline Blue is useful in staining of collagen [31–33]. In combination with Eosin-Y, Aniline Blue can also be used to assess sperm viability [34]. On the other hand, Crystal Violet (i.e., Methyl Violet 10B) is the primary staining agent in the Gram staining of bacteria [35, 36], Malachite Green is used as a staining agent for bacterial spores [37, 38], and as a counterstain in the Gimenez staining of bacterial infection in tissues [39, 40]. It is widely used, paired with glutaraldehyde, as a fixative-stain for phospholipids as well [41].

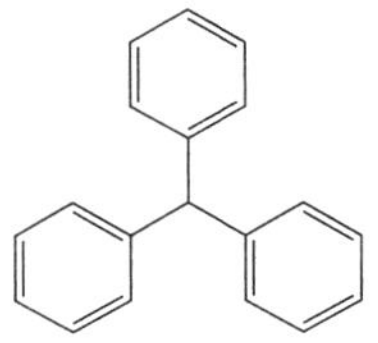

Triphenylmethane

Triarylmethine dyes	λ_{abs}(nm)	λ_{em}(nm)	Remarks
Methyl violet dyes	580–590	~590	/
Fuchsine dye	547–552	625	50% aqueous EtOH
Malachite green	446	614	/
Aniline blue	600		Mixture of methyl blue and water blue

FIGURE 11.8 The basic skeleton of staining agents belonging to the triarylmethine dye family.

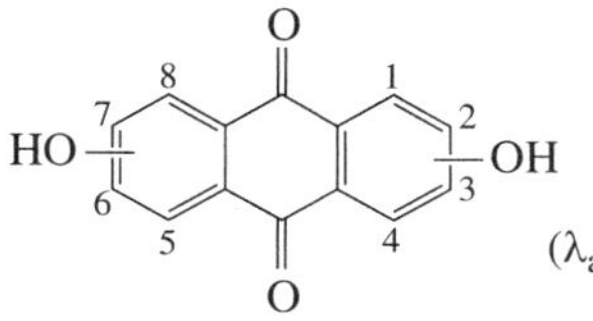

Alizarin: 1,2-Dihydroxyanthraquinone
($\lambda_{abs} = 438$ nm, $450-500$ nm after complexed with Ca^{2+})

FIGURE 11.9 Dihydroxyanthraquinone dyes.

TABLE 11.1 Various Azo and Diazo Staining Agents.

Important azo- and diazo staining agents	λ_{abs} (nm)	λ_{em} (nm)	Remarks
Janus Green B	630		50% aqueous EtOH
Congo Red	497	614	614 nm upon binding to amyloid
Sudan II	490		EtOH
Sudan III	503		EtOH
Sudan IV	520		EtOH
Oil Red O	518		toluene
Sudan Black B	598		EtOH
Tryphan Blue	588	600–670	600–670 when bound to proteins
Evans Blue	611	680	/

11.3.1.4 Dihydroxyanthraquinones The first dihydroxyanthraquinone dye, Alizarin (Figure 11.9), was introduced as a staining agent in 1874. It is still being used nowadays to reveal calcium in tissues and cells because it forms a distinct red precipitate with calcium ions [42, 43].

11.3.2 Azo- and Diazo-staining Agents

Toward the late 19th to early 20th century, other synthetic organic dyes began to emerge as useful histological staining agents. Amongst them are the azo- and diazo-dyes, and the reducible tetrazolium probes. Commonly used azo- and diazo-dyes in microscopy staining include Janus Green B, Congo Red, the Sudan lysochromes (Sudan II, III, and IV, Sudan Black B, and Oil Red O dyes, Table 11.1) and Trypan and Evans Blue. Janus Green B is a mitochondria-specific stain [43–45]. Congo Red is used as a cytoplasm and erythrocyte stain, as well as for the detection of amyloid plaques resulting from Alzheimer's disease [46, 47]. Besides being a colorimetric dye, Congo Red can also fluoresce to be used as a fluorescent stain for elastin [48] and polysaccharides [49–52].

The diazo dyes Trypan and Evans Blue were introduced as cell and tissue vitality stains, via exclusion, in 1915 [53, 54]. Both of them are colorimetric as well as fluorescent dyes. Most viable cells and tissues are able to exclude them, while the proliferated plasma membrane of dead cells cannot stop their penetration. Hence, they are used to stain the dead cells. Such an exclusion mechanism is different from the metabolic assessment mechanism of the MTT cell viability test using the tetrazolium compound methylthiaolyldophenyl tetrazolium (MTT) (to be discussed later). Tryphan Blue can also bind to proteins, especially albumin, give distinct red fluorescence at 600–670 nm, and is used to study exudation from blood vessels

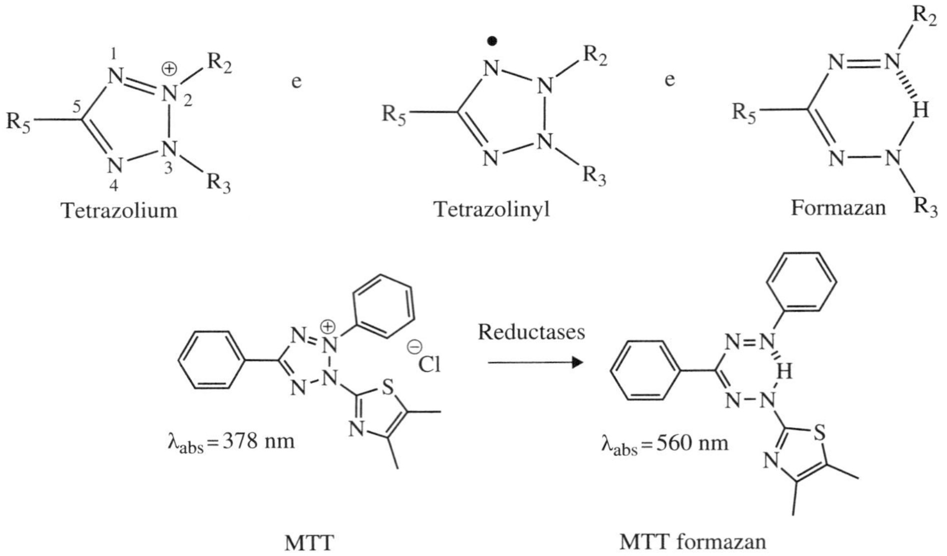

FIGURE 11.10 The tetrazolium-formazan transformation and the methylthiazolyldiphenyl tetrazolium assessment of metabolic viability of cells.

in the injured central nervous system [55, 56]. Similarly, the high affinity of Evans Blue for plasma albumin has made it a useful probe for the measurement of the blood volume of laboratory animals and humans [57–59]. It can also act as a fluorescent tag (fluoro-tag) to plasma albumin in the assessment of the integrity of the blood-brain barrier. If the blood-brain barrier has been compromised, the normally excluded serum albumin can penetrate into the brain. Such an event can be revealed by the albumin bound Evans Blue tag that fluoresces at 680 nm upon excitation at 470 nm [60].

11.3.2.1 Tetrazolium Compounds

Tetrazolium compounds, which possess an electron-deficient five-membered heterocyclic ring, can easily be reduced by reductase enzymes to the highly coloured and less water-soluble formazans (Figure 11.10) [61, 62]. Therefore, they can become useful colorimetric probes for general metabolic activities and the viability of cells. The most famous tetrazolium compound for this application is methylthiazolyldiphenyl tetrazolium (MTT) (Figure 11.10). MTT assay has been extensively adopted in the field of cell biology for the assessment of cell viability [63, 64]. The absorption maxima of MTT shifts from 378 nm (yellow) to 560 nm (magenta) upon reduction to MTT formazan.

With the rapid progress of fluorescent techniques in histochemistry, cell, and molecular biology since the early 20th century, a number of organic chromophores that were established centuries ago have found new utilities under the newly developed fluorescent microscopy and the subsequent confocal laser scanning microscopy. As highlighted in the previous section, these fluorescent organic dyes form the basic building blocks for the fluoro-tags for antibodies, proteins, and other biomolecules, as well as signalling units in chemosensors to report analyte-binding events. Some of their spectrofluorometric and photophysical properties have also been found to be extremely useful for the design of advanced probes to reveal complex histo- and cytochemical interactions. Here, we will focus our discussion on a number of important families of fluorophores, such as xanthenes, acridines and phenanthridines, and polymethines, as well as some miscellaneous fluoro-tags, such as NBD chloride, dansyl chloride, BODIPY, and its derivatives [65, 66].

11.3.3 Xanthene-based Luminophores

As introduced in the previous section, xanthene dyes are important fluoro-tags because of their outstanding photophysical properties and biocompatibility (Figure 11.11). Both anionic hydroxyxanthenes, for example, Fluorescein, and cationic aminoxanthenes, for example, Rhodamine B, are commonly used luminophores for the labelling of biomolecules and as signal-transducers in the chemosensing of specific analytes. For example, the famous Calcium-Green™-1, for real-time in vitro and in vivo calcium flux imaging, makes use of a fluorescein moiety as the signal transducers to report the binding of calcium ions [67, 68]. Given their relatively low energy fluorescence (emission maxima >500 nm), they are also widely used as energy acceptors in Fluorescent Resonance Energy Transfer (FRET) processes. In fact, a number of xanthene derivatives that carry reactive substituents are routinely being used as biolabelling agents in biomedical and cell biology research. These include Fluorescein isothiocyanate (FITC), tetramethylrhodamine isothiocyanate (TRITC), Oregon Green 514 Ester, and Texas Red.

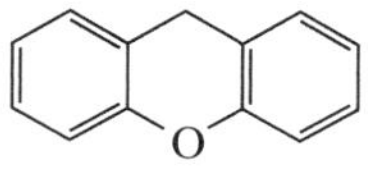

Xanthene

Important xanthene-based fluorescent biolabelling agents	λ_{abs}(nm)	λ_{em}(nm)	Remarks
Rhodamine B (aminoxanthene dye)	548	627	Acidic aqueous EtOH
Fluorescein isothiocyanate (FITC)	494	519	pH 9.0
Tetramethylrhodamine isothiocyanate (TRITC)	537–555	564–580	pH 8.0
Oregon green 514 ester	506	526	pH 9.0
Texas red	587	602	Chloroform

FIGURE 11.11 Structure of xanthene.

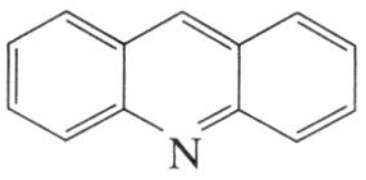

Acridine

Acridine-based dyes	λ_{abs}(nm)	λ_{em}(nm)	Remarks
Acridine orange (AO)	492	535	502 nm when bound with DNA, 650 nm when bound with RNA
Acriflavine	452	510	/
Quinacrine	445	500	/

FIGURE 11.12 The acridine dyes.

The isothiocyanate, succinimidyl ester, and sulfonyl chloride moieties of these 'reactive dyes' enable them to form covalent linkages with the free, and exposed, amino moieties in peptides, proteins, and antibodies for their fluoro-tagging.

11.3.4 Acridine- and Phenanthridine-based Luminophores

Both acridines and phenanthridines contain a central pyridine ring fused with two benzo rings, one on each side (Figures 11.12 and 11.13). They are small cationic and planar dyes that are able to interact with DNA and RNA via intercalation and coulombic attraction [69, 70]. This makes acridine and phenanthridine luminophores useful nucleic acid selective stains for cell cycle determination [71–74]. Important acridine dyes include Acridine Orange, Acriflavine, and Quinacrine. Acridine Orange (AO) was first reported as a fluorescent microscopy stain in 1940 [75–77]. It can enter both live and dead cells and has unique properties in differentiating between DNA and RNA as it gives green fluorescence (λ_{max} at 502 nm) when bound to DNA, and orange fluorescence (λ_{max} at 650 nm) when bound to RNA. Acriflavin is used as a general oversight stain in fluorescent microscopy in entomological specimens [78]. It can also be used in the *in vivo* imaging of rat brain by laser scanning confocal microscopy [79]. Quinacrine is a decent stain for lysosome and nucleic acid [80, 81]. It can also stain platelets as they store the dye in dense granules. This is useful for the assessment of platelet adhesion and aggregation [82–84].

Perhaps the most important phenanthridine dye for bioimaging is Ethidium Bromide (EB). It is useful for the detection of nucleic acids and the staining of cell nuclei because it can intercalate into DNA helices, similar to that of acridine dyes, to give strong orange colour fluorescence. However, it only interacts weakly with RNA to give a weak red fluorescence. It is often used in combination with AO to reveal the viability status of cells [85, 86]. Both live and dead cells can take up AO, while EB can only enter dead cells and their nuclei via their proliferated plasma and nuclear membranes. Therefore, under the co-staining of AO and EB with blue light excitation, viable cells will show bright green nuclei and red cytoplasm, while dead cells will show bright orange nuclei and cytoplasm (as fluorescence of EB overwhelms that of AO). Any remaining RNA will appear dark red. Nuclei of both live and dead cells have well-defined euchromatin and heterochromatin and will appear as distinct structures under the co-staining of AO and EB. However, apoptotic nuclei possessing highly condensed chromatin will appear as uniformly stained bodies. Also, in advanced apoptosis, cell nuclei will only be weakly stained, because the DNA in the nuclei is gone.

Another phenanthridine dye (Figure 11.13) that has similar properties to EB is Propidium Iodide (PI). It is often used in combination with a benzimidazole dye, Hoechst 33342 (Hoe 33342), to report the cell viability status [87, 88]. Nuclei of the viable cells will be stained blue (by Hoe 33342), while those of the dead cells will appear bright pink (a blend of fluorescence from Hoe and PI). Cytoplasm of the viable cells is free from staining (all Hoe enters cell nuclei and PI cannot penetrate the plasma membrane), while that of the dead cells will be stained bright pink (same as their nuclei).

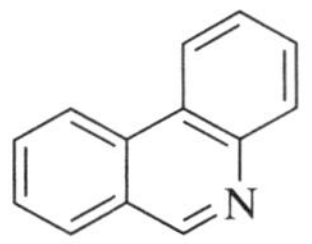

Phenanthridine

Phenanthridine-based dyes	λ_{abs}(nm)	λ_{em}(nm)	Remarks
Ethidium bromide (EB)	480	605	620 nm when bound with DNA
Propidium iodide (PI)	493	636	617 nm when bound with DNA

FIGURE 11.13 The phenanthridine dyes.

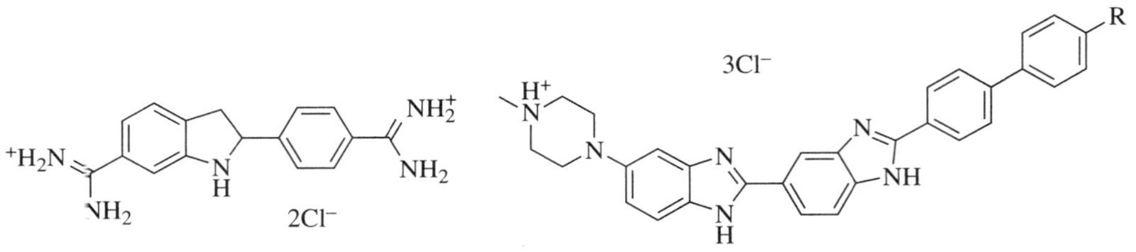

DAPI (λ_{abs} = 344 nm; λ_{em} = 450 nm, 461 nm when bound to DNA)

R = OEt $\Longrightarrow$ Hoechst 33342 (λ_{abs} = 350 nm; λ_{em} = 461 nm, 461 nm when bound to DNA)

R = OH $\Longrightarrow$ Hoechst 33258 (λ_{abs} = 352 nm; λ_{em} = 461 nm, 461 nm when bound to DNA)

FIGURE 11.14 The indolenine- and benzimidazole-based DNA probes DAPI, Hoechst 33342, and Hoechst 33258.

11.3.5 Polymethine-based Luminophores

The basic structure of polymethine luminophores consists of an electron donor and an electron acceptor moiety separated by a conjugated carbon chain or spacer group. Polymethines can be classified into a number of sub-families, such as indolenines and benzimidazoles, cyanines, styryls, and coumarins. Two sets of famous indolenine- and benzimidazole-based polymethine fluorescent probes for bioimaging are 4',6-diamino-2-phenylindole dichloride (DAPI) and Hoechst (Hoe) stains (Figure 11.14). Both are live-cell compatible DNA probes that can be used to stain cell nuclei. DAPI is an indolenine dye that gives bright blue colour fluorescence when bound to DNA [89]. It has a strong affinity for the A-T rich regions of DNA strands. However, it can also bind to RNA to give weak fluorescence at around 500 nm.

The Hoechst stains are benzimidazole dyes that also possess high affinity for the A-T rich regions of DNA. They are relatively less toxic and more cell permeable than DAPI. Fluorescence from Hoechst stains can be quenched by bromodeoxyuridine (BrdU). The latter can be used by cells as a substituent for thymidine in DNA synthesis. The Hoechst-BrdU pair can, therefore, be used to detect dividing cells. During the S-Phase of the cell cycle before mitosis, DNA in the cell nuclei is replicated, and chromosomes are duplicated. If cells have been given BrdU instead of thymidine, they will incorporate the BrdU into their newly synthesised DNA. These BrdUs quench the fluorescence of Hoechst dyes, which results in the absence of the blue fluorescence from cell nuclei that are about to undergo mitosis [90, 91]. Another use of Hoechst stains is for the assessment of cell viability, in combination with the live cell impermeable Propidium Iodide (PI) dye. This technique has already been outlined in the previous section.

Cyanine stains are polymethine dyes that possess nitrogen-based electron donors and acceptors. Famous examples of these fluorescent probes in bioimaging applications include Merocyanine 540, CellTracker™ CM-DiI, and Cy3 & Cy5 fluoro-tagging agents. Merocyanine 540 (Figure 11.15) is the first electrochromic dye used for the imaging of membrane potential. It is a slow-response membrane-potential probe that binds to the surface of polarised membranes in a perpendicular orientation, but dimerises into non-fluorescent dimers upon membrane depolarisation [92–94].

The CellTracker™ CM-DiI (Figure 11.16) is a symmetrical cyanine i.e., both the electron donor and acceptor end-groups on the conjugated carbon chain are identical, with excellent cell retention and minimal cytotoxicity. This cyanine membrane stain can be used for long-term labelling and tracking of live cells, intracellular membranes, liposomes, viruses, and lipoproteins [95–97]. The stain can be retained on the cell membranes throughout fixation and permeabilisation steps. It has two lipophilic C_{18}-carbon chains and a mildly thiol-reactive chloromethyl moiety for the labelling of membranes and thiol-containing peptides and proteins.

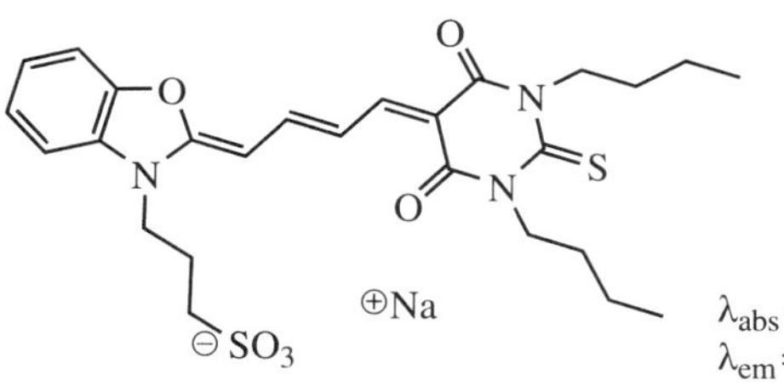

$\lambda_{abs} = 500, 534$ nm;
$\lambda_{em} = 577$ nm at pH = 7.0, 500 nm at pH>7.6

FIGURE 11.15   Structure of Merocyanine 540.

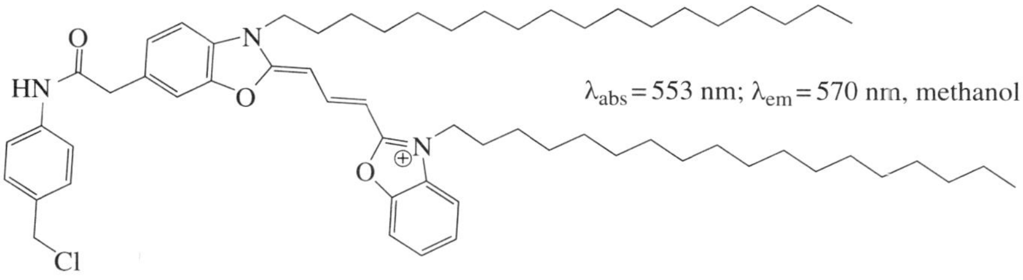

$\lambda_{abs} = 553$ nm; $\lambda_{em} = 570$ nm, methanol

FIGURE 11.16   Structure of Celltrack™ CM-DiI.

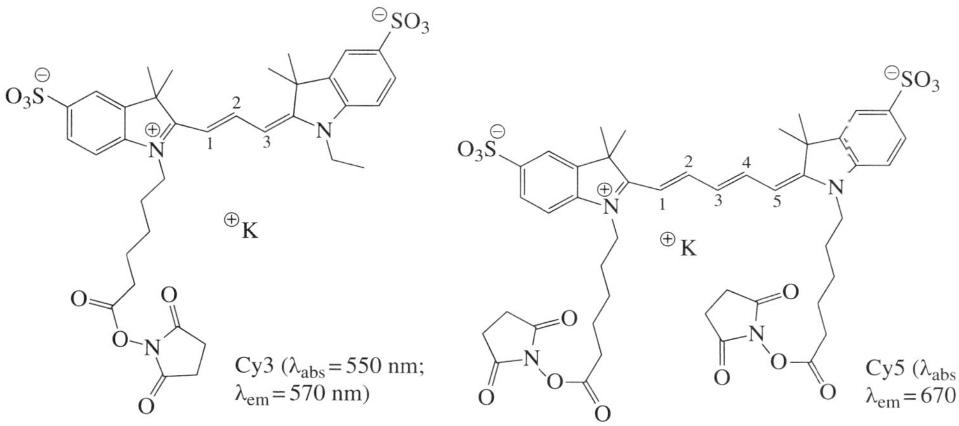

Cy3 ($\lambda_{abs} = 550$ nm; $\lambda_{em} = 570$ nm)

Cy5 ($\lambda_{abs} = 649$ nm; $\lambda_{em} = 670$ nm)

FIGURE 11.17   The reactive cyanine stains Cy3 and Cy5.

The other two symmetrical cyanine polymethine stains that are widely used as fluoro-tagging agents for nucleic acids, peptides, proteins, and antibodies are Cy3 and Cy5 (Figure 11.17). They are water soluble and possess the reactive succinimidyl ester groups that can form covalent linkages with the amino groups of their substrates. Cy3 is red fluorescing, while Cy5 fluoresces in the far-red region.

Styryl polymethine dyes generally contain a styrene core with electron donor and acceptor attached at each end. Interestingly, other polymethine stains that possess longer methane chains, such as a naphthyl or pyridyl ring, are also referred to as styryl dyes. Important members of the group include DASPMI and the FM series of lipophilic membrane stains. The 4-(4-(dimethylamino)styryl)-*N*-methylpyridinium iodide (DASPMI) is a hemicyanine whose pyridine and phenyl rings respectively contain and carry the acceptor and donor nitrogens (Figure 11.18). It is a very useful stain for metabolically active mitochondria in live cells [98–100]. It can also be used to assess the activity of membrane cation transporters [101].

The FM series of styryl polymethine stains are water-soluble lipophilic fluorescent staining agents commonly used in the imaging of plasma membranes and vesiculation. They show minimal cytotoxicity and are non-fluorescent in aqueous media. The strong green fluorescence from the FM1-43 and red fluorescence from the FM4-64 are observable only when these staining agents are inserted into the outer leaflet of the surface membranes of live cells [102, 103]. They are commonly used in synaptosomal studies, because they are internalised, during the recycling of synaptic vesicles, into neuron terminals that have been actively releasing neurotransmitters [104, 105].

Coumarin dyes possess a benzo-2-pyrone fluorophore with a carbonyl moiety as an electron acceptor (Figure 11.19). It is a large dye family with most members having very high photoluminescent quantum efficiency. Some of the family members

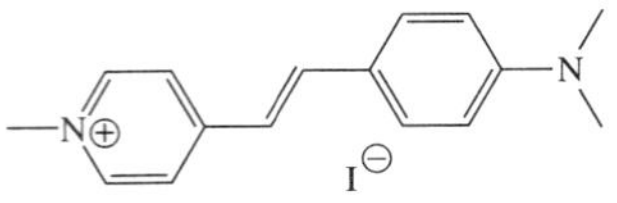

4-(4-(Dimethylamino)styryl)-N-methylpyridinium iodide (DASPMI)

Styryl-based dyes	λ_{abs}(nm)	λ_{em}(nm)	Remarks
DASPMI	475	605	MeOH
FM1-43	512	626	MeOH, emit at 598 nm when bound to biomembranes
FM4-64	560	767	MeOH

FIGURE 11.18   DASPMI and the FM series of plasma membrane stains.

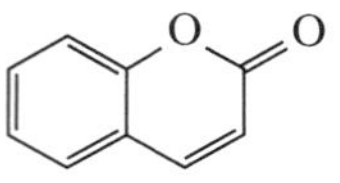

Coumarin

Coumarin-based dyes	λ_{abs}(nm)	λ_{em}(nm)	Remarks
Calcein blue	360	449	pH 9.0
DACM	383	463	MeOH
AMCA succinimidyl ester	354	440	MeOH

FIGURE 11.19   Structure of coumarin.

are laser dyes. They are also useful luminophores for bioimaging because of their biocompatibility and their higher energy fluorescence ($\lambda_{max} < 500$ nm), which makes them complementary energy donors to fluorescein-based acceptors in FRET processes [106–109]. Calcein Blue, consisting of a coumarin luminophore linked with a calcium-chelating methyliminodiacetic acid functionality, is used as a marker of bone growth and to identify microcracks in bones [110, 111]. The N-(7-dimethylamino-4-methylcoumarin-3-yl) maleimide (DACM) and 7-hydroxycoumarin-3-carboxylic acid succinimidyl ester (AMCA) are two commonly used fluoro-tagging agents for peptides, proteins, antibodies, and other biomolecules.

11.3.6   Reactive Chloride Dyes

4-Chloro-7-nitrobenz-2-oxa-1,3-diazole (NBD chloride) is a benzofurazan heterocycle with a reactive chloro group (Figure 11.20). It is a useful fluoro-tag for a wide range of biomolecules and xenobiotics because it is able to form covalent linkages with exposed amino, hydroxyl, and thiol moieties. It is biocompatible with live cells. The fluoro-tagged phospholipids, NBD-phospholipids, have been used to trace lipid metabolism in yeast [112]. The fluoro-tagged peptides, NBD-peptides, have been used to study the adsorptive endocytosis process of small peptides in live cells [113]. Dansyl chloride is a widely adopted luminophore for *in vitro* and *in vivo* fluoro-tagging of peptides, proteins, and antibodies. It has a naphthalene ring and a reactive sulfonyl chloride substituent [114–116] (Figure 11.20).

11.3.7   BODIPY and Its Derivatives

Any discussion of organic-based luminophores for biomedical and imaging applications would not be complete without an introduction to the BODIPY luminophore and its derivatives. BODIPY is the acronym of boron dipyrromethine; its full nomenclature is 4,4-difluoro-4-bora-3a,4a-diaza-*s*-indacene. It is a very popular luminophore for fluoro-tagging and bioimaging applications due to its unique advantageous photophysical properties, such as near 100% photoluminescent quantum yield in aqueous media, environment-independent fluorescence, ease of derivatisation, and the ease of fine-tuning of fluorescence wavelength by the introduction of different substituents [117–119].

In fact, in spectroscopic and spectrofluorometric terms, commercially available BODIPY derivatives (Figure 11.21) have already covered the full range of the visible spectrum. Reactive functionalities have been grafted onto the luminophore in order to generate fluoro-tags, which are capable of labelling biomolecules and xenobiotics. Special substituents have also been incorporated onto the luminophore to bring about specific responsive properties toward targeted stimulations and enzymatic/analyte activities (Figure 11.22) [120–122].

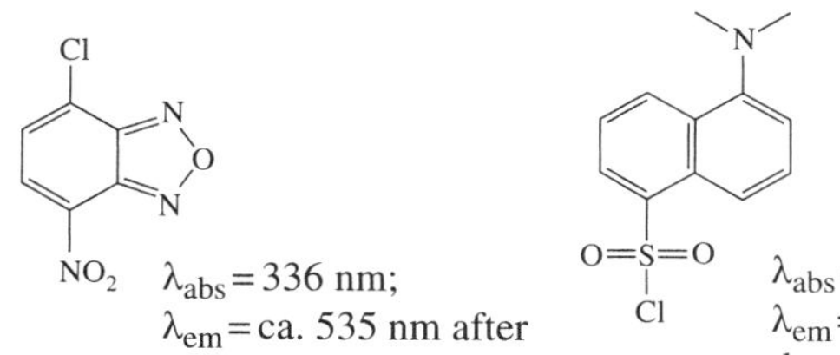
Cl
N
O
N
NO_2
λ_{abs} = 336 nm;
λ_{em} = ca. 535 nm after derviatisation with aliphatic amines, methanol
N
O=S=O
Cl
λ_{abs} = 372 nm;
λ_{em} = ca. 429 nm after derviatisation with aliphatic amines, chloroform

FIGURE 11.20 Structures of NBD chloride (left) and dansyl chloride (right).

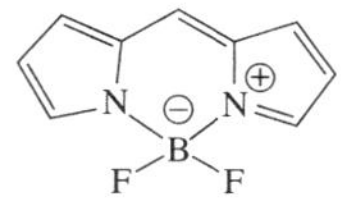
N
N
B
F
F

The BODIPY luminophore

BODIPY-based dyes	λ_{abs}(nm)	λ_{em}(nm)	Remarks
BODIPY FL	503	512	/
BODIPY R6G	528	547	/
BODIPY TMR	543	569	/
BODIPY TR	592	618	/
BODIPY 630/650	625	640	/
BODIPY 650/665	646	660	/
BODIPY 581/591		~510 to ~590	Oxidative responsive probe with λ_{em} shifts from ~590 nm to ~510 nm upon oxidation of the unsaturated butadienyl portion

FIGURE 11.21 Structure of BODIPY.

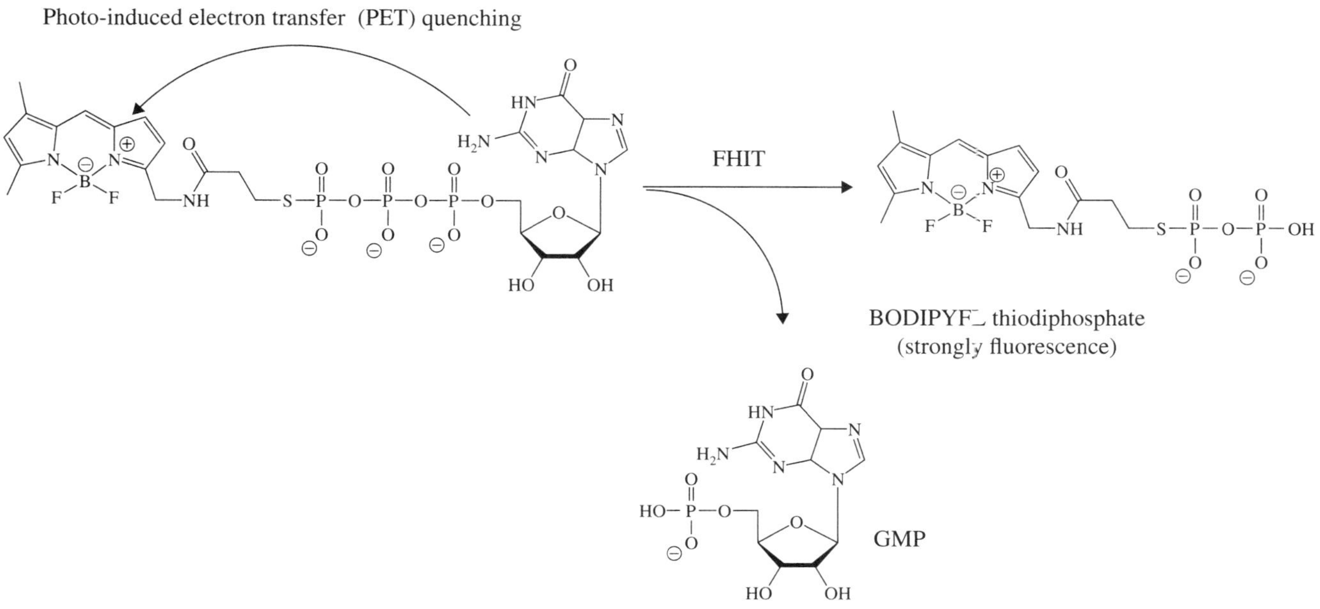
Photo-induced electron transfer (PET) quenching
FHIT
BODIPYFL thiodiphosphate
(strongly fluorescence)
GMP
HN
H_2N
N
O
NH
S
P
HO
OH

FIGURE 11.22 Specific responsive properties of BODIPY FL GTP-γ-*S*.

11.4 MECHANISMS OF PHOTOPHYSICAL PROCESSES AND THEIR APPLICATIONS IN MOLECULAR IMAGING AND CHEMOSENSING

Photophysical properties of fluorophores are strongly influenced by a variety of factors such as the viscosity, polarity, and rigidity of the media and the bio/macromolecules on which they are tagged; the presence of quenchers that interact with the fluorophores at either their ground or excited states; and their proximity to energy donors. Detailed information about the chemical nature of the surroundings of a fluorescent probe can be obtained by thorough interpretation of its luminescent behaviour. Fluorescent probes with specially incorporated features, such as analyte-specific receptors, energy/electron donor-acceptor pairs, can be utilised to selectively reveal certain important characteristics of the system under investigation. Therefore, fluorescent techniques making use of various kinds of fluoro-tags, fluorescent probes, and chemosensors have been becoming increasingly eminent to life science and biomedical studies ever since the advent of fluorescent microscopy and subsequently laser scanning confocal microscopy to the research communities. The following is a brief introduction to the fundamental principles of various fluorescent techniques that are commonly applied in bioimaging and *in vitro* and *in vivo* bio- and chemosensing.

11.4.1 Fluorescence Resonance Energy Transfer

The extra energy possessed by an excited fluorophore in a more energetic excited state can be transferred to and, hence, excites another fluorophore in a lower energy excited state. There are generally two possible pathways for such energy transfer—the *radiative* and *non-radiative* decay. In radiative energy transfer, an emitted photon from a donor molecule is reabsorbed by an acceptor. In this case, energy is transferred directly through long-range (1–10 nm) dipole interactions between a donor and an acceptor. This can occur if the emission spectrum of the donor overlaps with the absorption spectrum of the acceptor, given that several vibronic transitions in the donor have practically the same energy as the corresponding transitions in the acceptor (Figure 11.23) [123]. Because fluorescence resonance energy transfer is highly sensitive to the distance between the donor fluorophore and the acceptor fluorophore, it can be used as a spectroscopic ruler to measure intermolecular interactions in the range of 10–100 Å.

11.4.2 Photo-induced Electron Transfer (PET)

Photo-induced electron transfer (PET) is often responsible for the quenching of fluorescence from excited fluorophores. This can be mediated by either oxidants or reductants. As illustrated in Figure 11.24, reductive electron transfer quenching involves the transfer of an electron from the HOMO of an electron-rich quencher (usually a lone pair of electrons on a hetero atom) to the "hole" left in the HOMO of an excited fluorophore. Similarly, the excited electron of an excited fluorophore can be transferred to the LUMO of an electron-deficient quencher, resulting in an oxidative electron transfer quenching. Like FRET, PET is another photophysical process that leads to the variation of fluorescence efficiency by distance-dependent fluorescence quenching between a fluorophore and a quenching moiety [124]. This PET mechanism offers extraordinary

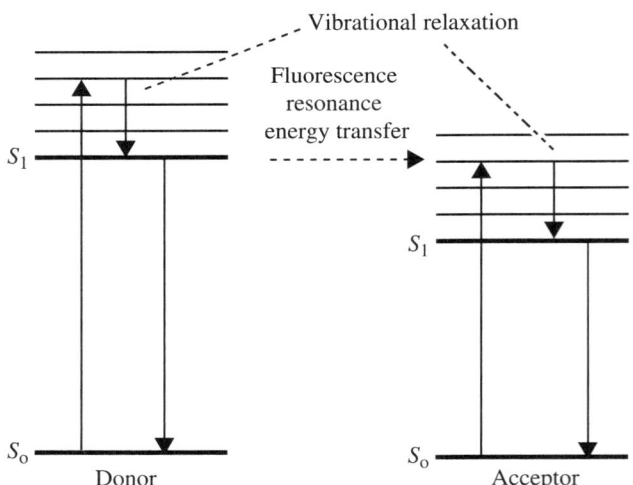

FIGURE 11.23 Illustration of fluorescence resonance energy transfer.

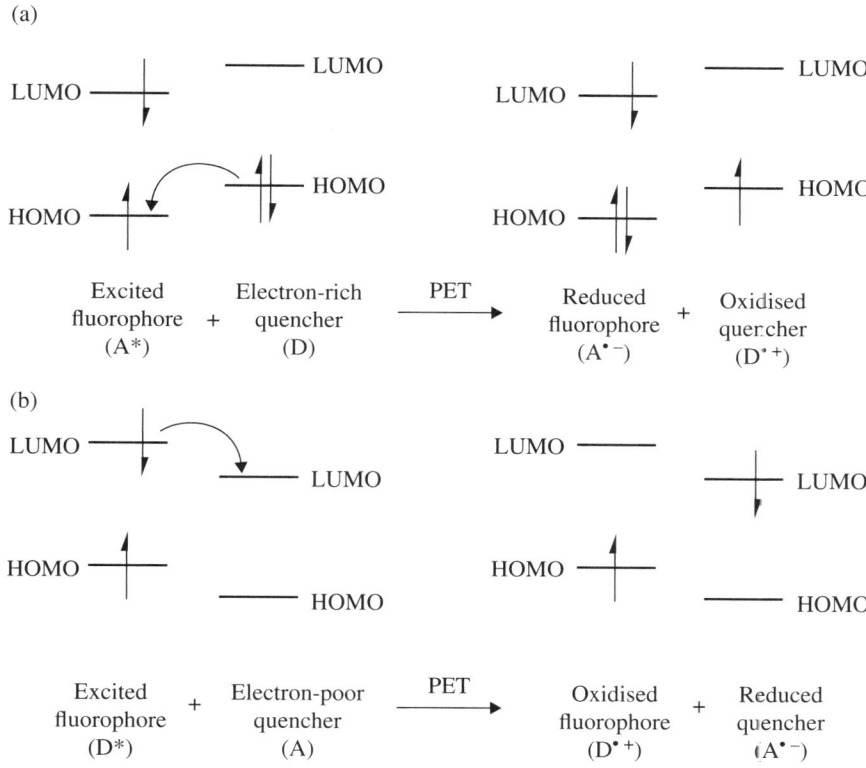

FIGURE 11.24 The PET mechanism: (a) reductive and (b) oxidative electron transfer.

sensitivity for fluorescence-based chemosensing of minute amount of target analytes. A typical PET fluorescent probe consists of a fluorophore and a quencher moiety connected together through a spacer. The intramolecular PET process keeps the probe in the 'off' status at rest. In the presence the targeted analyte, its specific interaction with the intramolecular quencher moiety turns off the PET process and 'turns on' the fluorophore to produce a strong fluorescent signal. Figure 11.25 shows a PET chemosensor Zinpyr-1 for Zn^{2+} ions that contains a $2',7'$-dichlorofluorescein fluorophore and two dipicolylamine moieties for Zn^{2+} binding [125]. The lone pairs of electrons on the tertiary amines of dipicolylamine act as the PET quencher. Fluorescence of the chemosensor is 'turned on' upon binding with Zn^{2+} or H^+ by the dipicolylamines.

The PET process is also useful in biosensing, where specific fluorescence quenching of fluorophores by selected quenchers such as naturally occurring DNA nucleotides (e.g., guanine) and amino acids (e.g., tryptophan) is adopted to probe the presence of target molecules (Figure 11.26). With careful design of these conformationally flexible biosensors, efficient PET-probes with single molecule detection sensitivity can be produced [126].

11.4.3 Excimer/Exciplex Formation

Excimers are excited state dimers, which are formed by the collision between an excited molecule and a ground state molecule of the same species. Exciplex are excited-state complexes that are formed by the collision of an excited molecule with a ground state molecule of another species, forming excited-donor/acceptor or excited-acceptor/donor complexes (Figure 11.27). These excimer/exciplex formations are diffusion-controlled processes. In polar media, de-excitation of exciplexes can lead to fluorescence emission of ion pairs and subsequently 'free' solvated ions. Therefore, exciplexes can be considered as intermediate species in electron transfer from a donor to an acceptor in some cases. The lifetime of excimers/exciplexes is very short, on the order of nanoseconds, and their corresponding fluorescence are usually broad and red-shifted. This mechanism normally occurs at high concentration due to the increased probability of forming the excimer [127].

11.4.4 Internal Charge Transfer and Twisted Internal Charge Transfer

Excitation of a fluorophore promotes an electron from one molecular orbital to another one of higher energy. If the initial and the final molecular orbitals are located on different parts of the fluorophore, the electronic transition will be accomplished by an almost instantaneous change in the molecular dipole moment. In such a manner, excitation gives rise to a

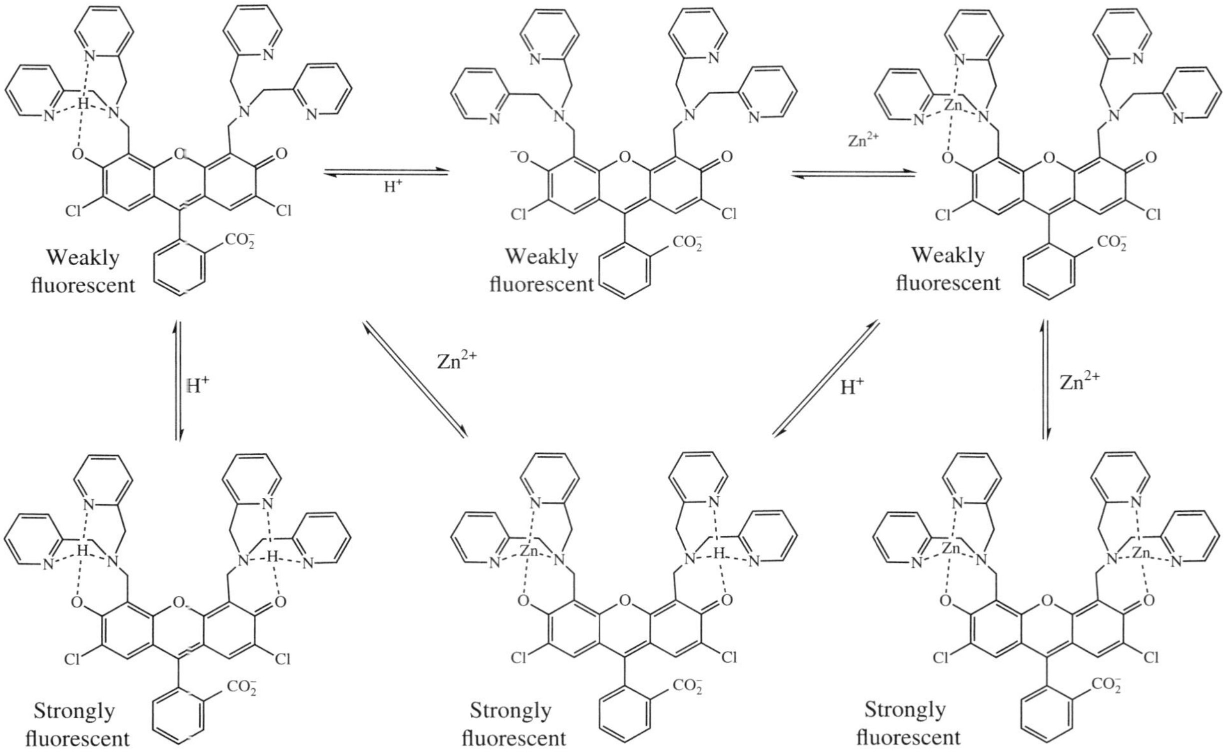

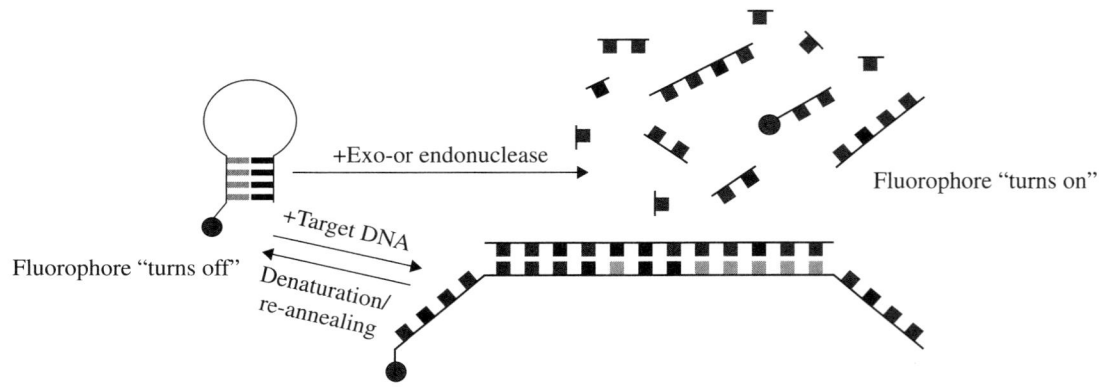

FIGURE 11.25   The PET chemosensor Zinpyr-1 for Zn^{2+} [56].

FIGURE 11.26   An application of the PET mechanism for smart biosensors. A fluorophore is attached to 5′ end of the oligonucleotide and quenched by the guanine residues of the complementary stem. Upon hybridisation to the target sequence, or exo-/endo-nucleolytic digestion, fluorescence of the fluorophore is restored.

$$M^* \; + \; M \; \rightleftharpoons \; (MM)^* \quad \text{(Excimer)}$$
$$D^* \; + \; A \; \rightleftharpoons \; (DA)^* \quad \text{(Exciplex)}$$
$$A^* \; + \; D \; \rightleftharpoons \; (AD)^* \quad \text{(Exciplex)}$$

FIGURE 11.27   Excimer and exciplex formation.

redistribution of electron density and the creation of a substantial dipole. Consequently, the excited state, upon excitation (known as the Frank-Condon state or the locally excited state) is not in equilibrium with the surrounding polar solvent molecules. For a fluorophore with two co-planar π-conjugated moieties joined together by a σ-bond, photo-excitation and subsequent intramolecular transfer of an electron from one moiety to the other generates an electron-rich and an electron-deficient conjugated region in the excited molecule. According to the Franck-Condon principle, these two conjugated regions can remain co-planar in the locally excited state. Solvent relaxation then takes place with a concomitant rotation around the sigma bond connecting the electron-rich and electron-poor parts of the excited fluorophore until they are orthogonal to each other. This results in a twisted internal charge transfer (TICT) state with complete charge separation (Figure 11.28). In addition to the fluorescence band due to the emission from the locally excited state ('normal' band), an additional emission band corresponding to emission from the TICT state can be observed at longer wavelengths ('anomalous' band) [128].

11.4.5 Excited-state Proton Transfer

In comparison with its ground state, a photo-excited molecule may possess different acid-base properties. The redistribution of the electronic density upon photo-excitation is one of the reasons for such a phenomenon. If an excited species is a strong acid or base compared with its ground state, photo-induced proton transfer may take place. The acidic character of a proton donor (e.g., the hydroxyl substituent of an aromatic ring) can be enhanced upon excitation so that its pK* becomes much lower than that of its group state (Figure 11.29a). In the same way, the pK* of an acceptor group (e.g., a heterocyclic nitrogen atom or carboxylate) in the excited state can be much higher than that of its ground state (Figure 11.29b) [129].

The excited state proton transfer (ESIPT) process can be easily recognised in steady-state spectra: The absorbance is generally similar to that of the parent chromophore, but the fluorescence is significantly different. ESIPT dyes generally show large Stokes' shifts and are ideal candidates for being used as fluorescence labels to avoid interference from other fluorescent materials present in the specimen. An advantage of the large Stokes' shift is the almost complete lack of spectral overlapping between absorption and emission, which makes ESIPT dyes promising for their use in fluorescent chemosensing. Very recently, a sensor based on a *bis*(benzoxazole) derivative (Figure 11.30) in which metal binding enables the ESIPT

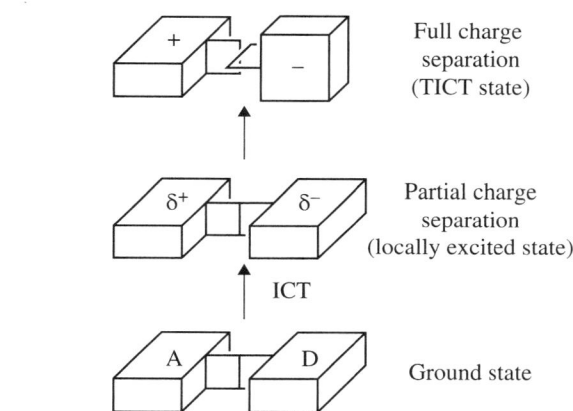

FIGURE 11.28 Charge separation in the ICT state and the TICT state.

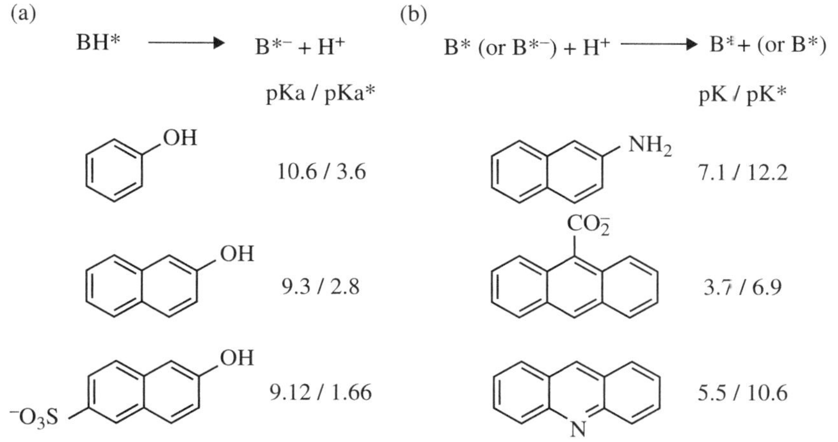

(a)   (b)

$$BH^* \longrightarrow B^{*-} + H^+ \qquad\qquad B^* \text{ (or } B^{*-}) + H^+ \longrightarrow B^* + \text{ (or } B^*)$$

pKa / pKa*   pK / pK*

10.6 / 3.6   7.1 / 12.2

9.3 / 2.8   3.7 / 6.9

9.12 / 1.66   5.5 / 10.6

FIGURE 11.29 Photo-induced changes in the acid-base properties of chromophores: (a) photo-induced enhancement in acidity resulting in excited state deprotonation, and (b) photo-induced enhancement in basicity resulting in excited state protonation.

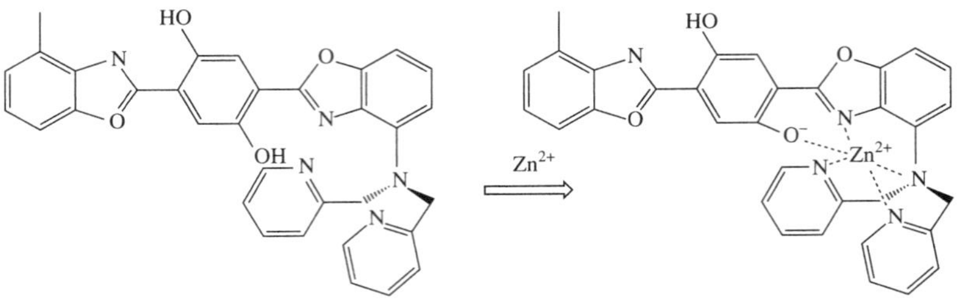

FIGURE 11.30   A Zn^{2+} sensor based on excited-state intramolecular proton transfer (ESIPT) [130].

bis(benzoxazole) derivative to give a new emission band in the NIR region with a large Stokes' shift (ca. 230 nm) has been reported [130]. The fluorescence of the free ligand is quite weak ($\lambda_{em}=543$ nm; $\Phi_{fl}=0.0067$), possibly due to the presence of its deprotonated form which exists in a small amount in equilibrium with its neutral form in the polar media. Upon addition of Zn^{2+}, the intensity of the emission signal at 543 nm increased gradually, and a new emission band appeared in the NIR region at 712 nm. Other metal ions (Cd^{2+} and Hg^{2+}) give rise to different changes in the fluorescent behaviour as they perturbed the enol and keto emission of *bis*(benzoxazole) derivative to different degrees.

11.5   TWO/MULTI-PHOTON INDUCED EMISSION AND *IN VITRO/IN VIVO* IMAGING

11.5.1   Principles of Two-photon Induced Emission and Imaging

In the previous sections, we have looked at a variety of organic-based colorimetric and fluorescent staining agents, probes and chemosensors and their applications in bioimaging. Most of these bio-bioimaging/biolabelling agents are single-photon fluorophores. Their operational spectral range is generally in the UV and visible region, which has low tissue penetration ability. Also, their relatively short fluorescent lifetimes sometimes makes it difficult for researchers to distinguish their signals from background autofluorescence in cells and tissues. This situation was changed in 1990 by the pioneering work of Webb and his co-workers on two-photon laser scanning fluorescence microscopy that made use of the lower energy, but much greater tissue-penetrating near-infrared (NIR) radiation as the excitation source [131]. Advantageous features of multiphoton bioimaging include a reduction in photobleaching and photodamage to the imaging probes and cellular structures, the capability to penetrate thick tissues, and the ability to bring about precise three-dimensional localised photosensitisation, photolysis, ablation, and cutting at the subcellular level. Even though the two-photon absorption in some organic materials can be as high as 100,000 GM (GM $= 10^{-50}$ cm^4 s photon^{-1}) (Figure 11.31), there are very few materials that can be used for *in vitro/in vivo* imaging. It is only in the last decade that a few two-photon excitable organic dyes become commercially available.

The elementary processes of linear and two-photon absorptions are both induced by a single laser beam. The schematic diagram for these processes in the regime of quantum theory is shown in Figure 11.32. An intermediate state (schematically represented by a dashed line) is introduced between the two real eigenstates (i.e., the ground and excited states) of a molecule. The occurrence of two-photon absorption, inducing the molecular transition between its two real states, can be visualised as a 'two-step' event: (i) In the first step, one photon is absorbed while the electron of molecule leaves its initial state E_g and be promoted to an intermediate state; (ii) in the second step, another photon is absorbed for the same electron to complete its transition from the intermediate state to the final real state E_f. The key connection between these two steps is the intermediate state in which the molecular status is not certain in the sense that the molecule may stay in all of its possible eigenstates (except E_g and E_f) with a certain probability of distribution. When a two-photon fluorophore absorbs light energy, it is often excited to a higher vibrational energy level in the first excited state, S(1), before rapidly relaxing to the lowest vibrational energy level. This event, depicted as a 'stair-step' transition from the upper to lower vibrational energy levels in S(1), is termed vibrational relaxation or internal conversion and takes place within a picosecond or less [134, 135]. Typically, fluorescence lifetimes are of more or less four orders of magnitude shorter than vibrational relaxation, giving the molecules sufficient time to achieve a thermally equilibrated lowest-energy excited state prior to fluorescence emission. The two-photon absorption cross-section is one of the parameters for the evaluation of the efficiency of a two-photon process. Z-scan and comparison with relevant fluorescence standards are the two commonly adopted measurements for the evaluation of it [136].

Since 1990, many classes of materials have been found capable of participating in direct two- or multi-photon absorption processes and become potential candidates for *in vitro* and *in vivo* bioimaging [137, 138]. Here, we will limit our discussion

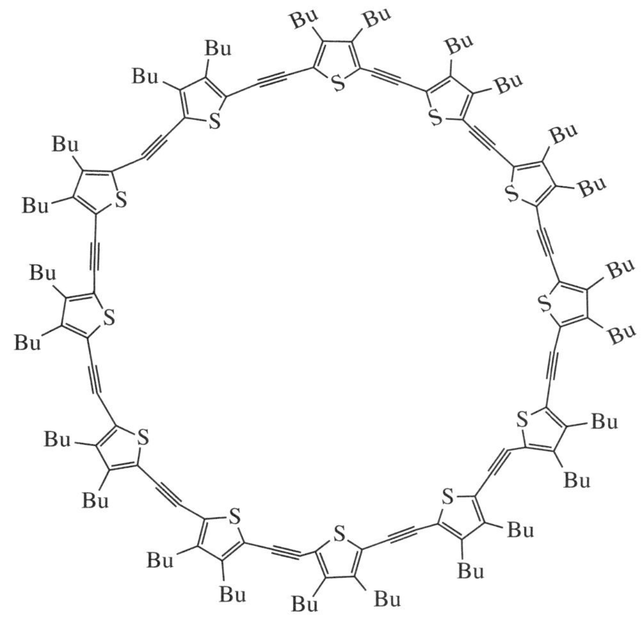

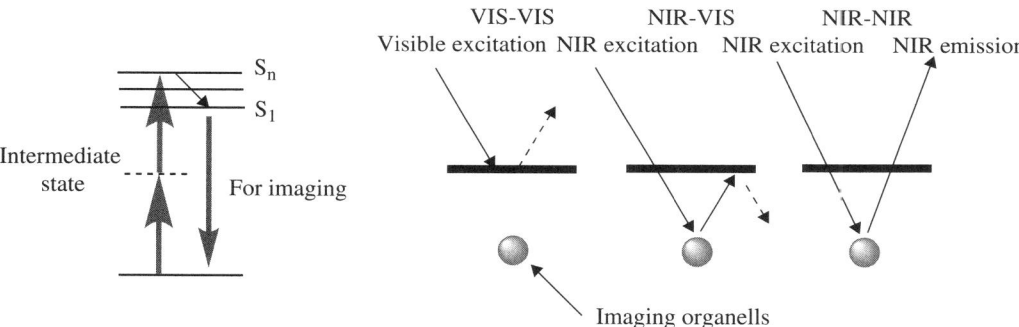

Macrocyclic thiophene 30-mers 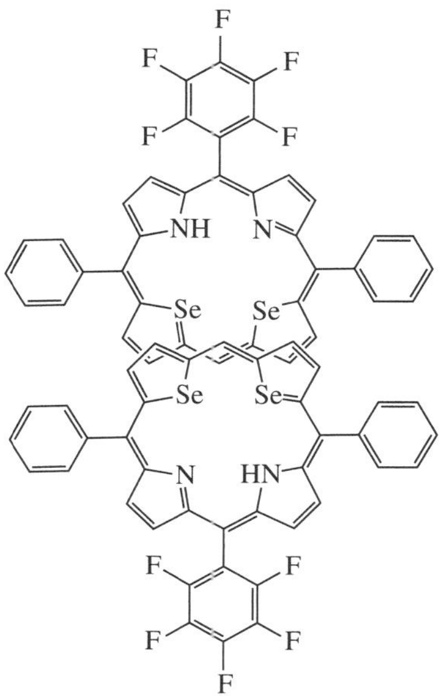   34π octaphyrin

FIGURE 11.31 Two of the highest two-photon absorption cross-section (~100,000 GM) organic materials [132, 133].

FIGURE 11.32 Schematic representations of the mechanism of the two-photon induced emission (left) and advantages of NIR excitation and NIR emission as *in vitro/in vivo* imaging agents.

to recently reported multi-photon materials in the period 2009 to 2011, with an emphasis on their two-photon absorption cross-sections for *in vitro* imaging. One needs to notice that even though some of the molecular and dendrimeric two-photon absorbers possess very high two-photon absorption cross-sections, their poor water solubility and cell permeability render them unsuitable for *in vitro* applications. The current available cell permeable materials for two-photon bioimaging generally have their two-photon absorption cross-sections of >20 GM. Five organic molecules with impressive two-photon absorption cross-section from 600 GM to 2500 GM are shown in Figure 11.33.

11.5.2 Two-photon Induced Visible *In Vitro* Organelle-specific Imaging

The study of cellular organelles and their associated signalling pathways in normal and malignant human cells lead to a gradual understanding of how normal cells transform into malignant cells. There have been already a lot of organelle-specific organic-based fluorescent staining agents in the market [144]. However, the majority of them are comprised of ordinary linear fluorophores that can only be excited in the ultraviolet-visible region. Figure 11.34 depicts some examples of cell permeable, two-photon excitable organic fluorescent probes that have been used in the imaging of biologically meaningful species in various cancer and normal cell lines.

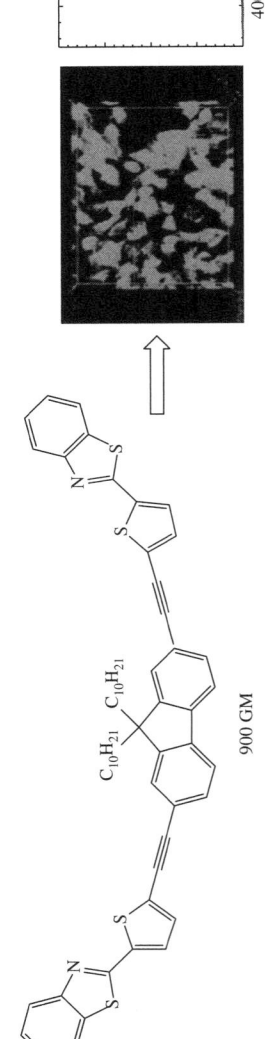

1200 GM

4-(2-(7-(2-(4-(dimethylamino)phenyl)ethynyl)-9,9-bis(2-methoxyethyl)-9H-fluoren-2-yl)ethynyl)-N,N-dimethylbenzenamine

2517 GM

N-(4-(2-(7-(4-(phenylsulfonyl)phenethyl)-9,9-didecyl-9H-fluoren-2-yl)ethyl)phenyl)-N-phenylbenzenamine

4-(2-(2,5-dibromo-4-(4-(4-(dihexylamino)phenyl)ethynyl)phenyl)ethynyl)-N,N-dihexylbenzenamine
570 GM

(E)-2-(4-(4-(pyridin-2-ylmethyl)piperazin-1-yl)styryl)-5-methylterephthalonitrile
600 GM

900 GM

2,2'-(5,5'-(9,9-didecyl-9H-fluorene-2,7-diyl)bis(ethyne-2,1-diyl))bis(thiophene-5,2-diyl))-dibenzo[d]-thiazole

Intensity, a.u.

Wavelength, nm

Lasing

Fluor×100

Superfluor×10

FIGURE 11.33 Structures of five fluorescence compounds available for *in vitro* imaging with two-photon absorption cross section greater than 500 GM and one example of the superfluorescence properties of 2,2'-(5,5'-(9,9-didecyl-9H-fluorene-2,7-diyl)bis(ethyne-2,1-diyl))bis(thiophene-5,2-diyl))-dibenzo[d]-thiazole with their two-photon induced green emission *in vitro* [139–143].

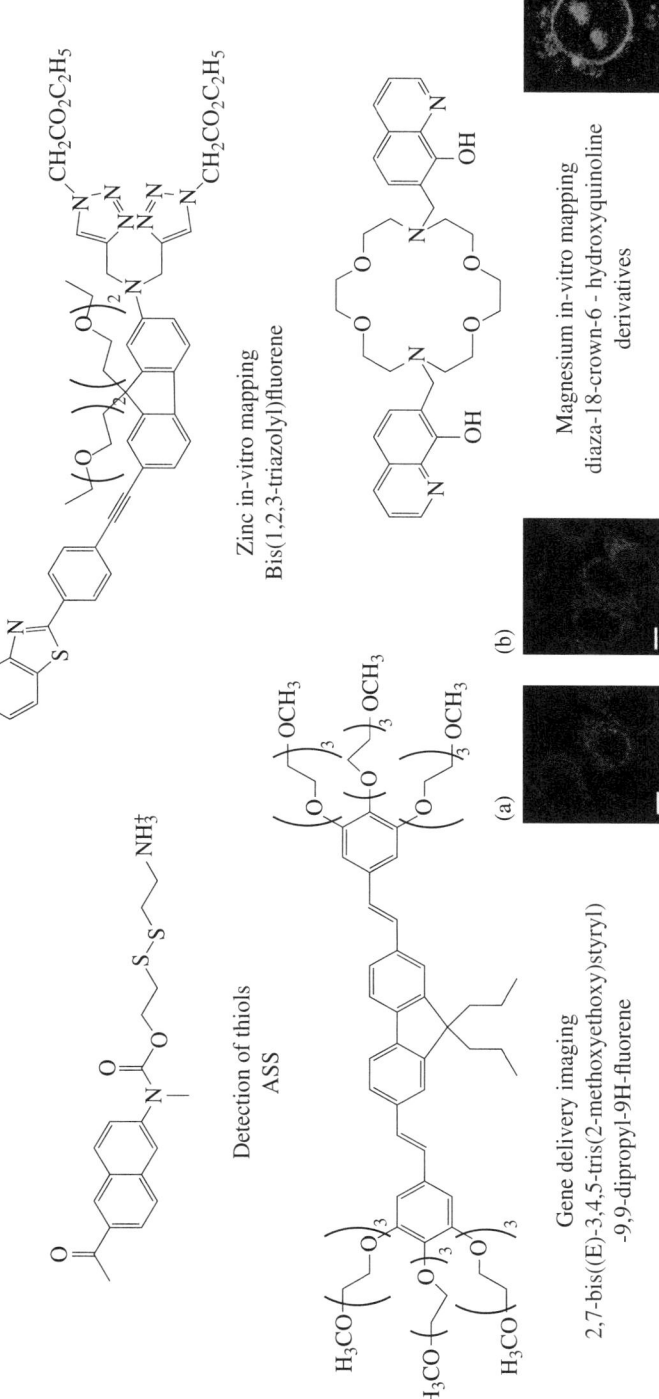

Detection of thiols
ASS

Zinc in-vitro mapping
Bis(1,2,3-triazolyl)fluorene

Magnesium in-vitro mapping
diaza-18-crown-6 - hydroxyquinoline
derivatives

(a)

(b)

Gene delivery imaging
2,7-bis((E)-3,4,5-tris(2-methoxyethoxy)styryl)
-9,9-dipropyl-9H-fluorene

FIGURE 11.34   Structures of some two-photon excitable fluorescence probes responsible for *in vitro* imaging of biologically meaningful small molecules [144–146].

11.5.3   Two-photon Induced *In Vivo* Imaging

With the advantages of the deep tissue penetration by NIR radiation, two-photon fluorescence imaging (especially combined with fluorescence lifetime imaging to remove interferences from the background autofluorescence of biological specimens) has contributed significantly to the *ex* (tissues) and *in vivo* (whole body) diagnosis of cancers, ranging from the unveiling of the basic mechanisms of cancer to clinical diagnostics. These include the search for cancer biomarkers in cells, assessing histological architecture of the cancer tissue, early cancer diagnosis, and the delineation of the tumour from normal tissues for staging and surgical removal.

Two-photon imaging of spine dynamics has been carried out on transgenic mice, showing that yellow fluorescent protein (YFP) or green fluorescent protein (GFP) are overexpressed predominantly in a subset of layer V pyramidal neurons driven by the Thy-1 promoter [145–153]. Although many mouse lines expressing fluorescent proteins have been generated [154], some might be too faint for imaging or too dense for the spines to be distinguished from one another. In addition to neuronal labelling, pretreatments where the skulls of the animals need to be thinned to ~20 µm (known as thinned-skull preparations) [155] are required before the imaging of fluorescently labelled dendritic spines in the cortex.

11.6   TIME-RESOLVED IMAGING

Fluorescence lifetime imaging can be performed either directly by measuring the fluorescence lifetime for each pixel of the image to generate a lifetime map of the specimen, or via time-gated experiments, where the fluorescence intensity for each pixel is determined after a short time-lap of photo-excitation to generate an intensity map (Figure 11.35). The former method is generally used in the monitoring of functional changes caused by environmental factors, while the latter offers the potential to eliminate background fluorescence and enhance imaging contrast [156].

It should be noted that both linear and two-photon induced fluorescence imaging and time-resolved imaging are also affected by numerous factors that were mentioned earlier, such as FRET, excimer formation, intersystem crossing, and the presence of internal quenchers. Compared with lanthanide or transition metal systems, the lifetime of organic fluorophores is much shorter. This, in fact, makes them more suitable for time-resolved imaging because commercial available FLIM systems nowadays are only capable of monitoring emission lifetime from nanoseconds to a few microseconds, given that coordination complexes with very long lifetimes are not suitable for fast scanning devices used in fluorescent lifetime imaging microscopy [157]. Furthermore, the quantum efficiency of organic fluorophores, which are often >50 %, is much higher than lanthanide/transition metal systems (~30%) [158].

11.6.1   Fluorescent Lifetime Imaging with Endogenous Probes via Autofluorescence

Autofluorescence lifetime imaging is an attractive modality because it does not involve *in vitro* staining procedures, and it purely relies on endogenous fluorophores, such as tyrosine, phenylalanine, riboflavin, and NADH (Figure 11.36). It provides rich information detailing the morphological and organisational structure of cells and tissues. In general, autofluorescence lifetime imaging is capable of differentiating one type of tissue from another, healthy from pathologic tissues.

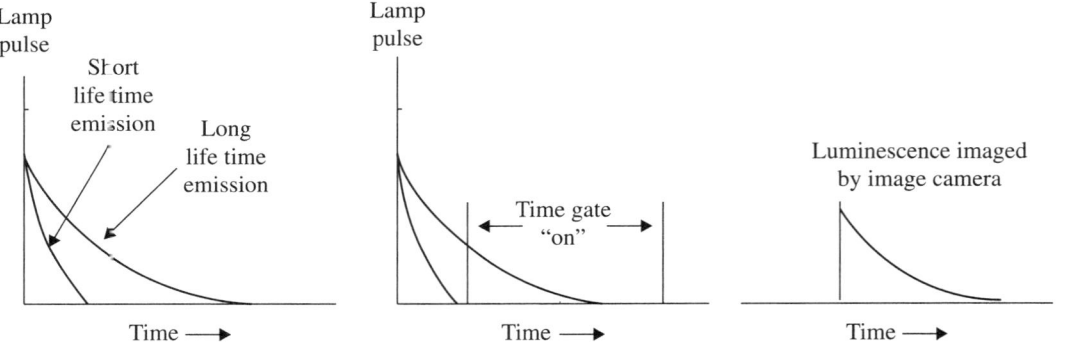

FIGURE 11.35   Illustrations of the time-resolved detection. In the experiment the short-lived fluorescence from the matrix (for endogenous probe: eliminate the signal from all parts inside the cell except endogenous probe; for exogenous probe: eliminate the entire emission signal *in vitro*) has decayed before the gate is turned on and the photoluminescence from the emissive probe is exclusively monitored. This microscope detection enables us to study the single cell.

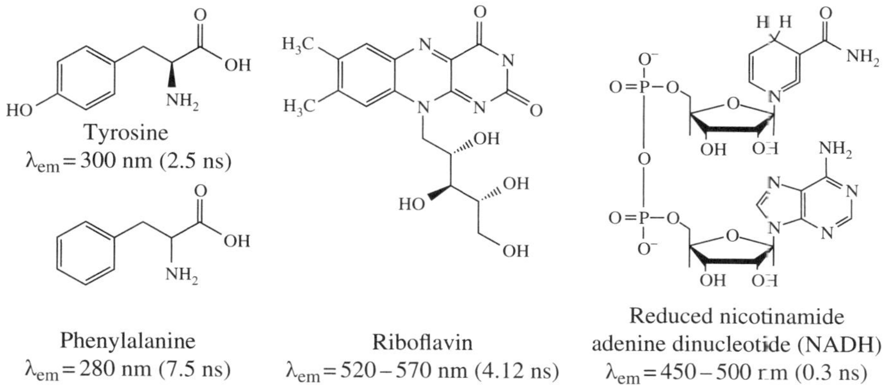

FIGURE 11.36 Examples of endogenous fluorophores responsible for *in vitro/in vivo* autofluorescence lifetime imaging [159–162].

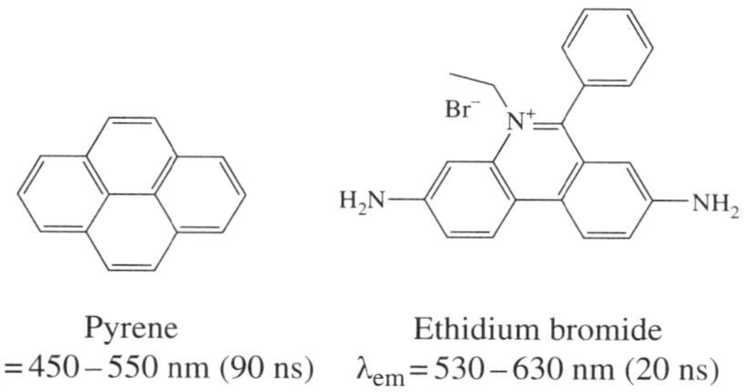

FIGURE 11.37 Examples of exogenous fluorophores (>90 ns emission lifetime) responsible for *in vitro/in vivo* fluorescence lifetime imaging [159–162].

11.6.2 Fluorescent Lifetime Imaging with Exogenous Probes

Although autofluorescence lifetime imaging does not require complicated staining procedures, it has a number of limitations due to its weak and nonspecific fluorescence signals originating from a variety of endogenous fluorophores. Since the emergence of fluorescence lifetime as an imaging modality, many commercially available and specially designed dyes have been tested for their applications in time-resolved bioimaging. For efficient fluorescence lifetime imaging with exogenous fluorophores, several factors should be considered. These include the use of fluorescent molecular probes (i) with longer lifetimes than those of their endogenous counterparts, (ii) in the spectral window with minimal fluorescence from endogenous fluorophores, and (iii) with dominant fluorescence that overwhelms the contribution of autofluorescence to the measured lifetime. Figure 11.37 shows some exogenous organic probes of the emission lifetime greater than 90 ns for *in vitro* and *in vivo* imaging [163, 164].

Pyrenes, polyaromatic compounds with exceptionally long fluorescence lifetimes, have been used for rapid and sensitive detection of the plated-derived growth factor and mRNA [165], whilst pyrenes with sufficiently long emission lifetimes have been developed as oxygen sensors. However, as the partial pressure of oxygen (P_{O2}) *in vitro* is in the range of 0.7–2.3 kPa (depending on organelles, such as mitochondria, which have a P_{O2} of 0.7 to 1.3 kPa), the sensitivity of pyrenes for oxygen (1 kPa) may not be sufficient to be used as quantitative *in vitro* oxygen sensors [166].

11.7 BIOLUMINESCENCE IN MOLECULAR IMAGING

Naturally occurring bioluminescence is a process, in line with chemiluminescence, by which living organisms emit visible light either inside or outside the cell through chemical reactions. It is one of the luminescence processes exhibiting the highest quantum efficiency, nearly up to 95% luminescence quantum yield [167]. The basic principle of bioluminescence is that two types of substances combine together in a light-producing reaction: luciferin and luciferase. Luciferin is a light-producing pigment that reacts with oxygen to yield oxyluciferins and light, while luciferase is a catalysing enzyme that triggers

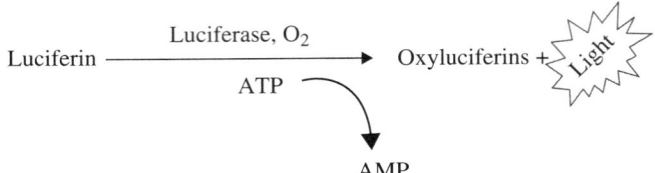

FIGURE 11.38 The general mechanism for bioluminescence.

the oxidation of luciferin. In some cases, luciferin is a protein known as photoprotein, which upon binding to metal ions, such as calcium(II) ions or magnesium(II) ions, will activate the reaction during a light-making process. Such binding to co-factors leads to a conformational change of the protein, giving the living organisms a way to precisely control the light production and emission. Adenosine triphosphates (ATP) are often consumed during the processes. Of course, there are also neurological, mechanical, chemical, or as-yet-undiscovered triggers that can start the reactions to release energy in form of light (Figure 11.38) [168]. In bacteria, luminescence involves the oxidation of $FMNH_2$ along with a long-chain aldehyde and a two-subunit luciferase, which is controlled by an operon called Lux operon.

Over the last decade, bioluminescence imaging has become a powerful methodology for molecular imaging of small laboratory animals, which allows the study of ongoing biological processes *in vivo*. This can be achieved by detecting luciferase in genetically modified cells, bacteria, or animals. Although bioluminescent probes (luciferases) are not so popular because they are exceedingly dim in comparison with fluorescence, their unique and advantageous features render them promising for wide applications. For example, these probes do not require exogenous illumination, and there is no phototoxicity or photobleaching of the emitting molecules or artificial perturbations of light-sensitive cells. Even though bioluminescence is much dimmer, it can be up to 50 times more sensitive than fluorescence, and images can be generated with remarkably high signal-to-noise ratios because there are low background noises, with the tissues having rather low intrinsic bioluminescence [169]. The reasons for such low background are that compared with endogenous fluorescence (autofluorescence), which can be as bright as the signal itself, endogenous bioluminescence (autoluminescence) of most cells is extremely low, and that there are no excitation photons that contribute greatly to background in fluorescence imaging. Therefore, the background levels in bioluminescence imaging are especially low and the signal-to-noise ratio can be very high, although the signals are very dim.

Biological events with fast kinetics can also be detected by adapted instruments, where *in vitro* luminescence intensities are quantitative to the amount of labelled cells. However, the absence of background signals in bioluminescence makes the localisation of the contour of the animal impossible, and imaging on moving animals requires an additional tracking system to follow their motion. Until recently, bioluminescence imaging has been harnessed to detect ATP because the luciferase reaction depends on the presence of oxygen and ATP, and this has had a much larger range of applications. Luciferase activity can now be used to detect protein-protein interactions, tracking cells *in vivo*, measuring levels of calcium and other signalling molecules, detecting protease activity, and even reporting circadian clock gene expression [170].

Among many luminescent species existing in nature, only three luciferases have been studied in detail and are being used in biomedical research: the *Photinus pyralis* (firefly) luciferase (Fluc), the sea pansy *Renilla reniformis* luciferase (Rluc), and the marine copepod *Gaussia princeps* (Gluc) (Figure 11.39). Gluc and Rluc give off blue luminescence (480 nm), which is strongly absorbed by pigmented molecules such as hemoglobin and melanin. However, blue emission is easily scattered by tissues and hence makes them less suitable for *in vivo* imaging. In contrast, Fluc is a better candidate for *in vivo* imaging, because it emits green light (562 nm). The sensitivity of bioluminescence imaging in deep tissues can be enhanced by the use of red emitting luciferases from *Pyrophorus plagiophthalamus, Lhotinus pyralis, Luciola italic,* and the railroad worm. Firefly luciferase was first cloned in 1985 and soon it had developed into an assay to measure luciferase in mammalian cell lysates. This development enabled the luciferase gene to become a useful tool for *in vivo* studies of gene regulation [171].

Bioluminescence has become indispensable for noninvasive monitoring of biologic phenomena *in vitro/vivo* by various approaches, such as genetic receptor assays (measure changes in luciferase production), ATP assays (measure changes in ATP levels) and Luciferin Substrate Assays (measure enzymatic release of free luciferin). Inside the nucleus, bioluminescence is available to monitor miRNA expression and evaluate siRNA delivery to specific tissue by fusing luciferase to the miRNA target site at the 3′ UTR and silencing siRNA of luciferase expression respectively. In the cytoplasm, protein folding and secretion can be monitored by chaperone-mediated protein-folding using secreted luciferase [171].

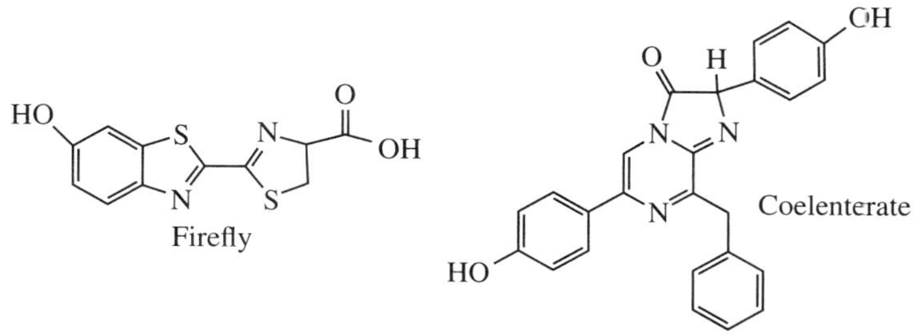

FIGURE 11.39 Structures of two commonly used luciferins for molecular imaging from different types of bioluminescent organisms.

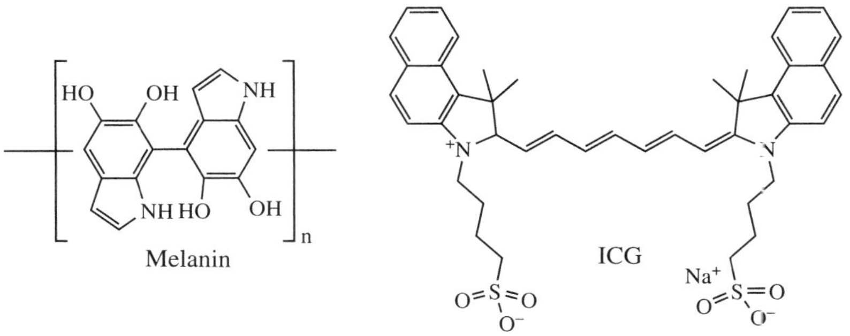

FIGURE 11.40 The structures of two commonly used organic probes for molecular photoacoustic imaging.

11.8 PHOTOACOUSTIC IMAGING

Photoacoustic imaging is a new hybrid biomedical imaging technique that has been developed based on photoacoustics, where the absorption of electromagnetic energy transforms into acoustic waves (pressure wave or sound). RF or short laser pulses are used for photoacoutic imaging excitation in soft tissues, usually pulsed on a nanosecond timescale, because the waves are non-ionising in these regions, which is safer for humans and can provide high contrast and enough penetration depths. The RF properties of biological tissues can provide the information of the physiological nature of their electrical properties [172].

In biomedical applications, photoacoustic imaging offers a number of advantages. For example, it combines ultrasonic resolution and high contrast from light or RF and does not rely on the photons for excitations. Also, photoacoustic imaging can overcome the problems existing in the conventional optical such as micro-absorption spectrometry and dark-field microscopy and ultrasound imaging. It produces images of high electromagnetic contrast at high ultrasonic resolution at large depth. The resolution and imaging depth of photoacoustic imaging is tunable by controlling the frequency of the ultrasound transducer used. When a tissue is irradiated by an incident laser pulse, absorption of the laser energy will cause a rapid temperature rise, consequently leading to transient thermoelastic expansion and the concomitant generation of broadband ultrasound waves inside the tissue. Optical properties (e.g., the optical absorption) of the tissue that is used to characterise biological tissues can then be estimated by the waves generated.

In general, there are two types of contrast agents for photoacoustic imaging: endogenous (e.g., haemoglobin and melanin) and exogenous (e.g., indocyanine green [ICG], various gold nanoparticles, single-walled carbon nanotubes [SWNTs], quantum dots [QDs], and fluorescent proteins). Applications of haemoglobin in photoacoustic imaging can greatly facilitate brain-and-blood dynamics-related research, such as measuring microvascular blood flow, visualising brain structure and lesions, monitoring haemodynamics, delineating tumour vasculature, and imaging small animals. Recently, haemoglobin has been found useful in monitoring burn recovery by using multi-wavelength photoacoustic measurements. Haemoglobin-based photoacoustic imaging for tumour angiogenesis is another case in point.

The second main endogenous contrast agent is melanin. It is employed mainly for diagnosis, prognosis, and treatment planning of melanotic melanoma (>90% of all melanomas). Furthermore, more parameters for the detection of melanomas can now be obtained by the mixed use of haemoglobin and melanin. Noninvasively three-dimensional images of subcutaneous melanomas and their surrounding vasculature in living nude mice can now be obtained by dual-wavelength reflection-mode photoacoustic imaging, because there is a vast difference between the absorption coefficients of blood and melanin-pigmented melanomas when two wavelengths at 584 nm and 764 nm are used for haemoglobin and melanin respectively (Figure 11.40) [174].

ICG has an absorption peak at around 800 nm, and it is employed as a diagnostic aid for measuring blood volume, cardiac output, or hepatic function. ICG is also applied in noninvasive photoacoustic angiography of animal brains, whereupon injecting ICG into the circulation system of a rat, the absorption contrast between the blood vessels and the background tissues will be enhanced significantly. Photoacoustic imaging allows the reconstruction of the vascular distribution in the rat brain with high spatial resolution and low background as a result of the ability of near-infrared light penetrating deep into brain tissues through the skin and the skull. However, the resolution was found to deteriorate slowly with increasing imaging depth [175].

11.9 CONCLUSION AND FUTURE PERSPECTIVES

An outline of the development of molecular imaging has been summarised in this chapter, with an array of representative organic probes. With prominent advantages of luminescent organic materials in aqueous solutions—their ability, availability, and applicability to be used as imaging agents to provide accurate and detailed diagnostic information, they have been manipulated to understand the processes involved in certain therapeutic treatments on a molecular level, thus acting as a tool to evaluate the efficacy of such treatments. More importantly, to modify and complement these luminescent organic materials, a large variety of tracers or probes can be devised and developed and be equipped themselves with a more precise vector to bolster their effectiveness when used as imaging agents. Most importantly, foresight tells us that the next breakthrough in this aspect, perhaps, lies elsewhere—time-resolved near-infrared imaging agents and the new epoch involving the coupled benefits of time-resolved microscopy and NIR-emissive materials that will demonstrate a profound capacity to improve the imaging sensitivity. Recently, metal-based organic systems emissive in the NIR region have been being studied extensively, and a decent profile of their luminescent properties has been established while technicality issues, in particular instrumentation for NIR emission imaging, has yet to be mature, or even, prove possible.

REFERENCES

[1] W. J. Turner, *Anat. Physiol.* **24**, 253–287 (1890).

[2] M. Tavassoli, *Am. J. Pathol.* **98**, 44–48 (1980).

[3] S. Bradbury, *An Introduction to the Optical Microscopy*, Revised Ed., Royal Microscopical Society, OUP, Oxford (1989).

[4] W. J. Croft, *Under the Microscope: A Brief History of Microscopy*, World Scientific, Hackensack, N J (2006).

[5] B. R. Masters, P. T. C. So and E. Gratton, *Laser. Medical Sci.* **13**, 196–203 (1998).

[6] J. V. Rocheleau, W. S. Haed and D. W. J. Piston, *Biol. Chem.* **279**, 31780–31787 (2004).

[7] E. P. W. Kable and A. K. Kiemer, *Immunol. Lett.* **96**, 33–42 (2005).

[8] E. Beaumont, J. C. Lambry, A. C. Robin, P. Martasek, M. Blanchard-Desce and A. Slama-Schwok, *Chem. Phys. Chem.* **9**, 2325–2331 (2008).

[9] O. Masihzadeh, D. A. Ammar, T. C. Lei, E. A. Gibon and M. Y. Kahook, *Experiment. Eye Res.* **93**, 316–320 (2011).

[10] P. M. Conn Ed., *Green Fluorescent Protein*, Academic Press, San Diego, CA, London, UK (1999).

[11] B. W. Hicks Ed., *Green Fluorescent Protein: Applications and Protocols*, Humana Press, Totowa, NJ (2002).

[12] K. L. Tucker, *Histochem. Cell Biol.* **115**, 31–39 (2001).

[13] J. C. March, G. Rao and W. E. Bentley, *Appl. Microbiol. Biotechnol.* **62**, 303–315 (2003).

[14] R. N. Day, *Mol. Cellular Endocrinol.* **230**, 1–2 (2005).

[15] F. Bohmer, *Aerzlich Intelligenzblatt (Munchen).* **12**, 539–550 (1865).

[16] P. Mayer, *Mitteilungen aus dem Zoologische Station zu Neapel.* **10**, 170–186 (1891).

[17] F. N. Egerton, *J. History Biol.* **1**, 1–22 (1967).

[18] F. N Egerton, *A History of the Ecological Sciences, Part 19, Bull. Ecol. Soc. Am.* **87**, 47–58 (2006).

[19] R. W. Horobin and J. A. Kiernan, Eds., *Conn's Biological Stains: A Handbook of Dyes, Stains and Fluorochromes for Use in Biology and Medicine*, 10th ed., Biological Stain Commission, BIOS Scientific Publishers, Oxford, U K (2002).

[20] W. H. Perkin, *J. Chem. Soc., Trans.* **35**, 717–732 (1879).

[21] K. Modha, J. P. Whiteside and R. E. Spier, *Cytotechnol.* **13**, 227–232 (1993).

[22] K. H. Sit, B. H, Bay R. Paramanantham, H. M. Tana and K. P. Wong, *Cancer Lett.* **104**, 63–69 (1996).

[23] R. Seeley, P. J. Vancemark and J. J. Lee, *Microbes in Action. A Laboratory Manual of Microbiology*, 4th Ed., Freeman, New York (1991).

[24] C. C. Churukian, *J. Histotechnol.* **22**, 309–311 (1999).

[25] P. Ehrlich and A. Lazarus, *Die Anamie. I. Normale und pathologische Histologie des Blutes*, Alfred Holder, Wien and Leipzig (1909).

[26] D. Wittekind, *Clin. Lab. Haematol.* **1**, 247–262 (1979).

[27] R. W. Horobin and K. J. Wlater, *Histochem.* **86**, 331–336 (1987).

[28] H. Zollinger, *Color: A Multidisciplinary Approach,* Wiley-VCH, Weinheim (1999).

[29] J. A. Kiernan, *Histological and Histochemical Methods: Theory and Practice*, 3rd Ed., Butterworth-Heinemann, Oxford (1999).

[30] F. DiAmico, *Biotechn. & Histochem.* **80**, 207–210 (2005).

[31] P. Masson, *J. Technical Method. Bull. Inter. Assoc. Med. Museum.* **12**, 75–90 (1929).

[32] R. D. Lillie, R. E. Tracy, P. Pizzolato, P. T. Donaldson and C. Reynolds, *Virchows Arch. A. Path. Anal. Histol.* **386**, 153–159 (1980).

[33] S. R. Slivka and R. L. Bartel, *In Vitro Cell., Dev. Biol.*, **28A**, 11–12 (1992).

[34] K. Borg, B. Colenbrander, A. Fazeli, J. Parevliet, and L. Malmgren, *Theriogenology.* **48**, 531–536 (1997).

[35] A. Popescu and R. J. Doyle, *Biotechn. Histochem.* **71**, 145–151 (1996).

[36] G. Jones and P. Landsman, *J. Photochem. Photobiol.* **189**, 147–152 (2007).

[37] D. S. King, V. A. Luna, A. C. Cannons and P. T. Amuso, J. Appl. *Microbiol.* **108**, 1817–1827 (2010).

[38] P. Bruneval, J. Choucair, F. Paraf, J.-P. Casalta, D. Raoult, F. Scherchen and J.-L. Mainardi, *J. Clin. Pathol.* **54**, 238–240 (2001).

[39] M. A. Hayat, *Stains and Cytochemical Methods*, Plenum Press, New York (1993).

[40] H. Puchtler, S. N. Melon and M. S. Terry, *J. Histochem. Cytochem.* **17**, 110–124 (1969).

[41] D. Bilgic, S. Karaderi and I. Bapli, *Rev. Anal. Chem.* **26**, 99–108 (2007).

[42] R. R. Bensley, *Am. J. Anatomy.* **12**, 297–388 (1911).

[43] M. Urata, Y. Ishikawa, S. Seno and M. Miyahara, *Acta Histochem. Cytochem.* **7**, 100–106 (1974).

[44] J. L. Yang, L. Ma, Y. Zhang, F. Fang and L. J. Li, *Protoplasma.* **231**, 249–258 (2007).

[45] M. Wolman and J. J. Bubis, *Histochem.* **4**, 351–356 (1965).

[46] K. Hsiao, P. Chapman, S. Nilsen, C. Eckman, Y. Harigaya, S. Younkin, F. S. Yang and G. Cole, *Science* **274**, 99–102 (1996).

[47] A. C. Hospelhorn, B. Faris, P. J. Mogayzel, O. T. Tan and C. J. Franzblau, *Histochem. Cytochem.* **36**, 1353–1358 (1988).

[48] J. Webster and B. A. Stone, *Aquat. Botany* **47**, 185–189 (1994).

[49] B. Downie, H. W. M. Hilhorst and J. D. Bewlwy, *Phytochem.* **36**, 829–835 (1994).

[50] J. Nitschke, H. Modick, E. Busch, R. W. von Rekowski, H. J. Altenbach and H. Molleken, *Food Chem.* **127**, 791–796 (2011).

[51] J. B. Rollefson, C. S. Stephen, M. Tien and D. R. Bond, *J. Bacteriol.* **193**, 1023–1033 (2011).

[52] G. Clark and F. H. Kasten, *History of Staining*, 3rd Ed., Williams and Wilkins, Baltimore (1983).

[53] S. A. Altman, L. Randers and G. Rao, *Biotechnol. Prog.* **9**, 671–674 (1993).

[54] J. H. Lang and E. C. Lasser, *Biochem.* **6**, 2403–2409 (1967).

[55] M. K. Baskaya, A. M. Rao, A. Dogan, D. Donaldson and R. J. Dempsey, *Neurosci. Lett.* **226**, 33–36 (1997).

[56] J. G. Gibson and W. A. Jr. Evans, *J. Clin. Invest.* **16**, 301–316 (1937).

[57] J. K. Thomsen, L. B. Fogh-Anderson, N. Bulow, K. Devantier and A. Scand. *J. Clin. Lab. Invest.*, **51**, 185–190 (1991).

[58] L. B. Johansen, N. Foldager and C. Stadeager, *J. Appl. Physicol.* **73**, 539–544 (1992).

[59] B. T. Hawkin and R. D. Egleton, *J. Neurosci. Method.* **151**, 262–267 (2006).

[60] F. P. Altman, *Histochem.* **38**, 155–171 (1974).

[61] F. P. Altman, *Prog. Histochem. Cytochem.* **9**, 1–56 (1976).

[62] D. A. Scudiero, R. H. Shoemaker and K. D. Paull, *Cancer Res.* **48**, 4827–4833 (1988).

[63] J. D. Bancroft, H. C. Cook, *Manual of Histological Techniques,* Churchill-Livingstone, Edinburgh (1994).

[64] F. H. Kasten, *The Origin of Modern Fluorescence Microscopy. In Cell Structure and Function by Microspectrofluorometry*, Academic Press, San Diego (1989).

[65] W. B. Amos and J. G. White, *Biol. Cell.* **95**, 335–342 (2003).

[66] S. K. Lee, J. Y. Lee, M. Y. Lee, S. M. Chung and J. H. Chung, *Anal. Biochem.* **273**, 186–191(1999).

[67] A. Takahashi, P. Camacho, J. D. Lechletter and B. Herma, *Physicol. Rev*, **79**, 1089–1125 (1999).

[68] M. Gilbert and P. Claverie, *J. Theoret. Biol.* **18**, 330–349 (1968).

[69] K. Fukul and K. Tanaka, *Nucl. Acid. Res.* **24**, 3962–3967 (1996).

[70] A. Adams, *Curr. Med. Chem.* **9**, 1667–1675 (2002).

[71] Z. Darzynkiewicz, F. Traganos and M. R. Melamed, *Cytometry* **1**, 98–108 (1980).

[72] R. Baserga, *New Engl. J. Med.* **304**, 453–459 (1981).

[73] D. W. Coats and J. F. Heinbokel, *Mar. Biol.* **67**, 71–79 (1982).

[74] K. D. Bauer, C. V. Clevenger, T. J. Williams and A. L. Epstein, *J. Histochem. Cytochem.* **34**, 245–250 (1986)

[75] S. Strugger and Z. Jena, *Naturwiss.* **73**, 97–134 (1940).

[76] F. Bukatsch and M. Haitinger, *Protoplasma* **34**, 515–523 (1940).

[77] S. Strugger, *Fluoreszenmicroscopie und Microbiologie*, Shaper, M. and Shaper H. Eds., Hanover, Germany (1949).

[78] R. L. Metcalf and R. L. Patton, *Stain Technol.* **19**, 11–27 (1944).

[79] D. E. Becker, H. Ancin, D. G. Szarowski, J. N. Turner and B. Roysam, *Cytometry* **25**, 235–245 (1996).

[80] A. Bastos, A. M. Terrinha, J. D. Vigário, J. F. Moura Nunes and J. L. Nunes Petisca, *Exp. Cell Res.* **42**, 84–88 (1966).

[81] A. T. Summer, *Histochem.* **84**, 566–586 (1986).

[82] A. A. Chan, E. T., Pritchard, J. M. Gerrard, R. Y. K. Man and P. C. Choy, *Biochim. Biophys. Acta – Lipids Lipid Metabol.***713**, 170–172 (1982).

[83] Y. Usui, Y. Ohshima and K. Yoshida, *J. Gen. Microbiol.* **133**, 1593–1600 (1987).

[84] E. F. Grabowski, *Blood* **15**, 390–398 (1990).

[85] J. J. Chen, *Immunol. Today* **14**, 126–128 (1993).

[86] J. E. Coligan, A. M. Kruisbeek, D. H. Margulies, E. M. Shevach and W. Strober, *Current Protocols in Immunology. In Related Isolation Procedures and Functional Assays*, Volume I, Coico, R. Ed., John Wiley & Sons, Inc., New York (1995).

[87] H. Lecoeur, *Exp. Cell Res.* **277**, 1–14 (2002).

[88] T. Suzuki, K. Fujikura, T. Higashiyama and K. Takata, *J. Histochem. Cytochem.* **45**, 49–53 (1997).

[89] J. Kapuscinski, *Biotech. Histochem.* **70**, 220–233 (1995).

[90] M. Kubbies and P. S. Rabinovitch, *Cytometry* **3**, 276–281 (1983).

[91] S. Y. Breusegem, R. M. Clegg and F. G. Loontiens, *J. Mol. Biol.* **315**, 1049–1061 (2002).

[92] A. S. Waggoner, *J. Membrane Biol.* **27,** 7117–7122 (1976).

[93] A. S. Waggoner, *J. Ann. Rev. Biophys. Bioenerg.* **8**, 47–68 (1979).

[94] A. Kalenak, R. J. McKenzie and T. E. Conover, *J. Membrane. Biol.* **123**, 23–31 (1991).

[95] M. Terasaki and L. A. Jaffe, *J. Cell Biol.* **114**, 929–940 (1991).

[96] T. Wang and H R Petty, *Cytometry.* **14**, 16–22 (1993).

[97] W. Y. Leung, F. Mao, R. P. Haugland and D. H. Klaubert, *Bioorg. Med. Chem. Lett.* **6**, 1479–1482 (1996).

[98] J. Bereiter-Hahn, *Biochim. Biophys. Acta.* **423**, 1–14 (1976).

[99] J. Bereiter-Hahn, K. H. Seipel, M. Vöth and J. S. Ploem, *Cell Biochem. Funct.* **1**, 147–155 (1983).

[100] R. Ramadass and J. Bereiter-Hahn, *Biophys. J.* **95**, 4068–4076 (2008).

[101] N. V. Gohad, G. H. Dickinson, B. Orihuela, D. Rittschof and A. S. Mount, *J. Exp. Mar. Biol. Ecol.* **380**, 88–98 (2009).

[102] T. Whalley, M. Teresaki, M. S. Cho and S. S. Vogel, *J. Cell Biol.* **131**, 1183–1192 (1995).

[103] A. Wiederkehr, S. Avaro, C. Prescianotto-Baschong, R. Haguenauer-Tsapis and H. Riezman, *J. Cell Biol.* **149**, 397–410 (2000).

[104] V. N. Murthy and C. F. Stevens, *Nature* **392**, 497–501 (1998).

[105] R. Rea, J. Li, A. Dharia, E. S. Levitan, P. Sterling and R. H. Kramer, *Neuron.* **41**, 755–766 (2004).

[106] T. Mitsui, H. Nakano and K. Yamana, *Tetrahed. Lett.* **41**, 2605–2608 (2000).

[107] M. Akhter Hossain, H. Mihara and A. Ueno, *J. Am. Chem. Soc.* **125**, 11178–11179 (2000).

[108] G. Zheng, Y.-M. Guo and W.-H. Li, *J. Am. Chem. Soc.* **129**, 10616–10617 (2007).

[109] D. Ho, M. R. Lugo, A. L. Lomiz, I. D. Pogozheva, S. P. Singh, A. L. Schwan and A. R. Merrill, *Biochem.* **50**, 4830–4842 (2011).

[110] T. C. Lee, T. L. Arthur, L. J. Gibson and W. C. Hayes, *J. Orthopaedic Res.* **18**, 322–325 (2000).

[111] T. C. Lee, F. J. O'Brien and D. Taylor, *Inter. J. Fatigue* **22**, 847–853 (2000).

[112] P. J. Trotter, *Traffic* **1**, 425–434 (2000).

[113] Y. Sai, M. Kajita, I. Tamai, J. Wakam, T. Wakamiya and A. Tsuji, *Am. J. Physicol.* **275**, G514–G520 (1998).

[114] N. M. Abuharfeil, R. F. Atmeh, M. N. Abo-Shehada and S. N. el-Sukhon. *Electrophoresis* **12**, 683–684 (1991).

[115] S. Kouvroukoglou, K. C. Dee, R. Bizios, L. V. McIntire and K. Zygourakis, *Biomater.* **21**, 1725–1733 (2000).

[116] L. You and G. W. Gokel, *Chem. Eur. J.* **14**, 5861–5870 (2008).

[117] A. Loudet and K. Burgess, *Chem. Rev.* **107**, 4891–4930 (2007).

[118] R. Ziessel, G. Ulrich and A. Harriman, *New J. Chem.* **31**, 496–501 (2007).

[119] X. H. Qian, Y. Xiao, Y. F. Xu, X. F. Guo, J. H. Qian and W. P. Zhu, *Chem. Commun.* **46**, 6418–6436 (2010).

[120] S. Arttamangkul, V. Alvarez-Maubecin, G. Thomas, J. T. Williams and D. K. Grandy, *Mol. Pharmacol.* **58**, 1570–1580 (2000).

[121] S. A. Farber, M. Pack, S. Y. Ho, I. D. Johnson, D. S. Wagner, R. Dosch, M. C. Mullins, H. S. Hendrickson, E. K. Hendrickson and M. E. Halpern, *Science* **292**, 1385–1388 (2001).

[122] J.-S. Lee, N.-Y.; Kang, Y.-K.; Kim, A. Samanta, S. Feng, H. K. Kim, M. Vendrell, J. H. Park and Y. -T. Chang, *J. Am. Chem. Soc.* **131**, 10077–10082 (2009).

[123] K. V. Mikkelsen and M. A. Ratner, *Chem. Rev.* **87**, 113–153 (1987).

[124] H. Tian, T. Xu, Y. Zhao and K. Chen, J. Chem. Soc., Perkin Trans. 2545–2549 (1999).

[125] B. A. Wong, S. Friedle and S. J. Lippard, *J. Am. Chem. Soc.* **131**, 7142–7152 (2009).

[126] I. Trkulja and R. Häner, *Bioconjugate Chem.* **18**, 289–292 (2007).

[127] C. Cazeau-Dubroca, S. Ait Lyazidi, P. Cambou, A. Peirigua, P. Cazeau and M. Pesquer, *J. Phys Chem.* **93**, 2347–2358 (1989).

[128] G. P. Kushto1 and P. W. Jagodzinski, *J. Mol. Struct.* **516**, 215–250 (2000).

[129] H. Shizuka, *Acc. Chem. Res.* **18**, 141–147 (1985).

[130] Y. Xu and Y. Pang, *Dalton Trans.* **40**, 1503–1509 (2011).

[131] W. Denk, J. H. Strickler and W. W. Webb, *Science* **248**, 73–76 (1990).

[132] M. W. Harry, A. Bhaskar, G. Ramakrishna, T. Goodson, III, M. Imamura, A. Mawatari, K. Nakao, H. Enozawa, T. Nishinaga and M. Iyoda, *J. Am. Chem. Soc.* **130**, 3252–3253 (2008).

[133] H. Rath, J. Sankar, V. PrabhuRaja, T. K. Chandrashekar, A. Nag and D. Goswami, *J. Am. Chem Soc.* **127**, 11608–11609 (2005).

[134] W. M. McClain, *Acc. Chem. Res.* **7**, 129–135 (1974).

[135] K.-L. Wong, G.-L. Law, W.-M. Kwok, W.-T. Wong and D. L. Phillips. *Angew. Chem. Int. Ed.* **44**, 3436–3439 (2005).

[136] M. Sheik-Bahae, A. A. Said T.-H.; Wei, D.-J.; Hagan and E. W. Van Stryland, *IEEE J. Quantum Electron.* **26**, 760–769 (1990).

[137] W. Denk, J. H. Stricker and W. W. Webb, *Science* **248**, 73–76 (1990).

[138] M. Albota, C. Xu and W. W. Webb, *Applied Optics.* **37**, 7352–7356 (1998).

[139] L .-S. Tan, Q. Zheng and P.N. Prasad, *Chem. Rev.* **108**, 1245–1330 (2008).

[140] T. Gallavardin, M. Maurin, S. Marotte, T. Simon, A.-M. Gabudean, Y. Bretonnière, M. Lindgren, F. Lerouge, P. L. Baldeck, O. Stéphan, Y. Leverrier, J. Marvel, S. Parola, O. Maury and C. Andraud, *Photochem. Photobiol. Sci.* **10**, 1216–1225 (2011).

[141] C. Huang, J. Qu, J. Qi, M. Yan and G. Xu, *Org. Lett.* **13**, 1462–1465 (2011).

[142] K. D. Belfield, C. D. Andrade, C. O.Yanez, M. V. Bondar MV, F. E. Hernandez and O. V. Frzhonska, *J. Phys. Chem. B.* **114**, 14087–14095 (2010).

[143] X. Wang, D. M. Nguyen, C. O. Yanez, L. Rodriguez, H.-Y. Ahn, M. V. Bondar and K. D. Belfield, *J. Am. Chem. Soc.* **132**, 12237–12239 (2010).

[144] J. H. Lee, C. S. Lim, Y. S. Tian, J. H. Han and B. R. Cho, *J. Am. Chem. Soc.* **132**, 1216–1217 (2010).

[145] D. M. Nguyen, X. H. Wang, H. Y. Ahn, L. Rodriguez, M. V. Bondar and K. D. Belfield, *ACS Appl Mater Interfaces.* **2**, 2978–2981 (2010).

[146] A. Hayek, S. Ercelen, X. Zhang, F. Bolze, J. F. Nicoud, E. Schaub, P. L. Baldeck and Y. Mély, *Bioconjugate Chem.* **18**, 844–851 (2007).

[147] J. Grutzendler, N. Kasthuri and W. B. Gan, *Nature* **420**, 812–816 (2002).

[148] J. T. Trachtenberg, B. E. Chen, G. W. Knott, G. Feng, J. R. Sanes, E. Welker and K. Svoboda, *Nature* **420**, 788–794 (2002).

[149] A. J. Holtmaat, J. T. Trachtenberg, L. Wilbrecht, G. M. Shepherd, X. Zhang, G. W. Knott and K. Svoboda, *Neuron.* **45**, 279–291 (2005).

[150] R. Yuste and A. Konnerth, Eds. *Imaging in Neuroscience and Development: A Laboratory Manual,* Cold Spring Harbor, NY: Cold Spring Harbor Laboratory Press, pp. 627–638 (2007).

[151] S. Zhang, J. Boyd, K. Delaney and T. H. Murphy, *J. Neurosci.* **25**, 5333–5338 (2005).

[152] Y. Zuo, A. Lin, P. Chang and W. B. Gan, *Neuron* **46**, 181–189 (2005).

[153] Y. Zuo, G. Yang, E. Kwon and W. B. Gan, *Nature* **436**, 261–265 (2005).

[154] G. Feng, R. H. Mellor, M. Bernstein, C. Keller-Peck, Q. T. Nguyen, M. Wallace, J. M. Nerbonne, J. W. Nerbonne, J. W. Lichtman and J. R. Sanes, *Neuron* **28**, 41–51 (2000).

[155] H. T. Xu, F. Pan, G. Yang and W. B. Gan. *Nat. Neurosci.* **10**, 549–551 (2007).

[156] J. R. Lakowicz, H. Szmacinski, K. Nowaczyk, K. W. Berndt and M. Johnson, *Anal. Biochem.* **202**, 316–330 (1992).

[157] M. Y. Berezin and S. Achilefu, *Chem. Rev.* **110**, 2641–2681(2010).

[158] L. Murphy, A. Congreve, L.-O. Pålsson and J. A. G. Williams, *Chem. Commun.* **46**, 8743–8745 (2010).

[159] H. E. Rajapakse, N. Gahlaut, S. Mohandessi, D. Yu, J. R. Turner and L. W. Miller. *Proc. Nat. Acad. Sci.* **107**, 13582–13587 (2010).

[160] I. Ashikawa, Y. Nishimura, M. Tsuboi, K. Watanabe and K. Iso. *J. Biochem.* **91**, 2047–2055 (1982).

[161] B. Koziol, M. Markowicz, J. Kruk and B. Plytycz, *Photochem. Photobiol.* **82**, 570–573 (2006)

[162] K. König, *J. Biophotonics.* **1**, 13–23 (2008).

[163] J.R. Lakowicz, H. Szmacinski, K. Nowaczyk, W. J. Lederer, M.S. Kirby and M. L. Johnson, *Cell Calcium* **15**, 7–27 (1994).

[164] H. Szmacinski, J. R. Lakowicz and M. L. Johnson, *Methods Enzymol.* **240**, 723–748 (1994).

[165] J. B. Birks, D. J. Dyson and I. H. Munro, *Proc. R. Soc. London Ser. A.* **275**, 575–588 (1963).

[166] B. L. Sailer, J. G. Valdez, J. A. Steinkamp and H. A. Crissman, *Cytometry* **31**, 208–216 (1998).

[167] R. T. Sadikot and T. S. Blackwell, *Proc. Am. Thorac. Soc.* **2**, 537–540 (2005).

[168] R. S. Negrin and C. H. Contag, *Nat. Rev. Immun.* **6**, 484–490 (2006).

[169] K. E. Luker and G. D. Luker, *Antivir. Res.* **78**, 179–187 (2008).

[170] M. Brock, *Internctional Journal of Microbiology,* Article ID 956794 (2012).

[171] A. Kheirolomoom, D. E. Kruse, S. Qin, K E. Watson, C.-Y. Lai, L. J.T. Young, R. D. Cardiff and K. W. Ferrara, *J. Control. Releas.* **141**, 128–136 (2010).

[172] C. Li and L. V. Wang, *Phys. Med. Biol.* **54**, R59–R97 (2009).

[173] S. Mallidi, G. P. Luke and S. Emelianov, *Trends in Biotechnology* **29**, 213–221 (2011).

[174] J. Yao and L. V. Wang, *Contrast Media and Molecular Imaging* **6**, 332–345 (2011).

[175] P. Beard, *Interface Focus* **1**, 602–631 (2011).

12

APPLICATION OF *d*- AND *f*-BLOCK FLUORESCENT CELL IMAGING AGENTS

Michael P. Coogan

Department of Chemistry, Lancaster University, Lancaster, UK

Simon J. A. Pope

School of Chemistry, Cardiff University, Cardiff, UK

ABBREVIATIONS

bpy	2,2′-bipyridine
CHO	Chinese Hamster Ovarian
Cyclen	1,4,7,10-tetraazacyclododecane
DO3A	1,4,7,10-tetraazacyclododecane-1,4,7-triacetic acid
DOTA	1,4,7,10-tetraazacyclododecane-1,4,7,10-tetraacetic acid
DPPZ	dipyrido-[3,2-a:2′,3′-c]-phenazine
FLIM	Fluorescence Lifetime Imaging Microscopy
LMCT	Ligand to Metal Charge Transfer
MLCT	Metal to Ligand Charge Transfer
NADPH	Nicotinamide adenine dinucleotide phosphate
NHC	Nitrogen Heterocyclic Carbene
NIR	Near Infra Red
phen	1,10-phenanthroline
PNA	Peptide Nucleic Acid
ppy	2-phenylpyridine
pybz	2-pyridyl-benzimidazole
ROS	reactive oxygen species
terpy	2,2′;6′,2″-terpyridine
TRLM	Time-resolved Luminescence Microscopy
U.V.	Ultra Violet

12.1 INTRODUCTION

Because the previous chapter focused on the principles of fluorescent techniques (e.g., PET, FRET, FLIM) for the *in vitro* and *in vivo* studies of interactions among biomolecules and the chemosensing of important analytes, whilst this chapter will focus heavily on applications in fluorescence microscopy, a short overview of the techniques involved may be useful. Fluorescence

The Chemistry of Molecular Imaging, First Edition. Edited by Nicholas Long and Wing-Tak Wong.
© 2015 John Wiley & Sons, Inc. Published 2015 by John Wiley & Sons, Inc.

microscopy relies upon the use of fluorescent or phosphorescent agents, or those parts of a sample that are naturally emissive, to generate an emitted light signal upon excitation. This allows greater contrast between sections of the specimen and greater signal-to-noise ratios than conventional microscopy, which uses detection of reflected or transmitted light. Fluorescence imaging in the life sciences is now usually performed in either epifluorescence or confocal microscope apparatus. Epifluorescence microscopy, the simpler technique, refers to an arrangement in which the excitation light is focused down onto the sample through the objective lens, meaning that most of this 'noise' passes on through the sample and is lost, with only the small proportion that is reflected being mixed with the emitted light signal passed back up through the objective to the detector. The unwanted excitation light can be further removed by optical filtering, which is standard in microscopy, and can be achieved either with bandpass filters, or by monochromation. Confocal microscopy refers to a more refined technique in which the excitation light is passed through a small aperture before focusing through the objective lens on a particular point and at a particular focal plane of the sample. The emitted light is also refocused through the same lens and another pinhole aperture before detection. In this way, only a small section of sample in the focal plane is illuminated with highly focused light, and any light emitted by the sample without the focal plane is eliminated because it is not focused through the final pinhole before detection. This setup gives a greatly enhanced signal-to-noise ratio and resolution, but much reduced intensity, thus long signal acquisition is often required. Because images are only detected from a small section of the sample at a time, scanning of rows of points is required to build up a 2D image. Additionally, shifting the focal plane in the z-dimension (z-stacking) can allow the acquisition and reconstruction of 3D images of thicker samples. A popular refinement is confocal laser scanning microscopy, which uses mirrors to scan an excitation laser across the x- and y-planes of the sample. The limitations of fluorescence microscopy as an imaging technique (as opposed to those defined by the choice of fluorophore, which are discussed individually in the sections on *d*- and *f*- block imaging agents) are the fundamental limitation of resolution to approximately half the wavelength of the light involved in the experiment and the limited depth of tissue penetration of the light used in the experiment. This limits visible light fluorescence microscopy to sample depths of a few millimetres and near IR microscopy to a few centimetres.

The application of metal-based lumophores in fluorescent cell imaging is a relatively new and currently growing area. While organic fluorophores (discussed in the previous chapter) dominate the commercially available agents, there are numerous advantages to the use of metal complexes in this field, which has led to a rapid growth of interest in their development. The advantages of metal-based systems are largely a result of the differences of the photophysical mechanism of the photoluminescence processes. Photoluminescence is the phenomenon in which a molecule absorbs a photon of light with a concomitant transfer of an electron to a higher energy orbital to generate an electronically excited state. This excited state loses some energy non-radiatively through vibrational and other processes, and the electron then returns to the original orbital (the electronic ground state). Energy can be emitted radiatively as a photon of light of lower energy than that which was absorbed, the difference in wavelength of light absorbed and emitted being the Stokes' shift. Most purely organic fluorophores absorb light to give singlet electronically excited states, in which the electron spins are formally paired, although the electrons occupy different orbitals. Typically, they then emit without any change in spin-state, although in some cases structural geometric changes radically alter the energetics of the excited states giving large Stokes' shifts. In contrast, most of the *d*-block lumophores that have been developed emit from triplet states in which the electron spins are parallel, leading to long excited state lifetimes due to the forbidden nature of the emission. The associated Stokes' shifts are therefore large, due to the energy difference between the singlet excitation and the emissive triplet state. The Stokes' shifts from *f*-block lumophores can also be advantageously large, although this is due to the specific mechanism for populating the metal-centred *f* excited state (the details of these different photophysical processes will be discussed in the sections on *d*- and *f*-block imaging agents). Long luminescence lifetimes and large Stokes' shifts are desirable properties for optical imaging agents because they aid in differentiating signal from background noise derived from endogenous materials such as flavones and NADPH (autofluorescence) [1]. Autofluorescence is typically characterised by short lifetimes, so time-gated techniques (TRLM) can allow its removal, and small Stokes' shifts, so selection of detection wavelength allows its removal. As well as being easy to distinguish from autofluorescence, metal-based systems also typically show less photobleaching than organic fluorophores. Photobleaching, the photo-induced destruction of fluorophores, typically involves the *in vivo* reaction of the excited states of organic heterocycles with oxygen, or ROS, to give non-emissive compounds, thus inducing a loss of signal intensity over prolonged image acquisitions. Metal-based lumophores are generally less prone to this phenomenon than purely organic compounds due to the different electronic structures of the excited states leading to lower reactivity with the medium. A further advantage of metal-based systems is the opportunity to exploit the longer luminescence lifetimes in imaging techniques, which rely upon a change in the lifetime, as opposed to the wavelength or intensity of emission for the detection of intracellular analytes [2]. Lifetime-based techniques, such as FLIM, overcome the classic limitation of responsive probes that rely upon changes in signal intensity in the presence of an analyte; such probes can work well *in vitro*, but are difficult to interpret *in vivo* because the concentration of the probe in a certain cellular compartment is usually difficult to assess, making direct correlation between intensity and analyte concentration impossible. It is important to note that many of the *d*- and *f*- block metals that have been applied in imaging are highly toxic as the free ions and often suffer from poor cell uptake. To overcome both of these problems, ligand systems have been designed to control their toxicity and endow membrane permeancy on the imaging agents, but these are best treated by class of metal as the design criteria of the *d*- and *f*- block cases are markedly different.

12.1.1 *d*- and *f*-Block Imaging Agents

The *d*-block (the transition metals) offers a vast range of molecular complexes that have suitable properties for application as agents in fluorescence cell imaging, with luminescent examples of complexes of almost all metals having been reported. However, as yet there are only a small number of types of complex that have been so used, and even fewer that have gained popularity. The most widely studied families of *d*-block imaging agents are those based around second- and third-row transition metals complexes bearing conjugated aromatic ligands in which the metal is in the *d*⁶ configuration [3]; however, there are also a number of examples of other types of complex, notably *d*⁸ and *d*¹⁰ complexes of the platinum group metals [4]. The trivalent lanthanide ions (Ln^{III}) of the *f*-block also provide many opportunities [5] for consideration as optical cellular imaging agents. Unlike the *d*⁶ transition metal ion complexes, which benefit from an inherent and advantageous kinetic inertness, a significant challenge with Ln^{III} based systems is addressing the issue of kinetic stability in aqueous, and more pertinently, physiologically relevant environments: typically this requires encapsulation of the Ln^{III} within a multidentate ligand array. Unlike the more rigid coordination geometries imposed by the *d*⁶ electronic arrangement, the Ln^{III} ions typically experience much weaker ligand field preferences and higher coordination numbers. In this context, the design of effective ligand systems for Ln^{III} reflects the predominantly electrostatic nature of bonding between Lewis acidic Ln^{III} ions and hard ligand donor moieties.

12.2 *d*⁶ METAL COMPLEXES IN FLUORESCENT CELL IMAGING

Several common features are shared by the three main families of *d*⁶ lumophores, that is, iridium cyclometallates, rhenium *fac* tricarbonyl polypyridyls, and ruthenium trisbipyridyls (Figure 12.1a–c, respectively) that make them attractive for cell imaging:

(a) **Photophysics:** All show excitation in the visible region of the spectrum, large (100 s of nm) Stokes' shifts and long (100–1000 ns) luminescence lifetimes, resulting from emission from triplet excited states with variable degrees of MLCT and IL character.

(b) **Photostability:** All show reduced photobleaching compared to common organic fluorophores.

(c) **Kinetic stability toward ligand exchange:** Heavy metal toxicity is largely associated with interactions of biomolecules with the metal centres, and so these coordinatively saturated complexes with very slow rates of ligand exchange typical of the *d*⁶ low-spin configuration prevent these interactions and thus reduce toxicity.

12.2.1 Photophysics of *d*⁶ Polypyridyl Lumophores

The photophysical properties of *d*⁶ lumophores are complex with variation between specific examples, but there is a general theme of triplet metal-to-ligand-charge-transfer (³MLCT) that is important in the majority of examples. Briefly, this involves (Figure 12.2a) excitation by a photo-induced electron transfer from metal-based orbitals to a conjugated π-system usually located on an aromatic heterocyclic ligand (often pyridyl). This generates a singlet excited state (¹MLCT) that is rapidly converted *via* intersystem crossing (Figure 12.2b) to the triplet excited state, from which emission (Figure 12.2c) occurs. It is important that other metal-based orbitals are not available for electron transfer from the excited state, because this can lead to energy loss through pathways that do not involve emission of light. For this reason the complexes usually include *high-field* ligands that will ensure that the vacant *d*-orbitals are of too high an energy to become involved in any of the deactivation processes. Because the excited state therefore involves charge transfer from metal to ligand, the metal should be easily oxidised, and the ligand

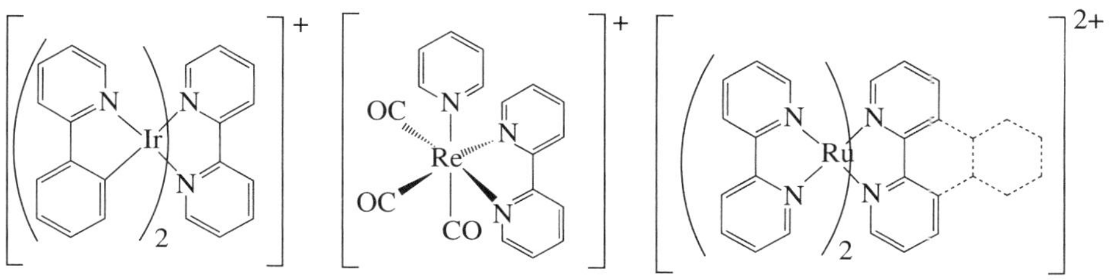

FIGURE 12.1 Common structural types in *d*⁶ imaging agents.

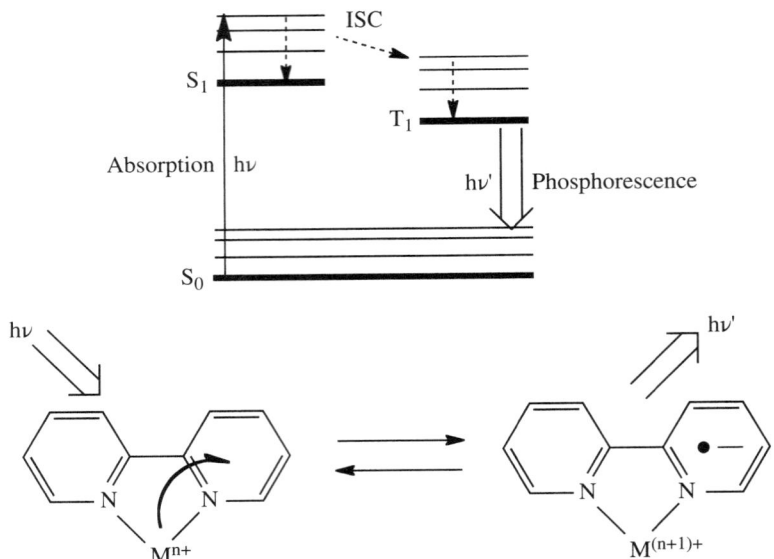

FIGURE 12.2   The electronic origin of luminescence from d^6 polypyridyls.

easily reduced. Thus low oxidation state metal complexes (Ir^{III}, Re^{I}, Ru^{II}) combined with highly conjugated ligands that can easily accept an electron are good combinations for this application. A degree of tuning of the excitation and emission wavelengths is possible by structural changes to the ligand involved in accepting the electrons, or other ligands, because these change the energy gaps involved. This can be used in sensing of intracellular species, but a detailed discussion of these photophysical phenomena (which are in fact much more complex than this simplistic model suggests) is beyond the scope of this chapter [3].

12.2.2   Iridium and Rhodium Cyclometallates as Imaging Agents

Complexes of the general formula [Ir(C^N)$_2$(N^N)], where C^N represents a cyclometallating ligand such as 2-phenyl pyridine, and N^N represents a chelating diimine such as 2,2'-bipyridine, have been the most widely applied iridium imaging agents, although the neutral [Ir(C^N)$_3$] analogues have been studied too. There have been many studies of the luminescence of a large range of examples of these species including examples that show responses to analytes and thus represent sensors [6]. These complexes are attractive for cell imaging applications due to their visible excitation and emission wavelengths, which are highly tunable. In some cases these complexes show extremely long (ms) lifetimes arising from ligand-centred triplet (^{3}LC) emissive states. The bis-cyclometallated complexes [Ir(C^N)$_2$(N^N)] are monocationic, which is likely to assist in their cell uptake due to the membrane potential-driven preferential uptake of cations.

12.2.2.1   Synthesis   Typically, iridium imaging agents are prepared by the reaction of the cyclometallated dimer [(C^N)$_2$IrCl]$_2$ with the chelating diimine ligand, either under forcing thermal conditions or in the presence of an activating agent such as a silver salt (Scheme 12.1).

12.2.2.2   Development of Iridium Imaging Agents   While the useful photophysical properties of iridium cyclometallates have been known for many years, it was only in 2008 that the first report of the application of such species in cell imaging appeared. Li designed a series of cyclometallated iridium species with fluorous substituents (to increase lipophilicity, Figure 12.3) and showed that they were taken up by cells and that the Ir-based luminescence was retained *in vivo*, allowing imaging by fluorescence microscopy [7].

There has now been a large range of Ir^{III} complexes synthesised and applied to cell imaging, particularly by the Lo and Li groups. A range of biologically relevant substituents have been attached to these complexes (e.g., biotin [8], indoles [9], Figure 12.4), and extensive studies of their uptake and localisation in a range of mammalian cells have been carried out. In general, iridium cyclometallates show reasonable-to-good cell uptake, which is attributed to their lipophilicity and (usually) cationic nature. Predominantly, these complexes show accumulation in lipophilic structures of the cytoplasm, which again is attributed to the extended hydrocarbon skeleton intrinsic to the cyclometallated structures. This predominance of cytoplasmic localisation means that the design of targeted iridium cyclometallated probes is a challenging area, although there are some examples of specific localisation in organelles. For instance, complex **5** (Figure 12.5) has been shown by co-localisation techniques to localise in the nucleoli (although there is also cytoplasmic staining) [10].

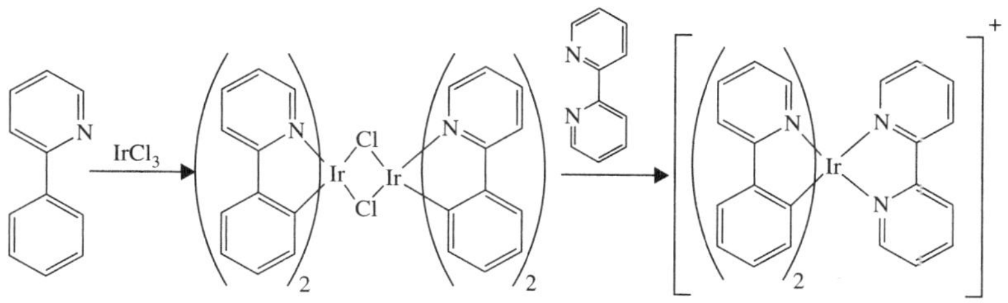

SCHEME 12.1 Typical synthesis of iridium imaging agents.

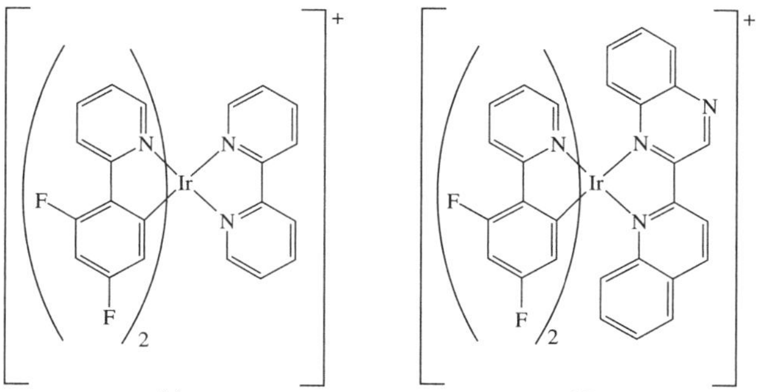

4A   4B

FIGURE 12.3 Li's fluorous iridium cyclometallates.

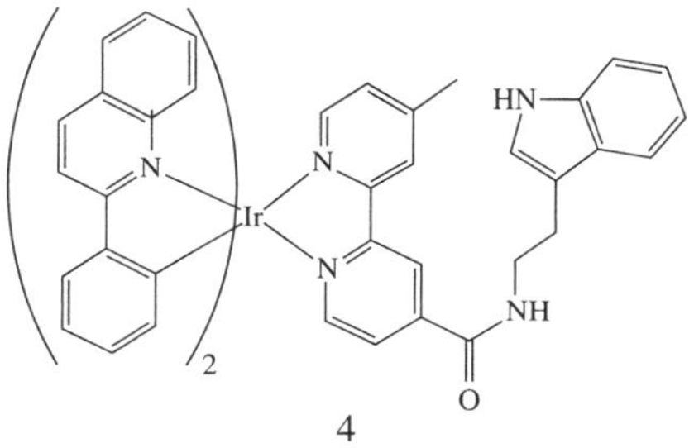

4

FIGURE 12.4 Indolated iridium cyclometallate **4**.

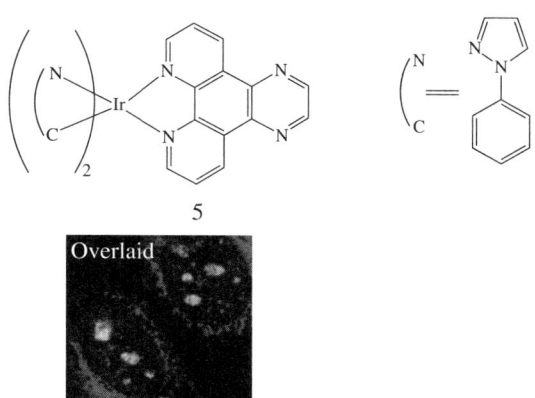

5

FIGURE 12.5 Nucleolar-targeting iridium complex **5** and cell image showing nucleolar localisation (Reproduced with permission from Ref. [10]). (*See insert for colour representation of the figure.*)

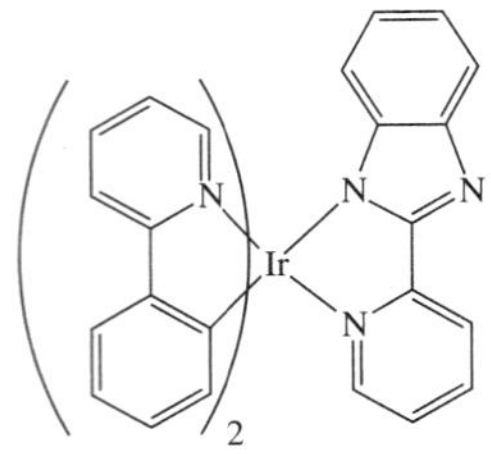

FIGURE 12.6 Neutral, high pH form of imaging probe **6**.

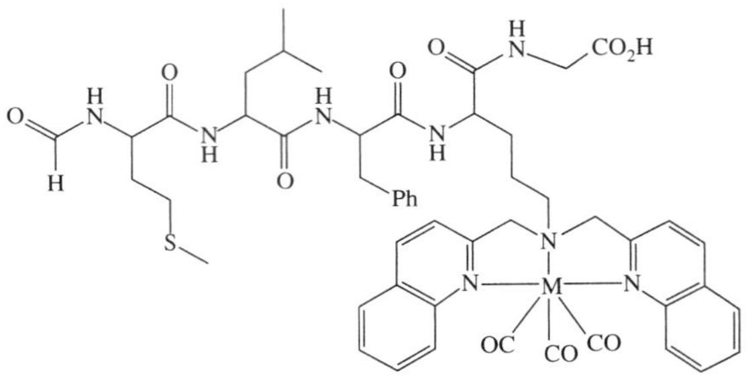

7    $M = Re/^{99m}Tc$

FIGURE 12.7 Bisquinoline complex **7**.

There has been one report of the use of time-gated techniques with iridium cyclometallates in which a pH-responsive iridium probe [Ir(ppy)$_2$(pybz)] **6** (Figure 12.6) with a long lifetime was co-incubated in CHO cells with an organic nuclear stain [11]. With no time-delay, the emission of the organic dye dominates, but with a 10 ns time gate before image acquisition, the cytoplasmic staining of the iridium complex is visible, because the short-lived organic fluorescence has decayed away by this point. There are a small number of rhodium analogues of the iridium cyclometallates, which have been applied in cell imaging and appear to show very similar patterns of uptake and localisation to the iridium cases [12].

12.2.2.3 Summary of Iridium/Rhodium Cell Imaging

Iridium cyclometallates have been established as promising candidates for live cell imaging, with numerous examples showing good uptake, ideal photophysics with highly tunable emission characteristics, and very long lifetimes. However, a propensity for widespread distribution throughout lipophilic organelles may hinder the design of agents that target specific sites in the cell, which will be required for genuine biomedical applications. Many examples also suffer from low aqueous solubility and significant cytotoxicity.

12.2.3 Rhenium *fac*-Tricarbonyl Heterocyclic Complexes as Imaging Agents

12.2.3.1 Bisquinoline Complexes and Related Species

The first report of a rhenium *fac*-tricarbonyl heterocyclic complex in cell imaging appeared in 2004 when Zubieta used the luminescence of a bisquinoline rhenium complex **7** (Figure 12.7) conjugated to fMLF a peptide, which targets the formyl peptide receptor (FPR), to verify that conjugation of this unit to a rhenium or technetium complex did not interfere with the receptor targeting [13]. Similar bisquinolines have since been applied with other vectors, such as cobalamin B$_{12}$ [14], and the rhenium complexes of the dipicolyl amine analogues are often used as cold analogues of their ^{99m}Tc complexes, which find use in SPECT imaging [15]. Many examples of similar complexes based on the dipicolyl amine core with additional organic fluorophores have been used in cell imaging experiments to determine the cellular behaviour of Re/Tc complexes designed for radioimaging/therapy, but because the emission of these species is not metal-based, they are better considered as examples of organic fluorophores [16]. Although the emissive bisquinoline complexes were the first rhenium species reported as lumophores in cell imaging, typically their photophysical properties are not as amenable to the technique as those of the bipyridyl and related complexes that have come to dominate the area.

12.2.3.2 Rhenium *fac*-Tricarbonyl Bipyridine Complexes

The most studied rhenium lumophores are based on a *fac*-tricarbonyl Re^I core with chelating N^N polypyridyl ligands. These species show photophysics dominated by emission from a ^{3}MLCT state localised on the diimine ligand. The neutral complexes [ReX(bipy)(CO)$_3$] (X=Cl/Br) typically show maximum absorption at around 350 nm and emission at 500–600 nm. The luminescence lifetimes are around 100 s of ns, with typical quantum yields around 0.1%. However, upon replacement of the halogen with a neutral (L=Py, PR$_3$) ligand, the cations [Re(bipy)(CO)$_3$(L)]$^+$ show longer lifetimes (up to ms) and larger quantum yields (1–10%) [17]. Although the excitation maxima are still in the UV region (~380 nm), the characteristic broad excitation bands associated with ^{1}MLCT absorption mean that excitation with visible light (often a 405 nm laser line) is effective, and the large Stokes' shift (>100 nm, emission maxima ~550 nm) allows easy detection that is conveniently free from autofluorescence.

The ease of synthesis of a range of complexes of the general formula [Re(CO)$_3$(bipy)(PyX)]$^+$ in which bipy refers to 2,2′-bipyridine or a similar system, and PyX a substituted pyridine has allowed a systematic variation of both the photophysical properties (by variation in the bipy unit) and the cellular behaviour (by variation in the pyridine). There are now cellular imaging studies of over 50 complexes reported in the literature that demonstrate that variations of the bipy unit have relatively minor effects on the uptake and localisation of the complexes (at least for the examples so far reported), whereas variations in the pyridine substituents can be used to give a degree of control of the uptake and cellular localisation of the agents.

12.2.3.3 Synthesis

Typically, the synthesis of rhenium-based imaging agents follows the approach outlined in Scheme 12.2. The lumophoric [Re(CO)$_3$(N^N)] core is first prepared by the reaction of rhenium halide pentacarbonyl with the chelating bisimine N^N. This species is then activated to ligand exchange by abstraction of the halide with silver salts, typically to generate the labile acetonitrile adduct, which is then reacted further with a substituted pyridine to give the cationic probe.

12.2.3.4 Development of Rhenium Imaging Agents

The first report of application of [Re(CO)$_3$(N^N)PyX] species in cell imaging appeared in 2007, applying a range of cationic, anionic, and neutral complexes in the imaging of parasitic flagellates (*Spironucleus Vortens*) [18]. This preliminary study demonstrated that variations in the charge and lipophilicity of the complexes were effective in controlling uptake and cellular localisation with wide variations observed. A series of subsequent studies by a number of groups has shown that many complexes of this general structure are taken up well in a variety of mammalian cells and that cellular localisation can be controlled by substitution patterns [19]. For instance, a chloromethyl-substituted complex **8** (Figure 12.8) targets mitochondria, in line with the known propensity of chloromethyl units to react with reduced thiols, in which mitochondria are particularly rich [20]. Biotin-appended complexes have been explored as a method of assisting cell uptake, and a biotin appended complex **9** (Figure 12.9) shows preferential accumulation in the Golgi apparatus [21]. Rhenium complexes based on the [Re(CO)$_3$(N^N)PyX] structure have now been reported that target a wide range of cellular organelles including the nucleolus, Golgi apparatus, ER, internal and plasma membranes, as well as mitochondria. Complexes of this general structure are particularly amenable to the control of localisation and uptake as the basic core [Re(CO)$_3$(bipy)Py]$^+$ **10** (Figure 12.10) appears to be poorly taken up by cells, with additional lipophilic substituents on the pyridine required for good uptake (at least by passive diffusion), and the nature of these substituents determines localisation. Small variations in the structure of the pyridine substituent can have a large effect on uptake and localisation. For example, a

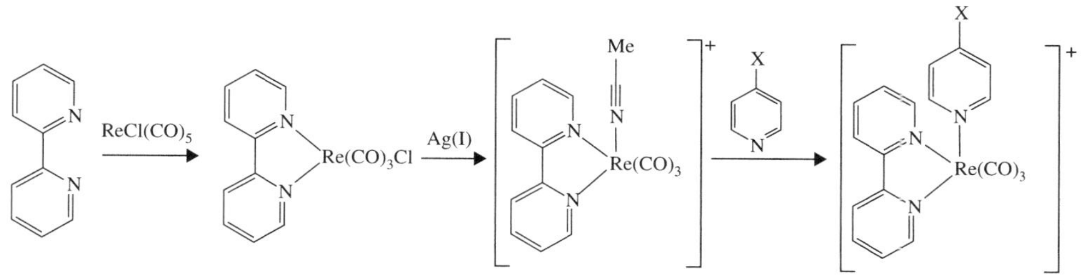

SCHEME 12.2 Synthesis of typical rhenium-based imaging probes.

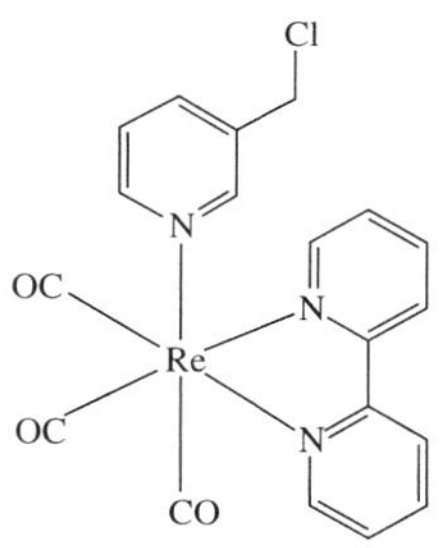

FIGURE 12.8 Mitochondrial targeting chloromethyl complex **8**.

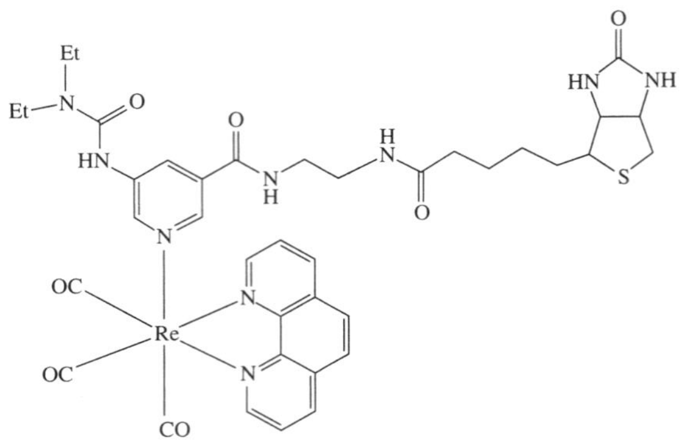

FIGURE 12.9   Biotin-appended rhenium complex **9**.

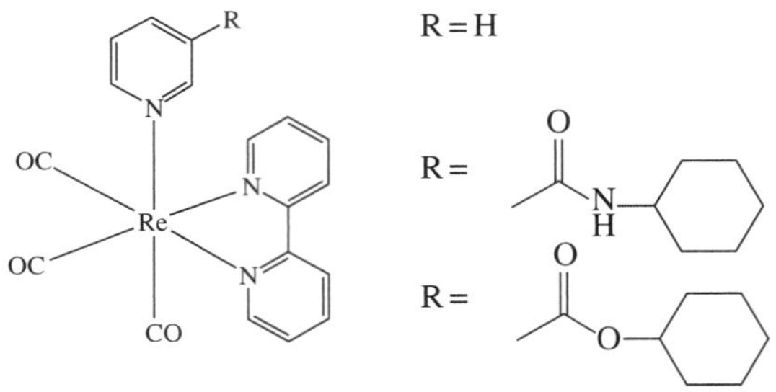

FIGURE 12.10   Structure of complexes **10–12**.

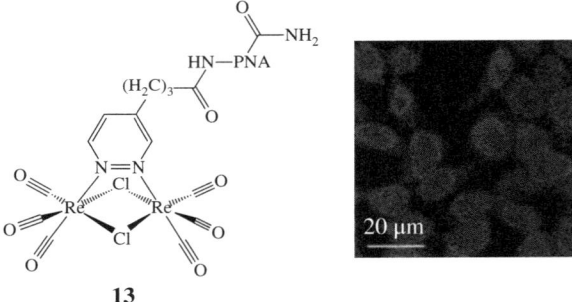

FIGURE 12.11   Thymine-rich PNA appended complex **13** and cell imaging showing blue-shifted nuclear emission (Reproduced with permission from Ref. [23]).

cyclohexyl amide-substituted complex **11** is not taken up by healthy cells, while the analogous ester **12** is taken up well by passive diffusion [22]. In this sense, the lower lipophilicity of the rhenium core is more attractive than the highly lipophilic bis-cyclometallated [Ir(ppy)$_2$(N^N)] core, because small variations can be used to tune localisation, while it seems more difficult to overcome the affinity of the [Ir(ppy)$_2$(N^N)] unit to show nonspecific distribution throughout lipophilic regions of cells.

A single example of an unusual complex based on a pyridazine-bridged di-rhenium core **13** (Figure 12.11) appended with a thymine PNA unit has been shown to be taken up well by cells and localise in the nucleus, directed by the PNA unit [23]. The emission from the nucleus is blue-shifted compared to that from other cellular regions, indicative of an interaction with DNA.

12.2.3.5   Summary of Rhenium Cell-Imaging Studies   Rhenium complexes have proven to be highly amenable to tuning of uptake and localisation properties by making small changes to the pyridine substituent, although the photophysics are not as easy to tune as some other classes of complex. The lower lipophilicity of the rhenium core is more attractive than the highly lipophilic bis-cyclometallated [Ir(ppy)$_2$(N^N)] core, because these small variations can be used to tune localisation, while it seems more difficult to overcome the affinity of the [Ir(ppy)$_2$(N^N)] unit to show nonspecific distribution throughout lipophilic regions of cells.

12.2.4 Ruthenium and Osmium Complexes in Cell Imaging

Ruthenium complexes of the general structure [Ru(N^N)$_3$]$^{2+}$, where N^N represents a chelating polypyridine unit, (bipy, phen) are the archetypal transition metal lumophores and have been widely studied in a variety of applications. Emission emanates from a ^{3}MLCT excited state localised on the ligand framework, with all of the N^N ligands contributing to the excited state (unlike the rhenium complexes in which only one ligand is involved). Typically, these complexes show excitation maxima around 450 nm and emission maxima around 610 nm, with luminescence lifetimes of 0.6–6 µs and quantum yields typically around 0.1, but up to 0.6%, making their photophysical properties ideal for cell imaging applications [24]. The sensitivity of the excited state to quenching by triplet dioxygen has led to the development of a range of ruthenium-based oxygen sensors, with lifetime-based methods widely used to correlate lifetime and [O$_2$] [25]. One major focus of the studies of Ru^{II} polypyridyls in biological systems has been the detection of DNA and other poly/oligo-nucleotides. Complexes such as [Ru(bipy)$_2$(dppz)]$^{2+}$ show only weak luminescence in aqueous solutions, but, upon intercalation of the highly conjugated aromatic dppz ligand into DNA, enhancements of 10^4-fold are observed [26]. While complexes with these intercalating ligands have been shown to be excellent DNA sensors *in vitro*, uptake and localisation issues complicate the issue *in vivo* [27].

12.2.4.1 Synthesis Typically, ruthenium-based imaging agents are synthesised in a two-step sequence, with ruthenium trichloride reacting with two equivalents of a bipyridine or similar ligand to give the neutral [RuCl$_2$(N^N)$_2$] species, which are then converted by reaction with a different chelating polypyridine to the cationic [Ru(N^N)$_2$(N'^N')]$^{2+}$ agents (Scheme 12.3). Usually, the unique N'^N' ligand is chosen for the specific localisation or sensing properties desired, and the two N^N ligands are chosen to tune the overall photophysical properties.

12.2.4.2 Development of Ruthenium Imaging Agents The uptake of simple ruthenium complexes is low, and in several cases it has been found that although they are taken up by active transport, the luminescence is only evident in endosomes from which it appears that the complexes do not escape. The first reports [28] of the application of ruthenium in cell imaging concentrated on mapping oxygen concentration in cells using [Ru(bipy)$_3$]$^{2+}$ in FLIM. Because this technique relies only on lifetime, and not intensity, the relative concentrations of the complex in different cellular compartments are irrelevant. Subsequent studies [29] highlighted problematic uptake, with complexes such as [Ru(phen)$_3$]$^{2+}$ being unable to cross healthy, intact membranes, leading to encapsulation in endosomes. In order to overcome problems, a range of complexes bearing substituents known to assist in uptake such as polyarginines have been prepared [30].

Barton has conducted a systematic study of complexes bearing dppz and related ligands as cellular DNA probes. In many cases the problems of uptake have been overcome by appending more lipophilic units to assist membrane transport, and other units such as oligoarginines and fluorescein have also been introduced [31]. A series of complexes of the general formula [Ru(dppz)(N^N)$_2$]$^{2+}$, **15** allowed a correlation between the lipophilicity of the N^N unit and the degree of uptake to be established [32]; this has been shown to be by a passive mechanism [33]. Highly lipophilic analogues of simple dppz complexes based around alkyl-ether substituted dppz ligands (Figure 12.12) developed by Svensson all showed good uptake.

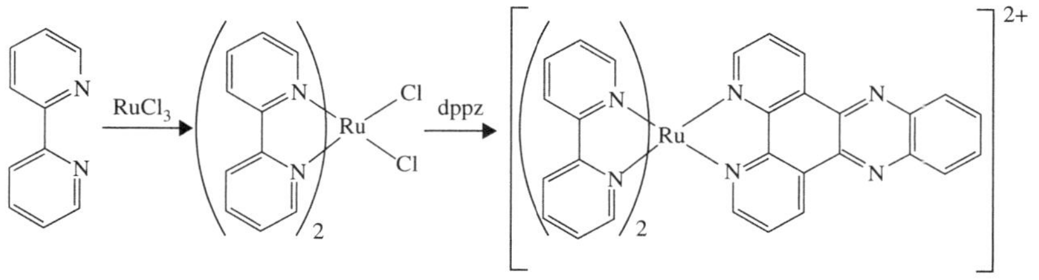

SCHEME 12.3 Synthesis of ruthenium imaging agents.

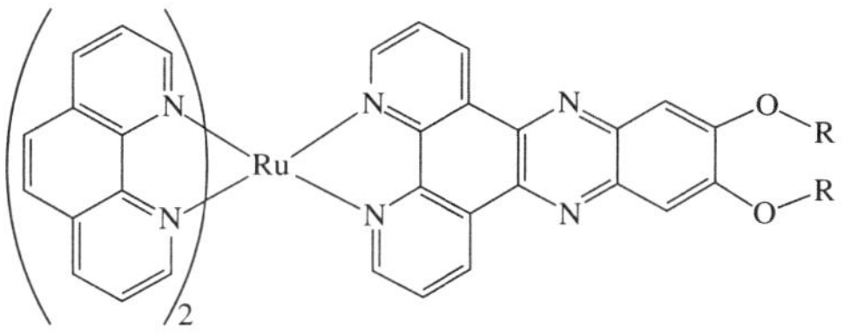

FIGURE 12.12 Lipophilic Ru dppz analogues **15a** (R = Et) b (R = Bu) c (R = hex).

Interestingly, the patterns of localisation were a function of the chain-length, with nuclear, nucleolar, and cytoplasmic staining all being available by changing the alkyl chain lengths from ethyl to hexyl [34].

Lo has reported cell imaging ruthenium complexes conjugated with estradiol, which is known to target the estrogen receptor-α (ER-α) [35]. These species show good uptake; however, studies indicate that regardless of the lipophilic structure of estradiol, the uptake is mainly by an energy-dependent mechanism.

While osmium is generally regarded as too toxic to use in biological applications, an example of an emissive Os^{II} complex with a chelating diphosphine ligand capping the osmium bis-phenanthroline core has been embedded in nanoparticles and applied in cell imaging. Os^{II} polypyridyls are particularly attractive from a photophysical standpoint because they can exhibit NIR emission, although lifetimes are consequently shorter, which gives good tissue penetration [36].

Ruthenium complexes have shown great promise in cell imaging applications, with attractive photophysics and sensing ability, particularly as DNA and O$_2$ probes. It is likely that their progress will depend upon the development of general and predictable methods to overcome problems associated with uptake and cellular distribution.

12.2.5   *d*^{8} Metal Complexes in Cell Imaging

The luminescence of platinum(II) and iridium(I) complexes is well established, but there are relatively few examples of their applications in cell imaging. A time-gated study of a platinum(II) complex **16** with a lifetime of 580 ns allowed complete separation of the Pt-derived signal, localised in the nucleus, from a fluorescein ($\tau=4$ ns) co-stain [37]. Studies of the related complexes **17** [38] and **18** (Figure 12.13) [39] showed preferential uptake in nuclei and cytoplasm respectively, indicating that platinum may well be a promising candidate for the development of organelle-specific agents. Platinum complexes have also been studied in cell imaging using 2-photon excitation, which avoids issues of tissue penetration of high energy excitation [37].

12.2.6   *d*^{10} Complexes in Cell Imaging

12.2.6.1   Au^I Complexes in Cell Imaging   Gold(I) complexes show emission from a variety of photophysical pathways, typically LMCT, or those relying upon dimeric species showing aurophilic interactions. There have been relatively few examples of gold lumophores applied in cell imaging, but the medicinal properties of certain gold complexes have led to some effort in this area. Gold phosphine complexes of thionaphthalimide have been shown to target the nucleus [40], while a dinuclear gold NHC complex, **19** which emits due to aurophilic interactions, targets lysosomes (Figure 12.14) [41].

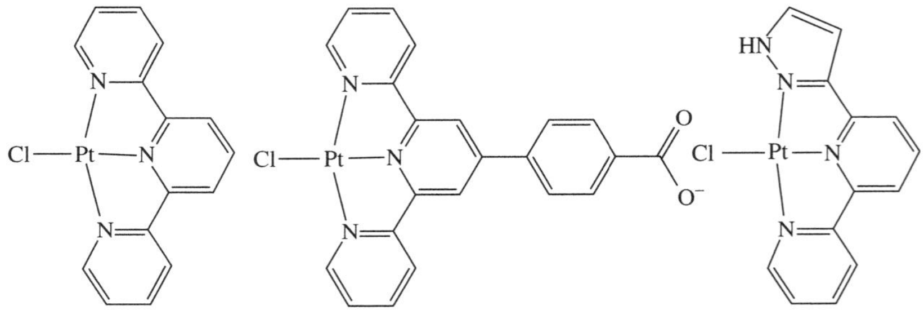

FIGURE 12.13   Pt complexes **16–18**.

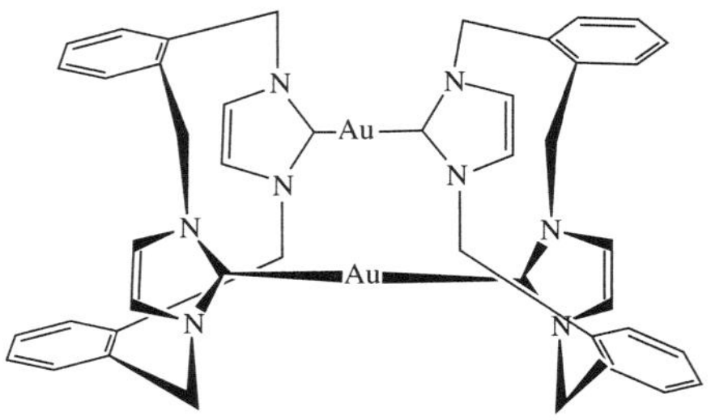

FIGURE 12.14   Au-NHC dimer **19**.

12.2.7 Zinc Complexes in Cell Imaging

There are a range of zinc species related to or based upon the ATSM (diacetyl-bis(N4-methylthiosemicarbazone) ligand that are fluorescent and have been applied in cell imaging, mainly as models for the non-emissive copper analogues [42]. The copper analogues are applied as PET imaging agents with the positron-emitting radioisotope ^{64}Cu. While these complexes are effective cell imaging agents, they are beyond the scope of this chapter. Regardless of the semantics of definition of the *d*-block, the electronic nature of the fluorescence of zinc ATSM (and zinc and other porphyrin complexes) makes them more akin to organic fluorophores than the phosphorescent species that are the focus of this section.

12.2.8 Summary of *d*-Block Lumophores in Cell Imaging

The work described in this chapter demonstrates that *d*-block lumophores have now been developed as viable cell imaging agents and molecular probes, with examples from a range of metals and complex types illustrating successful methods to overcome issues of uptake, localisation, and toxicity. The d^6 family have been most widely studied, and amongst these it appears that while iridium and ruthenium complexes have the most attractive photophysics in terms of tunable and/or responsive emission and lifetime profiles, in terms of controlling uptake and localisation, rhenium has, to date, been most successful.

12.3 *f*-BLOCK IMAGING AGENTS

12.3.1 Photophysical Properties of Ln^{III} Luminophores

Although the majority of the trivalent lanthanide (Ce to Lu) ions (Ln^{III}) (electronic configuration of [Xe]$4f^n$ where n = 0–14) are luminescent, the most important ions in the context of optical imaging are currently Sm^{III}, Eu^{III}, Tb^{III}, Dy^{III}, and Yb^{III}. The luminescence from emissive Ln^{III} originates from inner shell $4f$-$4f$ transitions, which are observably sharp in appearance and characteristic of the specific ion: Ln^{III} ions can possess emission bands that usefully address the UV, visible, or NIR regions. The associated Stokes' shifts can be large, but are a function of the particular pathway for populating the *f*-centred excited state. Lifetimes can be even longer (from micro- to millisecond domain) than those described earlier for the phosphorescent *d*-block lumophores due to the forbidden nature of the *f*-*f* relaxation. Because the $4f$-$4f$ transitions are symmetry (parity) forbidden, they possess very low molar absorptivities for direct excitation, so the established strategy for overcoming this is to incorporate a sensitising chromophore (commonly known as an antenna group), which absorbs light and transfers energy to the $4f$ excited state (the origin of the apparent Stokes' shift) via a mechanism that often, but not exclusively, involves the triplet excited state of the sensitiser (Scheme 12.4). The antenna group can be covalently attached to the ligand architecture or introduced via other means (as a co-ligand, for example). Although there is a vast range of sensitising chromophores that have been studied for various Ln^{III}, for Eu^{III} and Tb^{III} the choice is perhaps more limited. For these ions, the antenna group must absorb light very effectively, but it must also possess a triplet state that is of sufficient energy (i.e., >2000 cm^{-1}) above the accepting Ln^{III} state to allow sensitisation and prevent back energy transfer (which generally results in low emission intensity from the Ln^{III}). Chromophore types that have proved to be effective sensitisers of the accepting states of Eu^{III} ^{5}D$_0$ (~17200 cm^{-1}) and Tb^{III} ^{5}D$_4$ (~20400 cm^{-1}) are generally based upon (poly) aromatic, often heterocyclic antennae, which ideally (for the applications described here) absorb between 350–410 nm, with intrinsically small singlet-triplet energy gaps. From an energetic perspective, it is worth noting that consideration of the NIR emitting lanthanides such as Yb^{III} (accepting state of ^{2}F$_{5/2}$ at *ca.* 10200 cm^{-1}) broadens chromophoric options significantly [43]. The quantum yields of emissive Ln^{III} complexes vary dramatically due to the extremely sensitive nature of the $4f$-centred excited states to O–H, N–H, and C–H vibrational manifolds, which provide efficient, non-radiative deactivation pathways; the efficiency of energy transfer between the antenna and lanthanide ion also determines overall quantum yields. A classical approach to maximising the emissivity of Ln^{III} complexes is to therefore inhibit the approach of water solvent to the inner coordination sphere; high denticity, metal ion encapsulating ligands with hydrophobic peripheries can achieve this very effectively [44].

A unique and advantageous attribute of luminescent Ln^{III} complexes is the dependence of the emission spectral form and lifetime on the coordination environment. Eu^{III} is particularly valuable in this regard with sharp emission bands ^{5}D$_0 \rightarrow$ ^{7}F$_J$ (J = 0, 2, 3, 4) that are subtly sensitive to the nature/type of the donor and the coordination geometry at Eu^{III} [45]. When compared to the *d*-block lumophores described earlier, such behaviour presents unique opportunities in the design of responsive probes; thus, binding events at Ln^{III} can be interrogated directly using luminescence methods and ratiometric analyses (i.e., independent of probe concentration) [46]. For example, the affinity of Ln^{III} for various anions is dictated by electrostatics and the steric demand of the metal centre, in turn governed by the polydentate ligand, especially the arm substituents of

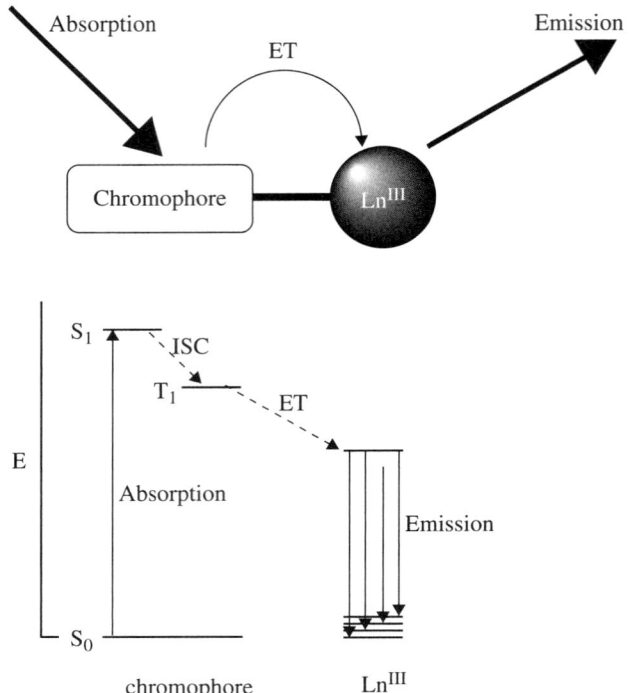

SCHEME 12.4 Top: Simplified model for sensitised lanthanide luminescence. Bottom: Typical energy level diagram for an emissive chromophore-appended lanthanide complex sensitised via a ligand-centred triplet excited state.

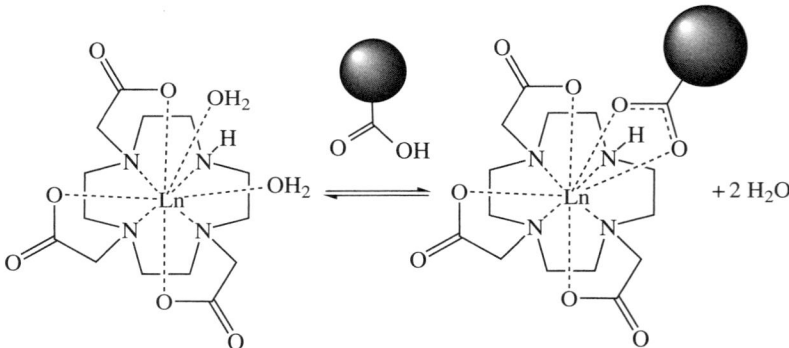

SCHEME 12.5 An example of reversible anion binding that induces a change in Ln^{III} hydration.

the ligand core (i.e., aromatic/heterocyclic sensitiser). Anion binding to the metal ion typically occurs through a reversible intermolecular process (Scheme 12.5), inducing reversible displacement of coordinated water molecules from the Ln^{III} centre. Therefore, for a heptadentate ligand the resultant Ln^{III} complex would typically expect to have 1 or 2 coordinated water molecules; coordination of mono- or bi-dentate anions will likely liberate one or both of these water molecules, resulting in measurable changes in luminescent output (e.g., intensity and lifetime). An understanding of the anion-binding affinities and the resultant perturbation of the Ln^{III} luminescence are important in a biological context because various anionic residues are available for binding [47]. In fact, this behaviour has allowed the development of responsive luminescent probes, which can relay information on anion concentrations [48]. Finally, luminescing Ln^{III} ions should be considered 'spherical' emitters and therefore avoid anisotropy; as a consequence, enantiopure Eu^{III}/Tb^{III} complexes can give intense CPL signals in the visible region. The implications of this observation are important because binding to a bio-macromolecule such as serum albumin can invert helicity, switching the sign of the CPL signal. Certain enantiopure complexes have been shown, through competition studies, to demonstrate selective binding to specific drug sites on proteins; CPL may be a very useful tool for tracking protein association through the use of chiral probes [49, 50].

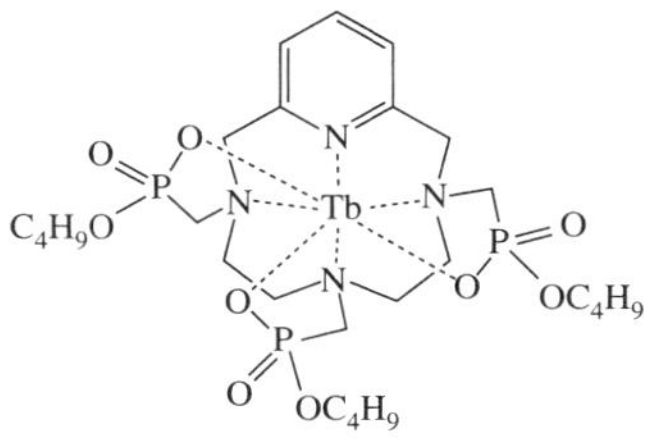

FIGURE 12.15 Bornhop's macrocyclic terbium complex.

12.3.2 Examples of *f*-Block Lumophores as Imaging Labels

The use of luminescent lanthanide complexes as optical labels in various fluoroimmunoassays has been established for over 30 years and was developed in parallel with their use as labels for cytochemistry and histochemical studies. In a biological imaging context, specific design criteria must be addressed to ensure that highly emissive, long-lived Ln^{III} probes persist in aqueous environments: The last 10 to 15 years of development has seen this challenge met, and the application of such species to confocal fluorescence microscopy and time-resolved luminescence microscopy is now rapidly developing [51].

One of the earliest examples of optical imaging using an Ln^{III} complex in a biological context was reported in 1999 by Bornhop. The development of a polyazamacrocyclic Tb^{III} complex (Figure 12.15) that incorporates a rigidifying ring-pyridine donor and phosphonic acid ester arms demonstrated promise as an abnormal tissue marker. Both *in vivo* and *in vitro* imaging showed that the complex, which has low cytotoxicity, possessed useful optical properties and demonstrated fluorescence imaging capability at picomolar concentrations in tissues. Adenocarcinoma cells showed good affinity for the complex, enabled by the lipophilic nature of the ligand periphery and the overall neutral charge of the complex, as demonstrated through *in vitro* fluorescence detection and histological assessments [52].

More recently Gunnlaugsson has extended the utility of emissive Ln^{III}-based complexes to the imaging assessment of damaged bone structure (specified as microcracks). An emissive, DO3A-type Eu^{III} complex (Figure 12.16), functionalised with an amido-naphthalene antenna and peripheral iminodiacetate groups, can effectively target exposed Ca^{II} sites of the hydroxyapatite lattice of the bone. Steady state luminescence spectroscopy was used to identify the binding of the Eu^{III} complex to a scratched (i.e., damaged) bone surface, through observed changes in the ratiometric emission of the Eu^{III}. The application of confocal fluorescence microscopy (Figure 12.16) demonstrated the potential for revealing far greater fine detail of the bone surface morphology through the observation of Eu-based visible emission and improved signal contrast [53].

An assessment of binding events at cell surfaces can be probed with luminescent labels. Both Tb^{III} and Eu^{III} complexes of chromophore-appended DO3A-type ligands have been designed to target the lipophilic plasma membrane of cells by incorporating long alkyl chains into the ligand framework. The amphiphilic complexes (Figure 12.17) appear to bind to the cell surface, presumably via the alkyl chains, and can be imaged using confocal fluorescence microscopy. The use of both Eu^{III} and Tb^{III} complexes to dual-label the cell provides further possibilities for interrogating the cell surface by studying the distance-dependent intermolecular energy transfer processes that can occur between the Ln^{III} ions. In this case, the energetic pathway favours Tb^{III}-sensitised-Eu^{III} emission, through observation of the relative quenching of the donor component (Tb^{III}) versus the acceptor (Eu^{III}) [54].

12.3.3 Examples of Time-Resolved Imaging with *f*-Metal Labels

As discussed earlier, time-resolved luminescence microscopy and/or FLIM allow for very effective removal of autofluorescence and improve the imaging sensitivity of the microscopy study. For maximal resolution, the lifetime of the lumophore should preferably reside in the microsecond-to-millisecond domain; emissive Ln^{III} complexes represent ideal candidates for this purpose. The application of Ln^{III} complexes to time-resolved luminescence microscopy has been demonstrated through labelling of small silica particles with cationic, chromophore-functionalised DO3A-type complexes of Eu^{III} (Figure 12.18). These particles were compared to those labelled with rhodamin 6G (i.e., a short-lived fluorophore), which absorbs and emits at coincident wavelengths to that of the Eu^{III} complex. The microscopy studies demonstrated that employing a microsecond time-delay allowed the Eu-labelled silica particles to be easily distinguished from those labelled with rhodamin 6G [55].

A demonstration of these optical advantages has been shown in a biological context, where imaging experiments on a fluorescein-labelled antibody and a Tb^{III}-labelled BSA conjugate were conducted, comparing the prompt fluorescence (in essence, total emission) with time-resolved (0.5 ms time delay) luminescence imaging. The latter revealed a 1000-fold increase in the Tb^{III}/fluorescein emission ratio, demonstrating the utility of biospecific probes based on emissive Ln^{III} chelates [56].

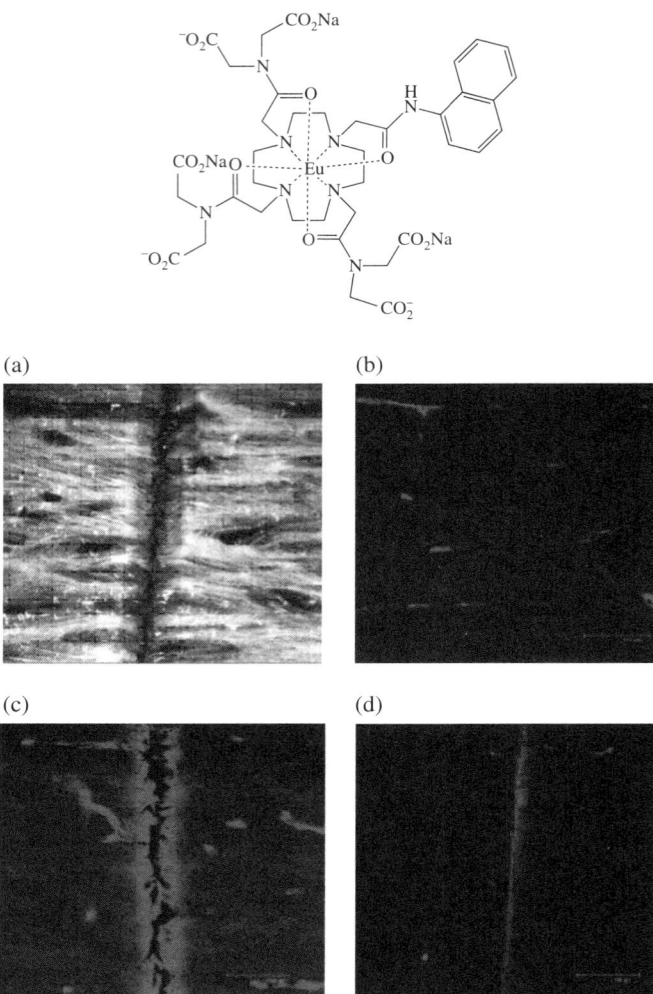

FIGURE 12.16 A polycarboxylate terminated Eu^{III} complex (left) and microscopy images of bone sample immersed in 10^{-3} M solution of the complex. (a) Reflected light image: 0 h (b) control (c) 4 h (d) 24 h. (*See insert for colour representation of the figure.*)

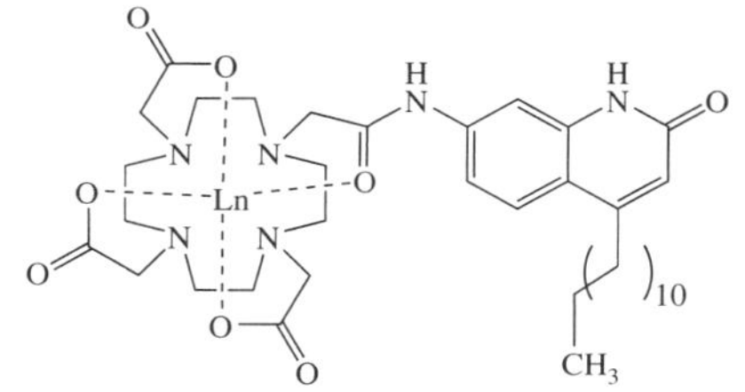

FIGURE 12.17 Structure of the amphiphilic Ln(III) complexes for labelling cell surfaces.

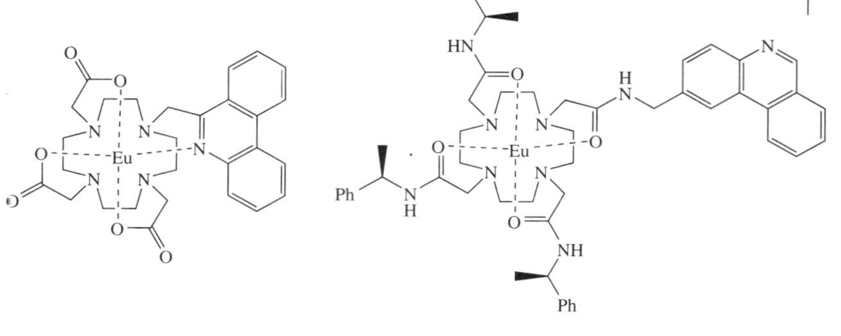

FIGURE 12.18 Structures of the Ln(III) complexes for labelling silica particles.

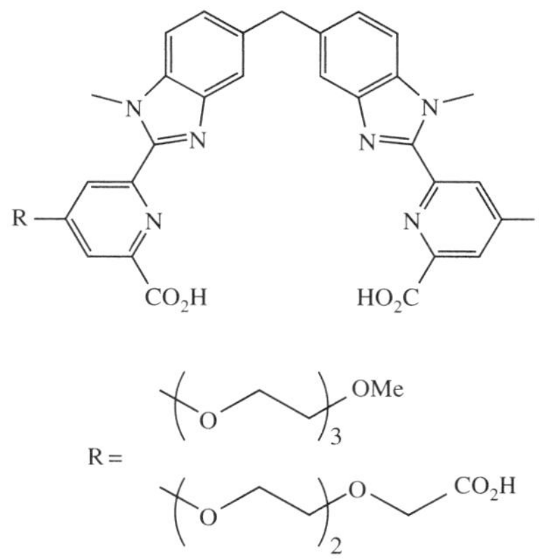

FIGURE 12.19 An example of a ditopic ligand used for the formation of bimetallic helical complexes.

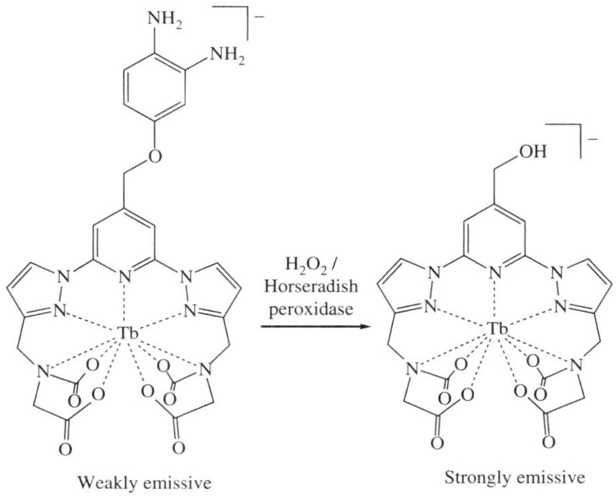

FIGURE 12.20 The two forms of the Tb^{III} complex.

In tandem with the diverse development of myriad cyclen-derived Ln^{III} complexes, an important, alternative class of species conceived by Bünzli are luminescent bimetallic, triple-stranded helical complexes of the general formula [Ln$^{III}_2$L$_3$]. Based on chromophoric, hexadentate ditopic ligands (Figure 12.19), the complexes are formed through a self-assembly process at room temperature. These emissive Ln^{III} complexes can also be bioconjugated with targeting vectors such as avidin, or monoclonal antibodies, allowing recognition of proteins expressed on the surface of breast cancer cells. The use of TRLM improves the signal-to-reference ratios, allowing high performance screening of cells and tissues; the diagnostic implications of this approach are significant to pathology [57].

The development of a responsive Tb^{III}-based probe for imaging hydrogen peroxide evolution in plant tissue (Figure 12.20) further highlights the utility of time-resolved imaging experiments. The starting Tb^{III} complex is essentially non-emissive, due to intramolecular quenching pathways within the terpyridine-based ligand; reaction with H$_2$O$_2$, in the presence of hydrogen peroxidase, cleaves a diaminophenylether unit, switching on visible luminescence from the Tb^{III} ion. The responsiveness of the complex was assessed *in vitro* using TRLM studies on tobacco leaf epidermal tissues. The obtained images are background free, with autofluorescence from the tobacco cells effectively suppressed, and show remarkable detail and spatial resolution [58].

12.3.4 Cellular Imaging with *f*-Metal Complexes

The following discussion focuses on the utility of emissive Ln^{III} complexes as intracellular optical imaging agents and is subdivided according to ligand classification. It is important to note that many of the imaging studies described below are accompanied by additional investigations, which are not dealt with in detail here, including the physiological relevance of the probes, such as protein affinity (secondary confirmation can be alternatively obtained through companion relaxivity studies with Gd^{III} complex analogues) and excited state quenching (e.g., Stern-Volmer plots show that urate is a very effective quencher of DO3A-type complexes) phenomena.

12.3.4.1 Macrocyclic Complexes
The cyclen ligand framework has provided one of the ideal structural scaffolds upon which to build multidentate ligands suitable for Ln^{III} encapsulation, providing complexes with high kinetic and thermodynamic stability. The reaction chemistry of the cyclen ring-nitrogens allows functional groups to be added in a stepwise manner, thereby allowing a huge variety of ligands to be synthesised. A significant body of work has been undertaken by Parker that investigated the cellular imaging potential of a diverse range of monometallic Ln^{III}-based optical probes (visibly emissive Eu^{III} Φ up to 10% and Tb^{III} Φ up to 40%) complexes), all of which are based on the cyclen ligand framework [59]. Generally, each of the complexes possesses a sensitising chromophore (Figure 12.21; e.g. tetraazatriphenylene, azaxanthone, azathiaxanthone, pyrazolyl-azaxanthone [60], acridone [61]), which is covalently linked to the macrocyclic framework and can also coordinate to the Ln^{III} if desired. The remainder of the coordination sphere comprises either phosphinate, amide, or carboxylate donors; these groups, and the associated periphery of the ligand architectures, can be designed to control overall charge and influence lipophilicity.

The types of cell used in these studies were NIH-T3, CHO, and HeLa, and flow cytometry quantified cellular uptake. Typical incubations were conducted at 50 μM and cell loading correlated with approximately 10^8 complexes per cell; the dominant mechanism of uptake for these complexes is macropinocytosis (the formation of large endocytotic vesicles of irregular shape and size). The key discovery from these far-ranging studies is that the nature and linkage of the sensitising chromophore is the most important factor in determining cellular uptake and localisation; unlike the *d*-metal complexes described earlier, charge, lipophilicity, and donor group substituents are not a critical factor. The nature of the aromatic planar chromophore is often implicated in protein binding and thus, presumably, trafficking. The intracellular localisation profile that is observed for the majority of these macrocyclic Ln^{III} complexes is endosomal-lysosomal (confirmed through co-staining experiments with LysoTracker); generally the rates of uptake and egress are fast. A smaller number of complexes show fast uptake and slow egress and rapid shuttling between endosomal/lysosomal compartments and mitochondria (confirmed through co-staining experiments with MitoTracker), and such behaviour does not compromise the mitochondrial membrane potential (i.e., the complexes are nontoxic). Those complexes (Figure 12.22) that do localise in the mitochondria for long periods of time (up to 10 hrs) demonstrate lower IC$_{50}$ values [62].

Ln^{III} complexes that incorporate an *N*-coordinated azathiaxanthone sensitiser show localised preferences for protein rich domains (ribosomes, nucleoli). Unlike the majority of the complexes, these species are characterised by slow uptake, slow egress, and moderate toxicity (IC$_{50}$ 40–90 μM), which is attributed to the specific chemical nature of the antenna (i.e., the product of oxidative metabolism at sulphur, giving a sulfoxide or sulfone, the latter resulting in significant cytotoxicity; the azaxanthone structural analogues were shown not to be toxic in control experiments). The known DNA intercalating ability of the chromophore may determine the fate of the complexes, which appear only to penetrate compromised membranes, ultimately localising in areas of chromosomal DNA. In fact, monocationic complexes incorporating two azaxanthone-type chromophores in the 1- and 7-positions of the cyclen ring (Figure 12.23) have demonstrated selective staining of chromosomal DNA in dividing cells [63]. However, it is interesting to note that although the complexes possess low intrinsic

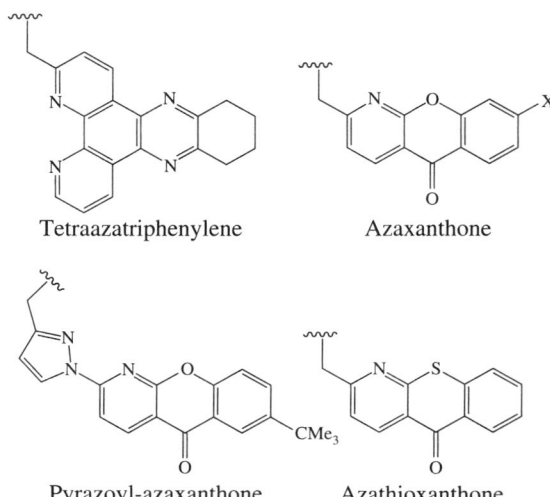

Tetraazatriphenylene   Azaxanthone

Pyrazoyl-azaxanthone   Azathioxanthone

FIGURE 12.21 Examples of the aromatic sensitising chromophores utilised in Parker's complexes.

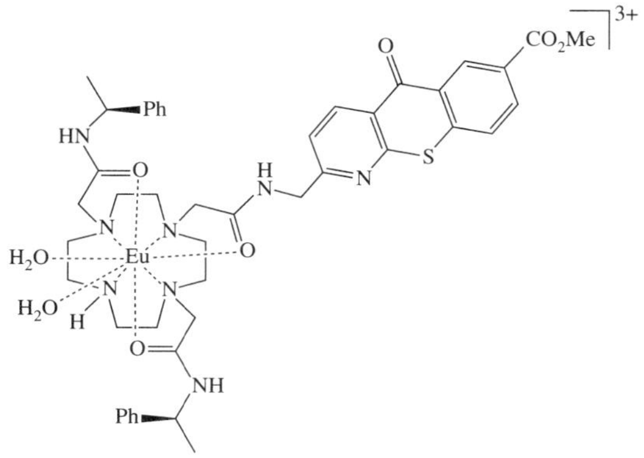

FIGURE 12.22 An example of a mitochondrial localising probe.

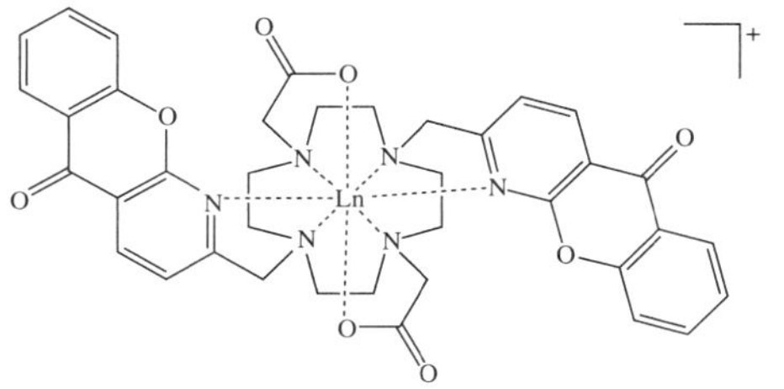

FIGURE 12.23 A 1,7-chromophorically substituted Ln(III) complex for staining chromosomal DNA.

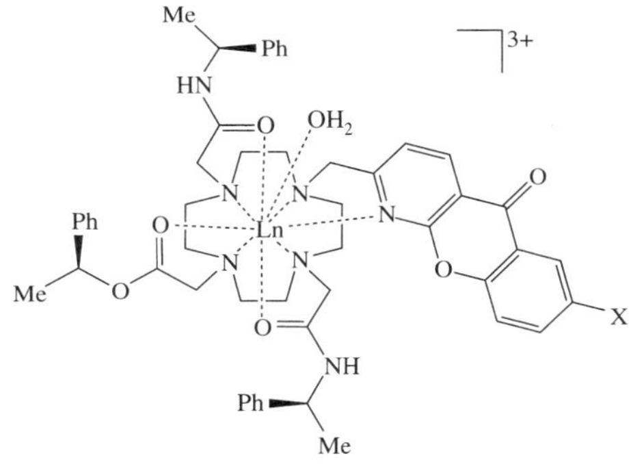

X = H; ^{t}Bu; CO_2Me; CO_2^-; CONHMe;
$CONHC_6H_{13}$; $CONHC_{12}H_{25}$;
CO-LysArg$_7$; CO-Arg$_8$;
CO-HSA; CO-guan.$_4$

FIGURE 12.24 Functionalising the periphery of the *N*-coordinated azaxanthone chomophore: hydrophilic (X = carboxylate and carboxamide), lipophilic (X = tertiary butyl, alkyl) and bio-inspired (X = LysArg, HSA, guan) variants.

cytotoxicity (IC$_{50}$ > 400 µM), single-photon illumination induces phototoxicity; it may be that for certain chromophores, two-photon absorption, to which these compounds are amenable, will reduce such phototoxic effects.

The importance of the appended chromophore in determining behaviour *in cellulo* was further demonstrated by a series of complex variants, wherein substituents added to the *N*-coordinated azaxanthone sensitiser dramatically influence the trafficking behaviour of the Ln^{III} probe. Relatively simple structural changes (Figure 12.24), that allow tuning of hydrophilic (carboxylate and carboxamide) through to lipophilic (tertiary butyl) character, demonstrate an element of control over cellular uptake, trafficking, localisation, and toxicity. More complex targeting vectors can also be conjugated to the azaxanthone antenna: Peptide conjugates promote rapid internalisation, cytosolic localisation, and lower toxicity; lipophilic oligo-guanidinium vectors induced apoptotic cell death (IC$_{50}$ 12 µM) following localisation within mitochondria [64].

Intriguingly, the ratiometric luminescence characteristics of Eu^{III} can be exploited using hyper-spectral analysis of microscopy images to signal changes in intracellular biochemical species in real time. Complexes with an amide-linked azaxanthone chromophore and two coordinated water molecules selectively and reversibly bind bicarbonate (for example, Scheme 12.5), intracellularly modulated by variation in external pCO$_2$; in a cellular context the complexes localise in the mitochondria of living cells and indicate a bicarbonate concentration of 10–30 mM [65].

Whilst numerous examples of dimetallic, bis-macrocyclic Ln^{III} complexes (Figure 12.25) have been reported for a variety of studies, the nature of the bridging unit can also be used advantageously in a cellular imaging context. Wong has reported the dimetallic Eu^{III} complex of a ligand, comprising two DO3A units linked via a bent biphenyl-methylene type bridging chromophore, that demonstrates binding to HSA (log K = 4.84) through an enhancement of sensitised (λ_{ex} = 350 nm) Eu^{III} emission. Altering the bridging chromophore to a linear dimethoxy-biphenyl bridge inhibited the binding ability of the complex. Confocal fluorescence microscopy imaging with HeLa cells (Figure 12.25) showed that both complexes are taken up by the cells and distributed in the cytoplasm, with the more lipophilic bent complex demonstrating much more rapid uptake and reduced cytotoxicity (IC$_{50}$ = 3 mM) [66].

12.3.4.2 Acyclic Complexes

The proclivity of cyclen-derived Ln^{III} complexes has not precluded the development of other ligand types for the biological exploitation of Ln^{III} complexes. Of the cellular imaging work conducted using Ln^{III} complexes of non-macrocyclic/cyclen type ligand frameworks, the most promising have been the developments by Bünzli of bimetallic, triple-stranded type helical complexes (discussed briefly above) of the general formula [Ln$_2$L$_3$], relatively high MW probes (>2500 Da) that are formally charge neutral. The bimetallic complexes can be formed through self-assembly under physiologically mimicked conditions [67]. Within each discrete complex the two Ln^{III} ions benefit from a nine-coordinate tri-capped trigonal prismatic coordination sphere, which encapsulates the ion tightly and prevents the approach of inner sphere water. Whilst these complexes would appear to be less water soluble than the more common (poly)amino-carboxylate type species, their hydrophilicity can be appropriately tuned through the addition of polyoxoethylene chains to the ligand periphery [68]. In fact, these ligands can be advantageously adorned (Figure 12.26) with a variety of substituents (e.g., positions R$^{1-3}$) that allow issues of solubility, optical/photophysical tuning, bioconjugation, and cell permeability to be effectively addressed, rendering this class of species particularly attractive in an imaging context. The Eu^{III} complexes benefit from good overall quantum yields and long millisecond lifetimes in water, whilst single-photon excitation wavelengths can be easily tuned toward 400 nm and thus biocompatibility with common 405 nm laser lines. The thermodynamic, photophysical, and biochemical attributes of homo-bimetallic helicates have proved to be sufficient for cellular imaging applications.

For the imaging studies, both cancerous (HeLa, MCF-7, HaCat) and non-cancerous (Jurkat) cell lines have been investigated. It was found that incubation concentrations could be as low as 5 µM, with uptake defined through endocytosis with the Eu^{III} complexes showing staining of the cytoplasm, where the helicates localise in lysosomes (which localise around the nucleus) and liposomes of the ER. Co-localisation experiments with known organic dyes (such as ER-Tracker, Blue-White DPX) suggest that the complex stains vesicles localised in the ER, rather than the Golgi apparatus [69]. Egress can be very slow and limited to around 30% over a 24 hr period, and the complexes have been shown to be non-cytotoxic. Grafting PEG chains (i.e., increasing hydrophilicity) does not appear to alter the uptake mechanism or localisation characteristics of the

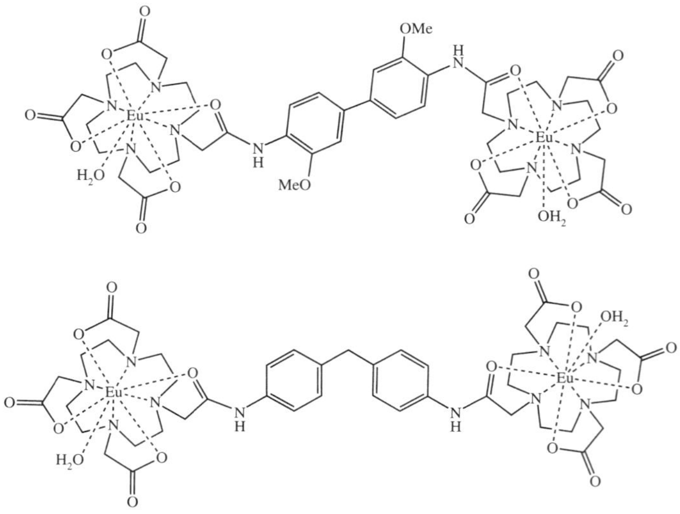

FIGURE 12.25 Dimetallic Eu^{III} complexes with bridging chromophores.

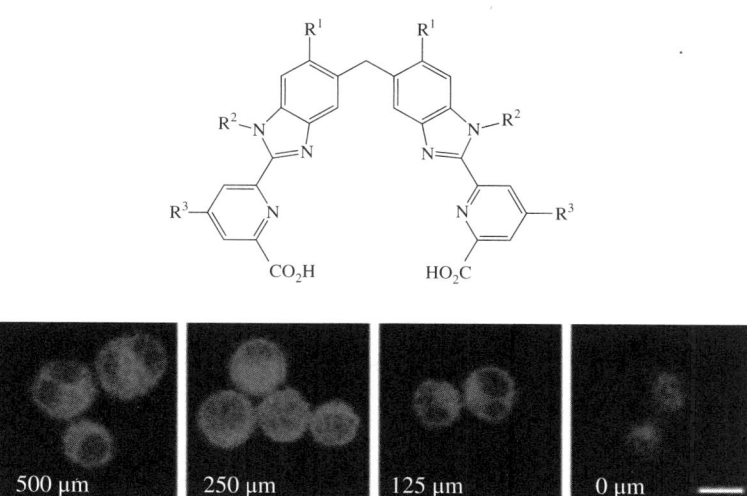

FIGURE 12.26 Ligand utilised for the bimetallic helicate [Eu$_2$L$_3$] (R^1 = H, R^2 = Me, R^3 = PEG chain). Cells were incubated in the presence of different concentrations of the helicate in RPMI-1640 for 24 h. The images were taken using a Zeiss LSM 500 META confocal microscope (λ_{ex} 405 nm). (*See insert for colour representation of the figure.*)

(a)   (b)   (c)

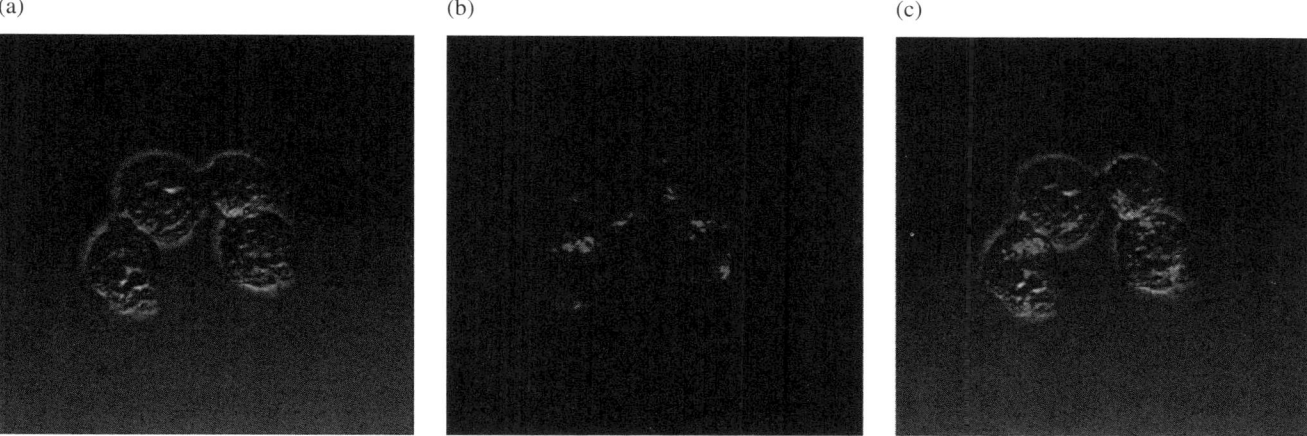

FIGURE 12.27 Two-photon microscopy images of HeLa cells incubated with 200 µM of a bimetallic Eu^{III} helicate [Eu$_2$(L)$_3$] in RPMI-1640 culture medium for 12 h at 37 °C, 5% CO$_2$: (a) bright field image; (b) luminescence (λ_{ex} = 750 nm, λ_{em} = 570 – 650 nm); (c) merged image. (*See insert for colour representation of the figure.*)

probes, suggesting that these complexes are less advantageous in terms of assessing cell functionality through organelle targeting. However, the intrinsic chirality of these bioprobes will also lend itself to potential application with CPL analyses. It should also be noted that the complexes are very amenable to multi-photon excitation, wherein both two-photon (Figure 12.27) and (uniquely) three-photon absorption have been demonstrated (i.e., excitation using NIR wavelengths) in an imaging context [70].

Two-photon imaging has also been shown with an architecturally simple Tb^{III} complex, based upon a tripodal ligand possessing a benzamide-type chromophore (Figure 12.33) with the formula [TbL(NO$_3$)$_3$]. Although the solid-state characterisation of the complex revealed a polymeric structure, in solution the complex was sufficiently emissive (Φ_{MeOH} = 0.11) and possessed a long millisecond lifetime (in HEPES) indicative of Tb^{III}-centred luminescence. Cellular studies were conducted with A549, HeLa, and HONE1, each of which was exposed to the complex at 20 µg/ml concentrations for different time durations; all showed good uptake of the complex. For the imaging studies, neither the free ligand nor the complex absorbs light at 400–800 nm, but excitation at 800 nm induces Tb^{III}–centred emission at 480 nm, indicative of a multi-photon process, which was subsequently confirmed via power-dependence experiments. The cellular imaging (Figure 12.28) showed that the green Tb^{III} signal was predominantly localised in the cytoplasm, but not the nuclei. Assessments of the cytotoxic effect of the complex (MTT assay) showed that at the concentration used for imaging the cell lines were viable over a 24 hr period [71].

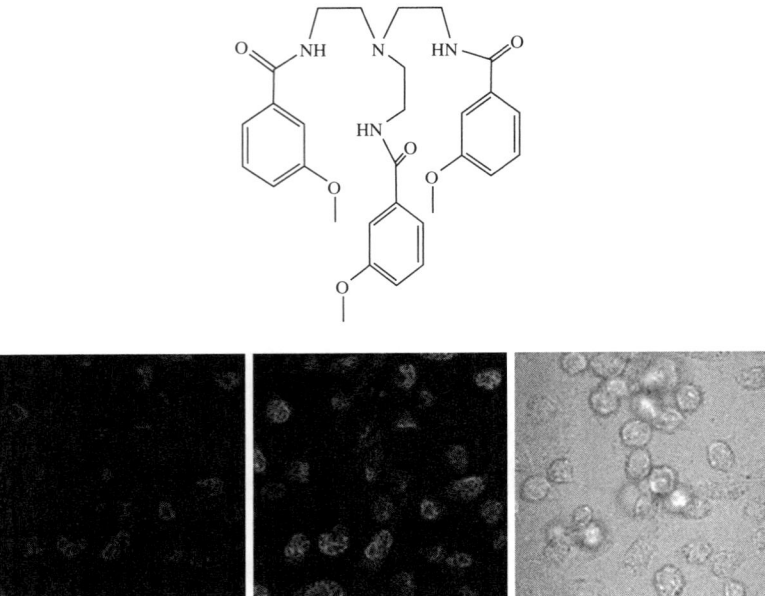

FIGURE 12.28 Top eft: Free ligand. Bottom: Multi-photon confocal microscopy images of the terbium complexes. Hoechst 33342-labelled nuclei (blue) in the HONE1 cells with excitation at 800 nm, Filter BP = 400–480 nm (left); (middle) terbium complexes loaded into the cell for 1 hour (Figure 2c) with nuclei labelled, filter BP = 400–615 nm; (right) bright field image of the HONE1 cells. (*See insert for colour representation of the figure.*)

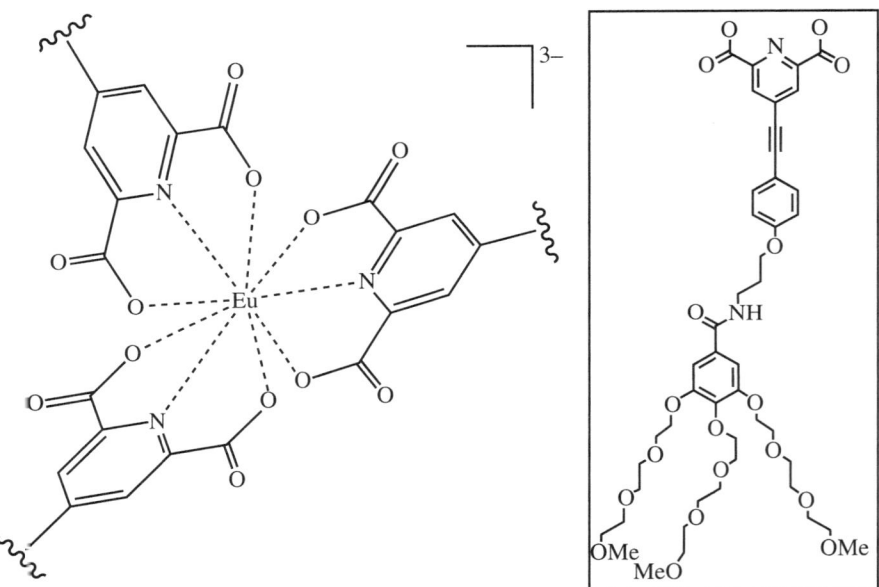

FIGURE 12.29 The tris picolinate Ln(III) complex core (left) and free ligand (inset).

The utility of two-photon sensitisation of water-soluble Eu^{III} complexes of the form [Eu(L)$_3$]$^{3-}$ has also been demonstrated by Maury, with ligands derived from 2,6-pyridine dicarboxylic acid (Figure 12.29). Additional functionality was achieved via an alkylphenylacetylene bridge, which serves dually to extend the π-conjugation and provide improved hydrophilicity through terminating tris(triethyleneglycol)phenyl groups. The absorption properties of the Eu^{III} complex are dominated by ligand-centred transitions, with the longest wavelength contribution (332 nm) assigned to a CT (alkoxy-to-pyridine) type electronic absorption. The complex was strongly emissive (Φ ~16%) with a long lifetime in water (1.06 ms) characteristic of encapsulated Eu^{III} and a maximal two-photon absorption at 700 nm. T24 cancer cells were incubated with the [Eu(L)$_3$]$^{3-}$ complex and imaged using two-photon excitation at 760 nm ($\sigma_{2PA}(760) = 19$ GM). Despite the trianionic nature of the complex, intracellular localisation was observed in the perinuclear region and distributed like ER; bright red spots were also observed in the nucleus (which could be indicative of nucleoli targeting) [72].

Acyclic derivatives based upon the DTPA ligand core have been utilised for targeting and imaging intracellular zinc (Figure 12.30). The design rationale for such a species requires a dual functioning binding site with good affinity for Zn^{II} and an ability to sensitise the chelated Eu^{III}. The DTPA core was appended with a bridging quinoline-type chromophore, which was further functionalised with a dipicolylamine unit. The quinoline unit provides a slightly longer wavelength of absorption in comparison to pyridyl, facilitating excitation at 320 nm. Although the emissivity associated with the free Eu^{III} complex was much less favourable (<1%) than those discussed above, binding Zn^{II} induced a favourable 8.2-fold increase in quantum yield. The use of luminescence lifetime measurements and water relaxivity measurements on the corresponding Gd^{III} analogue show that the lanthanide coordination sphere is not perturbed by the Zn-binding event. More recent developments have shown that it is now possible to design Zn-responsive probes perturbing inner sphere Ln^{III} hydration and thus induce more pronounced changes in optical output [73, 74]; similar concepts have now been adopted to target copper [75]. The utility of the Eu^{III} complex to image zinc in a cellular context was demonstrated using HeLa cells. The complex was injected into the cell, and under ambient levels of zinc, the cells showed no significant emission that could be attributed to the Eu^{III} complex. However, upon addition of a zinc ionophore (pyrithione) and ZnSO$_4$ (added up to 1 equivalent) the cell image brightened considerably (Figure 12.30), which was attributed to an increase in emission from the Zn-bound Eu^{III} complex. Further, the effect was reversible: Addition of *N,N,N,N*-tetra(2-picolyl)-ethylene diamine (a membrane permeable chelator) resulted in a subsequent loss of cellular fluorescence [76].

An emissive Eu^{III} complex has also demonstrated promise as a selective probe for imaging singlet oxygen over other ROS. The complex is based upon an aminocarboxylate-derived 2,2′:6′,2″-terpyridine ligand, which is substituted with a 9-anthryl unit (Figure 12.31). In its normal form, the complex is weakly luminescent; however, upon exposure to singlet oxygen, the complex becomes very emissive with long-lived luminescence characteristic of Eu^{III}-centred processes. The deactivation of the non-radiative quenching pathways (i.e., the 'switching on' of the emission) occurs via the anthryl moiety, which reacts with singlet oxygen, generating an endoperoxide. Methyl-substitution in the 10-position of the anthracene group significantly accelerates the rate of this reaction ($\sim K$ 10^{10} M^{-1} s^{-1}) and thus increases the sensitivity of the probe. The application to time-gated luminescence imaging microscopy has been demonstrated using HeLa cells co-incubated with the complex and ubstituted porphyrin as photosensitser, the latter providing *in cellulo* generation of singlet oxygen upon selective irradiation. The photosensitiser predominantly localises in the nuclei, and it is these regions that demonstrated the fastest increases in

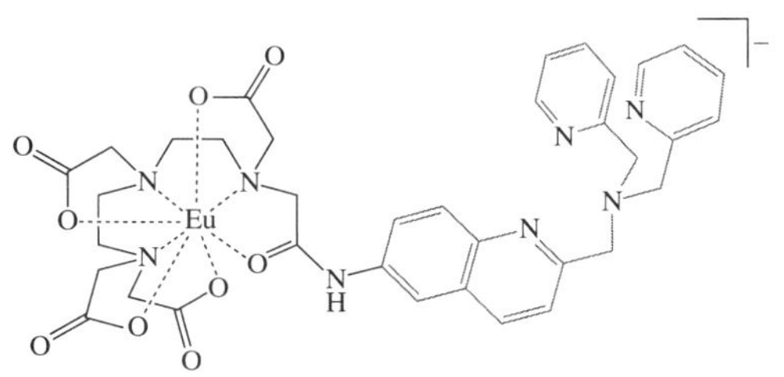

FIGURE 12.30 Eu^{III} complex structure of the zinc-responsive probe.

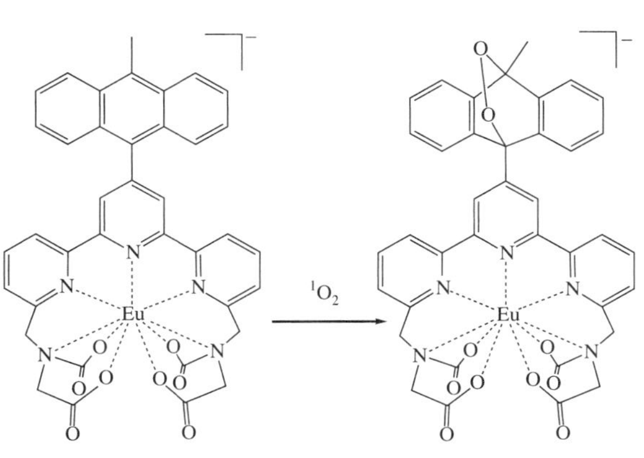

FIGURE 12.31 A singlet oxygen reactive Eu^{III} complex.

emission intensity following irradiation of the photosensitiser (and generation of singlet oxygen). It is noteworthy that irradiation did not induce apoptosis or necrosis, suggesting that the rapid reactivity of the Eu^{III} complex may actually inhibit reaction of singlet oxygen with other cell components [77].

12.4 CONCLUSIONS

Luminescent lanthanide complexes are applicable to a variety of optical imaging applications. The numerous opportunities afforded through targeted ligand design have demonstrated the wide utility of these complexes from luminescent labels of tissue to intracellular probes with defined localisation profiles. Perhaps the most intriguing attributes are revealed in the photophysically responsive nature of Ln^{III} complexes (particularly the emission of Eu^{III}) toward analytes of biological significance. Because the long-lived luminescent lifetimes are ideal for time-resolved luminescence microscopy, the application of lifetime imaging will be invaluable in a noninvasive biological context; combination of these optical attributes with related Gd^{III}-based contrast agents clearly has implications to dual modality imaging assessments [78]. All of the examples above have focused on the imaging ability of visibly emitting Ln^{III} ions, but the future development of probes for deeper tissue imaging requires consideration of the NIR-emitting lanthanides. Because Nd^{III} complexes are extremely sensitive to excited state quenching from C–H, as well as O–H and N–H oscillators, Yb^{III} might represent the greatest opportunity for development in this area: Engineering good emissivity from such complexes is an active area of research. Maximising the efficiency of energy transfer is a clear objective, whilst inducing fast radiative deactivation of Yb^{III} can yield brightly emissive complexes; elucidating the role of the radiative lifetime may well illuminate further developments in this area [79].

REFERENCES

[1] J. B. Pawlet (Ed.), *Handbook of Biological Confocal Microscopy*, Springer, New York (2006).

[2] J. R. Lakowicz, *Principles of Fluorescence Spectroscopy*, Springer, New York (2006).

[3] V. Fernández-Moreira, F. L. Thorp-Greenwood and M. P. Coogan, *Chem. Commun.* **46**, 186–202 (2010).

[4] Q. Zhao, C. H. Huang and F. Y. Li, *Chem. Soc. Rev.* **40**, 2508–2524 (2011).

[5] J-C. G. Bunzl, *Chem. Rev.* **110**, 2729–2755 (2010).

[6] K. K. W. Lo, S. P. Y. Li and K. Y. Zhang, *New J. Chem.* **35**, 265–287 (2011).

[7] M. X. Yu, Q. Zhao, L. X. Shi, F. Y. Li, Z. G. Zhou, H. Yang, T. Yi and C. H. Huang, *Chem. Commun.* 2115–2117 (2008).

[8] K. Y. Zhang and K. K-W. Lo, *Inorg. Chem.* **48**, 6011–6025 (2009).

[9] J. S.-Y. Lau, P.-K. Lee, K. H.-K. Tsang, C. H.-C. Ng, Y.-W. Lam, S.-H. Cheng and K. K.-W. Lo, *Inorg. Chem.* **48**, 708–718 (2009).

[10] K. Y. Zhang, S. P.-Y. Li, N. Zhu, I. W.-S. Or, M. S.-H. Cheung, Y.-W. Lam and K. K.-W. Lo, *Inorg. Chem.* **49**, 2530–2540 (2010).

[11] L. Murphy, A. Corgreve, L.-O. Pålsson and J. A. G. Williams, *Chem. Commun.* **46**, 8743–8745 (2010).

[12] S. K. Leung, K. Y. Kwok, K. Y. Zhang and K. K. W. Lo, *Inorg. Chem.* **49**, 4984–4995 (2010).

[13] K. A. Stephenson, S. R. Banerjee, T. Besenger, O. O. Sogbein, M. K. Levadala, N. McFarlane, J. A. Lemon, D. R. Boreham, K. P. Maresca, J. C. Brennan, J. W. Babich, J. Zubieta and J. F. Valliant, *J. Am. Chem. Soc.* **126**, 8598–8599 (2004).

[14] N. Viola-Villegas, A. E. Rabideau, M. Bartholoma, J. Zubieta and R. Doyle, *J. Med. Chem.* **52**, 5253–5261 (2009).

[15] S. R. Banerjee, J. W. Babich and J. Zubieta, *Chem. Commun.* 1784–1786 (2005).

[16] T. Esteves, C. Xavier, S. Gama, F. Mendes, P. D. Raposinho, F. Marques, A. Paulo, J. C. Pessoa, J. Rino, G. Viola and I. Santos, *Org. Biomol. Chem.* **8**, 4104–4116 (2010).

[17] D. J. Stufkens and A. Vlcek, Jr, *Coord. Chem. Rev.* **177**, 127–179 (1998).

[18] A. J. Amoroso, M. P. Coogan, J. E. Dunne, V. Fernández-Moreira, J. B. Hess, A. J. Hayes, D. Lloyd, C. Millet, S. J. A. Pope and C. Williams, *Chem. Commun.* 3066–3068 (2007).

[19] R. G. Balasingham, M. P. Coogan and F. L. Thorp-Greenwood, *Dalton Trans.* 2011, **40**, 11663–11674 (2011)

[20] A. J. Amoroso, R. J. Arthur, M. P. Coogan, J. B. Court, V. Fernandez-Moreira, A. J. Hayes, D. Lloyd, C. Millet, S. J. A. Pope and C. Williams, *New J. Chem.* **32**, 1097–1102 (2008).

[21] K. K.-W. Lo, M.-W. Louie, K.-S. Sze and J. S.-Y. Lau, *Inorg. Chem.* **47**, 602–611 (2008).

[22] V. Fernández-Moreira, F. L. Thorp-Greenwood, A. J. Amoroso, J. Cable, J. B. Court, V. Gray, A. J. Hayes, R. L. Jenkins, B. M. Kariuki, D. Lloyd, C. O. Millet, C. Ff. Williams and M. P. Coogan, *Org. Biomol. Chem.* **8**, 3888–3901 (2010).

[23] E. Ferri, D. Donghi, M. Panigati, G. Prencipe, L. D'Alfonso, I. Zanoni, C. Baldoli, S. Maiorana, G. D'Alfonso and E. Licandro, *Chem. Commun.* **46**, 6255–6257 (2010).

[24] A. Juris, V. Balzani, F. Barigellatti, S. Campagna, P. Belser and A. von Zelewsky, *Coord. Chem. Rev.* **84**, 85–277 (1988).

[25] M. Stucker, L. Schulze, G. Pott, P. Hartmann, D. W. Lubbers, A. Rochling and P. Altmeyer, *Sensors and Actuators* **51**, 171–175 (1998).

[26] C. V. Kumar, J. K. Barton, and N. J. Turro, *J. Am. Chem. Soc.* **107**, 5518–5523 (1985).

[27] F. L. Thorp-Greenwood, M. P. Coogan, L. Mishra, N. Kumari, G. Rai and S. Saripella, *New J. Chem.* **36**, 64–72 (2012).

[28] H. C. Gerritsen, R. Sanders, A. Draijer, C. Ince and Y. K. Levine, *J. Fluorescence* **7**, 11–15 (1997).

[29] J. W. Dobrucki, *J. Photochem. Photobiol. B* **65**, 136–144 (2001).

[30] U. Neugebauer, Y. Pellegrin, M. Devocelle, R. J. Forster, W. Signac, N. Morand and T. E. Keyes, *Chem. Commun.* 5307–5309 (2008).

[31] C. A. Puckett and J. K. Barton, *J. Am. Chem. Soc.* **131**, 8738–8739 (2009).

[32] C. A. Puckett and J. K. Barton, *J. Am. Chem. Soc.* **129**, 46–47 (2007).

[33] C. A. Puckett and J. K. Barton, *Biochemistry* **47**, 11711–11716 (2008).

[34] M. Matson, F. R. Svensson, B. Norden and P. Lincoln, *J. Phys. Chem. B* **115**, 1706–1711 (2011).

[35] K. K. W. Lo, T. K. M. Lee, J. S. Y. Lau, W. L. Poon and S. H. Cheng, *Inorg. Chem.* **47**, 200–208 (2008).

[36] T. S. Yang, A. Xia, Q. Liu, M. Shi, H. Z. Wu, L. Q. Xiong, C. H. Huang and F. Y. Li, *J. Mater. Chem.* **21**, 5360–5367 (2011).

[37] S. W. Botchway, M. Charnley, J. W. Haycock, A. W. Parker, D. L. Rochester, J. A. Weinstein and J. A. G. Williams, *Proc. Natl. Acad. Sci. U. S. A.* **105**, 16071–16076 (2008).

[38] P. Wu, E. L. M. Wong, D. L. Ma, G. S. M. Tong, K. M. Ng and C. M. Che, *Chem. Eur. J.* **15**, 3652–3656 (2009).

[39] C. K. Koo, K. L. Wong, C. W. Y. Man, Y. W. Lam, L. K. Y. So, H. L. Tam, S. W. Tsao, K. W. Cheah, K. C. Lau, Y. Y. Yang, J. C. Chen and M. H. W. Lam, *Inorg. Chem.* **48**, 872–878 (2009).

[40] C. P. Bagowski, Y. You, H. Scheffler, D. H. Vlecken, D. J. Schmitz and I. Ott, *Dalton Trans.* 10799–10805 (2009).

[41] P. J. Barnard, M. V. Baker, S. J. Berners-Price and D. A. Day, *J. Inorg. Biochem.* **98**, 1642–1647 (2004).

[42] S. I. Pascu, P. A. Waghorn, T. D. Conry, B. Lin, H. M. Betts, J. R. Dilworth, R. B. Sim, G. C. Churchill, F. I. Aigbirhio and J. E. Warren, *Dalton Trans.* 2107–2110 (2008).

[43] S. Faulkner, S. J. A. Pope and B. P. Burton-Pye, *Appl. Spec. Rev.* **40**, 1–31 (2005).

[44] J.-C. G. Bünzli and C. Piguet, *Chem. Soc. Rev.* **34**, 1048–1077 (2005).

[45] T. Gunnlaugsson, J. P. Leonard, K. Senechal and A. J. Harte, *J. Am. Chem. Soc.*, **125**, 12062–12063 (2003).

[46] C. M. G. dos Santos, A. J. Harte, S. J. Quinn and T. Gunnlaugsson, *Coord. Chem. Rev.* **252**, 2512–2527 (2008).

[47] S. Pandya, J. Yu and D. Parker, *Dalton Trans.* 2757–2766 (2006).

[48] T. Gunnlaugsson and J. P. Leonard, *Chem. Commun.* 3114–3131 (2005).

[49] E. J. New, D. Parker and R. D. Peacock, *Dalton Trans.* 672–679 (2009).

[50] C. P. Montgomery, B. S. Murray, E. J. New, R. Pal and D. Parker, *Acc. Chem. Res.* **42**, 925–937 (2009).

[51] E. G. Moore, A. P. S. Samuel and K. N. Raymond, *Acc. Chem. Res.* **42**, 542–552 (2009).

[52] D. J. Bornhop, D. S. Hubbard, M. P. Houlne, C. Adair, G. E. Kiefer, B. C. Pence and D. L. Morgan, *Anal. Chem.* **71**, 2607–2615 (1999).

[53] B. McMahon, P. Mauer, C. P. McCoy, T. C. Lee and T. Gunnlaugsson, *J. Am. Chem. Soc.* **131**, 17542–17543 (2009).

[54] M. Lee, M. S. Tremblay, S. Jockusch and N. J. Turro, *Org. Lett.* **13**, 2802–2805 (2011).

[55] A. Beeby, S. W. Botchway, I. M. Clarkson, S. Faulkner, A. W. Parker, D. Parker and J. A. G. Williams, *J. Photochem. Photobiol. B*, **57**, 83–89 (2000).

[56] N. Weibel, L. J. Charbonnière, M. Guardigli, A. Roda and R. Ziessel, *J. Am. Chem. Soc.* **126**, 4888–4896 (2004).

[57] V. Fernández-Moreira, B. Song, V. Sivagnanam, A-S. Chauvin, C. D. B. Vandevyver, M. Gijs, I. Hemmilä, H-A. Lehr and J-C. G. Bünzli, *Analyst* **135**, 42–52 (2010).

[58] Z. Ye, J. Chen, G. Wang and J. Yuan, *Anal. Chem.* **83**, 4163–4169 (2011).

[59] C. P. Montgomery, B. S. Murray, E. J. New, R. Pal and D. Parker, *Acc. Chem. Res.* **42**, 925–937 (2009).

[60] P. Atkinson, K. S. Findlay, F. Kielar, R. Pal, D. Parker, R. A. Poole, H. Puschmann, S. L. Richardson, P. A. Stenson, A. L. Thompson and J. Yu, *Org. Biomol. Chem.* **4**, 1707–1722 (2006).

[61] A. Dadabhoy, S. Faulkner and P. G. Sammes, *J. Chem. Soc., Perkin Trans.* **2**, 348–357 (2002).

[62] C. P. Montgomery, B. S. Murray, E. J. New, R. Pal and D. Parker, *Acc. Chem. Res.* **42,** 925–937 (2009).

[63] G-L. Law, C. Man, D. Parker and J. W. Walton, *Chem. Commun.* **46**, 2391–2393 (2010).

[64] C. P. Montgomery, B. S. Murray, E. J. New, R. Pal and D. Parker, *Acc. Chem. Res.* **42**, 925–937 (2009).

[65] D. G. Smith, G-L. Law, B. S. Murray, R. Pal, D. Parker and K-L. Wong, *Chem. Commun.* **47**, 7347–7349 (2011).

[66] Y. O. Fung, W. Wu, C-T. Yeung, H-K. Kong, K. K-C. Wong, W-S. Lo, G-L. Law, K-L. Wong, C-K. Lau, C-S. Lee and W-T. Wong, *Inorg. Chem.* **50**, 5517–5525 (2011).

[67] E. Deiters, B. Song, A-S. Chauvin, C. D. B. Vandevyver, F. Gumy and J-C. G. Bünzli, *Chem. Eur. J.* **15**, 885–900 (2009).

[68] C. D. B. Vandevyver, A-S. Chauvin, S. Comby and J-C. G. Bünzli, Chem. Commun. 1716—1718 (2007).

[69] B. Song, C. D. B. Vandevyver, A-S. Chauvin and J-C. G. Bünzli, *Org. Biomol. Chem.* **6**, 4125–4133 (2008).

[70] S. V. Eliseeva, G. Auböck, F. van Mourik, A. Cannizzo, B. Song, E. Deiters, A-S. Chauvin, M. Chergui and J.-C.G. Bünzli, *J. Phys. Chem. B* **114**, 2932–2937 (2010).

[71] G-A. Law, K-L. Wong, C.W-Y. Man, W-T. Wong, S-W. Tsao, M. H-W. Lam and P. K-S. Lam, *J. Am. Chem. Soc.* **130**, 3714–3715 (2008).

[72] A. Picot, A. D'Aleo, P. L. Baldeck, A. Grichine, A. Duperray, C. Andraud and O. Maury, *J. Am. Chem. Soc.* **130**, 1532–1533 (2008)

[73] S. J. A. Pope and R. H. Laye, *Dalton Trans.*, 3108–3113 (2006).

[74] J. L. Major, R. M. Boiteau and T. J. Meade, *Inorg. Chem.*, **47**, 10788–10795 (2008).

[75] E. L. Que and C. J. Chang, *J. Am. Chem. Soc.* **128**, 15942–15943 (2006)

[76] K. Hanaoka, K. Kikuchi, H. Kojima, Y. Urano and T. Nagano, *J. Am. Chem. Soc.* **126**, 12470–12476 (2004).

[77] B. Song, G. Wang, M. Tan and J. Yuan, *J. Am. Chem. Soc.* **128**, 13442–13450 (2006).

[78] E. J. New, D. Parker, D. G. Smith and J. W. Walton, *Curr. Opinion Chem. Biol.* **14**, 238–246 (2010).

[79] N. M. Shavaleev, R. Scopelliti, F. Gumy and J-C. G. Bünzli, *Inorg. Chem.* **48**, 7937–7946 (2009).

13

LANTHANIDE-BASED UPCONVERSION NANOPHOSPHORS FOR BIOIMAGING

FUYOU LI, WEI FENG, JING ZHOU AND YUN SUN

Department of Chemistry, Fudan University, Shanghai, China

13.1 INTRODUCTION

Lanthanide-based upconversion nanomaterials have attracted much attention in recent years because of their unique ability to generate visible or near-infrared (NIR) upconversion emission with excitation at near-infrared wavelengths [1]. For bioimaging applications, this unique upconversion luminescent property promises low background signals and deep penetration [2–4], which are key factors in determining the imaging quality in luminescence imaging techniques. Together with the high photostability and low phototoxicity originating from the inorganic nature of the lattice, lanthanide-based upconversion nanomaterials are considered candidates for a new generation of biolabels for luminescent imaging [5–8].

The upconversion process is different from the conventional luminescent process, as described in Scheme 13.1 [1]. After being excited to state *1* from ground state *0* by absorbing one excitation photon, the conventional luminescent process includes a non-radiative relaxation process to state *2*, followed by the radiative transition back to the ground state and the generation of an emission photon. Due to the law of conservation of energy, the energy of the emission photon is less than that of the absorbed excitation photon, which brings about a Stokes shift. However, in the upconversion process, the excited state *1* can absorb energy from another excitation photon to generate excited state *2*, followed by the radiative transition to generate an anti-Stokes shift emission photon.

The upconversion mechanism can be divided into three main classes: excited state absorption, energy transfer upconversion (ETU), and photon avalanche. Because all of these mechanisms require relatively high stability of the intermediate excited state (such as state *1* in Scheme 13.1b), lanthanide ions are the most commonly used luminescent centres for upconversion luminescence as a result of their abundant f-electron configurations and stable excitation state. Detailed discussion about these mechanisms can be found in the review articles and references therein [1, 9].

In light of the rapid development of the field concerning lanthanide-based upconversion nanophosphors (Ln-UCNPs), numerous works have described their synthesis, surface modification, and bioimaging applications [10, 11]; several review articles summarise recent progress [5–8]. Herein we will introduce the main methods and techniques developed in recent years that promote the bioimaging application of Ln-UCNPs. Some typical examples will be included to describe the methods or techniques in detail.

13.2 FABRICATION OF Ln-UCNPs SUITABLE FOR BIOIMAGING

An ideal luminescent biolabel must have the following unique properties: high luminescent efficiency to improve the sensitivity; uniform size, shape, and luminescent properties; suitable surface properties, which enable the dispersion and functionalisation of biomolecules in biological surroundings; and low cytotoxicity in living systems [5].

The Chemistry of Molecular Imaging, First Edition. Edited by Nicholas Long and Wing-Tak Wong.
© 2015 John Wiley & Sons, Inc. Published 2015 by John Wiley & Sons, Inc.

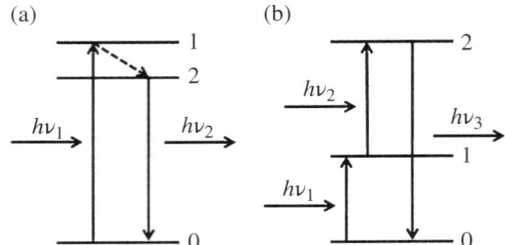

SCHEME 13.1   Schematic illustration of the conventional (a) and upconversion (b) luminescence processes.

In order to satisfy the above requirements, numerous techniques have been adopted in the fabrication of UCNPs for bioimaging applications, and these will be discussed herein.

13.2.1   General Composition for UCNPs

For nano-sized upconversion materials, the most commonly used system features Yb^{3+}-sensitised doping materials, with Er^{3+}, Tm^{3+}, or Ho^{3+} as emitters. Yb^{3+} can absorb energies from the 980 nm laser excitation, and the subsequent energy transfer upconversion (ETU) process populates the excitation state of emitters and generated upconversion emissions. In particular, the transitions of $^4F_{5/2} \rightarrow {}^4I_{15/2}$, $^2P_{3/2} \rightarrow {}^4I_{11/2}$, $^4F_{7/2} \rightarrow {}^4I_{15/2}$, $^2H_{11/2} \rightarrow {}^4I_{15/2}$, $^4S_{3/2} \rightarrow {}^4I_{15/2}$, and $^4F_{9/2} \rightarrow {}^4I_{15/2}$ of Er^{3+} give emissions at 450, 470, 495, 525, 550, and 660 nm, respectively. The transitions of $^3P_0 \rightarrow {}^3F_4$, $^1D_2 \rightarrow {}^3H_6$, $^1G_4 \rightarrow {}^3H_6$, $^3F_3 \rightarrow {}^3H_6$, and $^3H_4 \rightarrow {}^3H_6$ of Tm^{3+} generate emissions at 353, 368, 490, 695, and 800 nm, respectively, whilst the transitions of $^4F_3 \rightarrow {}^5I_8$, $^5F_4$, $^5S_2 \rightarrow {}^5I_8$, $^5F_5 \rightarrow {}^5I_8$, and $^5F_4$, $^5S_2 \rightarrow {}^5I_7$ of Ho^{3+} bring out emissions at 486, 541, 647, and 751 nm, respectively. The combination of these three doping ions can generate various colours for upconversion labelling (Figure 13.1) [5, 11, 12].

For the selection of the host matrix, rare earth-based materials are optimal because rare earth ions have similar properties, thus enabling the effective doping of sensitiser and emitter ions in a wide concentration range. Therefore, the optically inert rare-earth ions with full or half-full valence shells such as Y^{3+}, Gd^{3+}, and Lu^{3+} are commonly used cations in the host materials. In selecting anions for the host materials, the final materials should have low solubility in water and other solvents, which minimises the deliquescence effect, thus improving the stability of the materials. Until now, rare-earth halides, oxides, oxy-sulfides, phosphates, and vanadates have been successfully used as host materials for the upconversion luminescence process. Among them, rare-earth fluorides are considered to be one of the best choices because of their transparency in the NIR and visible range, low phonon energies (minimising the non-radiative process), and high stability compared to other halides [13].

13.2.2   Synthetic Technique

By virtue of the significant developments in nanotechnology in the past few decades, the synthesis of nano-sized upconversion materials has been achieved by various methods (Figure 13.2). The fine-tuning of these methods can control the composition, phase, size, and shape, as well as the surface ligands of UCNPs, which influence the upconversion properties and potential applications.

Because the commonly used host materials for Ln-UCNPs are insoluble in most solvents, co-precipitation is the easiest method to obtain such nanoparticles by mixing the cations and anions together in the solvent. Capping ligands are usually employed in the reaction to hinder the aggregation and control the growth of nanoparticles.

In 2004, $NaYF_4$:Yb,Er/Tm nanoparticles were synthesised by Yi and co-workers via a simple co-precipitation method [19]. NaF was used as the precipitator and ethylenediaminetetraacetic acid (EDTA) was employed as the capping agent to control the size of the products. An annealing process promoted the cubic-to-hexagonal phase transition and enhanced the upconversion luminescence emissions. Many other upconversion nanoparticles, such as Y_2O_3, Gd_2O_3, Lu_2O_3, $Gd_4O_3F_6$, $Gd_3Ga_5O_{12}$, GdOF, $BaYF_5$, $NaGdF_4$, $LuPO_4$, and $YbPO_4$ have also been synthesised via similar co-precipitation methods by the careful selection of precipitators [5].

The hydro(solvo)thermal method is also based on the principle of co-precipitation. The difference is that the relatively violent reaction conditions accelerate the crystallisation and recrystallisation rates, thus leading to kinetically stable products with high crystallinity. Taking $NaREF_4$, which is regarded as the most efficient host material for the upconversion luminescence (UCL) process as an example, hexagonal-phased Ln-UCNPs will be obtained after extended hydrothermal treatment, although cubic phase materials can be formed at the beginning of the reaction. In a common hydrothermal process, rare-earth ions and anion sources are added to the solvent, and cubic phase precipitations will immediately form. After hydrothermal treatment, the recrystallisation process occurs with the formation of hexagonal Ln-UCNPs. The size and shape can be fine-tuned by controlling reaction conditions such as temperature, time, and capping agents [14, 20].

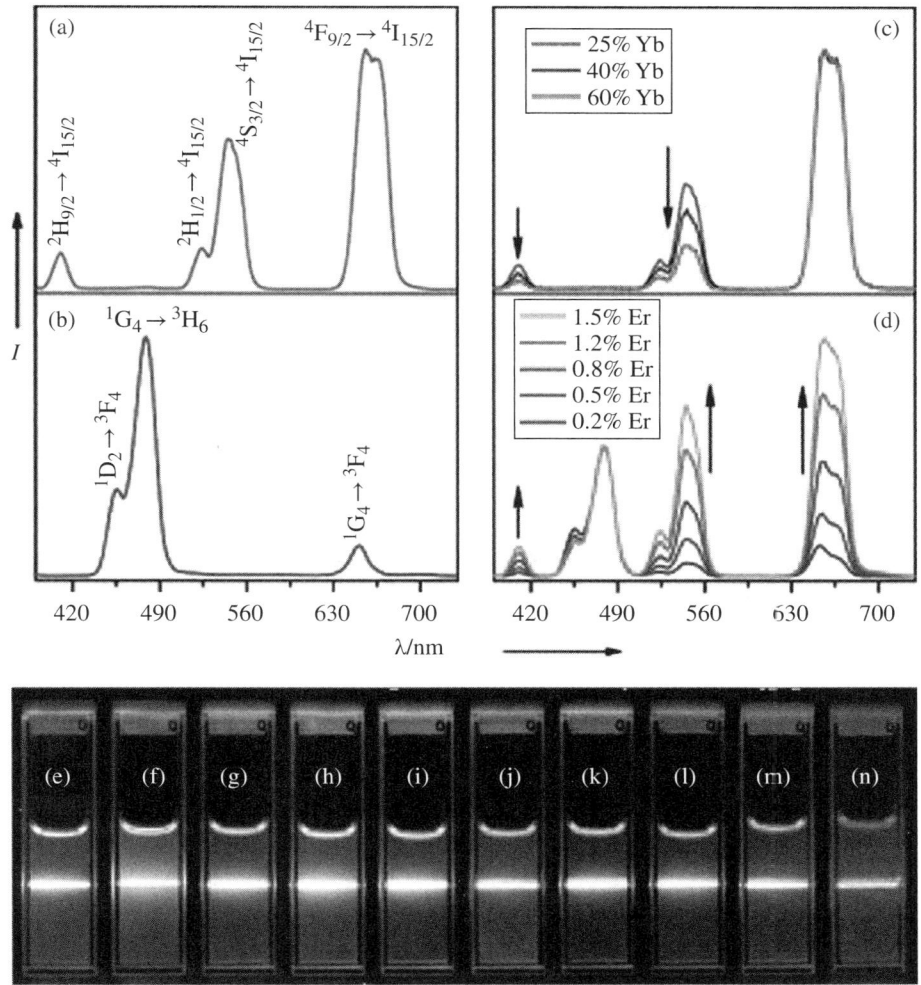

FIGURE 13.1 UCL emission spectra of (a) NaYF$_4$:Yb,Er (18/2 mol%), (b) NaYF$_4$:Yb,Tm (20/0.2 mol%), (c) NaYF$_4$:Yb,Er (25–60/2 mol%), and (d) NaYF$_4$:Yb,Tm/Er (20/0.2/0.2–1.5 mol%) particles in ethanol. (e)–(n) are compiled luminescent photos showing corresponding colloidal solutions of the samples shown in (a)–(c). Reprinted with permission from Ref. [12]. Copyright 2008 American Chemical Society. (*See insert for colour representation of the figure.*)

In order to obtain monodispersed Ln-UCNPs, a thermolysis method was recently developed to synthesise Ln-UCNPs. Organic solvents with high boiling points are employed to provide a high-temperature reaction medium. The concentration of capping agents can reach a high level in these hydrophobic solvents to form a metal ion buffer. For example, Yan and co-workers used 1-octadecene as the solvent, and oleic acid (OA) as the capping agent for the synthesis of series of rare earth compounds [21, 22]. To obtain NaREF$_4$–based UCNPs, the corresponding rare-earth trifluoroacetates can be employed as the precursors. After heating in the solvents, rare-earth ions are coordinated by the excess OA ligands to form a buffer system, in which the concentration of 'naked' metal ions is maintained at a stable level. The concentration of F$^-$ can also be controlled by adjusting the rate of the decomposition of trifluoroacetate ions. Careful control of the concentration of rare-earth and fluoride ions results in good uniformity of the products [16, 23]. Trioctylphosphine oxide (TOPO) and polyethylene (PEG) can be also employed as solvents, and oleylamine (OM) can be used as a capping agent. A similar method is used to synthesise various UCNPs such as NaGdF$_4$, NaYbF$_4$, LiYF$_4$, LiREF$_4$, KY$_3$F$_{10}$, BaGdF$_5$, Y$_2$O$_3$, ZrO$_2$, and GdOF [5].

As a variation, F$^-$ can be supplied as ions directly to the reaction system in the thermal decomposition protocol. Because the concentration of F$^-$ used in this method is much larger than that of the 'naked' rare earth ions, the variation will be minimal and will not affect the uniform growth of Ln-UCNPs. Using this method, Zhang and co-workers employed rare-earth chloride NaOH and NH$_4$F as rare earth, Na$^+$, and F$^-$ sources, respectively, to synthesise hexagonal-phased NaREF$_4$-

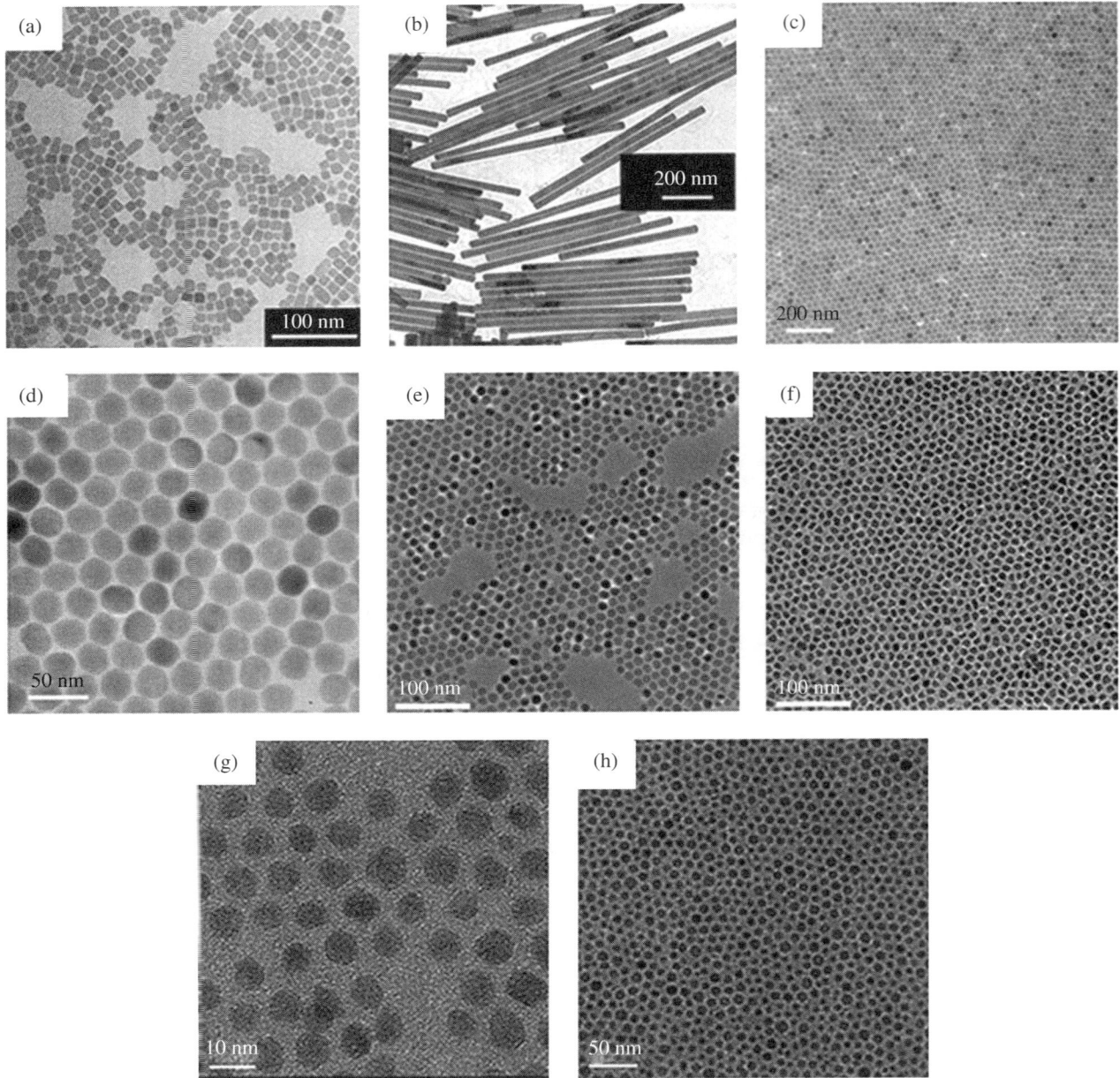

FIGURE 13.2 TEM images of the upconversion nanocrystals obtained by hydrothermal methods (a, b), thermal decomposition in high-boiling solvents (c, d), and thermal decomposition methods (e-h). Reprinted with permission from refs. [14–18]. Copyright 2007 American Chemical Society, 2008 IOP Publishing Ltd., 2006 American Chemical Society, 2010 American Chemical Society, 2006 Wiley-VCH Verlag GmbH&Co. KGaA., Weinheim.

based UCNPs in OA and 1-octadecence (ODE) [15]. The shape and size of UCNPs can also be controlled by fine-tuning the composition of the solvents.

In addition, Li and co-workers developed a general method to synthesise nano-sized materials by utilising the interfaces of the liquid, solid, and solution (LSS) phases [10, 24]. They used this LSS method to fabricate a series of inorganic nanocrystals with controllable size and shape. Ln-UCNPs can also be obtained by such a general method.

In addition to the hydrothermal and thermal decomposition methods, other methods such as sol-gel [25, 26], combustion, [27] ionic liquid-based synthesis [28–30], and microwave-assisted synthesis [31, 32] are also used for the synthesis of Ln-UCNPs. The principles and controlling methods are similar to those used in the hydrothermal and thermal decomposition methods.

13.2.3 Colour and Efficiency Tuning

In addition to the synthesis of uniform Ln-UCNPs, there is also the need to find an efficient way to optimise the upconversion properties of the as-prepared Ln-UCNPs. There are two main aims in the optimisation of Ln-UCNPs: One is the upconversion efficiency and the other is multicolour emissions for multiplexed labelling.

The factors that influence the luminescence efficiency of phosphors have been investigated for the conventional luminescent materials. These principles can also be applied for Ln-UCNPs. The most important factor restricting the luminescence efficiency of Ln-UCNPs is the abundant quenching route for nanoparticles. The small size of the nanoparticles results in poor crystallinity and a large surface-to-volume ratio, which generates more quenching centres. In this aspect, Ln-UCNPs with larger size will have better luminescence efficiency [33].

To achieve high luminescence efficiency in small-sized Ln-UCNPs, one effective way is to construct a core-shell structure to block the quenching route to the surface-quenching centres. Pioneering reports include the simple growth of a pure host layer around the Ln-UCNPs [34, 35]. The identical crystal lattice minimises the lattice strains to enable the growth of a core-shell structure. These as-prepared core-shell structured Ln-UCNPs are proven to have higher luminescence efficiency than the simple Ln-UCNPs. Recently, Yan and co-workers reported a core-shell structured Ln-UCNP employed CaF_2 as the shell (Figure 13.3) [36]. Benefitting from a lattice similar to that of the cubic-phased $NaREF_4$, CaF_2 shells can easily grow around $NaREF_4$-based UCNPs to obtain core-shell structures with high luminescent efficiency. Meanwhile, the CaF_2 shell can minimise the leakage of rare earth ions from Ln-UCNPs in the bio-surroundings, releasing Ca^{2+} instead, which has low biotoxicity.

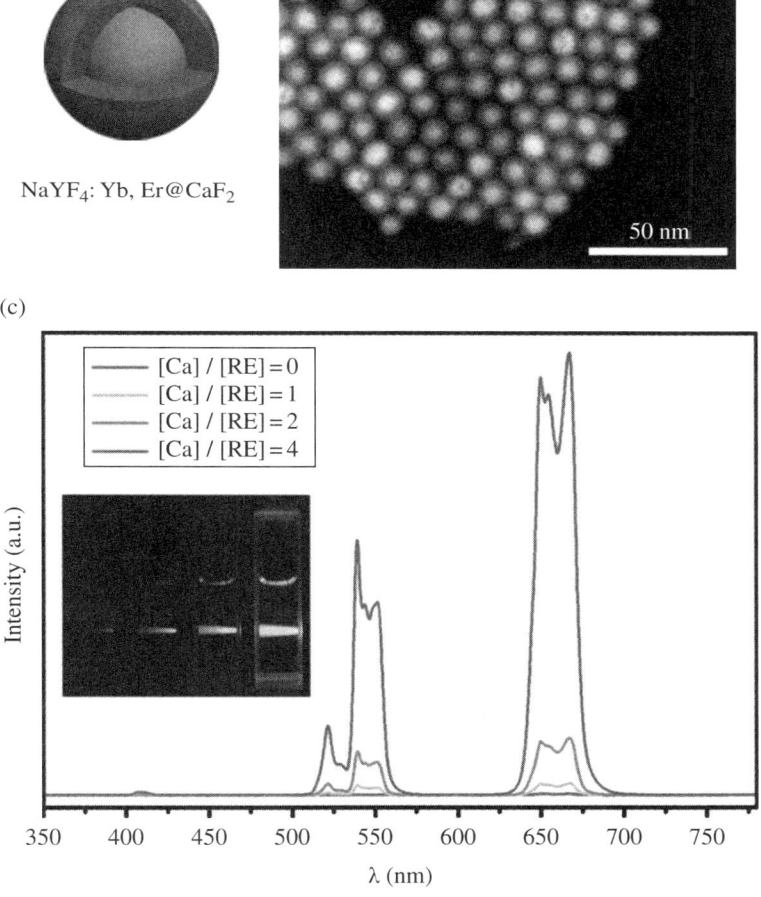

(a)

$NaYF_4$: Yb, Er@CaF_2

(b)

50 nm

(c)

— [Ca] / [RE] = 0
— [Ca] / [RE] = 1
— [Ca] / [RE] = 2
— [Ca] / [RE] = 4

Intensity (a.u.)

350 400 450 500 550 600 650 700 750

λ (nm)

FIGURE 13.3 Schematic illustration of Ln-UCNPs@CaF_2 core-shell structures (a), HAADF-STEM images of the nanoparticles and the corresponding upconversion emission spectra of Ln-UCNPs@CaF_2 nanoparticles with different thicknesses of shell layer. Reprinted with permission from Ref. [36]. Copyright 2012 Wiley-VCH Verlag GmbH&Co. KGaA, Weinheim.

Another method for achieving high luminescence efficiency is to reduce the amount of surface ligands in order to minimise the quenching effect. Besides the annealing method to remove surface ligands [37], Li and co-workers have reported the synthesis of small-sized $NaLuF_4$-based UCNPs [4]. The efficiency is much higher than the $NaYF_4$-based analogues, because of fewer surface-capping ligands, probably resulting from the different coordination abilities of OA molecules to Y^{3+} and Lu^{3+} ions.

A recently developed route to enhance the UCL is the metal surface-enhanced luminescence method. Although such metal-enhanced fluorescence effects were found early in 1980, the use of this method for Ln-UCNPs was only reported a few years ago. Because the surface plasmon resonance of noble metals have the ability to change the spatial distribution of the excitation energy field, the emitters located at the hot spot will gain more excitation energy, thus generating intense upconversion luminescence. Yan and co-workers reported the surface-enhanced upconversion luminescence of UCNPs on silver nanowires [38]. Duan and co-workers fabricated a $NaYF_4$:Yb,Tm@Au heterostructure and observed a more than 150% enhancement of the blue upconversion emission [39]. Recently, Kennedy [40] and Qin [41] reported the similar surface-enhanced upconversion luminescence phenomenon, employing Au nanoparticles to the surface of UCNPs, respectively. This kind of enhancement was further confirmed by a single particle experiment carried out on a combined optical and atomic force microscope setup by Schietinger and coworkers [42].

Relatively speaking, the colour tuning of Ln-UCNPs is easier. The common method is to dope multiple kinds of activator ions within the lattice [43]. Controlling the kinds and amounts of activators will lead to different upconversion emission colours [44, 45]. For example, Er^{3+} usually gives green and red emissions, Tm^{3+} usually gives blue emission, while Ho^{3+} usually gives green emissions. The mixture of these three kinds of activators can produce a series of upconversion emission colours [46–48].

Another method of colour tuning is to employ a quencher to selectively quench one emission in order to change the emission ratios of Ln-UCNPs. For example, gold nanoparticles have absorption in the visible range. In a heterostructure composed of Au NPs and Ln-UCNPs, the emission of Ln-UCNP will be quenched by the Au NPs [49]. Varying the ratio of these two NPs can adjust the emission ratio in the visible range.

In all, tuning the efficiency and colour of Ln-UCNPs can be achieved by controlling the structure and composition of the nanoparticles. However, how to obtain small Ln-UCNPs with enough efficiency and tunable upconversion emissions is still the main challenge for their bioimaging application.

13.3 SURFACE MODIFICATION OF Ln-UCNPs

Even though many groups have obtained Ln-UCNPs with the desired size and upconversion properties, the materials are not ready for bioimaging use. The important requirements for ideal biolabels include biocompatibility and some functional groups, which are convenient for linking to biological molecules in order to achieve selective targeting. Therefore, it is necessary to modify the surface of the Ln-UCNPs in order to endow them with these functions.

As mentioned in section 2.2, most of the successful synthetic methods use organic species as capping ligands to obtain uniform Ln-UCNPs. However, these organic species make the surface of Ln-UCNPs hydrophobic. Therefore, surface modification protocols need to be developed to transform the surface to a hydrophilic nature to suit the biological surroundings. Until now, several strategies have been developed to successfully modify the surface properties of Ln-UCNPs (Scheme 13.2) [5].

The first surface modification strategy is to directly alter the surface ligands. Ligand exchange is the most commonly used protocol in this strategy. Because these ligands are usually small molecules, the exchange process does not affect the crystallinity and morphology of Ln-UCNPs. The detailed exchange process is driven by the higher affinity between rare-earth ions and the additional ligands than those for the original ligands or by just the high concentration of the additional ligands. In particular, OA-coated Ln-UCNPs usually have strong interactions between rare-earth ions and OA ligands because of the high affinity of rare-earth ions to the oxygen terminated species. Thus the exchange of OA molecules usually requires excess additional ligands or the application of a chelating agent. Up to now, poly(ethyleneglycol) (PEG)-phosphate, polyacrylic acid (PAA), polyethylenimine (PEI), polyvinylpyrrolidone, hexanedioic acid, 3-mercaptopropionic acid, dimercaptosuccinic acid, mercaptosuccinic acid, citrate, 1,10-decanedicarboxylic acid, 11-mercaptoundecanoic acid, and poly(amidoamine) have been used to replace surface OA ligands to achieve hydrophilic properties [5].

Another way to modify the surface ligand is to perform some chemical reactions. Li and co-workers have developed a versatile protocol to oxidise surface OA molecules with Lemieux – von Rudloff reagent [50]. The oxidation agents break the double bond in the OA molecule to form azelaic acid products. The terminal carboxylic groups in the products endow the Ln-UCNPs with hydrophilic properties. Yan and co-workers have used ozone as an oxidation agent, in which the terminal group can be controlled as –OH, –CHO, or –COOH units by employing specific hydrolysis conditions [51]. The hydrophilic terminal functional groups can also be reacted directly to the biofunctional species, which provides a direct route for the biofunctionalisation.

Another strategy is to introduce some amphiphilic species outside the hydrophobic Ln-UCNPs. The hydrophobic groups of amphiphilic molecules will interact with the hydrophobic chains of the surface ligands in the Ln-UCNPs. The hydrophilic

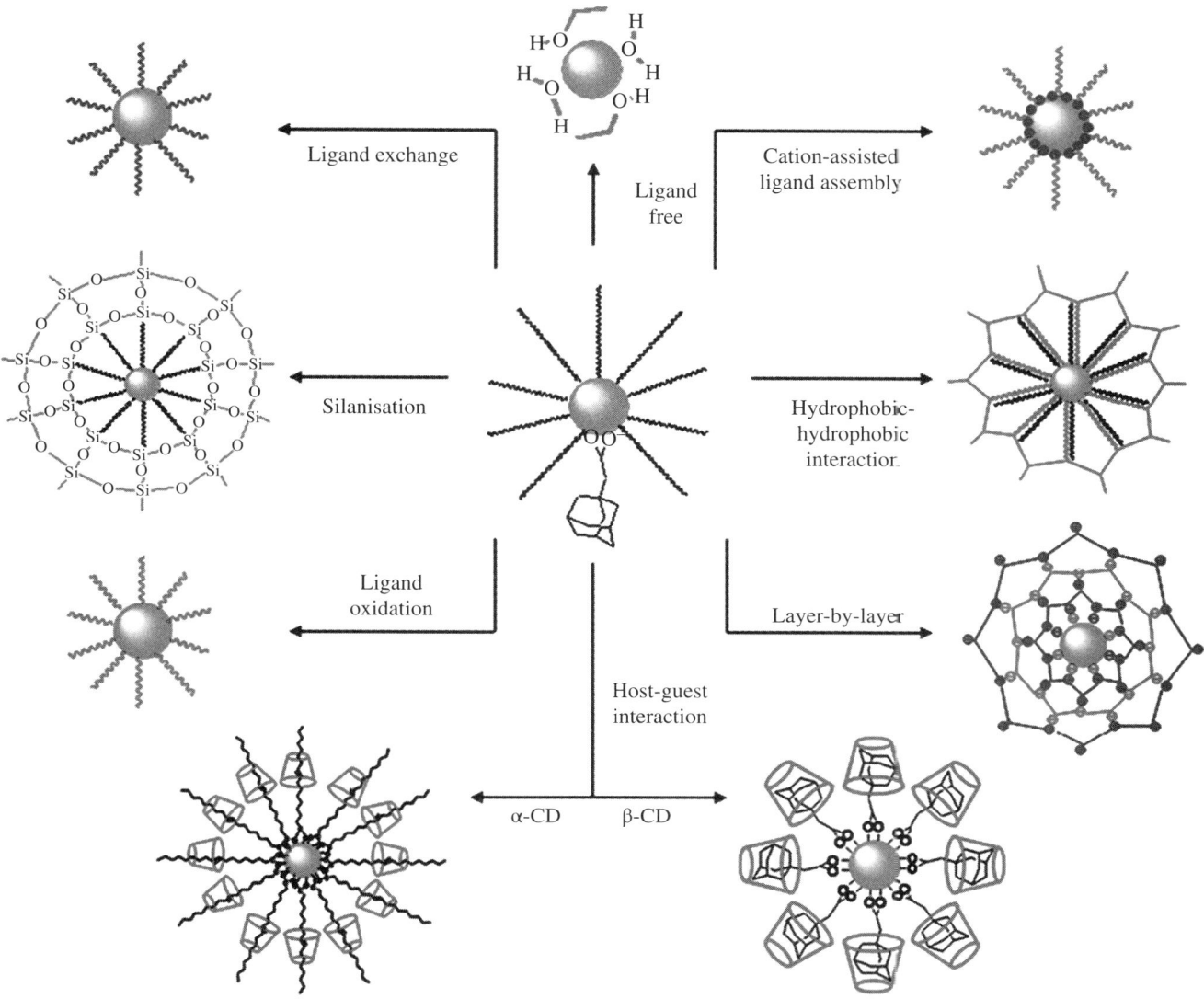

SCHEME 13.2 General strategies to modify the surface properties of Ln-UCNPs [5].

groups then point outward to form a hydrophilic layer. OA-PAA-PEG [52], CTAB, anionic AOT, and nonionic PEG *tert*-octylphenyl ether [53] have been shown to serve as amphiphilic molecules to achieve this goal.

A layer-by-layer assembly method is another route to construct a hydrophilic surface. Li and co-workers have used poly-electrolytes, such as PAH and PSS deposited on the surface of Ln-UCNPs [54]. Due to the repulsion between charges of the same kind, only one layer of polyelectrolytes can be deposited on the Ln-UCNPs at a time. The alternate usage of negative and positive polyelectrolytes will result in the layer-by-layer assemblies. Thus, the size of the assembled NPs can be con-trolled accurately.

Host-guest interaction is another force to enable the effective assembly of ligands. Li and co-workers developed an efficient method to draw OA-coated Ln-UCNPs into water by introducing *alpha*-cyclodextrin as the host to interact with OA (guest) molecules [55]. A similar method is also used for the adamantaneacetic acid-capped Ln-UCNPs, using *beta*-cyclodextrin instead of *alpha*-cyclodextrin.

The last but perhaps most important method is the silanisation of Ln-UCNPs. This protocol is the easiest one and gives the best flexibility to link Ln-UCNPs with biomolecules. To date, a series of silanisation methods have been developed for Ln-UCNPs with various surface ligands. Ln-UCNPs with hydrophilic capping ligands can be directly coated by silica via the common Stöber method; therefore, the abovementioned products can all be coated with a silica shell after transitioning to the hydrophilic surface [56]. The thickness of the silica shell can be controlled by the ratio of TEOS and Ln-UCNPs.

Reverse microemulsion systems can be used to obtain a silica shell outside the hydrophobic Ln-UCNPs [57]. Typically, polyoxyethylene 5-nonylphenylether, branched (Igepal CO520) is used as a surfactant to form the microemulsion. Ln-UCNPs are located in the microemulsions, while the silica shell forms at the interface between the aqueous solution and organic solvent (usually cyclohexane). The silica shells obtained in microemulsions are usually uniform, but the size of the entire Ln-UCNPs@SiO$_2$ is limited by the size of the droplets. When the size of Ln-UCNPs@SiO$_2$ is larger than the droplets, the hydrolysis of TEOS cannot be well controlled to occur at the interface, resulting in non-uniform products.

After silica coating, functional groups such as –NH$_2$ can be directly induced to the Ln-UCNPs by applying silica sources containing –NH$_2$ groups (APS). Researchers have also developed a method of constructing mesoporous silica shells outside Ln-UCNPs, which enables a larger loading amount of the structure.

In addition to the post-surface modification method, there are also some reports on the direct synthesis of surface functional Ln-UCNPs. In these methods, functional groups are introduced to the Ln-UCNPs during the synthesis procedure, usually acting as moieties of the capping ligands.

13.4 *IN VIVO* IMAGING APPLICATIONS

13.4.1 Setup of Imaging Instruments

Because most commercial microscopes and cameras are based on the detection of visible emissions, imaging instruments should be designed and modified to obtain upconversion imaging.

For microscopes, upconversion has luminescent characteristics similar to two-photon luminescence. Therefore, the instrument can be set up on a similar microscope. Li and co-workers chose a commercial inverted microscope (Olympus IX81) with a confocal scanning unit (FV1000, Olympus, Japan) as the base to construct the imaging instrument (Scheme 13.3a) [2]. A continuous-wave laser emitting at 980 nm was introduced as the excitation source. The laser beam was

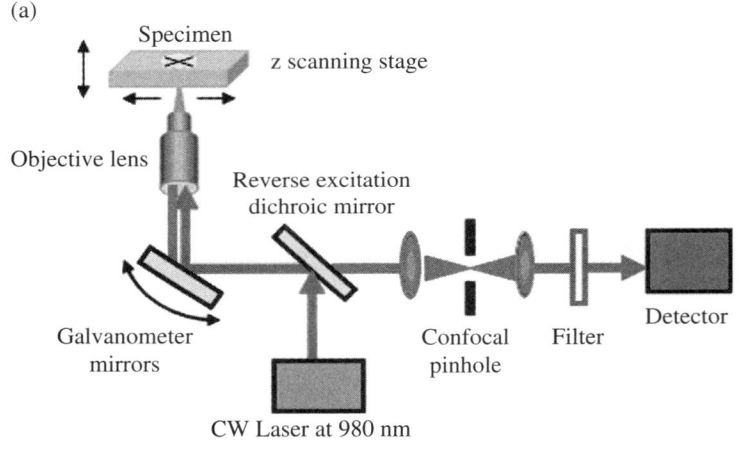

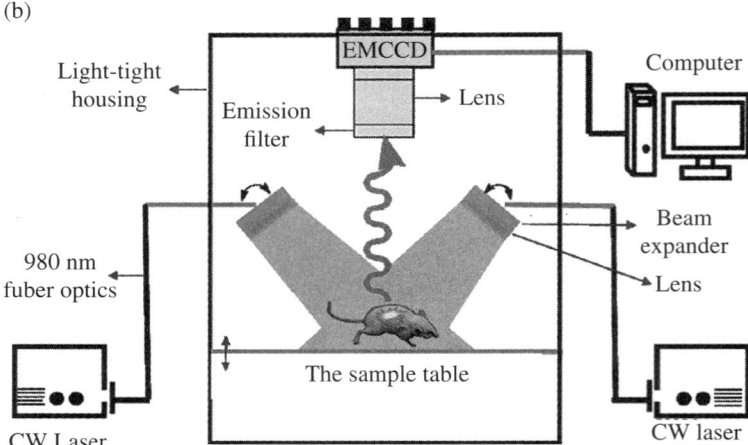

SCHEME 13.3 Schematic illustration of the instruments for cell imaging (a) and animal imaging (b).

reflected by the dichroic mirror and directed by the galvanometer mirrors to a position on the samples. The upconversion luminescence signals were collected by an objective lens and deflected by the galvanometer mirrors again to the dichroic mirror. Different from the excitation laser beam, upconversion luminescence signals can transmit through the dichroic mirror to the spectroscopical system and be detected by a photomultiplier tube (R6357 Enhanced model, HAMAMATSU, Japan). A filter was placed before the detector to further filter the excitation laser line and select the detection wavelength. A confocal pinhole was set behind the dichroic mirror and in front of the filter to improve the resolution in the z-direction. This *in vitro* upconversion imaging system is known as laser scanning upconversion luminescence microscopy (LSUCLM), which is ready for the upconversion imaging characterisation of cells and micro-samples.

Li and co-workers also developed an *in vivo* upconversion imaging system. Two 0-10 W output-tunable 980 nm lasers were introduced to the commercial imaging system (Scheme 13.3b) [3]. Optical filters were employed before the detector to block the excitation laser line and select the desired UCL signals. An electron multiplying charge coupled device (EMCCD, Andor DU987) was used as the detector. The most used UCL signals for small-animal imaging are located at around 800 nm from the emission of Tm^{3+} to enable large penetrating depth.

13.4.2 Experimental Proof Showing the Superiority of Upconversion Bioimaging

Using the above mentioned imaging instrument, upconversion images were obtained from biosamples to verify this upconversion imaging technique.

13.4.2.1 Low Background and High Signal-to-Noise Ratio (SNR)
The main noise problem in bioimaging experiments results from autoluminescence and bioluminescence, widely found in biotissues. These luminescent signals can be excited by common excitation sources, such as ultraviolet and visible light. Upconversion luminescence uses NIR photons as excitation sources. These photons have low energy, which do not excite the lumino-groups in biotissues, thus the background luminescent signals can, in theory, be blocked.

An experimental example has been provided by Li and co-workers (Figure 13.4) [2]. The confocal image of a live HeLa cell dual-labelled with 1,1'-dioctadecyl-3,3,3',3'-tetramethylindocarbocyanine perchlorate (DiI) and Ln-UCNPs showed no fluorescence signals from DiI or cell tissues under excitation at 980 nm, while the emission intensity from Ln-UCNPs exceeded the saturation value of the detection.

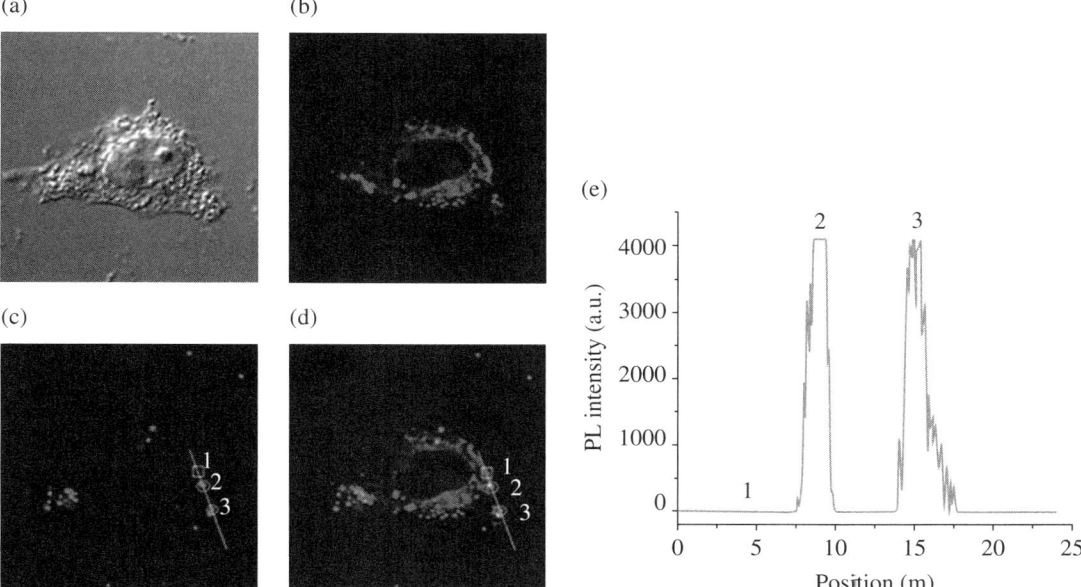

FIGURE 13.4 Bright field (a) and luminescent images (b–d) of a live HeLa cell dual-labelled with DiI and UCNPs. (b) Image of DiI when emission was collected at 560–600 nm (λ_{ex} = 543 nm). (c) UCL image of UCNPs when emission was collected at 500–600 nm (λ_{ex} = 980 nm). (d) The image of DiI and UCNPs when overlaid. (e) The detected upconversion signals along the line in (c) and (d). Reprinted with permission from Ref. [2]. Copyright 2009 American Chemical Society. (*See insert for colour representation of the figure.*)

13.4.2.2 High Photostability   Traditional organic dyes and quantum dots both have obvious photo-oxidative degradation, which leads to low photostability, and are not suitable for continuous imaging. Ln-UCNPs usually have high photostability due to their inorganic nature and can be excited by high energy lasers.

Using 4,6-diamino-2-phenyl indole (DAPI), DiI, and Ln-UCNPs multilabelled HeLa cells as an example, Li and co-workers have carried out the following experiment to directly prove that Ln-UCNPs have higher photostability than common organic dyes (Figure 13.5) [2]. The signals from all these labels can be collected in an initial image. After illumination by the lasers at 405, 543, and 980 nm for 400 s, the fluorescence signals from DAPI and DiI disappeared, while the signals from Ln-UCNPs were maintained under the excitation from the corresponding lasers.

13.4.2.3 Large Penetrating Depth   The excitation of Yb^{3+}-based UCNPs is usually located at 980 nm. This NIR excitation has a large penetrating depth in biological tissues. Taking Tm^{3+} as the doping ion, upconversion emissions at 800 nm can also be located in the 'optical transmission window' (650~1000 nm) of the biological tissues. Therefore, the Yb^{3+}-Tm^{3+} co-doped UCNPs are used to demonstrate the high penetrating ability of upconversion imaging techniques.

Zhang and co-workers employed $NaYF_4$:Yb,Tm@SiO_2 nanoparticles for in-depth imaging and showed that the largest imaging depth of cells can reach 3 mm [58]. Liu and co-workers reported a penetration depth of 0.8 cm in packing pork (a material commonly used to act as a biotissue) for $NaYF_4$:Yb,Tm nanoparticles [59]. Recently, using $NaLuF_4$:Yb,Tm as a

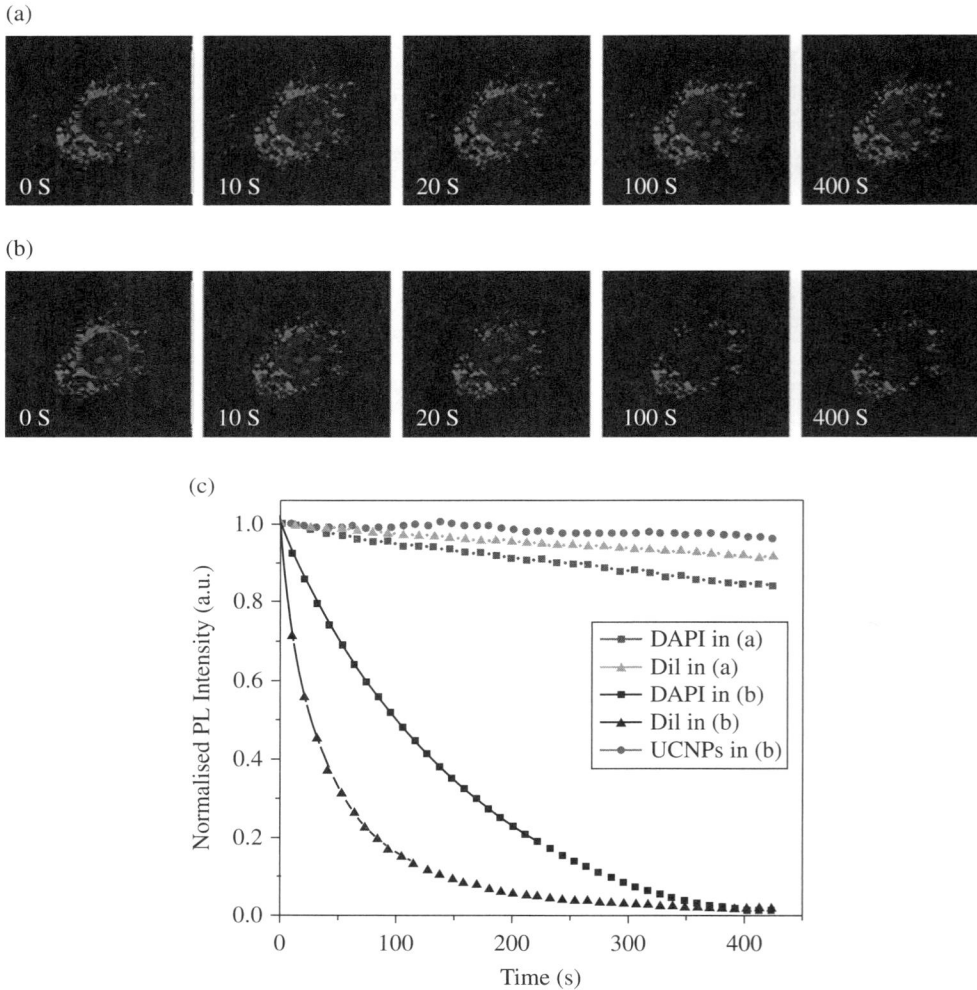

FIGURE 13.5   (a–b) Comparison of photobleaching of DAPI, DiI, and Ln-UCNPs in LSUCLM and conversional confocal microscopy imaging. The luminescence signals of DAPI, DiI, and Ln-UCNPs are shown in blue, red, and green, respectively. Simultaneous excitation was provided by CW lasers at 405, 543, and 980 nm with powers of approximately 19 mW, 15 μW, and 0.8 μW in (a) and 1.6, 0.13, and 19 mW in (b) in the focal plane, respectively. (c) Quantitative analysis of the signals in (a) and (b). Reprinted with permission from Ref. [2]. Copyright 2009 American Chemical Society. (*See insert for colour representation of the figure.*)

probe with higher quantum efficiency, Li and co-workers reported the imaging with penetration depth larger than 1.5 cm in a mouse [4] or packing pork [60].

13.4.2.4 *High Sensitivity*

Due to the rapid development in electro-optical devices, modern instruments can collect weak luminescent signals from minute amounts of samples. These techniques endow the luminescent imaging techniques with high sensitivity. Cohen and co-workers reported cell imaging stained with individual $NaYF_4$:Yb,Er nanocrystals with no measurable background signals and blinking [61]. This experiment proved that the sensitivity of the upconversion imaging technique can reach the single particle level in cell imaging. Recently, Li and co-workers tested the imaging sensitivity in animals [4]. KB cells stained by $NaLuF_4$:Yb,Tm UCNPs were injected into athymic nude mice to test the sensitivity. After subcutaneous injection of 50 KB cells, upconversion signals were detected with a signal-to-noise ratio of about 5. With vein injection of 1000 cells, upconversion signals can be collected in the region corresponding to the lung of the mice.

13.4.3 UCNPs for *In Vitro* Bioimaging

13.4.3.1 *UCNPs for Living Cells Imaging*

UCNPs with various surface modifications have been successfully used in living cell imaging. For example, Zhang and co-workers incubated skeletal myoblasts and bone marrow-derived mesenchymal stem cells with silica-coated Ln-UCNPs [57]. The UCL signals were collected in the confocal images, and Ln-UCNPs were mainly located in the cell cytoplasm. Li and co-workers also demonstrated that azelaic acid-coated Ln-UCNPs can be used for cell imaging via the endocytosis mechanism [2]. Ln-UCNPs coated by polymers can also be used for cell imaging. PEG, PEI, PAA, and PVP are the most commonly used polymers for the surface modification of Ln-UCNPs. Van Veggel's group [62] and Li's group [63] employed PEG-based polymer-coated Ln-UCNPs to successfully image ovarian cancer cells, respectively, whilst Capobianco and co-workers used PEI-coated Ln-UCNP to image living HeLa cells [64].

In addition, different surface properties will affect the cell uptake of Ln-UCNPs. Wong and co-workers studied the uptake of various charged Ln-UCNPs by mammalian cells [65]. PVP (neutral), PEI (positive), and PAA (negative)-coated Ln-UCNPs were used as models with different kinds of charges. As shown in confocal imaging and ICP-MS analysis, positively charged PEI-coated Ln-UCNPs can be internalised by the cells more easily than the neutral or negatively charged ones.

13.4.3.2 *Tumour Cell Targeting*

Tumour cell targeting is important in the detection of tumour cells and the labelling of cells for cell tracking. Some typical targeting species, such as folic acid (FA), peptides, and antibodies, have been used to link to the Ln-UCNPs for tumour cell targeting. Folic acid is one of the most commonly used functional molecules for tumour targeting because folate receptors (FR) are overexpressed in many human cancerous cells. Thus, FA-modified species have high affinity to these cells.

Zhang and co-workers were the first to conjugate FA molecules to the PEI-coated Ln-UCNPs [66]. These nanoparticles can target FR-positive human HT29 adenocarcinoma and human OVCAR3 ovarian carcinoma cell lines for upconversion imaging. Similarly, Li and co-workers have conjugated FA molecules on the surface of Ln-UCNPs@SiO$_2$ [67]. The carboxyl groups of FA are used to link with –NH$_2$ groups from the existing capping ligand (e.g., 6-aminohexanoic acid), activated by sulfo-NHS and EDC. As confirmed by confocal images and quantitative flow cytometric analysis, FA-labelled Ln-UCNPs were proven to have high efficiency to target tumour cells.

RGD peptide is another popular species for tumour targeting, based on its high affinity to the $\alpha_v\beta_3$ integrin receptor [68]. Li and co-workers developed a method to label Ln-UCNPs with RGD peptide by PEG linkage, for targeted imaging of tumours *in vitro* and *in vivo* [3]. Zako and co-workers have also reported RGD labelled Y_2O_3:Er for U87MG cells targeted imaging [69].

The interactions between antibody and antigen comprise another way to target tumour cells. Xu and co-workers applied anti-Her2 antibodies attached to the $NaYF_4$:Yb/Er@SiO$_2$ nanoparticles to target Her2 receptors of SK-BR-3 cells [70]. Similar strategies have been successfully applied for the corresponding tumour cell targeting imaging with Ln-UCNPs, using anti-claudin-4 [71], anti-mesothelin [71], or rabbit CEA8 antibody [72] as antibody, respectively.

13.4.4 UCNPs for *In Vivo* Bioimaging

The upconversion imaging technique has also been demonstrated at the animal level. Various kinds of animals, such as *C. elegans*, mouse, and rabbit have been employed as sample animals to evaluate their imaging qualities.

13.4.4.1 *Nematode* C. elegans *Imaging*

Ln-UCNPs can be introduced to label *C. elegans* by simple incubation, because *C. elegans* can eat Ln-UCNPs in the substrate. Thus the upconversion signals can be collected from inside *C. elegans*. Lim and co-workers first reported the *in vivo* upconversion imaging of *C. elegans* fed 150 nm Ln-UCNPs [73].

Recently, Yan et al. used a similar system to evaluate the metabolism of Ln-UCNPs through the digestive system. [74]. NaYF$_4$:Yb,Tm can be eaten by *C. elegans* and found in the gut from the pharynx to the anus. After being fed with *Escherichia coli*, *C. elegans* will excrete these Ln-UCNPs, and no upconversion signals can be found.

13.4.4.2 Animal Whole-Body Imaging

Apart from *C. elegans*, mice and rats are the most commonly used small animal models upon which upconversion imaging has been performed. Zhang and co-workers reported *in vivo* upconversion imaging near the body surface and deep in the body of rats [66]. The emission at 800 nm from PEI-coated NaYF$_4$:Yb,Tm nanoparticles can be used as a detection signal. Following this, many research groups have used Ln-UCNPs as luminescent labels for the *in vivo* bioimaging in animals, and the nude mouse is the common animal model because of the minimum scattering effect. However, upconversion imaging of mice with fur can also be achieved. For example, Prasad's group also applied NaYF$_4$:Yb,Tm nanoparticles for *in vivo* imaging [75]. They proved that this upconversion imaging provides deeper light penetration with almost no autofluorescence.

Liu and co-workers reported a penetration depth of 0.8 cm for upconversion imaging, which was evaluated in packing pork using NaYF$_4$:Yb,Tm as probes [59]. Li and co-workers applied NaLuF$_4$-based UCNPs with higher upconversion efficiency for imaging to achieve larger penetration depth (Figure 13.6) [4]. Black mice, in which the absorption of excitation and emission are larger than in nude mice, were employed as the animal model. By injecting subcutaneously with NaLuF$_4$:Yb,Tm nanoparticles, upconversion signals can be collected from both the front and back sides of the mouse, indicating the successful penetration of the whole mouse with a penetration depth of ~2 cm.

Furthermore, Li and co-workers applied NaLuF$_4$:Yb,Tm nanoparticles obtained by solvothermal methods for the upconversion imaging of a rabbit [60]. In this large animal model, upconversion imaging signals can also be collected with excellent signal-to-noise ratio.

As described above, upconversion imaging can be achieved for mice and rabbits for whole body imaging. However, because no target species were linked to the Ln-UCNPs in these examples, the imaging signals generally locate in the lung, liver, and spleen based on the biodistribution of Ln-UCNPs (which will be discussed in section 13.5).

13.4.4.3 Lymphatic Imaging

The lymph node is an important organ of the immune system in mammals; it has the ability to trap foreign particles and give the corresponding signals to the immune system. Lymph nodes are the most common place for cancers especially to be metastasised. Therefore, the imaging and position of lymph nodes have essential significance in clinical and biological applications. Because of the ability of lymph nodes to trap foreign particles, Ln-UCNPs can accumulate in lymph nodes easily after injection in the paw of a mouse. Kobayashi and co-workers first obtained two-colour

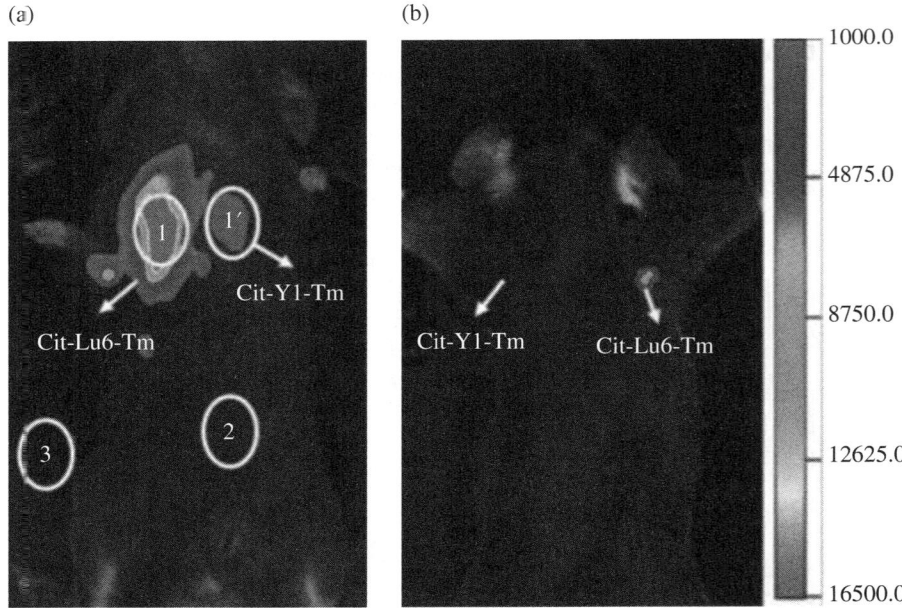

FIGURE 13.6 Upconversion imaging of a black mouse by subcutaneous injection of NaLuF$_4$ and NaYF$_4$-based UCNPs, (a) detection from the side of chest, (b) from the side of back. λ_{ex} = 980 nm, λ_{em} = 800 nm. Reprinted with permission from Ref. [4]. Copyright 2011 American Chemical Society.

lymphatic UCL imaging using $NaYF_4$:Yb,Er and $NaYF_4$:Yb,Tm nanoparticles [76]. Li and co-workers further applied amphiphilic LaF_3:Yb,Tm nanoparticles to image lymph nodes using NIR upconversion emission as the output signal (Figure 13.7) [77]. Liu and co-workers employed three different kinds of Ln-UCNPs ($NaY_{0.78}Yb_{0.2}Er_{0.02}F_4$, $NaY_{0.69}Yb_{0.3}Er_{0.01}F_4$, $NaY_{0.78}Yb_{0.2}Tm_{0.02}F_4$) with multicolour emissions for UCL imaging of lymph nodes at different positions [52]. These results demonstrate that the lymph nodes can be easily imaged by upconversion luminescence. Benefitting from the high sensitivity and signal-to-noise ratio, this imaging technique has potential in the luminescent imaging-guided surgery to remove the lymph nodes infected by cancers.

13.4.4.4 Vascular Imaging Vascellum is one of the most important organs in animals to transport blood around the body. Vascular imaging can provide information about the shape, position, spacing, permeability, for example. Ln-UCNPs can also be used for vascular imaging after proper surface modification to prolong their existence in the blood. Hilderbrand and co-workers first used PEG-coated Y_2O_3:Yb,Er nanoparticles for blood-pool UCL imaging via intravenous injection of the Ln-UCNPs. A NIR-emitting carbocyanine fluorophor was also used for co-localisation imaging (Figure 13.8) [78]. Zhang and co-workers also applied both $NaYF_4$:Yb,Er@SiO_2 nanoparticles and $NaYF_4$:Yb,Er@SiO_2-labelled cells for the UCL imaging of mouse ear blood vessels [79]. Furthermore, van Veggel et al. employed *in vivo* two-photon upconversion wide field microscopy to image brain blood vessels of a mouse after skull thinning [80].

13.4.4.5 Cell Tracking Another imaging application is to track cells in animals. Because some kinds of cells can be easily labelled by Ln-UCNPs as described in section 13.4.3.1, the upconversion luminescence from those labelled cells can help to track the motion and accumulation of these cells.

Zhang and co-workers first used $NaYF_4$:Yb,Er@SiO_2 nanoparticles to label myoblast cells, which were introduced to a mouse model with a cryo-injured hind limb [79]. The upconversion emission signals can be detected inside the limb of the

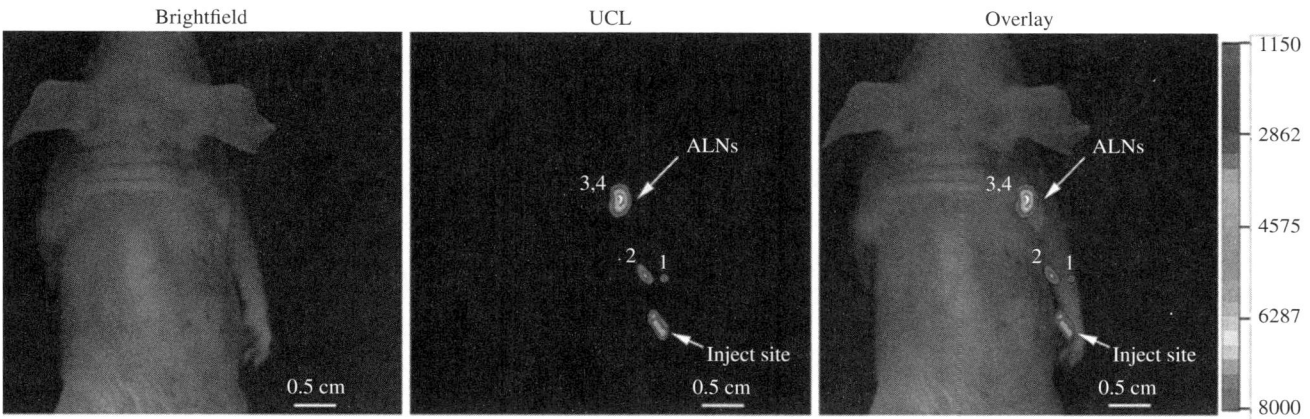

FIGURE 13.7 *In vivo* lymphatic drainage upconversion imaging at 800 nm to show four different draining lymph basins (1, 2, 3, 4) along the right antebrachium of the nude mouse. Reprinted with permission from Ref. [77]. Copyright 2011 Elsevier Ltd. (*See insert for colour representation of the figure.*)

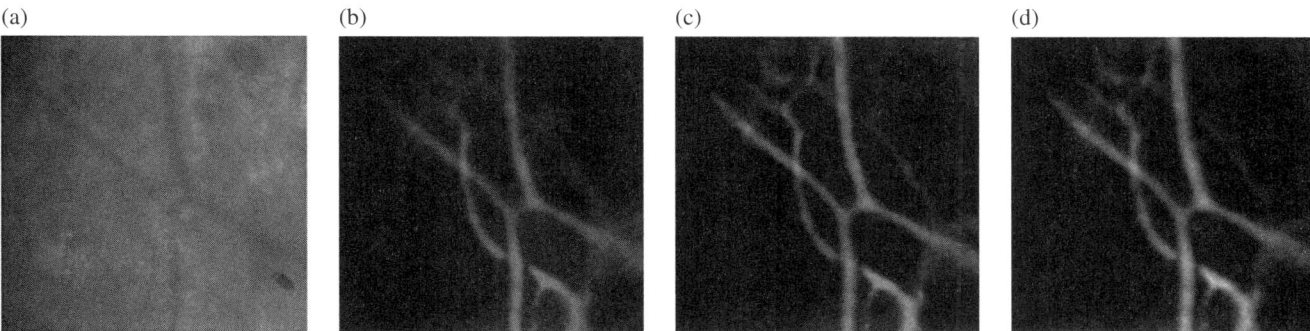

FIGURE 13.8 Optical imaging of blood vessels in the mouse ear following tail vein injection of the nanoparticles (10 mg); (a) blood vessels imaged with a blue light filter, (b) upconversion image with excitation at 980 nm, (c) fluorescence image of the carbocyanine dye with excitation at 737 nm, (d) merged image of the upconversion and fluorescence signals. Reprinted with permission from Ref. [78]. Copyright 2009 The Royal Society of Chemistry. (*See insert for colour representation of the figure.*)

mouse, demonstrating that upconversion imaging can be performed as a noninvasive technique to track cells *in vivo*. Liu et al. further applied $NaY_{0.78}Yb_{0.2}Er_{0.02}F_4$, $NaY_{0.69}Yb_{0.3}Er_{0.01}F_4$, and $NaY_{0.78}Yb_{0.2}Tm_{0.02}F_4$ with different upconversion emission properties to label cancer cells [52]. Seven days after subcutaneous injection of labelled cells, they also found the tumour at the location of injection, which agreed with the UCL cell-tracking imaging. This result indicates that the labelling protocol will not destroy the bioactivity of cancer cells. Further experiments are needed to track these cells during a long period to investigate the metastasis or other biological processes involving living cells.

13.4.4.6 *Tumour Targeting* **In Vivo**

Tumour-targeting techniques have attracted increasing attention, because this is a key point in tumour diagnosis and therapy. Ln-UCNPs have been successfully applied for tumour target imaging in a few cases, based on the ligand-acceptor or antigen-antibody interactions. However, the tumour-targeting properties originate from the active species linked to the Ln-UCNPs, rather than the nanoparticles themselves. Similar to the case for cell imaging, folic acid and peptides are still two kinds of specific active groups for tumour targeting.

On the basis of the high affinity between FA and FR, Li and co-workers reported the first example of imaging FR-overexpressed tumours in small animals model (Figure 13.9) [67]. FA-modified Ln-UCNPs were intravenously injected into a nude mouse with HeLa tumour-bearing athymic. Upconversion emission signals at 600–700 nm were collected from the position of the tumour. As a control experiment, amine-functionalised Ln-UCNPs were intravenously injected into the nude mouse via the same tumour, but no significant upconversion signals could be detected. A blocking dose of FA (10 mg/kg) was also introduced to the mouse to block the target effect.

Li and co-workers also used arginine-glycine-asparatic acid (RGD) peptide, which has a high affinity for the $\alpha_v\beta_3$ integrin receptor, to target tumour tissues [3]. Cyclopeptide c(RGDFK) was conjugated to $NaYF_4$:Yb,Er,Tm nanoparticles with the

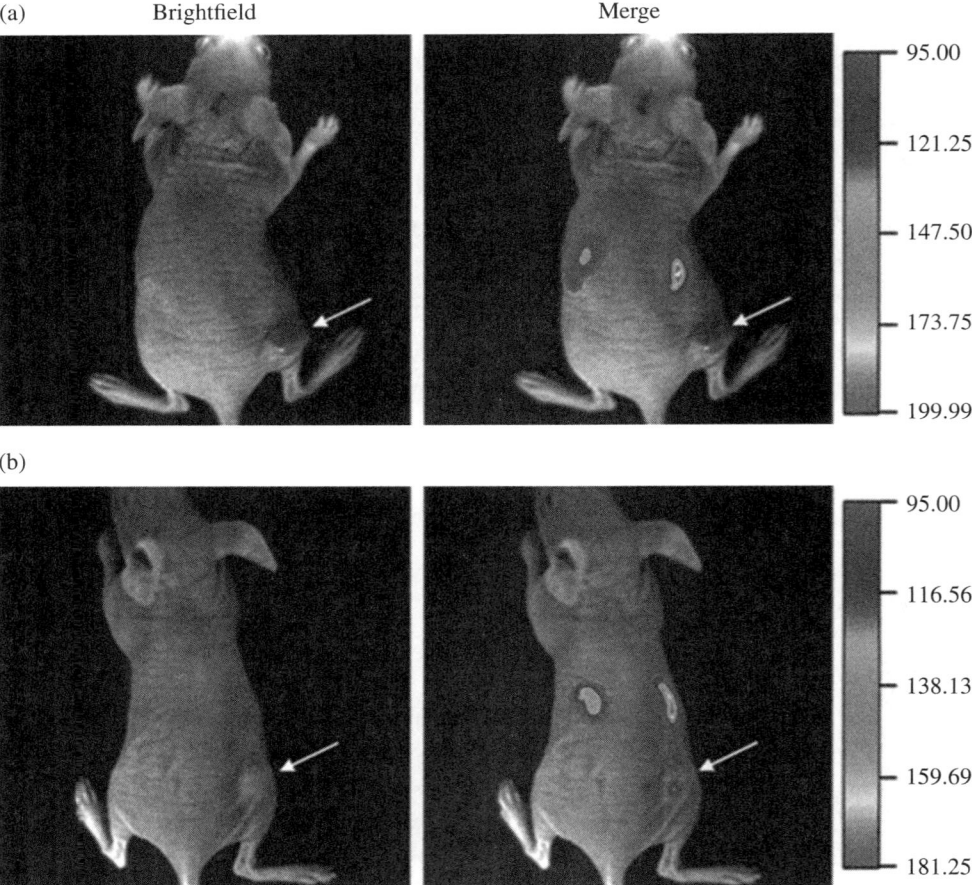

FIGURE 13.9 *In vivo* UCL imaging of subcutaneous HeLa tumour-bearing athymic nude mice (right hind leg, pointed by white arrows) after intravenous injection of UCNPs-NH$_2$ (a) and UCNPs-FA (b). All images were acquired under the same instrumental conditions (power density = 120 mW/cm^2 on the surface of mouse). Reprinted with permission from Ref. [67]. Copyright 2009 Elsevier Ltd.

PEG linkage. This kind of UCNP was applied for targeted imaging of mice bearing $\alpha_v\beta_3$-overexpressing U87MG tumours. Upconversion emission signals at 800 ± 10 nm can be collected from the position of the tumour. Comparing with the control groups, this study demonstrated the good target effect between RGD conjugated UCNPs and integrin-containing tumours.

These tumour target imaging examples demonstrate that Ln-UCNPs are promising candidates as luminescent labels for the cancer diagnosis and therapy.

13.4.4.7 Photo-Triggered Systems Because NIR excitation has large penetration depth in bioissues, Ln-UCNPs were also introduced to photo-triggered systems to manipulate the signals by NIR light. Zhao and co-workers used micelles composed of poly(ethylene oxide)-block-poly(4,5-dimethoxy-2-nitro-benzyl methacrylate) and NaYF$_4$:Yb,Tm nanoparticles to demonstrate this concept [81]. Under excitation at 980 nm, the micelles dissociated to release the co-loaded hydrophobic species. These light-responsive polymeric systems were suitable for potential biomedical applications such as drug delivery. Liu and co-workers further developed a system for *in vitro* and *in vivo* photo-triggered bioluminescence imaging [82]. The upconversion luminescence of NaYF$_4$:Yb,Tm nanoparticles were able to trigger the release of D-luciferin from D-luciferin-coujugated UCNPs. The bioluminescence of D-luciferin ($\lambda = 560$ nm) was then detectable after irradiation by 980 nm laser.

13.4.5 Multifunctional Imaging *In Vitro* and *In Vivo*

Multimodality imaging, including upconversion luminescent imaging techniques, can provide more information than single modal imaging techniques. Benefitting from the unique optical and magnetic properties of rare earth ions, conventional imaging techniques can be integrated to the upconversion luminescent imaging technique, employing UCNPs as the probes.

13.4.5.1 In Combination with Magnetic Resonance Imaging (MRI) Magnetic resonance imaging (MRI) is an important conventional imaging technique in clinical practice. The relaxation of protons gives three-dimensional information in the detected region. However, the poor spatial resolution limits the application of MRI technique. The combination of UCL and MRI gives the possibility of collecting 3D information by MRI, followed by highly sensitive UCL imaging to draw the outlines for the region of interest.

Because Gd^{3+} has seven unpaired electrons in its ground state, it has a large paramagnetic moment, which has strong interaction with surrounding protons to generate T$_1$ enhancement in MRI. Thus Gd^{3+}-containing species are usually used as T$_1$-enhanced MR contrast agents [83]. Rare earth ions have similar ionic radii and charges; therefore, Gd^{3+} can be easily introduced to the crystal lattice of Ln-UCNPs as host or doping ions. Heyon and co-workers first demonstrated the T$_1$-weighted MR images of SK-BR-3 cells with 20 nm NaGdF$_4$:Yb,Er nanoparticles [84]. Li and co-workers have used NaGdF$_4$ as host materials to synthesise Ln-UCNPs. The as-prepared Ln-UCNPs have an r$_1$ value of 5.6 s^{-1} mM^{-1} together with the intense UCL emission under CW excitation at 980 nm [47]. *In vivo* upconversion and T$_1$-weighted MR images confirm the existence of UCNPs in liver and spleen. Similarly, Gd$_2$O$_3$, [85] NaGdF$_4$, [86] BaGdF$_5$, [87] KGdF$_4$, [88] and GdVO$_4$ [89] can be used as host materials for UCL/MR two-modality imaging.

Because the interactions between Gd^{3+} and protons are affected by the distance between them, only the Gd^{3+} ions on the surface of Ln-UCNPs have the chance to be coordinated by the neighbouring protons from water molecules. van Veggel and co-workers synthesised a series of NaGdF$_4$ UCNPs with different sizes to investigate this effect [90]. The r$_1$ values increased with decreasing particle size. Shi et al. found that the Gd^{3+} buried deep within crystal lattices, larger than 4 nm, has nearly no T$_1$-enhancement effect (Figure 13.10) [91]. According to this principle, fabrication of NaYF$_4$:Yb,Er@NaGdF$_4$@SiO$_2$ with a diameter of 26.2 nm gave the highest r$_1$ value of 6.18 s^{-1} mM^{-1} (the shell thickness of NaGdF$_4$ is 0.2 nm).

Another approach for the introduction of Gd^{3+} is to use Gd^{3+} as a co-dopant as well as activator ions. Li and co-workers have reported Gd, Yb, and Er co-doped UCNPs with Gd^{3+} concentrations up to 60% [92]. The r$_1$ value is 0.41 s^{-1} mM^{-1} for these kinds of materials, which is significantly smaller than those using Gd-based materials as host because of the relatively low concentration of Gd^{3+}. Shi and co-workers also prepared NaYF$_4$:Yb,Tm,Gd@mSiO$_2$ for *in vivo* MRI and UCL imaging with injection of probes into the tumour issues directly [93]. Several materials, such as NaYF$_4$:Gd,Yb,Er, [94] NaLuF$_4$:Gd,Yb,Tm, [95] and BaF$_2$:Yb,Tm@SrF$_2$:Nd,Gd, [96] are synthesised to perform similar experiments.

T$_2$-enhancement MRI imaging is usually obtained with the assistance of Fe$_3$O$_4$ nanoprobes. Due to the similarities between Fe$_3$O$_4$ and Ln-UCNPs, probes for the combination of UCL and T$_2$-enhanced MRI imaging are often constructed by the fabrication of heterostructures composed of Ln-UCNPs and magnetic iron oxides. Shi and co-workers fabricated such a heterostructure via a 'neck-formation' strategy [97]. Silica shells are formed outside Fe$_3$O$_4$ and Ln-UCNPs, respectively, in the initial stage of the reaction. The growth of the silica shell will link two or more silica-coated NPs together to form heterostructures with random amounts of both Fe$_3$O$_4$ and Ln-UCNPs. T$_2$-weighted MR and UCL imaging of tumours in small animals was demonstrated by intra-tumour injection with such heterostructures.

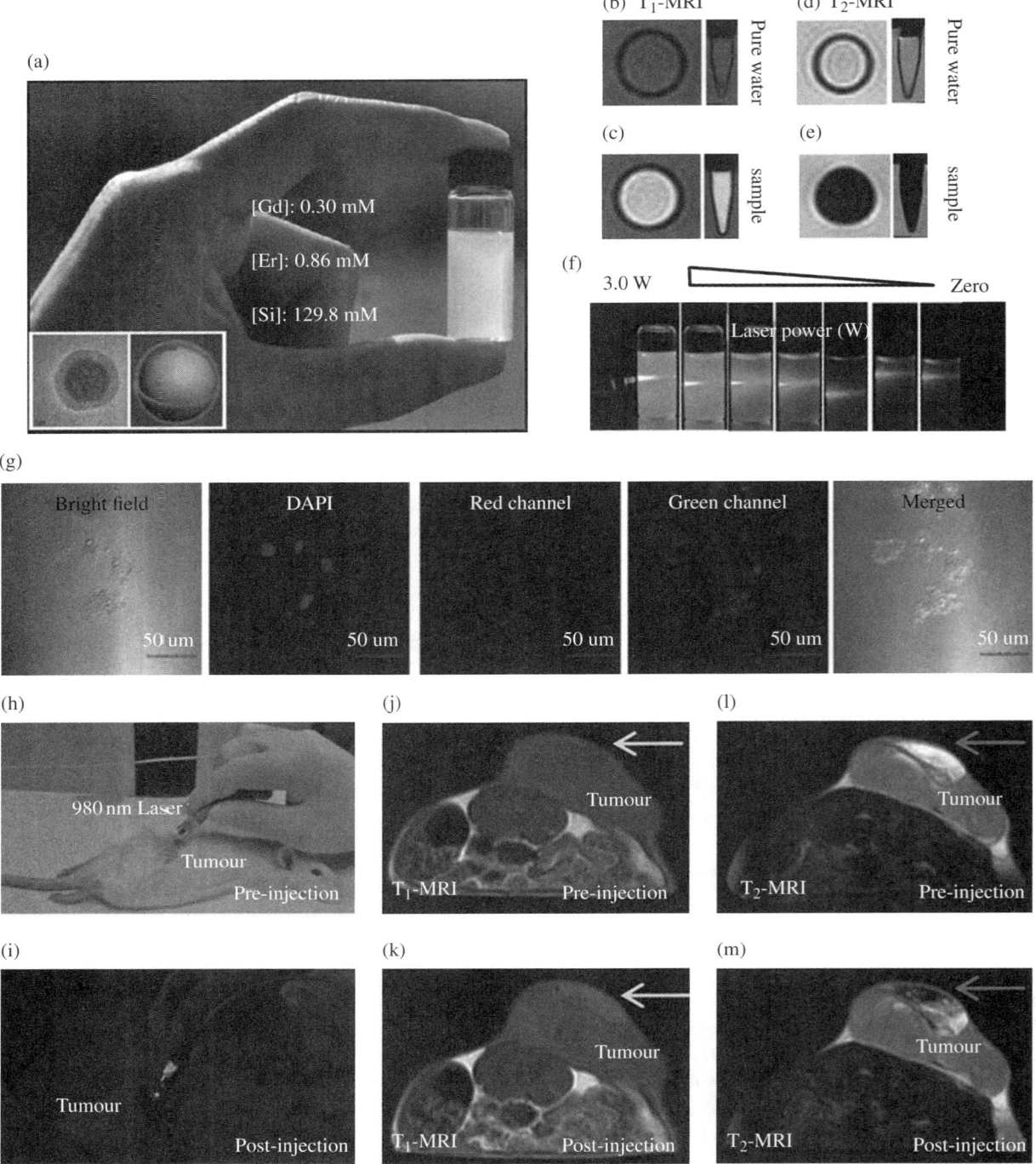

FIGURE 13.10   (a) Digital photograph of water soluble silica-protected NaGdF$_4$-based UCNPs. T$_1$-MR images of (b) pure water (r$_1$ = 0.35 1/s) and (c) UCNPs (0.2 nm) (r$_1$ = 2.03 1/s). T$_2$-MR images of (d) pure water (r$_2$ = 6.05 1/s) and (e) UCNPs (r$_2$ = 50 1/s). (f) Digital photos of samples under varied NIR laser powers. (g) Confocal images of MCF-7 cells incubated with the probe. *In vivo* optical imaging in tumour: (h) pre-injection, (i) post-injection. *In vivo* T$_1$-MR images of tumour: (j) pre-injection, (k) post-injection. *In vivo* T$_2$-MR images of tumour: (l) pre-injection, (m) post-injection. Tumour sites were marked with yellow and red arrows. Reprinted with permission from Ref. [91]. Copyright 2011 Wiley-VCH Verlag GmbH & Co. KgaA, Weinheim. (*See insert for colour representation of the figure.*)

Recently, Li et al. have synthesised NaYF$_4$:Yb,Tm@Fe$_x$O$_y$ core-shell nanocrystals directly via an epitaxial growth method [98]. A Fe$_x$O$_y$ shell with 5 nm thickness provides T$_2$-enhanced MRI characteristics. Lymphatic systems of small animals were chosen as the model to demonstrate UCL/MRI dual-modality imaging. In addition, Liu and coworkers have developed an electrostatic adsorption-based method to fabricate Fe$_3$O$_4$-UCNPs heterostructures. The UCL/MRI imaging was demonstrated by tumour targeting and lymphatic imaging in small animals.

Another novel multimodality imaging technique was recently demonstrated by Lin and coworkers. They fabricated a $Fe_3O_4@NaGdF_4:Yb,Er@NaGdF_4:Yb,Er$ core-multishell structure to combine UCL and T_1/T_2-weighted MRI together [99].

13.4.5.2 In Combination with X-ray CT Imaging

X-ray CT is another conventional imaging technique in the clinic. The CT image is based on the different X-ray absorption abilities of tissues. The common CT contrast agents consist of heavy elements, such as iodine and gold atoms. Lanthanide elements are located in the f-block (the sixth period) of the periodic table, and the large atomic numbers of lanthanide elements endow them with large absorption coefficiency for the X-ray techniques. Therefore, most lanthanide-based UCNPs can be used directly as CT contrast agents.

Cui and co-workers investigated the X-ray attenuation ability of Yb,Er/Ho/Tm co-doped $NaGdF_4$ UCNPs [100]. After subcutaneous injection into mice, UCNPs generated both UCL and CT signals. Lu and co-workers employed $NaYbF_4$ as the host material for the UCL and CT imaging [101]. The PEG-coated $NaYbF_4$:Gd,Yb,Er nanoparticles showed good CT contrast effects comparable to the commercial CT contrast agents. These works provide the evidence that UCNPs can serve as CT contrast agents and showed different metabolism characteristics compared to commercially used CT contrast agents.

13.4.5.3 In Combination with Positron Emission Tomography (PET) and Single-Photon Emission Computed Tomography (SPECT) Imaging

Positron emission tomography (PET) and single-photon emission computed tomography (SPECT) techniques can give accurate 3D information about the distribution of radioactive elements. ^{18}F is a commonly used radionuclide for PET imaging. Due to the high affinity between rare earth and fluorine ions, fluoride ions can directly bind to the surface of UCNPs. Li and co-workers proved that >90% of $^{18}F^-$ was bound to the UCNPs after mixing them in an aqueous solution and shaking for 5 min (Figure 13.11) [102]. This convenient method provides a powerful tool to investigate the biodistributions of UCNPs via PET imaging.

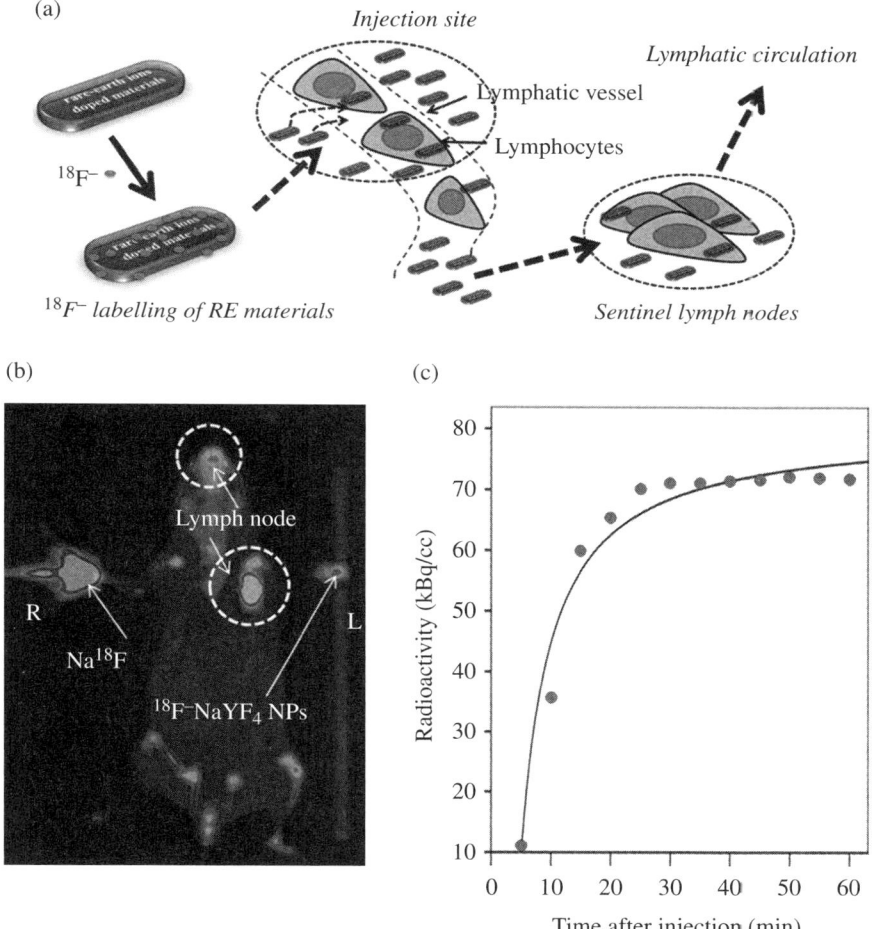

FIGURE 13.11 (a) Schematic representation of the preparation of ^{18}F-labelled UCNPs and the lymph node imaging mechanism. (b) PET/CT imaging of lymph node 30 min. after subcutaneous injection of 740 kBq/0.05 mL $^{18}F^-$ UCNPs into the left paw footpad. $Na^{18}F$ was injected into the right paw as a control experiment. (c) Curvilinear regression of real-time PET quantification detection of lymph node signals. Reprinted with permission from Ref. [102]. Copyright 2011 Elsevier Ltd. (*See insert for colour representation of the figure.*)

However, the short lifetime of ^{18}F limits its application in long-term imaging. ^{153}Sm is another kind of radionuclide that can be used in SPECT imaging. Because rare earth ions have similar chemical properties, ^{153}Sm can also label UCNPs directly by simple mixing. The relatively long half-lifetime of ^{153}Sm (1.929 day) enables the biodistribution of the labelled UCNPs to be conveniently analysed by SPECT imaging.

13.5 BIODISTRIBUTION AND TOXICITY OF UCNPs

Although UCNPs have been considered as one of the best candidates for *in vivo* luminescent imaging labels, the possible toxicity of UCNPs needs to be investigated in detail before any clinical application. Some work has been carried out to determine the toxicity of UCNPs against cells, but there are relatively few examples that have discussed the biodistribution and the possible bioeffects of UCNPs in animals.

13.5.1 *In Vitro* Toxicity of UCNPs

MTT (methyl thiazolyl tetrazolium), MTS ((3-(4,5-dimethylthiazol-2-yl)-5-(3-carbo-xymethoxyphenyl)-2-(4-sulfophenyl)-2H-tetrazolium, sodium salts), and CCK-8 mitochondrial metabolic activity assays are common techniques to evaluate the *in vitro* cytotoxicity of materials at the cell level. Numerous studies have been reported to test the cytotoxicity of UCNPs in different cell lines before the imaging applications. In these studies, cells were incubated with UCNPs for a certain period, and usually more than 75% cells were found to be viable when the concentration of UCNPs was in the range of 0.05 ~ 20000 µg/mL. These results demonstrated that UCNPs have low *in vitro* cytotoxicity and are biocompatible for the cell imaging.

Recently, Li and co-workers used the half-maximal inhibitory concentration (IC50) value as another parameter to evaluate the cytotoxicity of UCNPs [98]. They considered the water-soluble dopamine-modified NaYF$_4$:Yb,Tm@Fe$_x$O$_y$ core/shell nanoparticles in KB cells. After incubation for 24 and 48 hours, IC50 values of the nanoparticles were measured as approximately 295 and 190 µg/mL, respectively.

13.5.2 *In Vivo* Toxicity of UCNPs

13.5.2.1 *Biodistribution and Excretion of UCNPs* Before the clinical application of UCNPs, pharmacokinetic studies of UCNPs are necessary to evaluate their safety. Distribution, metabolism, and excretion are the main three parts in pharmacokinetics study. Plasma atomic emission spectroscopy (ICP-AES) is the most common technique to analyse the concentrations of UCNPs in different organs. For example, Zhang and co-workers investigated the biodistribution of SiO$_2$-coated [57] and PEI-modified NaYF$_4$:Yb,Er/Tm UCNPs [66], utilising the concentration of Y^{3+} as the parameter. A dose of 10 mg/kg wt SiO$_2$-coated UCNPs was intravenously injected into a Wistar rat. The concentration of UCNPs was found to be high in the lung and heart at the initial stage and gradually excreted by urine or faeces until completely cleared from body after 7 days. The PEI-modified UCNPs first accumulated in the lung. After 24 hours, the nanoparticles concentrated to the spleen, and after seven days, no UCNPs were detectable in the rats. Liu and co-workers also employed ICP-AES to analyse the biodistribution of UCNPs over a longer period of time [103]. They tracked PAA-UCNPs (35 nm) and PEG-UCNPs (~30 nm) for 90 days. The concentration of PAA-UCNPs in lung dropped significantly after three days, while the concentration remained at a certain level in all other organs.

Compared to the ICP-AES measurement, UCL imaging provides another route for the investigation of the biodistribution of UCNPs in animals *in vivo*. Li and coworkers employed NIR to NIR PAA-coated NaYF$_4$:Yb,Tm nanoparticles (~10 nm) for this [48], and the upconversion imaging results showed that most of the PAA-UCNPs were in the liver and the spleen at initial stages. As time increased, the upconversion signals in the spleen increased, while the signals in the liver decreased over 24 hours. After seven days, the upconversion signals were significantly reduced both in the liver and spleen. The *ex vivo* upconversion imaging analysis showed that after 14 days, almost no upconversion signal can be found in the liver and spleen, while obvious signals were observed in the intestine, indicating hepatobiliary excretion. At the 115th day, no upconversion signal could be found in any organs of the mouse, indicating complete excretion and clearance. These results were also confirmed by ICP-AES analysis.

Liu et al. observed upconversion signals from the liver, spleen, lungs, and bones one day after injection with PAA-UCNPs and PEG-UCNPs [103]. The signals decreased over the following days until no signals could be detected after seven days.

In these reports, SiO$_2$-coated, PEI-modified, PAA-coated, and PEG-coated UCNPs have different biodistribution characteristics in small animals after injection. The differences may come from the dosage, size, aggregation, and surface properties of Ln-UCNPs, as well as the animal type used in the experiments.

13.5.2.2 Long-Term Toxicity of Ln-UCNPs   To date, no obvious toxicity has been reported for Ln-UCNPs. The research work evaluated the behaviour, weight, and histological assessment of tissues. Li and co-workers reported that after injection of 15 mg/kg wt PAA-UCNPs (~11.5 nm), no changes in the common behaviour were observed for mice [48]. In other work, mice injected with 25 nm to 60 nm $NaGdF_4$:Yb,Er,Tm nanoparticles could also survive for more than one month without showing abnormal behaviour [92]. Body weight is another parameter to evaluate the living conditions of the mouse. Zhang and co-workers reported that injection with a dose of 10 mg/kg wt of silica-coated $NaYF_4$:Yb,Er UCNPs showed no obvious effect on the body weight of a mouse [57]. Li and co-workers observed tiny weight differences between a mouse injected with 15 mg/kg wt PAA-UCNPs (~11.5 nm) and the control group [48]. Neither death nor a significant body weight drop were observed for PAA-UCNPs and PEG-UCNPs treated mice [103].

Other analysis methods include the histological assessment of tissues and the routine analysis of blood. Li's group [48] and Liu's group [103] reported related results for PAA-UCNPs and PEG-UCNPs, and at general dose levels, they observed no abnormal data in these analysis.

13.6 FUTURE DIRECTIONS

As described in the previous sections, UCNPs can now be treated as candidates suitable for bioimaging. The fabrication of the labels and the imaging techniques are founded both in principle and in practice. However, along with the detailed investigation of these nanomaterials, some disadvantages need to be overcome before moving toward actual applications in experimental or clinical practice.

The most important disadvantage is the problem of the upconversion efficiency of UCNPs. Due to the large surface-to-volume ratio of small-sized nanoparticles, surface-quenching centres limit the upconversion efficiency. Although core-shell structures and some other techniques have been developed to overcome this problem, current materials still need significant improvement in their absolute efficiencies.

Another important question is the evaluation of the biosecurity of UCNPs. Some efforts have been made to investigate this point, but new techniques still need to be developed to monitor the metabolism, biodistribution, and possible toxicity of UCNPs. Furthermore, the long-term bioeffects of UCNPs need to be considered before the widespread clinical application of these materials.

REFERENCES

[1] F. Auzel, *Chem. Rev.* **104**, 139–174 (2004).

[2] M. X. Yu, F. Y. Li, Z. G. Chen, H. Hu, C. Zhan, H. Yang and C. H. Huang, *Anal. Chem.* **81**, 930–935 (2009).

[3] L. Q. Xiong, Z. G. Chen, Q. W. Tian, T. Y. Cao, C. J. Xu and F. Y. Li, *Anal. Chem.* **81**, 8687–8694 (2009).

[4] Q. Liu, Y. Sun, T. S. Yang, W. Feng, C. G. Li and F. Y. Li, *J. Am. Chem. Soc.* **133**, 17122–17125 (2011).

[5] J. Zhou, Z. Liu and F. Y. Li, *Chem. Soc. Rev.* **41**, 1323–1349 (2012).

[6] F. Wang, D. Banerjee, Y. S. Liu, X. Y. Chen and X. G. Liu, *Analyst* **135**, 1839–1854 (2010).

[7] D. K. Chatterjee, M. K. Gnanasammandhan and Y. Zhang, *Small* **6**, 2781–2795 (2010).

[8] H. S. Mader, P. Kele, S. M. Saleh and O. S. Wolfbeis, *Curr. Opin. Chem. Biol.* **14**, 582–596 (2010).

[9] D. R. Gamelin and H. U. Gudel, *Trans. Metal Rare Earth Comp.* **214**, 1–56 (2001).

[10] G. F. Wang, Q. Peng and Y. D. Li, *Acc. Chem. Res.* **44**, 322–332 (2011).

[11] F. Wang and X. G. Liu, *Chem. Soc. Rev.* **38**, 976–989 (2009).

[12] F. Wang and X. G. Liu, *J. Am. Chem. Soc.* **130**, 5642–5643 (2008).

[13] K. W. Krämer, D. Biner, G. Frei, H. U. Güdel, M. P. Hehlen and S. R. Lüthi, *Chem. Mater.* **16**, 1244–1251 (2004).

[14] L. Y. Wang and Y. D. Li, *Chem. Mater.* **19**, 727–734 (2007).

[15] Z. Q. Li and Y. Zhang, *Nanotechnology* **19**, 345606 (2008).

[16] H. X. Mai, Y. W. Zhang, R. Si, Z. G. Yan, L. D. Sun, L. P. You and C. H. Yan, *J. Am. Chem. Soc.* **128**, 6426–6436 (2006).

[17] G. Y. Chen, T. Y. Ohulchanskyy, R. Kumar, H. Agren and P. N. Prasad, *ACS Nano* **4**, 3163–3168 (2010).

[18] G. S. Yi and G. M. Chow, *Adv. Funct. Mater.* **16**, 2324–2329 (2006).

[19] G. S. Yi, H. C. Lu, S. Y. Zhao, G. Yue, W. J. Yang, D. P. Chen and L. H. Guo, *Nano Lett.* **4**, 2191–2196 (2004).

[20] F. Zhang, J. Li, J. Shan, L. Xu and D. Y. Zhao, *Chem. Eur. J.* **15**, 11010–11019 (2009).

[21] W. Feng, L. D. Sun, Y. W. Zhang and C. H. Yan, *Coord. Chem. Rev.* **254**, 1038–1053 (2010).

[22] Y. W. Zhang, X. Sun, R. Si, L. P. You and C. H. Yan, *J. Am. Chem. Soc.* **127**, 3260–3261 (2005).

[23] J. C. Boyer, F. Vetrone, L. A. Cuccia and J. A. Capobianco, *J. Am. Chem. Soc.* **128**, 7444–7445 (2006).

[24] X. Wang, J. Zhuang, Q. Peng and Y. D. Li, *Nature* **437**, 121–124 (2005).

[25] V. Mahalingam, F. Mangiarini, F. Vetrone, V. Venkatramu, M. Bettinelli, A. Speghini and J. A. Capobianco, *J. Phys. Chem. C* **112**, 17745–17749 (2008).

[26] M. Daldosso, D. Falcomer, A. Speghini, M. Bettinelli, S. Enzo, B. Lasio and S. Polizzi, *J. Alloy. Compd.* **451**, 553–556 (2008).

[27] R. Naccache, F. Vetrone, A. Speghini, M. Bettinelli and J. A. Capobianco, *J. Phys. Chem. C* **112**, 7750–7756 (2008).

[28] C. Chen, L. D. Sun, Z. X. Li, L. L. Li, J. Zhang, Y. W. Zhang and C. H. Yan, *Langmuir* **26**, 8797–8803 (2010).

[29] X. M. Liu, J. W. Zhao, Y. J. Sun, K. Song, Y. Yu, C. A. Du, X. G. Kong and H. Zhang, *Chem. Commun.* **43**, 6628–6630 (2009).

[30] T. Zhang, H. Guo and Y. M. Qiao, *J. Lumin.* **129**, 861–866 (2009).

[31] H. Q. Wang and T. Nann, *ACS Nano* **3**, 3804–3808 (2009).

[32] C. C. Mi, Z. H. Tian, C. Cao, Z. J. Wang, C. B. Mao and S. K. Xu, *Langmuir* **27**, 14632–14537 (2011).

[33] J. C. Boyer and F. C. J. M. van Veggel, *Nanoscale* **2**, 1417–1419 (2010).

[34] G. S. Yi and G. M. Chow, *Chem. Mater.* **19**, 341–343 (2007).

[35] H. X. Mai, Y. W. Zhang, L. D. Sun and C. H. Yan, *J. Phys. Chem. C* **111**, 13721–13729 (2007).

[36] Y. F. Wang, L. D. Sun, J. W. Xiao, W. Feng, J. C. Zhou, J. Shen and C. H. Yan, *Chem. Eur. J.* **18**, 5558–5564 (2012).

[37] A. Bednarkiewicz, D. Wawrzynczyk, A. Gagor, L. Kepinski, M. Kurnatowska, L. Krajczyk, M. Nyk, M. Samoc and W. Strek, *Nanotechnology* **23**, 145705 (2012).

[38] W. Feng, L. D. Sun and C. H. Yan, *Chem. Commun.* **29**, 4393–4385 (2009).

[39] H. Zhang, Y. J. Li, I. A. Ivanov, Y. Q. Qu, Y. Huang and X. F. Duan, *Angew. Chem. Int. Ed.* **49**, 2865–2868 (2010).

[40] L. Sudheendra, V. Ortalan, S. Dey, N. D. Browning and I. M. Kennedy, *Chem. Mater.* **23**, 2987–2993 (2011).

[41] N. Liu, W. P. Qin, G. S. Qin, T. Jiang and D. Zhao, *Chem. Commun.* **47**, 7671–7673 (2011).

[42] S. Schietinger, T. Aichele, H. Q. Wang, T. Nann and O. Benson, *Nano Lett.* **10**, 134–138 (2010).

[43] F. Vetrone, J. C. Boyer, J. A. Capobianco, A. Speghini and M. Bettinelli, *Chem. Mater.* **15**, 2737–2743 (2003).

[44] H. Q. Liu, L. L. Wang and S. G. Chen, *Mater. Lett.* **61**, 3629–3631 (2007).

[45] A. X. Yin, Y. W. Zhang, L. D. Sun and C. H. Yan, *Nanoscale* **2**, 953–959 (2010).

[46] O. Ehlert, R. Thomann, M. Darbandi and T. Nann, *ACS Nano* **2**, 120–124 (2008).

[47] J. Zhou, Y. Sun, X. X. Du, L. Q. Xiong, H. Hu and F. Y. Li, *Biomaterials* **31**, 3287–3295 (2010).

[48] L. Q. Xiong, T. S. Yang, Y. Yang, C. J. Xu and F. Y. Li, *Biomaterials* **31**, 7078–7085 (2010).

[49] Z. Q. Li, L. Wang, Z. Y. Wang, X. G. Liu and Y. J. Xiong, *J. Phys. Chem. C* **115**, 3291–3296 (2011).

[50] Z. G. Chen, H. L. Chen, H. Hu, M. X. Yu, F. Y. Li, Q. Zhang, Z. G. Zhou, T. Yi and C. H. Huang, *J. Am. Chem. Soc.* **130**, 3023–3029 (2008).

[51] H. P. Zhou, C. H. Xu, W. Sun and C. H. Yan, *Adv. Funct. Mater.* **19**, 3892–3900 (2009).

[52] L. Cheng, K. Yang, S. Zhang, M. W. Shao, S. T. Lee and Z. Liu, *Nano Res.* **3**, 722–732 (2010).

[53] K. Song, X. G. Kong, X. M. Liu, Y. L. Zhang, Q. H. Zeng, L. P. Tu, Z. Shi and H. Zhang, *Chem. Commun.* **48**, 1156–1158 (2012).

[54] L. Y. Wang, R. X. Yan, Z. Y. Hao, L. Wang, J. H. Zeng, H. Bao, X. Wang, Q. Peng and Y. D. Li, *Angew. Chem. Int. Ed.* **44**, 6054–6057 (2005).

[55] Q. Liu, C. Y. Li, T. S. Yang, T. Yi and F. Y. Li, *Chem. Commun.* **46**, 5551–5553 (2010).

[56] S. Sivakumar, P. R. Diamente and F. C. van Veggel, *Chem. Eur. J.* **12**, 5878–5884 (2006).

[57] R. A. Jalil and Y. Zhang, *Biomaterials* **29**, 4122–4128 (2008).

[58] S. Nagarajan and Y. Zhang, *Nanotechnology* **22**, 395101 (2011).

[59] C. Wang, H. Q. Tao, L. Cheng and Z. Liu, *Biomaterials* **32**, 6145–6154 (2011).

[60] T. S. Yang, Y. Sun, Q. Liu, W. Feng, P. Y. Yang and F. Y. Li, *Biomaterials* **33**, 3733–3742 (2012).

[61] S. W. Wu, G. Han, D. J. Milliron, S. Aloni, V. Altoe, D. V. Talapin, B. E. Cohen and P. J. Schuck, *Proc. Natl. Acad. Sci. U. S. A.* **106**, 10917–10921 (2009).

[62] J. C. Boyer, M. P. Manseau, J. I. Murray and F. C. J. M. van Veggel, *Langmuir* **26**, 1157–1164 (2010).

[63] H. Hu, M. X. Yu, F. Y. Li, Z. G. Chen, X. Gao, L. Q. Xiong and C. H. Huang, *Chem. Mater.* **20**, 7003–7009 (2008).

[64] F. Vetrone, R. Naccache, A. J. de la Fuente, F. Sanz-Rodriguez, A. Blazquez-Castro, E. M. Rodriguez, D. Jaque, J. G. Sole and J. A. Capobianco, *Nanoscale* **2**, 495–498 (2010).

[65] J. F. Jin, Y. J. Gu, C. W. Y. Man, J. P. Cheng, Z. H. Xu, Y. Zhang, H. S. Wang, V. H. Y. Lee, S. H. Cheng and W. T. Wong, *ACS Nano* **5**, 7838–7847 (2011).

[66] D. K. Chatterjee, A. J. Rufalhah and Y. Zhang, *Biomaterials* **29**, 937–943 (2008).

[67] L. Q. Xiong, Z. G. Chen, M. X. Yu, F. Y. Li, C. Liu and C. H. Huang, *Biomaterials* **30**, 5592–5600 (2009).

[68] Y. Sun, S. Cressman, N. Fang, P. R. Cullis and D. D. Y. Chen, *Anal. Chem.* **80**, 3105–3111 (2008).

[69] T. Zako, H. Nagata, N. Terada, A. Utsumi, M. Sakono, M. Yohda, H. Ueda, K. Soga and M. Maeda, *Biochem. Bioph. Res. Co.* **381**, 54–58 (2009).

[70] M. Wang, C. C. Mi, W. X. Wang, C. H. Liu, Y. F. Wu, Z. R. Xu, C. B. Mao and S. K. Xu, *ACS Nano* **3**, 1580–1586 (2009).

[71] R. Kumar, M. Nyk, T. Y. Ohulchanskyy, C. A. Flask and P. N. Prasad, *Adv. Funct. Mater.* **19**, 853–859 (2009).

[72] M. Wang, C. C. Mi, Y. X. Zhang, J. L. Liu, F. Li, C. B. Mao and S. K. Xu, *J. Phys. Chem. C* **113**, 19021–19027 (2009).

[73] S. F. Lim, R. Riehn, W. S. Ryu, N. Khanarian, C. K. Tung, D. Tank and R. H. Austin, *Nano Lett.* **6**, 169–174 (2006).

[74] J. C. Zhou, Z. L. Yang, W. Dong, R. J. Tang, L. D. Sun and H. Yan, *Biomaterials* **32**, 9059–9067 (2011).

[75] M. Nyk, R. Kumar, T. Y. Ohulchanskyy, E. J. Bergey and P. N. Prasad, *Nano Lett.* **8**, 3834–3838 (2008).

[76] H. Kobayashi, N. Kosaka, M. Ogawa, N. Y. Morgan, P. D. Smith, C. B. Murray, X. C. Ye, J. Collins, G. A. Kumar, H. Bell and P. L. Choyke, J. Mater. *Chem.* **19**, 6481–6484 (2009).

[77] T. Y. Cao, Y. Yang, Y. Gao, J. Zhou, Z. Q. Li and F. Y. Li, *Biomaterials* **32**, 2959–2968 (2011).

[78] S. A. Hilderbrand, F. W. Shao, C. Salthouse, U. Mahmood and R. Weissleder, *Chem. Commun* **28**, 4188–4190 (2009).

[79] N. M. Idris, Z. Q. Li, L. Ye, E. K. W. Sim, R. Mahendran, P. C. L. Ho and Y. Zhang, *Biomaterials* **30**, 5104–5113 (2009).

[80] J. Pichaandi, J. C. Boyer, K. R. Delaney and F. C. J. M. van Veggel, *J. Phys. Chem. C* **115**, 19054–19064 (2011).

[81] B. Yan, J. C. Boyer, N. R. Branda and Y. Zhao, *J. Am. Chem. Soc.* **133**, 19714–19717 (2011).

[82] Y. M. Yang, Q. Shao, R. R. Deng, C. Wang, X. Teng, K. Cheng, Z. Cheng, L. Huang, Z. Liu, X. G. Liu and B. G. Xing, *Angew. Chem. Int. Ed.* **51**, 3125–3129 (2012).

[83] S. Aime, D. D. Castelli, S. G. Crich, E. Gianolio and E. Terreno, *Acc. Chem. Res.* **42**, 822–831 (2009).

[84] Y. I. Park, J. H. Kim, K. T. Lee, K. S. Jeon, H. Bin Na, J. H. Yu, H. M. Kim, N. Lee, S. H. Choi, . I. Baik, H. Kim, S. P. Park, B. J. Park, Y. W. Kim, S. H. Lee, S. Y. Yoon, I. C. Song, W. K. Moon, Y. D. Suh and T. Hyeon, *Adv. Mater.* **21**, 4467–4471 (2009).

[85] S. Jiang, Y. Zhang, K. M. Lim, E. K. W. Sim and L. Ye, *Nanotechnology* **20**, 155101 (2009).

[86] L. Cheng, K. Yang, Y. G. Li, X. Zeng, M. W. Shao, S. T. Lee and Z. Liu, *Biomaterials* **33**, 2215–2222 (2012).

[87] M. E. Lim, Y. L. Lee, Y. Zhang and J. J. H. Chu, *Biomaterials* **33**, 1912–1920 (2012).

[88] L. Cheng, K. Yang, Y. G. Li, J. H. Chen, C. Wang, M. W. Shao, S. T. Lee and Z. Liu, *Angew. Chem. Int. Ed.* **50**, 7385–7390 (2011).

[89] S. S. Cui, H. Y. Chen, H. Y. Zhu, J. M. Tian, X. M. Chi, Z. Y. Qian, S. Achilefu and Y. Q. Gu, *J. Mater. Chem.* **22**, 4861–4873 (2012).

[90] N. J. J. Johnson, W. Oakden, G. J. Stanisz, R. Scott Prosser and F. C. J. M. van Veggel, *Chem. Mater.* **23**, 3714–3722 (2011).

[91] F. Chen, W. B. Bu, S. J. Zhang, X. H. Liu, J. N. Liu, H. Y. Xing, Q. F. Xiao, L. P. Zhou, W. J. Peng, L. Z. Wang and J. L. Shi, *Adv. Funct. Mater.* **21**, 4285–4294 (2011).

[92] J. Zhou, M. X. Yu, Y. Sun, X. Z. Zhang, X. J. Zhu, Z. H. Wu, D. M. Wu and F. Y. Li, *Biomaterials* **32**, 1148–1156 (2011).

[93] J. N. Liu, W. B. Bu, S. J. Zhang, F. Chen, H. Y. Xing, L. M. Pan, L. P. Zhou, W. J. Peng and J. L. Shi, *Chem. Eur. J.* **18**, 2335–2341 (2012).

[94] F. Liu, E. Ma, D. Q. Chen, Y. L. Yu and Y. S. Wang, *J. Phys. Chem. B* **110**, 20843–20846 (2006).

[95] H. Lin, D. Q. Chen, Y. L. Yu, Z. F. Shan, P. Huang, Y. S. Wang and J. L. Yuan, *J. Appl. Phys.* **107**, 103511 (2010).

[96] D. Q. Chen, Y. L. Yu, P. Huang, F. Y. Weng, H. Lin and S. Wang, *Appl. Phys. Lett.* **94**, 041909 (2009).

[97] F. Chen, S. J. Zhang, W. B. Bu, X. H. Liu, Y. Chen, Q. J. He, M. Zhu, L. X. Zhang, L. P. Zhou, W. J. Peng and J. L. Shi, *Chem. Eur. J.* **16**, 11254–11260 (2010).

[98] A. Xia, Y. Gao, J. Zhou, C. Y. Li, T. S. Yang, D. M. Wu, L. M. Wu and F. Y. Li, *Biomaterials* **32**, 7200–7208 (2011).

[99] C. N. Zhong, P. P. Yang, X. B. Li, C. X. Li, D. Wang, S. L. Gai and J. Lin, *RSC Adv.* **2**, 3194–3197 (2012).

[100] M. He, P. Huang, C. L. Zhang, H. Y. Hu, C. C. Bao, G. Gao, R. He and D. X. Cui, *Adv. Funct. Mater.* **21**, 4470–4477 (2011).

[101] Y. L. Liu, K. L. Ai, J. H. Liu, Q. H. Yuan, Y. Y. He and L. H. Lu, *Angew. Chem. Int. Ed.* **51**, 1437–1442 (2011).

[102] Y. Sun, M. X. Yu, S. Liang, Y. J. Zhang, C. G. Li, T. T. Mou, W. J. Yang, X. Z. Zhang, B. Li, C. H. Huang and F. Y. Li, *Biomaterials* **32**, 2999–3007 (2011).

[103] L. Cheng, K. Yang, M. W. Shao, X. H. Lu and Z. Liu, *Nanomedicine* **6**, 1327–1340 (2011).

14

MICROBUBBLES: CONTRAST AGENTS FOR ULTRASOUND AND MRI

APRIL M. CHOW AND ED X. WU

Laboratory of Biomedical Imaging and Signal Processing and Department of Electrical and Electronic Engineering, The University of Hong Kong, Pokfulam, Hong Kong SAR, China

14.1 INTRODUCTION

Gas-filled microbubbles offer new approaches in diagnostic imaging techniques and therapeutic interventions. Current applications of microbubbles include morphological enhancement and microvascular perfusion, which have been widely used clinically. Apart from conventional diagnostic purposes, microbubbles have also demonstrated potential in several exciting areas through various imaging modalities, including molecular imaging and therapeutic intervention. In this chapter, we will review the commonly used gas-filled microbubbles and their characteristics. We will summarise applications with several imaging modalities and future uses. Finally, we will discuss the bioeffects and safety issues related to microbubble applications.

The capability of gas-filled microbubbles in enhancement ultrasound (US) imaging contrast was first reported by Raymond Gramiak and Pravin Shah in 1968 [1]. During cardiac catheterisation, contrast enhancement was observed in the aorta following injection of agitated saline. Since the intervention of the gas-filled microbubbles for clinical applications, its development has been active and exciting in radiology. Gas-filled microbubbles then emerged as an essential adjunct to US imaging technology, bringing insights and breakthroughs toward therapeutic development and medical diagnosis.

Gas-filled microbubbles were originally developed as an intravascular contrast agent to enhance acoustic backscattering in US imaging. They are generally administered intravenously or in a cavity for better anatomy, measurement of tissue perfusion, precise drug delivery mechanisms, and determination of elastic properties of the tissue in US imaging [2–4]. Due to their unique cavitation and sonoporation properties [5–9], gas-filled microbubbles also play an expanding role in therapeutic applications. The relatively short *in vivo* lifetime of microbubbles may pose challenges in some of these applications. Nevertheless, the microbubble fabrication technology is advancing for increased *in vivo* lifetimes using surfactant molecules with multiphase mixing technique and molecular targeting capability by microbubble surface modification [10, 11].

14.2 CLASSIFICATION OF MICROBUBBLES

Microbubbles are composed of a shell of biocompatible materials, such as proteins, lipids, or polymers, with air, perfluorocarbon, or sulphur hexafluoride as the filling gas. In order to be a transpulmonary contrast agent, their size normally ranges from 3 to 10 μm. Figure 14.1 shows a representative light micrograph of SonoVue® microbubble suspensions and its histogram showing diameter distribution.

The Chemistry of Molecular Imaging, First Edition. Edited by Nicholas Long and Wing-Tak Wong.
© 2015 John Wiley & Sons, Inc. Published 2015 by John Wiley & Sons, Inc.

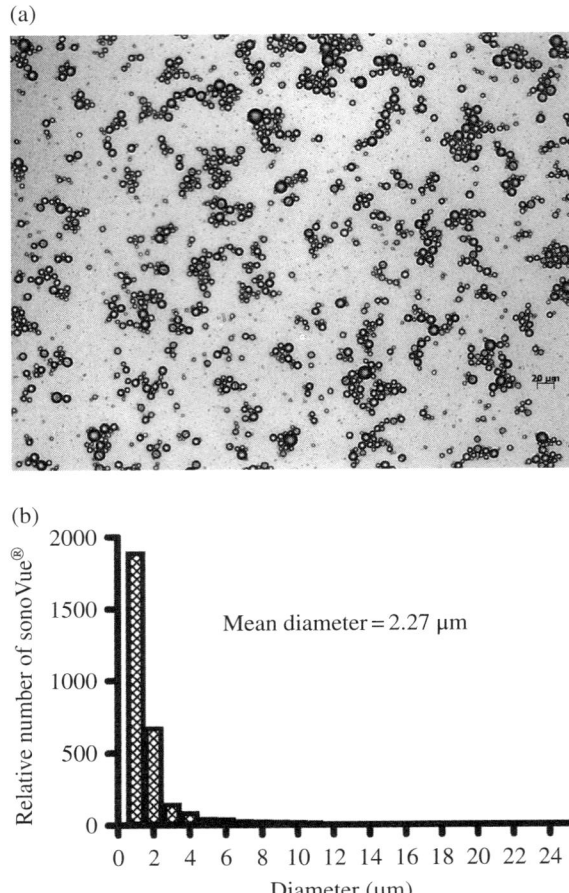

FIGURE 14.1 (a) Representative light micrograph of SonoVue® microbubble suspensions. (b) Histogram showing diameter distribution for a representative batch of SonoVue. The mean diameters of SonoVue microbubbles are 2.95 μm.

An ideal microbubble agent would have several essential characteristics:

- High echogenicity
- Inert
- Stable during bolus injection, infusion, cardiac and pulmonary passage
- Narrow size distribution
- Well-defined response toward incident US

Early versions of microbubbles were not optimised and yielded uneven sizes and poor stability. Advances in microbubble stabilisation technology and preparation processes have enhanced its *in vivo* stability and improved its size distribution to a certain extent [12–14]. To increase microbubble stability and persistence in the peripheral blood circulation, one can manipulate the shell materials or filling gases. The shell materials stabilise the microbubble itself and reduce gas diffusion into the surrounding blood. The shell can be stiff (denatured proteins or polymers) or flexible (phospholipids). Its thickness ranges from 10 nm to 200 nm with thinner shells typically used by protein and lipid microbubbles and thicker shells by polymeric microbubbles [15]. On the other hand, low solubility and low diffusibility gases, including perfluorocarbons and sulphur hexafluoride gas, have been found to enhance microbubble stability *in vivo*. The ideal filling gas should be inert and should present a high vapour pressure and the lowest solubility in blood. A limited solubility in blood produces an elevated vapour concentration in the microbubble relative to the surrounding blood and establishes an osmotic gradient that opposes the gas diffusion out of the microbubble. Recent microbubbles combined both approaches to yield longer plasma half-life.

Gas-filled microbubbles can be classified into non-transpulmonary and transpulmonary types. Non-transpulmonary microbubbles cannot pass through the lung, and are usually composed of air gas core. Transpulmonary microbubbles are

TABLE 14.1 Various Types of Microbubbles For Ultrasound (US) Imaging.

Microbubble	Shell	Filling Gas
Albunex®	Human albumin	Air
BR14	Phospholipid	Perfluorobutane
Definity®	Phospholipid	Octafluoropropane
Echogen®	-	Dodecafluoropentane
Echovist®	Galactose	Air
Imagent®	Phospholipid	Perfluorohexane
Myomap®	Recombinant albumin	Air
Optison®	Human albumin	Octafluoropropane
Quantison®	Recombinant albumin	Air
Perflubron®	-	Perfluorooctyl bromide
Sonavist®	Polymer	Air
Sonazoid®	Phospholipid	Perfluorobutane
SonoVue®	Phospholipid	Sulphur hexafluoride

usually composed of high molecular weight gas and with a stable lipid or polymer shell. Transpulmonary microbubbles offer higher diagnostic potential compared to those that cannot pass the pulmonary capillary bed after intravenous injection. A comparison between different types of microbubbles is shown in Table 14.1 with their coating materials and core gas types.

14.2.1 Air-Filled Microbubbles

Air-filled microbubbles were the earliest commercially available US contrast agent. While air presents high solubility in blood, the plasma half-life of air-filled microbubbles is very limited. The air quickly leaks from the microbubble shell into peripheral blood. In addition, microbubbles that were transpulmonary did not persist long enough for imaging because the filling air quickly dissolved into the blood.

14.2.1.1 Air-Filled Microbubbles with a Galactose Shell Echovist® (Schering AG, Berlin, Germany) is an air-filled microbubble stabilised with a galactose matrix. Its mean diameter is about 2 μm with 97% smaller than 6 μm; however, Echovist is not stable enough to pass the lung. They are administered intravenously for echocardiography to opacify right heart cavities and to detect cardiac shunts [16]. Levovist® (Schering AG, Berlin, Germany) is another air-filled microbubble with a galactose shell for imaging to enhance echostructures of liver, kidney, and heart. With the addition of palmitic acid, its stability increases such that it can pass the lung capillary and give systemic enhancements for up to 5 min. post-injection [17]. After plasma clearance, Levovist can accumulate in the reticuloendothelial system for up to 20 min. for hepatosplenic-specific enhancement [18].

14.2.1.2 Air-Filled Microbubbles with an Albumin Shell Albunex® (Molecular Biosystems Inc., San Diego, CA) is the first transpulmonary microbubble. It consists of air-filled microbubbles coated with human albumin of 30 to 50 nm thick. The mean diameter is 3.8 μm with standard deviation of 2.5 μm. Albunex can reach the left ventricle despite its relatively short half-life of less than 1 min. Quantison® (Quadrant Healthcare, Nottingham, United Kingdom) is similar to Albunex, but with a relatively thick and rigid shell of recombinant albumin of 200-300 nm. In a biodistribution study of Quantison, the highest uptake was observed in liver ($41.8 \pm 10.4\%$) at 1 h following intravenous administration [19], enabling liver-specific imaging. Myomap® (Quadrant Healthcare, Nottingham, United Kingdom) is another air-filled microbubble stabilised with recombinant albumin, which is 600-1000 nm thick.

14.2.1.3 Air-Filled Microbubbles with a Cyanoacrylate Shell Sonavist® (Schering AG, Berlin, Germany) is an air-filled microbubble synthesised by emulsion polymerisation. The shell material is a biodegradable n-butyl-2-cyanoacrylate polymer. Sonavist can stay intact in the circulation for up to 10 min. post-injection and provides passive targeting for the reticuloendothelial system [20, 21].

14.2.2 Perfluorocarbon-Filled Microbubbles

A significant improvement in the *in vivo* stability of microbubbles was achieved when perfluorocarbon gas began to be used as the core gas. Perfluorocarbon gases have high molecular weight, low solubility, and low diffusivity, thus they are good candidates for core gas that can keep the microbubble size stable *in vivo* [22].

14.2.2.1   Perfluorochemicals Perflubron® (Alliance Pharmaceutical Corporation, San Diego, CA) is liquid perfluoro-carbon emulsion with size ranges from 0.06 to 0.25 μm composed of carbon and bromine; it is inert with low surface tension and immiscible with water. Upon emulsification, it accumulates in the reticuloendothelial system and leaks from inflammatory or tumoural capillaries into the interstitial space for US imaging after intravenous injection [23, 24].

14.2.2.2   Perfluorocarbon-Filled Microbubbles with a Phospholipid Shell BR14 (Bracco Diagnostics, Geneva, Switzerland) is a perfluorobutane-filled microbubble stabilised using phospholipid. It can produce persistent contrast enhancement of tissue perfusion for liver and spleen imaging [25, 26]. Definity® (Bristol-Myers Squibb Medical Imaging, North Billerica, MA) is a phospholipid-coated microbubble filled with octafluoropropane with a mean diameter of 2.5 μm. Other phospholipid microbubbles include perfluorohexane-filled Imagent® (Imcor Pharmaceutical, San Diego, CA) and perfluorobutane-filled Sonazoid® (Amersham Health, Osla, Norway). Both Imagent and Sonazoid show a late liver-specific enhancement post-injection [27–29].

14.2.2.3   Perfluorocarbon-Filled Microbubbles with an Albumin Shell Optison® (Amersham Health, Princeton, NJ) is very similar to Albunex except the filling gas is octafluoropropane, with diameter ranges from 2 to 4.5 μm. When the micro-bubble shell collapses, the relatively inert octafluoropropane gas is released into the plasma and vents out via the lung. It has been reported that the Kupffer cells in the liver uptake albumin that includes both intact microbubbles and deflated albumin shells or fragments [30, 31]. It is approved for clinical use to provide opacification of cardiac chambers and to improve left ventriclar endocardial border delineation for cardiac imaging.

14.2.2.4   Phase Shift Perfluorocarbon-Filled Microbubbles EchoGen® (Sonus Pharmaceuticals, Bothell, WA) is a liquid/liquid emulsion with size of 0.4 μm [17], containing dodecafluoropentane in a dispersed phase. Dodecafluoropentane is a low diffusibility and low solubility perfluorocarbon gas, with a low boiling point of 28.5°C. Upon administration, dode-cafluoropentane shifts to gas phase at body temperature, forming microbubbles with a diameter of 2 to 5 μm. The phase transition from liquid to gas state is achieved by producing a hypobaric pressure followed by an intense shock within the syringe immediately prior to administration [32].

14.2.3   Sulphur Hexafluoride-Filled Microbubbles

Sulphur hexafluoride presents a low diffusibility through the phospholipid layer and a low solubility in blood, extending microbubble circulation time *in vivo*. SonoVue® (Bracco Diagnostics, Milan, Italy) consists of sulphur hexafluoride gas stabilised in an aqueous dispersion by a phospholipid monolayer. With its long plasma half-life of 6 min. and a narrow size distribution [33], SonoVue can improve the acoustic backscattering at low power insonation [34].

14.3   APPLICATIONS IN ULTRASOUND IMAGING

US imaging utilises typical frequency ranges from 2 to 18 MHz to visualise internal structures for diagnosis and therapeutic applications. Image contrast in US imaging is determined by the tissue acoustic impedance and the mode of imaging. The contrast mechanism of gas-filled microbubbles comes from the difference in acoustic impedance between the filling gas and the surrounding materials, as well as the compressed gas of the microbubbles compared to the incompressible mate-rials being displaced. Compared to the wavelength of the US beam, microbubbles are small and hence act as a point source for reflections. Moreover, the compressible microbubbles resonate in the US field, generating a nonlinear acoustic response that can be detected using harmonic imaging [35]. Under higher acoustic power, microbubble destruction occurs with the emission of an irregular nonlinear signal [36]. Behaviours of lipid and polymer microbubbles under different acoustic powers are depicted in Figure 14.2. Current applications of microbubbles in US imaging include, but are not limited to, morphological enhancement, microvascular perfusion, as well as molecular imaging and therapeutic intervention.

14.3.1   As an US Intravascular Contrast Agent

Due to the relatively large size of gas-filled microbubbles, they do not pass through the vascular endothelium and hence serve as pure intravascular contrast agents [37]. They have been used to improve the endocardial border of left ventricles for better delineation of left ventricular function and myocardial thickening [38]. Strong echoes were produced within the heart, due to the acoustic mismatch between microbubbles and the surrounding blood. Recent applications include

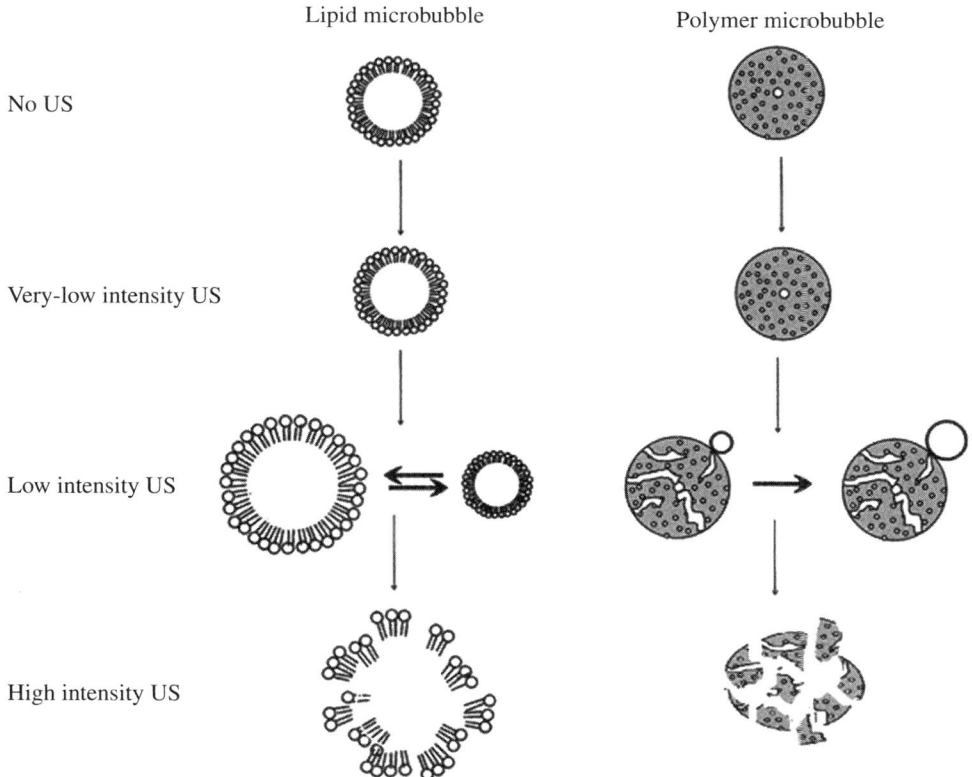

FIGURE 14.2 Interactions of microbubbles with US of increasing power. (Reproduced from Ref. [36], with permission from Elsevier).

vascular imaging to determine occlusion and blood flow and to detect atherosclerotic plaques. The enhancement of microbubbles is very promising in carotid artery imaging [39], nontargeting plaque visualisation [40, 41], and accurate stenosis assessment in carotid angiography compared to X-ray [41]. Contrast-enhanced US imaging has a unique application in microvascular perfusion. It depicts the nonperfused regions clearly and allows the detection of perfusion insufficiency during ischemia [42], vascular occlusion [43], and tissue infarction [44] and allows the detection of vascular insufficiency [45, 46]. Other than cardiac applications, contrast-enhanced US imaging has been widely used to characterise lesions in abdominal regions including liver [47–50], pancreas [51, 52], and gastrointestinal tract [53, 54]. The *in vivo* measurement of microvascular blood flow and blood volume also allows the detection of lesions/tumours from normal tissue [55, 56] and for the assessment of angiogenesis in tumour [57]. In addition to their vascular phase, some microbubbles can exhibit a tissue- or organ-specific phase for improved lesion conspicuity during late enhancement [21, 27, 29, 58–60]. Figure 14.3 illustrates the capability of identifying hepatocellular carcinoma using contrast-enhanced US imaging, which offers a unique approach for intravascular imaging noninvasively without ionising radiation.

14.3.2 As an Orally Administered US Contrast Agent

Diagnostic performance of abdominal US imaging is often limited by artefacts resulting from adjacent gas in the stomach and intestines. Water, simethicone, and methylcellulose were used to displace and disperse stomach and intestinal gas for better visualisation of upper abdomen [61]. However, it was prone to inter-subject variation and therefore results were inconsistent. SonoRx (Bracco Diagnostics Princeton, NJ), a simethicone-coated cellulose suspension, was then developed to improve visualisation of abdominal anatomy with reduction of gas artefacts [62, 63].

14.3.3 As a US Molecular Probe

Microbubbles can also be used in molecular imaging as a molecular probe [64]. The structure of targeted microbubbles consists of a lipid/polymer shell that can become a molecular probe by incorporating different antibodies, peptides, disintegrins, or other ligands [65, 66] (Figure 14.4).

(a)   (b)   (c)

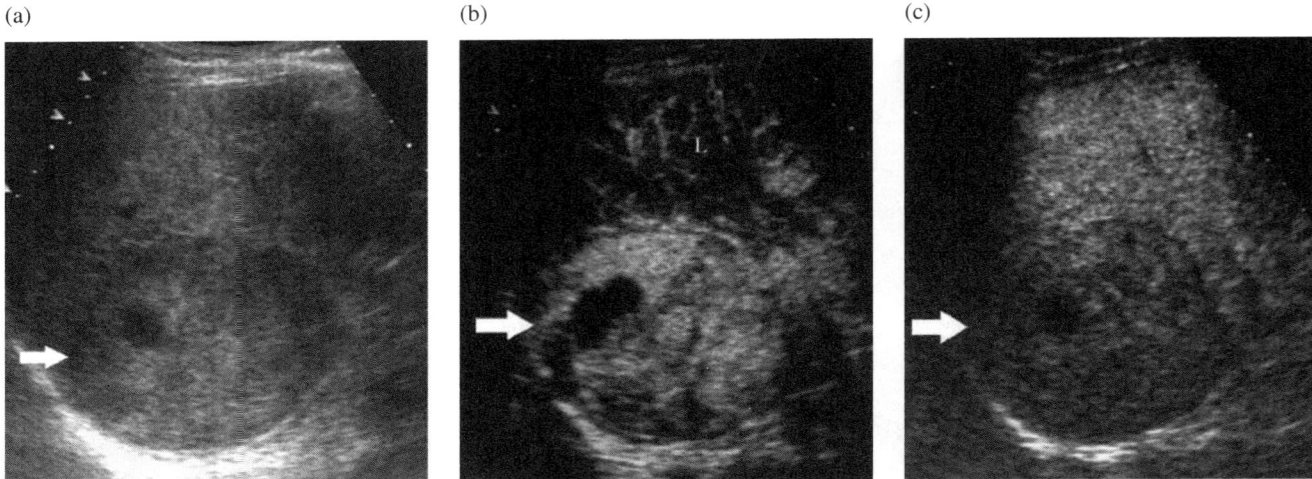

FIGURE 14.3 Hepatccellular carcinoma (arrow) presents heterogeneous appearance at (a) conventional US imaging. In contrast-enhanced US imaging, diffuse and heterogeneous contrast enhancement with hypervascular appearance (arrow) in comparison to the adjacent liver at (b) arterial phase, (c) 30 s after microbubble injection. Lesion appears hypovascular at (d) late phase, 120 s after microbubble injection. (Reproduced from Ref. [15], with permission from Springer).

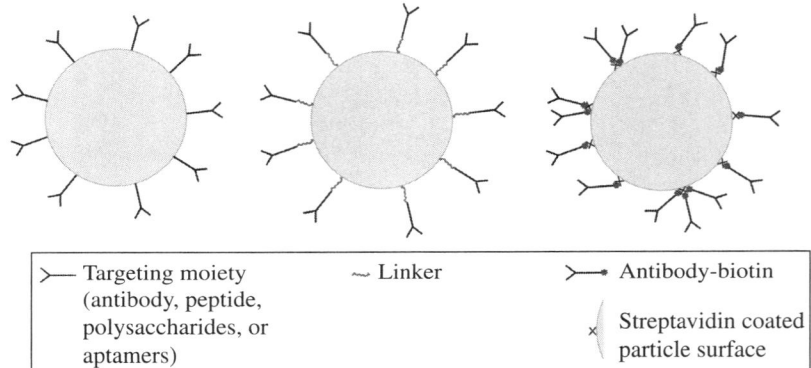

FIGURE 14.4 Targeted microbubbles can be produced by different coupling strategies depending on the targeting moiety (e.g., antibodies, peptides, polysaccharides, or aptamers).

Retention of targeted ligands and contrast agents leads to detection of molecular events through amplification effects [67]. Potential applications include detection of atherosclerotic plaques [68, 69], thrombi [70–74], inflammation [75–77], tumours, [78] and angiogenesis [79].

14.3.4 As a Therapeutic Agent

Local gene transfer and drug delivery have been another key area of application. Microbubble-mediated sonoporation can dramatically increase cell permeability and intracellular uptake with no apparent tissue damage and toxicity [80, 81]. Local microbubble cavitation by spatially focused US can be applied in achieving site-specific release of incorporated drugs or genes inside microbubbles [36, 82] (Figure 14.5).

Microbubble-mediated therapy has been used to deliver genes or drugs to specific tissues using microbubble cavitation and sonoporation effects, including neural tissues, skeletal muscles, myocardium, kidneys, vessels, and tumours [83–86]. Development of the improved uptake of vascular endothelial growth factor using microbubble sonication may provide coronary artery disease patients with new options such as therapeutic angiogenesis [87, 88].

Furthermore, the microbubble cavitation phenomenon has been put into practical use in achieving several therapeutic interventions. Sonothrcmbolysis, which employs the local shock waves produced by microbubble cavitation to fragment clots on a microscopic scale and restores blood flow, has been developed as a minimally invasive recanalisation technique in treating vascular thrombosis [89, 90]. In addition, transient opening of the blood-brain barrier through microbubble-enhanced

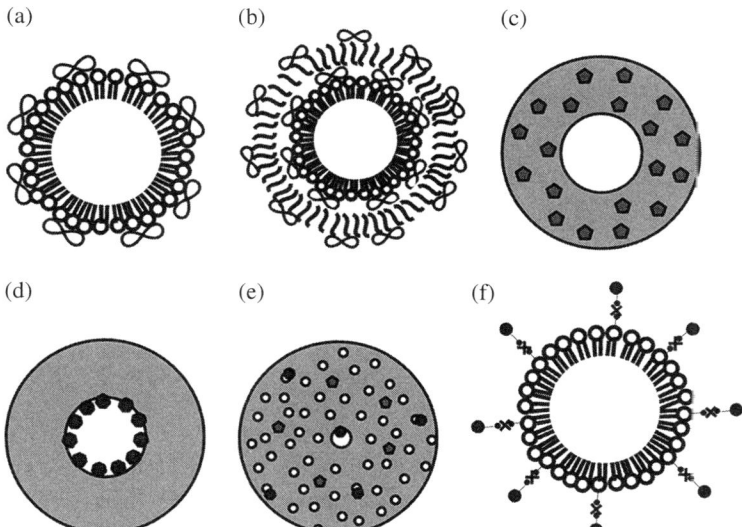

FIGURE 14.5 Schematic representation of loading strategies of drugs and genes on microbubbles. (a) Non-covalent binding of DNA to the surface of cationic lipid microbubbles. (b) Multilayered structure based on a lipid microbubble sequentially coated with DNA and poly-L-lysine layers. (c) Polymeric microbubble loaded with hydrophobic drug loaded in the shell phase. (d) Polymeric microbubble with hydrophilic drug loaded in the internal void. (e) Internal structure of polymeric microbubbles: Water-phase is dispersed through the polymer matrix, forming upon lyophilisation a plurality of cavities distributed over the particle volume. (f) Attachment of liposomes or nanoparticles to the surface of microbubbles through biotin–avidin–biotin bridging system. (Reproduced from Ref. [36], with permission from Elsevier).

sonoporation can be accomplished by applying transcranial US with intravenous injection of microbubbles without damaging the neurons [91]. Delivery of both low and high molecular weight therapeutic compounds to the central nervous system can be potentially attained through this noninvasive approach. Microbubbles are also used to enhance the noninvasive high intensity-focused US therapy by increasing the local heating rate [92].

14.3.5 Future Applications

The potential of molecular imaging and therapeutic interventions can be compromised by the limited sensitivities. Advanced microbubble fabrication technology could be sought to optimise size distribution, microbubble shell structure, ligand attachment, and drug/gene incorporation. Efforts could also be made in advancing US beam technique in order not to interfere with microbubbles.

14.4 APPLICATIONS IN MAGNETIC RESONANCE IMAGING

Relying on the differential decay and recovery characteristics of the nuclear magnetic resonance signals, magnetic resonance imaging (MRI) provides superb soft tissue contrast with high spatial resolution when compared with other imaging modalities. While MR image contrast can be flexibly controlled by varying pulse sequences and parameters, it is determined by the intrinsic tissue properties such as proton density, longitudinal relaxation time (T_1), and transverse relaxation time (T_2). At present, the exogenous contrast agents available for MRI mainly fall into three categories: gadolinium chelates, manganese chelates, and superparamagnetic iron oxide particles. Their effects are usually described by longitudinal relaxation rate (R_1) and transverse relaxation rate (R_2/R_2^*), where R_1, R_2 and R_2^* are defined as $1/T_1$, $1/T_2$, and $1/T_2^*$ respectively. Susceptibility contrast agents exhibit large R_2^*/R_1 ratios and predominantly induce signal loss through spin dephasing by strong magnetic susceptibility effects. Their T_2^* shortening effects are usually much stronger than the baseline T_2 effects.

14.4.1 As an MR Susceptibility Contrast Agent

Microbubbles can potentially be used as an intravascular MR susceptibility contrast agent *in vivo* due to the induction of large local magnetic susceptibility difference by the gas-liquid interface. Microbubble-induced signal perturbation depends on the microbubble radius, volume fraction, overall magnetic susceptibility difference between the microbubble and the blood plasma, and amplitude of the static field [93, 94].

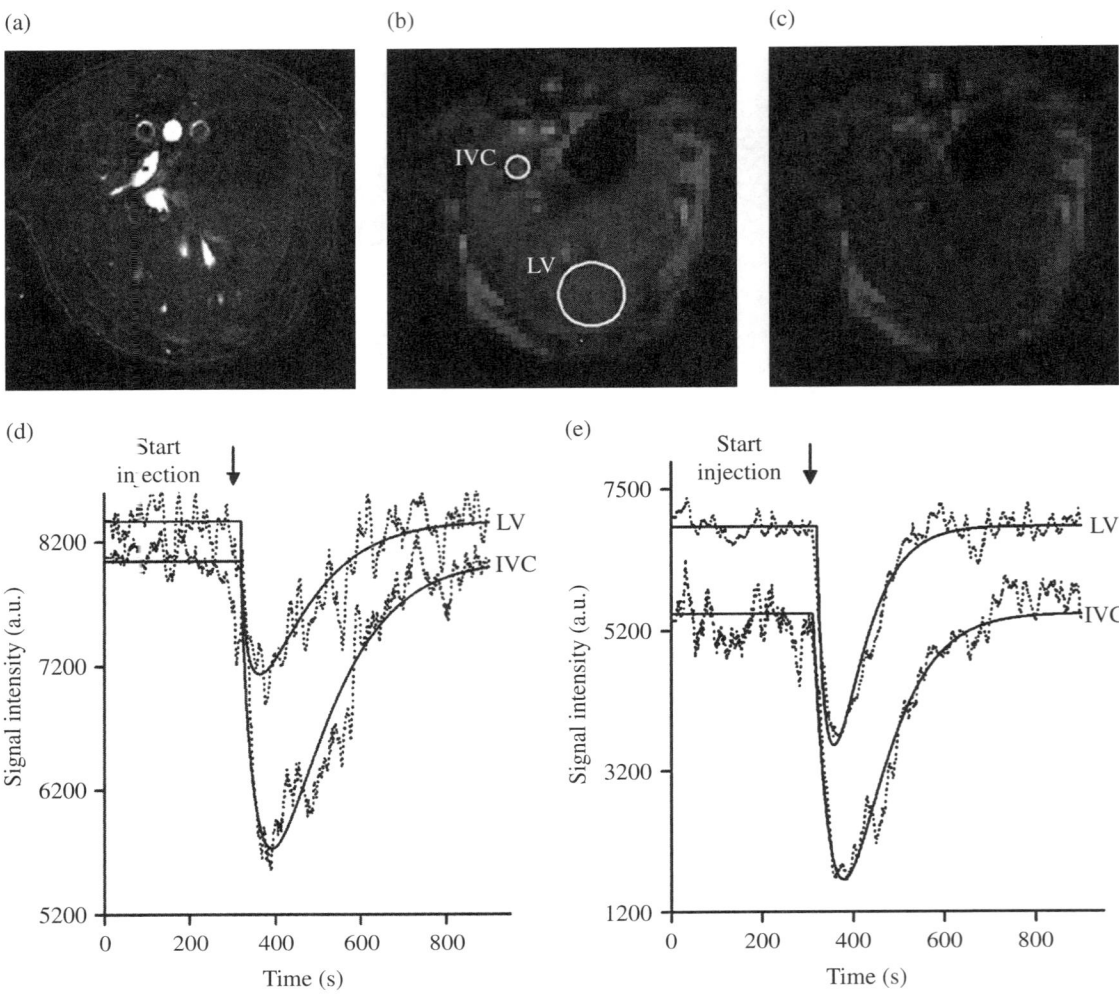

FIGURE 14.6 Corresponding images from the same rat at two adjacent slice locations during Optison® injection: (a) anatomical images, (b) preinjection T_2^*-weighted images, (c) maximum susceptibility contrast images after injection, (d) T_2^*-weighted images at ~12 min after injection, and (e) ΔR_2^* map computed from (b) and (c). Strong negative enhancement was seen in the venous branches in the liver. (Reproduced from Ref. [97], with permission from Wiley).

Early experiments with Albunex illustrated the potential of air-filled microbubbles as an MR susceptibility contrast agent [95]. Another experiment considered the potential of lipid-coated microbubbles as an MR susceptibility contrast agent for tumour detection, based on the accumulation of the microbubbles inside the tumours [96]. Linear relationship between apparent transverse relaxation rate ($R_2^* = 1/T_2^*$) and volume fraction for Optison was reported at 7 T [97]. R_2^* dependency on microbubble volume fraction was also reported for Levovist through an *in vitro* phantom study at 1.5 T [98]. Magnetic susceptibility enhancement induced by the gas-liquid interface was also demonstrated by simulations and MR experiments using air-filled cylinders in water [99], consolidating the feasibility of gas-filled microbubbles as a MR susceptibility contrast agent.

The first *in vivo* investigation of susceptibility contrast induced was performed in rat liver at 7 T [97] (Figure 14.6). Recently, the susceptibility effect of lipid-based microbubble SonoVue and air-filled custom-made albumin-coated microbubbles was also investigated in rat brain [100]. These results indicate that microbubbles could be used as a unique intravascular MR contrast agent. Such capability has the potential to lead to real-time MRI guidance in various microbubble-mediated drug delivery and therapeutic applications.

14.4.2 As an MR Pressure Sensor

Other than conventional intravascular contrast agents, the feasibility of microbubbles as an MR pressure sensor, based on the susceptibility change caused by pressure-induced microbubble size change, has also been explored through theoretical and phantom studies [94, 101–103].

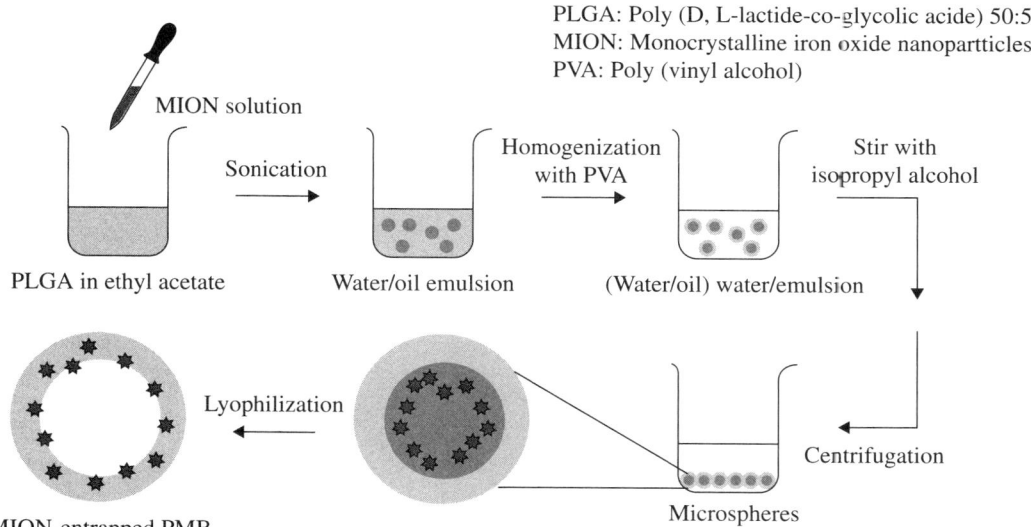

FIGURE 14.7 Flow diagram representing the method for synthesising iron oxide nanoparticles entrapped polymeric microbubbles to increase microbubble MR detectability. (Reproduced from Ref. [104], with permission from Wiley).

14.4.3 Increasing Microbubble MR Detectability

Microbubble susceptibility effects are relatively weak when compared with other intravascular MR susceptibility contrast agents. Given the possible microbubble toxicity at high dosage and the filtering of microbubbles larger than 10 μm by the lung capillary bed [97, 102], microbubble volume fraction and radius cannot be freely chosen. Without increasing static field strength, enhancement of overall magnetic susceptibility difference between the microbubble and the blood plasma is a preferred way to increase microbubble-induced signal perturbation.

Effective microbubble magnetic susceptibility can be manipulated by changing the shell thickness and the magnetic susceptibility of the shell or filling gas [102]. Because the type of filling gas is chosen mainly to improve the microbubble stability *in vivo*, modifying the magnetic susceptibility of the shell is a preferred means to increase the microbubble magnetic susceptibility. Theoretical studies have indicated that, by embedding or coating magnetic nanoparticles, the magnetic susceptibility of the shell can be increased [94, 102], thus enhancing the microbubble susceptibility effect and alleviating the dosage requirement for MRI applications (Figure 14.7). It was experimentally demonstrated that gas-filled polymeric microbubble susceptibility effects can be substantially enhanced by incorporating iron oxide nanoparticles into microbubble shells through phantom and rat studies [104, 105]. Such microbubbles loaded with oxide nanoparticles were also reported to possess increased acoustic backscattering and longer MR contrast enhancement [106, 107], suggesting its potential application as a dual contrast agent in US and MRI.

Microbubbles filled with hyperpolarised ^{3}He gas provide an alternative to increase MR signal, allowing acquisition of angiography and perfusion images with higher signal-to-noise ratio [108, 109]. However, it requires hardware tuning for signal reception at ^{3}He resonance frequency and may possibly hinder its clinical applications.

14.4.4 Future Applications

As discussed previously, microbubbles can be locally cavitated and destroyed by focused US [5]. Therefore, their MR signals can be temporally and spatially manipulated because microbubble disappearances will diminish the susceptibility effect, as confirmed in a phantom study [110]. Without causing microbubble sonoporation and cavitation, MRI visualisation may provide the most effective imaging guidance for microbubble-mediated genes or drug delivery. Incorporating gadolinium/manganese chelates into microbubble shells may also improve their susceptibility effect. Upon cavitation, the entrapped paramagnetic chelates could be released and in contact with surrounding water molecules, thus shortening the T_1 relaxation time and producing increased signal in T_1-weighted imaging. Such dark-to-bright change could be a potentially attractive way to monitor microbubble-based drug delivery and therapeutic applications.

Moreover, due to the capability of being locally cavitated, microbubbles possess potential blood-labelling applications with perfusion analysis [111]. Due to the hyperpermeable nature of tumour neo-vessels as compared with normal vessels [112], microbubbles may be able to pass through tumour vessels and therefore accumulate in the tumour tissues, producing differential enhancement. Microbubbles may serve as a MR contrast agent for imaging tumour angiogenesis. The limited *in vivo* lifetime and stability can be a challenge in various microbubble applications, but optimisation of physiological properties should be sought for future study.

14.5 APPLICATIONS BEYOND US IMAGING AND MRI

14.5.1 Positron Emission Tomography

Positron emission tomography (PET) detects annihilation photons emitted from radionuclides administered as positrons interact with electrons, hence generating images depicting distributions of radionuclides in the body. It is often coupled with computed tomography to acquire anatomical and functional images simultaneously. In PET, suitable radionuclides are generally short half-lived positron-emitting isotopes, including ^{11}C, ^{13}N, ^{15}O, ^{18}F, ^{62}Cu, ^{68}Ga, and ^{82}Rb. These radioisotopes can replace atoms in molecules that are essential for metabolism or in molecules that bind to receptors or other sites of drug action, enabling tracing the biological pathway of radiolabelled compounds for studies of functional processes in the body. To date, most clinical PET applications utilise ^{18}F-labelled fluorodeoxyglucose, an analogue of glucose, to probe regional glucose uptake for diagnosis, staging, and monitoring treatment of cancers.

The use of ^{18}F-labelled lipids, incorporated into microbubble shells, for measuring biodistribution of nontargeted microbubbles in rats was reported [113]. With the use of therapeutic US, microbubbles could be destructed at a specific site for characterisation of microbubble-mediated therapy. *Ex vivo* studies also confirmed that an increased deposition of lipid shells occurs locally after US sonication [113]. *In vivo* monitoring of biodistribution of targeted microbubbles in mice was also performed using ^{18}F-labelled lipids [114]. In general, PET provides a sensitive and quantitative means for evaluation of microbubble biodistribution via radiolabelling.

14.5.2 Diffraction-Enhanced Imaging (DEI)

Diffraction-enhanced imaging (DEI) is an X-ray phase contrast technique using diffraction contrast rather than absorption contrast. DEI is based on the use of a perfect crystal analyser placed between the region of interest and the imaging detector. After irradiating the sampling tissues with monochromatic X-rays, the analyser acts as an angular filter that selectively accepts those X-rays that have traversed the sample by satisfying the Bragg law for diffraction, thereby producing a two-dimensional spatial intensity mapping of the diffracted X-ray. Compared with conventional computed tomography, DEI can provide better soft tissue contrast with less radiation, and it has shown promise in breast lesion and bone and vascular imaging [115–117].

The feasibility of microbubbles Levovist and Optison as DEI scattering contrast agents was investigated [118]. Imaging contrast produced by microbubbles can be manipulated by varying the reflectivity of the analyser. Studies are warranted to further explore the values of microbubbles in enhancing DEI scattering for clinical applications.

14.6 CONCLUSIONS: LIMITATIONS, BIOEFFECTS, AND SAFETY

Despite the advantages of the unique capability of gas-filled microbubbles, there are several limitations. Microbubble-based applications are limited by the inherent short plasma half-lives, primarily as a result of their destruction in the alveoli due to gaseous exchange [15]. Possible toxicity is another issue that needs to be taken into consideration.

The potential hazards usually involve cavitation of microbubbles. At low acoustic pressures, microbubbles exhibit low amplitude volumetric oscillations following the changes in the local pressure, which is known as stable cavitation [119]. While exposed to higher pressures, the microbubble oscillation becomes increasingly asymmetric and eventually the shell may be broken, leading to the release of filling gas, which may lead to further cavitations. This nonlinear behaviour is called inertial cavitation [119]. Therefore, its potential risks increase with the acoustic pressure, which is defined by the mechanical index. Caution should be made in choosing the mechanical index for US to avoid undesired cavitation effects.

Generally, gas-filled microbubbles are relatively safe to use with no specific renal, liver, or cerebral toxicity [17]. The adverse reactions are rare, usually transient, and of mild intensity [17, 120]. Most frequently, local pain, warmth or cold, and tissue irritation may occur at the injection site or along the draining vein [121–123]. Recently, deaths and serious cardiopulmonary reactions following administration of microbubbles were reported [124]. However, the associated risk is 1:500000, which is advantageous with the 1:1000 mortality risk associated with diagnostic coronary angiography and the

1:2500 risk of myocardial infarction or death with treadmill exercise testing [125, 126]. Further research is required to clarify the biological effects of gas-filled microbubbles. In general, effects of microbubbles are reduced or eliminated through the use of lower pressure amplitudes, higher frequencies, and lower doses of contrast agent.

REFERENCES

[1] R. Gramiak and P. M. Shah, *Invest. Radiol.* **3**, 356–366 (1968).

[2] A. Kabalnov, D. Klein, T. Pelura, E. Schutt and J. Weers, *Ultrasound Med. Biol.* **24**, 739–749 (1998).

[3] A. Kabalnov, J. Bradley, S. Flaim, D. Klein, T. Pelura, B. Peters, S. Otto, J. Reynolds E. Schutt and J. Weers, *Ultrasound Med. Biol.* **24**, 751–760 (1998).

[4] N. de Jong, L. Hoff, T. Skotland and N. Bom, *Ultrasonics.* **30**, 95–103 (1992).

[5] E. C. Unger, E. Hersh, M. Vannan, T. O. Matsunaga and T. McCreery, *Prog. Cardiovasc. Dis.* **44**, 45–54 (2001).

[6] Y. Liu, H. Miyoshi and M. Nakamura, *J. Control. Release* **114**, 89–99 (2006).

[7] S. Mehier-Humbert, T. Bettinger, F. Yan and R. H. Guy, *J. Control. Release* **104**, 213–222 (2005).

[8] J. Wu, J. Pepe and M. Rincon, *Ultrasonics* **44 Suppl** 1, e21–25 (2006).

[9] K. W. Ferrara, *Adv. Drug Deliv. Rev.* **60**, 1097–1102 (2008).

[10] E. Dressaire, R. Bee, D. C. Bell, A. Lips and H. A. Stone, *Science* **320**, 1198–1201 (2008).

[11] S. M. Stieger, P. A. Dayton, M. A. Borden, C. F. Caskey, S. M. Griffey, E. R. Wisner and K. W. Ferrara, *Contrast Media Mol. Imaging* **3**, 9–18 (2008).

[12] S. B. Feinstein, F. J. Ten Cate, W. Zwehl, K. Ong, G. Maurer, C. Tei, S. Meerbaum and E. Corday, *J. Am. Coll. Cardiol.* **3**, 14–20 (1984).

[13] A. H. Myrset, H. Nicolaysen, K. Toft, C. Christiansen and T. Skotland, *Biotechnol. Appl. Biochem.* **24** (Pt 2), 145–153 (1996).

[14] C. Christiansen, H. Kryvi, P. C. Sontum, and T. Skotland, *Biotechnol. Appl. Biochem.* **19** (Pt 3), 307–320 (1994).

[15] E. Quaia, *Eur. Radiol.* **17**, 1995–2008 (2007).

[16] D. W. Droste, H. Lakemeier, M. Ritter, R. Dittrich, J. Stypmann, T. Wichter, E. B. Ringelstein, *Neurol. Res.* **26**, 325–330 (2004).

[17] J. M. Correas, L. Bridal, A. Lesavre, A. Mejean, M. Claudon and O. Helenon, *Eur. Radiol.* **11**, 1316–1328 (2001).

[18] M. Blomley, T. Albrecht, D. Cosgrove, V. Jayaram, J. Butler-Barnes and R. Eckersley, *Lancet* **351**, 568 (1998).

[19] A. C. Perkins, M. Frier, A. J. Hindle, P. E. Blackshaw, S. E. Bailey, J. M. Hebden, S. M. Middleton and M. L. Wastie, *Br. J. Radiol.* **70**, 603–611 (1997).

[20] A. Bauer, M. Blomley, E. Leen, D. Cosgrove and R. Schlief, *Eur. Radiol.* **9 Suppl** 3, S349–S352 (1999).

[21] F. Forsberg, B. B. Goldberg, J. B. Liu, D. A. Merton, N. M. Rawool and W. T. Shi, *Radiology* **210**, 125–132 (1999).

[22] S. Podell, C. Burrascano, M. Gaal, B. Golec, J. Maniquis and P. Mehlhaff, *Biotechnol. Appl. Biochem.* **30** (Pt 3), 213–223 (1999).

[23] R. F. Mattrey, G. R. Leopold, E. vanSonnenberg, B.B. Gosink, F. W. Scheible and D. M. Long. *J. Ultrasound Med.* **2**, 173–176 (1983).

[24] R. F. Mattrey, F. W. Scheible, B. B. Gosink, G. R. Leopold, D. M. Long and C. B. Higgins, *Radiology* **145**, 759–762 (1982).

[25] R. Basilico, M. J. Blomley, D. O. Cosgrove, J. B. Liull, A. Broillet, A. Bauer and L. Bonomo, *Acad. Radiol.* **9 Suppl** 2, S380–S381 (2002).

[26] N. G. Fisher, J. P. Christiansen, H. Leong-Poi, A. R. Jayaweera, J. R. Lindner and S. Kaul , *J. Am. Coll. Cardiol.* **39**, 530–537 (2002).

[27] Y. Kono, G. C. Steinbach, T. Peterson, G. W. Schmid-Schonbein and R. F. Mattrey, *Radiology* **224**, 253–257 (2002).

[28] C. Marelli, *Eur. Radiol.* **9 Suppl** 3, S343–S346 (1999).

[29] M. S, Girard, Y. Kono, C. B. Sirlin, K. G. Baker, L. H. Deiranieh and R. F. Mattrey, *Acad. Radiol.* **8**, 734–740 (2001).

[30] G. M. Kindberg, H. Tolleshaug and T. Skotland, *Cell Tissue Res.* **300**, 397–400 (2000).

[31] P. Walday, H. Tolleshaug, T. Gjoen, G. M. Kindberg, T. Berg, T. Skotland and E. Holtz, *Biochem. J.* **299** (Pt 2), 437–443 (1994).

[32] F. Forsberg, R. Roy, D. A. Merton, N. M. Rawool, J. B. Liu, M. Huang, D. Kessler and B. B. Goldberg, *Ultrasound Med.Biol.* **24**, 1143–1150 (1998).

[33] D. R. Morel, I. Schwieger, L. Hohn, J. Terrettaz, J. B. Llull, Y. A. Cornioley and M. Schneider, *Invest. Radiol.* **35**, 80–85 (2000).

[34] J. M. Gorce, M. Arditi and M. Schneider, *Invest. Radiol.* **35**, 661–671 (2000).

[35] D. H. Simpson, C. T. Chin and P. N. Burns, *IEEE Trans. Ultrason. Ferroelectr. Freq. Control* **46**, 372–382 (1999).

[36] S. Hernot and A. L. Klibanov, *Adv. Drug Deliv. Rev.* **60**, 1153–1166 (2008).

[37] S. R. Wilson and P.N .Burns, *Radiology* **257**, 24–39 (2010).

[38] N. C. Nanda, D. C. Wistran, R. P. Karlsberg, T. C. Hack, W. B. Smith, D. A. Foley, M. H. Picard and B. Cotter, *Echocardiography* **19**, 27–36 (2002).

[39] N. C. Nanda, A. P. Miller, R. Nekkanti and S. Aaluri, *Echocardiography* **18**, 711–716 (2001).

[40] C. B. Sirlin, Y. Z. Lee, M. S. Girard, T. M. Peterson G. C. Steinbach, K. G. Baker and R. F. Mattrey, *Acad. Radiol.* **8**, 162–172 (2001).

[41] Y. Kono, S. P. Pinnell, C. B. Sirlin, S. R. Sparks, B. Georgy, W. Wong and R. F. Mattrey, *Radiology* **230**, 561–568 (2004).

[42] M. Scherrer-Crosbie, W. Steudel, R. Ullrich, P. R. Hunziker, N. Liel-Cohen, J. Newell , J. Zaroff, W. M. Zapol and M. H. Picard, *Am. J. Physiol.* **277**, H986–H992 (1999).

[43] D. Rovai, V. Lubrano, C. Vassalle, C. Paterni, C. Marini, M. Kozakova, M. Castellari, L. Taddei, M. G. Trivella, A. Distante, A. N. DeMaria and A. L Abbate, *J. Am. Soc. Echocardiogr.* **11**, 169–180 (1998).

[44] M. P. Coggins, J. Sklenar, D. E. Le, K. Wei, J. R. Lindner and S. Kaul, *Circulation* **104**, 2471–2477 (2001).

[45] K. Wei, A. R. Jayaweera, S. Firoozan, A. Linka, D. M. Skyba and S. Kaul, *Circulation* **97**, 473–483 (1998).

[46] K. Ohmori, B. Cotter, O.L. Kwan, K. Mizushige and A. N. DeMaria, *Am. Heart J.* **134**, 1066–1074 (1997).

[47] R. Konopke, S. Kersting, H. Bergert, A. Bloomenthal, J. Gastmeier , H. D. Saeger and A. Bunk, *Int. J. Colorectal Dis.* **22**, 201–207 (2007).

[48] Z. Z. Song, *Hepatology* **47**, 2145–2146; *author reply* 2146–2147 (2008).

[49] M. Koda, Y. Matsunaga, M. Ueki, Y. Maeda, K. Mimura, K. Okamoto, K. Hosho and Y. Murawaki, *Eur. Radiol.* **14**, 1100–1108 (2004).

[50] C. B. Sirlin, M. S Girard, K. G. Baker, G. C. Steinbach, L. H. Deiranieh and R. F. Mattrey, *Ultrasound Med.Biol.* **25**, 331–338 (1999).

[51] N. Faccioli, S. Crippa, C. Bassi and M. D'Onofrio, *Pancreatology* **9**, 560–566 (2009).

[52] C. Recaldini, G. Carrafiello, E. Bertolotti, M. G. Angeretti and C. Fugazzola, *Int. J. Med. Sci.* **5**, 203–208 (2008).

[53] S. Odegaard, L. B. Nesje, D. A. Hoff, O. H. Gilja and H. Gregersen, *World J. Gastroenterol.* **12**, 2858–2863 (2006).

[54] N. Lassau, M. Lamuraglia, L. Chami, J. Leclere, S. Bonvalot, P. Terrier, A. Roche, A. Le Cesne, *AJR Am. J. Roentgenol.* **187**, 1267–1273 (2006).

[55] M. J. Blomley and R. J. Eckersley, *Eur. J. Cancer* **38**, 2108–2115 (2002).

[56] J. Hohmann, T. Albrecht, C. W. Hoffmann and K. J. Wolf, *Eur. J. Radiol.* **46**, 147–159 (2003).

[57] D. Strohmeyer, F. Frauscher, A. Klauser, W. Recheis, G. Eibl, W. Horninger, H. Steiner, H. Volgger and G. Bartsch, *Anticancer Res.* **21**, 2907–2913 (2001).

[58] T. Albrecht, C. W. Hoffmann, S. A. Schmitz, S. Schettler, A. Overberg, C. T. Germer and K.J. Wolf , *AJR Am. J. Roentgenol.* **176**, 1191–1198 (2001).

[59] M. Bertolotto, E. Quaia, R. Zappetti, G. Cester and A. Turoldo, *Radiol. Med.* **114**, 42–51 (2009).

[60] D. Patel, P. Dayton, J. Gut, E. Wisner and K. W. Ferrara, *IEEE Trans. Ultrason. Ferroelectr. Freq. Control.* **49**, 1641–1651 (2002).

[61] P. S. Warren, W. J. Garrett and G. Kossoff, *J. Clin. Ultrasound.* **6**, 315–320 (1978).

[62] P. J. Lund, T. A. Fritz, E. C. Unger, R. K. Hunt and E. Fuller, *Radiology* **185**, 783–788 (1992).

[63] A. S. Lev-Toaff, J. E. Langer, D. L. Rubin, J. V. Zelch, W. K. Chong, A. E. Barone and B. B. Goldberg, *AJR Am. J. Roentgenol.* **173**, 431–436 (1999).

[64] A. L. Klibanov, *Ernst Schering Res. Found. Workshop.* 171–191 (2005).

[65] A. L. Klibanov AL. *Med. Biol. Eng. Comput.* **47**, 875–882 (2009).

[66] P. Hauff, M. Reinhardt and S. Foster S, *Handb. Exp. Pharmacol.* 223–245 (2008).

[67] A. J. Sinusas, F. Bengel, M. Nahrendorf, F. H. Epstein, J. C. Wu, F. S. Villanueva, Z.A. Fayad and R. J. Gropler, *Circ. Cardiovasc Imaging* **1**, 244–256 (2008).

[68] J. R. Lindner, *Am. J. Cardiol.* **90**, 32L–35L (2002).

[69] S. B. Feinstein, *Am. J. Physiol. Heart Circ. Physiol.* **287**, H450–H457 (2004).

[70] E. C. Unger, T. P. McCreery, R. H. Sweitzer, D. Shen and G. Wu, *Am. J. Cardiol.* **81**, 58G–61G (1998).

[71] W. H. Wright, Jr., T. P. McCreery, E. A. Krupinski, P. J. Lund, S. H. Smyth, M. R. Baker, R. L. Hulett and E. C. Unger, *Acad. Radiol.* **5 Suppl** 1, S240–S242 (1998).

[72] Y. Wu, E. C. Unger, T. P. McCreery, R. H. Sweitzer, D. Shen, G. Wu, M. D. Vielhauer, *Invest. Radiol.* **33**, 880–885 (1998).

[73] P. A. Schumann, J. P. Christiansen, R. M. Quigley, T. P. McCreery, R. H. Sweitzer, E. C. Unger, J. R. Lindner and T. O. Matsunaga, *Invest. Radiol.* **37**, 587–593 (2002).

[74] M. Takeuchi, K. Ogunyankin, N. G. Pandian, T. P. McCreery, R. H. Sweitzer, V. E. Caldwell, E. C. Unger, E. Avelar, M. Sheahan and R. J. Connolly, *J. Am. Soc. Echocardiogr.* **12**, 1015–1021 (1999).

[75] J. R. Lindner, M. P. Coggins, S. Kaul, A. L. Klibanov, G. H. Brandenburger and K. Ley, *Circulation* **101**, 668–675 (2000).

[76] J. R. Lindner, J. Song, F. Xu, A. L. Klibanov, K. Singbartl, K. Ley and S. Kaul, *Circulation* **102**, 2745–2750 (2000).

[77] J. R. Lindner, J. Song, J. Christiansen, A. L. Klibanov, F. Xu and K. Ley, *Circulation* **104**, 2107–2112 (2001).

[78] A. N. Bian, Y. H. Gao, K. B. Tan, P. Liu, G. J. Zeng, X. Zhang and Z. Liu, *World J. Gastroenterol.* **10**, 3424–3427 (2004).

[79] H. Leong-Poi, J. Christiansen, A. L. Klibanov, S. Kaul and J. R. Lindner, *Circulation* **107**, 455–460 (2003).

[80] N. Kudo, K. Okada and K. Yamamoto, *Biophys. J.* **96**, 4866–4876 (2009).

[81] E. C. Unger, T. O. Matsunaga, T. McCreery, P. Schumann, R. Sweitzer and R. Quigley, *Eur. J. Radiol.* **42**, 160–168 (2002).

[82] A. L. Klibanov, *Adv. Drug Deliv. Rev.* **37**, 139–157 (1999).

[83] R. Bekeredjian, S. Chen, P. A. Frenkel, P. A. Grayburn and R. V. Shohet, *Circulation* **108**, 1022–1026 (2003).

[84] P. Hauff , S. Seemann, R. Reszka, M. Schultze-Mosgau, M. Reinhardt, T. Buzasi, T. Plath, S. Rosewicz and M. Schirner, *Radiology* **236**, 572–578 (2005).

[85] M. Shimamura, N. Sato, Y. Taniyama, S. Yamamoto, M. Endoh, H. Kurinami, M. Aoki, T. Ogihara, Y. Kaneda and R. Morishita, *Gene Ther.* **11**, 1532–1539 (2004).

[86] Y. Taniyama, K. Tachibana, K. Hiraoka, T. Namba, K. Yamasaki, N. Hashiya, M. Aoki, T. Ogihara, K. Yasufumi and R. Morishita, *Circulation* **105**, 1233–1239 (2002).

[87] D. Mukherjee, J. Wong, B. Griffin, S. G. Ellis, T. Porter, S. Sen and J. D.Thomas, *J. Am. Coll. Cardiol.* **35**, 1678–1686 (2000).

[88] D. Mukherjee and S.G. Ellis, *Cleve. Clin. J. Med.* **67**, 577–583 (2000).

[89] M. Daffertshofer and M. Hennerici, *Lancet Neurology.* **2**, 283–290 (2003).

[90] W. C. Culp, T. R. Porter, J. Lowery, F. Xie, P.K. Roberson and L. Marky, *Stroke* **35**, 2407–2411 (2004).

[91] N. Sheiko, N. McDannold, N. Vykhodtseva, F. Jolesz and K. Hynynen, *Ultra. Med. Biol.* **30**, 979–989 (2004).

[92] Y. Kaneko, T. Maruyama, K. Takegami, T. Watanabe, H. Mitsui, K. Hanajiri, H. Nagawa and Y. Matsumoto, *Eur. Radiol.* **15**, 1415–1420 (2005).

[93] A. L. Alexander, T. T. McCreery, T. R. Barrette, A. F. Gmitro and E. C. Unger, *Magn. Reson. Med.* **35**, 801–806 (1996).

[94] R. Dharmakumar, D. B. Plewes and G. A. Wright, *Magn. Reson. Med.* **47**, 264–273 (2002).

[95] M. E. Moseley, M. F. Wendland, I. Rampil and J. Barnhart, *Proceedings of the 10th Annual Meeting of the ISMRM*, 1020 (1991).

[96] W. Huang, J. C. Grecula, T. M. Button, D. P. Harrington, M. A. Davis, J. S. D'Arrigo, B. H. Laster and C. S. Springer, *Proceedings of the 12th Annual Meeting of the ISMRM*, 757 (1993).

[97] K. K. Wong, I. Huang, Y. R. Kim, H. Tang, E. S. Yang, K. K. Kwong and E. X. Wu, *Magn. Reson. Med.* **52**, 445–452 (2004).

[98] T. Ueguchi, Y. Tanaka, S. Hamada, R. Kawamoto, Y. Ogata, M. Matsumoto, H. Nakamura and T. Johkoh, *Magn. Reson. Med. Sci.* **5**, 147–150 (2006).

[99] F. De Guio, H. Benoit-Cattin and A. Davenel, *Magma.* **21**, 261–271 (2008).

[100] J. S. Cheung, A. M. Chow, H. Guo and E. X. Wu, *Neuroimage* **46**, 658–664 (2009).

[101] A. L. Alexander, T. T. McCreery, T. R. Barrette, A. F. Gmitro and E. C. Unger, *Magn. Reson. Med.* **35**, 801–806 (1996).

[102] R. Dharmakumar, D. B. Plewes and G. A. Wright, *Phys. Med. Biol.* **50**, 4745–4762 (2005).

[103] R. H. Morris, M. Bencsik, N. Nestle, P. Galvosas, D. Fairhurst, A. Vangala, Y. Perrie and G. McHale, *J Magn. Reson.* **193**, 159–167 (2008).

[104] A. M. Chow, K. W. Chan, J. S. Cheung and E. X. Wu , *Magn. Reson. Med.* **63**, 224–229 (2010).

[105] A. M. Chow, K. W. Chan and E. X. Wu, *Proceedings of the 18th Annual Meeting of the ISMRM*, 698 (2008).

[106] F. Yang, Y. Li, Z. Chen, Y. Zhang, J. Wu and N. Gu, *Biomaterials* **30**, 3882–3890 (2009).

[107] J. I. Park, D. Jagadeesan, R. Williams, W. Oakden, S. Chung, G. J. Stanisz and E. Kumacheva, *ACS Nano* **4**, 6579–6586 (2010).

[108] M.S. Chawla, X. J. Chen, H. E. Moller, G. P. Cofer, C. T. Wheeler, L. W. Hedlund and G.A. Johnson, *Proc. Natl. Acad. Sci. U. S. A.* **95**, 10832–10835 (1998).

[109] V. Callot, E. Canet, J. Brochot, H. Humblot, A. Briguet, H. Tournier and Y. Cremillieux, *Acad. Radiol.* **9 Suppl** 2, S501–S503 (2002).

[110] A. M. Chow, C. T. Chiu, K. X. Cai and E. X. Wu, *Proceedings of the 16th Annual Meeting of the ISMRM*, 72 (2008).

[111] G. Seidel, K. Meyer, V. Metzler, D. Toth, M. Vida-Langwasser and T. Aach, *Ultrasound Med. Biol.* **28**, 183–189 (2002).

[112] L. E. Benjamin, D. Golijanin, A. Itin, D. Pode and E. Keshet, *J. Clin. Invest.* **103**, 159–165 (1999).

[113] M. S. Tartis, D. E. Kruse, H. Zheng, H. Zhang, A. Kheirolomoom, J. Marik and K. W. Ferrara, *J. Cont. Rel.* **131**, 160–166 (2008).

[114] J. K. Willmann, Z. Cheng, C. Davis, A. M. Lutz, M. L. Schipper, C. H. Nielsen and S. S. Gambhir, *Radiology* **249**, 212–219 (2008).

[115] T. Kao, D. Connor, F. A. Dilmanian, L. Faulconer, T. Liu, C. Parham, E. D. Pisano and Z. Zhong, *Phys. Med. Biol.* **54**, 3247–3256 (2009).

[116] D. M. Connor, H. D. Hallen, D. S. Lalush, D. R. Sumner and Z, Zhong, *Phys. Med. Biol.* **54**, 6123–6133 (2009).

[117] L. Zhang, C. Hu, T. Zhao and S. Luo, *Eur. J. Radiol.* **80**, 158–162 (2010).

[118] F. Arfelli, L. Rigon and R. H. Menk, *Phys. Med. Biol.* **55**, 1643–1658 (2010).

[119] C. C. Coussios, C. H. Farny, G. T. Haar and A. A. Roy, *Int. J. Hyperthermia* **23**, 105–120 (2007).

[120] M. Claudon, F. Plouin, G. M. Baxter, T. Rohban and D. M. Devos, *Radiology* **214**, 739–746 (2000).

[121] J. L. Cohen, J. Cheirif, D. S. Segar, L. D. Gillam, J. S. Gottdiener, E. Hausnerova and D. E. Bruns, *J. Am. Coll. Cardiol.* **32**, 746–752 (1998).

[122] Y. Myreng, P. Molstad, K. Ytre-Arne, M. Aas, L. Stoksflod, J. O. Nossen and B. Oftedal, *Heart* **82**, 333–335 (1999).

[123] M. Kaps, G. Seidel, D. Bokor, B. Modrau and C. Algermissen, *J. Neuroimaging.* **9**, 150–154 (1999).

[124] G. ter Haar, *Med. Biol. Eng. Comput.* **47**, 893–900 (2009).

[125] L. W. Johnson, E. C. Lozner, S. Johnson, R. Krone, A. D. Pichard, G. W. Vetrovec and T. J. Noto, *Cathet. Cardiovasc. Diagn.* **17**, 5–10 (1989).

[126] R. J. Stuart Jr. and M. H. Ellestad, *Chest* **77**, 94–97 (1980).

15

NON-NANOPARTICLE-BASED DUAL-MODALITY IMAGING AGENTS

REINIER HERNANDEZ

Department of Medical Physics, University of Wisconsin-Madison, Madison, WI, USA

TAPAS R. NAYAK AND HAO HONG

Department of Radiology, University of Wisconsin-Madison, Madison, WI, USA

WEIBO CAI

Departments of Radiology and Medical Physics, University of Wisconsin-Madison, Madison, WI, USA

15.1 INTRODUCTION

In the first decade of the 21st century, molecular imaging, which embraces a variety of techniques such as positron emission tomography (PET), single-photon emission computed tomography (SPECT), optical bioluminescence, optical fluorescence, targeted ultrasound, magnetic resonance imaging (MRI), magnetic resonance spectroscopy (MRS), and many others that are under active development, has been an extremely dynamic interdisciplinary field [1–3]. Each of these imaging modalities has its intrinsic advantages and disadvantages in terms of spatial/temporal resolution, tissue penetration, sensitivity, cost, dependence on infrastructure, as well as other parameters. For example, nuclear medicine techniques (PET and SPECT) have very high sensitivity and excellent tissue penetration that can allow the acquisition of whole-body images. In addition, accurate quantification is another hallmark of nuclear medicine imaging techniques, in particular PET. However, PET and SPECT imaging have rather poor resolution, in the order of a few millimetres (mm), which is inadequate in many scenarios.

On the other hand, real-time images with resolution in the sub-mm range can be acquired using optical techniques in superficial tissue; however, the resolution and sensitivity deteriorates very rapidly with increasing depth due to many factors such as tissue absorption, scattering, and autofluorescence [4]. The near-infrared (NIR; 700–900 nm) region is optimal for *in vivo* optical imaging because the absorbance spectra for all biomolecules reach a minimum, thus providing a clear optical window for small animal studies and limited clinical scenarios (e.g., breast imaging, endoscopy, surgical guidance) [5]. In addition to better tissue penetration of light when compared to optical imaging in the visible range, there is also significantly less background signal from tissue autofluorescence in the NIR window. MRI can offer high resolution (sub-mm), three-dimensional (3D) imaging of anatomical structures with exquisite soft tissue contrast. In addition, it does not involve ionising radiation. However, the sensitivity of MRI is many orders of magnitude lower than the nuclear medicine and optical imaging techniques. Clearly, the combination of two or more molecular imaging modalities is synergistic and can overcome many of the abovementioned limitations for a given technique.

The Chemistry of Molecular Imaging, First Edition. Edited by Nicholas Long and Wing-Tak Wong.
© 2015 John Wiley & Sons, Inc. Published 2015 by John Wiley & Sons, Inc.

To date, most dual-modality and multimodality imaging agents are based on certain nanoparticles [6, 7], which will be the focus of Chapter 16. In the current chapter, a comprehensive overview of the dual-modality imaging approaches reported to date that are not based on nanoparticles will be presented. Because molecular imaging with ultrasound is rare and targeted ultrasound studies almost exclusively use microbubbles as the contrast agents [8, 9] (see Chapter 14), dual-modality imaging with ultrasound will not be discussed in this chapter.

Molecular imaging relies on the specific delivery of a contrast agent to achieve imaging contrast of the target of interest over the background; therefore, the design and synthesis of the imaging agents are critical for successful dual-modality imaging. These agents are typically composed of several different moieties, which include a targeting ligand (e.g., small molecule, peptide, protein, antibody) and the imaging tags that can be detected by multiple imaging modalities (e.g., radioisotopes, fluorescent dyes, Gd^{3+}). Although direct conjugation of different imaging tags to conform the dual-modality probe has been reported, a much more common approach is to use a linker to connect them. In many cases, the function of the linker is not restricted to simply serve as a bridge between the imaging tags, but also to improve the targeting efficacy and/or pharmacokinetics of the imaging agent. After a thorough literature survey, the dual-modality imaging agents are divided into the following categories: PET/optical, SPECT/optical, MRI/optical, and PET/MRI agents.

15.2 PET/OPTICAL AGENTS

PET imaging has been widely used in clinical oncology for cancer staging and monitoring the therapeutic response to various anticancer therapies [10, 11]. The success of PET is attributed to many factors, such as superb tissue penetration, which allows noninvasive visualisation of deep tissues/structures, excellent quantitation capability, wide availability of ^{18}F-FDG (a tracer for imaging glucose metabolism), among others. However, imaging with PET alone is far from ideal since it has relatively poor spatial resolution (a few mm) and does not provide adequate anatomical information.

Although optical imaging has only limited use in the clinical setting, such as imaging tissues close to the surface of the skin (e.g., breast imaging), tissues accessible by endoscopy (such as within the esophagus and colon), and intraoperative visualisation (typically image-guided surgery), the combination of PET and optical imaging is highly beneficial [12]. Clinically, dual-modality PET/optical agents may be particularly useful in cancer patient management by employing the whole-body PET scan to identify the location of tumour(s) and optical imaging to guide tumour resection. From a regulatory perspective, the need for comprehensive toxicity/dosimetry studies in multiple animal species for only one imaging agent instead of two separate agents (one for each imaging modality) can significantly reduce the development cost and facilitate future clinical translation of novel imaging agents, which is the bottleneck for moving state-of-the-art molecular imaging technology into clinical practice and day-to-day patient management. A tabulated summary of PET/optical agents is provided in Table 15.1 and discussed in detail below.

15.2.1 Small Molecule-Based Agents

A compound called PS-2, synthesised through electrophilic aromatic iodination with Na[124]I in the presence of iodogen beads, has been investigated for both PET/optical imaging and photodynamic therapy (Figure 15.1) [13].

TABLE 15.1 A Tabulated Summary of Dual-modality PET/optical Agents.

Radioisotope	Fluorophore	Targeting Ligand	Target	References
[124]I	*Porphyrin*	-	-	[13]
[11]C	Styryl dye	-	RNA	[17]
[18]F	BODIPY	-	-	[18]
[64]Cu	Cy5.5	Knottin 2.5D peptide	Integrin $\alpha_v\beta_3/\alpha_v\beta_5$	[29]
[68]Ga	800CW	KKAHWGFTLD peptide	MMP-2/9	[36]
[64]Cu	Cypate	Octreotate peptide	Somatostatin receptor	[46]
[64]Cu	LS-276	DEVD peptide	Caspase-3	[49]
[64]Cu	AlexaFluor 750	NuB2 mAb	CD20	[52]
[64]Cu	800CW	Bevacizumab	VEGF	[54]
[64]Cu/[89]Zr	800CW	TRC105 mAb	CD105/endoglin	[66, 67]
[18]F	C7-Cy	Mannosyl-dextran	Mannose receptor	[70]

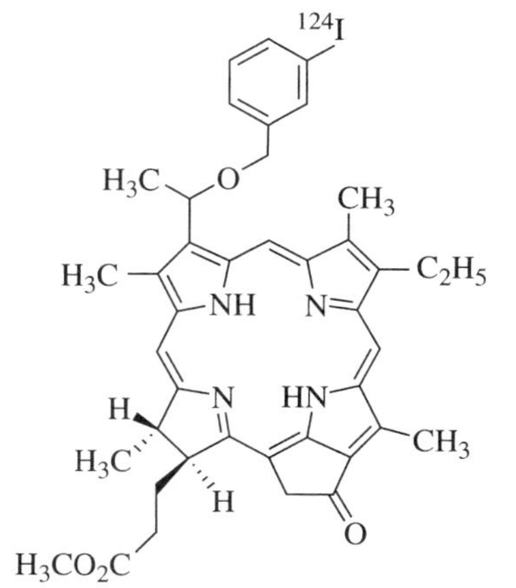

FIGURE 15.1 A small molecule probe for PET/optical imaging of cancer in living mice.

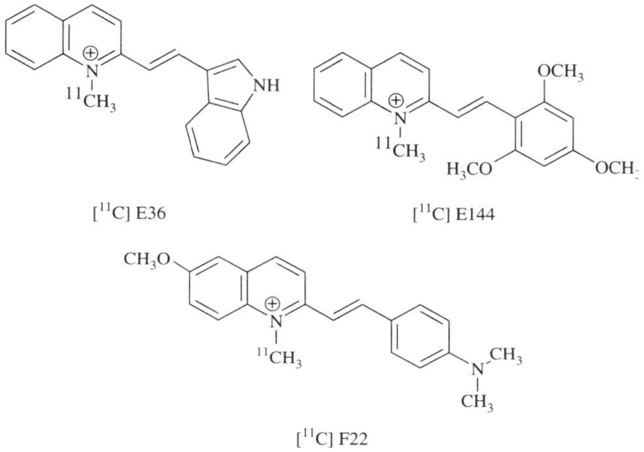

[11C] E36   [11C] E144

[11C] F22

FIGURE 15.2 ^{11}C-labelled styryl dyes.

PS-2 is a porphyrin-based photosensitiser that exhibits marked fluorescence at 720 nm upon excitation at 540 nm. The use of a long-lived PET isotope, ^{124}I ($t_{1/2}$: 4.2 days), was necessary because maximum tumour accumulation of the agent occurs at 24 h post-injection in mice. *In vivo* investigation in tumour-bearing mice clearly demonstrated the effectiveness of the probe for PET/optical imaging, which can be used for guiding photodynamic therapy in the future. Although ^{11}C ($t_{1/2}$: 20.3 min.) has not been used as widely as ^{18}F ($t_{1/2}$: 109.8 min.), due to its much shorter half-life and limited radiochemistry, recent efforts have been made to synthesise PET/optical imaging agents using ^{11}C as the positron emitter. Specifically, ^{11}C-labelling of styryl dyes constitutes an intriguing field to explore because these molecules can target RNA and have been widely used for visualising RNA metabolism in living cells [14–16]. The inclusion of PET imaging capability may open the possibilities of new ways to elucidate RNA-related biological processes such as gene expression. In one report, three styryldye-based PET probes were successfully synthesised (Figure 15.2), in which the quaternary amino groups allowed for facile ^{11}C-labelling using [^{11}C]methyltriflate as the methylating agent [17]. Good radiochemical yield was achieved, and the final products could be easily purified with solid phase extraction columns because of the presence of quaternary amino groups.

Recently, an ^{18}F-labelled BODIPY dye has been reported where direct radiolabelling of the dye was carried out with KH^{18}F$_2$ [18]. *In vivo* imaging in normal mice showed liver and kidney accumulation, corresponding to nonspecific uptake and metabolic clearance of the probe. No accumulation of ^{18}F was found in the bone, suggesting good stability of the B-F bond *in vivo*. Successful correlation between the images obtained by PET and optical scans was validated *ex vivo*. Although the fluorescence emission of BODIPY (~580 nm) makes it unsuitable for *in vivo* imaging, it may have a potential role in intraoperative surgical guidance.

15.2.2 Peptide-Based Agents

A number of peptides have been used as the targeting ligands for the development of dual-modality PET/optical agents [19]. One of the most intensively studied molecular targets using peptide-based imaging probes is integrin $\alpha_v\beta_3$, which binds to arginine-glycine-aspartic acid (RGD)-containing peptides/proteins [20, 21]. As a key protein involved in tumour angiogenesis and metastasis [21–23], integrin $\alpha_v\beta_3$ is overexpressed in a variety of solid tumour types (e.g., melanoma, late stage glioblastoma, breast, prostate, and ovarian cancer) but is not readily detectable in resting endothelial cells or most normal organs [24], which makes it a universally applicable target for molecular imaging and therapy of cancer. Because one of the key requirements during tumour development is angiogenesis [25–27], tremendous effort has been devoted to angiogenesis-related research over the last decade, particularly those involving noninvasive molecular imaging techniques [28].

An engineered Cystine knot peptide (knottin 2.5D), which contains the RGD motif and exhibits high binding affinity for integrin $\alpha_v\beta_3$ and $\alpha_v\beta_5$, has been used to develop a PET/optical imaging agent for noninvasive monitoring of integrin expression in tumour models (Figure 15.3) [29]. Synthesis of this agent was achieved by stoichiometric conjugation of the NIR fluorescent (NIRF) dye Cy5.5, DOTA (1,4,7,10-tetraazacyclododecane-N,N',N'',N'''-tetraacetic acid, for ^{64}Cu-labelling), and the N-terminus of knottin 2.5D to a peptide linker. Solid-phase peptide synthesis using appropriate protective groups was employed to assure consistency in the 3D structure and target binding affinity of the knottin 2.5D peptide. The advantage of dual-labelled knottin 2.5D over the analogous ^{64}Cu-labelled single-modality probe was demonstrated by decreased washout and significantly better retention of the PET/optical dual-modality probe in the tumour. Good correlation between the two imaging modalities was achieved, suggesting stability of the dual-modality probe *in vivo*. Future studies with different combinations of imaging labels may yield additional information through multimodal imaging.

Disrupted expression and activity of matrix metalloproteinases (MMPs) is present in many diseases such as rheumatoid arthritis, atherosclerosis, heart failure, pulmonary emphysema, and tumour development/metastasis [30–35]. A cyclic peptide that contains a peptide sequence cleavable by MMP-2/9 (i.e., HWGF) was dual-labelled with ^{68}Ga and the NIRF dye IRDye800CW [36]. Solid-phase synthesis of the peptide KKAHWGFTLD, followed by DOTA conjugation, yielded the DOTA-KKAHWGFTLD conjugate. 800CW was then attached to the lysine residue to render the dual-modality probe. Although fluorescence properties of the dye survived the harsh conditions associated with ^{68}Ga-labelling, which was carried out at 95°C and pH 4, the binding affinity of the peptide to MMP-2/9 was significantly compromised. The use of other PET isotopes is warranted in future investigation of this class of imaging probes. For example, ^{64}Cu can be labelled with DOTA at mild conditions (37°C) in close to neutral pH values [37–39].

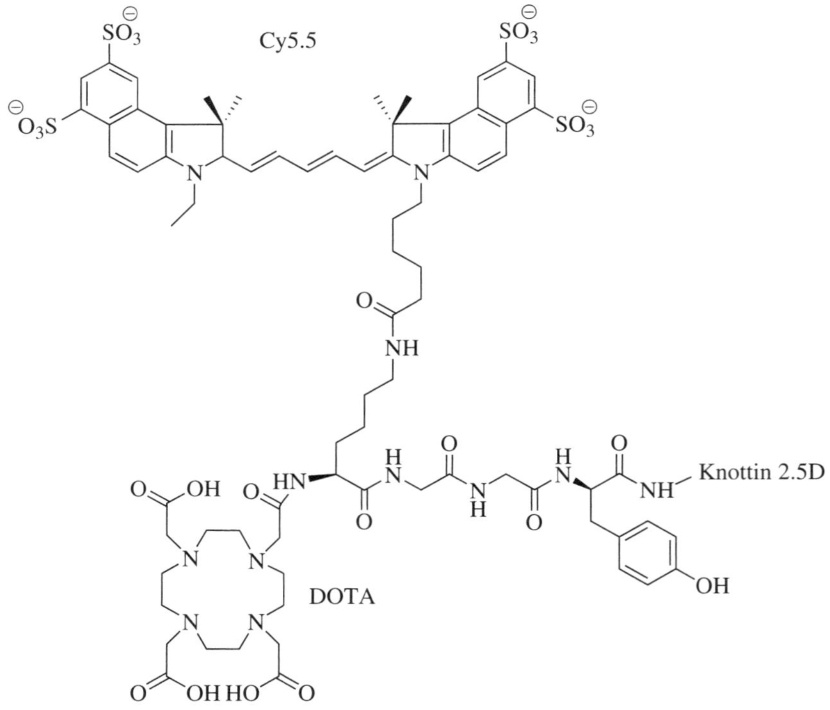

FIGURE 15.3 A knottin-based probe for *in vivo* NIRF and PET imaging in mouse tumour models.

Somatostatin receptors, in particular subtype 2 (sst2), are usually upregulated in endocrine tumours [40, 41]. Analogues of somatostatin such as the octreotide and octreotate peptides, which have improved biological half-lives over somatostatin itself [42, 43], have been extensively studied for cancer imaging in both animal models and clinical settings [44, 45]. In one report, the peptide octreotate was conjugated to both DOTA and the NIRF dye Cypate [46]. In addition to optical methods, radiolabelling with ^{64}Cu and ^{177}Lu enables *in vivo* imaging with PET and SPECT respectively. However, although target specificity of the agents was demonstrated *in vitro*, tumour uptake of the tracer was very low in AR42J tumour-bearing rats, which exhibited predominantly hepatobiliary clearance.

Bombesin, a peptide with 14 amino acid residues that binds to the gastrin-releasing peptide receptor (GRPR) [47], has also been dual-labelled with the dye FITC (fluorescein isothiocyanate) and DTPA (diethylenetriaminepentaacetic acid), or DOTA, by the same research group [48]. It was found that a spacer between the peptide and the two labels was necessary to maintain the target binding affinity of the probe. Furthermore, positioning of the label also had a significant impact on the target-binding properties. However, no PET studies were carried out on these potential PET/optical dual-modality agents.

Different from the direct targeting strategies mentioned above, an activatable multimodality probe LS498 has been constructed [49]. This agent contains DOTA (for ^{64}Cu-labelling), a NIRF dye LS-276, a fluorescence-quenching molecule QC-1, as well as a peptide sequence Asp-Glu-Val-Asp (DEVD) that can be cleaved by Caspase-3 [50], an enzyme upregulated in apoptotic cells [51]. Enzyme kinetics assay showed that LS498 could be readily cleaved by Caspase-3, which was also successfully tested in living mice using fluorescence imaging (Figure 15.4). It was suggested that the 'always on' nuclear signal can be useful for quantifying and localising the distribution of the probe, while optical imaging can report the functional status of a molecular event (e.g., enzymatic activity).

15.2.3 Macromolecule-Based Agents

Dual-modality imaging using a small molecule- or peptide-based probe is very challenging due to the limited number of attachment points and the potential interference with its receptor binding affinity. Therefore, various nanoparticles [6, 7] and macromolecules (e.g., antibodies and polymers), which have more functional groups available to attach the image labels for dual-modality molecular imaging, have also been investigated. A dual-modality probe for PET and NIRF imaging has been reported that employs a monoclonal antibody (mAb) that binds to CD20 [52]. The mAb NuB2 was conjugated to DOTA and an AlexaFluor 750 dye, where no perturbation of the absorption and fluorescence spectra of the dye was observed upon conjugation to the mAb. The imaging characteristics of ^{64}Cu-DOTA-NuB2-AlexaFluor 750 were assessed in a CD20-positive Raji lymphoma-bearing mouse, which had much higher uptake than the nonconjugated ^{64}Cu-DOTA-AlexaFluor 750. Furthermore, robust correlation between the biodistribution ratios obtained by radioactivity and fluorescence measurement was achieved, demonstrating the effectiveness of this dual-modality agent in imaging CD20-expressing lymphomas.

The pivotal role of vascular endothelial growth factor (VEGF) in cancer is underscored by the approval of bevacizumab (Bev, a humanised anti-VEGF mAb) as first-line treatment of cancer patients [53]. We recently reported dual-labelled Bev (with 800 CW and ^{64}Cu) for both PET and NIRF imaging of VEGF [54]. Because it has been generally recognised that NOTA (1,4,7-triazacyclononane-1,4,7-triacetic acid) is one of the most suitable chelators for ^{64}Cu-labelling [55], NOTA was used as the chelator for ^{64}Cu in this study instead of DOTA. Flow cytometry analysis of U87MG human glioblastoma cells revealed no difference in VEGF binding affinity/specificity between Bev and NOTA-Bev-800CW. Serial PET imaging of U87MG tumour-bearing female nude mice revealed excellent tumour contrast and high tumour uptake of ^{64}Cu-NOTA-Bev-800CW, corroborated by *in vivo/ex vivo* NIRF imaging and biodistribution studies (Figure 15.5). Furthermore, tumour uptake as measured by *ex vivo* NIRF imaging had a good linear correlation with the percentage injected dose per gram of tissue (%ID/g) values obtained from PET, which warrants further investigation and future clinical translation of such Bev-based imaging agents for many applications such as disease diagnosis, patient stratification, treatment monitoring, and image-guided surgery.

In addition to the integrin $\alpha_v\beta_3$ and VEGF/VEGF receptors (VEGFRs) mentioned above, for which several tracers have entered clinical investigation [28, 56–59], another attractive target related to tumour angiogenesis that deserves more investigation is CD105 (also called endoglin), a 180 kDa disulphide-linked homodimeric transmembrane protein [60–62]. Immunohistochemistry of CD105 on tumour tissue is now the accepted standard approach for determining tumour microvessel density (MVD), a quantitative measure of tumour angiogenesis and an independent prognostic factor for survival in patients with many types of solid tumours [61, 62]. Noninvasive imaging of CD105 expression (essentially noninvasive measurement of whole-body MVD) has promising clinical potential in cancer diagnosis, anti-angiogenic drug development, personalised medicine, and ultimately the day-to-day cancer patient management.

After demonstrating the feasibility of PET or NIRF imaging of CD105 using an anti-CD105 mAb called TRC105 [39, 63–65], we are currently optimising the development of dual-modality PET/NIRF agents for noninvasive imaging of CD105 [66, 67]. In both studies, good correlation of tracer uptake based on PET and NIRF measurements was achieved, confirming that imaging in the NIR window in small animal models can give semi-quantitative to quantitative information in superficial (tumour)

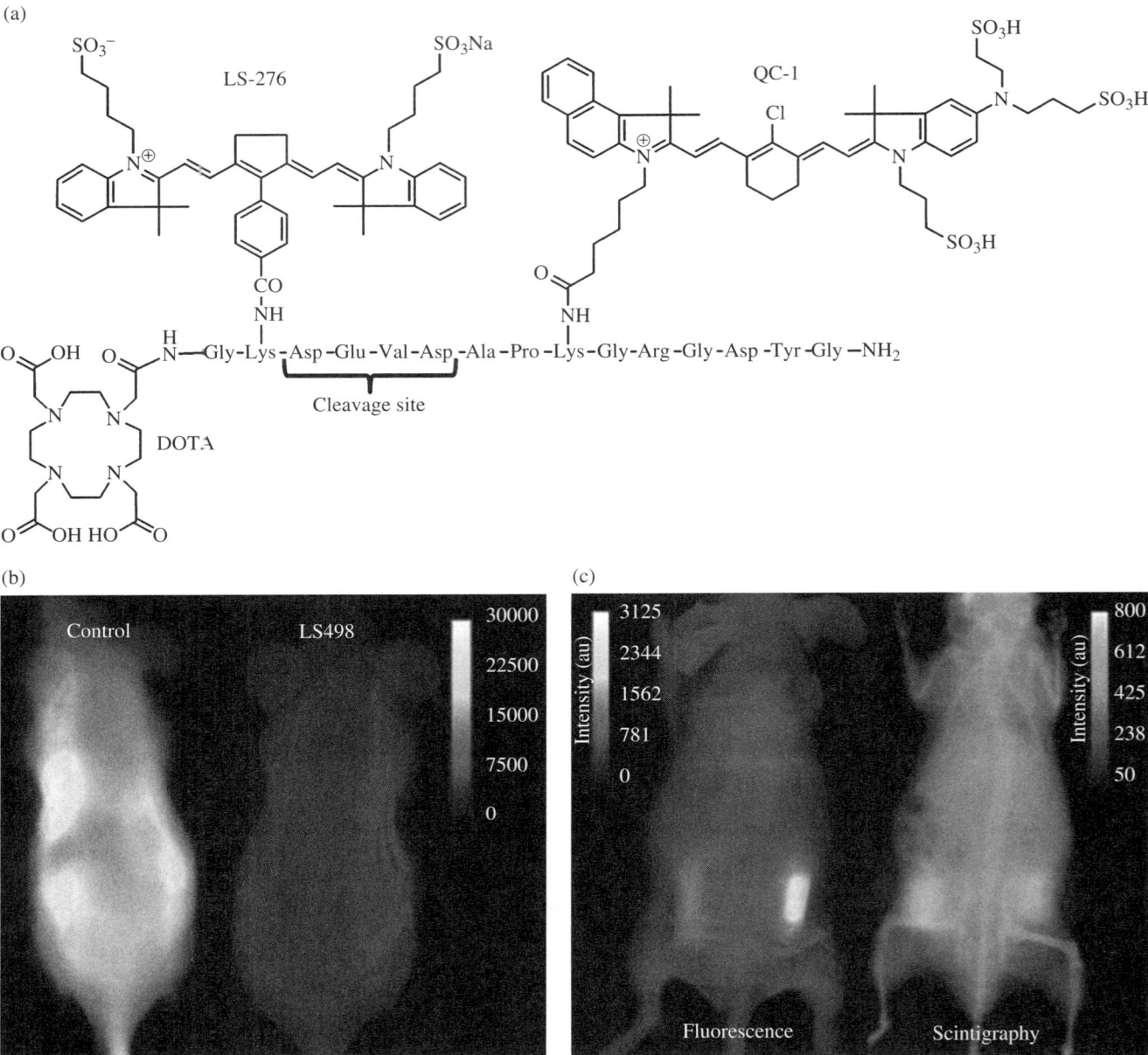

FIGURE 15.4 Noninvasive imaging of Caspase-3 activity with an activatable probe. (a) Chemical structure of the probe. (b) Activated (left) versus quenched (right) fluorescence signal of the probe in mice. (c) Fluorescence (left) and scintigraphy (right) images of subcutaneously implanted capsules containing ^{64}Cu-labelled activatable probe (i.e., LS498). Tubes were implanted subcutaneously below the surface of the skin. Comparable signal intensity was observed in scintigraphy ('always on'). However, the fluorescence signal was much stronger in the presence of Caspase-3. Adapted with permission from Ref. [49]. (*See insert for colour representation of the figure.*)

tissue. Upon further optimisation and development, such dual-labelled PET/NIRF agents can be translated into the clinic for both disease diagnosis and image-guided surgery in multiple solid tumour types, because all solid tumours depend on angiogenesis.

DTPA-mannosyl-dextran (Lymphoseek), a dextran polymer that contains several mannose groups for targeting the mannose receptor on macrophages, has been tested for sentinel lymph node (SLN) mapping in clinical trials [68, 69]. Recently, SLN mapping with both PET and NIRF imaging was reported in a preclinical study using modified Lymphoseek [70]. A boronate trap for F$^-$ and a NIR fluorophore was combined into a single molecule, which was conjugated to the free amino groups in the dextran chain. The fact that both imaging labels are combined in one molecule is highly desirable, because only one attachment site is required on the targeting ligand, which leads to much easier synthesis and more facile purification of the imaging probe. In addition, the boron trap can conjugate three ^{18}F atoms, which can give high specific activity of the imaging agent. It was suggested that due to the relatively short half-life of ^{18}F, hydrolysis of the B-F bond (less stable than the C-F bond) can be neglected.

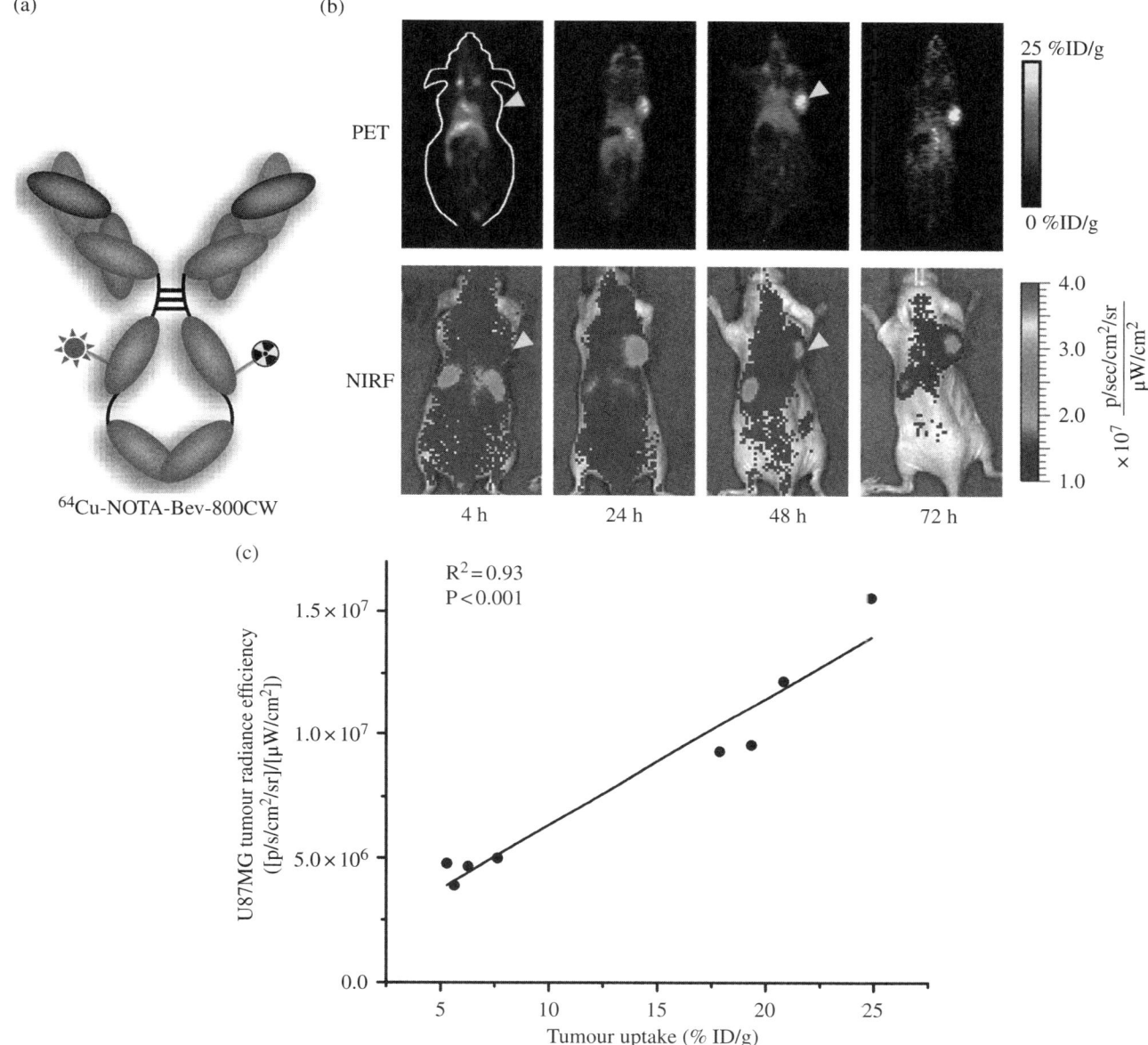

FIGURE 15.5 PET/NIRF imaging of VEGF. (a) Schematic representation of the dual-modality probe ^{64}Cu-NOTA-Bev-800CW. (b) Serial *in vivo* PET/NIRF imaging of U87MG tumour-bearing mice at 4, 24, 48 and 72 h post-injection of ^{64}Cu-NOTA-Bev-800CW. Arrowheads indicate the U87MG tumours. (c) Linear correlation of the *ex vivo* NIRF signal intensity with the percentage injected dose per gram of tissue (%ID/g) values based on PET in all U87MG tumour-bearing mice at 72 h post-injection [54]. (*See insert for colour representation of the figure.*)

In vivo studies of the dual-modality agent revealed rapid, receptor-specific, and positive identification of the SLN in both the NIRF and PET imaging modes, while maintaining little breakthrough to distal lymph nodes.

15.3 SPECT/OPTICAL AGENTS

The source of SPECT images are gamma ray emissions [71, 72]. The first object that an emitted gamma photon encounters after exiting the body is the collimator, which allows it to travel only along certain directions to reach the detector to ensure that the signal position on the detector accurately represents the source of the gamma ray. Because of the use of collimators, the sensitivity of SPECT is much lower than PET. The most common radioisotopes used for SPECT imaging include ^{99m}Tc ($t_{1/2}$: 6.0 h), ^{111}In ($t_{1/2}$: 2.8 days), and radioiodine (e.g. ^{131}I, $t_{1/2}$: 8.0 days). Similar to the abovementioned PET/optical agents, many dual-modality SPECT/optical agents have also been developed. A tabulated summary of SPECT/optical agents is provided in Table 15 2 and discussed in detail below.

15.3.1 Small Molecule-Based Agents

Possessing high binding affinity for hydroxyapatite (HA) that is present in bone surfaces, bisphosphonates (BP) have been extensively utilised to diagnose bone lesions and to treat breast cancer bone metastasis [73]. To target the microcalcification in metastatic tumours, a BP-based dual-modality imaging agent (Pam-Tc-800) was developed [74]. In this probe, a lysine residue was used as a linker to connect the three functional entities: 3-amino tetramethyl 1-hydroxy-propylidenebis-bisphosphonate (Me-Pam) for HA targeting [75], S-acetylmercaptoacetyltriserine (MAS$_3$) for ^{99m}Tc-labelling [76], and the NIRF dye 800CW (Figure 15.6a). *In vivo* studies of the agent in a breast cancer rat model showed high sensitivity detection

TABLE 15.2   A Tabulated Summary of Dual-modality SPECT/optical Agents.

Radioisotope	Fluorophore	Targeting Ligand	Target	References
^{99m}Tc	800CW	Bisphosphonate	Hydroxyapatite	[74]
^{99m}Tc	Acridine orange	Bombesin peptide	GRPR	[77]
^{99m}Tc	Acridine derivatives	Pyrazolyl-diamine ligands	DNA	[78]
^{99m}Tc	2-(4'-Aminophenyl)benzothiazole derivatives	2-(4'-Aminophenyl)benzothiazole derivatives	Cancer (target unclear)	[79]
^{99m}Tc	Flavone	Flavone	β-Amyloid	[80]
^{99m}Tc	Rhodamine 110	DEVD peptide	Caspase-3	[86]
^{111}In	800CW	c(RGDfK) peptide	Integrin $\alpha_v\beta_3$	[87, 88]
^{111}In	Cypate	c(RGDyK) peptide	Integrin $\alpha_v\beta_3$	[89]
^{111}In	IR-783	c(CGRRAGGSC)NH$_2$ peptide	IL-11Rα	[95]
^{111}In	800CW	Trastuzumab	HER2	[103]
^{111}In	Cy5.5	Trastuzumab	HER2	[104]
^{111}In	Cy5.5/Cy7	Trastuzumab/Cetuximab	HER2/HER1	[105]

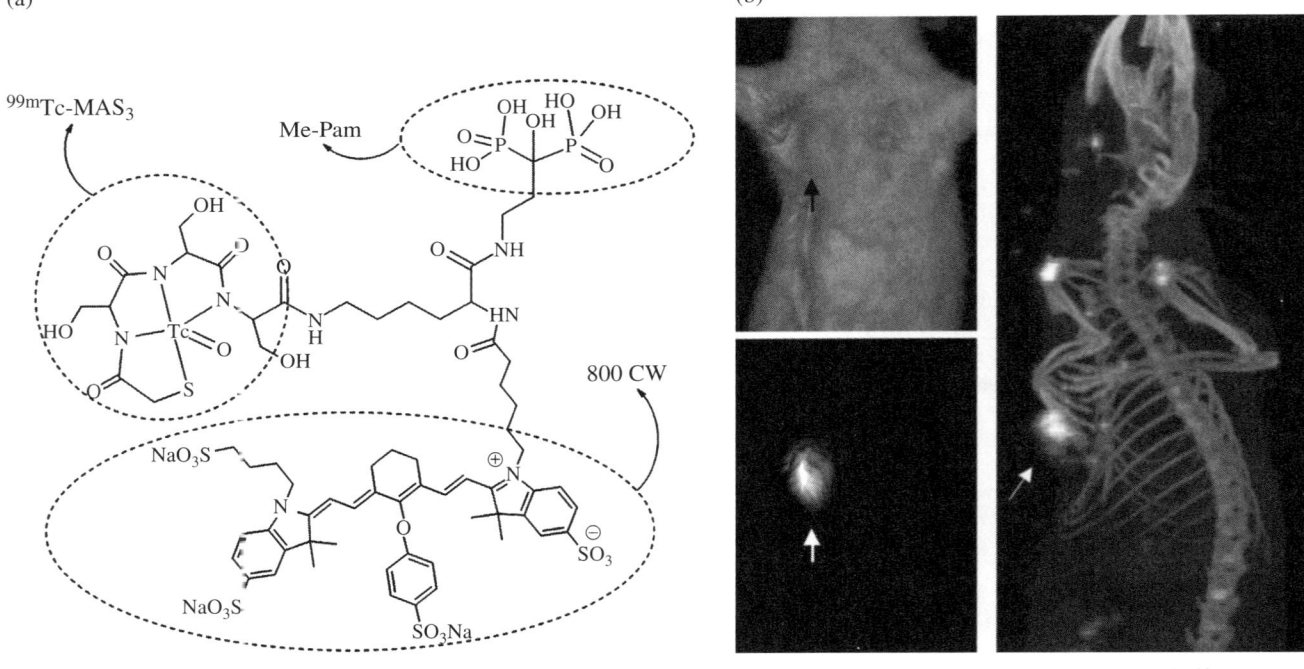

FIGURE 15.6 *In vivo* SPECT/NIRF imaging of breast cancer microcalcification. (a) Chemical structure of the Pam-Tc-800 probe. (b) NIRF and SPECT/CT images after administration of the probe in a breast cancer rat model. Arrows indicate the tumour. Adapted with permission from Ref. [74]. (*See insert for colour representation of the figure.*)

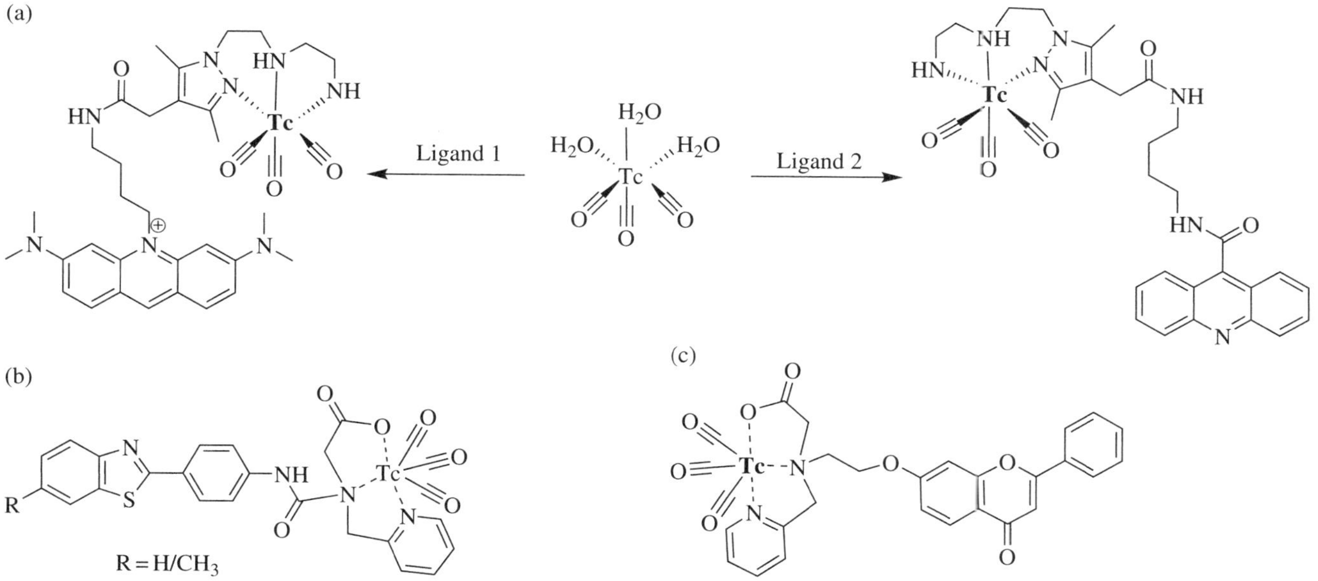

(a)

(b)

R = H/CH$_3$

(c)

FIGURE 15.7 Representative dual-modality agents that combine naturally fluorescent molecules with ^{99m}Tc. (a) Synthesis of acridine-based agents. (b) Aminophenylbenzothiazole-based agents. (c) A flavone-based agent.

of tumour microcalcifications (Figure 15.6B). In addition, rapid clearance of the probe from soft tissue, along with high tumour uptake of the agent, makes it an attractive imaging agent for bone metastasis detection.

A series of dual-modality agents that combine naturally fluorescent molecules with ^{99m}Tc have been recently synthesised [77–80]. Figure 15.7a shows an example reaction where a $[^{99m}Tc(CO)_3(H_2O)_3]^+$ complex is used as the precursor for the reaction with a pyrazolyldiamine chelator that was functionalised with an acridine moiety. Several other representative compounds of this class are also shown in Figure 15.7b and c [77–80]. Although these agents have been used for cell-based studies, they are not ideal for *in vivo* imaging applications.

Aiming toward chelation of Re and Tc, single amino acid chelates (SAACs) based on lysine derivatives have been developed [81, 82]. The construction of tridentate SAAC ligands containing two pyridine/quinoline and one tertiary amine groups as donors was achieved by modifying the ε-NH$_2$ group of lysine with two pyridine/quinoline units, which readily reacts with Re/Tc to give the [M(CO)$_3$(ligand)]$^+$ complex in high yield. One of the major advantages of these SAACs is their easy incorporation into peptide sequences by solid phase peptide synthesis. Although still in the preliminary stage with no animal studies reported, several reports exist in the literature regarding the development of potential dual-modality SPECT/optical agents using this strategy [83–85].

In one study, SAAC was employed for ^{99m}Tc-labelling to create a Caspase-3-sensitive SPECT/optical imaging agent [86]. The design of this agent is similar to the activatable PET/optical agent mentioned above [49]. The fluorescent molecule rhodamine 110 was linked to both a SAAC and a DEVD peptide, which was then labelled with ^{99m}Tc to form the dual-modality SPECT/optical imaging probe. Being nonfluorescent in the bonded state, rhodamine 110 can be activated in the presence of Caspase-3, which cleaves the DEVD sequence. After *in vitro* validation in various cell types using fluorescence techniques, *in vivo* imaging in mice revealed significantly higher uptake of the radiolabel in apoptosis-induced liver than in normal liver [86]. This study represented the first demonstration of dual-modality SPECT/optical imaging of enzymatic activity using a single imaging probe. The use of NIRF dyes is expected to give better *in vivo* performance in future studies because of better tissue penetration of NIR light.

15.3.2 Peptide-Based Agents

As previously discussed in PET/optical agents, several RGD peptides have also been reported for SPECT/optical imaging. The cyclic peptide c(RGDfK) was labelled with both ^{111}In (through DTPA) and the NIRF dye 800CW for gamma scintigraphy and continuous-wave imaging of integrin $\alpha_v\beta_3$-positive tumours xenografts, respectively [87]. Twenty-four hours after administration of the dual-modality agent at a dose equivalent to 90 μCi of ^{111}In and 5 nmol of 800CW, whole-body gamma scintigraphy and optical imaging was conducted. It was found that the target-to-background ratios of nuclear and optical imaging were similar for surface regions of interest, consistent with the origin of gamma and NIRF signal from a common targeted peptide. Furthermore, an analysis of signal-to-noise ratio versus contrast showed greater sensitivity of NIRF over nuclear imaging for subcutaneous tumour targets, while gamma scintigraphy allowed for more sensitive detection of deeper structures, which was corroborated by a follow-up study [88]. Although accurate comparison between the two modalities was not feasible because fluorescence imaging was not very quantitative, such proof-of-principle study did demonstrate for the first time the direct comparison of molecular optical and planar nuclear imaging for subcutaneous and deeper tumours.

The c(RGDyK) peptide has also been labelled with ^{111}In and the NIRF dye cypate and investigated in a 4T1 murine breast cancer model [89]. With tumours visualised by both optical and nuclear imaging methods in a receptor-specific manner, it was suggested that such a dual-modality imaging approach could provide important complementary diagnostic information for improving patient management.

The cytokine interleukin (IL)-11 and its receptor, IL-11Rα, have been linked to breast cancer development and progression, as well as bone metastasis [90–92]. A cyclic peptide that mimics the receptor binding motif within IL-11, c(CGRRAGGSC), has been identified through phage display [93, 94]. A dual-modality imaging agent, ^{111}In-DTPA-Bz-NH-SA-K(IR-783-S-Ph-CO)-c(CGRRAGGSC)NH$_2$, was constructed in which DTPA and a fluorescence dye (IR-783-S-Ph-CO) were linked to the peptide via a lysine linker [95]. The selection of IR-783-S-Ph-CO-NHS (NHS denotes N-hydroxysuccinimide) as the fluorophore over its analogues was due to its superior chemical stability [96]. Cell-based studies revealed that the cyclic peptide maintained the targeting capability to IL-11Rα after conjugation of both the optical and radioisotope labels. Cross-validation and direct comparison of optical and nuclear imaging of the agent in a small animal tumour model was achieved using a single injection, which demonstrated the tumour-targeting capability of the dual-modality probe *in vivo* (Figure 15.8).

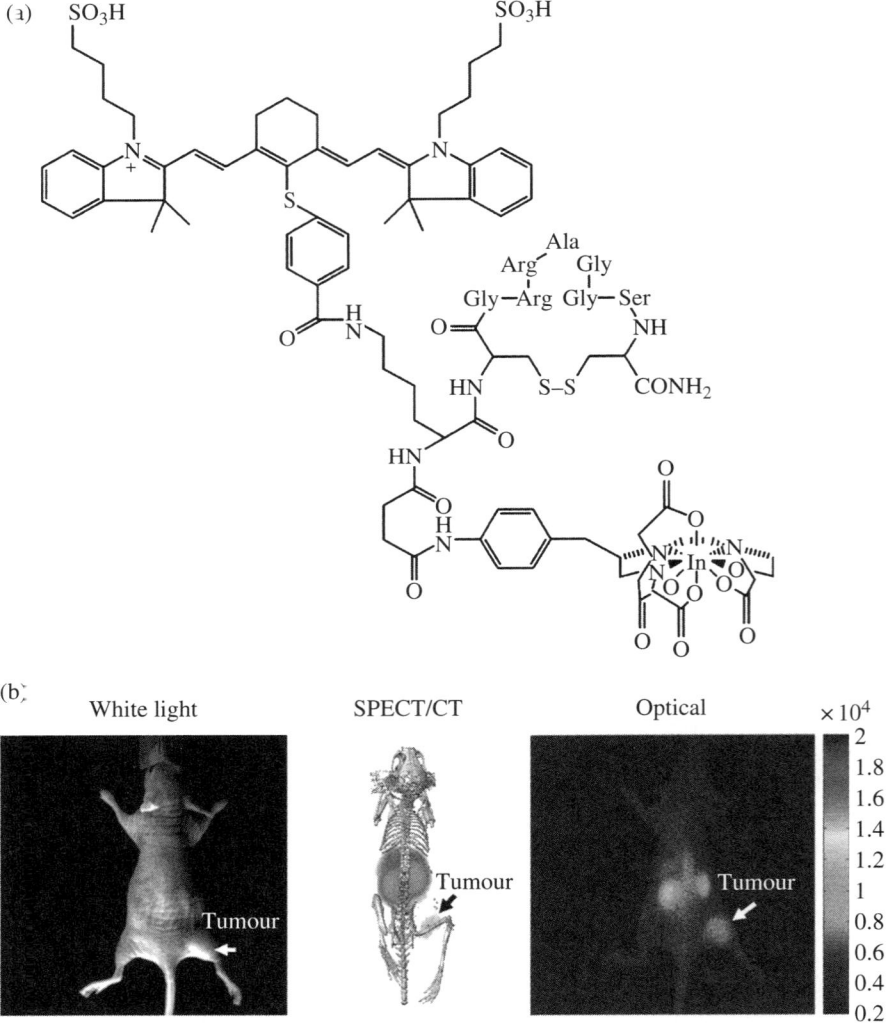

FIGURE 15.8 Dual-modality SPECT/optical imaging of IL-11Rα. (a) Chemical structure of the dual-modality probe. (b) SPECT/CT and optical images of a tumour-bearing mouse at 24 h after probe injection. Adapted with permission from Ref. [95]. (*See insert for colour representation of the figure.*)

In several of the abovementioned studies, the reason why imaging was carried out at 24 h after injection for such low molecular weight agents (which are typically cleared from the circulation in minutes) was unclear. When compared to the size of either cyclic RGD peptides or c(CGRRAGGSC), the ^{111}In-DTPA complex and the fluorescent dyes are relatively bulky and may significantly affect receptor binding. Recognising such limitations, several groups have also investigated the use of dual-labelled antibodies for SPECT/optical imaging using a similar approach.

15.3.3 Antibody-Based Agents

Overexpression of the human epidermal growth factor receptor (HER) family has been implicated in many cancer types, because of its multiple roles in signalling pathways that regulate cellular proliferation, differentiation, motility, and survival [97, 98]. Approximately one-third of breast cancer patients overexpress HER2, a member of the HER family [99]. Trastuzumab (Herceptin; Genentech, Inc.) is an inhibitory mAb developed against the extracellular domain of HER2, which has been used for cancer therapy in HER2-positive patients [100–102].

In one report, $(^{111}$In-DTPA$)_n$-trastuzumab-(IRDye800)$_m$ was synthesised, and fluorescence confocal microscopy was used to determine the molecular specificity of (DTPA)$_n$-trastuzumab-(IRDye800)$_m$ in vitro in SKBr3 (HER2-positive) and MDA-MB-231 (HER2-negative) breast cancer cells [103]. It was found that (DTPA)$_n$-trastuzumab-(IRDye800)$_m$ had significantly greater binding to SKBr3 than to MDA-MB-231 cells, and the binding occurred predominantly within the cell membrane, because HER2 is a membrane-bound protein. In vivo SPECT and optical imaging of SKBr3 xenografts injected with $(^{111}$In-DTPA$)_n$-trastuzumab-(IRDye800)$_m$ revealed significantly more uptake in the tumour region than in the contralateral muscle. This dual-modality agent, $(^{111}$In-DTPA$)_n$-trastuzumab-(IRDye800)$_m$, may be an effective diagnostic biomarker capable of tracking HER2 overexpression in breast cancer patients. Several other reports also investigated the potential of trastuzumab for the development of dual-modality SPECT/optical or PET/optical agents using similar approaches [104, 105]. However, no in vivo studies were reported using these agents.

15.4 MRI/OPTICAL AGENTS

MRI is a noninvasive diagnostic technique based on the interaction of protons (or other MRI-active nuclei) with each other and with surrounding molecules in a tissue of interest [106]. Different tissues have different relaxation times, which can result in endogenous MRI contrast. The major advantages of MRI over radionuclide-based imaging are the absence of radiation, higher spatial resolution, and exquisite soft tissue contrast. The major disadvantage of MRI is its inherent low sensitivity, which can be partially compensated for by working at higher magnetic fields (4.7-14 T in small animal models), acquiring data for a much longer time period, and using exogenous contrast agents. Traditionally, Gd^{3+}-chelates have been used to enhance the T_1 contrast [107], and iron oxide nanoparticles have been used to increase the T_2 contrast [108]. Generally, the design of non-nanoparticle-based dual-modality MRI/optical imaging agents consists of Gd^{3+}-chelates that are covalently conjugated to a fluorescent dye. A tabulated summary of MRI/optical agents is provided in Table 15.3 and discussed in detail below.

In an early study of dual-modality MRI/optical imaging agents, several groups of compounds with different cell permeability were investigated for such applications [109]. One of these agents, gadolinium rhodamine dextran, was further studied for its effect on neural stem cells [110]. Although no significant effect on cell viability was observed, a decrease in

TABLE 15.3 A Tabulated Summary of Dual-modality MRI/optical Agents.

Paramagnetic Moiety	Fluorophore	Targeting Ligand	Target	References
	Rhodamine	-	-	[109, 110]
	Dicarbocyanine	LDL	LDL receptor	[111]
	Oregon Green 488	Cyclic NGR peptide	CD13	[112]
	Fluorescein	-	-	[115]
	Cyanine dyes	-	-	[116]
Gd^{3+}-chelates	Porphyrin	-	-	[121]
	Fluorescein derivative	Albumin	Caveolae	[124, 125]
	Congo red	Albumin	Amyloid plaque	[126]
	NIR813	-	-	[128]
	Cy5.5	Bombesin peptide	GRPR	[129]
	HPPH	-	-	[130]

proliferation was evident in cells that underwent 24 h of labelling. Further *in vivo* testing will be required to ensure that labelling with these dual-modality MRI/optical agents does not perturb the cells' function.

A lipophilic agent composed of Gd-DTPA, a fluorescent dye, and a 16-carbon alkyl chain for intercalative labelling of low-density lipoprotein (LDL) particles was reported for *in vivo* detection of LDL receptors (LDLR) by MRI and *in vitro* monitoring of cellular localisation by confocal fluorescence microscopy [111]. While no significant membrane-associated fluorescence signal was observed, differential shortening of T_1 relaxation time was found in normal liver (>70%) when compared to LDLR$^{-/-}$ liver (12%) in mice. In addition, uptake of labelled LDL particles in subcutaneously implanted B16 melanoma tumours in mice led to a modest decrease in T_1 relaxation time of the tumour.

In another report, a chemoselective reaction was employed to synthesise an agent that contains a peptidic targeting ligand, Gd-DTPA, and a fluorescent dye Oregon Green 488 [112]. The cyclic peptide used in this study contains the asparagine-glycine-arginine motif (NGR), which can bind to an angiogenesis-related protein CD13 [113, 114]. However, *in vitro* and *in vivo* behaviour of this agent was not reported.

A bifunctional agent that contains a Gd chelate and fluorescein has been developed and investigated [115]. *In vitro* studies in NIH-3T3 mouse fibroblast cells demonstrated a concentrated-dependent increase in fluorescence as well as significant increase in T_1, suggesting cell binding and uptake of the agent. However, the use of fluorescein made it only suitable for cell-based assays but not *in vivo* imaging, due to the short excitation/emission wavelength. Cyanine dyes have been covalently attached to Gd-DTPA-polylysine of an extended, uncoiled conformation for fluorescence imaging of preclinical subcutaneous and orthotopic mammary gland tumours [116]. When a wide-field illumination camera system was used, this agent could allow for intraoperative delineation of the tumour margin.

Gadophrin-2, composed of a porphyrin ring and two covalently linked Gd-DTPA, has been investigated as a MRI contrast agent in various studies [117–120]. In one report, it was also used to label and trace intravenously injected human hematopoietic cells in athymic mice [121]. Unlike the use of covalently linked fluorophores in other dual-modality MRI/optical imaging agents described above, the fluorescence properties of the porphyrin ring within gadophrin-2 was used for optical imaging. After intravenous injection, the distribution of gadophrin-2-labelled cells in nude mice was visualised by MRI, optical imaging, and fluorescence microscopy (Figure 15.9).

In these abovementioned reports, the ratio of Gd:dye is quite low, in many cases 1:1. Due to the very low sensitivity of Gd-based MRI, such composition is not optimal for *in vivo* imaging applications. A class of polymer-based MRI/optical agents has been reported that contains multiple copies of Gd in each molecule [122, 123]. However, since these agents fall into the category of nanoparticle-based dual-modality imaging agents, they will not be discussed in this chapter.

Albumin has been conjugated to multiple imaging tags for MRI/optical tracking of labelled fibroblasts in animal tumour models to understand the role of fibroblasts in tumour differentiation, tumour angiogenesis, and vessel maturation [124]. It was demonstrated that this contrast agent can be internalised into intracellular granules via caveolae-mediated endocytosis, which leads to MRI signal enhancement [125]. Interestingly, similar Gd^{3+}-containing albumin-based agents that contain Congo Red as the fluorophore have also been reported for detection of amyloid plaques in Alzheimer disease (AD) [126]. Because the amyloid deposits detected by the probe were found to be comparable to the amyloid deposits detected by an amyloid-specific mAb it was suggested that this agent may be useful for detection of individual senile plaques and diagnosis of AD.

In the NIR region, the absorbance of all biomolecules reaches a minimum and provides a clear window for *in vivo* optical imaging [127]. Furthermore, there is also significantly less tissue autofluorescence in this region. Poly(L-glutamic acid) conjugated with Gd-DTPA and a NIRF dye NIR813 (emission maximum: 813 nm) has been used for SLN mapping in normal and tumour-bearing mice [128]. It was reported that there were more than 50 Gd-DTPA units per polymer chain. After subcutaneous injection, axillary and branchial lymph nodes were clearly visualised with both T_1-weighted MRI and optical imaging within 3 minutes, even at the lowest dose tested (2 µmol of Gd/kg and 4.8 nmol of NIR813). After intralingual injection in tumour-bearing mice, both MRI and NIRF imaging identified most of the superficial cervical lymph nodes. Histopathologic examination of the SLNs resected under NIRF imaging guidance revealed micrometastases in all SLNs identified.

A MRI/optical agent has been developed and evaluated for imaging of GRPR in prostate cancer [129]. This agent is consisted of Gd^{3+} complexed with a chelator abbreviated as TTDA-NP, the NIR dye Cy5.5, and a bombesin peptide. Both MRI and NIRF imaging displayed good tumour contrast of the agents (i.e., Gd-TTDA-NP-BN and Gd-TTDA-NP-BN-Cy5.5; Figure 15.10) in PC-3 tumour-bearing mice.

In a recent report, an MRI/optical agent has been developed for both imaging and photodynamic therapy of tumours [130]. The fluorophore used in this agent, 3-(1′-hexyloxyethyl)pyropheophorbide-a (abbreviated as HPPH), is a photosensitiser that displayed the best photodynamic therapy efficacy *in vivo* when it was conjugated with multiple Gd^{3+}-DTPA chelates. It was suggested that this multifunctional agent has the potential for screening, diagnosis, treatment monitoring, as well as image-guided intervention of cancer.

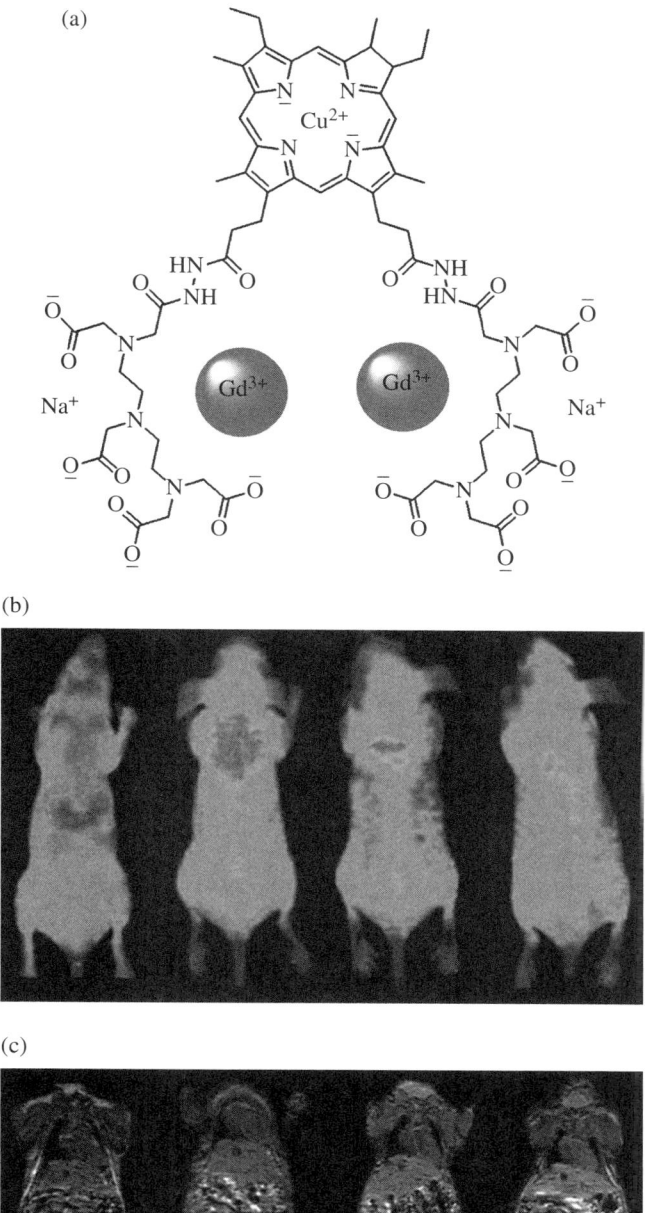

(a)

(b)

(c)

0 h   1 h   4 h   24 h

FIGURE 15.9 Dual-modality MRI/optical imaging with gadophrin-2. (a) Chemical structure of gadophrin-2. (b) Optical images and MRI images (c) of a mouse before and after injection of gadophrin-2-labelled hematopoietic cells, which showed a progressive cell accumulation in various organs (e.g., the lung). Adapted with permission from Ref. [121]. (*See insert for colour representation of the figure.*)

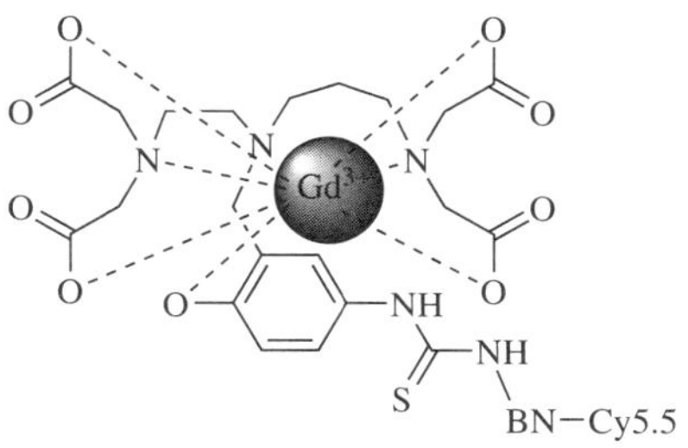

FIGURE 15.10 Chemical structure of Gd-TTDA-NP-BN-Cy5.5, which was used for MRI/optical imaging of prostate cancer in tumour-bearing mice.

When compared to the abovementioned dual-modality PET/optical and SPECT/optical imaging agents, where most of the agents developed were specifically directed to a certain protein or other targets of interest, most of the dual-modality MRI/optical agents are not molecularly targeted. In many cases, these MRI/optical agents take advantage of the enhanced permeability and retention effect [131] for tumour accumulation or certain lymphatic drainage patterns for SLN mapping. One major reason for this phenomenon is that Gd-based MRI has low sensitivity (in the millimolar range); therefore, it is very challenging to achieve high enough local concentration of a MRI contrast agent for noninvasive imaging using a targeted approach.

15.5   PET/MRI AGENTS

The dual-modality PET/CT scanner, already being used on a routine basis in clinical oncology, greatly facilitated pinpointing the regions of increased activity on PET, which does not provide anatomical information by itself [132, 133]. However, accurate localisation of PET probe uptake, even with PET/CT, can be very difficult in some cases due to the absence of identifiable anatomical structures, particularly in the abdomen [134–136]. Because of the exquisite soft tissue contrast of MRI, combination of PET and MRI can have many synergistic effects. For example, highly accurate image registration can offer the possibility of using the MRI image to correct for PET partial volume effect (i.e., image blurring introduced by the finite spatial resolution of the imaging system [137]) and aid in PET image reconstruction. In addition, PET/MRI also has greatly reduced radiation exposure than PET/CT.

PET/MRI systems have been developed for small animal imaging as well as human studies [136, 138]. PET/MRI, acquired in one measurement, has the potential to become the imaging modality of choice for various clinical applications such as neurological studies, certain types of cancer, stroke, and the emerging field of stem cell therapy. The future of PET/MRI scanners will greatly benefit from the use of dual-modality PET/MRI agents, which are currently being investigated [7]. Recently, a PET/MRI probe was synthesised through partial exchange of Gd^{3+} for ^{64}Cu in EP-2104R [139], a fibrin-specific MRI contrast agent for thrombus detection [140]. This agent was successfully utilised for simultaneous MRI and PET imaging of thrombus in a rat arterial thrombus model.

15.6   CONCLUSIONS

The synergy of combining multiple imaging modalities is well recognised by the imaging community, and one of the best examples is the development of PET/MRI scanners that recently became commercially available. This chapter aims to give a comprehensive summary of the dual-modality imaging agents that have been reported to date, which are not based on nanoparticles. When compared to other dual-modality agents reported in the literature, dual-modality MRI/optical agents are among the most intensively studied, which can be partly attributed to the wide availability of the imaging agents (a representative list of fluorescent dyes that have been used for the development of dual-modality imaging agents is shown in Figure 15.11) as well as the imaging systems. However, whether MRI/optical combination is the most optimal dual-modality approach is questionable because neither modality is highly quantitative. Because PET has significantly higher sensitivity than SPECT (at least an order of magnitude) [10], dual-modality PET/optical and PET/MRI agents deserve the dedication of significant research effort in the future. We envision that PET/MRI/optical agents may find the most widespread use for future biomedical/clinical applications because such a combination provides extremely high sensitivity (PET), quantitation capability (PET), excellent anatomical information and soft tissue contrast (MRI), as well as a means for *ex vivo* validation (optical), which itself can also be useful for highly sensitive imaging in certain sites of the human body.

Robust chemistry for both radiolabelling and targeting ligand conjugation is of critical importance to the potential clinical applications of dual-modality imaging agents. For example, one major challenge in antibody labelling is to minimise the potential interference with its antigen binding affinity/specificity, hence overconjugation of the image tags should be avoided. One of the key requirements for accurate PET/SPECT imaging with radiolabelled imaging probes is that the tracer should be sufficiently stable during the imaging period, because the PET/SPECT scanner detects the distribution of the radioisotopes instead of the targeting ligand itself (e.g., peptide, antibody). A number of chelators can be used for complexation of certain radiometals for PET/SPECT applications (Figure 15.12).

In preclinical studies, ^{64}Cu is one of the most widely used radioisotopes for PET, and a variety of chelators have been explored for ^{64}Cu-labelling [141]. Recently, it has been generally recognised that NOTA is one of the most suitable chelators for ^{64}Cu-labelling [55, 65], even though several other chelators also exhibit similar *in vitro* stability and comparable tumour uptake in animal models.

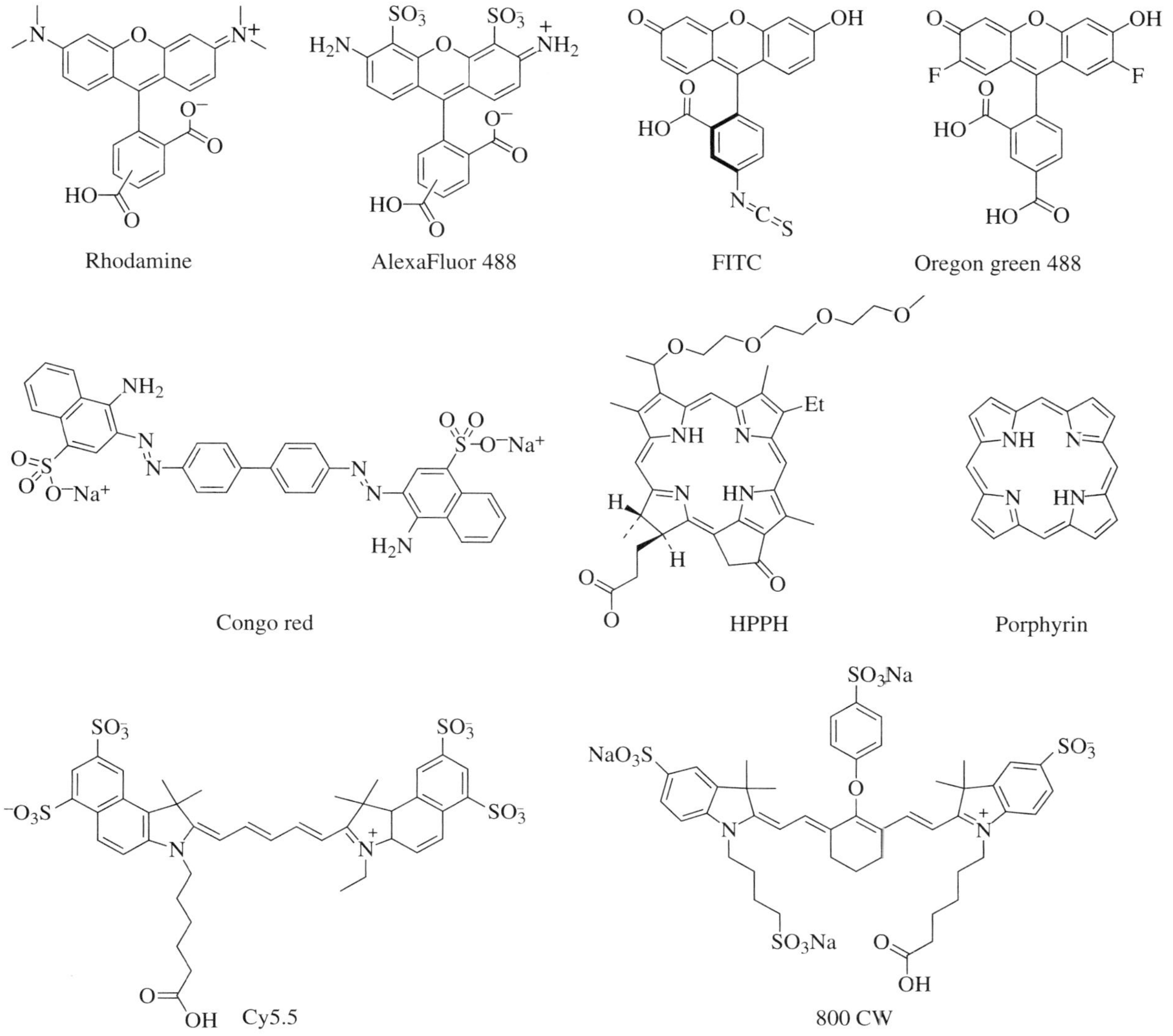

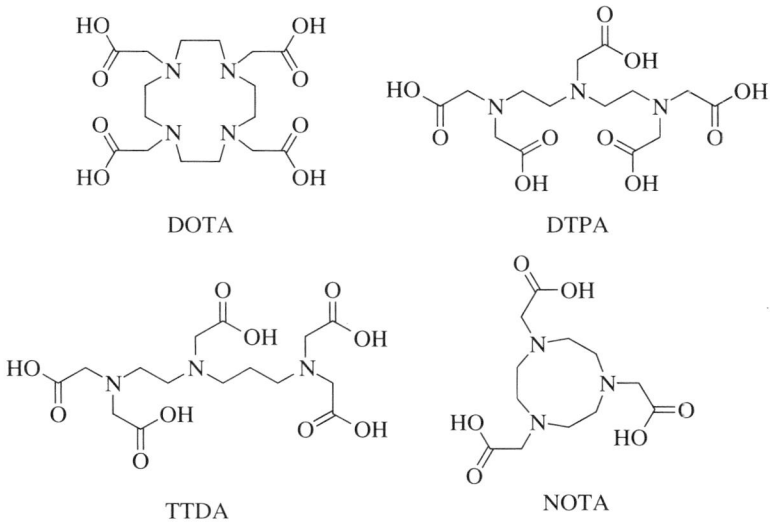

Rhodamine

AlexaFluor 488

FITC

Oregon green 488

Congo red

HPPH

Porphyrin

Cy5.5

800 CW

FIGURE 15.11 Representative fluorescent dyes that have been used for the development of dual-modality imaging agents.

DOTA

DTPA

TTDA

NOTA

FIGURE 15.12 A selected list of chelators that have been used for the development of dual-modality imaging agents.

Most of the dual-modality imaging agents developed to date are for oncology applications. Noninvasive imaging of cancer has clinical applications in many aspects including, but not limited to, lesion detection, patient stratification, new drug development/validation, treatment monitoring, and dose optimisation [142–144]. The complete sequencing of the human genome has ushered in a new era of systems biology referred to as '-omics.' [145] Genomic and proteomic molecular profiling technologies are transforming cancer research [146, 147]. New technologies such as microarray analysis, genomics, proteomics, and high-throughput screening may lead to the discovery of new molecular targets/targeting ligands for multimodality imaging of cancer. With the development of new imaging agents with better targeting efficacy and desirable pharmacokinetics, clinical translation of the probes will be critical for the maximum benefit of cancer patients.

REFERENCES

[1]   R. Weissleder and M. J. Pittet, *Nature* **452,** 580–589 (2008).

[2]   T. F. Massoud and S. S. Gambhir, *Genes Dev.* **17,** 545–580 (2003).

[3]   M. A. Pysz, S. S. Gambhir and J. K. Willmann, *Clin. Radiol.* **65,** 500–516 (2010).

[4]   F. Leblond, S. C. Davis, P. A. Valdes and B. W. Pogue, *J. Photochem. Photobiol. B* **98,** 77–94 (2010).

[5]   W. Cai, A. R. Hsu, Z. B. Li and X. Chen, *Nanoscale Res. Lett.* **2,** 265–281 (2007).

[6]   W. Cai and X. Chen, *Small* **3,** 1840–1854 (2007).

[7]   H. Hong, Y. Zhang, J. Sun and W. Cai, *Nano Today* **4,** 399–413 (2009).

[8]   S. H. Bloch, P. A. Dayton and K. W. Ferrara, *IEEE Eng. Med. Biol. Mag.* **23,** 18–29 (2004).

[9]   P. A. Dayton and J. J. Rychak, *Front. Biosci.* **12,** 5124–5142 (2007).

[10]   S. S. Gambhir, *Nat. Rev. Cancer* **2,** 683–693 (2002).

[11]   S. S. Gambhir, J. Czernin, J. Schwimmer, D. H. Silverman, R. E. Coleman and M. E. Phelps, *J. Nucl. Med.* **42,** 1S–93S (2001).

[12]   F. L. Thorp-Greenwood and M. P. Coogan, *Dalton Trans.* **40,** 6129–6143 (2011).

[13]   S. K. Pandey, A. L. Gryshuk, M. Sajjad, X. Zheng, Y. Chen, M. M. Abouzeid, J. Morgan, I. Charamisinau, H. A. Nabi, A. Oseroff and R. K. Pandey, *J. Med. Chem.* **48,** 6286–6295 (2005).

[14]   A. A. Bogdanov, Jr., C. P. Lin, M. Simonova, L. Matuszewski and R. Weissleder, *Neoplasia* **4,** 228–236 (2002).

[15]   B. Ballou, G. W. Fisher, J. S. Deng, T. R. Hakala, M. Srivastava and D. L. Farkas, *Cancer Detect. Prev.* **22,** 251–257 (1998).

[16]   Q. Li, Y. Kim, J. Namm, A. Kulkarni, G. R. Rosania, Y. H. Ahn and Y. T. Chang, *Chem. Biol.* **13,** 615–623 (2006).

[17]   M. Wang, M. Gao, K. D. Miller, G. W. Sledge, G. D. Hutchins and Q. H. Zheng, *Eur. J. Med. Chem.* **44,** 2300–2306 (2009).

[18]   Z. Li, T. P. Lin, S. Liu, C. W. Huang, T. W. Hudnall, F. P. Gabbai and P. S. Conti, *Chem. Commun.* **47,** 9324–9326 (2011).

[19]   J. Kuil, A. H. Velders and F. W. B. van Leeuwen, *Bioconjug. Chem.* **21,** 1709–1719 (2010).

[20]   J. P. Xiong, T. Stehle, R. Zhang, A. Joachimiak, M. Frech, S. L. Goodman, M. A. Arnaout, B. Diefenbach, R. Dunker and D. L. Scott, *Science* **296,** 151–155 (2002).

[21]   W. Cai, G. Niu and X. Chen, *Curr. Pharm. Des.* **14,** 2943–2973 (2008).

[22]   J. D. Hood and D. A. Cheresh, *Nat. Rev. Cancer* **2,** 91–100 (2002).

[23]   G. Bergers and L. E. Benjamin, *Nat. Rev. Cancer* **3,** 401–410 (2003).

[24]   W. Cai and X. Chen *Anti-Cancer Agents Med. Chem.* **6,** 407–428 (2006).

[25]   D. Hanahan and R. A. Weinberg, *Cell* **144,** 646–674 (2011).

[26]   P. Carmeliet, *Nature* **438,** 932–936 (2005).

[27]   J. Folkman, *Nat. Med.* **1,** 27–31 (1995).

[28]   W. Cai and X. Chen, *J. Nucl. Med.* **49 Suppl** 2, 113S–128S (2008).

[29]   R. H. Kimura, Z. Miao, Z. Cheng, S. S. Gambhir and J. R. Cochran, *Bioconjug. Chem.* **21,** 436–444 (2010).

[30]   Y. H. Chang, I. L. Lin, G. J. Tsay, S. C. Yang, T. P. Yang, K. T. Ho, T. C. Hsu and M. Y. Shiau, *Clin. Biochem.* **41,** 955–959 (2008).

[31]   E. Lancelot, V. Amirbekian, I. Brigger, J. S. Raynaud, S. Ballet, C. David, O. Rousseaux, S. Le Greneur, M. Port, H. R. Lijnen, P. Bruneval, J. B. Michel, T. Ouimet, B. Roques, S. Amirbekian, F. Hyafil, E. Vucic, J. G. Aguinaldo, C. Corot and Z. A. Fayad, *Arterioscler. Thromb. Vasc. Biol.* **28,** 425–432 (2008).

[32]   M. E. Muroski, M. D. Roycik, R. G. Newcomer, P. E. Van den Steen, G. Opdenakker, H. R. Monroe, Z. J. Sahab and Q. X. Sang, *Curr. Pharm. Biotechnol.* **9,** 34–46 (2008).

[33]   E. I. Deryugina and J. P. Quigley, *Cancer Metastasis Rev.* **25,** 9–34 (2006).

[34]   P. Manduca, A. Castagnino, D. Lombardini, S. Marchisio, S. Soldano, V. Ulivi, S. Zanotti, C. Garbi, N. Ferrari and D. Palmieri, *Bone* **44,** 251–265 (2009).

[35]   M. Egeblad and Z. Werb, *Nat. Rev. Cancer* **2,** 161–174 (2002).

[36] A. Azhdarinia, N. Wilganowski, H. Robinson, P. Ghosh, S. Kwon, Z. W. Lazard, A. R. Davis, E. Olmsted-Davis and E. M. Sevick-Muraca, *Bioorg. Med. Chem.* **19,** 3769–3776 (2011).

[37] W. Cai, K. Chen, K. A. Mohamedali, Q. Cao, S. S. Gambhir, M. G. Rosenblum and X. Chen, *J. Nucl. Med.* **47,** 2048–2056 (2006).

[38] W. Cai, Y. Wu, K. Chen, Q. Cao, D. A. Tice and X. Chen, *Cancer Res.* **66,** 9673–9681 (2006).

[39] H. Hong, Y. Yang, Y. Zhang, J. W. Engle, T. E. Barnhart, R. J. Nickles, B. R. Leigh and W. Cai, *Eur. J. Nucl. Med. Mol. Imaging* **38,** 1335–1343 (2011).

[40] C. Scarpignato and I. Pelosini, *Chemotherapy* **47 Suppl** 2, 1–29 (2001).

[41] I. Virgolini, T. Pangerl, C. Bischof, P. Smith-Jone and M. Peck-Radosavljevic, *Eur. J. Clin. Invest.* **27,** 645–647 (1997).

[42] M. Werle and A. Bernkop-Schnurch, *Amino Acids* **30,** 351–367 (2006).

[43] J. C. Reubi, J. C. Schar, B. Waser, S. Wenger, A. Heppeler, J. S. Schmitt and H. R. Macke, *Eur. J. Nucl. Med.* **27,** 273–282 (2000).

[44] W. A. Breeman, M. de Jong, D. J. Kwekkeboom, R. Valkema, W. H. Bakker, P. P. Kooij, T. J. Visser and E. P. Krenning, *Eur. J. Nucl. Med.* **28,** 1421–1429 (2001).

[45] J. S. Lewis and C. J. Anderson, *Methods Mol. Biol.* **386,** 227–240 (2007).

[46] W. B. Edwards, B. Xu, W. Akers, P. P. Cheney, K. Liang, B. E. Rogers, C. J. Anderson and S. Achilefu, *Bioconjug. Chem.* **19,** 192–200 (2008).

[47] D. B. Cornelio, R. Roesler and G. Schwartsmann, *Ann. Oncol.* **18,** 1457–1466 (2007).

[48] S. Achilefu, H. N. Jimenez, R. B. Dorshow, J. E. Bugaj, E. G. Webb, R. R. Wilhelm, R. Rajagopalan, J. Johler and J. L. Erion, *J. Med. Chem.* **45,** 2003–2015 (2002).

[49] H. Lee, W. J. Akers, P. P. Cheney, W. B. Edwards, K. Liang, J. P. Culver and S. Achilefu, *J. Biomed. Opt.* **14,** 040507/040501–040507/040503 (2009).

[50] Q. P. Peterson, D. R. Goode, D. C. West, R. C. Botham and P. J. Hergenrother, *Nat. Protoc.* **5,** 294–302 (2010).

[51] A. G. Porter and R. U. Janicke, *Cell Death Differ.* **6,** 99–104 (1999).

[52] P. Paudyal, B. Paudyal, Y. Iida, N. Oriuchi, H. Hanaoka, H. Tominaga, T. Ishikita, H. Yoshioka, T. Higuchi and K. Endo, *Oncol. Rep.* **22,** 115–119 (2009).

[53] N. Ferrara, K. J. Hillan, H. P. Gerber and W. Novotny, *Nat. Rev. Drug Discov.* **3,** 391–400 (2004).

[54] Y. Zhang, H. Hong, J. W. Engle, Y. Yang, T. E. Barnhart and W. Cai, *Am. J. Nucl. Med. Mol. Imaging* **2,** 1–13 (2012).

[55] J. L. J. Dearling, S. D. Voss, P. Dunning, E. Snay, F. Fahey, S. V. Smith, J. S. Huston, C. F. Meares, S. T. Treves and A. B. Packard, *Nucl. Med. Biol.* **38,** 29–38 (2011).

[56] A. J. Beer and M. Schwaiger, *Cancer Metastasis Rev.* **27,** 631–644 (2008).

[57] W. Cai and X. Chen, *Front. Biosci.* **12,** 4267–4279 (2007).

[58] I. Dijkgraaf and O. C. Boerman, *Cancer Biother. Radiopharm.* **24,** 637–647 (2009).

[59] E. S. Mittra, M. L. Goris, A. H. Iagaru, A. Kardan, L. Burton, R. Berganos, E. Chang, S. Liu, B. Shen, F. T. Chin, X. Chen and S. S. Gambhir, *Radiology* **260,** 182–191 (2011).

[60] N. A. Dallas, S. Samuel, L. Xia, F. Fan, M. J. Gray, S. J. Lim and L. M. Ellis, *Clin. Cancer Res.* **14,** 1931–1937 (2008).

[61] E. Fonsatti, H. J. Nicolay, M. Altomonte, A. Covre and M. Maio, *Cardiovasc. Res.* **86,** 12–19 (2010).

[62] B. K. Seon, A. Haba, F. Matsuno, N. Takahashi, M. Tsujie, X. She, N. Harada, S. Uneda, T. Tsujie, H. Toi, H. Tsai and Y. Haruta, *Curr. Drug Deliv.* **8,** 135–143 (2011).

[63] H. Hong, G. W. Severin, Y. Yang, J. W. Engle, Y. Zhang, T. E. Barnhart, G. Liu, B. R. Leigh, R. J. Nickles and W. Cai, *Eur. J. Nucl. Med. Mol. Imaging* **39,** 138–148 (2012).

[64] Y. Yang, Y. Zhang, H. Hong, G. Liu, B. R. Leigh and W. Cai, *Eur. J. Nucl. Med. Mol. Imaging* **38,** 2066–2076 (2011).

[65] Y. Zhang, H. Hong, J. W. Engle, J. Bean, Y. Yang, B. R. Leigh, T. E. Barnhart and W. Cai, *PLoS One* **6,** e28005 (2011).

[66] Y. Zhang, H. Hong, G. W. Severin, J. W. Engle, Y. Yang, S. Goel, A. J. Nathanson, G. Liu, R. J. Nickles, B. R. Leigh, T. E. Barnhart and W. Cai, *Am. J. Transl. Res.* **5,** 291–302, (2013).

[67] Y. Zhang, H. Hong, J. W. Engle, Y. Yang, C. P. Theuer, T. E. Barnhart and W. Cai, *Mol. Pharm.* **9,** 645–653 (2012).

[68] A. M. Wallace, C. K. Hoh, S. J. Ellner, D. D. Darrah, G. Schulteis and D. R. Vera, *Ann. Surg. Oncol.* **14,** 913–921 (2007).

[69] A. M. Wallace, C. K. Hoh, D. R. Vera, D. D. Darrah and G. Schulteis, *Ann. Surg. Oncol.* **10,** 531–538 (2003).

[70] R. Ting, T. A. Aguilera, J. L. Crisp, D. J. Hall, W. C. Eckelman, D. R. Vera and R. Y. Tsien, *Bioconjug. Chem.* **21,** 1811–1819 (2010).

[71] K. Peremans, B. Cornelissen, B. Van Den Bossche, K. Audenaert and C. Van de Wiele, *Vet. Radiol. Ultrasound* **46,** 162–170 (2005).

[72] A. Kjaer, *Adv. Exp. Med. Biol.* **587,** 277–284 (2006).

[73] A. Lipton, *Cancer* **88,** 3033–3037 (2000).

[74] K. R. Bhushan, P. Misra, F. Liu, S. Mathur, R. E. Lenkinski and J. V. Frangioni, *J. Am. Chem. Soc.* **130,** 17648–17649 (2008).

[75] K. R. Bhushan, E. Tanaka and J. V. Frangioni, *Angew. Chem. Int. Ed. Engl.* **46,** 7969–7971 (2007).

[76] F. Chang, T. Qu, M. Rusckowski and D. J. Hnatowich, *Appl. Radiat. Isot.* **50,** 723–732 (1999).

[77] N. Agorastos, L. Borsig, A. Renard, P. Antoni, G. Viola, B. Spingler, P. Kurz and R. Alberto, *Chemistry* **13**, 3842–3852 (2007).

[78] T. Esteves, C. Xavier, S. Gama, F. Mendes, P. D. Raposinho, F. Marques, A. Paulo, J. C. Pessoa, J. Rino, G. Viola and I. Santos, *Org. Biomol. Chem.* **8**, 4104–4116 (2010).

[79] S. Tzanopoulou, M. Sagnou, M. Paravatou-Petsotas, E. Gourni, G. Loudos, S. Xanthopoulos, D. Lafkas, H. Kiaris, A. Varvarigou, I. C. Pirmettis, M. Papadopoulos and M. Pelecanou, *J. Med. Chem.* **53**, 4633–4641 (2010).

[80] Y. Yang, L. Zhu, M. Cui, R. Tang and H. Zhang, *Bioorg. Med. Chem. Lett.* **20**, 5337–5344 (2010).

[81] K. A. Stephenson, J. Zubieta, S. R. Banerjee, M. K. Levadala, L. Taggart, L. Ryan, N. McFarlane, D. R. Boreham, K. P. Maresca, J. W. Babich and J. F. Valliant, *Bioconjug. Chem.* **15**, 128–136 (2004).

[82] S. R. Banerjee, M. K. Levadala, N. Lazarova, L. Wei, J. F. Valliant, K. A. Stephenson, J. W. Babich, K. P. Maresca and J. Zubieta, *Inorg. Chem.* **41**, 6417–6425 (2002).

[83] S. James, K. P. Maresca, J. W. Babich, J. F. Valliant, L. Doering and J. Zubieta, *Bioconjug. Chem.* **17**, 590–596 (2006).

[84] P. Kurz, B. Probst, B. Spingler and R. Alberto, *Eur. J. Inorg. Chem.* 2966–2974 (2006).

[85] Y. Tooyama, A. Harano, T. Yoshimura, T. Takayama, T. Sekine, H. Kudo and A. Shinohara, *J. Nucl. Radiochem. Sci.* **6**, 153–155 (2005).

[86] C. Xiong, W. Lu, R. Zhang, M. Tian, W. Tong, J. Gelovani and C. Li, *Chemistry* **15**, 8979–8984 (2009).

[87] J. P. Houston, S. Ke, W. Wang, C. Li and E. M. Sevick-Muraca, *J. Biomed. Opt.* **10**, 054010 (2005).

[88] C. Li, W. Wang, Q. Wu, S. Ke, J. Houston, E. Sevick-Muraca, L. Dong, D. Chow, C. Charnsangavej and J. G. Gelovani, *Nucl. Med. Biol.* **33**, 349–358 (2006).

[89] W. B. Edwards, W. J. Akers, Y. Ye, P. P. Cheney, S. Bloch, B. Xu, R. Laforest and S. Achilefu, *Mol. Imaging* **8**, 101–110 (2009).

[90] Y. Kang, W. He, S. Tulley, G. P. Gupta, I. Serganova, C. R. Chen, K. Manova-Todorova, R. Blasberg, W. L. Gerald and J. Massague, *Proc. Natl. Acad. Sci. U.S.A.* **102**, 13909–13914 (2005).

[91] M. Lacroix, B. Sivek, P. J. Marie and J. J. Body, *Cancer Lett.* **127**, 29–35 (1998).

[92] Y. Morinaga, N. Fujita, K. Ohishi, Y. Zhang and T. Tsuruo, *J. Cell Physiol.* **175**, 247–254 (1998).

[93] A. J. Zurita, P. Troncoso, M. Cardo-Vila, C. J. Logothetis, R. Pasqualini and W. Arap, *Cancer Res.* **64**, 435–439 (2004).

[94] W. Arap, M. G. Kolonin, M. Trepel, J. Lahdenranta, M. Cardo-Vila, R. J. Giordano, P. J. Mintz, P. U. Ardelt, V. J. Yao, C. I. Vidal, L. Chen, A. Flamm, H. Valtanen, L. M. Weavind, M. E. Hicks, R. E. Pollock, G. H. Botz, C. D. Bucana, E. Koivunen, D. Cahill, P. Troncoso, K. A. Baggerly, R. D. Pentz, K. A. Do, C. J. Logothetis and R. Pasqualini, *Nat. Med.* **8**, 121–127 (2002).

[95] W. Wang, S. Ke, S. Kwon, S. Yallampalli, A. G. Cameron, K. E. Adams, M. E. Mawad and E. M. Sevick-Muraca, *Bioconjug. Chem.* **18**, 397–402 (2007).

[96] S. A. Hilderbrand, K. A. Kelly, R. Weissleder and C. H. Tung, *Bioconjug. Chem.* **16**, 1275–1281 (2005).

[97] W. Cai, G. Niu and X. Chen, *Eur. J. Nucl. Med. Mol. Imaging* **35**, 186–208 (2007).

[98] G. Niu, W. Cai and X. Chen, *Front. Biosci.* **13**, 790–805 (2008).

[99] D. J. Slamon, W. Godolphin, L. A. Jones, J. A. Holt, S. G. Wong, D. E. Keith, W. J. Levin, S. G. Stuart, J. Udove, A. Ullrich, et al., *Science* **244**, 707–712 (1989).

[100] M. Harries and I. Smith, *Endocr. Relat. Cancer* **9**, 75–85 (2002).

[101] C. H. Yeon and M. D. Pegram, *Invest. New Drugs* **23**, 391–409 (2005).

[102] E. Tokunaga, E. Oki, K. Nishida, T. Koga, A. Egashira, M. Morita, Y. Kakeji and Y. Maehara, *Int. J. Clin. Oncol.* **11**, 199–208 (2006).

[103] L. Sampath, S. Kwon, S. Ke, W. Wang, R. Schiff, M. E. Mawad and E. M. Sevick-Muraca, *J. Nucl. Med.* **48**, 1501-1510 (2007).

[104] H. Xu, K. Baidoo, A. J. Gunn, C. A. Boswell, D. E. Milenic, P. L. Choyke and M. W. Brechbiel, *J. Med. Chem.* **50**, 4759–4765 (2007).

[105] H. Xu, P. K. Eck, K. E. Baidoo, P. L. Choyke and M. W. Brechbiel, *Bioorg. Med. Chem.* **17**, 5176–5181 (2009).

[106] A. P. Pathak, B. Gimi, K. Glunde, E. Ackerstaff, D. Artemov and Z. M. Bhujwalla, *Methods Enzymol.* **386**, 3–60 (2004).

[107] A. de Roos, J. Doornbos, D. Baleriaux, H. L. Bloem and T. H. Falke, *Magn. Reson. Annu.* 113–145 (1988).

[108] D. L. Thorek, A. K. Chen, J. Czupryna and A. Tsourkas, *Ann. Biomed. Eng.* **34**, 23–38 (2006).

[109] M. M. Huber, A. B. Staubli, K. Kustedjo, M. H. Gray, J. Shih, S. E. Fraser, R. E. Jacobs and T. J. Meade, *Bioconjug. Chem.* **9**, 242–249 (1998).

[110] C. Brekke, S. C. Morgan, A. S. Lowe, T. J. Meade, J. Price, S. C. Williams and M. Modo, *NMR Biomed.* **20**, 77–89 (2007).

[111] H. Li, B. D. Gray, I. Corbin, C. Lebherz, H. Choi, S. Lund-Katz, J. M. Wilson, J. D. Glickson and R. Zhou, *Acad. Radiol.* **11**, 1251–1259 (2004).

[112] A. Dirksen, S. Langereis, B. F. de Waal, M. H. van Genderen, E. W. Meijer, Q. G. de Lussanet and T. M. Hackeng, *Org. Lett.* **6**, 4857–4860 (2004).

[113] F. Curnis, G. Arrigoni, A. Sacchi, L. Fischetti, W. Arap, R. Pasqualini and A. Corti, *Cancer Res.* **62**, 867–874 (2002).

[114] R. Pasqualini, E. Koivunen and E. Ruoslahti, *Nat. Biotechnol.* **15**, 542–546 (1997).

[115] A. Mishra, J. Pfeuffer, R. Mishra, J. Engelmann, A. K. Mishra, K. Ugurbil and N. K. Logothetis, *Bioconjug. Chem.* **17**, 773–780 (2006).

[116] E. E. Uzgiris, A. Sood, K. Bove, B. Grimmond, D. Lee and S. Lomnes, *Technol. Cancer Res. Treat.* **5**, 301–309 (2006).

[117] B. Hofmann, A. Bogdanov, Jr., E. Marecos, W. Ebert, W. Semmler and R. Weissleder, *J. Magn. Reson. Imaging* **9**, 336–341 (1999).

[118] T. K. Kim, B. I. Choi, S. W. Park, W. Lee, J. K. Han, M. C. Han and H. J. Weinmann, *AJR Am. J. Roentgenol.* **175**, 227–234 (2000).

[119] S. W. Young, M. K. Sidhu, F. Qing, H. H. Muller, G. Neuder, G. Zanassi, T. D. Mody, G. Hemmi, W. Dow, J. D. Mutch, et al., *Invest. Radiol.* **29**, 330–338 (1994).

[120] S. W. Young and Q. Fan, *Invest. Radiol.* **31**, 280–283 (1996).

[121] H. E. Daldrup-Link, M. Rudelius, S. Metz, G. Piontek, B. Pichler, M. Settles, U. Heinzmann, J. Schlegel, R. A. Oostendorp and E. J. Rummeny, *Eur. J. Nucl. Med. Mol. Imaging* **31**, 1312–1321 (2004).

[122] V. S. Talanov, C. A. Regino, H. Kobayashi, M. Bernardo, P. L. Choyke and M. W. Brechbiel, *Nano Lett.* **6**, 1459–1463 (2006).

[123] Y. Koyama, V. S. Talanov, M. Bernardo, Y. Hama, C. A. Regino, M. W. Brechbiel, P. L. Choyke and H. Kobayashi, *J. Magn. Reson. Imaging* **25**, 866–871 (2007).

[124] H. Dafni, T. Israely, Z. M. Bhujwalla, L. E. Benjamin and M. Neeman, *Cancer Res.* **62**, 6731–6739 (2002).

[125] D. Granot, L. A. Kunz-Schughart and M. Neeman, *Magn. Reson. Med.* **54**, 789–797 (2005).

[126] S. Li, H. He, W. Cui, B. Gu, J. Li, Z. Qi, G. Zhou, C. M. Liang and X. Y. Feng, *Anat. Rec. (Hoboken)* **293**, 2136–2143 (2010).

[127] J. V. Frangioni, *Curr. Opin. Chem. Biol.* **7**, 626–634 (2003).

[128] M. P. Melancon, Y. Wang, X. Wen, J. A. Bankson, L. C. Stephens, S. Jasser, J. G. Gelovani, J. N. Myers and C. Li, *Invest. Radiol.* **42**, 569–578 (2007).

[129] Y. H. Lin, K. Dayananda, C. Y. Chen, G. C. Liu, T. Y. Luo, H. S. Hsu and Y. M. Wang, *Bioorg. Med. Chem.* **19**, 1085–1096 (2011).

[130] L. N. Goswami, W. H. White, 3rd, J. A. Spernyak, M. Ethirajan, Y. Chen, J. R. Missert, J. Morgan, R. Mazurchuk and R. K. Pandey, *Bioconjug. Chem.* **21**, 816–827 (2010).

[131] H. Maeda, J. Wu, T. Sawa, Y. Matsumura and K. Hori, *J. Control. Release* **65**, 271–284 (2000).

[132] T. Beyer, D. W. Townsend, T. Brun, P. E. Kinahan, M. Charron, R. Roddy, J. Jerin, J. Young, L. Byars and R. Nutt, *J. Nucl. Med.* **41**, 1369–1379 (2000).

[133] D. W. Townsend and T. Beyer, *Br. J. Radiol.* **75 Spec No,** S24–S30 (2002).

[134] J. Ruf, E. Lopez Hanninen, H. Oettle, M. Plotkin, U. Pelzer, C. Stroszczynski, R. Felix and H. Amthauer, *Pancreatology* **5**, 266–272 (2005).

[135] H. K. Pannu, C. Cohade, R. E. Bristow, E. K. Fishman and R. L. Wahl, *Abdom. Imaging* **29**, 398–403 (2004).

[136] M. S. Judenhofer, H. F. Wehrl, D. F. Newport, C. Catana, S. B. Siegel, M. Becker, A. Thielscher, M. Kneilling, M. P. Lichy, M. Eichner, K. Klingel, G. Reischl, S. Widmaier, M. Rocken, R. E. Nutt, H. J. Machulla, K. Uludag, S. R. Cherry, C. D. Claussen and B. J. Pichler, *Nat. Med.* **14**, 459–465 (2008).

[137] M. Soret, S. L. Bacharach and I. Buvat, *J. Nucl. Med.* **48**, 932–945 (2007).

[138] C. Catana, Y. Wu, M. S. Judenhofer, J. Qi, B. J. Pichler and S. R. Cherry, *J. Nucl. Med.* **47**, 1968–1976 (2006).

[139] R. Uppal, C. Catana, I. Ay, T. Benner, A. G. Sorensen and P. Caravan, *Radiology* **258**, 812–820 (2011).

[140] K. Overoye-Chan, S. Koerner, R. J. Looby, A. F. Kolodziej, S. G. Zech, Q. Deng, J. M. Chasse, T. J. McMurry and P. Caravan, *J. Am. Chem. Soc.* **130**, 6025–6039 (2008).

[141] T. J. Wadas, E. H. Wong, G. R. Weisman and C. J. Anderson, *Chem. Rev.* **110**, 2858–2902 (2010).

[142] W. Cai, J. Rao, S. S. Gambhir and X. Chen, *Mol. Cancer Ther.* **5**, 2624–2633 (2006).

[143] M. Rudin and R. Weissleder, *Nat. Rev. Drug Discov.* **2**, 123–131 (2003).

[144] J. K. Willmann, N. van Bruggen, L. M. Dinkelborg and S. S. Gambhir, *Nat. Rev. Drug Discov.* **7**, 591–607 (2008).

[145] E. S. Lander, L. M. Linton, B. Birren, C. Nusbaum, M. C. Zody, J. Baldwin, K. Devon, K. Dewar, M. Doyle, W. FitzHugh, R. Funke, D. Gage, K. Harris, A. Heaford, J. Howland, L. Kann, J. Lehoczky, R. LeVine, P. McEwan, K. McKernan, J. Meldrim, J. P. Mesirov, C. Miranda, W. Morris, J. Naylor, C. Raymond, M. Rosetti, R. Santos, A. Sheridan, C. Sougnez, N. Stange-Thomann, N. Stojanovic, A. Subramanian, D. Wyman, J. Rogers, J. Sulston, R. Ainscough, S. Beck, D. Bentley, J. Burton, C. Clee, N. Carter, A. Coulson, R. Deadman, P. Deloukas, A. Dunham, I. Dunham, R. Durbin, L. French, D. Grafham, S. Gregory, T. Hubbard, S. Humphray, A. Hunt, M. Jones, C. Lloyd, A. McMurray, L. Matthews, S. Mercer, S. Milne, J. C. Mullikin, A. Mungall, R. Plumb, M. Ross, R. Shownkeen, S. Sims, R. H. Waterston, R. K. Wilson, L. W. Hillier, J. D. McPherson, M. A. Marra, E. R. Mardis, L. A. Fulton, A. T. Chinwalla, K. H. Pepin, W. R. Gish, S. L. Chissoe, M. C. Wendl, K. D. Delehaunty, T. L. Miner, A. Delehaunty, J. B. Kramer, L. L. Cook, R. S. Fulton, D. L. Johnson, P. J. Minx, S. W. Clifton, T. Hawkins, E. Branscomb, P. Predki, P. Richardson, S. Wenning, T. Slezak, N. Doggett, J. F. Cheng, A. Olsen, S. Lucas, C. Elkin, E. Uberbacher, M. Frazier, R. A. Gibbs, D. M. Muzny, S. E. Scherer, J. B. Bouck, E. J. Sodergren, K. C. Worley, C. M. Rives, J. H. Gorrell, M. L. Metzker, S. L. Naylor, R. S. Kucherlapati, D. L. Nelson, G. M. Weinstock, Y. Sakaki, A. Fujiyama, M. Hattori, T. Yada, A. Toyoda, T. Itoh, C. Kawagoe, H. Watanabe,

Y. Totoki, T. Taylor, J. Weissenbach, R. Heilig, W. Saurin, F. Artiguenave, P. Brottier, T. Bruls, E. Pelletier, C. Robert, P. Wincker, D. R. Smith, L. Doucette-Stamm, M. Rubenfield, K. Weinstock, H. M. Lee, J. Dubois, A. Rosenthal, M. Platzer, G. Nyakatura, S. Taudien, A. Rump, H. Yang, J. Yu, J. Wang, G. Huang, J. Gu, L. Hood, L. Rowen, A. Madan, S. Qin, R. W. Davis, N. A. Federspiel, A. P. Abola, M. J. Proctor, R. M. Myers, J. Schmutz, M. Dickson, J. Grimwood, D. R. Cox, M. V. Olson, R. Kaul, N. Shimizu, K. Kawasaki, S. Minoshima, G. A. Evans, M. Athanasiou, R. Schultz, B. A. Roe, F. Chen, H. Pan, J. Ramser, H. Lehrach, R. Reinhardt, W. R. McCombie, M. de la Bastide, N. Dedhia, H. Blocker, K. Hornischer, G. Nordsiek, R. Agarwala, L. Aravind, J. A. Bailey, A. Bateman, S. Batzoglou, E. Birney, P. Bork, D. G. Brown, C. B. Burge, L. Cerutti, H. C. Chen, D. Church, M. Clamp, R. R. Copley, T. Doerks, S. R. Eddy, E. E. Eichler, T. S. Furey, J. Galagan, J. G. Gilbert, C. Harmon, Y. Hayashizaki, D. Haussler, H. Hermjakob, K. Hokamp, W. Jang, L. S. Johnson, T. A. Jones, S. Kasif, A. Kaspryzk, S. Kennedy, W. J. Kent, P. Kitts, E. V. Koonin, I. Korf, D. Kulp, D. Lancet, T. M. Lowe, A. McLysaght, T. Mikkelsen, J. V. Moran, N. Mulder, V. J. Pollara, C. P. Ponting, G. Schuler, J. Schultz, G. Slater, A. F. Smit, E. Stupka, J. Szustakowski, D. Thierry-Mieg, J. Thierry-Mieg, L. Wagner, J. Wallis, R. Wheeler, A. Williams, Y. I. Wolf, K. H. Wolfe, S. P. Yang, R. F. Yeh, F. Collins, M. S. Guyer, J. Peterson, A. Felsenfeld, K. A. Wetterstrand, A. Patrinos, M. J. Morgan, P. de Jong, J. J. Catanese, K. Osoegawa, H. Shizuya, S. Choi and Y. J. Chen, *Nature* **409,** 860-921 (2001).

[146] P. Workman, *Ann Oncol.***13 Suppl** **4,** 115–124 (2002).

[147] J. Wulfkuhle, V. Espina, L. Liotta and E. Petricoin, *Eur. J. Cancer* **40,** 2623–2632 (2004).

16

CHEMICAL STRATEGIES FOR THE DEVELOPMENT OF MULTIMODAL IMAGING PROBES USING NANOPARTICLES

Amanda L. Eckermann, Daniel J. Mastarone and Thomas J. Meade

Departments of Chemistry, Molecular Biosciences, Neurobiology, Biomedical Engineering and Radiology, Northwestern University, Evanston, IL, USA

16.1 INTRODUCTION

The ability to interrogate *in vivo* anatomical features and biochemical processes has stimulated growth in the relatively new field of molecular imaging. In a sense, molecular imaging is the intersection of classical diagnostic imaging and molecular biology with the goal of allowing molecular processes to be visualised within a living organism. There are a number of physical techniques (or modalities) frequently employed in molecular imaging experiments, and each has its own inherent strengths and weaknesses regarding resolution, sensitivity, and temporal limitations. These techniques include, but are not limited to, positron emission tomography (PET), single photon emission computed tomography (SPECT), X-ray computed tomography (CT), magnetic resonance imaging (MRI), 2-photon optical imaging, near-IR fluorescence imaging (NIRFI), and ultrasound (US).

It has been evident for many years that combining one or more modalities to investigate a problem allows researchers to capture the strengths of more than one of these techniques. Therefore, the development of instrumentation capable of coupling more than one modality has been achieved (PET-CT-SPECT is an excellent example). Further, in order to differentiate specific anatomical features, individual cells, and subcellular domains, imaging agents (or probes) are frequently employed.

While molecular multimodal agents exist, the chemistry of combing multiple probes in one molecule, macromolecule, or supramolecule is complex [1–3]. Nanoparticles are ideal scaffolds for combining two or more imaging modalities. Multiple moieties may be attached to the nanoparticle surface for amplification while the nanoparticle itself may consist of material suitable for imaging. This chapter focuses on the most recently reported (2010–2012) nanoparticle agents that have been demonstrated *in vivo*. Wherever possible, we include the details of the chemistry used to assemble the multimodal components.

16.1.1 General Chemical Methods

While a variety of bioconjugate techniques for labelling nanoparticles have been developed, we have chosen to focus on the most widely used. This section describes the general methods of chemical conjugations (Scheme 16.1), silanisation (Scheme 16.2), and electrostatic adsorption (Figure 16.1). Nanoparticles can be generated with thiol-, carboxylic acid-, or amine-functionalised surfaces. Peptide coupling is the formation of an amide bond between a carboxylic acid and an amine. Dyes or other species with carboxylic acids are often pre-activated using 1-ethyl-3-(3-dimethylaminopropyl) carbodiimide (EDC) and N-hydroxyl succinimide (NHS) to form a succinimidyl ester that is more reactive toward the amine than the acid alone. This approach has been used, for example, to attach Gd(III) chelates to nanodiamonds [4]. Another common

The Chemistry of Molecular Imaging, First Edition. Edited by Nicholas Long and Wing-Tak Wong.
© 2015 John Wiley & Sons, Inc. Published 2015 by John Wiley & Sons, Inc.

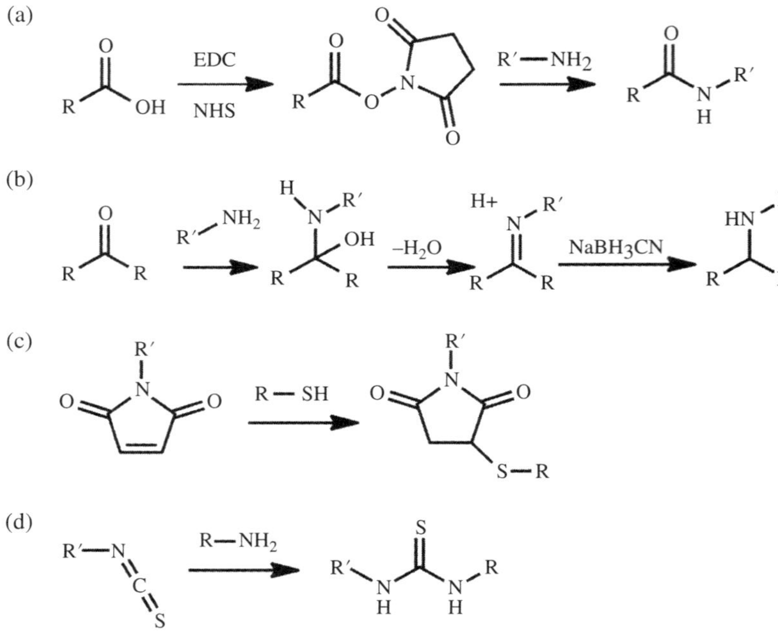

(a)

(b)

(c)

(d)

SCHEME 16.1 Common conjugation techniques include (a) peptide coupling, (b) reductive amination, (c) thiol-maleimide cross-linking, and (d) the reaction between isothiocyanate and an amine.

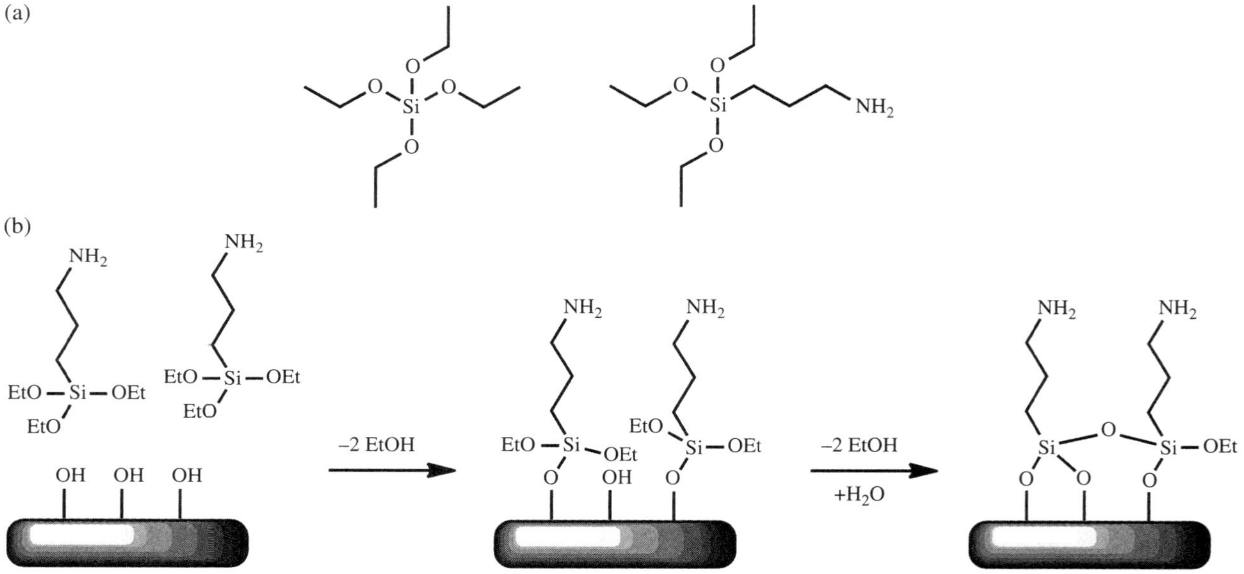

(a)

(b)

−2 EtOH

−2 EtOH
+H$_2$O

SCHEME 16.2 (a) Structures of tetraethyl orthosilicate (TEOS) and (3-aminopropyl) triethoxysilane (APTES). (b) Scheme showing how APTES may be used to functionalise a surface with amine groups.

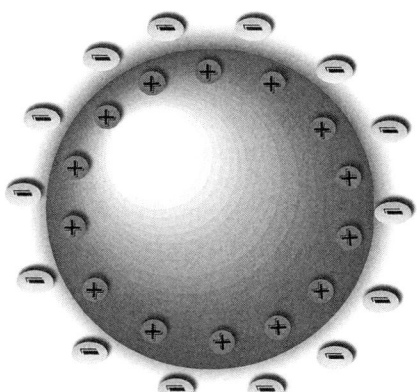

FIGURE 16.1 Electrostatic interactions can be used to build layers or adsorb molecules to a nanoparticle surface.

technique is reductive amination. Ketones or aldehydes react with amines to form imines that are further stabilised through reduction with sodium cyanoborohydride to provide a stable conjugation product. This procedure has been shown to be effective in aqueous conditions expanding its usefulness for biological systems [5, 6]. Further, amines can react with isothiocyanate groups [7–9]. Another common approach utilises thiols, either to form a strong gold-thiol bond [10] or a thiolmaleimide reaction that results in a thioether bond.

Silanisation is the controlled hydrolysis of siloxane species such as tetraorthosilicate (TEOS) resulting in the formation of a silica (SiO_2) coating on the surface of the nanoparticle [11]. This process lends itself to the addition of (3-aminopropyl) triethoxysilane(APTES), which is frequently used to incorporate an amine on the surface of the silica layer (Scheme 16.2). Finally, the use of charged polymers such as polylysine can improve the electrostatic adsorption of species on the surface of a nanoparticle or can be used to modify the surface with polymeric layers.

16.1.2 Early Work

Early reports combine multiple modalities on nanoparticle platforms. The combination of ultrasound and MRI was used to evaluate the effectiveness of targeting groups, stem-cell tracking, visualisation of angiogenesis, and imaging transplant rejection [12]. The combination of optical, CT, and MRI in the form of paramagnetic silica nanoparticles was proposed to be useful for preoperative diagnosis and intraoperative surgical resection of brain tumours or other surgical targets [13].

One of the first *in vivo* applications of multimodal nanoparticles was to assist the determination of brain tumour margins [14]. SPIONs with cross-linked dextran coating were labelled with Cy5.5 and injected into mice bearing GFP-expressing 9L glioma tumours. The T_2-weighted MRI showed accumulation of the SPIONs in the tumour after 24 hours. Optical imaging was performed and showed co-localisation of GFP with the Cy5.5 dye. At a cellular level, using CD11b staining, microglia were found to extend slightly beyond the GFP-defined tumour border. This finding was consistent with the estimation of tumour area by Cy5.5 fluorescence. The combined optical and magnetic properties of this probe allowed delineation of the brain tumour both by pre-operative magnetic resonance imaging and by intra-operative optical imaging.

Atherosclerosis is a chronic, progressive inflammatory disease, and there is significant need for diagnostic tools to noninvasively assess the therapeutic efficacy in the clinic as well as to identify molecular changes for basic research purposes. Vascular adhesion molecule-1 (VCAM-1) is upregulated in endothelial cells under inflammatory conditions such as those associated with atherosclerosis. In order to image inflammation, SPIONs were conjugated with a fluorescent dye and a peptide designed to target inflamed tissue [15]. The SPION effectively targeted VCAM-1 and *in vivo* accumulated in the vessel wall. The SPIONs were detectable for at least 24 hours by both optical and MRI modalities, proving that this nanoparticle system is an effective amplification strategy to achieve high target-to-background ratios.

16.2 FLUORESCENCE-MRI

Nanoparticles are frequently and widely used for the labelling of cells and targeting of tissues in theranostic medical studies [16–22]. The main applications for multimodal therapy are to (1) visualise the effectiveness of targeting cancerous tumours *in vivo*, (2) directly observe the effectiveness of therapy, or (3) label cells for *in vivo* tracking. In this section we examine the recent use of nanoparticles that incorporate a fluorescent dye and a magnetic nanoparticle together.

The combination of MRI and fluorescent imaging can be accomplished in two different ways. One method is to covalently attach an organic fluorescent dye to the surface of a superparamagnetic iron oxide nanoparticle (SPION) [7]. The second approach is to use a nanoparticle of 'inert' material, such as silicon oxide, as a scaffold and decorate the surface with both the fluorescent dye and small SPIONs [23, 24].

16.2.1 Theranostic Applications of Fluorescent-MRI Bimodal Agents

Nanoparticles that carry both imaging capability and therapeutic agents are known as *theranostic agents*. One major concern is the biodistribution and toxicity of the nanoparticle platforms used to deliver the treatment. One means of addressing this question is to use nanoparticle systems that attach imaging agents that can be tracked *in vivo*. This approach provides real-time location monitoring of the treatment and the delivery system.

Harrison and co-workers have developed aPGMA (poly(glycidyl methacrylate)-encapsulated SPION with rhodamine B (Figure 16.2) for fluorescent imaging [23]. The particles were synthesised according to a previous report [25]. In brief, glycidylmethacrylate was polymerised using azobisisobutyronitrile as an initiator. The dye was incorporated to the polymer by heating a solution of rhodamine B with PGMA under nitrogen for 18 h. The SPIONs were incorporated by emulsification with the dye-labelled PGMA.

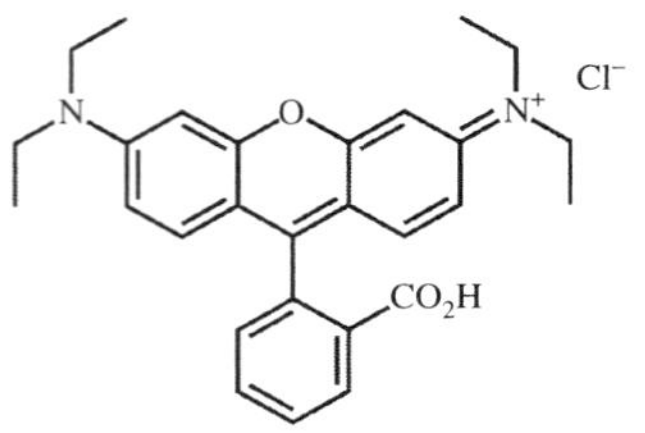

FIGURE 16.2   Structure of rhodamine B.

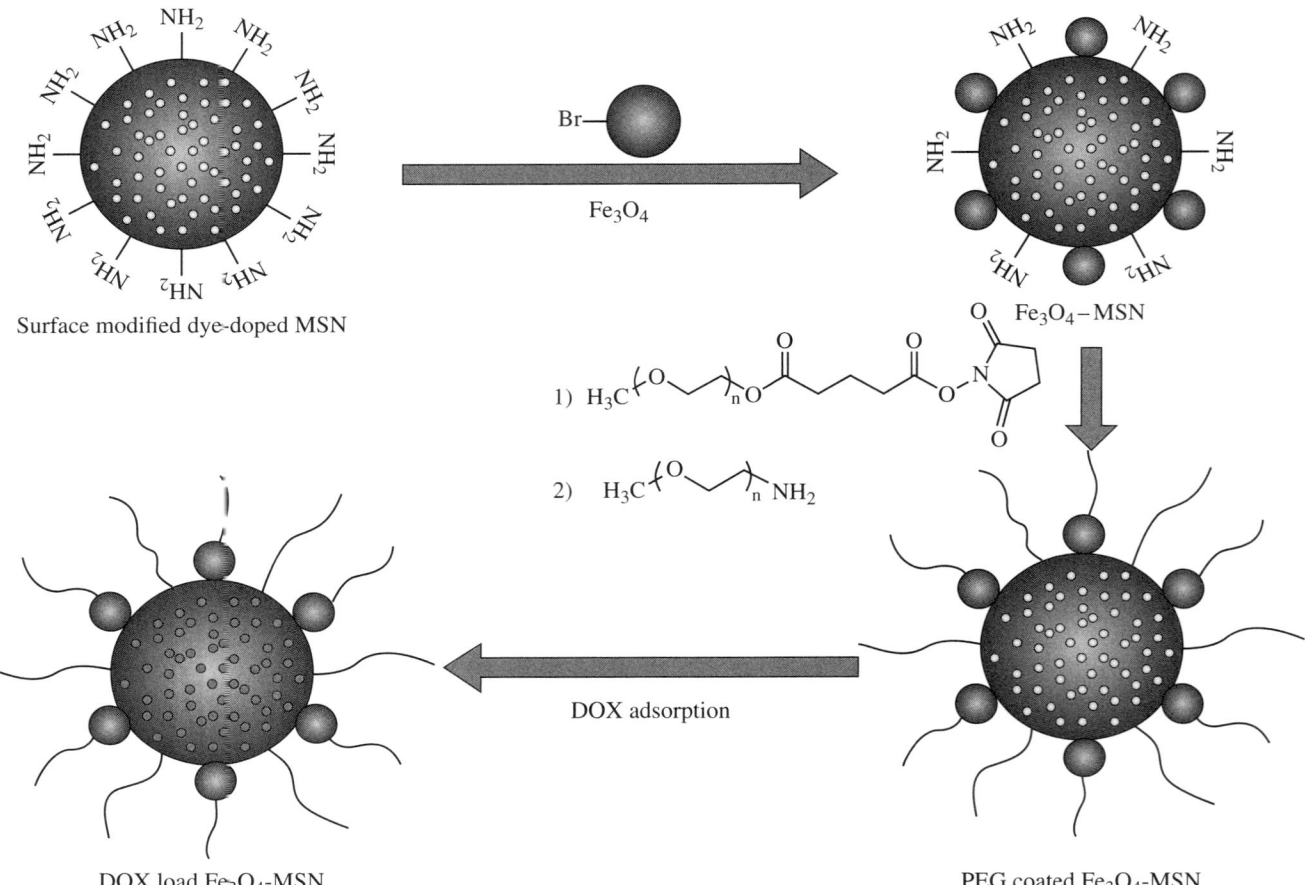

SCHEME 16.3   Functionalisation of MSNPs with SPIONs and DOX. Reproduced from Ref. [24] with permission from the American Chemical Society.

Injection into a damaged optic nerve and injection into the eye of the mouse model resulted in increased infiltration of activated microglia and macrophages but no peripheral toxicity in the subjects was observed. This result indicates that the nanoparticles are nontoxic despite the activation of the immune system.

Particles were tracked using MRI for 14 days and confirmed in *ex vivo* tissue slices through optical imaging of the fluorescent dye on the nanoparticles. The nanoparticles did not travel far from the injection site in either case and were not observed in the peripheral tissue samples. These results show the promise of these types of agents being used as a platform for sustained therapeutic delivery.

In a similar approach, Lee and co-workers have delivered chemotherapeutics to tumours in a mouse model using a mesoporous silica nanoparticle (MSNP) [24]. They modified the MSNP surface with fluorescent dyes as well as SPIONs to provide both MR and optical visualisation (Scheme 16.3). SPIONs synthesised in organic solution were modified with 2-bromo-2-methylpropionic acid. These particles were mixed with MSNPs that possess free amine groups. Reaction between the alkyl-bromide on the SPIONs and amine on the MSNPs provides a stable covalent bond. The porous nature of silica allows loading of the system with a chemotherapeutic, in this case, doxorubicin (DOX) (Figure 16.3).

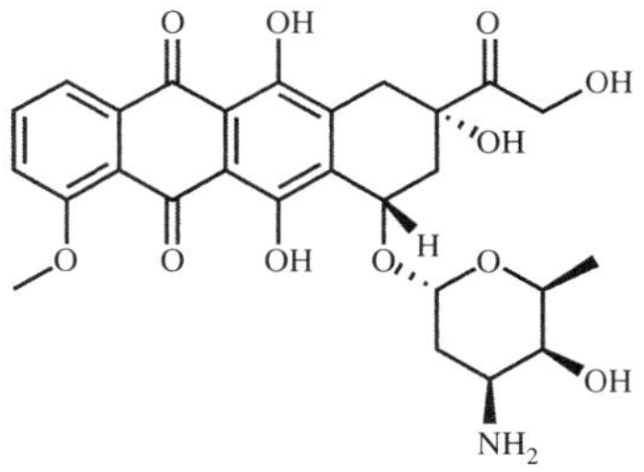

FIGURE 16.3 Structure of doxorubicin (DOX), a widely used cancer therapeutic. In biological conditions, the $-NH_2$ group is protonated to $-NH_3^+$, giving an overall charge of +1 to the molecule.

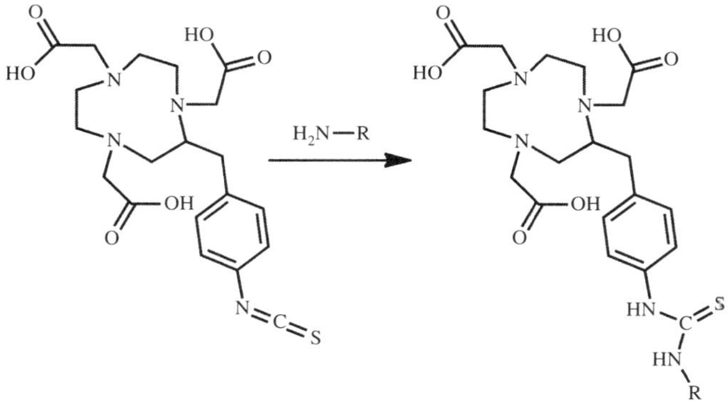

SCHEME 16.4 Reaction of *p*-NCS-Bn-NOTA, an empty chelate bearing an isothiocyanate group, with a generic amine results in formation of a thiourea group and a strong covalent bond.

For MR imaging, the relaxometric properties were determined with the labelled MSNP system having an r_2 of $76.2 \, mM^{-1} s^{-1}$ versus the free SPIONs with a r_2 of $26.8 \, mM^{-1} s^{-1}$. Three hours after injection, a hypointense signal at the tumour core was observed, indicating that the particles had migrated to the tumour. Fluorescence measurements of *ex vivo* tissue sections confirmed the delivery of the doxorubicin.

16.2.2 Targeting Applications of Fluorescent-MRI Bimodal Agents

Hwang and co-workers added a targeting functionality to a multimodal nanoparticle system by using an aptamer targeted to nucleolin, a cellular membrane protein highly expressed in cancer [7]. In this case, a silica-coated cobalt ferrite nanoparticle with an amine-functionalised surface has been generated for the attachment of a rhodamine dye and the targeting aptamer. Additionally, a backbone-modified chelate 2-(*p*-isothiocyanatobenzyl)-1,4,7-triazacyclonane-1,4,7-triacetic acid (p-NCS-Bn-NOTA) (Scheme 16.4) was covalently attached to the amine-functionalised surface to allow for radioisotope labelling with ^{67}Ga. Utilising MRI and scintillation images, they demonstrated successful targeting of the nanoparticle system to an *in vivo* mouse tumour model. In contrast, an NP synthesised with a mutant aptamer rapidly cleared from the model and did not accumulate in the tumours. *Ex vivo* optical images of the tumours confirmed the presence of the fluorescent NPs in the aptamer-targeted system and the absence in the mutant system.

A carbogenic nanocomposite has been described by Srivastava [26]. This nanocomposite was synthesised by a partial thermal decomposition method using citric acid and l-lysine as organic precursors along with preformed SPIONs [27]. Through tuning the decomposition and growth of the organic layer, they demonstrate the ability to tune the fluorescent response of the particle. The authors further report proof-of-principle MR and fluorescent images indicating that this non-toxic system is effective as both a T_2 and T_1 agent. Fluorescent images of the *ex vivo* tissue sections confirm the results of the MRI scans.

16.3 NEAR-INFRARED FLUORESCENCE/MRI

The combination of NIR and MRI modalities and *in vivo* characterisation has been executed in two ways. The most straight-forward method is to attach an organic NIR dye to a SPION (e.g., [28]). The most widely used commercially available organic NIR dyes are based on a cyanine core (e.g., Cy5, Figure 16.4). The second approach is to combine SPIONs with a materials-based NIR fluorescent species such as a quantum dot or multilayered gold/silica nanoparticles (e.g., [29]).

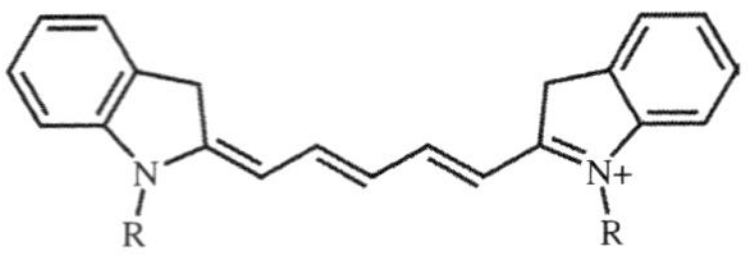

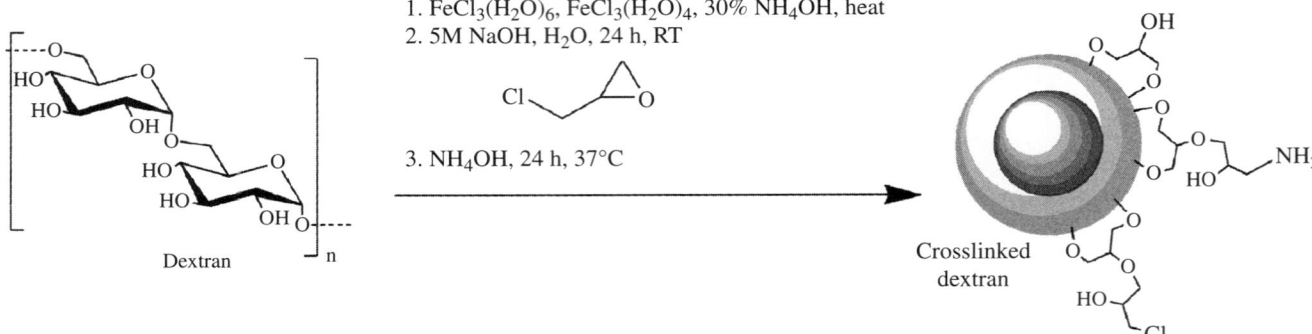

FIGURE 16.4 The conjugated structure of cyanine dyes imparts their photophysical attributes. The R groups are alkyl chains with or without functional groups such as an NHS ester or maleimide. The '5' in Cy5 refers to the number of carbon atoms bridging the two heterocyclic indole groups.

1. $FeCl_3(H_2O)_6$, $FeCl_3(H_2O)_4$, 30% NH_4OH, heat
2. 5M NaOH, H_2O, 24 h, RT

3. NH_4OH, 24 h, 37°C

Dextran

Crosslinked dextran

SCHEME 16.5 Dextran coating and cross-linking on SPION surface results in -OH, $-NH_2$, and -Cl reactive groups. Adapted from refs. [30, 31].

16.3.1 Cell Tracking Using NIR-MRI Bimodal Agents

The monitoring of B-cell depletion is a critical need in immunotherapy because it can be used to evaluate the biological response to and efficacy of immunotherapeutics. In this report, dextran-coated SPIONs (Scheme 16.5) were labelled with a NIR dye AlexaFluor680 (Invitrogen) [28]. B-cells were harvested from mice and labelled with the nanoparticles and a lipophilic dye, CellVue NIR 815 (Polysciences, Inc.).

NIR815-labelled B-cells with and without SPIONs were administered to mice and tracked using MR and NIR fluorescence imaging. The mice were treated with either PBS or antiCD79. NIRF imaging showed depletion of B cells treated with NIR815, but not the cells labelled only with the SPIONs. Although the expression levels of several cell surface markers, including CD40, CD80, CD86m, and MHC11, did not change when the cells were labelled with the nanoparticles, it appears that labelling with the SPIONs had an effect on the ability of antiCD79 to interact with the B-cells.

In some systems, NIR fluorescence is an intrinsic property. Kim et al. report the labelling of therapeutic dendritic cells with a nanogel comprised of cross-linked manganese ferrite nanoparticles [32]. The $MnFe_2O_4$ particles are coated with polyglutamic acid (PGA), then treated with polylysine (PLL), which electrostatically adsorbs to the PGA layer (Figure 16.5). Cross-linking to form the nanogel is performed using glutaraldehyde (Figure 16.5) that forms a series of imine bonds between the side chain amines of lysine. A consequence of this cross-linking is the formation of a Schiff base that produces a surprisingly strong autofluorescence due to the conjugated bonds.

Dendritic cells were labelled with this nanogel and injected into nude mice. The cells were found to migrate to the lymph nodes. A darkening in the MR image of the mouse was observed, indicating that the nanogel is an effective T_2 MRI contrast agent *in vivo*. The lymph nodes were detected using NIRFI, further indicating that the nanogel is effective for *in vivo* imaging. The excised organs were imaged and exhibited strong NIR fluorescence.

Quantum dots (QDs) are nanoparticles based on semiconducting materials. Due to their nanometer size, their electronic properties fall between those of bulk materials and those of discrete molecules. Semiconductor materials that are free of heavy metals, such as nanoscale silicon, are being investigated as replacement materials for the development of quantum dots due to the toxicity of cadmium, lead, and gallium commonly found in QDs [33, 34].

A magnetofluorescent probe with luminescence and superparamagnetic properties has been produced by combining silicon-based quantum dots (SiQDs) and SPIONs in micelles [35]. SiQDs were synthesised by CO_2 laser pyrolysis of silane (SiH_4) in an aerosol reactor and found to be ~4nm in diameter [36]. These particles were co-encapsulated in a lipid micelle of phospholipids (Figure 16.6, 1,2-distearoyl-sn-glycero-3-phosphoethanolamine-N-[amino(polyethylene glycol)-2000]; DSPE-PEG 2000; n ~ 142) with iron oxide particles (5–9nm) to give a ratio by mass of Si:Fe_3O_4=80:1. The micelle size varied from 50–100nm. If the amount of Fe_3O_4 was increased, quenching of the SiQD fluorescence was observed.

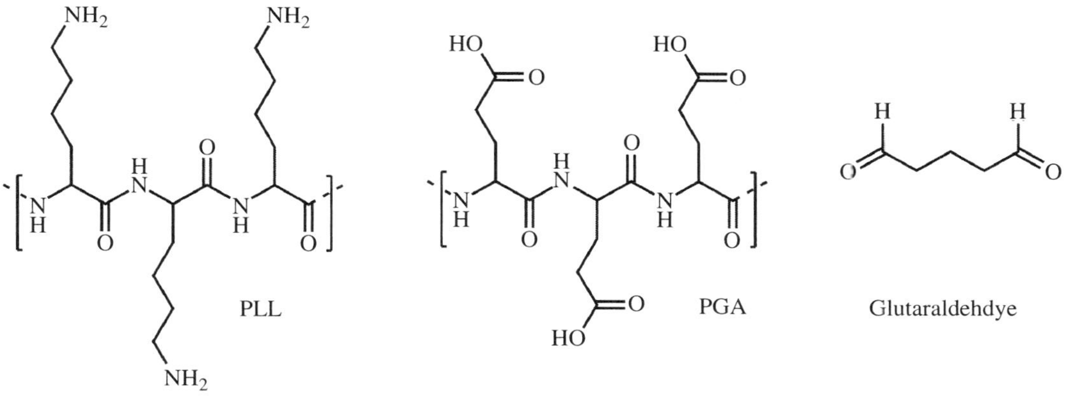

FIGURE 16.5   Structures of polylysine (PLL), polyglutamic acid (PGA), and glutaraldehyde.

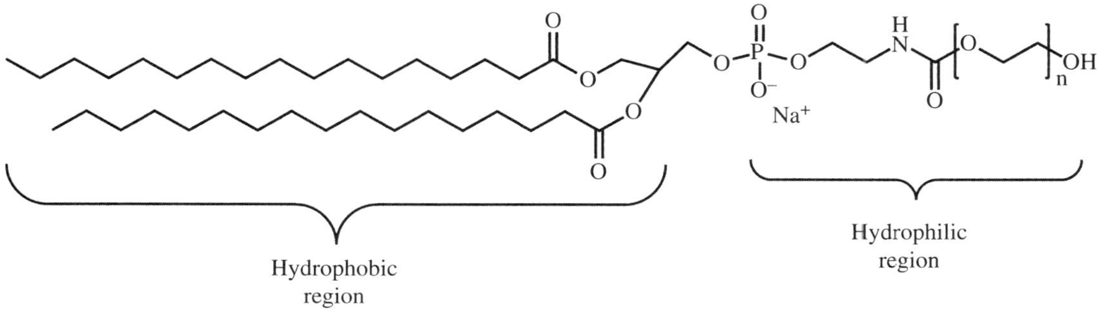

FIGURE 16.6   Structure of DSPE-PEG 2000, n ~ 142 showing the hydrophobic and hydrophilic regions of the molecule that allow it to form micelles.

Cell culture experiments with an external magnet suggest that the uptake by cells can be influenced by external magnetic forces and confirmed that uptake can be tracked with fluorescence microscopy. The micelles were injected directly into xenograft PC-3 tumours in athymic mice. The tumours were clearly observable and the overall signal intensity remained constant for 24 h. No MR imaging was reported.

16.3.2   Theranostic Applications of NIR-MRI Bimodal Agents

Graphene is a strong NIR absorber and has potential use in photothermal therapy [37]. SPIONs can be grown from ferrous chloride on reduced graphene oxide (RGO) by hydrothermal reaction [38]. The resulting nanocomposite can be further functionalised with PEG-grafted poly(maleicanhydride-alt-1-octadecene) (C18PMH–PEG) for biocompatibility (Scheme 16.6) [39]. The final product was found to be 50 nm in diameter with an r_2 of $108\,\text{mM}^{-1}\text{s}^{-1}$. Labelling the RGO-$Fe_3O_4$-PEG particles with ^{125}I showed that they have a long blood circulation time. MRI and photoacoustic tomography showed accumulation in tumours in mice. Irradiating the tumour with laser light (808 nm, 0.5 W/cm, 5 min) eradicated the tumour after one day.

Mesoporous silica coatings on nanoparticles have been widely investigated for biomedical applications due to biocompatibility. The porous nature of this material allows drugs to be adsorbed for slow release *in vivo*. Ma et al. synthesised oblong iron oxide particles coated with silica, followed by covalent attachment of gold nanorods (AuNRs) to make AuNR-SPIONs (Figure 16.7) [40]. The AuNRs are known to be effective photothermal therapeutics. Further, modification of gold nanoparticles is facile through the formation of a gold-thiol bond. Previous work has demonstrated the effectiveness of this approach by using DNA modified with Gd(III) chelates and fluorescent dyes to modify gold nanoparticles for cell labelling [10].

First, 200 nm hematite (Fe_2O_3) spindles were prepared by aging a solution containing $Fe(ClO_4)_3 \cdot 6H_2O$, urea, and NaH_2PO_4 in deionised water at 100 °C for 24 h. An amine-functionalised silica coating was added using TEOS and APTES, and the particles were reduced to Fe_3O_4 using a mixture of Ar and H_2 at high temperature. Separately, AuNRs were grown and modified with PEG and carboxylic acid disulfides to provide functionality. The AuNRs were covalently bound to the SPIONs using the EDC/NHS coupling procedure. Doxorubicin was loaded by nonspecifically adsorbing into the silica shell pores.

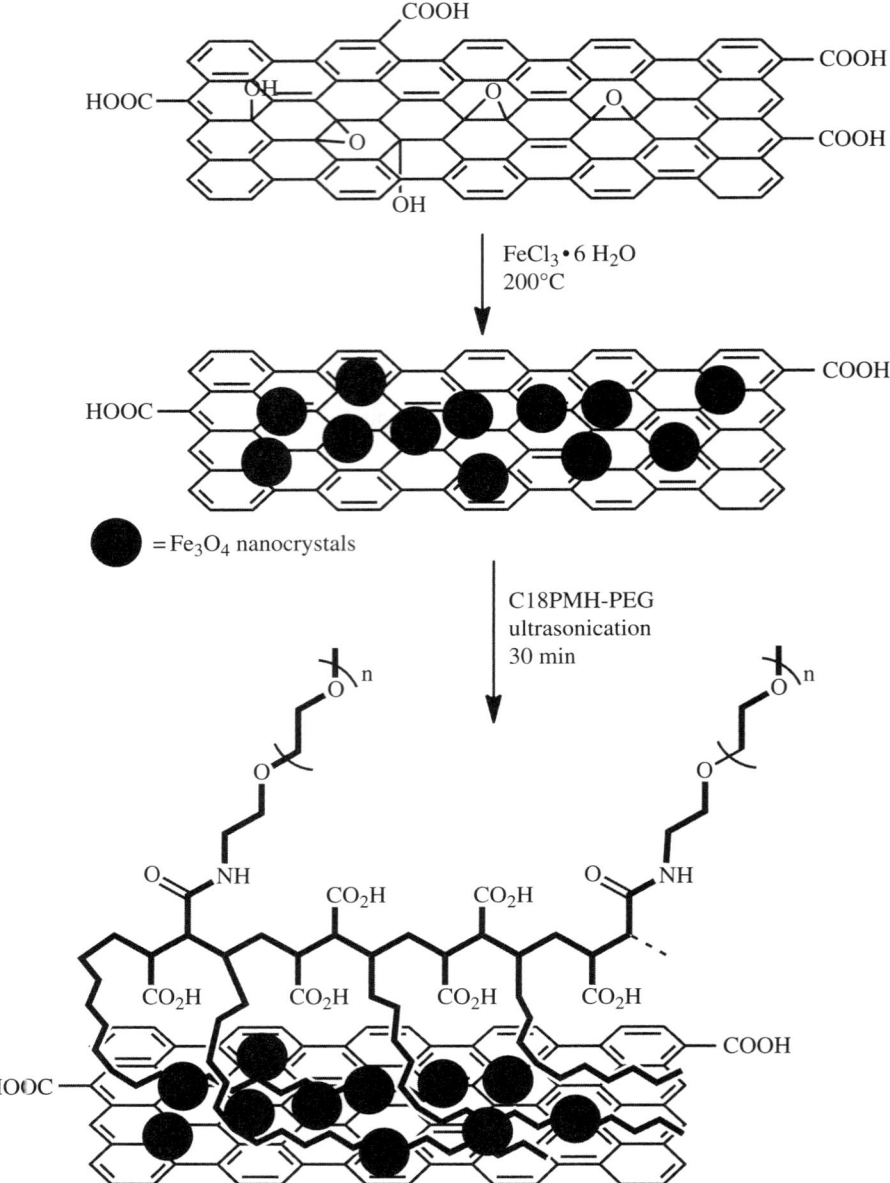

SCHEME 16.6   SPIONs are grown on RGO and modified with PEG for solubility and biocompatibility. Adapted from Ref. [39].

The particles possessed a T_2 relaxivity of approximately $400\,mM^{-1}\,s^{-1}$ and were shown to localise to the xenograft tumour through the use of an external magnet. MR imaging showed darkening of the tumour following injection of the AuNR-SPIONs. When the tumour was irradiated with 808 nm laser light, the toxicity increased. The combination of phototherapy with chemotherapy was most effective.

A novel theranostic multimodal platform based on a soybean oil nanoemulsion (Figure 16.8a) was designed to include the delivery of hydrophobic drugs [41]. The NIR dye Cy7 was used to label the periphery of the particle containing SPIONs and the therapeutic glucocorticoid prednisolone acetate valerate (PAV, Figure 16.8b) [41]. A solution of the components dissolved in chloroform was slowly dropped into an aqueous buffer at 70 °C followed by sonication and concentration using Vivaspin concentrators with a 100.000 MW cut-off size. *Ex vivo* NIRFI shows accumulation in the liver, kidneys, and tumour. No difference in RGD-targeted versus nontargeted particles was noted. Tumour growth was inhibited by particles that delivered PAV. MR imaging showed accumulation in the tumours *in vivo*.

In another NIRFI-MRI system, chlorin e6 was chosen as the NIR dye and as a photosensitiser for photodynamic therapy (PDT) [42]. Upon irradiation, photosensitisers generate reactive oxygen species (ROS), which wreak havoc on cells in the immediate proximity, but cause minimal damage to adjacent healthy tissue. Chlorin has a structure similar to that of a porphyrin (Figure 16.9) and emits at 660–670 nm [43]. SPIONs ($r_2 = 118\,mM^{-1}\,s^{-1}$) were functionalised with APTES followed by peptide coupling to chlorin. The formation of ROS by chlorin was found to be efficient after conjugation to the nanogel.

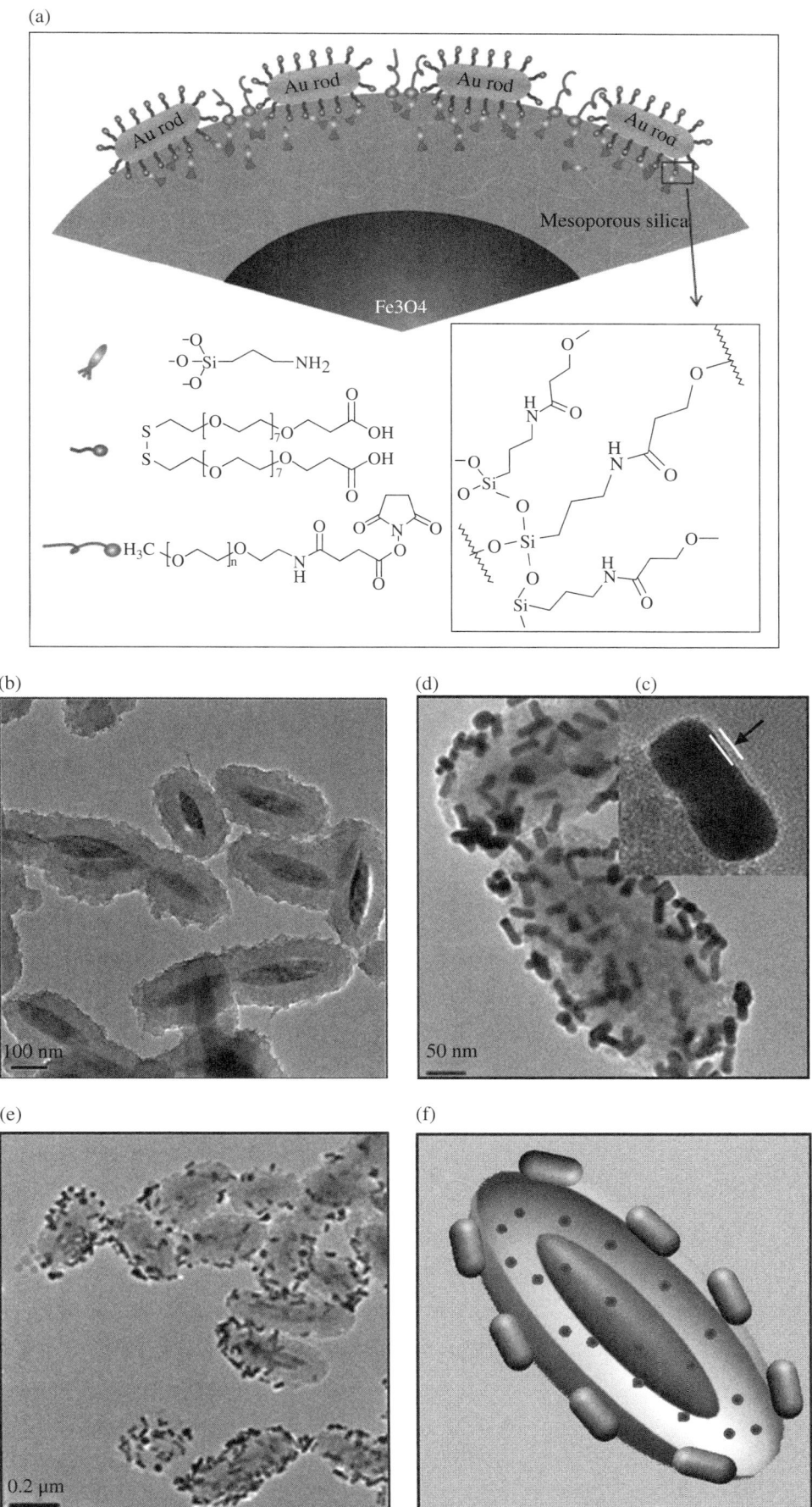

FIGURE 16.7 (a) Attachment chemistry and (b–e) TEM images of the AuNR-SPIONs. Reproduced with permission from Ref. [40].

(a)

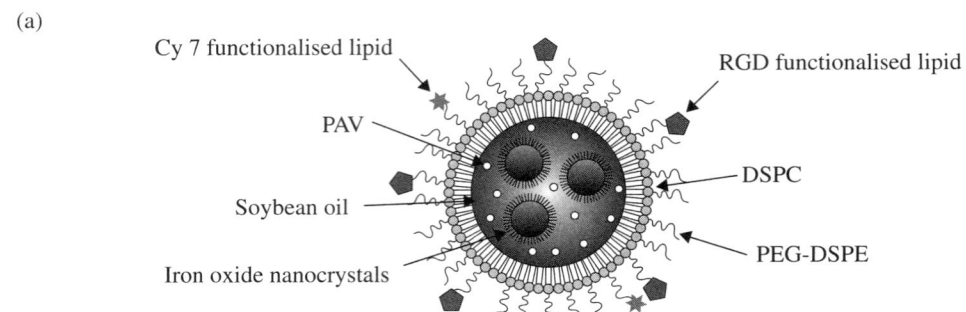

(b)

	Soy b.oil (mg)	PEG-DSPE (mmole)	DSPC (mmole)	Mal-PEG-DSPE (mmole)	PEG:DSPE:Mal (molar ratio)	PAV (mg)	FeO (mg)
CTRL nano	80	0.013	0.0138	-	0.94:1	-	-
PAV nano	80	0.013	0.0138	-	0.94:1	3.75	-
RGD-PAV nano	80	0.0015	0.0138	0.0014	0.83:1:0.1	3.75	-
FeO nano	80	0.013	0.0138	-	0.94:1	-	7.5
PAV+FeO nano	80	0.013	0.0138	-	0.94:1	3.75	7.5

(c)

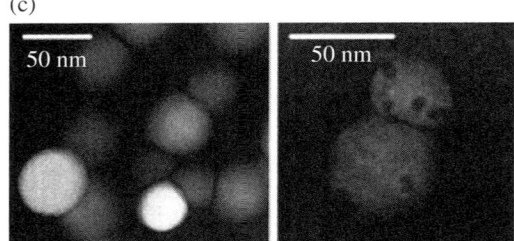

(d)

DLS size (nm) and polydispersity				
PAV nano	RGD-PAV nano	PAV+FeO nano	FeO nano	CTRL nano
50.5 ± 0.14	48.4 ± 0.14	52.5 ± 0.13	50.7 ± 0.15	49.27 ± 0.14
0.137	0.138	0.128	0.150	0.143

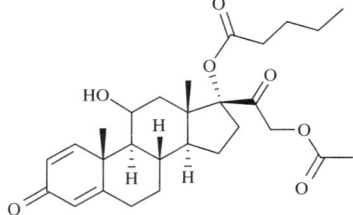

FIGURE 16.8 (a) Functionalised soybean oil emulsion containing PAV and SPIONs. (b) Structure of prednisolone acetate valerate. Reproduced from Ref. [41] with permission from the American Chemical Society.

FIGURE 16.9 The macrocyclic structure of chlorin is very similar to that of a porphyrin.

In nude mice, the nanogel was localised to a gastric cancer tumour using an external magnet. MRI shows a darkening of the tumour, indicating the presence of the particles after 8 h. NIRF imaging shows that the agent localises in the tumour between 0 and 6 h. Over the period of one month, the tumour was treated with a combination of the chlorin nanogel, magnetic targeting, and irradiation, and did not grow in size in contrast to the control groups.

16.3.3 Tumour Targeting Using NIR-MRI Bimodal Agents

In order to improve diagnostics and therapeutics, the efficiency of tumour targeting must be optimised. Various strategies have been employed, including using targeting peptides, antibodies, and reliance on the EPR effect. Tumour targeting can be evaluated using multimodal imaging of nanoparticle platforms *in vivo*.

High-density lipoprotein (HDL) is a biologically endogenous nanoparticle that can be readily modified [44, 45]. In this report, HDL was reconstituted with a hydrophobic Gd chelate (gadolinium diethylenetriaminepentaacetate-di(stearylamide), or Gd-DTPA-DSA) and 1,1-dioctadecyl-3,3,3′,3′-tetramethylindotricarbocyanine iodide (DiR), a NIR dye (Figure 16.10) [46]. Finally, either the targeting cyclic peptide cRGD or the control peptide cRAD was conjugated to the particle through formation of a disulfide bond (Scheme 16.7). The r_1 was found to be 8.5 mM^{-1} s^{-1} with the particles being approximately 12 nm in diameter.

In vitro tests with human umbilical vascular endothelial cells show the RGD targeting is specific [46]. *In vivo* tests, however, show that rHDL, rHDL-RGD, and rHDL-RAD all accumulate in the tumour over 24 h, likely due to the EPR effect. However, confocal microscopy of tumour sections shows that the accumulation patterns are different. NIRFI also showed that the binding kinetics *in vivo* were different for the three HDL platforms.

Gliosarcoma is a particularly difficult cancer to treat and the subject of increasing investigation [47]. SPIONs with a silica shell, PEG coating, and amine functional groups were modified to conjugate chlorotoxin (CTX), a targeting

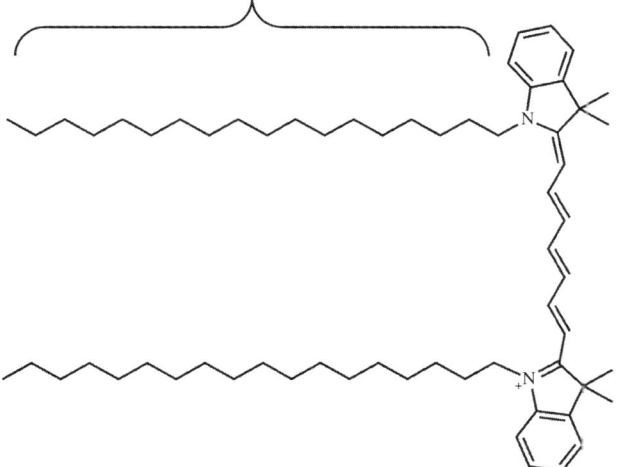

Hydrophobic regions for membrane labelling

FIGURE 16.10 Structures of GdDTPA-DSA and DiR. The long hydrophobic alkyl chains are designed to optimise membrane labelling.

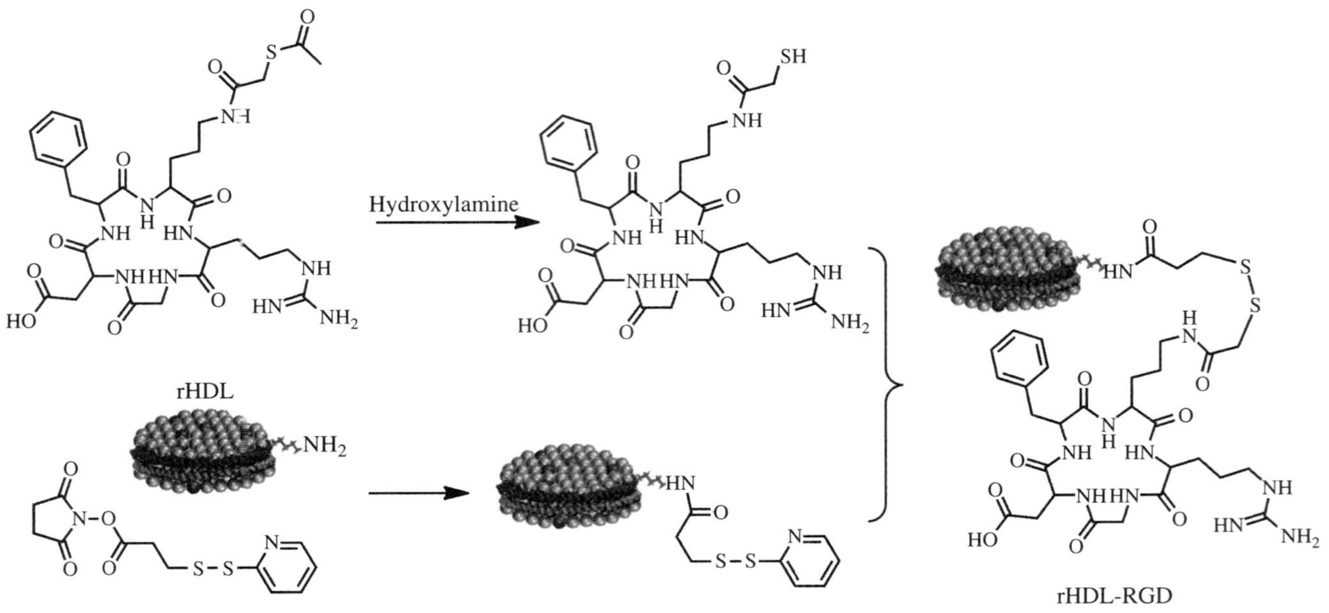

SCHEME 16.7   rHDL was modified with RGD via formation of a disulfide bond. Adapted from Ref. [46].

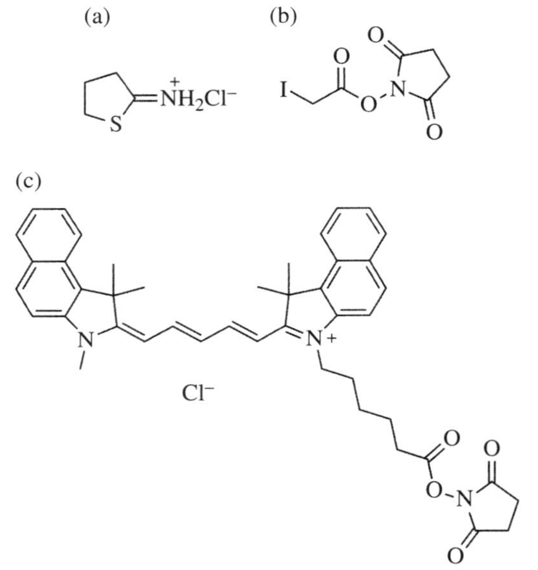

FIGURE 16.11   (a) Structure of Traut's reagent, which reacts with amine groups and provides a thiol group. (b) structure of the cross-linker succinimidyliodoacetate, which can react with thiol groups. (c) structure of the NHS ester of Cy5.5.

group derived from the venom of the Leiurus quinquestriatus scorpion. CTX is a 36-mer peptide (sequence: MCMPCFTTDHQMARKCDDCCGGKGRGKCYGPQCLCR) with a high affinity for neuroectodermal cancer cells [48–50]. CTX was pre-functionalised with Traut's reagent and covalently attached to the particle surface via a succinimidyli-odoacetate cross-linking (Figure 16.11, Scheme 16.8) [47]. To attach an NIR dye, Cy5.5 NHS ester was used. The particles have a mean diameter of 13.5 nm and an r_2 of 126 mM^{-1} s^{-1}.

In vitro studies showed the functionalised nanoparticles are taken up by gliosarcoma cells after 1 h incubation. *In vivo*, MRI shows heterogeneous darkening of the gliosarcoma xenograft tumour in a mouse model (Figure 16.12). *Ex vivo* NIRFI confirmed the localisation of the nanoparticles in the tumour 48 h after injection. Biodistribution studies showed that at 72 h after injection, the liver, spleen, and kidneys exhibited significant NIRF, but none was observed in the lungs or heart. Liver enzyme serum levels and blood counts of the mice showed the particles are nontoxic to the liver.

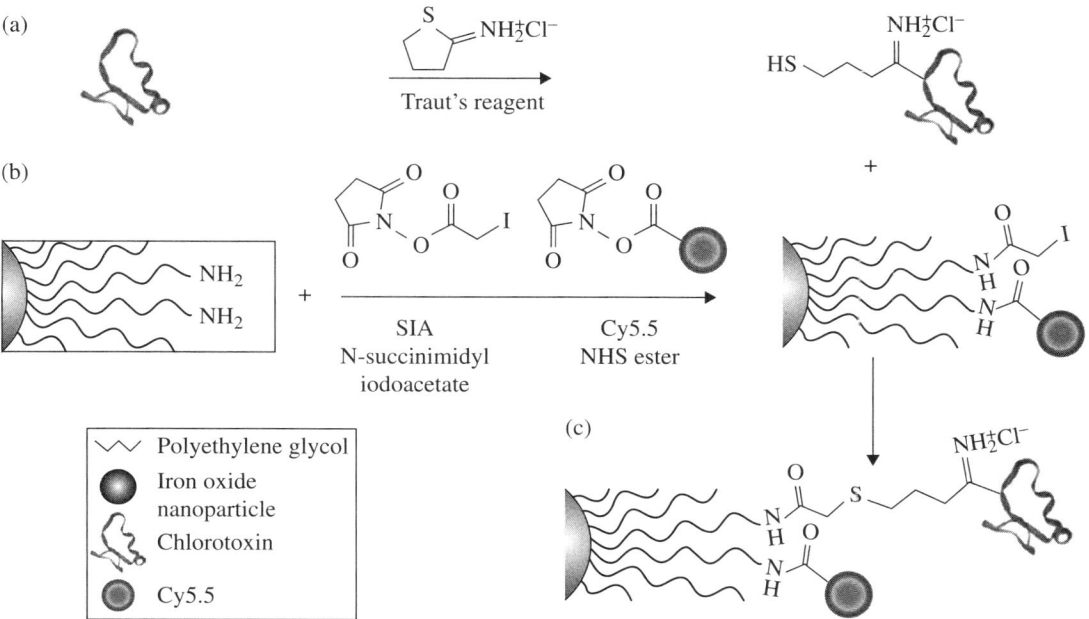

SCHEME 16.8 Covalent attachment of CTX to nanoparticle surface using a cross-linking strategy. Reproduced from Ref. [47] with permission from the American Chemical Society.

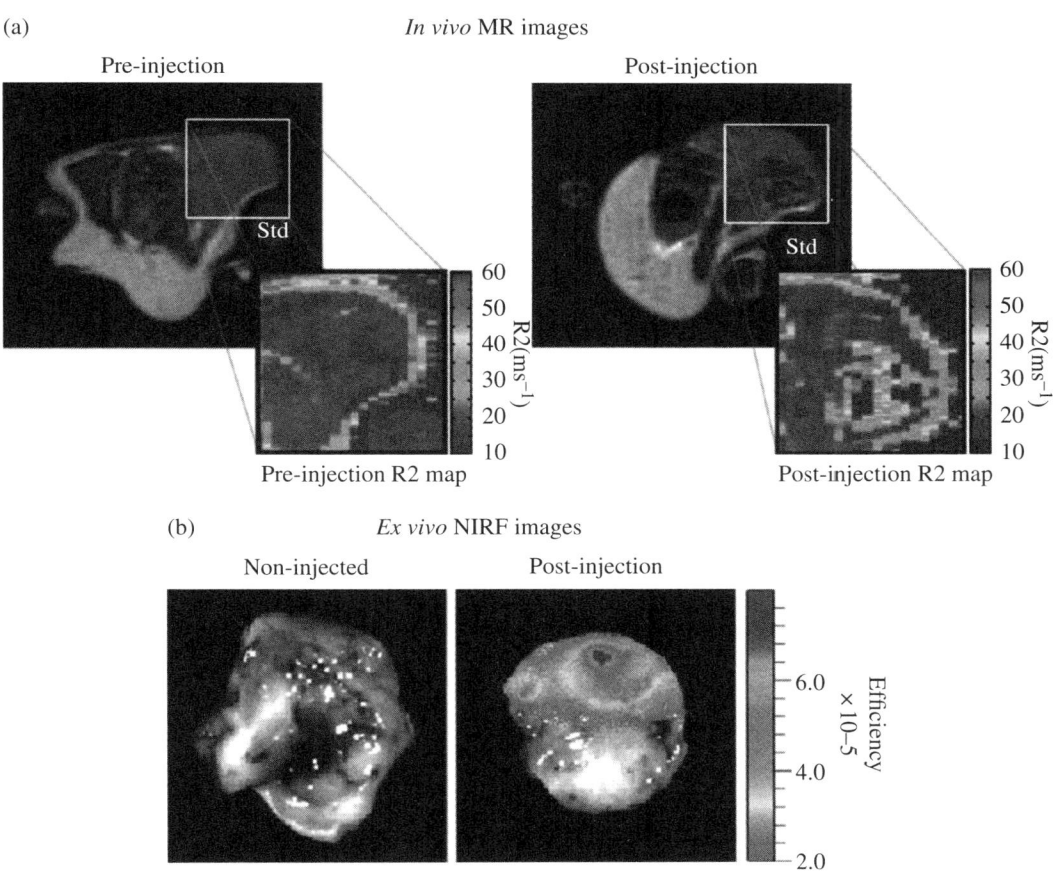

FIGURE 16.12 (a) MRI of the mouse bearing a 9 L tumour pre- and post-injection (48 h) of the CTX-Cy5.5 nanoparticles. Heterogeneous accumulation of the nanoparticles is observed by negative contrast enhancement and changes in the R_2 measurements in the region of interest. (b) *Ex vivo* NIRF image of xenograft tumours acquired from a control mouse (left) and the mouse injected with CTX-Cy5.5 nanoparticles (right). Reprinted with permission from Ref. [47]. Copyright 2010 American Chemical Society. (*See insert for colour representation of the figure.*)

Bioactivatable or bioresponsive agents can be used to report on the biochemical status of tissues, such as the level of enzyme activity near tumours. In one design, a NIR Cy5.5 dye is attached to the surface of a ~43 nm SPION via peptide coupling. The Cy5.5 dye is attached via a peptide linker, Cy5.5-GPLGVRG, which can be cleaved by a matrix metalloprotease [51]. The NIRF of Cy5.5 is efficiently quenched when conjugated to the SPION, but when the MMP cleaves the MMP-Cy5.5 substrate peptide, the dye is freed from the particle and NIRF is restored.

The proposed mechanism of enzyme activation of NIRF was confirmed by *in vitro* studies with the enzyme. *In vivo*, a SCC7 tumour xenograft in mice was used. This cell line was chosen for its high expression of MMP-2. No NIRF was observed in the control mice. In tumour mice, a NIRF image was obtained after 12 h post injection. Administration of an MMP-2 inhibitor 30 min. before injection reduced the observed NIR fluorescence. MR imaging showed highest relative signal enhancement in the tumour at 12 h. *Ex vivo* imaging showed NIRF in the liver, heart; the highest levels were noted in the tumour. The NIRF observed in the heart is assumed to be due to hydrophilic Cy5.5 dyes that circulated after decomposition of the peptide.

16.3.4 Multilayered Nanoparticle Agents

By changing the size and aspect ratio of nanostructures, the plasmon-derived optical resonance of some particles can be tuned to the NIR region for *in vivo* applications [52–54]. Nanoshells consist of a silica core surrounded by a thin metallic (e.g., Au) layer. The plasmonic properties of nanoshells have been shown to further enhance fluorescence of some organic NIR dyes that are covalently attached and are within 10 nm of the surface [55].

Nanoshell (Au) SPIONs have been used to investigate the efficiency of HER2 antibody targeting [29]. To generate the Au-SPIONs, water-soluble SPIONs were treated with APTES. Gold nanoshells were coated with these amine-functionalised SPIONs followed by a layer of silica. The fluorophoreindocyanine green (ICG) was included and trapped within the silica layer to make the magnetofluorescent Au-SPION. HER2 antibodies were attached on the surface via biotin-avidin conjugation.

Mice were injected with either BT474AZ or MDAMB231 cells to grow tumours. The BT474AZ cell line overexpresses HER2 while MDAMB231 only expresses HER2 at low levels. The comparison of these two cell lines is designed to show the targeting ability of the HER2 antibody. In both cases, injection of the Au-SPIONs was followed by NIRFI and MRI, which showed that the Au-SPIONs cleared from the animals within 72 h. Imaging showed higher intensity for the BT474AZ animals [29]. Biodistribution investigations showed the presence of the NIR agent in the tumour, kidneys, and liver of these animals after 72 h, which the authors presume is due to the higher levels of accumulation due to targeting.

A similar multilayer approach was reported where SPIONs were surrounded by a silica layer entrapping visible and NIR-fluorescent quantum dots and finished with HER2 antibodies [56]. The final product had a mean diameter of 150 nm and the r_2 was found to be 1914 mg^{-1} L^{-1} s^{-1}, or 105 mM^{-1} s^{-1}. As a control, nanocomplex BSA was conjugated to the particles rather than the HER2 antibodies.

Xenograft human breast cancer tumours were grown in mice, and the mice were injected with the modified nanoparticles [56]. T_2-weighted MR imaging shows that the nanocomplexes are found largely in the peripheral regions around blood vessels of the tumour. NIRF imaging of the excised organs shows the HER2 nanocomplex localises in the tumour and lungs whereas the BSA-conjugated control does not.

16.4 NIR-PET

16.4.1 Biodistribution

Silica and organically modified silica (ORMOSIL) nanoparticles have been shown to be highly biocompatible. The nature of silica allows dyes and other materials such as quantum dots or iron oxide particles to be encapsulated. The synthesis of ultrafine ORMOSIL particles and biodistribution of these particles labelled with *either* a NIR dye *or* ^{124}I for PET is reported [57]. ORMOSIL nanoparticles were synthesised in the nonpolar core of an oil-in-water microemulsion [57, 58]. A surfactant docusate sodium, butanol, and DMSO were dissolved in water. To this microemulsion, DY776-silane and VTES were added, and polymerisation was initiated by APTES. PEG silane was added to coat the surface of the particles with PEG groups for water solubility. The mixture was dialysed against distilled water and sterile filtered. DY776-silane was generated from the NHS ester of NIR dye DY776 and APTES in order to incorporate a silane reactive group on the dye (Figure 16.13). The ^{124}I label was introduced via the Bolton-Hunter reagent (Scheme 16.9).

PET imaging using ^{124}I-labelled particles shows that most of the particles accumulate in the liver and spleen. The accumulation in the spleen of the ^{124}I-labelled particles differs from the DY776-labelled particles, most likely due to

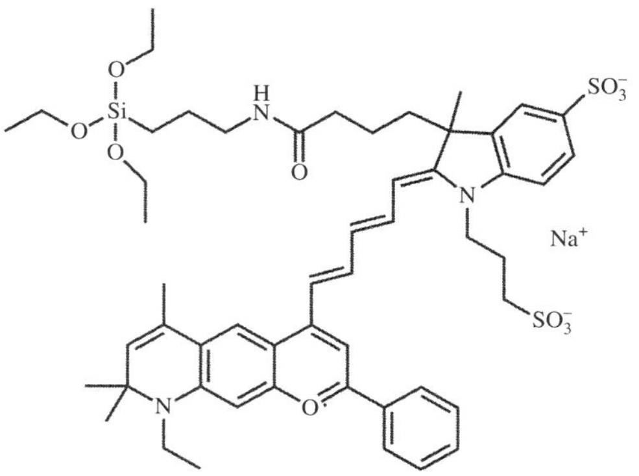

FIGURE 16.13 Structure of DY776-silane.

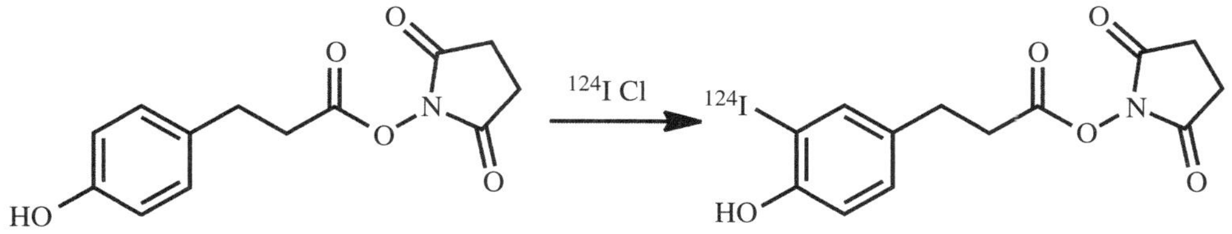

SCHEME 16.9 The iodination reaction of the Bolton-Hunter reagent.

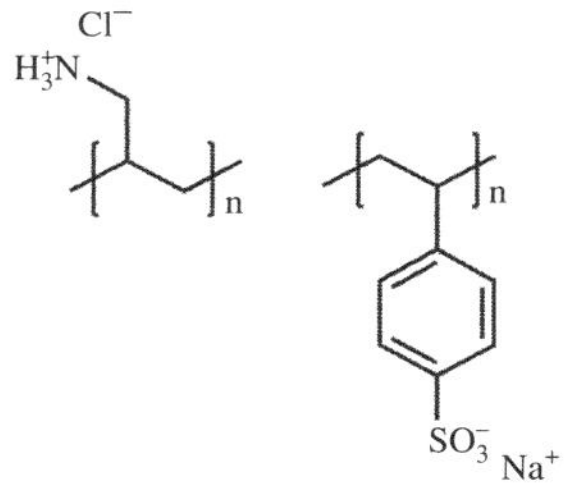

FIGURE 16.14 Structures of polyallylamine and polystyrenesulfonate.

difference in the surface charges incurred by the different labels. *Ex vivo* NIRFI of the individual organs shows accumulation in liver, stomach/intestines, and skin, with minimal intensity observed for kidney, heart, lungs, and spleen [57]. These results suggest accumulation in the RES organs and that the clearance is achieved effectively by hepatobiliary excretion with no signs of organ toxicity as determined by Prussian Blue staining [57].

Understanding how nanoparticles cross the blood-brain barrier presents one of the most important challenges to current biomedical research. AuNPs coated with multilayers of polyelectrolytes have been shown to cross the blood-brain barrier [59, 60]. However, previous reports indicated that these AuNPs were toxic. In order to reduce toxicity in this study, HSA was applied as a coating [61].

Polyallylamine and polystyrenesulfonate (Figure 16.14) were used as the multilayers. *In vivo* NIRFI shows that the particles localise in the head, and X-ray microtomography show boundaries between regions that are not typically obvious in this imaging technique. The particles were observed in the thalamus and hypothalamus. *Ex vivo* NIRFI shows the particles in the brain. Confocal microscopy of brain slices indicates that the AuNPs localised in the hippocampus, hypothalamus, thalamus, and the cortex. FITC on the AuNPs co-localised with Nissl or DAPI stain shows that the particles are in the cells but not in the nucleus. These areas of the brain are of interest due to their relevance to Parkinson's, Alzheimer's, and prion diseases.

16.4.2 Theranostic Applications of NIR-PET Bimodal Nanoparticle Systems

Polymeric nanoparticles based on the self-assembly of glycol chitosan (Figure 16.15) have been previously reported as drug delivery vehicles, and N-acetyl histidine can be used to modify the polymer to control the hydrophobicity [62–64]. In this report, particles of 240–290 nm were prepared using three different weight percentage amounts of NAcHis and labelled with Cy5.5 [65].

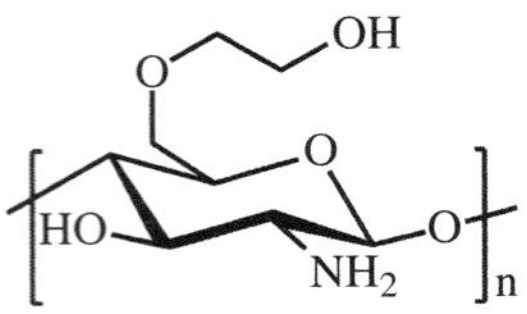

FIGURE 16.15 Glycol chitosan (GC) is a polysaccharide composed of β-(1-4)-linked D-glucosamine subunits where one alcohol has been modified with a glycol (-CH$_2$CH$_2$OH) group.

These GC particles were loaded with DOX. The drug release was evaluated at pH 7.4 and 6.4. At pH 6.4, drug release reached a plateau within 6 h, whereas at pH 7.4, complete drug release was not reached for 48 h.

To track the particles *in vivo*, ^{131}I-labelled nanoparticles were used. Scintigraphic images of tumour-bearing mice show GC3 particles, or those prepared with the highest wt% NAcHis targeted the tumour well while the other formulations (GC1 and GC2) did not. This result is in contrast to the NIRF images, which were comparable for all three compositions, but highest intensity in the tumour region was found for GC3. *Ex vivo* imaging showed high fluorescence in the tumour tissue and some fluorescence in the liver, spleen, and kidneys.

The *in vivo* images for the DOX-loaded GCs varied dramatically. Although the scintigraphic images were very similar for GC1 and GC2, the GC1 DOX formulation localised to a much larger extent in the tumour according NIRFI. *Ex vivo* NIRF imaging shows fluorescence in the lung, liver, kidney, with the highest intensity in the tumour.

In order to test the drug delivery capability, HT29 tumour-bearing mice were treated with DOX, GC1, and DOX-GC1 every 3 days (5 mg dox/kg). Tumour volume was monitored over 16 days and growth was inhibited best by DOX-GC1.

To generate a PET-NIRFI-MRI trimodal agent, liposomes were formed with gadolinium labels (GdDOTA-DSPE) for MRI, IRDye for NIRFI, and empty chelates (DOTA-DSPE) for PET nuclei (Figure 16.16). The liposomes were formed by dissolving all the components in chloroform, followed by lipid film hydration and extrusion [66]. The authors demonstrated that the DOTA-DSPE chelates could bind ^{64}Cu ions for PET imaging. ^{99m}Tc BMEDA was formed by reduction of NaTcO$_4$ with SnCl$_2$ in the presence of BMEDA and loaded into the preformed Gd-liposomes.

In vivo NIRFI showed superior accumulation in the tumour of the liposomes when compared to the free NIR dye. The excised tumour showed significantly more uptake than the other organs. Planar gamma camera images were obtained for mice injected with ^{99m}Tc-labelled liposomes and showed accumulation in the tumour and fast clearance of the remainder. 3D tomography and projection images from the microPET imaging of the ^{64}Cu labelled liposomes shows local retention and micro-distribution of the radioactivity within the tumour.

16.4.3 Tumour Targeting Using NIR-PET Bimodal Nanoparticle Systems

Glycol chitosan was modified with 5-β-cholanic acid (Figure 16.17) and Cy5.5 and the bladder cancer targeting peptide CSNRDARRC [68]. This polymer was sonicated in the presence of iron oxide nanocubes, which became non-covalently incorporated within the polymer matrix. The average diameter of the final particles was found to be 480 nm.

In vitro cellular targeting was investigated using K9TCC bladder cancer cells incubated with the nanocubes for 3 h. Confocal microscopy showed nanoparticles bound to the K9TCC cells. Lower concentration of nanocubes corresponds to lower NIRF and decreased negative MR contrast, as expected.

NIRF imaging was utilised to evaluate the contrast effect of the modified nanocubes *in vivo*. Images of tumour-bearing mice clearly indicated strong NIRF intensity at the tumour 24 h post administration of the nanoparticles. *Ex vivo* images showed much brighter NIRF in the tumour than liver, spleen, and kidneys. The lung and heart showed negligible NIRF. This result supports active targeting and retention of the nanocubes at tumours in mice. The results might be explained by both passive accumulation of nanoparticles through EPR effects and active targeting of nanoparticles by specific binding effect of CSNRDARRC peptide to bladder cancer. No MR imaging of the mice was reported.

Ferritin is a protein composed of 24 subunits that self-assemble into a nanostructure 12 nm in size with an internal cavity of 8 nm [69]. The subunit proteins may be chemically or genetically modified and may be assembled to encapsulate metal ions within the hollow cavity [70–72]. In this report, ferritin was genetically modified with RGD, the integrin-targeting peptide, and chemically modified with Cy5.5 [73]. The RGD-ferritin and Cy5.5-ferritin assemblies were combined, dissociated at pH 2, and then reconstituted at pH 7.4 in the presence of ^{64}Cu^{2+}. This procedure scrambles the ferritin components, and re-assembly results in ferritin species with both RGD and Cy5.5 on the outer surface and a payload of ^{64}Cu^{2+} ions encapsulated within the cavity.

PET and NIRF images show the labelled ferritin is taken up by the tumour *in vivo*. A decrease in levels is observed between 24 and 40 h in the PET images, while NIRF images show a decrease between 4 and 24 h. Administration of a blocking

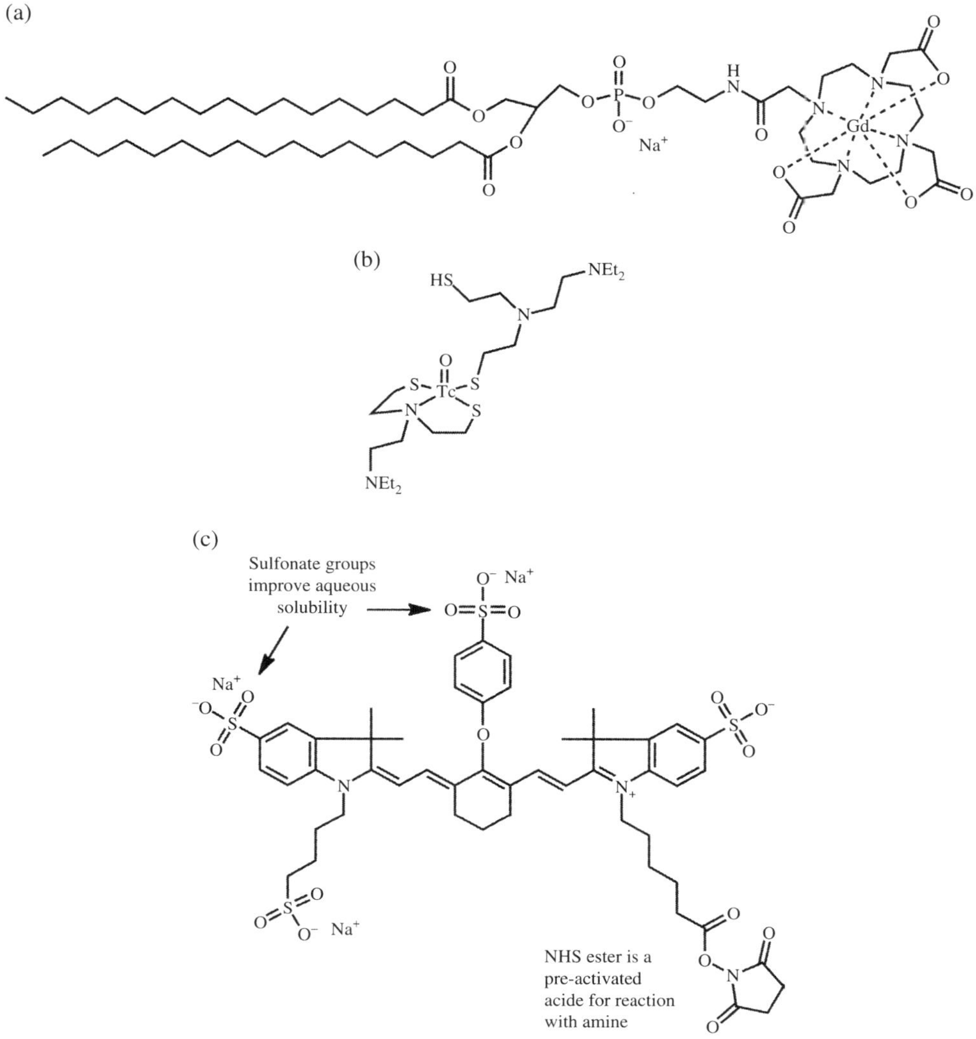

(a)

(b)

(c)

Sulfonate groups
improve aqueous
solubility

NHS ester is a
pre-activated
acide for reaction
with amine

FIGURE 16.16   Structures of (a) GdDOTA-DSPE, (b) ^{99m}Tc(V)BMEDA [67], and (c) IRDye.

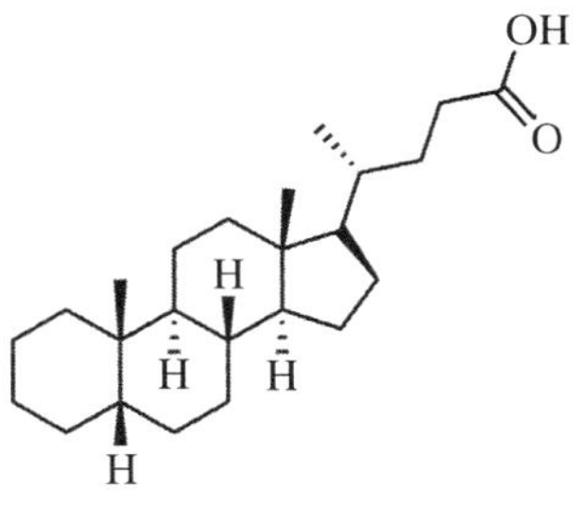

FIGURE 16.17   Structure of 5-β-cholanic acid.

peptide that interferes with the RGD-integrin interaction 30 min. prior to injection of the labelled ferritin results in decreased ferritin uptake by the tumour. *Ex vivo* images of organs from animals sacrificed at 40 h show a lower amount in the animal that was treated with the blocking peptide. The authors conclude that accumulation in the tumour is due to both RGD targeting and the EPR effect.

Changes in the lymph nodes are closely related to tumour metastasis. Sentinel lymph node (SNL) imaging is important because it can give information on the metastatic status of tumours, which can be useful for tumour staging and therapeutic decision making. In this work, NIR dye ZW800 (Figure 16.18a), Gd^{3+}, and ^{64}Cu were incorporated into 60 nm mesoporous

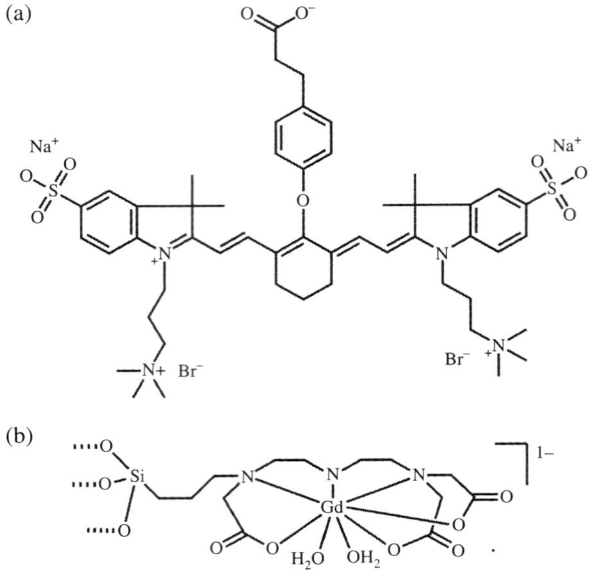

FIGURE 16.18 Structures of (a) ZW800 and (b) Si-Gd(DTTA) as it might be bound to the nanoparticle surface (hashed bonds).

silica nanoparticles for trimodal imaging of tumour metastatic SNLs [74]. The ZW800 dye was modified with APTES and incorporated into the silica matrix during the particle formation from TEOS [75, 76]. Si-DTTA (Figure 16.18b) was prepared from bromoacetic acid and 3-(trimethoxysilylpropyl)diethylenetriamine and reacted with the silica particle surface followed by metallation with GdCl$_3$. Finally, DOTA-NHS was conjugated to the surface and metallated with ^{64}Cu from ^{64}Cu(OAc)$_2$. The relationship of $1/T_1$ to Gd concentration was linear and $r_1 \sim 15\,\mathrm{mM^{-1}\,s^{-1}}$.

A tumour metastasis model was established by injecting 4T1 cells into mice. The tumour metastatic SLN (T-SLN) was compared to the normal SLN from the non-injected side of the mouse. NIRF imaging showed that the T-SLN was visible for up to 21 days. *Ex vivo* NIRF imaging showed localisation of the probe in the liver and spleen. After two weeks, the organs still possessed a detectable optical signal.

PET imaging showed immediate accumulation in the T-SLN. Uptake by the T-SLNs was faster than by the normal SLNs. Between 6 and 24 h, accumulation in the liver began. MRI showed an increase in contrast in the area of the SLNs that maximised at 1 day, although contrast was observable for up to 21 days. Further, MRI tomographically shows a single layer of T-SLNs in contrast with the 2D projection from the optical imaging results.

The authors suggest the following clinical orchestration of modalities: PET imaging can be used to first identify areas of interest. MRI can be used to refine the anatomic localisation. Finally, optical imaging can be used to guide surgery. Overall, the three modes of imaging give consistent results and are complementary.

16.5 UPCONVERSION LUMINESCENCE

Upconversion luminescence (UCL) is the process by which continuous wave low-energy (near-infrared) light is converted to higher energy (visible) light through multiple absorptions or energy transfers [77]. This unique luminescence process has been observed in nanoparticles of compositions such as NaYF$_4$ doped with sensitiser ions such as Yb^{3+} and activator ions such as Er^{3+}. UCL often has sharp emission lines, superior photostability, no autofluorescence, good tissue penetration, and low toxicity. The topic of UCL is discussed in greater detail in Chapter 13.

16.5.1 Bimodal UCL-CT Nanoparticles

NaYF$_4$ nanoparticles were combined with 5-amino-2,4,6-triiodoisophthalic acid (AIPA) to make the first bimodal UCL/CT contrast agent [78]. A silica shell on the UCL particles was introduced by the hydrolysis of TEOS using a standard microemulsion procedure. The surface was further functionalised with amines using APTES. AIPA, a standard iodine-containing CT agent, was attached to the amines by peptide coupling. Finally, PEG groups were covalently attached using O-[2-(N-Succinimidyloxycarbonyl)-ethyl]-O'-methylpolyethyleneglycol (SEMG, Figure 16.19). The final diameter of the particles was found to be 54 nm. Cytotoxicity studies using MTT showed the cells to be >90% viable at the highest dose of 1000 μg/mL.

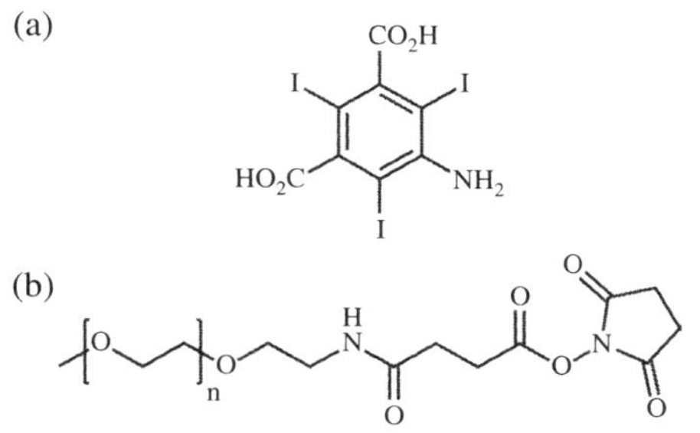

FIGURE 16.19 Structures of (a) 5-amino-2,4,6-triiodoisophthalic acid (AIPA) and (b) SEMG.

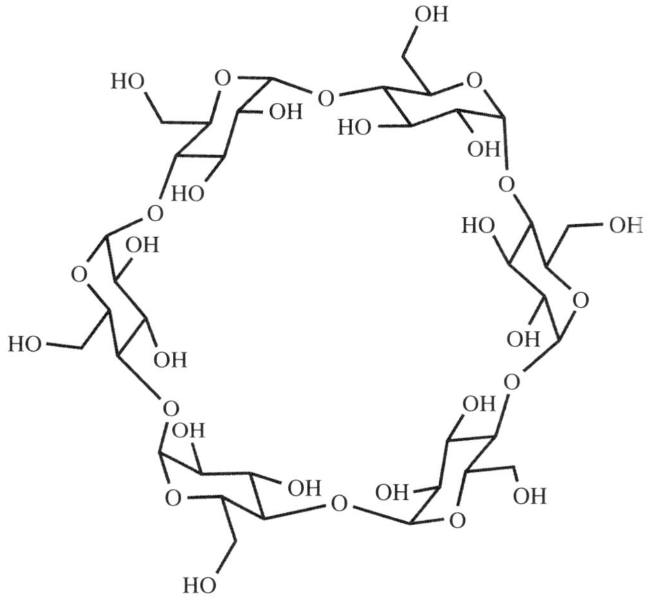

FIGURE 16.20 Structure of α-cyclodextrin, commonly used to host hydrophobic guests.

Solutions of the particles were injected into the subcutaneous tissue and intramuscular tissue of rats and visualised using both UCL and CT. The UCL signal was weak but observable for the particles injected subcutaneously. Further, using control solutions *in vitro*, it was found that the rare earth elements of the UCL particles contributed to the CT contrast.

16.5.2 Biodistribution (UCL-PET)

UCL nanoparticles (UCL-NPs) can be synthesised from rare earth salts by solvothermal, thermal decomposition, and hydrothermal methods [79]. The resulting hydrophobic UCL-NPs can be solubilised in aqueous solutions using *alpha*-cyclodextrin (α-CD, Figure 16.20) to make CD-UCL-NPs [79]. Further, the probes can be used for PET imaging by incorporating ^{18}F for biodistribution and *in vivo* imaging.

The size of the particles prepared by the solvothermal and thermal decomposition methods was approximately ~18 nm. Addition of α-CD combined with shaking solubilised the particles. Larger particles prepared using the hydrothermal method (400 nm) were also solubilised using this method.

In vivo UCL imaging of tail vein-injected CD-UCL-NPs showed the nanoparticles are taken up largely by the liver and spleen, as would be expected for nanoparticles of this size. For lymphatic imaging, UCL-NPs were injected into the mouse claw with expected lymphatic drainage to the oxter nodes as observed by UCL imaging.

^{18}F-labelled particles were generated by ligand exchange on the particle surface and used for *in vivo* microPET imaging and *ex vivo* biodistribution. The particles were found to quickly accumulate in the liver and spleen. Uptake by other organs was minimal. Because bone can take up free F$^-$ ions, the amount of ^{18}F in the bone was investigated and found to be very low, supporting claims of the biostability of these particles.

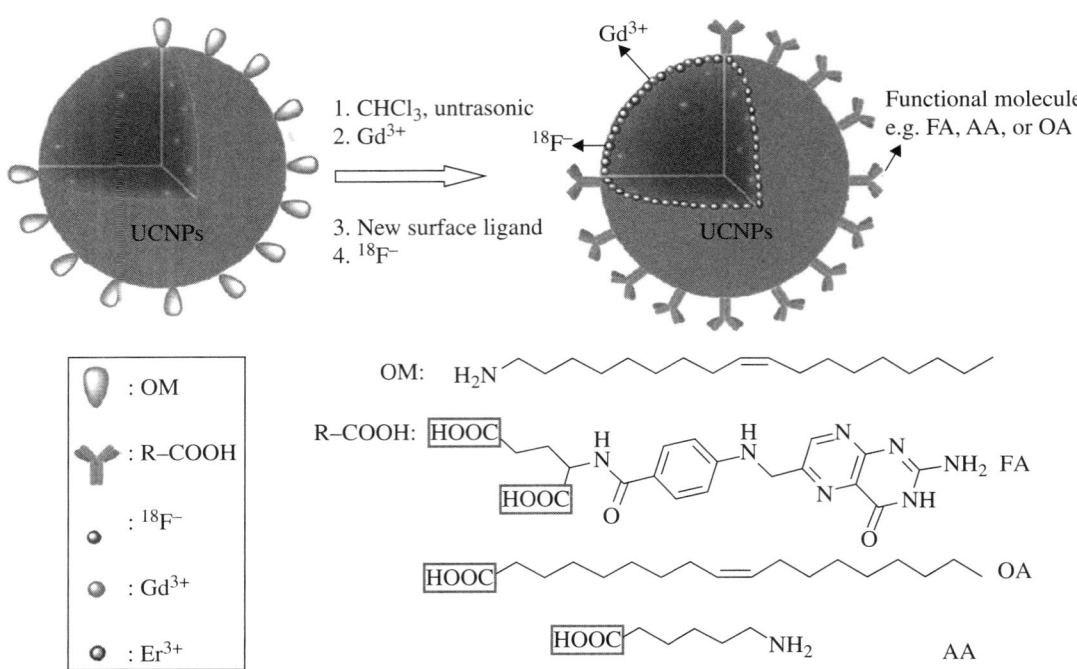

SCHEME 16.10   Modification of UCNPs with Gd^{3+} and F^- ions and carboxylic acid-bearing surface ligands. Reprinted with permission from Ref. [80]. Copyright 2011 American Chemical Society.

Surface ligand exchange can be used to solubilise $NaYF_4$ particles (UCNPs) as well as introduce other ions and targeting ligands. In this report, Gd^{3+} ions, $^{18}F^-$ ions, and aminocaproic acid (AA) are introduced during sonication [80]. The ions exchange on the surface and are subsequently stabilised by the acid ligand (Scheme 16.10). The r_1 per Gd(III) ion was found to be $28\,mM^{-1}s^{-1}$ at 3 T, 38 °C.

In vivo imaging in mice was undertaken using AA-Gd-UCL-NPs [80]. Rapid accumulation in the liver and spleen was observed as expected in the UCL image (Figure 16.21). T_1-weighted MR imaging showed 30% enhancements in the liver and spleen 10 min. after injection. ^{18}F-AA-Gd-UCL-NPs were used for *in vivo* PET imaging and showed rapid uptake by the liver and spleen and very low uptake by other organs, including bone.

16.5.3 Theranostic Applications of UCL-NPs

Two nanocomposites are described where UCL nanoparticles are combined with iron oxide particles in a polymeric shell and either a squaraine dye (SQ, Figure 16.22, synthesised as in Ref. [81]) or DOX was added [82]. The polymers used were poly(styrene block allyl alcohol) and polyvinyl alcohol. Iron oxide particles of 6-8 nm in size were incorporated into the nanocomposite using a microemulsion method. The overall r_2 of the particles was found to be $84\,mM^{-1}s^{-1}$, with the diameter of the nanoparticles being about ~200 nm. pH-dependent release of DOX was observed, with more DOX being released at lower pH. Cell toxicity studies showed that UCL-DOX was slightly less toxic than free DOX. Magnetic targeting was employed to localise the particles in cell culture. Uptake of UCL-DOX was higher in cells growing near the magnet and cell viability was lower.

4T1 cells were labelled with the UCL-SQ nanoparticles for 24 h and injected into mice. After 10 days, the UCL could still be observed as well as the down conversion fluorescence from the SQ. Between 0 and 5 days, the MR signal decreased, possibly due to tumour growth.

Biodistribution of the UCL-SQ particles showed that the particles localised quickly in the liver (within 2 h). T_2 weighted MR imaging showed darkening of the liver, spleen, and lung, but no change in the other major organs.

To take advantage of the photothermal therapeutic properties of Au nanoparticles, UCL nanoparticles were combined with iron oxide particles and coated with a thin shell of gold via a seed-mediated procedure [83]. The final particles consisted of cube-like structures with a diameter of approximately 200 nm, with ratio of Y/Fe/Au of 100:37:4. When irradiated with 808 nm laser light at $1\,W/cm^2$, the temperature of the solution increased rapidly, indicating that the UCL emission is unperturbed in this nanocomposite.

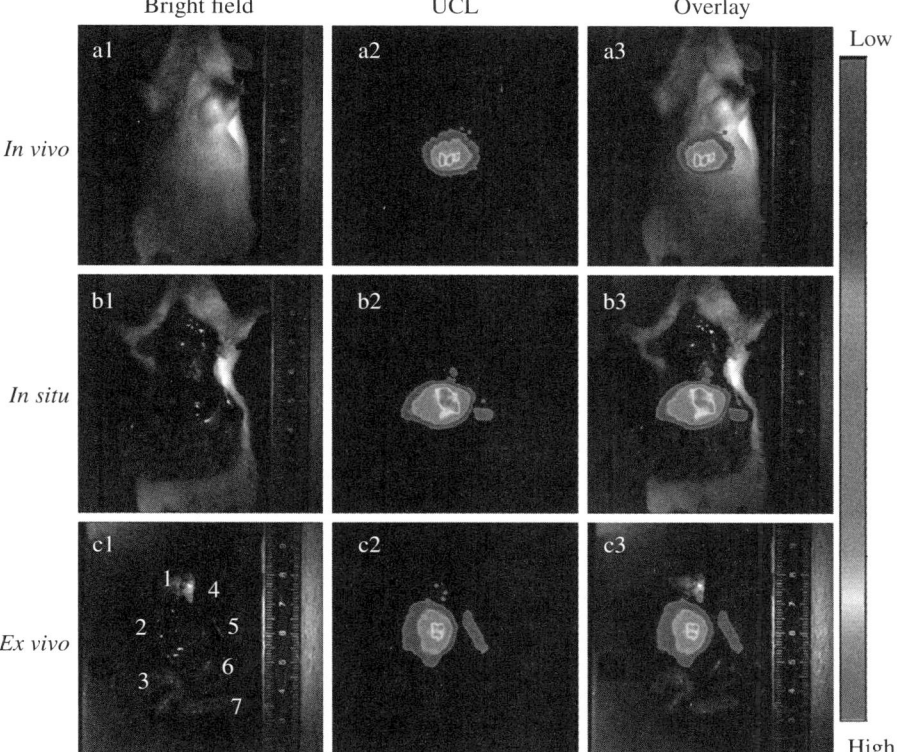

FIGURE 16.21 *In vivo* (a), *in situ* (b), and *in vitro* (c) upconversion imaging at 10 min. post-injection of AA-Gd-UCNPs. 1, lung; 2, liver; 3, stomach; 4, heart; 5, spleen; 6, kidney; 7, intestines. Reprinted with permission from Ref. [80]. Copyright 2011 American Chemical Society.

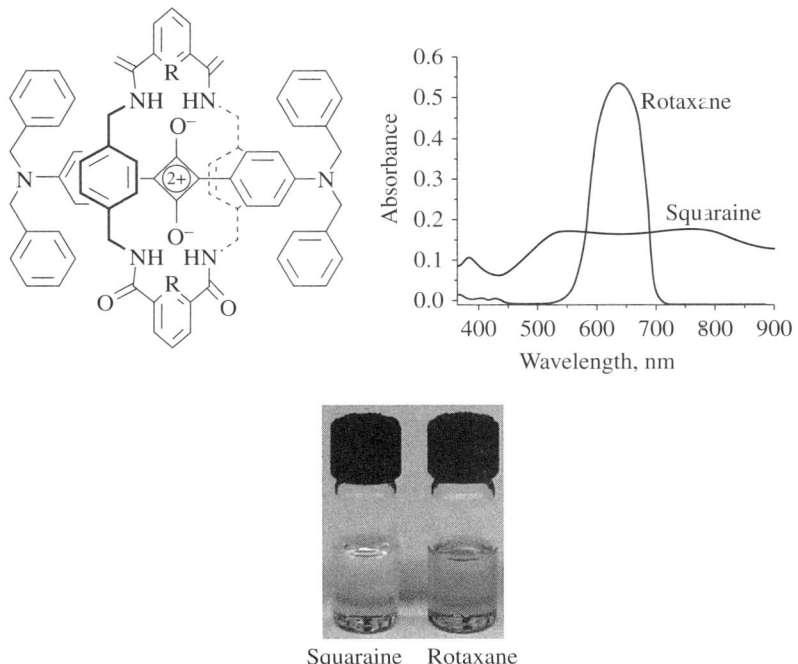

FIGURE 16.22 Structure of the squaraine dye. Reproduced from Ref. [81] with permission from the American Chemical Society.

In vivo studies showed that particles administered intravenously could be directed to a tumour using an external magnet. The UCL increased over time and the MR image of the tumour darkened. Biodistribution showed the particles could be found (detection of Y^{3+}) in the lung, spleen, and liver for up to 40 days after injection. However, histological analysis showed that the particles did not appear to induce any toxic effects, damage, or inflammation in these organs.

The particles were used as a photothermal therapeutic. Several control systems were studied to observe the effect of magnetic targeting and photothermal therapy (PTT). The systems studied were untreated, laser only, NPs + magnet, NPs + laser, NPs + magnet + laser. Tumour volume increased in all cases except the last, which were tumour-free after the treatment. Further, the mice survived over 40 days in contrast to the mice from the control groups which lived 14–18 days. Future work will involve reducing the nanoparticle size, varying the surface coating to reduce the circulation time and RES retention.

16.6 PET-SPECT-CT-MRI

16.6.1 Biodistribution

A top-down approach was used to generate silica-based particles for combined fluorescence/MR/SPECT/CT [84]. First, Gd_2O_3 nanoparticles were coated with SiO_2 and Cy5.5 was encapsulated during the formation of the SiO_2 layer. DOTA-GA anhydride (1,4,7,10-tetraazacyclododecane-1-glutaric anhydride-4,7,10-triacetic acid; CheMatech, Figure 16.23) was added, which effectively dissolves the Gd_2O_3 core and provides DOTA-chelated Gd ions on the surface of a collapsed polysiloxane platform. The result is a ratio of 10 DOTA:7Gd:27 Si. Free DOTA groups were used to chelate ^{111}In from ^{111}InCl$_3$ solutions. The size of the particles was determined to be approximately 3 nm.

Biodistribution studies in mice showed accumulation in the kidney and bladder, but not lungs or liver, and indicate a renal clearance pathway. The circulation lifetimes were characterised by a multi-exponential fit that indicated the presence of a slow-clearing component. This study illustrates the existence of an intermediate size range that is too large for efficient renal clearance yet too small for hepatic clearance.

A similar system for fluorescence/MR/SPECT/CT was prepared from Gd_2O_3 particles that were synthesised according to a modified polyol method [85]. TEOS/APTES were used to modify the surface via controlled hydrolysis and condensation. Cy5-NHS and DTPA-bisanhydride were attached to the surface amine groups from the APTES (Scheme 16.11). ^{111}In was chelated by coordinating to the surface DTPA.

Biodistribution showed renal excretion and no uptake by the RES. Evaluation of ^{111}In levels in brain, liver, heart, lung, muscles, bone, skin, spleen, and kidney was determined at 2 h, and at 1, 3, 10, and 18 days. By far the majority of the radioactive ^{111}In was detected in the kidneys at all time points. Similar results were obtained using SPECT/CT, MR, and optical imaging.

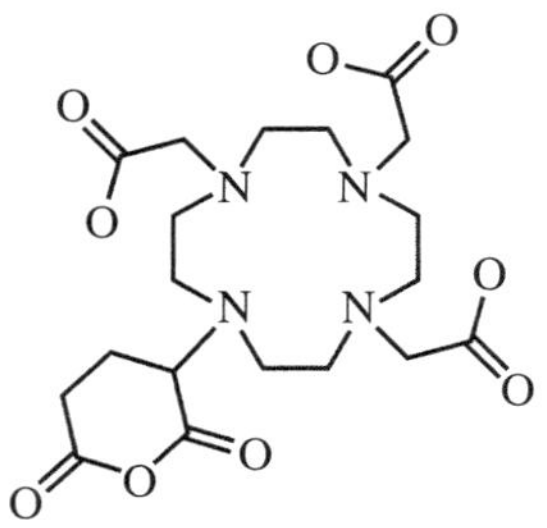

FIGURE 16.23   Structure of 1,4,7,10-tetraazacyclododecane-1-glutaric anhydride-4,7,10-triacetic acid.

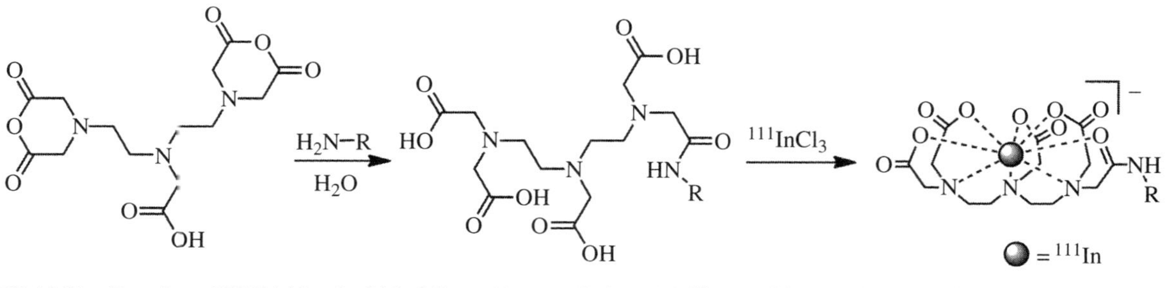

SCHEME 16.11   Reaction of DTPA bisanhydride followed by metallation with ^{111}In. In this case, the amine is on the nanoparticle surface.

The use of phosphonate groups has proven to be a stable method of nanoparticle surface modification [86]. In this work, the metal-binding chelate dipicolyl amine (DPA) was combined with alendronate (ale) in order to introduce ^{99m}Tc chelates onto the surface of SPIONS [87]. First, ^{99m}Tc –DPA-ale (Figure 16.24) was formed by mixing ^{99m}Tc(CO)$_3$(H$_2$O)$_3$$^+$ with DPA-ale in water with brief heating [88]. Finally, ^{99m}TcDPA-ale was heated with Endorem, a commercially available dextran-coated SPION. The transverse relaxivity was found to be r$_2$ ~ 11 at 9.4 T. SPECT, CT, and PET were used to show that the ^{99m}TcDPA-ale-Endorem particles accumulated in the liver and spleen, in contrast to free ^{99m}TcDPA-ale, which localised in the femur (Figures 16.25 and 16.26).

16.6.2 Targeting

Streptavidin (SA) is a tetrameric protein 66 kDa in size that binds to the small molecule biotin. Each of the four binding sites is identical with the highest binding constant of known non-covalent interactions. Streptavidin serves as the 'nanoparticle' for a NIRFI/SPECT/CTsystem [89]. First, Cy5.5-NHS was attached to biocytin and combined with SA in a 1:1 ratio. Biotin-DOTA was added to the Cy5.5-SA in a 1:1 ratio and added to a solution of ^{111}InCl$_3$. Finally, the HER2 antibody was biotinylated according to literature procedures and added to the SA. The agent was administered to mice intravenously. Fluorescence at 2, 15, and 40 h as well as SPECT and CT showed accumulation in the liver, kidneys, spleen, and tumour. SUM190 (HER2+) xenograft tumours accumulated the agent, whereas SUM149 (HER2-) tumours did not.

A multimodal system based on the biotin-avidin interaction has been developed for the detection of brain tumours. The Gaussia luciferase reporter gene was fused with a membrane anchor and BAP, biotin acceptor peptide. The biotin acceptor peptide was attached so that biotin ligase attaches biotin on the cell surface at a specific lysine residue [90]. This approach allows the cells to be detected with any modality attached to streptavidin (SA).

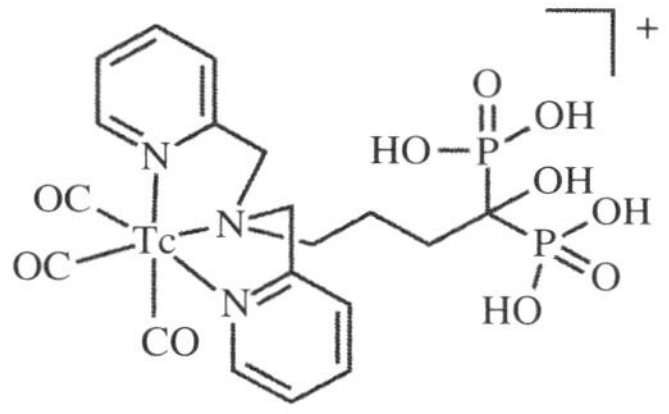

FIGURE 16.24 Structure of bisphosphonate chelate ^{99m}TcDPA-ale.

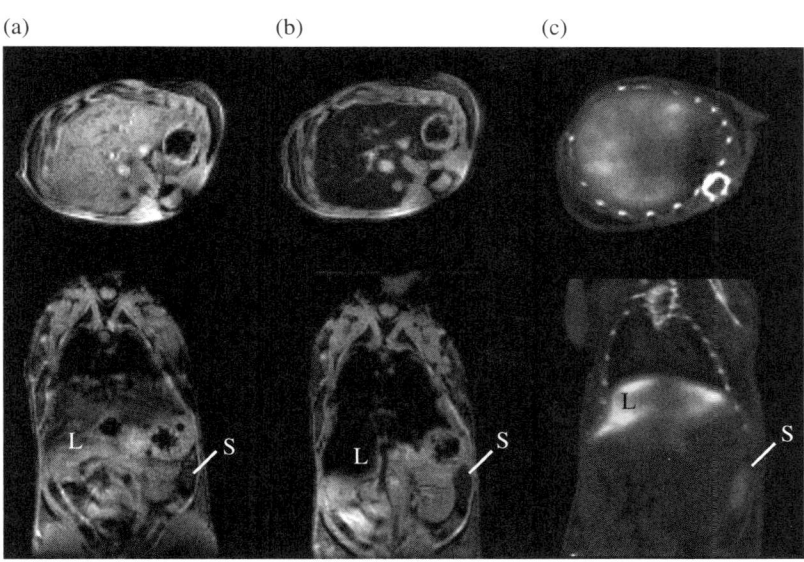

FIGURE 16.25 Short-axis view (top) and coronal view (bottom) images: (a) T_2*-weighted MR images before injection of ^{99m}TcDPA-ale-Endorem, (b) T_2*-weighted MR image 15 min. post injection, and (c) nanoSPECT-CT image of the same animal in a similar view 45 min. post injection. Contrast in the liver (L) and spleen (S) changes after injection due to accumulation of ^{99m}TcDPA-ale-Endorem, in agreement with the nanoSPECT-CT image, which shows almost exclusively liver and spleen accumulation of radioactivity. Reproduced with permission from Ref. [87]. (*See insert for colour representation of the figure.*)

(a)   (b)

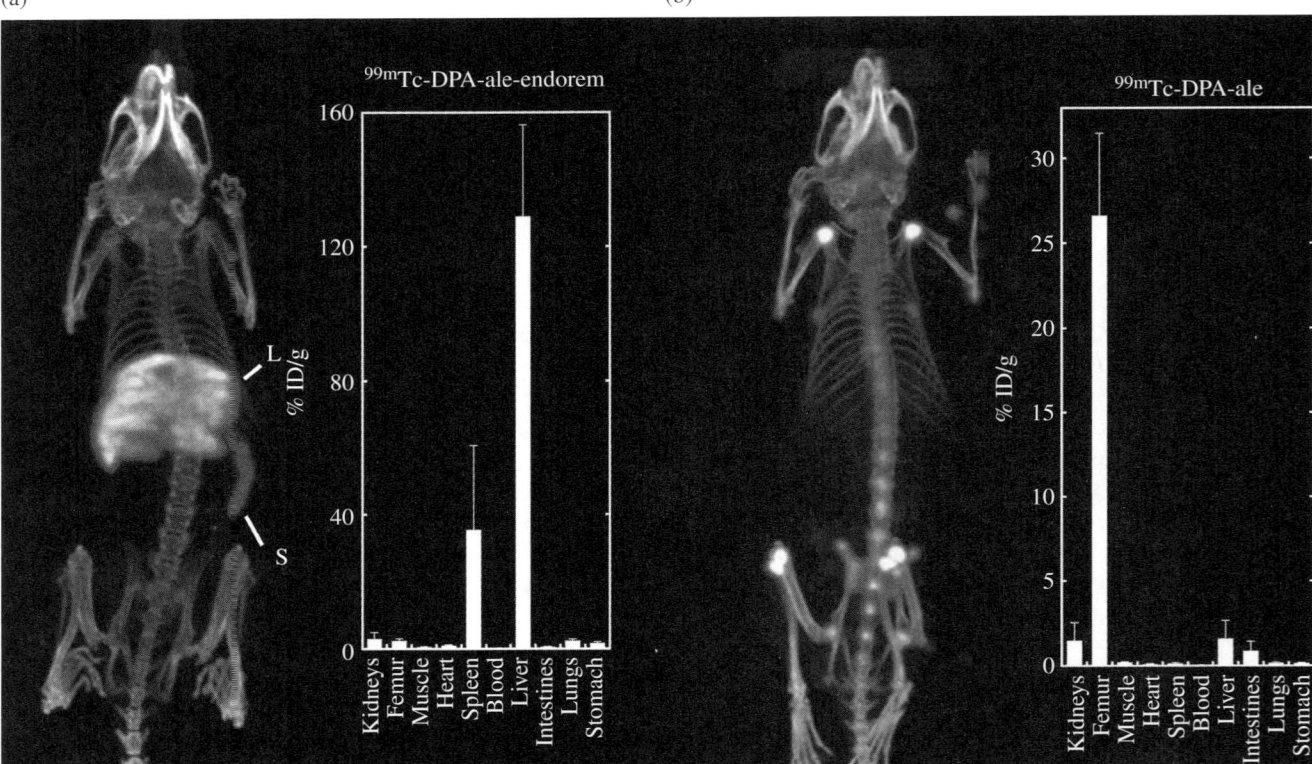

FIGURE 16.26 Whole-body SPECT-CT maximum intensity projection (left) and biodistribution studies (right) of [99m]TcDPA-ale-Endorem (a) and [99m]TcDPA-ale (b). Reprinted with permission from Ref. [87]. Copyright 2011 American Chemical Society. (*See insert for colour representation of the figure.*)

To form a tumour, Cli36-sshBirA cells were injected intracranially in mice. Streptavidin with AlexaFluor750 was used for fluorescence-mediated tomographic imaging. Subsequently, streptavidin labelled with either ^{111}In-DTPA-biotin or SPIONs were injected intravenously for SPECT imaging or MRI imaging respectively. The tumour was detected using bioluminescence (following injection of coelenterizine), fluorescence tomography, SPECT, and MRI.

A novel targeted MRI-CT nanoparticle system was investigated using FePt alloy particles and the HER2 antibody [91]. These nanoparticles were synthesised by heating $Pt(acac)_2$ and $Fe(CO)_5$ with dioctyl ether, hexadecandiol, and oleyl amine or oleic acid. The size was found to be 3, 6, or 12 nm upon varying the heating conditions [91]. The surface was passivated by coating with cysteamine, and the amine groups were used for peptide coupling to the HER2 antibody for targeting. The particles were cuboctahedral at 3 and 6 nm, and cubic at 12 nm. The ratios of Fe:Pt were 58:42, 51:49, and 33:67 respectively. Cell viability with MTT showed ~90% up to 10 mM Fe. Extensive biodistribution studies were undertaken at several time points. In general, the highest concentrations of the particles were found in the lungs and spleen. Importantly, the particles were found in the brain, indicating that these small particles may be useful for brain imaging.

Imaging using CT and MRI showed that only the 12 nm particles were appropriate for MR imaging, because they were comparable to Resovist at 3 T. 10 mM in Fe was needed to detect enhancement in the image. As a CT agent, the 12 nm particles showed similar contrast to the same concentration of commonly used iodine agents. In cell pellets, the MRI and CT showed greater labelling of MBT2 (HER2+) cells than MBT2-KD cells. *In vivo*, tail vein injection of the 12 nm particles in mice with transplanted MBT2 tumours allowed the visualisation of the tumour by both MRI and CT.

Nanochains of iron oxide particles were used for geometrically enhanced multivalent docking [92]. The nanochains were formed by a previously reported multistep process [93]. First, using solid phase chemistry, amine-functionalised SPIONs (N-SPIONs) were covalently bonded to a surface via DTSSP (Scheme 16.12). This attachment allows the SPIONS to be freed by reduction of the S-S bond, leaving the particles with amine (NH_2) groups on one side and thiol (SH) groups on the other side (N,S-SPIONs). In the second step, N,S-SPIONs were again combined with DTSSP beads. The amine groups react with the exposed SulfoNHS and leave the –SH groups exposed. These are reacted with Sulfo-SMCC, which contains a maleimide group and a sulfoNHS group. The thiol groups react with the maleimide to form a highly stable thioether bond. N,S-SPIONs are added and the amine groups react with the sulfoNHS group. This step is repeated n times to make a chain of n + 2 particles.

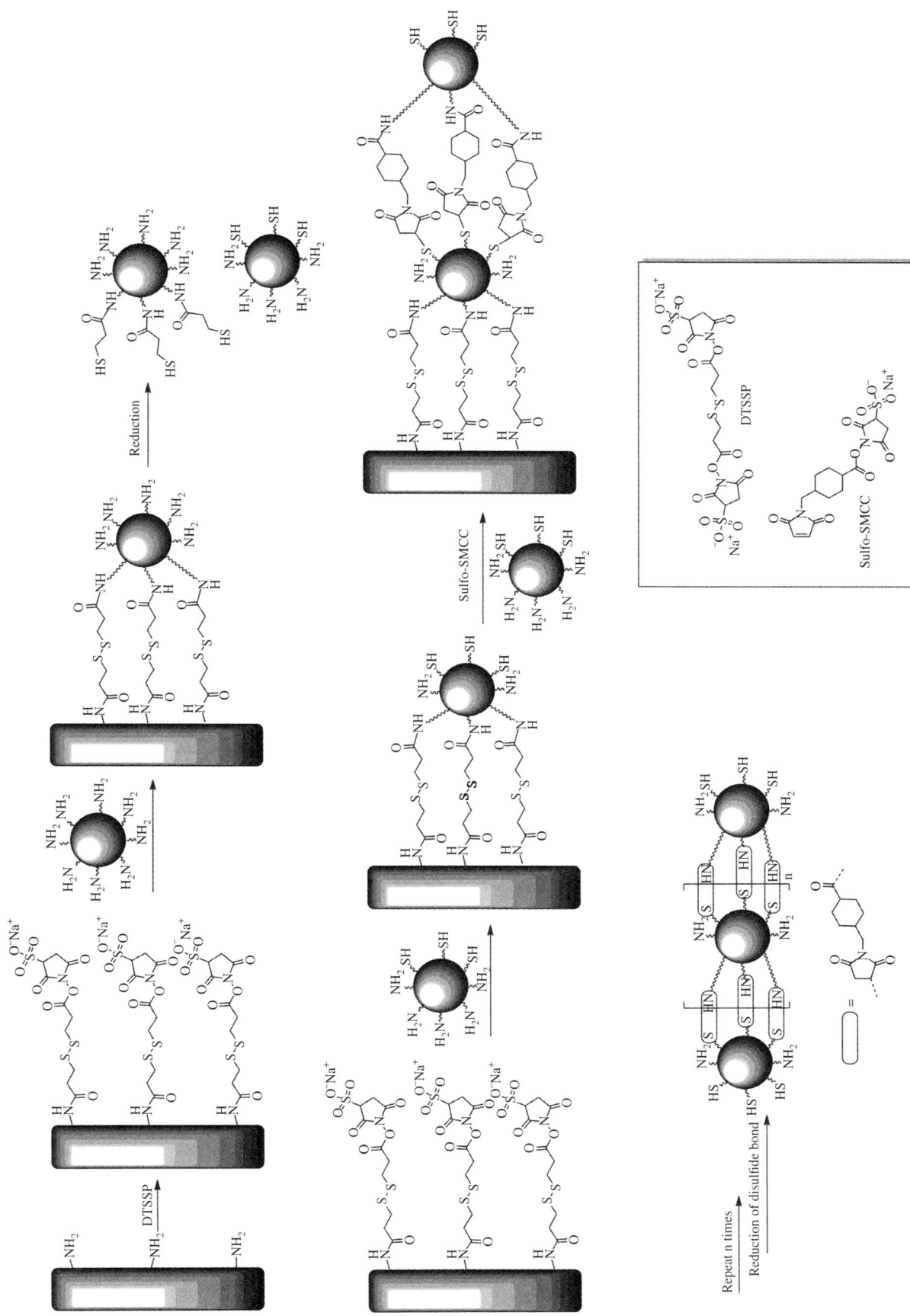

SCHEME 16.12 Solid phase chemistry for nanochain formation. Adapted from refs. [92] and [93].

The isolated nanochains were found to consist of three to five particles. After cleavage from the beads, the mixture was found to be 6% monomeric SPIONs, 72% four SPIONs, 12% three SPIONs, and 10% five SPIONs. The relaxivity of the nanochains was found to be 121 mM^{-1} s^{-1} at 1.4 T, which is approximately twice that of the individual nanoparticles. Vivotag 680 was included for fluorescence imaging.

These nanochains were further modified with cyclic RGD to target a$_v$B$_3$ integrin. The cyclic (Arg-GlyAsp-D-Phe-Cys) was attached to the nanochains via maleimide chemistry. Nanochains with cRGD targeted tumours *in vivo* more efficiently than either non-targeted nanochains or targeted nanoparticles, most likely due to the multivalent effect of the chain structure (Figure 16.27). Notably, MRI was used to detect the presence of the targeted nanochains in liver and lung metastases.

16.6.3 Medical Applications

Tantalum oxide nanoparticles were investigated for SLN mapping using fluorescence and CT [9]. A microemulsion was prepared using Igepal CO-520 (Figure 16.28) and tantalumethoxide was added. Rhodamine B isothiocyanate (RITC) was functionalised with a silane group by mixing with APTES. The RITC-silane and PEG-silane were added to the tantalum microemulsion to modify the surface of the nanoparticle.

Cell studies using MTT showed high viability. Histology showed low toxicity of injected particles. For SLN mapping, particles were injected in the paws of mice. 3D CT imaging showed the location of the SLNs, which were successfully removed using fluorescence-guided surgery.

Atherosclerotic disease is typically investigated using angiography. In order to identify areas of probable rupture, there is a need for the ability to image the vessel wall. In this study, PET and MRI were combined to address this need [8]. Dextran-coated SPIONs with amine functional groups were modified with *p*-NCS-Bn-DOTA-^{64}Cu (Figure 16.29). PET was used to screen for inflamed lesions while macrophages were visualised using MRI at high resolution. This 'non-invasive' endoscopy shows macrophage density and distribution, which can in turn be used to estimate the probability of rupture (Figure 16.30).

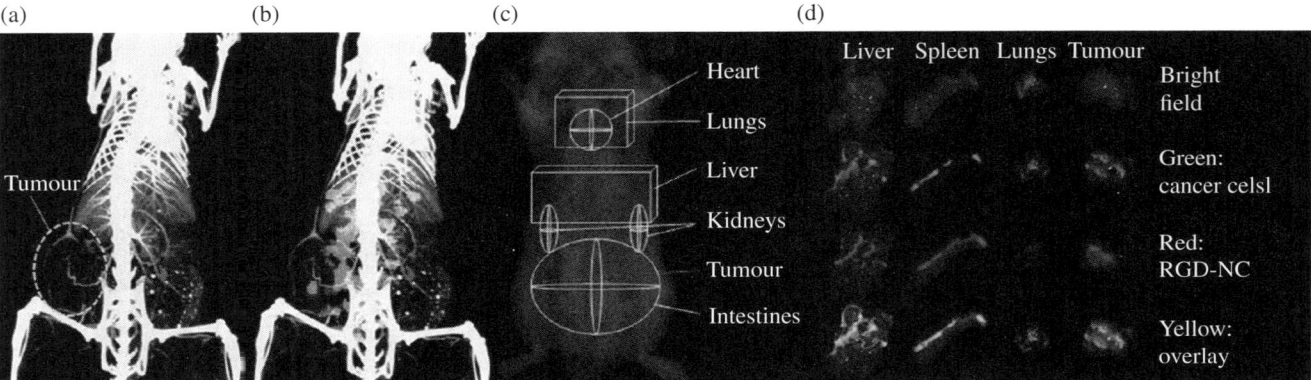

FIGURE 16.27 CT/fluorescence imaging of metastases. (a) Micromorphological imaging of normal and tumour vasculature at 99 μm resolution of a metastatic 4T1 tumour (week 5) using a liposome-based iodinated contrast agent. (b) Co-registration of the micro-CT image with the fluorescence image of the same animal injected with the RGD-NC nanoparticles. (c) Regions of interest indicate the location of the tumour and organs. (d) *ex vivo* imaging of organs indicates the co-localisation of RGD-NC particles and 4T1 metastatic cells expressing GFP. Reprinted with permission from Ref. [92]. Copyright 2012 American Chemical Society. (*See insert for colour representation of the figure.*)

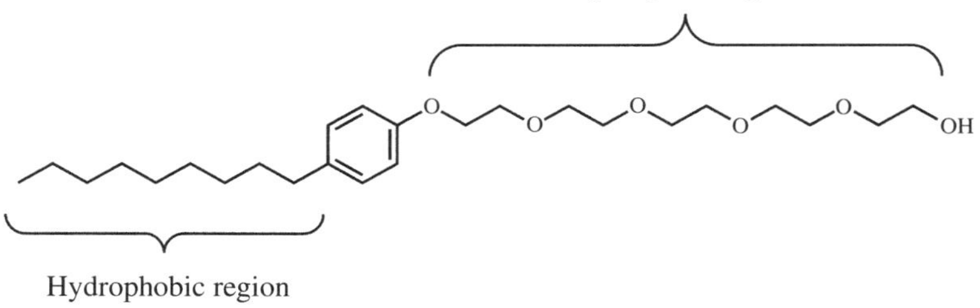

FIGURE 16.28 Structure of the surfactant Igepal CO-520.

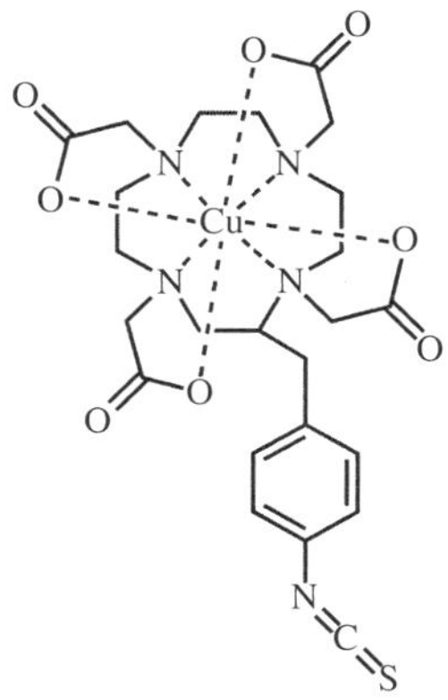

FIGURE 16.29 Structure of *p*-NCS-Bn-DOTA -⁶⁴Cu.

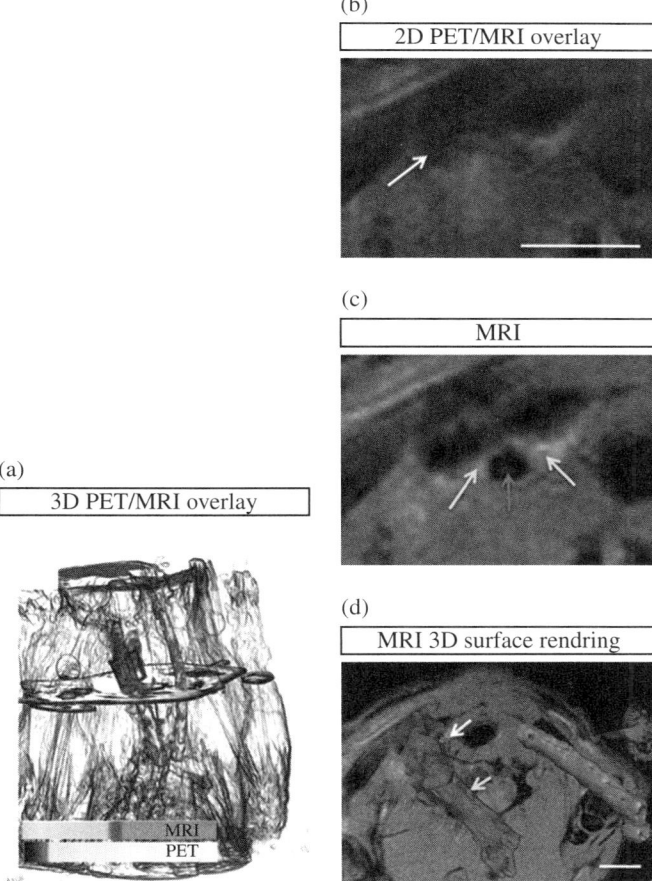

(b)

2D PET/MRI overlay

(c)

MRI

(a)

3D PET/MRI overlay

(d)

MRI 3D surface rendring

MRI
PET

FIGURE 16.30 Multimodal macromolecular probes localise to the injured vessel in the rat copper cuff model. (a) Overlay of PET with MRI 24 h post injection (head at top, out of field of view). The PET and MR signal from vessels was rendered in colour using the assignments shown in the scale bar. The plane through the image indicates the position of the image shown in (b). This co-registered MRI/PET image shows a diffuse cloud of PET signal in the region around the injured vessel and region of higher MR intensity on the right side of the vessel. The clavicle is the dark region indicated by the white arrow. Scale bar = 2.5 mm. (c) The MR image from the same plane clearly shows elevated MR contrast in the walls of the vessel. (d) 3D reconstruction of MRI and PET data in an oblique orientation shows the discrete accumulations of macrophages on the vessel wall. Scale bar = 5 mm. This view is zoomed cut from the FOV in panels (b) and (c) to include both vessels. Injured carotid artery is purple, increased MR signal intensity relative to vessel background is green, PET signal is orange, and contralateral vessel is gold. Reproduced with permission from Ref. [8]. (*See insert for colour representation of the figure.*)

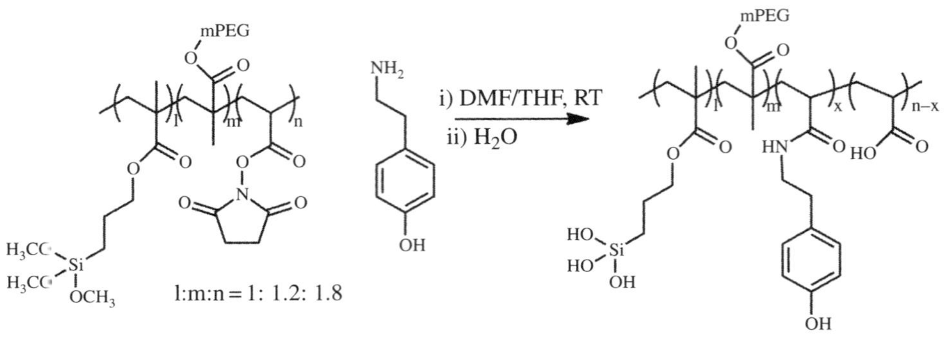

SCHEME 16.13   Tyramine conjugation to the TMSMSA-PEGMA-NAS polymer. Adapted from Ref. [94].

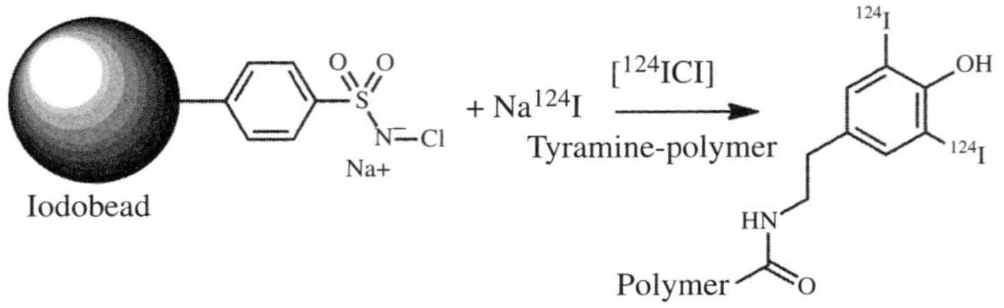

SCHEME 16.14   Iodination of the tyramine-conjugated polymer using IodoBeads.

Iodinated SPIONs were designed to utilise Cerenkov radiation in a trimodal imaging system [94]. Cerenkov radiation is a decay phenomenon in which optical photons are emitted when a charged particle moves faster than the speed of light in a particular medium [95]. Therefore, no external light source is needed to stimulate Cerenkov light emission, which is continuous. The intensity is dependent on the frequency and kinetic energy of the emitted particle. PET nuclides can be used with the requirement of the positron energy be >262 eV for aqueous media.

To generate the iodinated nanoparticles, SPIONs were coated with poly(TMSMSA-rPEGMA-rNAS), which was synthesised by combining 3-(trimethoxysilyl)propyl methacrylate (TMSMA), poly(ethylene glycol)methyl ether methacrylate (PEGMA), and N -acryloxysuccinimide (NAS) with 2,2′-azo-bisisobutyronitrile, a radical polymerisation initiator [94]. Tyramine was added to the purified polymer, where the amine group reacts with the succinimide of NAS (Scheme 16.13). Hydrolysis results in -Si(OH)$_3$ groups and free carboxylic acids. SPIONS were coated with the polymer and heated to form cross-linking bonds between the Si(OH)$_3$ groups and carboxylic acids of the polymer on the particle surface.

IodoBeads (Pierce) were used to iodinate the tyramine on the particles. The IodoBeads convert sodium iodide to iodo-monochloride (ICl), which readily incorporates iodine into tyrosyl groups next to the OH group on the phenolic ring (Scheme 16.14). ^{124}I was used because of its long half-life of four days and high energy positrons (819 keV). The authors found that Cerenkov imaging using ^{124}I has a resolution of 1.6-1.2 mm, while the PET resolution was limited to 2.4 mm. All three modalities, optical, PET, and MRI, showed that the SLNs near the tumour showed lower uptake due to metastasis of the tumour and subsequent destruction of the lymph nodes (Figure 16.31).

For a SPECT/MRI system, PEG-functionalised SPIONs were modified with ^{99m}Tc for bimodal imaging of sentinel lymph nodes [96]. ^{99m}TcO$_4$ was reduced using SnCl$_2$ in the presence of the SPIONs. Magnetic separation of the labelled SPIONs from the solution showed no free ^{99m}Tc was present.^{99m}Tc-SPIONs were detected in SLN by SPECT as well as MRI. Digital autoradiography showed the uptake by the SLN was not homogeneous, and the authors suggest that the particles were phagocytosed in the sub-central, central, and medullary sinuses.

16.7   ULTRASOUND

Microbubbles (MBs), a term loosely applied to microstructures with an internal cavity, have been widely implemented as ultrasound contrast agents [97–99]. The efficiency of MBs is dependent on the compressibility and density of the MB and the gas or liquid within it. The difference in these factors between the MB and the surrounding medium allows contrast to be observed. MBs have been made of lipids and polymers, and gases such as perfluorocarbons [100] and sulfurhexafluoride [101] have been used [100, 101]. The walls of polymeric MBs have the disadvantage of being thicker than lipid-based MBs,

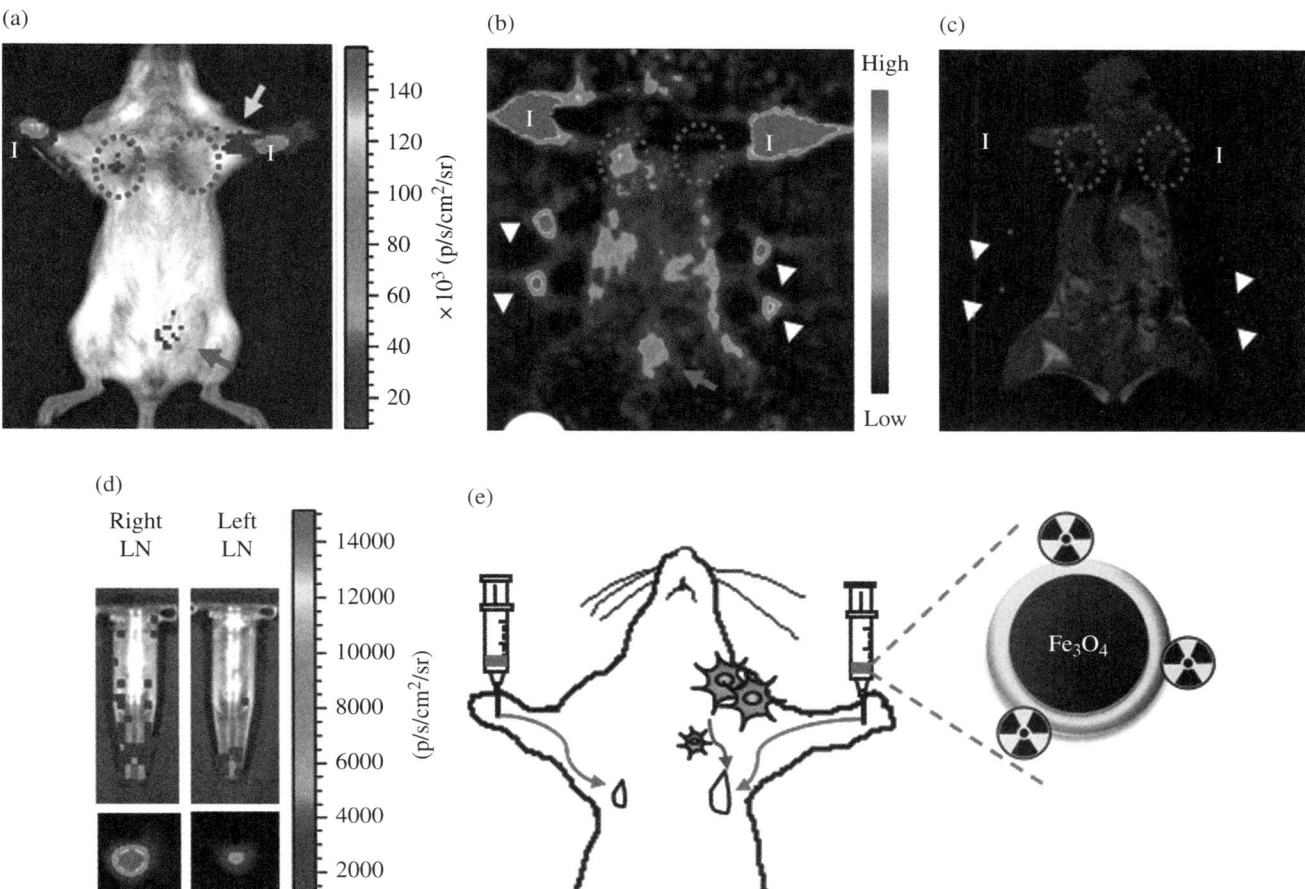

FIGURE 16.31 Triple-modality imaging of radiolabelled nanoparticles: (a) optical, (b) microPET, and (c) MRI of ^{124}I-labelled SPIONs injected into the front paws of a BALB/c mouse bearing a 4T1 tumour implanted on its shoulder. Tumour: arrow; sentinel lymph node: dotted circle; injection site: "I"; bladder: arrow; fiduciary markers: white arrow head. (d) *Ex vivo* luminescence (top) and microPET (bottom) images of the dissected lymph nodes. (e) Schematic diagram of tumour metastasis model and injection route of radiolabelled nanoparticles. Reprinted with permission from Ref. [94]. (*See insert for colour representation of the figure.*)

which dampens oscillations and reduces the power of the backscatter. However, the rigid base demonstrates a longer shelf life and allows for covalent conjugation of targeting groups or addition of other imaging modalities and drug delivery [97].

Ultrasound and MR imaging are complementary because they combine cost-effective 2D US imaging with the 3D soft tissue images from MR. In order to prepare MBs for use in US and MRI, SPIONs were combined with previously characterised MBs [102]. First, MBs were prepared from PVA as previously reported, and SPIONs of size 8–10 nm were prepared by a co-precipitation method. To covalently attach the SPIONs to the PVA microbubbles, amino groups were introduced on the surface of the Fe_3O_4 particles via silanisation with APTES. The amino groups reacted with aldehyde groups on the microbubble surface, followed by reductive amination at pH 5.0 with $NaBH_3CN$. Alternatively, unmodified SPIONs were added during the PVA shell formation to physically embed them within the shell. The MBs were found to be 3.8 +/−0.6 μm in diameter. Neither modification of the PVA MBs with iron oxide impacted the echogenicity of the MBs in a negative way [102]. The SPIONs enabled visualisation of the microbubbles both *in vitro* and *in vivo* using MRI. Modification of MBs in this way may be a new way of tuning the echogenicity and improving the US contrast.

16.8 MAGNETOMOTIVE OPTICAL COHERENCE TOMOGRAPHY (MM-OCT)

Optical coherence tomography is a relatively new imaging technique that is capable of mapping 3D structures of tissue based on optical scattering properties. The resolution is similar to that of histology. Magnetomotive optical coherence tomography (MM-OCT) is used to image dynamic magnetic nanoprobes in biological subjects because these probes alter the local optica. properties of the tissue. This approach makes it possible to detect extremely low concentrations of nanoprobes. Importantly, this method can take advantage of nanoprobes designed for other imaging modalities without need for alteration.

In the first study, *in vivo* imaging of a preclinical mammary tumour was undertaken using MM-OCT and MRI [103]. Dextran-coated SPIONs were synthesised and conjugated with Her2 neu antibodies for tumour targeting. Tumour-bearing mice were injected with targeted or untargeted SPIONs. MM-OCT showed accumulation of the targeted SPIONs in the tumour. A high signal observed in the spleen due to the high concentrations of ferritin; however, the control animals showed a similar signal. In the animal treated with non-targeted particles, high signals were observed in the liver, spleen, and lung.

MRI showed shifts toward lower T_2^* values in the tumour when targeted nanoparticles were used. In the cases of non-targeted particles or saline, the shifts were minimal (non-targeted) or to higher T_2^* values (saline). This result suggests that the targeting antibody successfully localises the SPIONs at the tumour.

In order to add additional modalities, protein microspheres were employed as the base for multimodal MM-OCT/MRI imaging agents [104]. The hydrophilic protein shell consists of BSA surrounding a core composed of vegetable oil. SPIONs (20–30 nm), Nile red, and vegetable oil were layered with an aqueous solution of BSA. The mixture was sonicated at 45°C with surfactants. The product was centrifuged, washed, and filtered through a 5 μm filter. The cross-linked protein shell was subjected to a layer-by-layer deposition of PDDA and silica, and functionalisation of the RGD peptide to target cells that over-express integrin receptors. The final product was 2–5 μm in diameter.

In vivo, the microspheres show negative T_2 contrast enhancement in the MR image. MM-OCT allowed the accumulation of targeted particles in the tumour to be observed in real time. *Ex vivo* MM-OCT confirmed that the RGD-targeted microspheres accumulated in the tumour. Histology used the fluorescence of Nile red dye to show presence of microspheres in the tumour tissue. These *ex vivo* results confirm that high-resolution, real-time *in vivo* imaging is possible using MM-OCT and labelled protein microspheres.

16.9   PHOTOACOUSTIC IMAGING

The number of multimodal agents that include photoacoustic imaging has burgeoned in the past 2 years [105–115]. Photoacoustic imaging (PAI), the combination of optical and ultrasound imaging, is an emerging modality for noninvasive detection of structural and functional anomalies in biological tissues, which can assist with image-guided therapy [108]. A wide variety of nanoparticulate agents have been developed to include PAI capability, including gold nanotripods [110], porphyrin-lipid-based supramolecular structures [111], single-walled and multi-walled carbon nanotubes [106, 109], as well as multifunctional microbubbles and nanobubbles [107]. PAI is a natural, advantageous modality because commonly used highly absorbing optical contrast agents, such as gold nanoparticles, gold-coated carbon nanotubes or Indocyanine green, can be utilised. PAI offers high spatial resolution, a perfect complement to other modalities that suffer from low spatial resolution. For example, MRI can reveal tumour location and time-dependent behaviour of the nanoparticles, while PAI allows delineation of tumour margin and vivid 3D visualisation of theranostic nanoparticles inside the tumour [112].

16.10   CONCLUSIONS

It is clear that the ability to simultaneously interrogate anatomical features and biochemical processes by molecular imaging has greatly increased our understanding of these important events. Visualising molecular processes *within* a living organism in real time has become a key element of both basic research and diagnostic radiology. Combining the most powerful imaging modalities to exploit their individual strengths overcomes the weaknesses of any individual technique. The advantages of obtaining higher resolution images, sensitivity, and temporal resolution are obvious. Further, with the advent of multimodal hardware, the development of multimodal imaging probes to facilitate co-registration has been critical for optimising experiments.

In this chapter we have attempted to provide a snapshot of the many advances that have been made in demonstrating the utility of multimodal nanoparticles for molecular imaging. While a great deal of progress has been achieved, *significant* hurdles remain. For example, a great deal of nanoparticle research is focused on controlling the size, stability, clearance, and immunogenic properties of nanoparticles. Translation of these probes to the clinic faces the well-known problems of long-term toxicity, pharmacokinetics, and pharmacodynamics of nanoparticle platforms. While many challenges remain, it has become increasingly apparent that the benefits of using multimodal nanoparticles in basic research are just beginning to be realised.

REFERENCES

[1] *Supramolecular chemistry in biological imaging in vivo*, E. A. Schultz-Sikma and T. J. Meade, Eds., 2012; Vol. 4.

[2] J. L. Major and T. J. Meade, *Acc. Chem. Res.* **42**, 893–903 (2009).

[3] L. Frullano and T. J. Meade, J. Biol. *Inorg. Chem.* **12**, 939–949 (2007).

[4] L. M. Manus, D. J. Mastarone, E. A. Waters, X. Q. Zhang, E. A. Schultz-Sikma, K. W. Macrenaris, D. Ho and T. J. Meade, *Nano Lett.* **10**, 484–489 (2010).

[5] B. P. Barnett, J. Ruiz-Cabello, P. Hota, R. Ouwerkerk, M. J. Shamblott, C. Lauzon, P. Walczak, W. D. Gilson, V. P. Chacko, D. L. Kraitchman, A. Arepally and J. W. M. Bulte, *Contrast Media Mol. Imaging* **6**, 251–259 (2011).

[6] P. A. Sukerkar, U. G. Rezvi, K. W. Macrenaris, P. C. Patel, J. C. Wood and T. J. Meade, *Magn. Reson. Med.* **65**, 522–530 (2011).

[7] D. W. Hwang, H. Y. Ko, J. H. Lee, H. Kang, S. H. Ryu, I. C. Song, D. S. Lee and S. Kim, *J. Nucl. Med.* **51**, 98–105 (2010).

[8] B. R. Jarrett, C. Correa, K. L. Ma and A. Y. Louie, *PLoS One* **5**, e13254 (2010).

[9] M. H. Oh, N. Lee, H. Kim, S. P. Park, Y. Piao, J. Lee, S. W. Jun, W. K. Moon, S. H. Choi and T. Hyeon, *J. Am. Chem. Soc.* **133**, 5508–5515 (2011).

[10] Y. Song, X. Xu, K. W. MacRenaris, X. Q. Zhang, C. A. Mirkin and T. J. Meade, *Angew. Chem. Int. Ed.* **48**, 9143–9147 (2009).

[11] M. Zhang, B. L. Cushing and C. J. O'Connor, *Nanotechnology* **19**, 085601/085601-085601/085605 (2008).

[12] A. M. Morawski, G. A. Lanza and S. A. Wickline, *Curr. Opin. Biotechnol.* **16**, 89–92 (2005).

[13] S. Santra, R. P. Bagwe, D. Dutta, J. T. Stanley, G. A. Walter, W. Tan, B. M. Moudgil and R. A. Mericle, *Adv. Mater.* **17**, 2165–2169 (2005).

[14] M. F. Kircher, U. Mahmood, R. S. King, R. Weissleder and L. Josephson, *Cancer Res.* **63**, 8122–8125 (2003).

[15] K. A. Kelly, J. R. Allport, A. Tsourkas, V. R. Shinde-Patil, L. Josephson and R. Weissleder, *Circ. Res.* **96**, 327–336 (2005).

[16] M. L. James and S. S. Gambhir, *Physiol. Rev.* **92**, 897–965 (2012).

[17] M. Janowski, J. W. M. Bulte and P. Walczak, *Adv. Drug Delivery Rev.* **64**, 1488–1507 (2012).

[18] P. A. Jarzyna, A. Gianella, T. Skajaa, G. Knudsen, L. H. Deddens, D. P. Cormode, Z. A. Fayad and W. J. M. Mulder, *Wiley Interdiscip. Rev.: Nanomed. Nanobiotechnol.* **2**, 138–150 (2010).

[19] J. V. Jokerst and S. S. Gambhir, *Acc. Chem. Res.* **44**, 1050–1060 (2011).

[20] A. Louie, *Chem. Rev.* **110**, 3146–3195 (2010).

[21] A. Taylor, K. M. Wilson, P. Murray, D. G. Fernig and R. Levy, *Chem. Soc. Rev.* **41**, 2707–2717 (2012).

[22] C. Xu, L. Mu, I. Roes, D. Miranda-Nieves, M. Nahremdorf, J. A. Ankrum, W. Zhao and J. M. Karp, *Nanotechnology* **22**, 494001/494001–494001/494017 (2011).

[23] J. Harrison, C. A. Bartlett, G. Cowin, P. K. Nicholls, C. W. Evans, T. D. Clemons, B. Zdyrko, I. A. Luzinov, A. R. Harvey, K. S. Iyer, S. A. Dunlop and M. Fitzgerald, *Small* **8**, 1579–1589 (2012).

[24] J. E. Lee, N. Lee, H. Kim, J. Kim, S. H. Choi, J. H. Kim, T. Kim, I. C. Song, S. P. Park, W. K. Moon and T. Hyeon, *J. Am. Chem. Soc.* **132**, 552–557 (2010).

[25] C. W. Evans, M. Fitzgerald, T. D. Clemons, M. J. House, B. S. Padman, J. A. Shaw, M. Saunders, A. R. Harvey, B. Zdyrko, I. Luzinov, G. A. Silva, S. A. Dunlop and K. S. Iyer, *ACS Nano* **5**, 8640–8648 (2011).

[26] S. Srivastava, R. Awasthi, D. Tripathi, M. K. Rai, V. Agarwal, V. Agrawal, N. S. Gajbhiye and R. K. Gupta, *Small* **8**, 1099–1109 (2012).

[27] S. Srivastava and N. S. Gajbhiye, *ChemPhysChem.* **12**, 2624–2632 (2011).

[28] D. L. J. Thorek, P. Y. Tsao, V. Arora, L. Zhou, R. A. Eisenberg and A. Tsourkas, *PLoS One* **5**, e10655 (2010).

[29] R. Bardhan, W. Chen, M. Bartels, C. Perez-Torres, M. F. Botero, R. W. McAninch, A. Contreras, R. Schiff, R. G. Pautler, N. J. Halas and A. Joshi, *Nano Lett.* **10**, 4920–4928 (2010).

[30] L. Josephson, C. H. Tung, A. Moore and R. Weissleder, *Bioconjugate Chem.* **10**, 186–191 (1999).

[31] E. A. Schultz-Sikma, PhD Thesis, Northwestern University (2010).

[32] H. M. Kim, Y.-W. Noh, H. S. Park, M. Y. Cho, K. S. Hong, H. Lee, D. H. Shin, J. Kang, M.-H. Sung, H. Poo and Y. T. Lim, *Small* **8**, 666–670 (2012).

[33] M. Longmire, P. L. Choyke and H. Kobayashi, *Nanomedicine* **3**, 703–717 (2008).

[34] M. R. Longmire, M. Ogawa, P. L. Choyke and H. Kobayashi, *Bioconjugate Chem.* **22**, 993–1000 (2011).

[35] F. Erogbogbo, K.-T. Yong, R. Hu, W.-C. Law, H. Ding, C.-W. Chang, P. N. Prasad and M. T. Swihart, *ACS Nano* **4**, 5131–5138 (2010).

[36] X. G. Li, Y. Q. He, S. S. Talukdar and M. T. Swihart, *Langmuir* **19**, 8490–8496 (2003).

[37] Z. M. Markovic, L. M. Harhaji-Trajkovic, B. M. Todorovic-Markovic, D. P. Kepic, K. M. Arsikin, S. P. Jovanovic, A. C. Pantovic, M. D. Dramicanin and V. S. Trajkovic, *Biomaterials* **32**, 1121–1129 (2011).

[38] H. M. Sun, L. Y. Cao and L. H. Lu, *Nano Res.* **4**, 550–562 (2011).

[39] K. Yang, L. Hu, X. Ma, S. Ye, L. Cheng, X. Shi, C. Li, Y. Li and Z. Liu, *Adv. Mater.* **24**, 1868–1872 (2012).

[40] M. Ma, H. Chen, Y. Chen, X. Wang, F. Chen, X. Cui and J. Shi, *Biomaterials* **33**, 989–998 (2012).

[41] A. Gianella, P. A. Jarzyna, V. Mani, S. Ramachandran, C. Calcagno, J. Tang, B. Kann, W. J. R. Dijk, V. L. Thijssen, A. W. Griffioen, G. Storm, Z. A. Fayad and W. J. M. Mulder, *ACS Nano* **5**, 4422–4433 (2011).

[42] P. Huang, Z. Li, J. Lin, D. Yang, G. Gao, C. Xu, L. Bao, C. Zhang, K. Wang, H. Song, H. Hu and D. Cui, *Biomaterials* **32**, 3447–3458 (2011).

[43] Y. Choi, R. Weissleder and C. H. Tung, *Cancer Res.* **66**, 7225–7229 (2006).

[44] A. G. Lacko, M. Nair, L. Prokai and W. J. McConathy, *Expert Opin. Drug Deliv.* **4**, 665–675 (2007).

[45] F. Tabet and K. A. Rye, *Clin. Sci.* **116**, 87–98 (2009).

[46] W. Chen, P. A. Jarzyna, G. A. F. van Tilborg, V. A. Nguyen, D. P. Cormode, A. Klink, A. W. Griffioen, G. J. Randolph, E. A. Fisher, W. J. M. Mulder and Z. A. Fayad, *Faseb J.* **24**, 1689–1699 1610.1096/fj.1609-139865 (2010).

[47] C. Sun, K. Du, C. Fang, N. Bhattarai, O. Veiseh, F. Kievit, Z. Stephen, D. Lee, R. G. Ellenbogen, B. Ratner and M. Zhang, *ACS Nano* **4**, 2402–2410 (2010).

[48] J. A. DeBin and G. R. Strichartz, *Toxicon.* **29**, 1403–1408 (1991).

[49] J. Deshane, C. C. Garner and H. Sontheimer, *J. Biol. Chem.* **278**, 4135–4144 (2003).

[50] S. A. Lyons, J. O'Neal and H. Sontheimer, *Glia* **39**, 162–173 (2002).

[51] E.-J. Cha, E. S. Jang, I.-C. Sun, I. J. Lee, J. H. Ko, Y. I. Kim, I. C. Kwon, K. Kim and C.-H. Ahn, J. Control. *Release* **155**, 152–158 (2011).

[52] A. M. Gobin, M. H. Lee, N. J. Halas, W. D. James, R. A. Drezek and J. L. West, *Nano Lett.* **7**, 1929–1934 (2007).

[53] D. P. O'Neal, L. R. Hirsch, N. J. Halas, J. D. Payne and J. L. West, *Cancer Lett.* **209**, 171–176 (2004).

[54] J. M. Stern, J. Stanfield, W. Kabbani, J. T. Hsieh and J. A. Cadeddu, *J. Urol.* **179**, 748–753 (2008).

[55] R. Bardhan, N. K. Grady, J. R. Cole, A. Joshi and N. J. Halas, *ACS Nano* **3**, 744–752 (2009).

[56] Q. Ma, Y. Nakane, Y. Mori, M. Hasegawa, Y. Yoshioka, T. M. Watanabe, K. Gonda, N. Ohuchi and T. Jin, *Biomaterials* **33**, 8486–8494 (2012).

[57] R. Kumar, I. Roy, T. Y. Ohulchanskky, L. A. Vathy, E. J. Bergey, M. Sajjad and P. N. Prasad, *ACS Nano* **4**, 699–708 (2010).

[58] R. Kumar, I. Roy, T. Y. Hulchanskyy, L. N. Goswami, A. C. Bonoiu, E. J. Bergey, K. M. Tramposch, A. Maitra and P. N. Prasad, *ACS Nano* **2**, 449–456 (2008).

[59] W. H. De Jong, W. I. Hagens, P. Krystek, M. C. Burger, A. J. Sips and R. E. Geertsma, *Biomaterials* **29**, 1912–1919 (2008).

[60] G. Sonavane, K. Tomoda and K. Makino, *Colloids Surf B Biointerfaces* **66**, 274–280 (2008).

[61] F. Sousa, S. Mandal, C. Garrovo, A. Astolfo, A. Bonifacio, D. Latawiec, R. H. Menk, F. Arfelli, S. Huewel, G. Legname, H.-J. Galla; S. Krol, *Nanoscale* **2**, 2826–2834 (2010).

[62] J. H. Kim, Y. S. Kim, K. Park, S. Lee, H. Y. Nam, K. H. Min, H. G. Jo, J. H. Park, K. Choi, S. Y. Jeong, R. W. Park, I. S. Kim, K. Kim and I. C. Kwon, *J. Control. Release* **127**, 41–49 (2008).

[63] K. Kim, S. Kwon, J. H. Park, H. Chung, S. Y. Jeong, I. C. Kwo and I. S. Kim, *Biomacromolecules* **6**, 1154–1158 (2005).

[64] J. S. Park, T. H. Han, K. Y. Lee, S. S. Han, J. J. Hwang, D. H. Moon, S. Y. Kim and Y. W. Cho, *J. Control. Release* **115**, 37–45 (2006).

[65] B. S. Lee, K. Park, S. Park, G. C. Kim, H. J. Kim, S. Lee, H. Kil, S. J. Oh, D. Chi, K. Kim, K. Choi, I. C. Kwon and S. Y. Kim, *J. Control. Release* **147**, 253–260 (2010).

[66] S. Li, B. Goins, L. Zhang and A. Bao, *Bioconjugate Chem.* **23**, 13221332 (2012).

[67] A. Bao, B. Goins, R. Klipper, G. Negrete and W. T. Phillips, *J. Pharmacol. Exp. Ther.* **308**, 419–425 (2004).

[68] J. Key, D. Dhawan, D. W. Knapp, K. Kim, I. C. Kwon, K. Choi and J. F. Leary, *Proc. SPIE* **8225**, 82251F/82251–82251F/82258 (2012).

[69] A. Maham, Z. Tang, H. Wu, J. Wang and Y. Lin, *Small* **5**, 1706–1721 (2009).

[70] T. Douglas and V. T. Stark, *Inorg Chem.* **39**, 1828–1830 (2000).

[71] M. Uchida, M. L. Flenniken, M. Allen, D. A. Willits, B. E. Crowley, S. Brumfield, A. F. Willis, L. Jackiw, M. Jutila, M. J. Young and T. Douglas, *J. Am. Chem. Soc.* **128**, 16626–16633 (2006).

[72] L. Zhang, L. Laug, W. Munchgesang, E. Pippel, U. Gosele, M. Brandsch and M. Knez, *Nano Lett.* **10**, 219–223 (2010).

[73] X. Lin, J. Xie, G. Niu, F. Zhang, H. Gao, M. Yang, Q. Quan, M. A. Aronova, G. Zhang, S. Lee, R. Leapman and X. Chen, *Nano Lett.* **11**, 814–819 (2011).

[74] X. Huang, F. Zhang, S. Lee, M. Swierczewska, D. O. Kiesewetter, L. Lang, G. Zhang, L. Zhu, H. Gao, H. S. Choi, G. Niu and X. Chen, *Biomaterials* **33**, 4370–4378 (2012).

[75] X. Huang, L. Li, T. Liu, N. Hao, H. Liu, D. Chen and F. Tang, *ACS Nano* **5**, 5390–5399 (2011).

[76] X. Huang, X. Teng, D. Chen, F. Tang and J. He, *Biomaterials* **31**, 438–448 (2010).

[77] F. Wang and X. G. Liu, *Chem. Soc. Rev.* **38**, 976–989 (2009).

[78] G. Zhang, Y. Liu, Q. Yuan, C. Zong, J. Liu and L. Lu, *Nanoscale* **3**, 4365–4371 (2011).

[79] Q. Liu, M. Chen, Y. Sun, G. Chen, T. Yang, Y. Gao, X. Zhang and F. Li, *Biomaterials* **32**, 8243–8253 (2011).

[80] Q. Liu, Y. Sun, C. Li, J. Zhou, C. Li, T. Yang, X. Zhang, T. Yi, D. Wu and F. Li, *ACS Nano* **5**, 3146–3157 (2011).

[81] E. Arunkumar, C. C. Forbes, B. C. Noll and B. D. Smith, *J. Am. Chem. Soc.* **127**, 3288–3289 (2005).

[82] H. Xu, L. Cheng, C. Wang, X. Ma, Y. Li and Z. Liu, *Biomaterials* **32**, 9364–9373 (2011).

[83] L. Cheng, K. Yang, Y. Li, X. Zeng, M. Shao, S.-T. Lee and Z. Liu, *Biomaterials* **33**, 2215–2222 (2012).

[84] F. Lux, A. Mignot, P. Mowat, C. Louis, S. Dufort, C. Bernhard, F. Denat, F. Boschetti, C. Brunet, R. Antoine, P. Dugourd, S. Laurent, L. Vander Elst, R. Muller, L. Sancey, V. Josserand, J.-L. Coll, V. Stupar, E. Barbier, C. Remy, A. Broisat, C. Ghezzi, G. Le Duc, S. Roux, P. Perriat and O. Tillement, *Angew. Chem. Int. Ed.* **50**, 12299–12303, S12299/12291–S12299/12216 (2011).

[85] D. Kryza, J. Taleb, M. Janier, L. Marmuse, I. Miladi, P. Bonazza, C. Louis, P. Perriat, S. Roux, O. Tillement and C. Billotey, *Bioconjugate Chem.* **22**, 1145–1152 (2011).

[86] D. Portet, B. Denizot, E. Rump, J. J. Lejeune and P. Jallet, *J. Coll. Interf. Sci.* **238**, 37–42 (2001).

[87] D. R. R. T. Martin, R. Tavare, A. Glaria, G. Varma, A. Protti and P. J. Blower, *Bioconjugate Chem.* **22**, 455–465 (2011).

[88] R. Torres Martin de Rosales, C. Finucane, S. J. Mather and P. J. Blower, *Chem. Commun.* 4847–4849 (2009).

[89] M. Liang, X. Liu, D. Cheng, G. Liu, S. Dou, Y. Wang, M. Rusckowski and D. J. Hnatowich, *Bioconjugate Chem.* **21**, 1385–1388 (2010).

[90] J. M. Niers, J. W. Chen, G. Lewandrowski, M. Kerami, E. Garanger, G. Wojtkiewicz, P. Waterman, E. Keliher, R. Weissleder and B. A. Tannous, *J. Am. Chem. Soc.* **134**, 5149–5156 (2012).

[91] S.-W. Chou, Y.-H. Shau, P.-C. Wu, Y.-S. Yang, D.-B. Shieh and C.-C. Chen, *J. Am. Chem. Soc.* **132**, 13270–13278 (2010).

[92] P. M. Peiris, R. Toy, E. Doolittle, J. Pansky, A. Abramowski, M. Tam, P. Vicente, E. Tran, E. Hayden, A. Camann, A. Mayer, B. O. Erokwu, Z. Berman, D. Wilson, H. Baskaran, C. A. Flask, R. A. Keri and E. Karathanasis, *ACS Nano* **6**, 8783–8795 (2012).

[93] P. M. Peiris, E. Schmidt, M. Calabrese and E. Karathanasis, *PLoS One* **6**, e15927(2011).

[94] J.-C. Park, M.-K. Yu, G.-I. An, S.-I. Park, J.-M. Oh, H.-J. Kim, J.-H. Kim, E.-K. Wang, I.-H. Hong, Y.-S. Ha, T.-H. Choi, K.-S. Jeong, Y.-M. Chang, M. J. Welch, S.-Y. Jon and J.-S. Yoo, *Small* **6**, 2863–2868 (2010).

[95] J. V. Jelley, *Br. J. Appl. Phys.* **6**, 227–232 (1955).

[96] R. Madru, P. Kjellman, F. Olsson, K. Wingaardh, C. Ingvar, F. Staahlberg, J. Olsrud, J. Laett, S. Fredriksson, L. Knutsson and S.-E. Strand, *J. Nucl. Med.* **53**, 459–463 (2012).

[97] K. Ferrara, R. Pollard and M. Borden, *Ann. Rev. Biomed. Eng.* **9**, 415–447 (2007).

[98] R. S. Meltzer, E. G. Tickner, T. P. Sahines and R. L. Popp, *J. Clin. Ultrasound* **8**, 121–127 (1980).

[99] R. Schlief, *Curr. Opin. Radiol.* **3**, 198–207 (1991).

[100] R. F. Mattrey, K. G. Baker, L. A. Hall, G. C. Steinbach and T. Peterson, *Acad Radiol* **3 Suppl** 2, S320–S321 (1996).

[101] A. K. Lim, N. Patel, R. J. Eckersley, S. D. Taylor-Robinson, D. O. Cosgrove and M. J. Blomley, *Radiology* **231**, 785–788 (2004).

[102] T. B. Brismar, D. Grishenkov, B. Gustafsson, J. Haermark, A. Barrefelt, S. V. V. N. Kothapalli, S. Margheritelli, L. Oddo, K. Caidahl, H. Hebert and G. Paradossi, *Biomacromolecules* **13**, 1390–1399 (2012).

[103] R. John, R. Rezaeipoor, S. G. Adie, E. J. Chaney, A. L. Oldenburg, M. Marjanovic, J. P. Haldar, B. P. Sutton and S. A. Boppart, *Proc. Natl. Acad. Sci. U. S. A.* **107**, 8085–8090, S8085/8081–S8085/8082 (2010).

[104] R. John, F. T. Nguyen, K. J. Kolbeck, E. J. Chaney, M. Marjanovic, K. S. Suslick and S. A. Boppart, *Mol. Imaging Biol.* **14**, 17–24 (2012).

[105] L.-S. Bouchard, M. S. Anwar, G. L. Liu, B. Hann, Z. H. Xie, J. W. Gray, X. Wang, A. Pines and F. F. Chen, *Proc. Natl. Acad. Sci. U. S. A.* **106**, 4085–4089 (2009).

[106] J.-W. Kim, E. I. Galanzha, E. V. Shashkov, H.-M. Moon and V. P. Zharov, *Nat. Nanotechnol.* **4**, 688–694 (2009).

[107] R. X. Xu, *Contrast Media Mol. Imaging* **6**, 401–411 (2011).

[108] Y. Huang, S. He, W. Cao, K. Cai and X.-J. Liang, *Nanoscale* **4**, 6135–6149 (2012).

[109] H. Gong, R. Peng and Z. Liu, *Adv. Drug Deliv. Rev.* **65**, 1951–1963 (2013).

[110] K. Cheng, S.-R. Kothapalli, H. Liu, A. L. Koh, J. V. Jokerst, H. Jiang, M. Yang, J. Li, J. Levi, J. C. Wu, S. S. Gambhir and Z. Cheng, *J. Am. Chem. Soc.* **136**, 3560–3571 (2014).

[111] E. Huynh and G. Zheng, *Nano Today* **9**, 212–222 (2014).

[112] Z. Li, S. Yin, L. Cheng, K. Yang, Y. Li and Z. Liu, *Adv. Funct. Mater.* **24**, 2312–2321 (2014).

[113] L.-S. Lin, Z.-X. Cong, J.-B. Cao, K.-M. Ke, Q.-L. Peng, J. Gao, H.-H. Yang, G. Liu and X. Chen, *ACS Nano* **8**, 3876–3883 (2014).

[114] C.-H. Wu, J. Cook, S. Emelianov and K. Sokolov, *Adv. Funct. Mater.* Ahead of Print (2014).

[115] Y. Zhang, M. Jeon, L. J. Rich, H. Hong, J. Geng, Y. Zhang, S. Shi, T. E. Barnhart, P. Alexandridis, J. D. Huizinga, M. Seshadri, W. Cai, C. Kim and J. F. Lovell, *Nat. Nanotechnol.* **9**, 631–638 (2014).

INDEX

Note: Tables are in *italics*; Figures are in **bold**.

The Chemistry of Molecular Imaging, First Edition. Edited by Nicholas Long and Wing-Tak Wong.
© 2015 John Wiley & Sons, Inc. Published 2015 by John Wiley & Sons, Inc.